ADVANCES IN SOUND LOCALIZATION

Edited by **Paweł Strumiłło**

Advances in Sound Localization
Edited by Paweł Strumiłło

Published by InTech
Janeza Trdine 9, 51000 Rijeka, Croatia

Publishing Process Manager Ivana Lorković

Technical Editor Teodora Smiljanic

Cover Designer InTech Design Team

Additional hard copies can be obtained from orders@intechopen.com

Advances in Sound Localization, Edited by Paweł Strumiłło
p. cm.
ISBN 978-953-307-224-1

For EU product safety concerns:
IN TECH d.o.o., Prolaz Marije Krucifikse Kozulić 3, 51000 Rijeka, Croatia,
info@intechopen.com or visit our website at intechopen.com.

Meet the editor

Pawel Strumillo received the Ph.D. degree in technical sciences from the University of Strathclyde, Glasgow, U.K. in 1993, and D.Sc. degree (habilitation) in electronics from the Technical University of Lodz (TUL), Poland in 2002. Currently, he holds the position of professor at TUL and is the head of the Medical Electronics Division. His research interests include: processing of biomedical and acoustic signal, computational intelligence and human-computer interaction systems. He has authored more than 100 publications in these fields. He has been principal and co-principal investigator of research grants on electronic systems aiding the blind in independent travel. Results of these projects were awarded at international innovation exhibitions. He has also received a number of industry grants from the Orange Telecommunications Group. He is a senior member of the Institute of Electrical and Electronics Engineers (IEEE) and member of board of the Polish Neural Network Society.
Sound source localization is an important research field that has attracted researchers' efforts from many technical and biomedical sciences. Sound source localization (SSL) is defined as the determination of the direction from a receiver, but also includes the distance from it.

Contents

Preface

Awareness of one's environment is important in everyday life situations for humans, animals and in various scientific and engineering applications. Living organisms can observe their surroundings using their senses, whereas man-made systems need to be equipped with different sensors (e.g. image, acoustic or touch). Whatever the nature of the signal acquisition system, be it technical or biological, an advanced processing of sensory data is needed in order to derive localization information.

Among the sources of physical modalities that can be localized from far distances are electromagnetic waves (that can propagate in vacuum) and sound waves that require some physical medium (air, water or a solid material) to propagate through. A consequence of the mechanical nature of sound propagation is the considerable dissipation of the carried energy and an a high dependence of the propagation speed on the medium type (e.g. 340m/s in air). Although, different techniques need to be engaged in locating electromagnetic and sound radiation sources, some of them are conceptually alike, e.g. processes used in radar and echolocation (also animal echolocation).

Sound source localization (SSL) is defined predominantly as the determination of the direction from a receiver, but also includes the distance from it. The direction can be expressed by two polar angles: the azimuth angle (i.e. horizontal bearings) and the elevation angle (i.e. vertical bearings). Determination of a sound source's distance can be achieved through measurements of sound intensity and/or its spectrum; however, a priori knowledge is needed about the source's radiation characteristic.

SSL is a complex computation problem. Because of the wave nature of sound propagation phenomena such as refraction, diffraction, diffusion, reflection, reverberation and interference occur. The wide spectrum of sound frequencies that range from infrasounds (lower than 20Hz) through acoustic sounds which are perceived by the human auditory system (nominally ~20Hz÷20kHZ) to ultrasounds (above 20kHz), also introduces difficulties, as different spectrum components have different penetration properties through the medium. Wide-band sound sources can be perceived differently (in terms of distance, direction and pitch) depending on the geometric characteristics of the sound propagation environment. Consequently, development of robust sound localization techniques calls for different approaches, including multisensor schemes, null-steering beamforming and time-difference arrival techniques.

SSL is an important research field that has attracted researchers' efforts from many technical and biomedical sciences. Sound localization techniques can be vital in rescue missions, medicine (ultrasonography), seismology (oil and gas exploration), as well as robotics, noise cancellation and improvement of immersion in virtual reality systems. Remarkable sound localization capabilities are featured by humans and other living organism who use them for communication, spatial orientation, wayfinding and also for locating prey or fleeing from predators.

Advances in Sound Localization is a collection of 30 contributions reporting up-to-date studies of different aspects of sound localization research, ranging from purely theoretical approaches to their implementation in specific applications. The contributions are organized in six major sections.

Part I provides state of the art exposition to a number of advanced concepts for SSL starting from a mathematical background of sensor arrays, binaural techniques (including the Head-Related Transfer Functions – HRTFs) to conceptually appealing methods that employ direction-selective filters and discrimination of acoustic transfer functions to achieve single-channel sound source localization.

Part II reports systems that implement signal processing techniques and sensor setups for robust SSL in real-life environments. It is shown that source localization can find application in robotics (e.g. for aiding environment mapping) and underwater acoustics. Techniques are proposed for considerable reduction of computing time required to run SSL algorithms. Also, approaches to generation of virtual acoustic space for studying SSL abilities in humans and animals are described. Finally, it is demonstrated how the use of SSL techniques can be applied for speech enhancement purposes.

In Part III applications of SSL techniques are covered that are aimed at enhancing human perception abilities. Applications include: aiding the blind in spatial orientation by means of auditory display systems and investigation on how bilateral hearing fittings improve spatial hearing. The part is concluded by studies underlining the importance of auditory information for environmental awareness in infants.

Applications of SSL in multimedia and teleconferencing systems are addressed in Part IV. The concept of an automatic cameraman is reported, in which a pan-tilt-zoom camera is driven by an SSL system to point in the direction of a speaker. Another communication deals with enriching video material that is projected onto large displays by spatialization of sounds using a novel loudspeaker setup. Finally, a technique employing a microphone array for spatial location of speakers in teleconferencing systems is described.

Part V is devoted to applications of SSL techniques in biomedical and diagnostic studies. First two contributions in this Section deal with studies of the human auditory cortex. The former attempts to identify characteristics of human binaural auditory filter by examining the activity of auditory evoked fields, whereas the latter explains how the binaural information is processed in the auditory cortex by using electroencephalography (EEG) and magnetoencephalography (MEG). In another interesting study it is postulated that sound stimuli (stochastic or deterministic) can facilitate perception of stimuli by other sensory modalities. This observation can be the basis for treatment of Parkinson and Alzheimer diseases. The Part is concluded by studies on detection of structural damage in materials using acoustic emission techniques.

Finally, Part VI focuses on the intriguing field of SSL in animal studies. Two lines of research are reported. The first addresses, how avian, terrestrial and aquatic animals excel in SSL by their extraordinary spatial hearing abilities. The second field of study is devoted to techniques used in practical application of SSL methods (e.g. matched filtering) for localizing animal groups or an individual animal within a group.

While preparing this preface I have become strongly convinced that this book will offer a rich source of valuable material on up-to-date advances on sound source localization that should appeal to researches representing diverse engineering and scientific disciplines.

March 2011

Paweł Strumiłło, Ph.D., D.Sc
Technical University of Lodz,
Poland

Part 1

Signal Processing Techniques for Sound Localization

The Linear Method for Acoustical Source Localization (Constant Speed Localization Method) - A Discussion of Receptor Geometries and Time Delay Accuracy for Robust Localization

Sergio R. Buenafuente and Carmelo M. Militello
University of La Laguna (ULL)
Spain

1. Introduction

One of the most widely used methodology for the passive localization of acoustic sources is based on the measurement of the time delay of arrival (TDOA) of the source signal to receptors pairs. In 2D, two pairs of receptors are necessary, implying the need of 3 receptors. In 3D, three pairs are needed, and a minimum of 4 receptors. The only data available to solve for the source spatial coordinates are the receptors spatial position and the best possible computation of TDOA between receptors pairs. In a 2D problem if we have two receptors and we compute a TDOA between them, it is a well known fact that the source capable to produce that delay must be placed over one of two symmetric hyperbolas, Figure 1. Because this is true for each pair, becomes clear that the source must be placed in the intersection of the hyperbolas of two different pairs. That is why this method is known as hyperbolic localization. HL for short.
The resulting system of equations is non linear. In 3D the hyperbolas become hyperboloids, a third coordinate appears as unknown, and one more pair of receptors is needed. This reasoning justifies the minimum number of receptors mentioned above. Of course, although the mathematical minimum is correct, in finite computations the pairs available can provide a numerically inadequate set of equations. To provide more pairs, and receptors, than necessary made available an ample set of equations from where to choose the adequate ones. Nevertheless, non linearity and equation redundancy are different issues that should not be confused.
For the sake of self consistency the equations of the HL problem are developed.
Be $s = \{x, y, z\}$ the unknown spatial position of the source. For each receptor m_i we have its position $\{xi, yi, zi\}$ and the vector $\vec{\mathbf{r_i}} = \vec{\mathbf{s}} - \vec{\mathbf{m_i}}$ that points from the receptor to the source. Assuming spherical sound propagation the following relationship is satisfied by each receptor pair:

$$r_i - r_j = d_{ij} = v\tau_{ij} \tag{1}$$

where d_{ij}, a signed quantity, is the difference between the distances of each receptor to the source, v is the sound propagation speed in the medium and τ_{ij} is the TDOA computed from the receptors registers. The τ_{ij} s are signed quantities too. Working over Equation 1, the

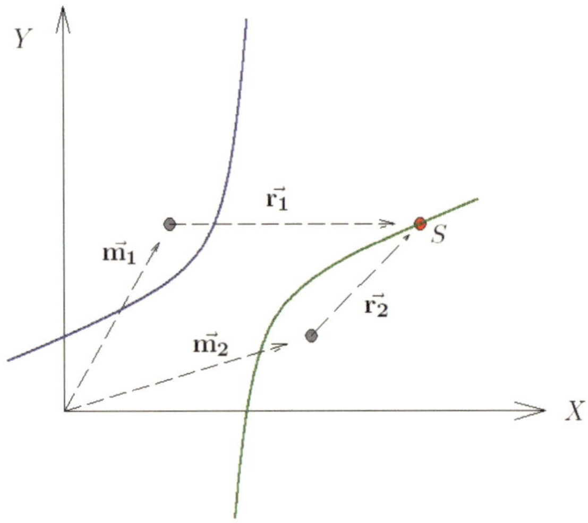

Fig. 1. A source positioned over the hyperbolas, irrespective of the distance, will produce the same TDOA absolute value. Which one is the involved hyperbola is determined by the TDOA sign.

following expression is obtained:

$$(x_i - x_j)x + (y_i - y_j)y + (z_i - z_j)z + d_{ij}r_j = \frac{m_i^2 - m_j^2 - d_{ij}^2}{2} \tag{2}$$

The same equation can be written for other two pairs. Assuming that the three pairs are constructed from three receptors the resulting system of equations is:

$$\begin{aligned}
(x_i - x_j)x + (y_i - y_j)y + (z_i - z_j)z + d_{ij}r_j &= 0.5(m_i^2 - m_j^2 - d_{ij}^2) \\
(x_k - x_l)x + (y_k - y_l)y + (z_k - z_l)z + d_{kl}r_l &= 0.5(m_k^2 - m_l^2 - d_{kl}^2) \\
(x_i - x_k)x + (y_i - y_k)y + (z_i - z_k)z + d_{ik}r_k &= 0.5(m_i^2 - m_k^2 - d_{ik}^2)
\end{aligned} \tag{3}$$

where

$$\begin{aligned}
r_q &= \sqrt{(x_q - x)^2 + (y_q - y)^2 + (z_q - z)^2} \\
m_q &= \sqrt{x_q^2 + y_q^2 + z_q^2} \text{ ; for } q = j, k, l
\end{aligned} \tag{4}$$

Equations 3 constitute a nonlinear system of equations and can be solved, iteratively, by traditional numerical methods. In 1987 many authors, in closely sequenced papers, presented a different way to obtain Equation 3 (Abel & Smith, 1987; Friedlander, 1987; H.C.Schau & Robinson, 1987). First they choose one of the receptors, for example receptor j, as a master receptor. This allows computing all the receptor-source distances as a function of the distance of the master receptor to the source. The values of d_{ij} are computed from the τ_{ij} and the medium propagation speed.

$$d_{jl} = r_j - r_l \Longrightarrow r_l = r_j - d_{jl} \tag{5}$$

Second, receptor m_j is renamed m_0 and r_j as r_0, obtaining

$$(x_i - x_j)x + (y_i - y_j)y + (z_i - z_j)z + d_{ij}r_0 = \frac{m_i^2 - m_j^2 - d_{ij}^2}{2} + d_{ij}d_{0j} \tag{6}$$

where r_0 is now the distance between the master receptor and the source, the so called range, computed as

$$r_0 = \sqrt{(x - x_0)^2 + (y - y_0)^2) + (z - z_0)^2} \tag{7}$$

In Equation 6, the unknowns still are $\{x, y, z\}$. One way to overcome the non linearity of the system was to introduce r_0 as a new unknown or parameter (Friedlander, 1987). The new unknown required the introduction of one more equation, expanding the original equations system. At that time nobody believed that the values of r_0 and $\{x, y, z\}$ obtained from the expanded system would satisfy Equation 7. It seems that nobody checked it either in the last 20 years. Because the clear non linear nature of Equation 7 many authors developed ways to solve the new expanded system by iterative methods (Chan & Ho, 1994).
The use of redundant pairs made it necessary to combine iterative methods with least square procedures, increasing the difficulty. In 2000, (Huang et al., 2000) found that the redundant system can be solved correctly in only one iteration. It was not noticed that it only can happen if the system is linear or if the initial guess in the nonlinear system is always coincident with the right solution.

2. The constant speed localization method, CSLM

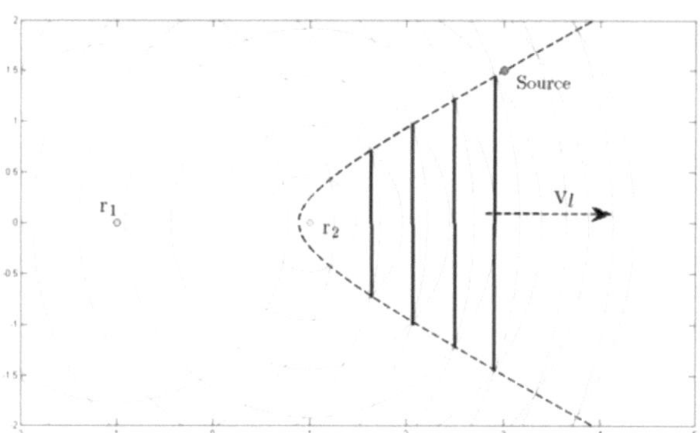

Fig. 2. Straight front propagation

In 2007 the authors (Militello & Buenafuente, 2007) presented a new way of interpreting the source localization problem, from now on CSLM (Constant Speed Localization Method). This allowed demonstrating that the problem could be transformed into a linear one by the mere fact of adding an additional receiver to the minimum required in the hyperbolic localization method. It was also shown that the work of Friedlander *et al.* and methods derived from it are special cases of the general case presented, making clear the linearity of the method. To explain the CSLM the receptors are considered to act as sources, each one emitting sound. But each one starts emitting in the inverse order they capture the sound from the source. In this way, all the wave fronts emitted will intersect the source at the same time.

Two receptors at a distance $2c$ from each other received the signal with a time delay t_a. For a sound speed v a spatial delay is defined as $2a = t_a v$. Now the two receptors start emitting with a time delay t_a. Both circles will intersect, and the successive intersections will describe a hyperbola. The hyperbola is symmetric with respect to the line joining the receptors and one of the branches will contain the source. But, if we join the successive intersection points with a straight line, as in Figure 2, a straight front can be identified. In (Militello & Buenafuente, 2007) it was proved that this front propagates with a constant speed $v_l = va/c$. Because of the straight front speed property the method is called Constant Speed Localization.
Each receptor pairs will produce one straight front propagating at a constant speed, and all the fronts will reach the source at the same time, i.e. all the constant speed traveling straight lines will intersect at the source position. In this way, a linear system of equations having as unknowns the source coordinates and the time of arrival can be constructed. The unknowns are clearly independent, and there is neither preferred coordinate system nor time origin. If one receptor position is considered as the coordinate centre, and the distance from this point to the source is called the range, the values of vt appearing in the equation can be substituted by r_0 and Friedlandert's equations are recovered. This is the only case where $R = vt = \sqrt{x^2 + y^2}$.
A detailed development of CSLM for 2D and 3D problems in its general form is presented in (Militello & Buenafuente, 2007). Here, for the sake of comparison, the equations are developed taking into account Friedlander's methodology and the following particular form is obtained:

$$(x_i - x_j)x + (y_i - y_j)y + (z_i - z_j)z + d_{ij}vt = \frac{m_i^2 - m_j^2 - d_{ij}^2}{2} + d_{ij}d_{0j} \tag{8}$$

To reach (8) the time origin is established as the time when receptor m_0 starts emitting. In the original CSLM method the time origin is the time when the furthest receptor starts emitting. Because the problem is linear in time and space, a time or a coordinate shift do not introduces changes in the solution nature.
Equations 6 and 8 are almost identical. The difference is that r_0 is replaced by vt. This replacement is consistent with the meaning of r_0 in Friedlandert's formulation and the meaning of the independent variable t in the CSLM formulation. Then r_0 is an independent variable because it can be obtained as the product of the independent variable t by the sound speed in the medium.
Now the linear nature of both methods and their equivalence has been established. Because a new independent variable appears, r_0 or t, one more equation is needed. The linear system can be solved by using a minimum of four sensors instead of three in a 2D problem and five sensors instead of four in a 3D problem. But the use of the correct number of sensors does not preclude the appearance of numerical errors when solving the system.
Something worth noting: in the CSLM method it is necessary to create a common time axis. It can only be done if the TDOA are not only computed between the active receptor pairs but also among one receptor, lets say a master one, and one of the receptors of each active pair. This is totally equivalent to Friedlandert's method when all the receptors positions are computed as a function of the position of the master receptor. Then, the computational work load involved in both methods is the same.

3. The design of the reception system

There are many variables and uncertainties in the design of a receptor system. To mention some of them the following list is proposed:
Uncertainties:

1. The error in TDOA estimations. This error depends on the ability to identify a specific perturbation introduced by the source in each sensor register and to assign a time to it. Or in the ability to compute the TDOA for a receptor pair.
2. The geometrical position of the receptor. Nowadays receptors are small in size and the pressure centre of a microphone can be determined with an error of the order of millimetres.

Design variables:

1. The spatial distribution of receptors.
2. The receptors chosen to constitute active pairs.

As it will be shown, the design variables will be responsible of the system performance. It will govern the way the effects of uncertainties are amplified in some detection scenarios and the quality of detection when the relative position of the source changes respect to our detection system.

3.1 Selecting the active pairs and the master receptor (time origin)

The study is focused in the way the design variables affects the source localization through the inevitable TDOA uncertainties. The superscript $^\circ$ is used to indicate the correct or exact values. They will be affected by an uncertainty value so that $\tau_{ij} = \tau_{ij}^\circ \pm e_{ij}$. By replacing it in (8) and rearranging terms:

$$(x_i - x_j)x^\circ + (y_i - y_j)y^\circ + (z_i - z_j)z^\circ + v\tau_{ij}^\circ vt^\circ - 0.5(m_i^2 - m_j^2 - v^2(d_{ij}^\circ)^2) = 0 \quad (9)$$

$$\pm v^2 e_{ij} t^\circ - 0.5 e_{ij}^2 \pm v\tau_{ij}^\circ e_{ij} + v^2(\tau_{ij}^\circ \tau_{0j}^\circ \pm \tau_{ij}^\circ e_{0j} \pm \tau_{0j}^\circ e_{ij} \pm e_{ij} e_{oj}) = \epsilon_{ij} \quad (10)$$

Equation 9 recasts Equation 8. Equation 10 is an error and can be seen as a contribution to the uncertainty value of the left hand side of the original equation system. Neglecting second order terms and adding up uncertainties an upper bound can be computed.

$$\epsilon_{ij} = v^2 \left(e_{ij}(t^\circ + \tau_{ij}^\circ + \tau_{0j}^\circ) + \tau_{ij}^\circ \tau_{0j}^\circ + \tau_{ij}^\circ e_{0j} \right) \quad (11)$$

This upper bound can be reduced if all the active pairs include the master receptor. In doing so $\tau_{00}^\circ = 0$. In this case Equation 11 can be further simplified to:

$$\epsilon_{i0} = v e_{i0}(vt^\circ + d_{i0}^\circ) \quad (12)$$

From this equation many conclusions can be drawn about the amplification of the TDOA inaccuracies. The main factors are:

1. The speed of sound in the medium.
2. The distance from the source.
3. The TDOA uncertainty.

In other words, for a given medium, the further the source the higher is the error. And, for a given set of receptors, it seems that the active pairs should be chosen so that one of the receptors appears in all the pairs and the distance between receptors is kept to a minimum.

4. Error propagation

Although the rules extracted in the preceding sections seems logical, they are not conclusive. This is due to the fact that in a linear problem the quality of the solution depends on the conditioning of the system of equations. In 3D the number of unknowns is four so that four pairs are needed. The system of equations gets the form $\mathbf{Mx} = \mathbf{b}$, where

$$\mathbf{M} = \begin{bmatrix} x_i - x_j & y_i - y_j & z_i - z_j & d_{ij} \\ x_k - x_l & y_k - y_l & z_k - z_l & d_{kl} \\ x_m - x_n & y_m - y_n & z_m - z_n & d_{mn} \\ x_p - x_q & y_p - y_q & z_p - z_q & d_{pq} \end{bmatrix} \quad (13)$$

$$\mathbf{x} = \begin{bmatrix} x & y & z & vt \end{bmatrix}^{\mathrm{T}} \quad (14)$$

$$\mathbf{b} = \frac{1}{2} \begin{bmatrix} m_i^2 - m_j^2 - d_{ij}^2 + 2d_{ij}d_{0j} \\ m_k^2 - m_l^2 - d_{kl}^2 + 2d_{kl}d_{0l} \\ m_m^2 - m_n^2 - d_{mn}^2 + 2d_{mn}d_{0n} \\ m_p^2 - m_q^2 - d_{pq}^2 + 2d_{pq}d_{0q} \end{bmatrix} \quad (15)$$

and the solution is

$$\mathbf{x} = \mathbf{M}^{-1}\mathbf{b} \quad (16)$$

provided that the inverse of $\mathbf{M}$ exists. Notice the use of eight different sensors, which is the most general case to construct the system. But, as one sensor can be part of many pairs, this number can be reduced to five. Because of the uncertainties pointed up before matrices $\mathbf{M}$ and $\mathbf{b}$ are perturbed. As before only TDOA uncertainties are considered. The real equation system becomes

$$(\mathbf{M} + \delta\mathbf{M})\,\hat{\mathbf{x}} = (\mathbf{b} + \delta\mathbf{b}) \quad (17)$$

being $\hat{\mathbf{x}}$ an approximation to the exact solution.

$$\hat{\mathbf{x}} = \mathbf{x}^{\circ} + \delta\mathbf{x} \quad (18)$$

Because the system is linear, perturbation theory can be applied in order to obtain a bound to the expected error in the system solution. The relative solution error will satisfy:

$$\frac{\|\delta\mathbf{x}\|}{\|\mathbf{x}^{\circ}\|} \leq \frac{\mathrm{cond}(\mathbf{M})}{1 - \mathrm{cond}(\mathbf{M})\frac{\|\delta\mathbf{M}\|}{\|\mathbf{M}\|}} \left(\frac{\|\delta\mathbf{M}\|}{\|\mathbf{M}\|} + \frac{\|\delta\mathbf{b}\|}{\|\mathbf{b}\|} \right) \quad (19)$$

where $\mathrm{cond}(\mathbf{M})$ is the matrix condition number defined as:

$$\mathrm{cond}(\mathbf{M}) = \|\mathbf{M}\| \left\|\mathbf{M}^{-1}\right\| \geq 1 \quad (20)$$

where $\|\cdot\|$ is a matrix norm, usually the l_2 norm. In a badly conditioned system the $\mathrm{cond}(\mathbf{M})$ is bigger than 1. If it is assumed that the perturbed matrices have a small norm and $\mathrm{cond}(\mathbf{M})$ is not a big number, (Moon & Stirling, 2000), the relative error in system solution can be approximated by

$$\frac{\|\delta\mathbf{x}\|}{\|\mathbf{x}^{\circ}\|} \leq \mathrm{cond}(\mathbf{M}) \left(\frac{\|\delta\mathbf{M}\|}{\|\mathbf{M}\|} + \frac{\|\delta\mathbf{b}\|}{\|\mathbf{b}\|} \right) + O(e^2) \quad (21)$$

Being e the order of magnitude of the TDOA uncertainty. From Equation 21 it can be seen that the relative error in the system solution can be approximated as the sum of the relative error in the matrix plus the relative error in the independent term, amplified by the condition number. In order to clarify the effect of this equation in the results two examples are presented.

4.1 Directivity of a given sensor configuration

In this context the term "directivity" is defined as $1/\text{cond}(\mathbf{M})$, having a maximum value of 1, and is used to point how a given sensor configuration will amplify the uncertainties from a source placed over a circle around the designed master receptor. Matrix **M** has three columns that can be evaluated from the receptors coordinates, but the fourth one depends on the relative positions of source and receptors pairs, the TDOA. Matrix **M** can be easily constructed from any expected source position and its condition evaluated. Following Equation 21 the value $1/\text{cond}(\mathbf{M})$ can be seen as a directivity property. A high value in a given direction indicates that direction as a preferred one with small uncertainty amplification.

Simulation A.

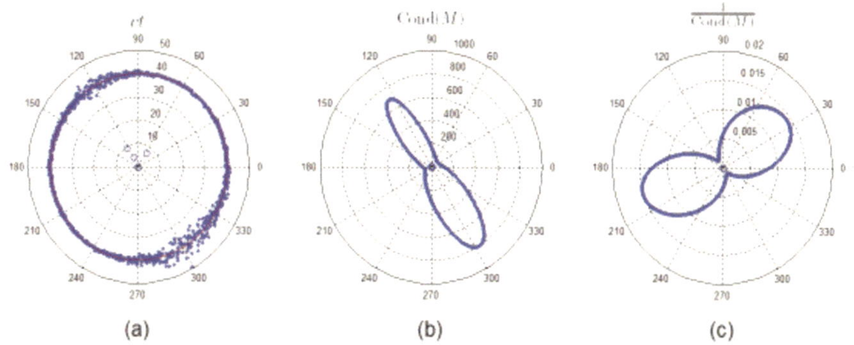

Fig. 3. **Simulación A**. (a) A starting receptors configuration and range computation with CSLM. (b) Matrix **M** condition showing the lobes responsible of error amplification. (c) Receptors array directivity, minimum directivity in the maximum error propagation direction.

A set of receptors are positioned: $m_0\{0,0\}$, $m_1\{-5,8\}$, $m_2\{4,6\}$, and $m_3\{-2,4\}$. The receptors pairs are $\{m_0, m_1\}$, $\{m_0, m_3\}$ and $\{m_0, m_2\}$. It must be noticed that receptors m_0, m_1 and m_3 seems to be over a straight line at 120° from the X axis but they are not. If they are over the same line the system is singular and can not be inverted. A circle of radius 40 m centered at m_0 is drawn and 1000 sources uniformly distributed over it. For each source exact, within machine precision, quantities are computed. The exact TDOAs are computed and perturbed with a random Gaussian error distribution. The error standard deviation is set to 10us. The values of vt computed for each source are plotted in Figure 3(a). Figure 3(b) plots the computed matrix condition and clearly shows the coincidence of big condition values with high source localization error. An amplification factor of 800 can be seen at 300°. Figure 3(c) is the directivity, showing a big value in the directions where the computed error will be low. From the traveling straight front point of view a wrong selection of receptors pairs will produce almost parallel lines, making it difficult to compute their intersection. Why the 120° direction produces less dispersion than the 300° one? It will be explained latter.

Simulation B

A robust configuration is defined as the one with not pronounced directivity lobes. Under this point of view the best one will be the one with no lobes and a directivity value near one. In order to achieve this receptors are placed in the vertex of an equilateral triangle and the master receptor is placed at the triangle centre of gravity, Figure 4. The triangle side is $4\sqrt{3}$

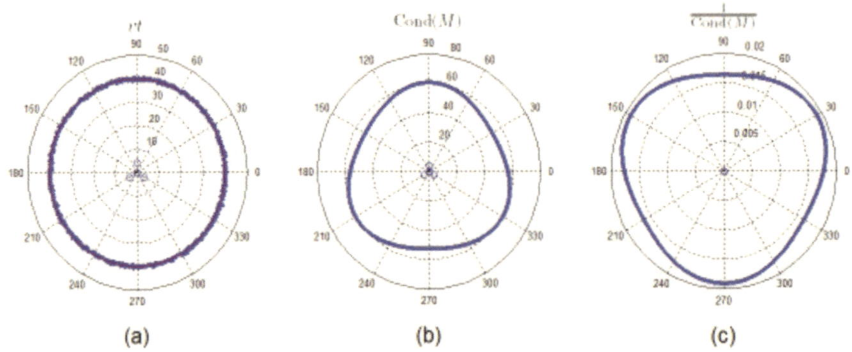

Fig. 4. **Simulación B**. A centred triangle configuration. a)Computed range with CSLM. b)Matrix M condition. c) Directivity.

m. The TDOA uncertainties are computed in exactly the same manner as in Simulation A. It can be seen that three lobes appear with a very uniform shape. The directivity is uniform too. It should be noticed that a directivity number better than 0.02 is not achieved for this configuration. Simulation B shows how with the same computational and hardware costs a better system can be constructed. The matrix condition number increases as the distance to the source increases. The ideal number of 1 is hard to get. For the triangular configuration of SIMULATION B a condition number of 1.4 is obtained for a source placed at the triangle centre, in top of the master receptor.

5. An upper bound for the solution error

When designing a reception system the effect of TDOA error in system performance is capital. All the electronics and computational effort used in reducing this uncertainty will have a direct impact in localization. Equation 21 provides an easy way to predict the value of uncertainty necessary for a desired performance. Assuming no error in receptors positions the perturbed matrix can be written as

$$\delta\mathbf{M} = \begin{bmatrix} 0 & 0 & 0 & e_{ij}v \\ 0 & 0 & 0 & e_{kl}v \\ 0 & 0 & 0 & e_{mn}v \\ 0 & 0 & 0 & e_{pq}v \end{bmatrix} \tag{22}$$

where e_{ij} is the error in computing the TDOA for each receptors pair. The maximum value for e_{ij} is set to $e_{\max}$. The l_1 norm is computed for this matrix obtaining a bound for the perturbed matrix:

$$\|\delta\mathbf{M}\| < nve_{\max} \tag{23}$$

In Equation 23 n is the number of receptor pairs.
To compute an upper bound to $\|\delta\mathbf{b}\|$ it must be recalled that $d_{ij} = d_{ij}^{\circ} + ve_{ij}$. The perturbed $\mathbf{b}$ can be written as:

$$\delta\mathbf{b} = -\frac{v^2}{2} \begin{bmatrix} e_{ij}^2 + 2\tau_{ij}e_{ij} \\ e_{kl}^2 + 2\tau_{kl}e_{kl} \\ e_{mn}^2 + 2\tau_{mn}e_{mn} \\ e_{pq}^2 + 2\tau_{pq}e_{pq} \end{bmatrix} \tag{24}$$

Now, if e_{ij} is neglected with respect to τ_{ij} (remember that τ_{ij} is the TDOA and e_{ij} the error in computing it. It is assumed that $e_{ij} << \tau_{ij}$), e_{ij} is bounded by $e_{\max}$ and d_{ij} is bounded by $D = d_{ij}^{\max}$, :

$$
\begin{aligned}
v^2 e_{ij}(e_{ij+2\tau_{ij}}) &\approx v^2 e_{ij} 2\tau_{ij} \\
&< 2ve_{\max} d_{ij} \\
&< 2ve_{\max} D
\end{aligned} \tag{25}
$$

Then, an upper bound for the perturbation is

$$\|\boldsymbol{\delta}\mathbf{b}\|_1 < nve_{\max} D \tag{26}$$

From (25) and (26) the relative error in source positioning can be bounded:

$$\frac{\|\boldsymbol{\delta}\mathbf{x}\|}{\|\mathbf{x}^{\circ}\|} < nve_{\max} \left(\frac{1}{\|\mathbf{M}\|} + \frac{D}{\|\mathbf{b}\|} \right) \operatorname{cond}(\mathbf{M}) \tag{27}$$

Finally the value of $e_{\max}$ can be computed from it:

$$e_{\max} = \frac{\Delta R}{Rnv \left(\frac{1}{\|\mathbf{M}\|} + \frac{D}{\|\mathbf{b}\|} \right) \operatorname{cond}(\mathbf{M})} \tag{28}$$

The values of the range R and its allowed uncertainty ΔR must be introduced and matrices $\mathbf{M}$ and $\mathbf{b}$ must be computed. The following examples will show how Equation 28 can be used.

Simulation C

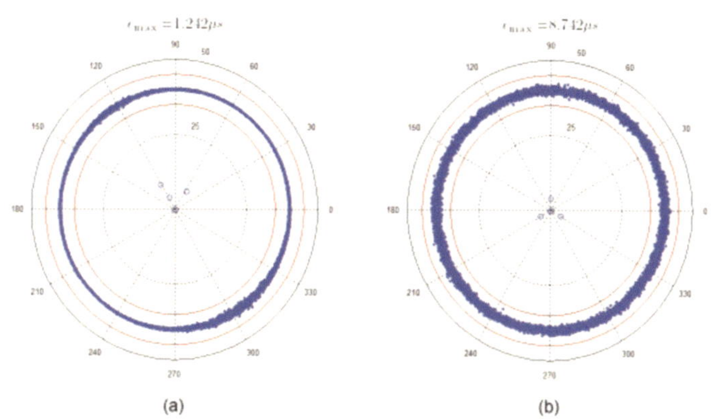

Fig. 5. **Simulation C**. 1000 sources are localized around the receptors. The red circles show the allowed error bound of ± 5 m.

For the examples the two configurations studied in simulations A and B are used. The range was 40 m and an uncertainty of $\pm$ 5 m ($\Delta R = 5$) is introduced. From (28) the values of $e_{\max}$ are computed for the 1000 sources equally spaced. The smallest value, $e_{\max}^{\min}$, imposes the hardware and software quality. Now the TDOAs are perturbed with a random Gaussian error with a standard deviation equal to $e_{\max}^{\min}$. The source position is computed. The results for both configurations are depicted in Figure 5. Configuration A needs an $e_{\max}^{\min}$ equal to 1.242 μs to keep positioning for the worst conditions within bounds. Configuration B can do the job with $e_{\max}^{\min}$ equal to 8.742 μs. All the sources are localized within bounds. It should be noticed the low values of $e_{\max}^{\min}$ needed to ensure an error of $\pm$ 5m in a 40 m range. To the best of the authors' knowledge is the first time this kind of quantification can be done a priori.

Simulation D

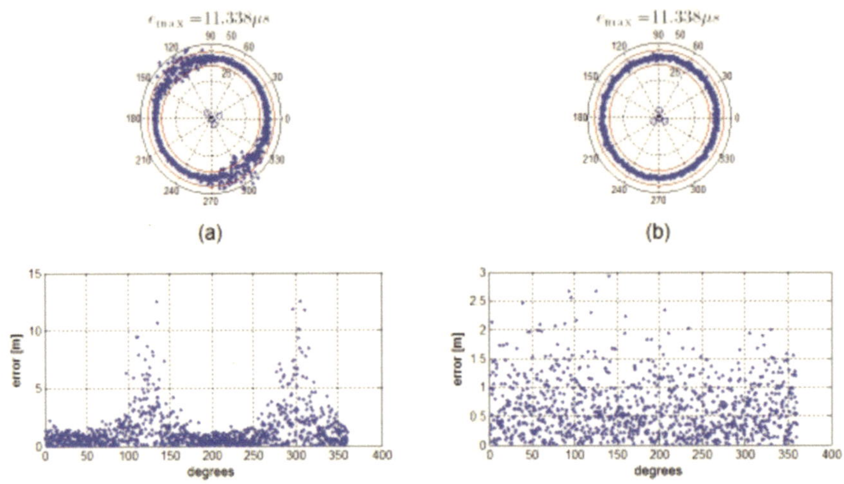

Fig. 6. Simulación D. Ascertaining whether the acquisition frequency is adequate or not. a) The acquisition frequency of 44100 Hz is not enough. b) 44100 Hz is enough.

One of the methods used for TDOA computation is the *Generalized Cross Correlation*, GCC (Knapp & Carter, 1976). In this method the uncertainty is bounded by the acquisition frequency used for the signal. For a given acquisition frequency a lower uncertainty can be achieved by the use of interpolation techniques (Tervo & Lokki, 2008), or regressive techniques (Brandstein & Silverman, 1997). With Equation 28 it can be established if interpolation is needed or not. For SIMULATION C the signal sampling frequency is 44100 Hz. A Gaussian noise with standard deviation equal to $\frac{1}{2}\Delta t = \frac{1}{2*44100}$ is added to the TDOA exact values. The resulting system is solved for each source position. The results can be seen in Figure 6. For configuration A, in order to keep the error within bounds it is necessary to use interpolation algorithms. Configuration B will do the job using the GCC algorithm alone.
It should be noticed that configuration A do not present noticeable differences at 120° and 300° as it did before. The only change is that receptor m_3 is used as coordinate centre instead of m_0.

6. 3D examples

If uniform directivity is considered a desirable property for a detection system the goal must be to achieve it with the minimum hardware and computational work, i.e: receptors and receptors pairs involved. A tetrahedron with a receptor at the geometrical centre is proposed as a guess, Figure 7(a). It is not a blind guess because of the properties shown by the centred equilateral triangle array (see Figure 4). The tetrahedron is a five receptors array, the minimum required. The receptor at the centre is designed as the master. All pairs include the master receptor. The distance from the centre to the corner is 1m. Figure 7(b) shows the directivity

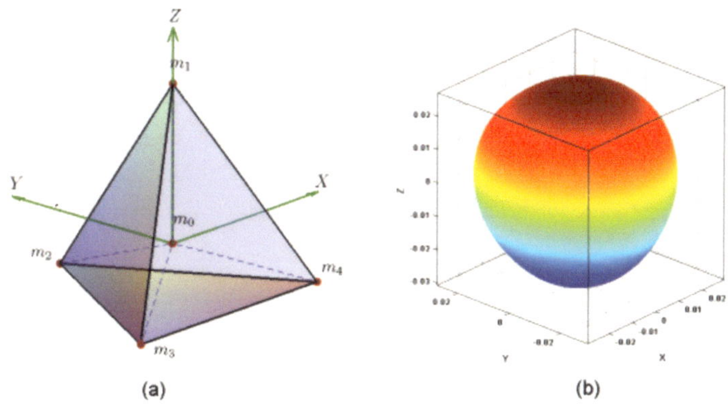

Fig. 7. 3D examples. (a) Receptors spatial configuration. All receptors pairs include the master receptor m0 (b) System directivity

pattern for sources located in a surrounding sphere of radius 10m. The maximum directivity value is 0.017.

The system is perturbed by changing the receptors pairs. The centre receptor is still used as the master. The receptors pairs are depicted by the solid lines in Figure 8(a–c). The corresponding computed directivity pattern is shown in Figure 9(a–c). The results are astonishing.

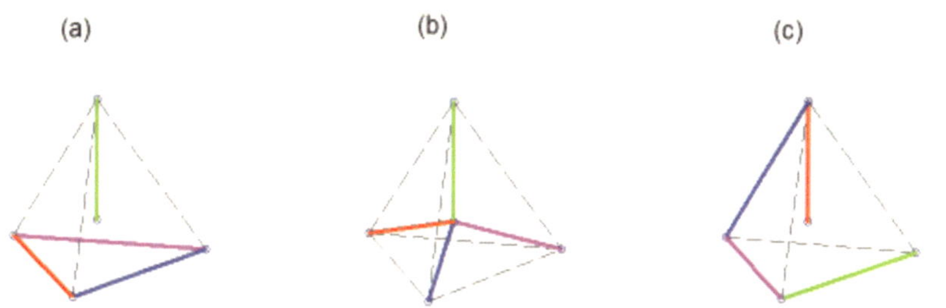

Fig. 8. Three detection systems made from the same receptors but choosing different pairs. The master receptor is always the one in the tetrahedron centre.

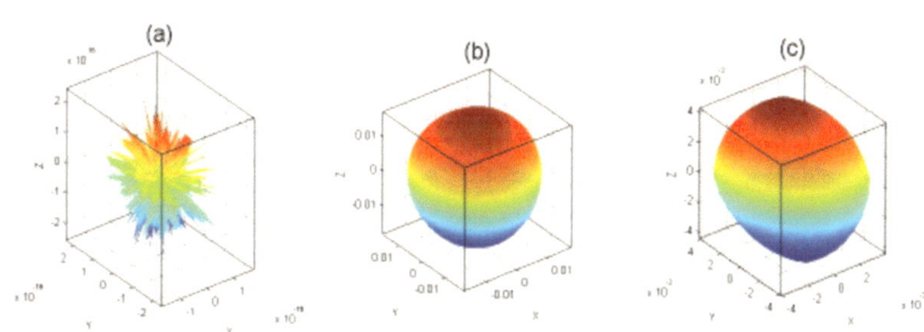

Fig. 9. Directivities computed for the three receptor systems from Figure 8.

The conclusion is that for the same hardware configuration receptors pairs are paramount to determine system directivity.

An alternative is the six receptors arrays of Figure 10(a). The distance of each receptor to the centre is 1 m. The master receptor is the one on top. Three pairs are constructed from the obvious on axis locations. A fourth pair is constructed with two corners from different axes. One more receptor is used. Eight directivity lobes in the axis direction can be seen, Figure 10(b). The maximum directivity is 0.01. Although the fourth receptor pair selection breaks symmetry the directivity pattern is symmetric.

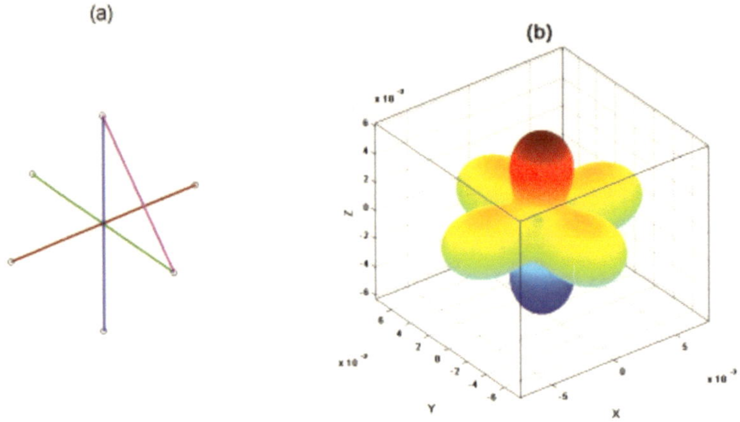

Fig. 10. Six receptors arrangement. a) Selected pairs, b) Computed directivities.

7. Experiment dimensions and effectiveness forecast

7.1 Localization errors as a function of TDOA uncertainties

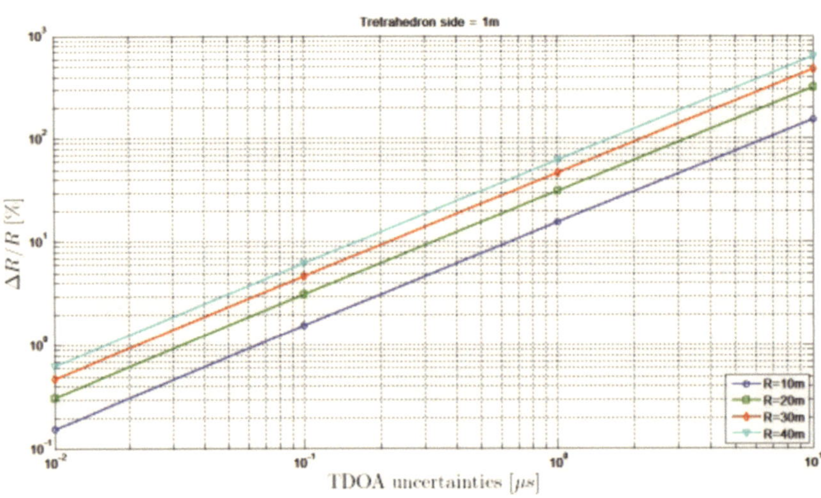

Fig. 11. Relative localization error for a 1 meter tetrahedron array side. TDOA uncertainties of 0.01, 0.1, 1 and 10 microseconds are considered for source distances of 10, 20, 30 and 40 m.

The starting experimental setting is a tetrahedron array with 1m side. For a source position at $r[\frac{1}{\sqrt{3}}, \frac{1}{\sqrt{3}}, \frac{1}{\sqrt{3}}]$ with $r = 10, 20, 30$ and 40 meters and TDOA uncertainties of 0.01, 0.1, 1 and 10 microseconds, by using Equation 28 the relative positioning error is computed. The results are plotted in Figure 11. For $1\mu s$ uncertainty the relative error in localizing a source at 20 m is 30%. That is 6 m. It can be seen that in order to reduce the localization uncertainty one order of magnitude, the TDOA uncertainty must be reduced one order of magnitude too.

7.2 The effect of receptors arrangement size

Assuming that the arrangement dimensions can be chosen freely Equation 28 is now computed for the same values of TDOA uncertainties but changing the tetrahedron sides to 0.01, 0.1, 1 and 10 meters. The source is placed at 20 metres. Results are plotted in Figure 12. For the same value of TDOA uncertainties, increasing the side one order of magnitude reduces the localization error in two orders of magnitude.

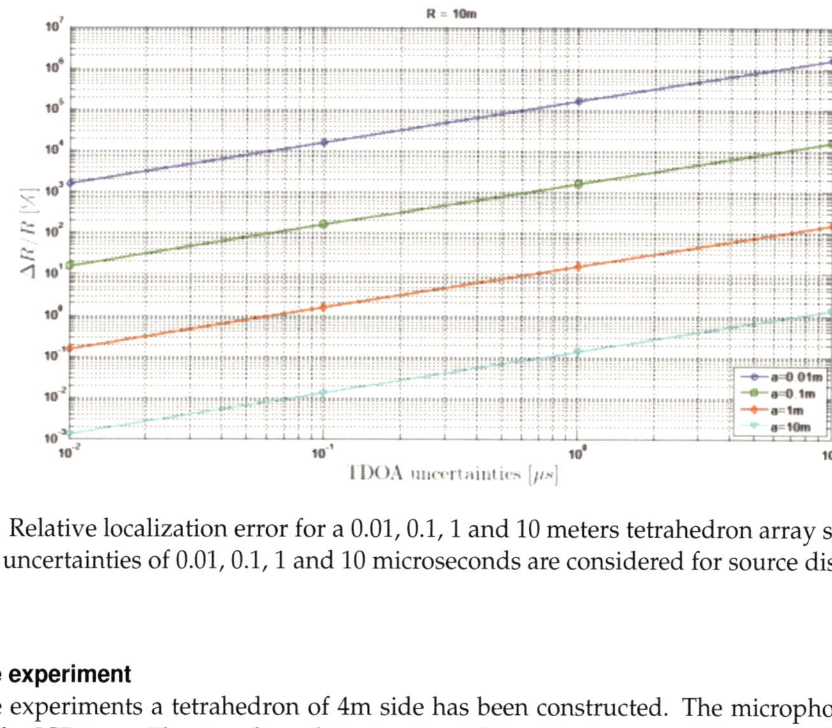

Fig. 12. Relative localization error for a 0.01, 0.1, 1 and 10 meters tetrahedron array side. TDOA uncertainties of 0.01, 0.1, 1 and 10 microseconds are considered for source distance of 10 m.

7.3 The experiment

For the experiments a tetrahedron of 4m side has been constructed. The microphones used are of the ICP type. The signal conditioning is a PCB with a low pass filter set at 10 Khz. A KHEITLEY USB ®, 16 bits, card attached to a portable PC is used as A/D converter.

An acoustic gun shot is used as the source. Acquisition frequency is set to 100 KHz. TDOA are computed by using the GCC algorithm.

From Figure 11, to localize a source with an upper bound of 17% relative error, a TDOA uncertainty of $10\mu s$, approximately, is needed. A GCC method will produce an uncertainty determined by the acquisition frequency, i.e., 10 us.

The complete experiment is mounted in a football stadium, Figure 13. The sources are placed over a 20m circle around the receptors arrangement. To install the receptors the following procedure is followed. One long vertical stick carries the central and the top microphones. Three short sticks with a microphone at the end are placed around the long one on a circle

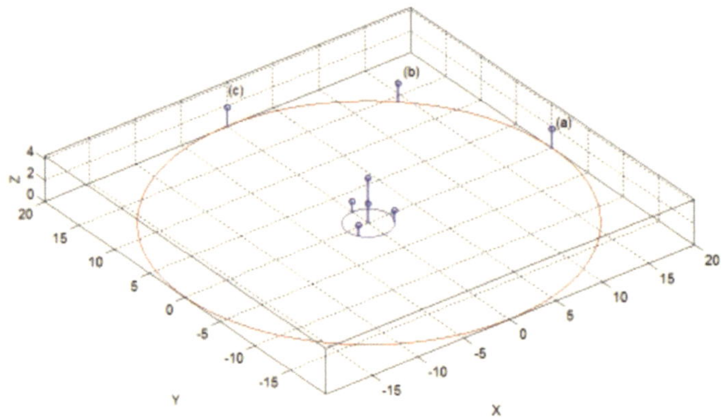

Fig. 13. Receptors array and source position for the experimental setting.

of radius $\frac{4}{\sqrt{3}}$. They position the lower microphones at 1m over the floor. The source height is coincident with the height of the centre receptor. Positioning of the source, end of gun barrel, with respect to the receptors centre is made with an estimated error of $\pm 5cm$, which is less than 0.6% of the radius. Positioning of the receptors is checked with a theodolite Model Wild-Leica-T2 of 1 second precision.

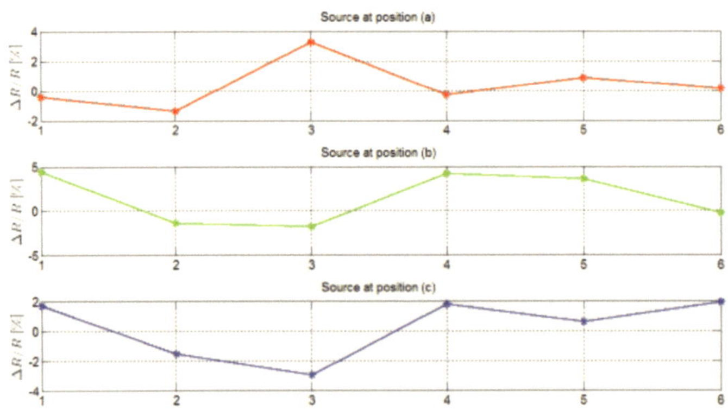

Fig. 14. Error computed for three gun shots at each source location. Notice that all of them are within the prescribed bound of 8%.

For each position the experiment is repeated six times. Figure 14 shows the results. It can be seen that the relative error remains within the 5% level, which is one third of the upper bound forecasted of 17%. The safety margin is in agreement with the ones that can be seen in Figure 5 (b).

8. Discussion, conclusions and future research

Nowadays a DSP can carry on thousands of operation per second. At first glance to solve the localization problem in one step or ten could be considered irrelevant. To add one more

receptor does not seem a big deal because redundancy is a common practice. To solve the HL non-linear original problem or the linear expanded one can be a matter of taste. But it is not. A linear system allows using well known, well established error propagation methods. Equation 28 is an invaluable tool for the one in charge to design a source localization system. For a given array a directivity pattern can be computed and observed easily with software like MATLAB®. Plots like the ones in Figures 9 and 10 will help in designing the acquisition system. The experiment shows that the upper bound computed is reliable.
Three points, among others, have not been reviewed in this work: the uncertainty in receptors position, the effect of using redundant pairs and adequate receptor pair selection for a given receptor geometry.
It is clear from this work that matrix **M** condition is important. It can be computed if the receptors pattern, receptors pairs and source position are known. The condition does not depend on geometrical or TDOA uncertainties. Geometrical uncertainties will add or will establish the upper bound for the $\boldsymbol{\delta}\mathbf{M}$ matrix norm. A rule of thumb is that receptors position uncertainties must be in the order of $v \cdot e_{max}$. For a time uncertainty of $10\mu s$ in air, the number is 0.35 cm. For a high frequency acquisition and very low errors in TDOA the ability to correctly position the receptors centre will impose the limits.
The use of redundant pairs seems plausible. At first glance it can be imagined that many of the selected pairs will produce a better problem conditioning or a more robust pseudoinverse for a given source location. But a meaningful error reduction can be obtained only if the condition is improved.
It has been shown that pair selection for a given receptors array is paramount. Recently (Gillette & Silverman, 2008) produced a redundant system by introducing more equations. The equations do not came from the introduction of more receptors but for arranging new pairs with the same existing receptors. In the author's opinion the same reasoning of the previous paragraph can be done.
This work does not give a guideline on receptors orientation and preferred receptors pairs. Research is carried on in order to develop a rationale to reduce the conditioning and is a matter of future research. It is worth noting the work of (Yang & Scheuing, 2005) as an effort to find a good receptors distribution geometry.
This work do shows a 2D and a 3D robust detection system and a simple way to validate them or any other configuration.

9. References

Abel, J. & Smith, J. (1987). The spherical interpolation method for closed–form passive source localization using range difference measurements, *ICASSP–87*, pp. 471–474.

Brandstein, M. S. & Silverman, H. F. (1997). A robust method for speech signal time-delay estimation in reverberant rooms, *in Proceedings of ICASSP*, pp. 375–378.

Chan, Y. & Ho, K. (1994). A simple and efficient estimator for hyperbolic location, *IEEE Trans. Signal Processing* 42: 1905–1915.

Friedlander, B. (1987). A passive localization algorithm and its accuracy analysis, *IEEE J.Ocean. Eng.* OE–12(No.): 234–245.

Gillette, M. D. & Silverman, H. F. (2008). A linear closed-form algorithm for source localization from time-differences of arrival, *IEEE Signal Processing Letters* 15: 1–4.

H.C.Schau & Robinson, A. (1987). Passive source localization employing intersecting spherical surfaces from time–or–arrival differences, *IEEE Trans. Acoust., Speech, Signal Processing* ASSP–35: 1223–1225.

Huang, Y., Benesty, J. & Elko, G. W. (2000). Passive acoustic source localization for video camera steering, *ICASSP '00: Proceedings of the Acoustics, Speech, and Signal Processing, 2000. on IEEE International Conference*, IEEE Computer Society, Washington, DC, USA, pp. II909–II912.

Knapp, C. H. & Carter, G. C. (1976). The generalized correlation method for estimation of time delay, *IEEE Trans. Acoust., Speech, Signal Processing* 24: 320–3327.

Militello, C. & Buenafuente, S. (2007). An exact noniterative linear method for locating sources based on measuring receiver arrival times, *J. Acoust. Soc. Am.* 121(6): 3595–3601.

Moon, D. K. & Stirling, W. C. (2000). *Mathematical Methods and Algorithms for Signal Processing*, Prentice Hall, New Jersey.

Tervo, S. & Lokki, T. (2008). Interpolation methods for the srp-phat algorithm, *In The 11th International Workshop on Acoustic Echo and Noise Control (IWAENC2008).*

Yang, B. & Scheuing, J. (2005). Cramer-rao bound and optimum sensor array for source localization from time difference of arrival, *ICASSP '05: Proceedings of the Acoustics, Speech, and Signal Processing, 2005. on IEEE International Conference*, IEEE Computer Society, Washington, DC, USA, pp. IV961–IV964.

2

Direction-Selective Filters for Sound Localization

Dean Schmidlin
El Roi Analytical Services
United States of America

1. Introduction

An important problem in sound localization is the determination of the polar and azimuthal angles of far-field acoustic sources. Two fundamental approaches to the solution can be identified: spatial filtering (beamforming) and the parameter estimation approach. Van Veen and Buckley (1988) and Krim and Viberg (1996) give comprehensive reviews of the first and second approaches, respectively. Spatial filtering was carried out by an array of pressure sensors. A serious drawback to the filtering approach is that its performance depends directly on the physical size of the array (aperture), regardless of the data gathering time and signal-to-noise ratio. This aperture dependence together with more demanding applications motivated a good number of researchers to develop parametric estimation techniques. These methods can be separated into two main categories, namely, spectral-based and parametric approaches. The most famous example of the first is MUSIC (MUltiple Signal Classification) algorithm developed by Schmidt (1981) and Bienvenu and Kopp (1980), and of the second is the Maximum Likelihood (ML) method developed by Kumaresan and Shaw (1985) and Bresler and Macovski (1986).

In contrast to beamforming techniques, a MUSIC estimate of arbitrary accuracy can be achieved if the data gathering time is sufficiently long, the SNR high enough, and the signal model sufficiently accurate. However, a significant limitation is the inability to resolve closely spaced signals with small sample sizes and low SNR. Further deterioration occurs for highly correlated signals and complete breakdown for coherent signals. The interested reader is referred to Krim and Viberg (1996) for discussions on how these limitations have been addressed.

All of the methods for localizing acoustic sources had one thing in common. They used arrays composed of pressure sensors. This continued until Nehorai and Paldi (1994) introduced a new type of sensor called the vector sensor. An acoustic vector sensor measures the acoustic pressure and all three components of the acoustic particle velocity at a single point in space. The extra information provided by the vector sensor opened the door to improved source localization accuracy without increase in array aperture. Vector-sensor models and fundamental processing techniques were developed by Nehorai and Paldi (1994) and Hawkes and Nehorai (2000) for the case of sensors located away from and in the presence of a reflecting boundary, respectively. Parametric techniques that had been designed for arrays of pressure sensors were adapted to vector sensors. For example, Wong and Zoltowski (1999), (2000) introduced Root-MUSIC-based and MUSIC-based source localization algorithms for vector sensors. Theoretical and technological development of the vector sensor also revitalized interest in using spatial filtering (beamforming) to localize acoustic sound sources (D'Spain *et al.*, 1992; Hawkes and Nehorai, 1998; Wong and Chi, 2002; Zou and Nehorai, 2009).

D'Spain et al. (2006) pointed out that the Taylor series expansion of the acoustic pressure field about a single point in space provides the theoretical basis for array processing with measurements at a single point in space. Since the particle velocity is proportional to the gradient of the pressure, the vector sensor provides information for the first two terms in the Taylor series. Silvia et al. (2001) used the Taylor series to define a general class of directional acoustic receivers based on the number of series terms measured by the receiver. Based on this definition, a pressure sensor is considered as a directional acoustic sensor of order zero, and a vector sensor is referred to as a directional acoustic sensor of order one. Silvia (2001) performed a theoretical and experimental investigation of an acoustic sensor of order two. It was given the name "dyadic sensor" because in addition to measuring the pressure and the gradient of the pressure, it also measures the dyadic of the pressure. Cray (2002) and Cray *et al.* (2003) presented theory for acoustic receivers of order greater than two. Schmidlin (2007) extended the multichannel filtering approach of Silvia (2001) to directional acoustic sensors of arbitrary order ν. It was shown that the maximum directivity index is $20\log(1+\nu)$, and explicit expressions were derived for the optimum weights.

The primary interest in "beamforming from a single point in space" is the achievement of high directivity with a sensor system occupying a smaller area of space than the conventional pressure array. However, it is very difficult to physically measure the higher-order spatial partial derivatives of the pressure. This led to indirect means for measuring these derivatives. Hines *et al.* (2000) used the method of finite differences to implement a superdirective line array and Schmidlin (2010a) introduced a distribution theory approach for implementing directional acoustic sensors. Another difficulty with highly directional receivers is sensitivity to uncorrelated system noise (Hines and Hutt, 1999; Hines *et al.*, 2000; Cray, 2001). System noise includes pre-amplifier voltage noise, inter-channel imbalance in gain and/or phase, sensor spacing errors, acoustic scatter and hydrophone self-noise due to hydrodynamic flow past the sensors.

In the theory of digital filters, causal FIR filters and IIR filters have transfer functions that are polynomial functions and rational functions, respectively, of the complex variable z^{-1}. The primary advantage of IIR filters over FIR filters is that they usually satisfy a particular set of specifications with a much lower filter order than a corresponding order FIR filter. This paper uses this advantage as the starting point for generating direction-selective filters. Directional acoustic sensors have beampatterns that are polynomial functions of the direction cosine $\cos\psi$. The direction-selective filters presented herein have beampatterns that are rational functions of $\cos\psi$. Section 2 analyzes a first-order filter prototype, develops the concept of a discriminating function, and derives an expression for its directivity index. In Section 3, prototypical filters are connected in parallel to realize rational discriminating functions, and a detailed example is presented. It is also shown that a discriminating function can be designed from the magnitude-squared response of a digital filter. Section 4 summarizes the contents of the paper and discusses future research.

2. Direction-selective filters tuned to the look direction

2.1 Vector sensor as a direction-selective filter

A plane wave is traveling towards the origin of a rectangular coordinate system. Located at the origin is a directional acoustic sensor. If this sensor is a vector sensor then the expression for the linear beamformer output for the look direction u_L is given by (D'Spain *et al.*, 2006)

$$p_o(t) = a_0 p(t) + \rho_0 c \sum_{j=1}^{3} a_j \left[\upsilon_j(t) + n_j(t) \right] \cos\beta_j \tag{1}$$

The components of the look direction are the direction cosines $\cos\beta_j, j = 1,2,3$ where

$$\begin{aligned} \cos\beta_1 &= \cos\theta_L \sin\phi_L \\ \cos\beta_2 &= \sin\theta_L \sin\phi_L \\ \cos\beta_3 &= \cos\phi_L \end{aligned} \tag{2}$$

where the angles θ_L is the azimuthal angle and ϕ_L the polar or zenith angle. The time function $p(t)$ is the acoustic pressure at the origin and $\upsilon_j(t), j = 1,2,3$ the three orthogonal components of the acoustic particle velocity. The function $n_j(t)$ represents the self-noise at the $j-$th velocity sensor and $\rho_0 c$ the characteristic impedance of the medium. Ignoring the self-noise at each velocity sensor and letting $a_j = -a_\upsilon$ for $j = 1,2,3$ simplifies Eq. (1) to

$$p_o(t) = a_0 p(t) - \rho_0 c a_\upsilon \mathbf{v}(t) \cdot \mathrm{u}_L \tag{3}$$

The particle velocity at time *t* and position ***r*** is related to the pressure as follows (Ziomek, 1995)

$$\mathbf{v}(t,\mathrm{r}) = -\frac{p(t + \mathrm{u} \cdot \mathrm{r}/c)}{\rho_0 c}\mathrm{u} \tag{4}$$

Setting ***r*** to 0 and placing the result into Eq. (3) results in

$$p_o(t) = (a_0 + a_\upsilon \mathrm{u} \cdot \mathrm{u}_L) p(t) \tag{5}$$

The unit-vector u points in the direction of the arriving plane wave and the unit-vector u_L points in the look direction. The scalar product $\mathrm{u} \cdot \mathrm{u}_L$ is equal to the cosine of the angle ψ between them. If

$$g_{\mathrm{u}_L}(\psi) \equiv a_0 + a_\upsilon \mathrm{u} \cdot \mathrm{u}_L = a_0 + a_\upsilon \cos\psi \tag{6}$$

then the output of the linear beamformer is expressed as

$$p_o(t) = g_{\mathrm{u}_L}(\psi) p(t) \tag{7}$$

The function $g_{\mathrm{u}_L}(\psi)$ has some selectivity with regards to the direction of the plane wave and is generally referred to as the beampattern of the vector sensor. If the pair of weights are given the assignments

$$a_0 = -\frac{b}{1-b}, a_\upsilon = \frac{1}{1-b} \tag{8}$$

then

$$g_{\mathrm{u}_L}(\psi) = \frac{\cos\psi - b}{1-b} \tag{9}$$

The angle ψ goes from 0 to π. When $\psi = 0$, the plane wave is arriving in the look direction and $g_{\mathrm{u}_L}(0) = 1$. When $\psi = \pi$, the plane wave is arriving in a direction opposite to the look direction and

$$g_{\mathrm{u}_L}(\pi) = -\frac{1+b}{1-b} \tag{10}$$

Since it is desired that $\left|g_{\mathrm{u}_L}(\pi)\right| < \left|g_{\mathrm{u}_L}(0)\right| = 1$, the value of b must be negative. If the magnitude of b is not greater than 1, then $g_{\mathrm{u}_L}(\psi) = 0$ at $\psi = \cos^{-1} b$ and the vector sensor will have nulls. It has been shown by the author (2010b) that the null directions are given by

$$\mathrm{u} = \left(\mathrm{u}_1 \cos\zeta + \mathrm{u}_2 \sin\zeta\right)\sqrt{1-b^2} + b\mathrm{u}_L \tag{11}$$

where $0 \le \zeta < 2\pi$ and

$$\mathrm{u}_L = \begin{bmatrix} \cos\theta_L \sin\phi_L \\ \sin\theta_L \sin\phi_L \\ \cos\phi_L \end{bmatrix} \tag{12}$$

$$\mathrm{u}_1 = \begin{bmatrix} \cos\theta_L \cos\phi_L \\ \sin\theta_L \cos\phi_L \\ -\sin\phi_L \end{bmatrix}, \mathrm{u}_2 = \begin{bmatrix} -\sin\theta_L \\ \cos\theta_L \\ 0 \end{bmatrix} \tag{13}$$

The unit-vectors $\mathrm{u}_1, \mathrm{u}_2, \mathrm{u}_L$ define the coordinate axes of a new rectangular coordinate system where u_L points in direction of the new z axis. The angles ψ and ζ are the polar and azimuthal angles, respectively. This new coordinate system was generated by making two positive coordinate frame rotations, the first a rotation through an angle θ_L about the original z axis and the second a rotation through an angle ϕ_L about the newly formed y axis. The maximum directivity index occurs at $b = -1/3$ and has the value 6.02 dB.

The input-output equation (7) together with Eq. (9) define a spatial filter. The filter is centered in the direction u_L. In this paper, the function $g_{\mathrm{u}_L}(\psi)$ will be called the discriminating function because it favors a plane wave traveling in the look direction while tending to discriminate against plane waves moving in other directions. The discriminating function is a function of only one variable, ψ. If the angle between a direction u of a plane wave and the look direction u_L is ψ_1, then the set of all u vectors that experience the same attenuation $g_{\mathrm{u}_L}(\psi_1)$ is specified by

$$\mathrm{u} = \left(\mathrm{u}_1 \cos\zeta + \mathrm{u}_2 \sin\zeta\right)\sin\psi_1 + \mathrm{u}_L \cos\psi_1 \quad 0 \le \zeta < 2\pi \tag{14}$$

Equation (14) follows from Eq. (11). Note that when $\psi_1 = 0$, $\mathrm{u} = \mathrm{u}_L$ and when $\psi_1 = \pi$, $\mathrm{u} = -\mathrm{u}_L$. Both cases consist of only a single vector in the set.
In the Introduction it was mentioned that the vector sensor is called a directional acoustic sensor of order one. It owes its filtering capability to the fact that its discriminating function contains the scalar product $\mathrm{u} \cdot \mathrm{u}_L$. One can extend the order of the directional acoustic sensor by beginning with the expression for the acoustic pressure at time t and position r, namely,

$$p(t,\mathrm{r}) = p\left(t + \frac{\mathrm{u} \cdot \mathrm{r}}{c}\right) \tag{15}$$

Setting r to $r\mathrm{u}_L$ in Eq. (15) yields

$$p(t,r) = p\left(t + \frac{\mathrm{u} \cdot \mathrm{u}_L}{c} r\right) \tag{16}$$

The pressure function was transformed from a four-dimensional function to a two-dimensional one by restricting the spatial points to lie on the radial line extending out from the origin in the look direction u_L. Consider next the two-dimensional integro-differential operator

$$\mathrm{L}\left[p(t,r)\right] \equiv c\int \frac{\partial p(t,r)}{\partial r} dt \tag{17}$$

The substitution of Eq. (16) into Eq. (17) results in

$$\mathrm{L}\left[p(t,r)\right] \equiv (\mathrm{u} \cdot \mathrm{u}_L) p(t,r) \tag{18}$$

The function $p(t,r)$ is an eigenfunction of the linear operator L and $\mathrm{u} \cdot \mathrm{u}_L$ the associated eigenvalue. A generalized directional acoustic sensor of order ν can be defined as one whose beamformer output is given by

$$p_o(t) = \sum_{n=0}^{\nu} a_n \mathrm{L}^n\left[p(t,r)\right] = g_{\mathrm{u}_L}(\psi) p(t,r) \tag{19}$$

$$g_{\mathrm{u}_L}(\psi) = \sum_{n=0}^{\nu} a_n \cos^n \psi \tag{20}$$

The discriminating function is a polynomial in $\cos\psi$ of degree ν. The optimum directivity index is $20\log(1+\nu)$ (Schmidlin, 2007). It is a very difficult matter to implement the operations $\mathrm{L}^n\left[p(t,r)\right]$ for $n \geq 2$. This accounts for the sparsity of work on higher-order directional acoustic receivers. This paper attempts to alleviate this problem by introducing a special type of spatial filter, one whose discriminating function is a rational function of $\cos\psi$. The prototype filter is presented in the next section.

2.2 First-order prototype filter

The temporal-spatial filter that is to serve as the prototype for the filters considered herein is represented by the linear first-order partial differential equation

$$a\frac{\partial p_o(t,\tau)}{\partial t}-\frac{\partial p_o(t,\tau)}{\partial \tau}+\gamma\, p_o(t,\tau)=K\, p(t,\tau) \tag{21}$$

The variable τ is equal to r/c. The general solution to Eq. (21) when the forcing function is equal to zero is given by (Kythe et al., 2003)

$$p_o(t,\tau)=f(t+a\tau)\exp(-\gamma t/a) \tag{22}$$

The function $f(\cdot)$ is arbitrary. The forcing function of interest is the harmonic plane wave function

$$p(t,\tau)=\exp(j\omega t)\exp(j\omega\tau\cos\psi) \tag{23}$$

The response to this input can be found by assuming a solution of the form

$$p_o(t,\tau)=B(\omega:\psi)\exp(j\omega t)\exp(j\omega\tau\cos\psi) \tag{24}$$

The substitution of Eqs. (23) and (24) into Eq. (21) results in

$$B(\omega:\psi)=\frac{K}{\gamma+j\omega(a-\cos\psi)} \tag{25}$$

The function $B(\omega:\psi)$ is called the beam pattern of the filter. The total solution of the partial differential equation is the sum of the functions of Eqs. (22) and (24). The total solution is made unique by introducing the initial condition $p_o(0,\tau)=0$. This creates the constraint

$$f(a\tau)+B(\omega:\psi)\exp(j\omega\tau\cos\psi)=0 \tag{26}$$

Solving for $f(\tau)$ and then $f(t+a\tau)$ gives

$$\begin{gathered} f(\tau)=-B(\omega:\psi)\exp(j\omega\tau\cos\psi/a)\\ f(t+a\tau)=-B(\omega:\psi)\exp(j\omega t\cos\psi/a)\exp(j\omega\tau\cos\psi) \end{gathered} \tag{27}$$

The output of the prototype filter in response to the harmonic plane wave input is

$$p_o(t,\tau)=B(\omega:\psi)\exp(j\omega\tau\cos\psi)\left[\exp(j\omega t)-\exp(-\gamma t/a)\exp(j\omega t\cos\psi/a)\right] \tag{28}$$

When $\gamma\neq 0$, the second component within the brackets of Eq. (28) decays to zero as time increases. One observes from Eq. (25) that the beam pattern's sensitivity to variations in the angle ψ decreases with increasing γ. Consequently, a very small γ is desirable. For the special case $\gamma=0$, Eq. (25) becomes

$$B(\omega:\psi)=\frac{g_{u_L}(\psi)}{j\omega} \tag{29}$$

$$g_{u_L}(\psi)=\frac{K}{a-\cos\psi} \tag{30}$$

The function $g_{u_L}(\psi)$ is the discriminating function of the prototype filter. If K is chosen to be $a-1$ then

$$g_{u_L}(0)=1,\ g_{u_L}(\pi)=\frac{a-1}{a+1} \tag{31}$$

Since it is desirable for the discriminating function at $\psi=\pi$ to be less than one in magnitude, the value of a must be positive. And Eq. (30) reveals that for the discriminating function to be finite for $0\le\psi\le\pi$, the value of a must be greater than 1. For $\gamma=0$, the output $p_o(t,\tau)$ becomes

$$p_o(t,\tau)=g_{u_L}(\psi)\frac{\exp(j\omega\tau\cos\psi)}{j\omega}\left[\exp(j\omega t)-\exp(j\omega t\cos\psi/a)\right] \tag{32}$$

Of special interest is the behavior of the filter towards a plane wave coming from the look direction ($\psi=0$). Equation (32) simplifies to

$$p_o(t,\tau)=\frac{\exp(j\omega\tau)}{j\omega}\left[\exp(j\omega t)-\exp\left(j\frac{\omega}{a}t\right)\right] \tag{33}$$

The output of the prototype filter contains two sinusoidal components. The frequency of the first component is equal to the input frequency ω. The frequency of the second component is equal to ω/a which is less than the input frequency since $a>1$. This frequency can be eliminated by a temporal bandpass filter. If $\omega_{\min}$ and $\omega_{\max}$ denote the minimum and maximum frequencies of interest, then a constraint on the parameter a is

$$\frac{\omega_{\max}}{a}<\omega_{\min}\Rightarrow a>\frac{\omega_{\max}}{\omega_{\min}} \tag{34}$$

2.3 Directivity index of prototype filter

In a receiving aperture, directivity serves to reject noise and other interference arriving from directions other than the look direction. The directive effect of a spatial filter has been summarized in a single number called the directivity, which is computed from (Ziomek, 1995)

$$D=\frac{P(\omega:0)}{\frac{1}{4\pi}\int_0^{2\pi}\int_0^{\pi}P(\omega:\psi)\sin\psi\,d\psi\,d\zeta} \tag{35}$$

where $P(\omega:\psi)$ is the filter's beam power pattern and for $\gamma=0$ is given by

$$P(\omega:\psi)=\left|B(\omega:\psi)\right|^2=\frac{(a-1)^2}{\omega^2(a-\cos\psi)^2} \tag{36}$$

Equation (35) can be simplified to

$$D = \frac{2P(\omega:0)}{\int_{-1}^{1} P(\omega:x)dx} \tag{37}$$

where $x = \cos\psi$. The substitution of Eq. (36) into Eq. (37) results in

$$D = \frac{2}{\int_{-1}^{1} \frac{(a-1)^2}{(a-x)^2} dx} = \frac{a+1}{a-1} \tag{38}$$

Equation (38) represents the directivity of the first-order prototype filter. The directivity index is defined as

$$\mathrm{DI} \triangleq 10\log_{10} D \text{ dB} \tag{39}$$

Equation (34) gives a constraint on the parameter a. Let $\omega_1 \le \omega_{\min}$ and $\omega_2 \ge \omega_{\max}$ denote the lower and upper cutoff frequencies of the temporal bandpass filter that is to filter out the undesirable frequency component in Eq. (33), and let $a = \omega_2/\omega_1$. The lower and upper cutoff frequencies are related to the center frequency ω_0 and the quality factor Q by

$$\begin{aligned} \omega_1 &= \omega_0\left(\sqrt{1+\frac{1}{4Q^2}} - \frac{1}{2Q}\right) \\ \omega_2 &= \omega_0\left(\sqrt{1+\frac{1}{4Q^2}} + \frac{1}{2Q}\right) \end{aligned} \tag{40}$$

From Eq. (40) one may write

$$\frac{\omega_2+\omega_1}{\omega_2-\omega_1} = \frac{a+1}{a-1} = \sqrt{1+4Q^2} \tag{41}$$

From Eqs. (38) and (39) the directivity index becomes

$$\mathrm{DI} = 10\log_{10}\sqrt{1+4Q^2} \tag{42}$$

For $Q >> 1/2$ the DI may be approximated as

$$\mathrm{DI} = 3 + 10\log_{10} Q \text{ dB} \tag{43}$$

If the input plane wave function fits within the pass band of the temporal filter, then the directivity index is given by Eq. (43). For $Q = 10$, the directivity index is 13 dB. It was noted in Section 2.1 that the maximum directivity index for a vector sensor is 6.02 dB. Using Eq. (41) to Solve for a yields

$$a = \frac{\sqrt{1+4Q^2}+1}{\sqrt{1+4Q^2}-1} \tag{44}$$

When the quality factor is 10, then the parameter a of the prototype filter is 1.105. The discriminating function of the filter is given by Eq. (30). The function has a value of 1 at $\psi = 0$. The beamwidth of the prototype filter is obtained by equating Eq. (30) to $1/\sqrt{2}$, solving for ψ, and multiplying by 2. The result is

$$\text{BW} = 2\psi_{3dB} = 2\cos^{-1}\left[a\left(1-\sqrt{2}\right)+\sqrt{2}\right] \tag{45}$$

For the case $a = 1.105$, the beamwidth is 33.9°. This is in sharp contrast to the beamwidth of the maximum DI vector sensor which is 104.9°. Figure 1 gives a plot of the discriminating function as a function of the angle ψ. Note that the discriminating function is a monotonic function of ψ. This is not true for discriminating functions of directional acoustic sensors (Schmidlin, 2007).

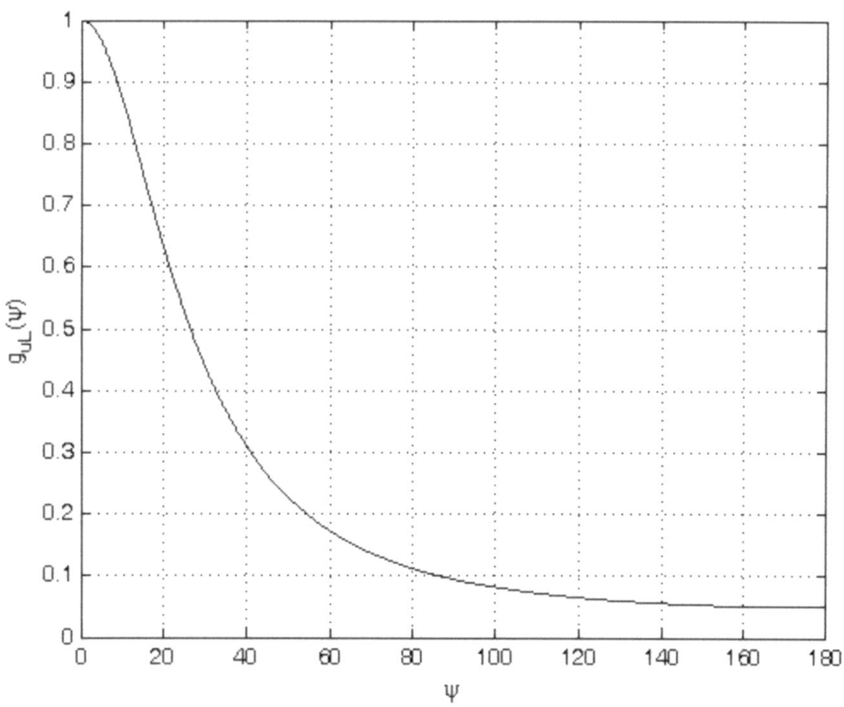

Fig. 1. Discriminating function for a = 1.105.

3. Direction-Selective filters with rational discriminating functions

3.1 Interconnection of prototype filters

The first-order prototype filter can be used as a fundamental building block for generating filters that have discriminating functions which are rational functions of $\cos\psi$. As an example, consider a discriminating function that is a proper rational function and whose denominator polynomial has roots that are real and distinct. Such a discriminating function may be expressed as

$$g_{\mathrm{u}_L}(\psi)=\frac{\sum_{j=0}^{\mu} d_j \cos^j \psi}{\sum_{j=0}^{\nu} c_j \cos^j \psi}=K\frac{\prod_{j=1}^{\mu}\left(b_j-\cos\psi\right)}{\prod_{j=1}^{\nu}\left(a_j-\cos\psi\right)} \tag{46}$$

where $c_\nu = 1$ and $\mu < \nu$. The discriminating function of Eq. (46) can expanded in the partial fraction expansion

$$g_{\mathrm{u}_L}(\psi)=\sum_{i=1}^{\nu}\frac{K_i}{a_i-\cos\psi} \tag{47}$$

The function specified by Eq. (47) may be realized by a parallel interconnection of ν prototype filters (with γ = 0). Each component of the above expansion has the form of Eq. (30). Normalizing the discriminating function such that it has a value of 1 at $\psi = 0$ yields

$$\sum_{i=1}^{\nu}\frac{K_i}{a_i-1}=1 \tag{48}$$

Similar to Eq. (36), the beam power pattern of the composite filter is given by

$$P(\omega:\psi)=\frac{\left|g_{\mathrm{u}_L}(\psi)\right|^2}{\omega^2} \tag{49}$$

Equations (47) and (49) together with Eq. (35) lead to the following expression for the directivity:

$$D^{-1} = \sum_{i=1}^{\nu}\sum_{j=1}^{\nu} K_i K_j g_{ij} \tag{50}$$

where

$$g_{ii}=\frac{1}{a_i^2-1} \tag{51}$$

$$g_{ij}=\frac{1}{a_i-a_j}\coth^{-1}\left(\frac{a_i a_j-1}{a_i-a_j}\right), i\neq j \tag{52}$$

For a given set of a_i values, the directivity can be maximized by minimizing the quadratic form given by Eq. (50) subject to the linear constraint specified by Eq. (48). To solve this optimization problem, it is useful to represent the problem in matrix form, namely,

$$\begin{gathered}\text{minimize } D^{-1} = \mathrm{K'GK} \\ \text{subject to } \mathrm{U'K} = 1\end{gathered} \tag{53}$$

where

$$\mathrm{K}' = \begin{bmatrix} K_1 & K_2 & \cdots & K_v \end{bmatrix} \tag{54}$$

$$\mathrm{U}' = \begin{bmatrix} \dfrac{1}{a_1 - 1} & \dfrac{1}{a_2 - 1} & \cdots & \dfrac{1}{a_v - 1} \end{bmatrix} \tag{55}$$

and $\boldsymbol{G}$ is the matrix containing the elements g_{ij}. Utilizing the Method of Lagrange Multipliers, the solution for K is given by

$$\mathrm{K} = \frac{\mathrm{G}^{-1}\mathrm{U}}{\mathrm{U}'\mathrm{G}^{-1}\mathrm{U}} \tag{56}$$

The minimum of D^{-1} has the value

$$D^{-1} = \mathrm{U}'\mathrm{G}^{-1}\mathrm{U} \tag{57}$$

The maximum value of the directivity index is

$$\mathrm{DI}_{\max} = -10\log_{10}\left(\mathrm{U}'\mathrm{G}^{-1}\mathrm{U}\right) \tag{58}$$

3.2 An example: a second-degree rational discriminating function

As a example of applying the contents of the previous section, consider the proper rational function of the second degree,

$$g_{\mathrm{u}_L}(\psi) = \frac{d_0 + d_1\cos\psi}{c_0 + c_1\cos\psi + \cos^2\psi} = \frac{K_1}{a_1 - \cos\psi} + \frac{K_2}{a_2 - \cos\psi} \tag{59}$$

where $a_2 > a_1$ and

$$\begin{gathered} d_0 = a_2K_1 + a_1K_2 \\ d_1 = -K_1 - K_2 \\ c_0 = a_1a_2, \quad c_1 = -a_1 - a_2 \end{gathered} \tag{60}$$

In the example presented in Section 2.3, the parameter a had the value 1.105. In this example let $a_1 = 1.105$, and let $a_2 = 1.200$. The value of the matrices $\boldsymbol{G}$ and $\boldsymbol{U}$ are given by

$$\mathrm{G} = \begin{bmatrix} 4.5244 & 3.1590 \\ 3.1590 & 2.227 \end{bmatrix} \tag{61}$$

$$\mathrm{U} = \begin{bmatrix} 9.5238 \\ 5.0000 \end{bmatrix} \tag{62}$$

If Eqs. (56) and (58) are used to compute $\boldsymbol{K}$ and $\mathrm{DI}_{\max}$, the result is

$$\mathrm{K} = \begin{bmatrix} 0.3181 \\ -0.4058 \end{bmatrix} \tag{63}$$

$$\mathrm{DI}_{\max} = 17.8289 \text{ dB} \tag{64}$$

From Eqs. (60), one obtains

$$\begin{aligned} d_0 &= -.0668, \quad d_1 = 0.0878 \\ c_0 &= 1.3260, \quad c_1 = -2.3050 \end{aligned} \tag{65}$$

Figure 2 illustrates the discriminating function specified by Eqs. (59) and (65). Also shown (as a dashed line) for comparison the discriminating function of Fig. 1. The dashed-line plot represents a discriminating function that is a rational function of degree one, whereas the solid-line plot corresponds to a discriminating function that is a rational function of degree two. The latter function decays more quickly having a 3-dB down beamwidth of 22.6° as compared to a 3-dB down beamwidth of 33.9° for the former function.

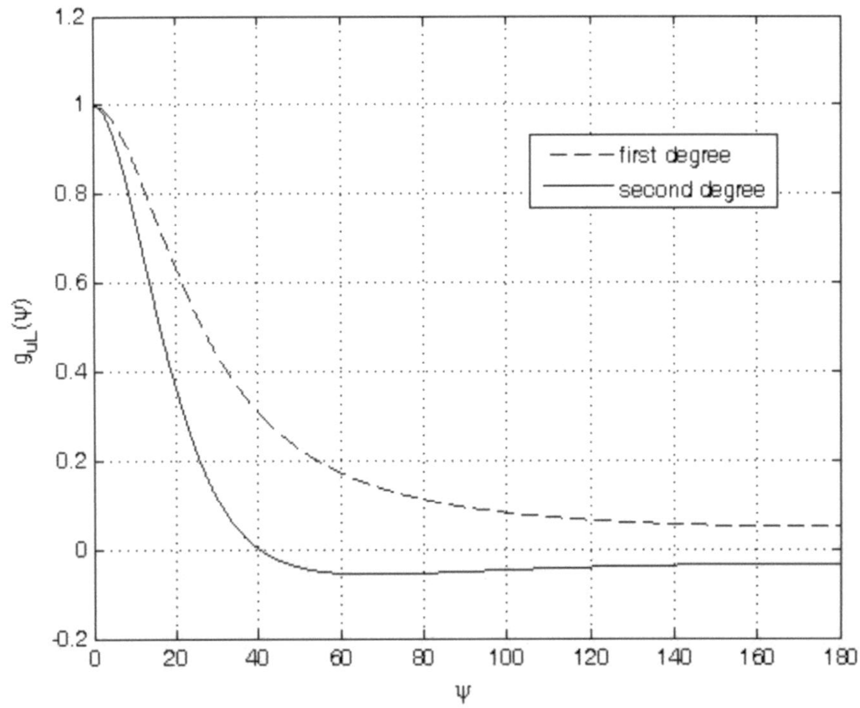

Fig. 2. Plots of the discriminating function of the examples presented in Sections 2.3 and 3.2.

In order to see what directivity index is achievable with a second-degree discriminating function, it is useful to consider the second-degree discriminating function of Eq. (59) with equal roots in the denominator, that is, $c_0 = a^2, c_1 = -2a$. It is shown in a technical report by the author (2010c) that the maximum directivity index for this discriminating function is equal to

$$D_{\max} = 4\frac{a+1}{a-1} \tag{66}$$

and is achieved when d_0 and d_1 have the values

$$d_0 = \frac{a-1}{4}(a-3) \tag{67}$$

$$d_1 = \frac{a-1}{4}(3a-1) \tag{68}$$

Note that the directivity given by Eq. (66) is four times the directivity given by Eq. (38). Analogous to Eqs. (42) and (43), the maximum directivity index can be expressed as

$$\mathrm{DI}_{\max} = 6 + 10\log_{10}\sqrt{1+4Q^2}\ \mathrm{dB} \approx 9 + 10\log_{10} Q\ \mathrm{dB} \tag{69}$$

For $a_1 = 1.105$, $Q = 10$ and the maximum directivity index is 19 dB which is a 6 dB improvement over that of the first-degree discriminating function of Eq. (30). In the example presented in this section, $a_1 = 1.105, a_2 = 1.200, \mathrm{DI}_{\max} = 17.8\ \mathrm{dB}$. As a_2 moves closer to a_1, the maximum directivity index will move closer to 19 dB. For a specified a_1, Eq. (69) represents an upper bound on the maximum directivity index, the bound approached more closely as a_2 moves more closely to a_1.

3.3 Design of discriminating functions from the magnitude response of digital filters

In designing and implementing transfer functions of IIR digital filters, advantage has been taken of the wealth of knowledge and practical experience accumulated in the design and implementation of the transfer functions of analog filters. Continuous-time transfer functions are, by means of the bilinear or impulse-invariant transformations, transformed into equivalent discrete-time transfer functions. The goal of this section is to do a similar thing by generating discriminating functions from the magnitude response of digital filters. As a starting point, consider the following frequency response:

$$H\left(e^{j\omega}\right) = \frac{1-\rho}{1-\rho e^{-j\omega}} \tag{70}$$

where ρ is real, positive and less than 1. Equation (70) corresponds to a causal, stable discrete-time system. The digital frequency ω is not to be confused with the analog frequency ω appearing in previous sections. The magnitude-squared response of this system is obtained from Eq. (70) as

$$\left|H\left(e^{j\omega}\right)\right|^2 = \frac{1-2\rho+\rho^2}{1-2\rho\cos\omega+\rho^2} \tag{71}$$

Letting $\rho = e^{-\sigma}$ allows one to recast Eq. (71) into the simpler form

$$\left|H\left(e^{j\omega}\right)\right|^2 = \frac{\cosh\sigma - 1}{\cosh\sigma - \cos\omega} \tag{72}$$

If the variable ω is replaced by ψ, the resulting function looks like the discriminating function of Eq. (30) where $a = \cosh\sigma$. This suggests a means for generating discriminating functions from the magnitude response of digital filters. Express the magnitude-squared response of the filter in terms of $\cos\omega$ and define

$$g_{u_L}(\psi) \triangleq \left|H\left(e^{j\psi}\right)\right|^2 \tag{73}$$

To illustrate the process, consider the magnitude-squared response of a low pass Butterworth filter of order 2, which has the magnitude-squared function

$$\left|H\left(e^{j\omega}\right)\right|^2 = \frac{1}{1+\left[\dfrac{\tan(\omega/2)}{\tan(\omega_c/2)}\right]^4} \tag{74}$$

where ω_c is the cutoff frequency of the filter. Utilizing the relationship

$$\tan^2\left(\frac{A}{2}\right) = \frac{1-\cos A}{1+\cos A} \tag{75}$$

one can express Eq. (74) as

$$\left|H\left(e^{j\omega}\right)\right|^2 = \frac{\alpha(1+\cos\omega)^2}{\alpha(1+\cos\omega)^2+(1-\cos\omega)^2} \tag{76}$$

where

$$\alpha = \tan^4\left(\frac{\omega_c}{2}\right) = \frac{(1-\cos\omega_c)^2}{(1+\cos\omega_c)^2} \tag{77}$$

The substitution of Eq. (77) into Eq. (76) and simplifying yields the final result

$$\left|H\left(e^{j\omega}\right)\right|^2 = \frac{1-\cos\theta}{2}\frac{1+2\cos\omega+\cos^2\omega}{1-2\cos\theta\cos\omega+\cos^2\omega} \tag{78}$$

where

$$\cos\theta = \frac{2\cos\omega_c}{1+\cos^2\omega_c} \tag{79}$$

By replacing ω by ψ in Eq. (78), one obtains the discriminating function

$$g_{u_L}(\psi) = \frac{1-\cos\theta}{2}\frac{1+2\cos\psi+\cos^2\psi}{1-2\cos\theta\cos\psi+\cos^2\psi} \tag{80}$$

where ω_c is replaced by ψ_c in Eq. (79). A plot of Eq. (80) is shown in Fig. 3 for $\psi_c = 10^\circ$. From the figure it is observed that $\psi_c = 10^\circ$ is the 6-dB down angle because the discriminating function is equal to the magnitude-squared function of the Butterworth filter. The discriminating function of Fig. 3 can be said to be providing a "maximally-flat beam" of order 2 in the look direction u_L. Equation (80) cannot be realized by a parallel interconnection of first-order prototype filters because the roots of the denominator of Eq. (80) are complex. Its realization requires the development of a second-order prototype filter which is the focus of current research.

4. Summary and future research

4.1 Summary

The objective of this paper is to improve the directivity index, beamwidth, and the flexibility of spatial filters by introducing spatial filters having rational discriminating functions. A first-order prototype filter has been presented which has a rational discriminating function of degree one. By interconnecting prototype filters in parallel, a rational discriminating function can be created which has real distinct simple poles. As brought out by Eq. (33), a negative aspect of the prototype filter is the appearance at the output of a spurious frequency whose value is equal to the input frequency divided by the parameter a of the filter where $a > 1$. Since the directivity of the filter is inversely proportional to $a-1$, there exists a tension as a approaches 1 between an arbitrarily increasing directivity D and destructive interference between the real and spurious frequencies. The problem was

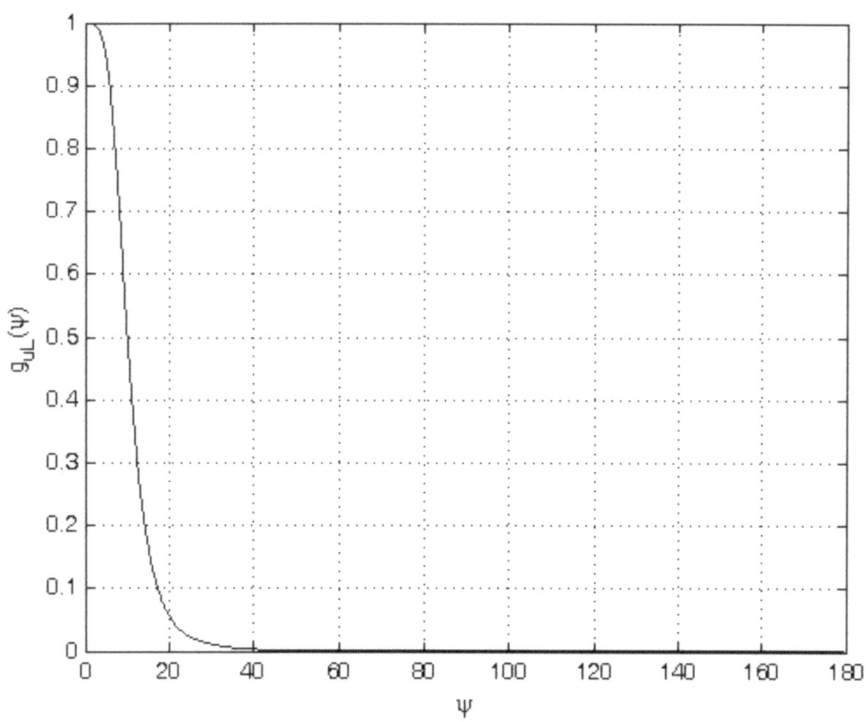

Fig. 3. Discriminating function of Eq. (80).

alleviated by placing a temporal bandpass filter at the output of the prototype filter and assigning a the value equal to the ratio of the upper to the lower cutoff frequencies of the bandpass filter. This resulted in the dependence of the directivity index DI on the value of the bandpass filter's quality factor Q as indicated by Eqs. (42) and (43). Consequently, for the prototype filter to be useful, the input plane wave function must be a bandpass signal which fits within the pass band of the temporal bandpass filter. It was noted in Section 2.3 that for $Q = 10$ the directivity index is 13 dB and the beamwidth is 33.9°. Directional acoustic sensors as they exist today have discriminating functions that are polynomials. Their processors do not have the spurious frequency problem. The vector sensor has a maximum directivity index of 6.02 dB and the associated beamwidth is 104.9°. According to Eq. (42) the prototype filter has a DI of 6.02 dB when $Q = 1.94$. The corresponding beamwidth is 87.3°. Section 3.2 demonstrated that the directivity index and the beamwidth can be improved by adding an additional pole. Figure 4 illustrates the directivity index and the beamwidth for the case of two equal roots or poles in the denominator of the discriminating function. As a means of comparison, it is instructive to consider the dyadic sensor which has a polynomial of the second degree as its discriminating function. The sensor's maximum directivity index is 9.54 dB and the associated beamwidth is 65°. The directivity index in Fig. 4 varies from 9.5 dB at $Q = 1$ to 19.0 dB at $Q = 10$. The beamwidth varies from 63.2° at $Q = 1$ to 19.7° at $Q = 10$. The directivity index and beamwidth of the two-equal-poles discriminating function at $Q = 1$ is essentially the same as that of the dyadic sensor. But as the quality factor increases, the directivity index goes up while the beamwidth goes down. It is important to note that the curves in Fig. 4 are theoretical curves. In any practical implementation, one may be required to operate at the lower end of each curve. However, the performance will still be an improvement over that of a dyadic sensor. The two-equal-poles case cannot be realized exactly by first-order prototype filters, but the implementation presented in Section 3.2 comes arbitrarily close. Finally, in Section 3.3 it was shown that discriminating functions can be derived from the magnitude-squared response of digital filters. This allows a great deal of flexibility in the design of discriminating functions. For example, Section 3.3 used the magnitude-response of a second-order Butterworth digital filter to generate a discriminating function that provides a "maximally-flat beam" centered in the look direction. The beamwidth is controlled directly by a single parameter.

4.2 Future research

Many rational discriminating functions, specifically those with complex-valued poles and multiple-order poles, cannot be realized as parallel interconnections of first-order prototype filters. Examples of such discriminating functions appear in Figs. 2 and 3. Research is underway involving the development of a second-order temporal-spatial filter having the prototypical beampattern

$$B(\omega:\psi) = \frac{g_{u_L}(\psi)}{(j\omega)^2} \tag{81}$$

where the prototypical discriminating function $g_{u_L}(\psi)$ has the form

$$g_{u_L}(\psi) = \frac{d_0 + d_1 \cos\psi}{1 + c_1 \cos\psi + c_2 \cos^2\psi} \tag{82}$$

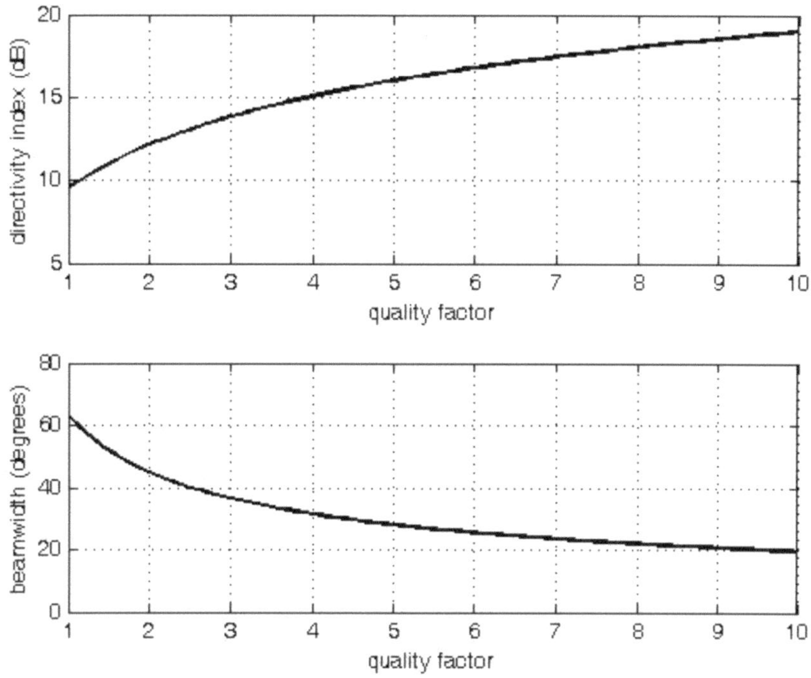

Fig. 4. DI and beamwidth as a function of Q.

With the second-order prototype in place, the discriminating function of Eq. (80), as an example, can be realized by expressing it as a partial fraction expansion and connecting in parallel two prototypal filters. For the first, $d_0 = (1-\cos\theta)/2$ and $d_1 = c_1 = c_2 = 0$, and for the second, $d_0 = 0, d_1 = \sin^2\theta, c_1 = -2\cos\theta, c_2 = 1$. Though the development of a second-order prototype is critical for the implementation of a more general rational discriminating function than that of the first-order prototype, additional research is necessary for the first-order prototype. In Section 2.2 the number of spatial dimensions was reduced from three to one by restricting pressure measurements to a radial line extending from the origin in the direction defined by the unit vector u_L. This allowed processing of the plane-wave pressure function by a temporal-spatial filter describable by a linear first-order partial differential equation in two variables (Eq. (21)). The radial line (when finite in length) represents a linear aperture or antenna. In many instances, the linear aperture is replaced by a linear array of pressure sensors. This necessitates the numerical integration of the partial differential equation in order to come up with the output of the associated filter. Numerical integration techniques for PDE's generally fall into two categories, finite-difference methods (LeVeque, 2007) and finite-element methods (Johnson, 2009). If q prototypal filters are connected in parallel, the associated set of partial differential equations form a set of q symmetric hyperbolic systems (Bilbao, 2004). Such systems can be numerically integrated using principles of multidimensional wave digital filters (Fettweis and Nitsche, 1991a, 1991b). The resulting algorithms inherit all the good properties known to hold for wave digital filters, specifically the full range of robustness properties typical for these filters (Fettweis, 1990). Of special interest in the filter implementation process is the length of the aperture. The goal is to achieve a particular directivity index and beamwidth with the smallest possible aperture length. Another important area for future research is studying the effect of noise (both ambient and system noise) on the filtering process. The fact that the prototypal filter tends to act as an integrator should help soften the effect of uncorrelated input noise to the filter. Finally, upcoming research will also include the array gain (Burdic, 1991) of the filter prototype for the case of anisotropic noise (Buckingham, 1979a,b; Cox, 1973). This paper considered the directivity index which is the array gain for the case of isotropic noise.

5. References

Bienvenu, G. & Kopp, L. (1980). Adaptivity to background noise spatial coherence for high resolution passive methods, *Int. Conf. on Acoust., Speech and Signal Processing*, pp. 307-310.

Bilbao, S. (2004). *Wave and Scattering Methods for Numerical Simulation*, John Wiley and Sons, ISBN 0-470-87017-6, West Sussex, England.

Bresler, Y. & Macovski, A. (1986). Exact maximum likelihood parameter estimation of superimposed exponential signals in noise, *IEEE Trans. ASSP*, Vol. ASSP-34, No. 5, pp. 1361-1375.

Buckingham, M. J. (1979a). Array gain of a broadside vertical line array in shallow water, *J. Acoust. Soc. Am.*, Vol. 65, No. 1, pp. 148-161.

Buckingham, M. J. (1979b). On the response of steered vertical line arrays to anisotropic noise, *Proc. R. Soc. Lond. A*, Vol. 367, pp. 539-547.

Burdic, W. S. (1991). *Underwater Acoustic System Analysis*, Prentice-Hall, ISBN 0-13-947607-5, Englewood Cliffs, New Jersey, USA.

Cox, H. (1973). Spatial correlation in arbitrary noise fields with application to ambient sea noise, *J. Acoust. Soc. Am.*, Vol. 54, No. 5, pp. 1289-1301.

Cray, B. A. (2001). Directional acoustic receivers: signal and noise characteristics, *Proc. of the Workshop of Directional Acoustic Sensors*, Newport, RI.

Cray, B. A. (2002). Directional point receivers: the sound and the theory, *Oceans '02*, pp. 1903-1905.

Cray, B. A.; Evora, V. M. & Nuttall, A. H. (2003). Highly directional acoustic receivers, *J. Acoust. Soc. Am.*, Vol. 13, No. 3, pp. 1526-1532.

D'Spain, G. L.; Hodgkiss, W. S.; Edmonds, G. L.; Nickles, J. C.; Fisher, F. H.; & Harris, R. A. (1992). Initial analysis of the data from the vertical DIFAR array, *Proc. Mast. Oceans Tech. (Oceans '92)*, pp. 346-351.

D'Spain, G. L.; Luby, J. C.; Wilson, G. R. & Gramann R. A. (2006). Vector sensors and vector sensor line arrays: comments on optimal array gain and detection, *J. Acoust. Soc. Am.*, Vol. 120, No. 1, pp. 171-185.

Fettweis, A. (1990). On assessing robustness of recursive digital filters, *European Transactions on Telecommunications*, Vol. 1, pp. 103-109.

Fettweis, A. & Nitsche, G. (1991a). Numerical Integration of partial differential equations using principles of multidimensional wave digital filters, *Journal of VLSI Signal Processing*, Vol. 3, pp. 7-24, Kluwer Academic Publishers, Boston.

Fettweis, A. & Nitsche, G. (1991b). Transformation approach to numerically integrating PDEs by means of WDF principles, *Multidimensional Systems and Signal Processing*, Vol. 2, pp. 127-159, Kluwer Academic Publishers, Boston.

Hawkes, M. & Nehorai, A. (1998). Acoustic vector-sensor beamforming and capon direction estimation, *IEEE Trans. Signal Processing*, Vol. 46, No. 9, pp. 2291-2304.

Hawkes, M. & Nehorai, A. (2000). Acoustic vector-sensor processing in the presence of a reflecting boundary, *IEEE Trans. Signal Processing*, Vol. 48, No. 11, pp. 2981-2993.

Hines, P. C. & Hutt, D. L. (1999). SIREM: an instrument to evaluate superdirective and intensity receiver arrays, *Oceans 1999*, pp. 1376-1380.

Hines, P. C.; Rosenfeld, A. L.; Maranda, B. H. & Hutt, D. L. (2000). Evaluation of the endfire response of a superdirective line array in simulated ambient noise environments, *Proc. Oceans 2000*, pp. 1489-1494.

Johnson, C. (2009). *Numerical Solution of Partial Differential Equations by the Finite-Element Method*, Dover Publications, ISBN-13 978-0-486-46900-3, Mineola, New York, USA

Krim, H. & Viberg, M. (1996). Two decades of array signal processing research, *IEEE Signal Processing Magazine*, Vol. 13, No. 4, pp. 67-94.

Kumaresan, R. & Shaw, A. K. (1985). High resolution bearing estimation without eigendecomposition , *Proc. IEEE ICASSP 85*, p. 576-579, Tampa, FL.

Kythe, P. K.; Puri, P. & Schaferkotter, M. R. (2003). *Partial Differential Equations and Boundary Value Problems with Mathematica*, Chapman & Hall/ CRC, ISBN 1-58488-314-6, Boca Raton, London, New York, Washington, D.C.

LeVeque, R. J. (2007). *Finite Difference Methods for Ordinary and Partial Differential Equations*, SIAM, ISBN 978-0-898716-29-0, Philadelphia, USA.

Nehorai, A. & Paldi, E. (1994). Acoustic vector-sensor array processing, *IEEE Trans. Signal Processing*, Vol. 42, No. 9, pp. 2481-2491.

Schmidlin, D. J. (2007). Directionality of generalized acoustic sensors of arbitrary order, *J. Acoust. Soc. Am.*, Vol. 121, No. 6, pp. 3569-3578.

Schmidlin, D. J. (2010a). Distribution theory approach to implementing directional acoustic sensors, *J. Acoust. Soc. Am.*, Vol. 127, No. 1, pp. 292-299.

Schmidlin, D. J. (2010b). Concerning the null contours of vector sensors, *Proc. Meetings on Acoustics*, Vol. 9, Acoustical Society of America.

Schmidlin, D. J. (2010c). The directivity index of discriminating functions, *Technical Report No. 31-2010-1*, El Roi Analytical Services, Valdese, North Carolina.

Schmidt, R. O. (1986). Multiple emitter location and signal parameter estimation, *IEEE Trans. Antennas and Propagation*, Vol. AP-34, No. 3, pp. 276-280.

Silvia, M. T. (2001). A theoretical and experimental investigation of acoustic dyadic sensors, *SITTEL Technical Report No. TP-4*, SITTEL Corporation, Ojai, Ca.

Silvia, M. T.; Franklin, R. E. & Schmidlin, D. J. (2001). Signal processing considerations for a general class of directional acoustic sensors, *Proc. of the Workshop of Directional Acoustic Sensors*, Newport, RI.

Van Veen, B. D. & Buckley, K. M. (1988). Beamforming: a versatile approach to spatial filtering, *IEEE ASSP Magazine*, Vol. 5, No. 2, pp. 4-24.

Wong, K. T. & Zoltowski, M. D. (1999). Root-MUSIC-based azimuth-elevation angle-of-arrival estimation with uniformly spaced but arbitrarily oriented velocity hydrophones, *IEEE Trans. Signal Processing*, Vol. 47, No. 12, pp. 3250-3260.

Wong, K. T. & Zoltowski, M. D. (2000). Self-initiating MUSIC-based direction finding in underwater acoustic particle velocity-field beamspace, *IEEE Journal of Oceanic Engineering*, Vol. 25, No. 2, pp. 262-273.

Wong, K. T. & Chi, H. (2002). Beam patterns of an underwater acoustic vector hydrophone located away from any reflecting boundary, *IEEE Journal Oceanic Engineering*, Vol. 27, No. 3, pp. 628-637.

Ziomek, L. J. (1995). *Fundamentals of Acoustic Field Theory and Space-Time Signal Processing*, CRC Press, ISBN 0-8493-9455-4, Boca Raton, Ann Arbor, London, Tokyo.

Zou, N. & Nehorai, A. (2009). Circular acoustic vector-sensor array for mode beamforming, *IEEE Trans. Signal Processing*, Vol. 57, No. 8, pp. 3041-3052.

3

Single-Channel Sound Source Localization Based on Discrimination of Acoustic Transfer Functions

Ryoichi Takashima, Tetsuya Takiguchi and Yasuo Ariki
Graduate School of System Informatics, Kobe University, Kobe
Japan

1. Introduction

Many systems using microphone arrays have been tried in order to localize sound sources. Conventional techniques, such as MUSIC, CSP, and so on (e.g., (Johnson & Dudgeon, 1996; Omologo & Svaizer, 1996; Asano et al., 2000; Denda et al., 2006)), use simultaneous phase information from microphone arrays to estimate the direction of the arriving signal. There have also been studies on binaural source localization based on interaural differences, such as interaural level difference and interaural time difference (e.g., (Keyrouz et al., 2006; Takimoto et al., 2006)). However, microphone-array-based systems may not be suitable in some cases because of their size and cost. Therefore, single-channel techniques are of interest, especially in small-device-based scenarios.

The problem of single-microphone source separation is one of the most challenging scenarios in the field of signal processing, and some techniques have been described (e.g., (Kristiansson et al., 2004; Raj et al., 2006; Jang et al., 2003; Nakatani & Juang, 2006)). In our previous work (Takiguchi et al., 2001; Takiguchi & Nishimura, 2004), we proposed HMM (Hidden Markov Model) separation for reverberant speech recognition, where the observed (reverberant) speech is separated into the acoustic transfer function and the clean speech HMM. Using HMM separation, it is possible to estimate the acoustic transfer function using some adaptation data (only several words) uttered from a given position. For this reason, measurement of impulse responses is not required. Because the characteristics of the acoustic transfer function depend on each position, the obtained acoustic transfer function can be used to localize the talker.

In this paper, we will discuss a new talker localization method using only a single microphone. In our previous work (Takiguchi et al., 2001) for reverberant speech recognition, HMM separation required texts of a user's utterances in order to estimate the acoustic transfer function. However, it is difficult to obtain texts of utterances for talker-localization estimation tasks. In this paper, the acoustic transfer function is estimated from observed (reverberant) speech using a clean speech model without having to rely on user utterance texts, where a GMM (Gaussian Mixture Model) is used to model clean speech features. This estimation is performed in the cepstral domain employing an approach based upon maximum likelihood. This is possible because the cepstral parameters are an effective representation for retaining useful clean speech information. The results of our talker-localization experiments show the effectiveness of our method.

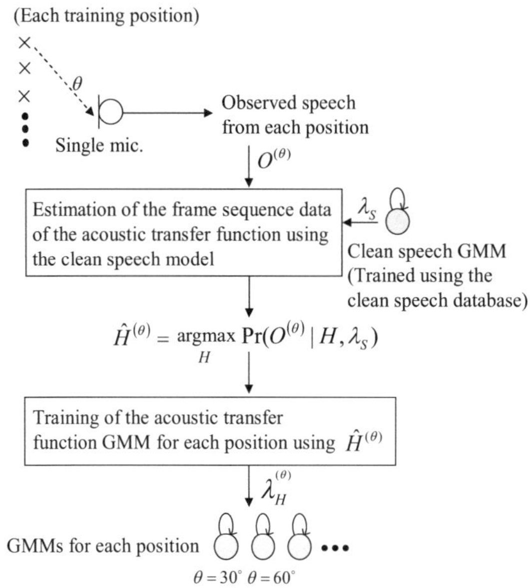

Fig. 1. Training process for the acoustic transfer function GMM

2. Estimation of the acoustic transfer function

2.1 System overview

Figure 1 shows the training process for the acoustic transfer function GMM. First, we record the reverberant speech data $O^{(\theta)}$ from each position θ in order to build the GMM of the acoustic transfer function for θ. Next, the frame sequence of the acoustic transfer function $\hat{H}^{(\theta)}$ is estimated from the reverberant speech $O^{(\theta)}$ (any utterance) using the clean-speech acoustic model, where a GMM is used to model the clean speech feature:

$$\hat{H}^{(\theta)} = \underset{H}{\text{argmax}}\, \Pr(O^{(\theta)}|H, \lambda_S). \tag{1}$$

Here, λ_S denotes the set of GMM parameters for clean speech, while the suffix S represents the clean speech in the cepstral domain. The clean speech GMM enables us to estimate the acoustic transfer function from the observed speech without needing to have user utterance texts (i.e., text-independent acoustic transfer estimation). Using the estimated frame sequence data of the acoustic transfer function $\hat{H}^{(\theta)}$, the acoustic transfer function GMM for each position $\lambda_H^{(\theta)}$ is trained.

Figure 2 shows the talker localization process. For test data, the talker position $\hat{\theta}$ is estimated based on discrimination of the acoustic transfer function, where the GMMs of the acoustic transfer function are used. First, the frame sequence of the acoustic transfer function $\hat{H}$ is estimated from the test data (any utterance) using the clean-speech acoustic model. Then, from among the GMMs corresponding to each position, we find a GMM having the maximum-likelihood in regard to $\hat{H}$:

$$\hat{\theta} = \underset{\theta}{\text{argmax}}\, \Pr(\hat{H}|\lambda_H^{(\theta)}), \tag{2}$$

where $\lambda_H^{(\theta)}$ denotes the estimated acoustic transfer function GMM for direction θ (location).

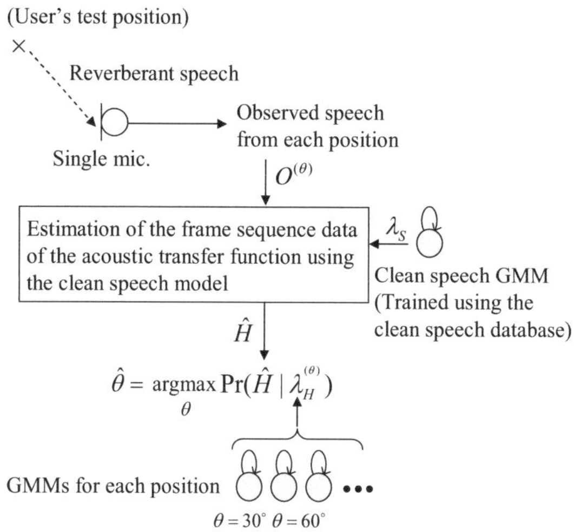

Fig. 2. Estimation of talker localization based on discrimination of the acoustic transfer function

2.2 Cepstrum representation of reverberant speech

The observed signal (reverberant speech), $o(t)$, in a room environment is generally considered as the convolution of clean speech and the acoustic transfer function:

$$o(t) = \sum_{l=0}^{L-1} s(t-l)h(l) \tag{3}$$

where $s(t)$ is a clean speech signal and $h(l)$ is an acoustic transfer function (room impulse response) from the sound source to the microphone. The length of the acoustic transfer function is L. The spectral analysis of the acoustic modeling is generally carried out using short-term windowing. If the length L is shorter than that of the window, the observed complex spectrum is generally represented by

$$O(\omega; n) = S(\omega; n) \cdot H(\omega; n). \tag{4}$$

However, since the length of the acoustic transfer function is greater than that of the window, the observed spectrum is approximately represented by $O(\omega; n) \approx S(\omega; n) \cdot H(\omega; n)$. Here $O(\omega; n)$, $S(\omega; n)$, and $H(\omega; n)$ are the short-term linear complex spectra in analysis window n. Applying the logarithm transform to the power spectrum, we get

$$\log |O(\omega; n)|^2 \approx \log |S(\omega; n)|^2 + \log |H(\omega; n)|^2. \tag{5}$$

In speech recognition, cepstral parameters are an effective representation when it comes to retaining useful speech information. Therefore, we use the cepstrum for acoustic modeling that is necessary to estimate the acoustic transfer function. The cepstrum of the observed signal is given by the inverse Fourier transform of the log spectrum:

$$O_{cep}(t; n) \approx S_{cep}(t; n) + H_{cep}(t; n) \tag{6}$$

where O_{cep}, S_{cep}, and H_{cep} are cepstra for the observed signal, clean speech signal, and acoustic transfer function, respectively. In this paper, we introduce a GMM (Gaussian Mixture Model) of the acoustic transfer function to deal with the influence of a room impulse response.

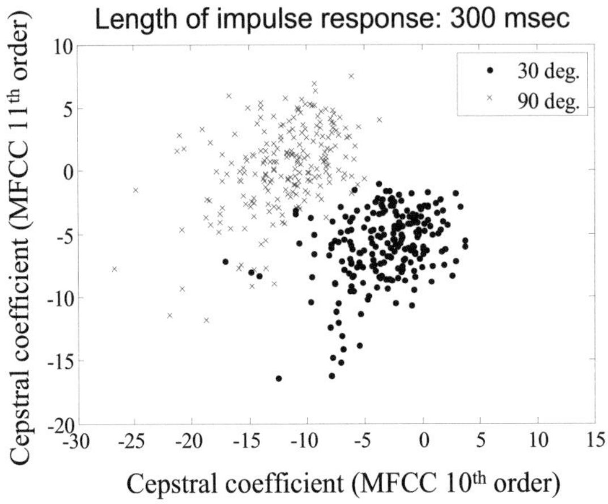

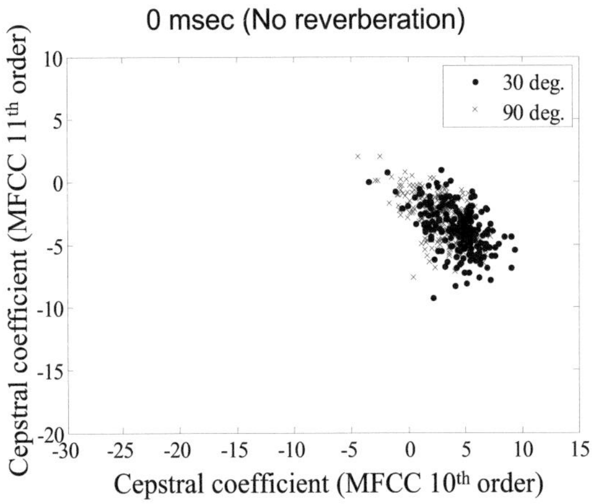

Fig. 3. Difference between acoustic transfer functions obtained by subtraction of short-term-analysis-based speech features in the cepstrum domain

2.3 Difference of acoustic transfer functions

Figure 3 shows the mean values of the cepstrum, H'_{cep}, that were computed for each word using the following equations:

$$H_{cep}(t;n) \approx O_{cep}(t;n) - S_{cep}(t;n) \tag{7}$$

$$H'_{cep}(t) = \frac{1}{N}\sum_{n}^{N} H_{cep}(t;n) \tag{8}$$

where t is the cepstral index. Reverberant speech, O, was created using linear convolution of clean speech and impulse response. The impulse responses were taken from the RWCP sound scene database (Nakamura, 2001), where the loudspeaker was located at 30 and 90 degrees from the microphone. The lengths of the impulse responses are 300 msec and 0 msec. The reverberant speech and clean speech were processed using a 32-msec Hamming window, and then for each frame, n, a set of 16 MFCCs was computed. The 10th and 11th cepstral coefficients for 216 words are plotted in Figure 3. As shown in this figure (300 msec) a difference between the two acoustic transfer functions (30 and 90 degrees) appears in the cepstral domain. The difference shown will be useful for sound source localization estimation. On the other hand, in the case of the 0 msec impulse response, the influence of the microphone and the loudspeaker characteristics are a significant problem. Therefore, it is difficult to discriminate between each position for the 0 msec impulse response.

Also, this figure shows that the variability of the acoustic transfer function in the cepstral domain appears to be large for the reverberant speech. When the length of the impulse response is shorter than the analysis window used for the spectral analysis of speech, the acoustic transfer function obtained by subtraction of short-term-analysis-based speech features in the cepstrum domain comes to be constant over the whole utterance. However, as the length of the impulse response for the room reverberation becomes longer than the analysis window, the variability of the acoustic transfer function obtained by the short-term analysis will become large, with acoustic transfer function being approximately represented by Equation (7). To compensate for this variability, a GMM is employed to model the acoustic transfer function.

3. Maximum-likelihood-based parameter estimation

This section presents a new method for estimating the GMM (Gaussian Mixture Model) of the acoustic transfer function. The estimation is implemented by maximizing the likelihood of the training data from a user's position. In (Sankar & Lee, 1996), a maximum-likelihood (ML) estimation method to decrease the acoustic mismatch for a telephone channel was described, and in (Kristiansson et al., 2001) channel distortion and noise are simultaneously estimated using an expectation maximization (EM) method. In this paper, we introduce the utilization of the GMM of the acoustic transfer function based on the ML estimation approach to deal with a room impulse response.

The frame sequence of the acoustic transfer function in (6) is estimated in an ML manner by using the expectation maximization (EM) algorithm, which maximizes the likelihood of the observed speech:

$$\hat{H} = \underset{H}{\operatorname{argmax}} \Pr(O|H, \lambda_S). \tag{9}$$

Here, λ_S denotes the set of clean speech GMM parameters, while the suffix S represents the clean speech in the cepstral domain. The EM algorithm is a two-step iterative procedure. In the first step, called the expectation step, the following auxiliary function is computed.

$$\begin{aligned} & Q(\hat{H}|H) \\ &= E[\log \Pr(O, c|\hat{H}, \lambda_S)|H, \lambda_S] \\ &= \sum_c \frac{\Pr(O, c|H, \lambda_S)}{\Pr(O|H, \lambda_S)} \cdot \log \Pr(O, c|\hat{H}, \lambda_S) \end{aligned} \tag{10}$$

Here c represents the unobserved mixture component labels corresponding to the observation sequence O.

The joint probability of observing sequences O and c can be calculated as

$$\Pr(O, c|\hat{H}, \lambda_S) = \prod_{n^{(v)}} w_{c_{n^{(v)}}} \Pr(O_{n^{(v)}}|\hat{H}, \lambda_S) \tag{11}$$

where w is the mixture weight and $O_{n^{(v)}}$ is the cepstrum at the n-th frame for the v-th training data (observation data). Since we consider the acoustic transfer function as additive noise in the cepstral domain, the mean to mixture k in the model λ_O is derived by adding the acoustic transfer function. Therefore, (11) can be written as

$$\begin{aligned} &\Pr(O, c|\hat{H}, \lambda_S) \\ &= \prod_{n^{(v)}} w_{c_{n^{(v)}}} \cdot N(O_{n^{(v)}}; \mu^{(S)}_{k_{n^{(v)}}} + \hat{H}_{n^{(v)}}, \Sigma^{(S)}_{k_{n^{(v)}}}) \end{aligned} \tag{12}$$

where $N(O; \mu, \Sigma)$ denotes the multivariate Gaussian distribution. It is straightforward to derive that (Juang, 1985)

$$\begin{aligned} &Q(\hat{H}|H) \\ &= \sum_k \sum_{n^{(v)}} \Pr(O_{n^{(v)}}, c_{n^{(v)}} = k|\lambda_S) \log w_k \\ &\quad + \sum_k \sum_{n^{(v)}} \Pr(O_{n^{(v)}}, c_{n^{(v)}} = k|\lambda_S) \\ &\quad \cdot \log N(O_{n^{(v)}}; \mu^{(S)}_k + \hat{H}_{n^{(v)}}, \Sigma^{(S)}_k) \end{aligned} \tag{13}$$

Here $\mu^{(S)}_k$ and $\Sigma^{(S)}_k$ are the k-th mean vector and the (diagonal) covariance matrix in the clean speech GMM, respectively. It is possible to train those parameters by using a clean speech database.

Next, we focus only on the term involving H.

$$\begin{aligned} &Q(\hat{H}|H) \\ &= \sum_k \sum_{n^{(v)}} \Pr(O_{n^{(v)}}, c_{n^{(v)}} = k|\lambda_S) \\ &\quad \cdot \log N(O_{n^{(v)}}; \mu^{(S)}_k + \hat{H}_{n^{(v)}}, \Sigma^{(S)}_k) \\ &= -\sum_k \sum_{n^{(v)}} \gamma_{k,n^{(v)}} \sum_{d=1}^{D} \left\{ \frac{1}{2} \log(2\pi)^D \sigma^{(S)2}_{k,d} \right. \\ &\quad \left. + \frac{(O_{n^{(v)},d} - \mu^{(S)}_{k,d} - \hat{H}_{n^{(v)},d})^2}{2\sigma^{(S)2}_{k,d}} \right\} \end{aligned} \tag{14}$$

$$\gamma_{k,n^{(v)}} = \Pr(O_{n^{(v)}}, k|\lambda_S) \tag{15}$$

Here D is the dimension of the observation vector O_n, and $\mu^{(S)}_{k,d}$ and $\sigma^{(S)2}_{k,d}$ are the d-th mean value and the d-th diagonal variance value of the k-th component in the clean speech GMM, respectively.

The maximization step (M-step) in the EM algorithm becomes "max $Q(\hat{H}|H)$". The re-estimation formula can, therefore, be derived, knowing that $\partial Q(\hat{H}|H)/\partial \hat{H} = 0$ as

$$\hat{H}_{n^{(v)},d} = \frac{\sum_k \gamma_{k,n^{(v)}} \dfrac{O_{n^{(v)},d} - \mu^{(S)}_{k,d}}{\sigma^{(S)2}_{k,d}}}{\sum_k \dfrac{\gamma_{k,n^{(v)}}}{\sigma^{(S)2}_{k,d}}} \tag{16}$$

After calculating the frame sequence data of the acoustic transfer function for all training data (several words), the GMM for the acoustic transfer function is created. The m-th mean vector and covariance matrix in the acoustic transfer function GMM ($\lambda_H^{(\theta)}$) for the direction (location) θ can be represented using the term $\hat{H}_n$ as follows:

$$\mu_m^{(H)} = \sum_v \sum_{n^{(v)}} \frac{\gamma_{m,n^{(v)}} \hat{H}_{n^{(v)}}}{\gamma_m} \tag{17}$$

$$\Sigma_m^{(H)} = \sum_v \sum_{n^{(v)}} \frac{\gamma_{m,n^{(v)}} (\hat{H}_{n^{(v)}} - \mu_m^{(H)})^T (\hat{H}_{n^{(v)}} - \mu_m^{(H)})}{\gamma_m} \tag{18}$$

Here $n^{(v)}$ denotes the frame number for v-th training data.
Finally, using the estimated GMM of the acoustic transfer function, the estimation of talker localization is handled in an ML framework:

$$\hat{\theta} = \underset{\theta}{\operatorname{argmax}} \Pr(\hat{H}|\lambda_H^{(\theta)}), \tag{19}$$

where $\lambda_H^{(\theta)}$ denotes the estimated GMM for θ direction (location), and a GMM having the maximum-likelihood is found for each test data from among the estimated GMMs corresponding to each position.

4. Experiments

4.1 Simulation experimental conditions

The new talker localization method was evaluated in both a simulated reverberant environment and a real environment. In the simulated environment, the reverberant speech was simulated by a linear convolution of clean speech and impulse response. The impulse response was taken from the RWCP database in real acoustical environments (Nakamura, 2001). The reverberation time was 300 msec, and the distance to the microphone was about 2 meters. The size of the recording room was about 6.7 m×4.2 m (width×depth). Figure 4 and Fig. 5 show the experimental room environment and the impulse response (90 degrees), respectively.
The speech signal was sampled at 12 kHz and windowed with a 32-msec Hamming window every 8 msec. The experiment utilized the speech data of four males in the ATR Japanese speech database. The clean speech GMM (speaker-dependent model) was trained using 2,620 words and has 64 Gaussian mixture components. The test data for one location consisted of 1,000 words, and 16-order MFCCs (Mel-Frequency Cepstral Coefficients) were used as feature vectors. The total number of test data for one location was 1,000 (words) $\times$ 4 (males). The number of training data for the acoustic transfer function GMM was 10 words and 50 words. The speech data for training the clean speech model, training the acoustic transfer function and testing were spoken by the same speakers but had different text utterances respectively. The speaker's position for training and testing consisted of three positions (30, 90, and 130 degrees), five positions (10, 50, 90, 130, and 170 degrees), seven positions (30, 50, 70, ..., 130 and 150 degrees) and nine positions (10, 30, 50, 70, ..., 150, and 170 degrees). Then, for each

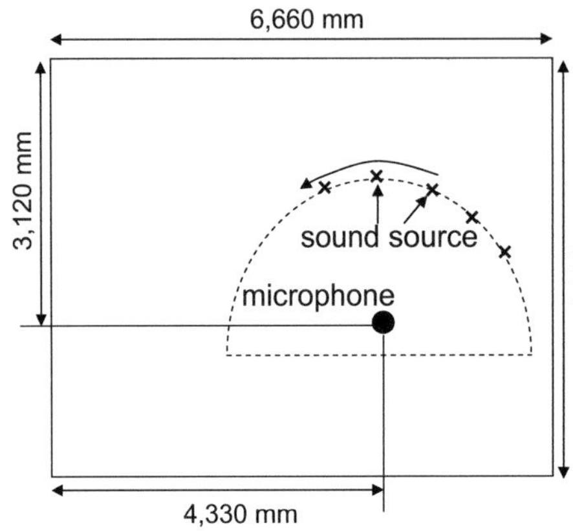

Fig. 4. Experiment room environment for simulation

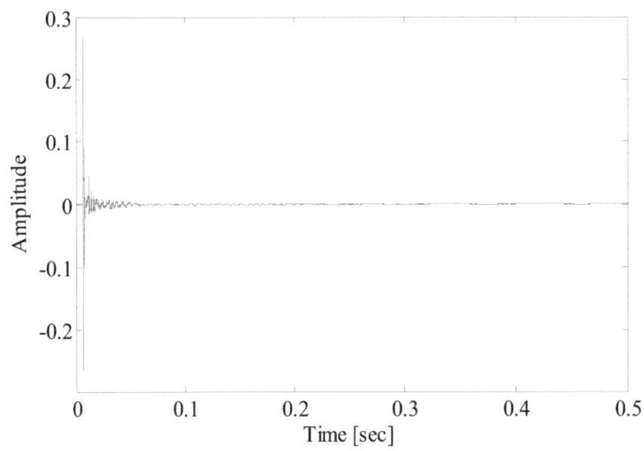

Fig. 5. Impulse response (90 degrees, reverberation time: 300 msec)

set of test data, we found a GMM having the maximum-likelihood from among those GMMs corresponding to each position. These experiments were carried out for each speaker, and the localization accuracy was averaged by four talkers.

4.2 Performance in a simulated reverberant environment

Figure 6 shows the localization accuracy in the three-position estimation task, where 50 words are used for the estimation of the acoustic transfer function. As can be seen from this figure, by increasing the number of Gaussian mixture components for the acoustic transfer function, the localization accuracy is improved. We can expect that the GMM for the acoustic transfer function is effective for carrying out localization estimation.

Figure 7 shows the results for a different number of training data, where the number of Gaussian mixture components for the acoustic transfer function is 16. The performance of the training using ten words may be a bit poor due to the lack of data for estimating the acoustic transfer function. Increasing the amount of training data (50 words) improves in the performance.

In the proposed method, the frame sequence of the acoustic transfer function is separated from the observed speech using (16), and the GMM of the acoustic transfer function is trained by (17) and (18) using the separated sequence data. On the other hand, a simple way to carry

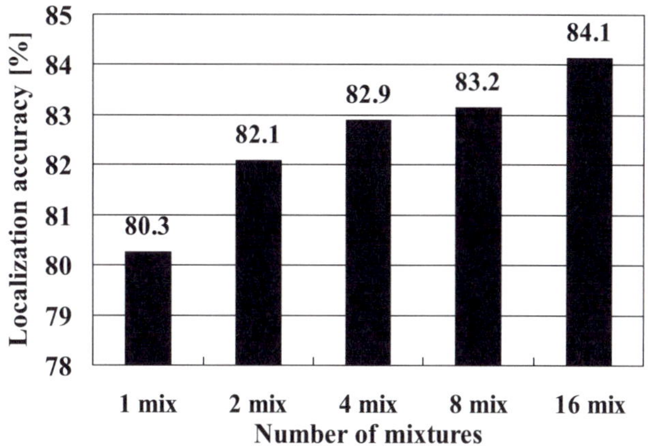

Fig. 6. Effect of increasing the number of mixtures in modeling acoustic transfer function Here, 50 words are used for the estimation of the acoustic transfer function.

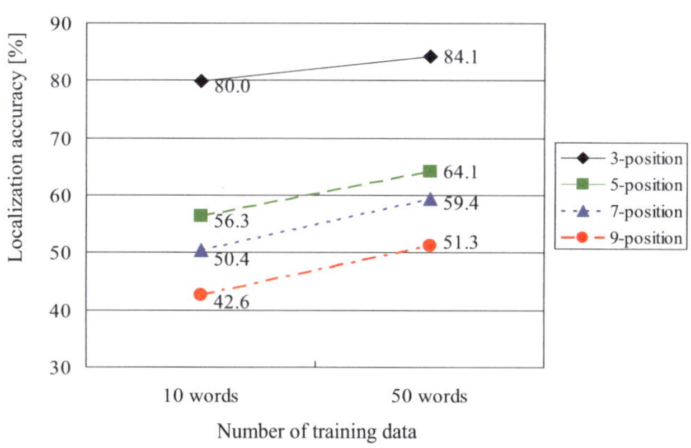

Fig. 7. Comparison of the different number of training data

out voice (talker) localization may be to use the GMM of the observed speech without the separation of the acoustic transfer function. The GMM of the observed speech can be derived in a similar way as in (17) and (18).

$$\mu_m^{(O)} = \sum_v \sum_{n^{(v)}} \frac{\gamma_{m,n^{(v)}} O_{n^{(v)}}}{\gamma_m} \tag{20}$$

$$\Sigma_m^{(O)} = \sum_v \sum_{n^{(v)}} \frac{\gamma_{m,n^{(v)}} (O_{n^{(v)}} - \mu_m^{(O)})^T (O_{n^{(v)}} - \mu_m^{(O)})}{\gamma_m} \tag{21}$$

The GMM of the observed speech includes not only the acoustic transfer function but also clean speech, which is meaningless information for sound source localization. Figure 8 shows the comparison of four methods. The first method is our proposed method and the second is the method using GMM of the observed speech without the separation of the acoustic transfer function. The third is a simpler method that uses the cepstral mean of the observed

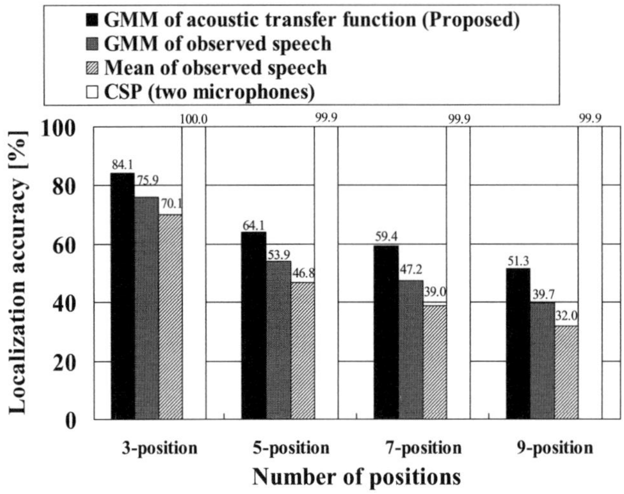

Fig. 8. Performance comparison of the proposed method using GMM of the acoustic transfer function, a method using GMM of observed speech, that using the cepstral mean of observed speech, and CSP algorithm based on two microphones

speech instead of GMM. (Then, the position that has the minimum distance from the learned cepstral mean to that of the test data is selected as the talker's position.) And the fourth is a CSP (Cross-power Spectrum Phase) algorithm based on two microphones, where the CSP uses simultaneous phase information from microphone arrays to estimate the location of the arriving signal (Omologo & Svaizer, 1996). As shown in this figure, the use of the GMM of the observed speech had a higher accuracy than that of the mean of the observed speech. And, the use of the GMM of the acoustic transfer function results in a higher accuracy than that of GMM of the observed speech. The proposed method separates the acoustic transfer function from the short observed speech signal, so the GMM of the acoustic transfer function will not be affected greatly by the characteristics of the clean speech (phoneme). As it did with each test word, it is able to achieve good performance regardless of the content of the speech utterance. But the localization accuracy of the methods using just one microphone decreases as the number of training positions increases. On the other hand, the CSP algorithm based on two microphones has high accuracy even in the 9-position task. As the proposed method (single microphone only) uses the acoustic transfer function estimated from a user's utterance, the accuracy is low.

4.3 Performance in simulated noisy reverberant environments and using a Speaker-independent speech model

Figure 9 shows the localization accuracy for noisy environments. The observed speech data was simulated by adding pink noise to clean speech convoluted using the impulse response so that the signal to noise ratio (SNR) were 25 dB, 15 dB and 5 dB. As shown in Figure 9, the localization accuracy at the SNR of 25 dB decreases about 30 % in comparison to that in a noiseless environment. The localization accuracy decreases further as the SNR decreases.

Figure 10 shows the comparison of the performance between a speaker-dependent speech model and a speaker-independent speech model. For training a speaker-independent clean speech model and a speaker-independent acoustic transfer function model, the speech data spoken by four males in the ASJ Japanese speech database were used. Then, the clean speech

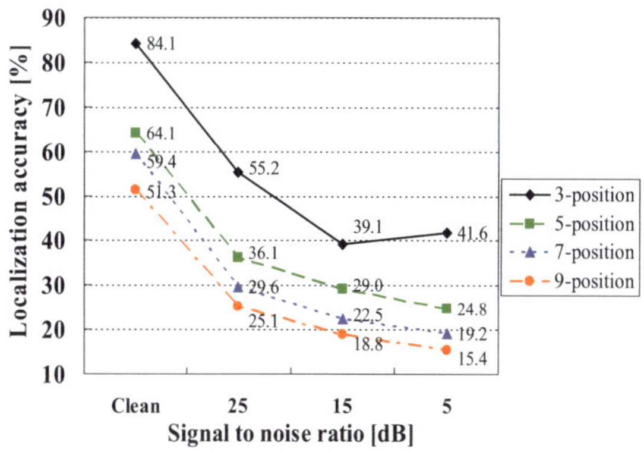

Fig. 9. Localization accuracy for noisy environments

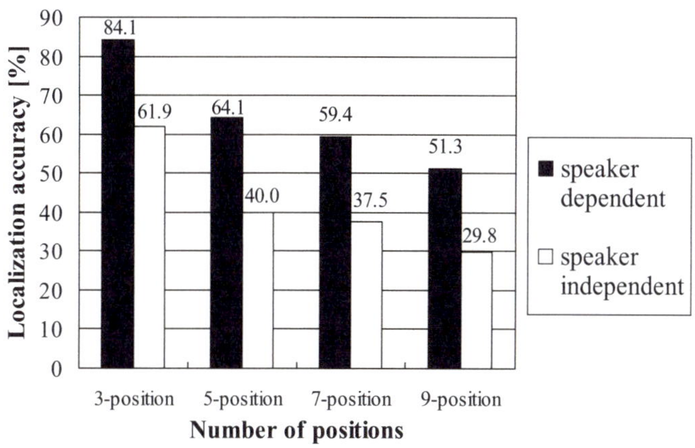

Fig. 10. Comparison of performance using speaker-dependent/-independent speech model (speaker-independent, 256 Gaussian mixture components; speaker-dependent,64 Gaussian mixture components)

GMM was trained using 160 sentences (40 sentences × 4 males) and it has 256 Gaussian mixture components. The acoustic transfer function for training locations was estimated by this clean speech model from 10 sentences for each male. The total number of training data for the acoustic transfer function GMM was 40 (10 sentences × 4 males) sentences. For training the speaker-dependent model and testing, the speech data spoken by four males in the ATR Japanese speech database were used in the same way as described in section 4.1. The speech data for the test were provided by the same speakers used to train the speaker-dependent model, but different speakers were used to train the speaker-independent model. Both the speaker-dependent GMM and the speaker-independent GMM for the acoustic transfer function have 16 Gaussian mixture components. As shown in Figure 10, the localization accuracy of the speaker-independent speech model decreases about 20 % in comparison to the speaker-dependent speech model.

4.4 Performance using Speaker-dependent speech model in a real environment

The proposed method, which uses a speaker-dependent speech model, was also evaluated in a real environment. The distance to the microphone was 1.5 m and the height of the microphone

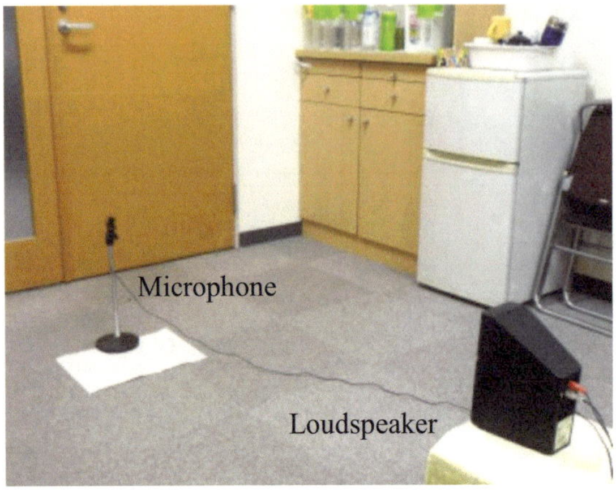

Fig. 11. Experiment room environment

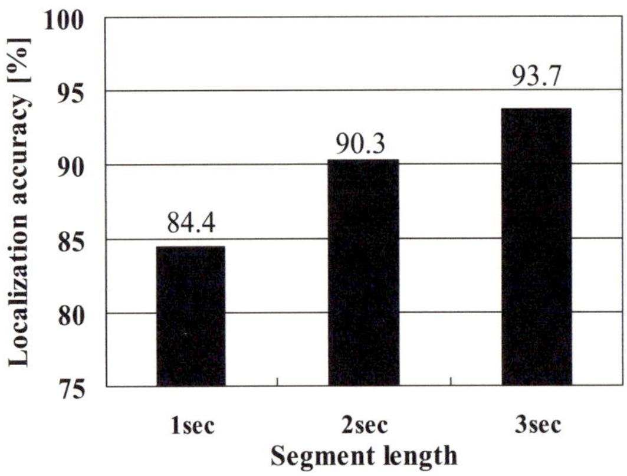

Fig. 12. Comparison of performance using different test segment lengths

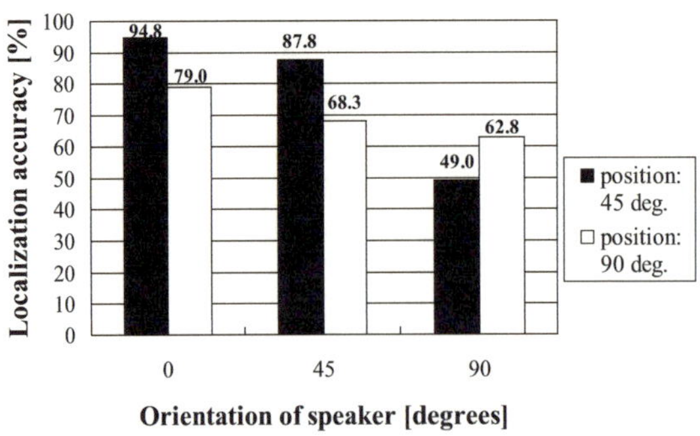

Fig. 13. Effect of speaker orientation

was about 0.45 m. The size of the recording room was about 5.5 m × 3.6 m × 2.7 m (width × depth × height). Figure 11 depicts the room environment of the experiment. The experiment used speech data, spoken by two males, in the ASJ Japanese speech database. The clean speech GMM (speaker-dependent model) was trained using 40 sentences and has 64 Gaussian

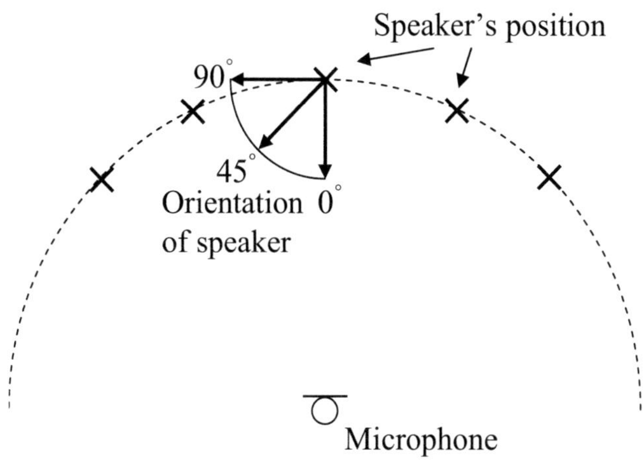

Fig. 14. Speaker orientation

mixture components. The test data for one location consisted of 200, 100 and 66 segments, where one segment has a length of 1, 2 and 3 sec, respectively. The number of training data for the acoustic transfer function was 10 sentences. The speech data for training the clean speech model, training the acoustic transfer function, and testing were spoken by the same speakers, but they had different text utterances respectively. The experiments were carried out for each speaker and the localization accuracy of the two speakers was averaged.

Figure 12 shows the comparison of the performance using different test segment lengths. There were three speaker positions for training and testing (45, 90 and 135 degrees) and one loudspeaker (BOSE Mediamate II) was used for each position. As shown in this figure, the longer the length of the segment was, the more the localization accuracy increased, since the mean of estimated acoustic transfer function became stable. Figure 13 shows the effect when the orientation of the speaker changed from that of the speaker for training. There were five speaker positions for training (45, 65, 90, 115 and 135 degrees). There were two speaker positions for the test (45 and 90 degrees), and the orientation of the speaker changed to 0, 45 and 90 degrees, as shown in Figure 14. As shown in Figure 13, as the orientation of speaker changed, the localization accuracy decreased. Figure 15 shows the plot of acoustic transfer function estimated for each position and orientation of speaker. The plot of the training data is the mean value of all training data, and that for the test data is the mean value of test data per 40 seconds. As shown in Figure 15, as the orientation of the speaker changed from that for training, the estimated acoustic transfer functions were distributed over the distance away from the position of training data. As a result, these estimated acoustic transfer functions were not correctly recognized.

5. Conclusion

This paper has described a voice (talker) localization method using a single microphone. The sequence of the acoustic transfer function is estimated by maximizing the likelihood of training data uttered from a position, where the cepstral parameters are used to effectively represent useful clean speech information. The GMM of the acoustic transfer function based on the ML estimation approach is introduced to deal with a room impulse response. The experiment results in a room environment confirmed its effectiveness for location

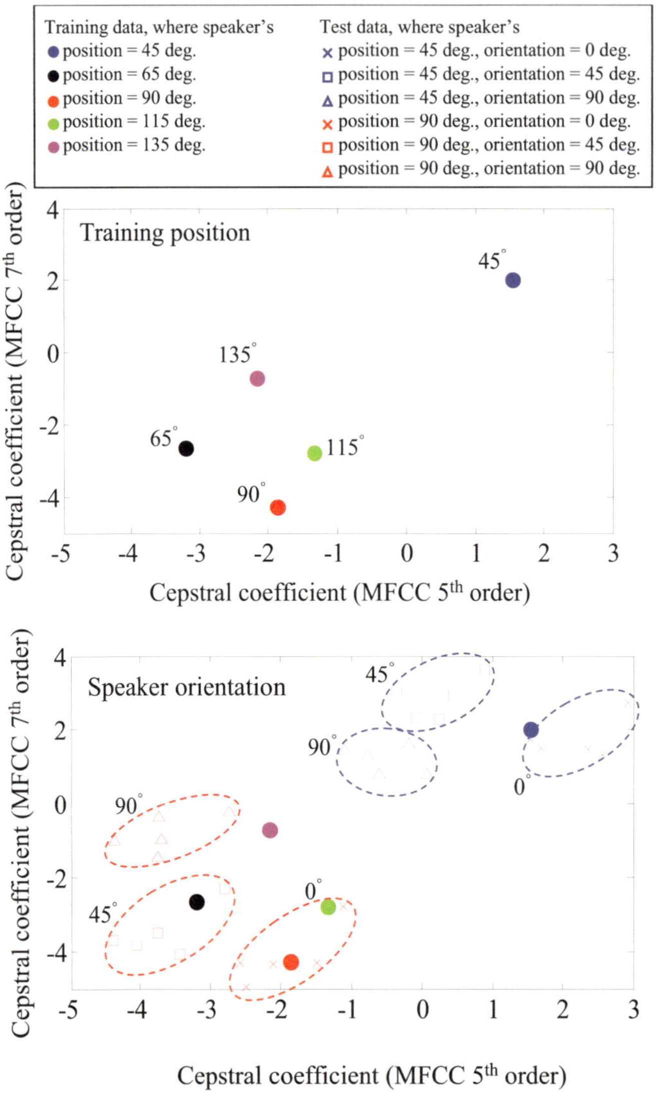

Fig. 15. Mean acoustic transfer function values for five positions (top graph) and mean acoustic transfer function values for three speaker orientations (0 deg, 45 deg, and 90 deg) at a position of 45 deg and 90 deg (bottom graph)

estimation tasks. But the proposed method requires the measurement of speech for each room environment in advance, and the localization accuracy decreases as the number of training positions increases. In addition, not only the position of speaker but also various factors (e.g., orientation of the speaker) affect the acoustic transfer function. Future work will include efforts to improve both localization estimation from more locations and estimation when the conditions other than speaker position change. We also hope to improve the localization accuracy in noisy environments and for speaker-independent speech models.

Also, we will investigate a text-independent technique based on HMM in the modeling of the speech content.

6. References

Johnson, D. & Dudgeon, D. (1996). *Array Signal Processing*, Prentice Hall, Englewood Cliffs, NJ.

Omologo, M. & Svaizer, P. (1996). Acoustic Event Localization in Noisy and Reverberant Environment Using CSP Analysis, *Proceedings of 1996 IEEE International Conference on Acoustics, Speech, and Signal Processing (ICASSP96)*, Institute of Electrical and Electronics Engineers (IEEE), Atlanta, Georgia, pp. 921-924.

Asano, F., Asoh, H. & Matsui, T. (2000). Sound Source Localization and Separation in Near Field, *IEICE Trans. Fundamentals* Vol. E83-A, No. 11, pp. 2286-2294.

Denda, Y., Nishiura, T. & Yamashita, Y. (2006). Robust Talker Direction Estimation Based on Weighted CSP Analysis and Maximum Likelihood Estimation, *IEICE Trans. on Information and Systems* Vol. E89-D, No. 3, pp. 1050-1057.

Keyrouz, F., Naous, Y. & Diepold, K. (2006) A New Method for Binaural 3-D Localization Based on HRTFs, *Proceedings of 2006 IEEE International Conference on Acoustics, Speech, and Signal Processing (ICASSP06)*, Institute of Electrical and Electronics Engineers (IEEE), Toulouse, France, pp. V-341-V-344.

Takimoto, M., Nishino, T. & Takeda, K. (2006). Estimation of a talker and listener's positions in a car using binaural signals, *The Fourth Joint Meeting Acoustical Society of America and Acoustical Society of Japan* Acoustical Society of America and Acoustic Society of Japan, Honolulu, Hawaii, 3pSP33, pp. 3216.

Kristjansson, T., Attias, H. & Hershey, J. (2004). Single Microphone Source Separation Using High Resolution Signal Reconstruction, *Proceedings of 2004 IEEE International Conference on Acoustics, Speech, and Signal Processing (ICASSP04)*, Institute of Electrical and Electronics Engineers (IEEE), Montreal, Quebec, Canada, pp. 817-820.

Raj, B., Shashanka, M. & Smaragdis, P. (2006). Latent Direchlet Decomposition for Single Channel Speaker Separation, *Proceedings of 2006 IEEE International Conference on Acoustics, Speech, and Signal Processing (ICASSP06)*, Institute of Electrical and Electronics Engineers (IEEE), Toulouse, France, pp. 821-824.

Jang, G., Lee, T. & Oh, Y. (2003). A Subspace Approach to Single Channel Signal Separation Using Maximum Likelihood Weighting Filters, *Proceedings of 2003 IEEE International Conference on Acoustics, Speech, and Signal Processing (ICASSP03)*, Institute of Electrical and Electronics Engineers (IEEE), Hong Kong, pp. 45-48.

Nakatani, T. & Juang, B. (2006). Speech Dereverberation Based on Probabilistic Models of Source and Room Acoustics, *Proceedings of 2006 IEEE International Conference on Acoustics, Speech, and Signal Processing (ICASSP06)*, Institute of Electrical and Electronics Engineers (IEEE), Toulouse, France, pp. I-821-I-824.

Takiguchi, T., Nakamura, S. & Shikano, K. (2001). HMM-separation-based speech recognition for a distant moving speaker, *IEEE Transactions on Speech and Audio Processing*, Vol. 9, No. 2, pp. 127-140.

Takiguchi, T. & Nishimura, M. (2004). Acoustic Model Adaptation Using First Order Prediction for Reverberant Speech, *Proceedings of 2004 IEEE International Conference on Acoustics, Speech, and Signal Processing (ICASSP04)*, Institute of Electrical and Electronics Engineers (IEEE), Montreal, Quebec, Canada, pp. 869-872.

Sankar, A. & Lee, C. (1996). A Maximum-Likelihood Approach to Stochastic Matching for Robust Speech Recognition, *IEEE Transactions on Speech and Audio Processing*, Vol. 4, No. 3, pp. 190-202.

Kristiansson, T., Frey, B., Deng, L. & Acero, A. (2001). Joint Estimation of Noise and Channel Distortion in a Generalized EM framework, *IEEE Automatic Speech Recognition and Understanding Workshop (ASRU)*, Institute of Electrical and Electronics Engineers (IEEE), Trento, Italy, pp. 155-158.

Juang, B. (1985). Maximum-likelihood estimation of mixture multivariate stochastic observations of Markov chains, *AT&T Tech. J.*, Vol. 64, No. 6, pp. 1235-1249.

Nakamura, S. (2001). Acoustic sound database collected for hands-free speech recognition and sound scene understanding, *International Workshop on Hands-Free Speech Communication*, International Speech Communication Association, Kyoto, Japan, pp. 43-46.

4

Localization Error: Accuracy and Precision of Auditory Localization

Tomasz Letowski[1] and Szymon Letowski[2]
[1]*U.S. Army Research Laboratory, Human Research and Engineering Directorate, Aberdeen Proving Ground, MD 21005-5425*
[2]*Evidence Based Research, Inc., Vienna, VA 22182*
USA

1. Introduction

The act of localization is the estimation of the true location of an object in space and is characterized by a certain amount of inherent uncertainty and operational bias that results in estimation errors. The type and size of the estimation errors depend on the properties of the emitted sound, the characteristics of the surrounding environment, the specific localization task, and the abilities of the listener.

While the general idea of localization error is straightforward, the specific concepts and measures of localization error encountered in the psychoacoustic literature are quite diverse and frequently poorly described, making generalizations and data comparison quite difficult. In addition, the same concept is sometimes described in different papers by different terms, and the same term is used by different authors to refer to different concepts. This variety of terms and metrics used with inconsistent semantics can easily be a source of confusion and may cause the reader to misinterpret the reported data and conclusions.

A fundamental property of localization estimates is that in most cases they are angular and thus represent circular (spherical) variables, which in general cannot be described by a linear distribution as assumed in classical statistics. The azimuth and elevation of the sound source locations define an ambiguous conceptual sphere, which can only be fully analyzed with the methods of spherical statistics. However, these methods are seldom used in psychoacoustic studies, and it is not immediately clear to what degree they should be utilized. In many cases, localization estimates may, in fact, be correctly analyzed using linear methods, but neither the necessary conditions for nor the limitations of linear methods have been clearly stated.

In sum, localization error is a widely used and intuitively simple measure of spatial uncertainty and spatial bias in the perception of sound source location, but both a common terminology for its description and a broad understanding of the implications of its circular character are lacking. Some of the issues related to these topics are discussed in the subsequent sections. The presented concepts and explanations are intended to clarify some existing terminological ambiguities and offer some guidance as to the statistical treatment of localization error data. The focus of the discussion is on issues related to localization judgments, with only marginal attention given to distance estimation judgments which deserve to be the object of a separate article.

2. Basis of auditory localization

Spatial hearing provides information about the acoustic environment; about its geometry and physical properties and about the locations of sound sources. Sound localization generally refers to the act or process of identifying the direction toward a sound source on the basis of sound emitted by the source (see discussion of this definition in Section 3). For living organisms, this is a sensory act based on the perceived auditory stimulation. In the case of machine localization, it is an algorithmic comparison of signals arriving at various sensors. The sound can be either the main product of the source or a by-product of its operation. The act of sound localization when performed by living organisms can also be referred to as *auditory localization,* and this term is used throughout this chapter.

The localization ability of humans depends on a number of anatomical properties of the human auditory system. The most important of these is the presence of two entry points to the auditory system (the external ears) that are located on opposite sides of the human head. Such a configuration of the auditory input system causes a sound coming at the listener from an angle to have a different sound intensity and time of arrival at each ear. The difference in sound intensity is mainly caused by the acoustic shadow and baffle effects of the head and results in a lower sound intensity at the ear located farther away from the sound source (Strutt, 1876; Steinhauser, 1879). The difference in time of arrival is caused by the difference in the distance the sound has to travel to each of the ears (Strutt, 1907; Wilson and Myers, 1908). These differences are normally referred to as the *interaural intensity difference* (IID) and the *interaural time difference* (ITD). In the case of continuous pure tones and other periodic signals the term *interaural phase difference* (IPD) is used in place of ITD since such sounds have no clear reference point in time. The IID and ITD (IPD) together are called the *binaural localization cues.* The IID is the dominant localization cue for high frequency sounds, while the ITD (IPD) is the dominant cue for low frequency sounds (waveform phase difference). The ITD (IPD) is additionally an important cue for high frequency sounds because of differences in the waveform envelope delay (group delay) (Henning, 1974; 1980; McFadden & Pasanen, 1976).

Binaural cues are the main localization mechanisms in the horizontal plane but are only marginally useful for vertical localization or front-back differentiation. This is due to spatial ambiguity caused by head symmetry and referred to as the *cone of confusion* (Wallach, 1939). The *cone of confusion* is the imaginary cone extending outward from each ear along the interaural axis that represents sound source locations producing the same interaural differences. Although asymmetry in ear placement on the head and in the shape of the pinnae provides some disambiguation, the sound source positions located on the surface of the cone of confusion cannot be identified using binaural cues and can only be resolved using spectral cues associated with the directional sound filtering of the human body. These cues are called monaural cues as they do not depend on the presence of two ears.

Monaural cues result from the shadowing and baffle effects of the pinna and the sound reflections caused by the outer ear (pinna and tragus), head, and upper torso (Steinhauser, 1879; Batteau, 1967; Musicant & Butler, 1984; Lopez-Poveda & Meddis, 1996). These effects and reflections produce peaks and troughs in the sound spectrum that are unique for each sound source location in space relative to the position of the listener (Bloom, 1977; Butler & Belendiuk, 1977; Watkins, 1978).

Monaural cues and the related Interaural Spectrum Difference (ISD) also help binaural horizontal localization (Jin et al., 2004; Van Wanrooij & Van Opstal, 2004), but they are most critical for vertical localization and front-back differentiation. The spectral cues that are the most important for accurate front-back and up-down differentiation are located in the 4-16 kHz frequency range (e.g., Langendijk & Bronkhorst, 2002). Spatial localization ability in both horizontal and vertical planes is also dependent on slight head movements, which cause momentary changes in the peak-and-trough pattern of the sound spectrum at each ear (Young, 1931; Wallach, 1940; Perrett & Noble, 1997; Iwaya et al., 2003), visual cues, and prior knowledge of the stimulus (Pierce, 1901; Rogers & Butler, 1992). More information about the physiology and psychology of auditory localization can be found elsewhere (e.g., Blauert, 1974; Yost & Gourevitch, 1987; Moore, 1989; Yost et al., 2008; Emanuel & Letowski (2009).

3. Terminology, notation, and conventions

The broad interest and large number of publications in the field of auditory localization has advanced our knowledge of neurophysiologic processing of spatial auditory signals, the psychology of spatial judgments, and environmental issues in determining the locations of sound sources. The authors of various experimental and theoretical publications range from physiologists to engineers and computer scientists, each bringing their specific expertise and perspective. The large number of diversified publications has also led to a certain lack of consistency regarding the meaning of some concepts. Therefore, before discussing the methods and measures used to describe and quantify auditory localization errors in Section 5, some key concepts and terminological issues are discussed in this and the following section.

Auditory spatial perception involves the perception of the surrounding space and the locations of the sound sources within that space on the basis of perceived sound. In other words, auditory spatial perception involves the perception of sound spaciousness, which results from the specific volume and shape of the surrounding space, and the identification of the locations of the primary and secondary (sound reflections) sound sources operating in the space in relation to each other and to the position of the listener.

In very general terms, auditory spatial perception involves four basic elements:

- Horizontal localization (azimuth, declination)
- Vertical localization (elevation)
- Distance estimation
- Perception of space properties (spaciousness)

The selection of these four elements is based on a meta-analysis of the literature on spatial perception and refers to the traditional terminology used in psychoacoustic research studies on the subject matter. It seems to be a logical, albeit obviously arbitrary, classification.

A direction judgment toward a sound source located in space is an act of localization and can be considered a combination of both horizontal and vertical localization judgments. Horizontal and vertical localization judgments are direction judgments in the corresponding planes and may vary from simple left-right, up-down, and more-less discriminations, to categorical judgments, to the absolute identifications of specific directions in space. A special form of localization judgments for phantom sound sources located in the head of the listener is called *lateralization*. Therefore, the terms lateralization and localization refer respectively to judgment of the internal and external positions of sound sources in reference to the listener's head (Yost & Hafter, 1987; Emanuel & Letowski, 2009).

Similarly to localization judgments, distance judgments may have the form of discrimination judgments (closer-farther), relative numeric judgments (half as far - twice as far), or absolute numeric judgments in units of distance. In the case of two sound sources located at different distances from the listener, the listener may estimate their relative difference in distance using the same types of judgments. Such relative judgments are referred to as *auditory distance difference* or *auditory depth* judgments.

Both distance and depth judgments are less accurate than angular localization judgments and show large intersubject variability. In general, perceived distance *PD* is a power function of the actual distance *d* and can be described as

$$PD = kd^{a}, \tag{1}$$

where *a* and *k* are fitting constants dependent on the individual listener. Typically *k* is close to but slightly smaller than 1 (k=0.9), and *a* is about 0.4 but varies widely (0.3-0.8) across listeners (Zahorik et al., 2005).

The above differentiation between localization and distance estimation is consistent with the common interpretation of auditory localization as the act of identifying the direction toward the sound source (White, 1987; Morfey, 2001; Illusion, 2010). It may seem, however, inconsistent with the general definition of localization which includes distance estimation (APA, 2007; Houghton Mifflin, 2007). Therefore, some authors who view distance estimation as an inherent part of auditory localization propose other terms, e.g., direction-of-arrival (DOA) (Dietz et al., 2010), to denote direction-only judgments and distinguish them from general localization judgments.

The introduction of a new term describing direction-only judgments is intended to add clarity to the language describing auditory spatial perception. However, the opposite may be true since the term *localization* has a long tradition in the psychoacoustic literature of being used to mean the judgment of direction. This meaning also agrees with the common usage of this term. Therefore, it seems reasonable to accept that while the general definition of localization includes judging the distance to a specific location, it does not mandate it, and in its narrow meaning, localization refers to the judgment of direction. In this context, the term localization error refers to errors in direction judgment, and the term distance estimation error to errors in distance estimation.

Spaciousness is the perception of being surrounded by sound and is related to the type and size of the surrounding space. It depends not only on the type and volume of the space but also on the number, type, and locations of the sound sources in the space. Perception of spaciousness has not yet been well researched and has only recently become of more scientific interest due to the rapid development of various types of spatial sound recording and reproduction systems and AVR simulations (Griesinger, 1997). The literature on this subject is very fragmented, inconsistent, and contradictory. The main reason for this is that unlike horizontal localization, vertical localization, and distance estimation judgments, which are made along a single continuum, spaciousness is a multidimensional phenomenon without well defined dimensions and one that as of now can only be described in relative terms or using categorical judgments.

The two terms related to spaciousness that are the most frequently used are *listener envelopment* (LEV) and *apparent source width* (ASW). Listener envelopment describes the degree to which a listener is surrounded by sound, as opposed to listening to sound that happens "somewhere else". It is synonymous to *spatial impression* as defined by Barron and

Marshall (1981). Some authors treat both these terms as synonymous to spaciousness, but spaciousness can exist without listener envelopment. The ASW is also frequently equated with spaciousness, but such an association does not agree with the common meanings of both *width* and *spaciousness* and should be abandoned (Griesinger, 1999). The concept of ASW relates more to the size of the space occupied by the active sound sources and should be a subordinate term to spaciousness. Thus, LEV and ASW can be treated as two complementary elements of spaciousness (Morimoto, 2002). Some other correlated or subordinate aspects of spaciousness are panorama (a synonym of ASW), perspective, ambience, presence, and warmth.

Depending on the task given to the listener there are two basic types of localization judgments:

- Relative localization (discrimination task)
- Absolute localization (identification task)

Relative localization judgments are made when one sound source location is compared to another, either simultaneously or sequentially. Absolute localization judgments involve only one sound source location that needs to be directly pointed out. In addition, absolute localization judgments can be made on a continuous circular scale and expressed in degrees (°) or can be restricted to a limited set of preselected directions. The latter type of judgment occurs when all the potential sound source locations are marked by labels (e.g., number), and the listener is asked to identify the sound source location by label. The actual sound sources may or may not be visible. This type of localization judgment, in which the identification data are later expressed as selection percentages, i.e., the percent of responses indicating each (or just the correct) location, is referred to throughout this chapter as *categorical localization*.

From the listener's perspective, the most complex and demanding judgments are the absolute localization judgments, and they are the main subject of this chapter. The other two types of judgments, discrimination judgments and categorization judgments, are only briefly described and compared to absolute judgments later in the chapter.

In order to assess the human ability to localize the sources of incoming sounds, the physical reference space needs to be defined in relation to the position of the human head. This reference space can be described either in the rectangular or polar coordinate system. The rectangular coordinate system x, y, z is the basis of Euclidean geometry and is also called the Cartesian coordinate system. In the head-oriented Cartesian coordinate system the x, y, and z axes are typically oriented as left-right (west-east), back-front (south-north), down-up (nadir-zenith), respectively. The east, front, and up directions indicate the positive ends of the scales.

The Euclidean planes associated with the Cartesian coordinate system are the vertical lateral (x-z), the vertical sagittal (y-z), and the horizontal (x-y) planes. The main reference planes of symmetry for the human body are:

- Median sagittal (midsagittal) plane: y-z plane
- Frontal (coronal) lateral plane: x-z plane
- Axial (transversal, transaxial) horizontal plane: x-y plane

The relative orientations of the sagittal and lateral planes and the positions of the median and frontal planes are shown in Figure 2. The virtual line passing though both ears in the frontal plane is called the *interaural axis*. The ear closer to the sound source is termed the ipsilateral ear and the ear farther away from the sound source is the contralateral ear.

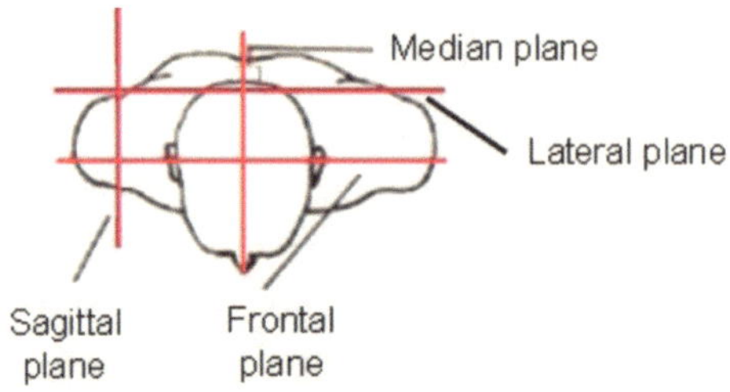

Fig. 1. Main reference planes of the human body. The axial plane is parallel to the page.

The median (midsagittal) plane is the sagittal plane (see Figure 1) that is equidistant from both ears. The frontal (coronal) plane is the lateral plane that divides the listener's head into front and back hemispheres along the interaural axis. The axial (transversal) plane is the horizontal plane of symmetry of the human body. Since the axial plane is not level with the interaural axis of human hearing, the respective plane, called the *visuoaural plane* by Knudsen (1982), is referred to here as the *hearing plane*, or as just the horizontal plane.
In the polar system of coordinates, the reference dimensions are d (distance or radius), θ (declination or azimuth), and φ (elevation). Distance is the amount of linear separation between two points in space, usually between the observation point and the target. The angle of declination (azimuth) is the horizontal angle between the medial plane and the line connecting the point of observation to the target. The angle of elevation is the vertical angle between the hearing plane and the line from the point of observation to the target. The Cartesian and polar systems are shown together in Figure 2.

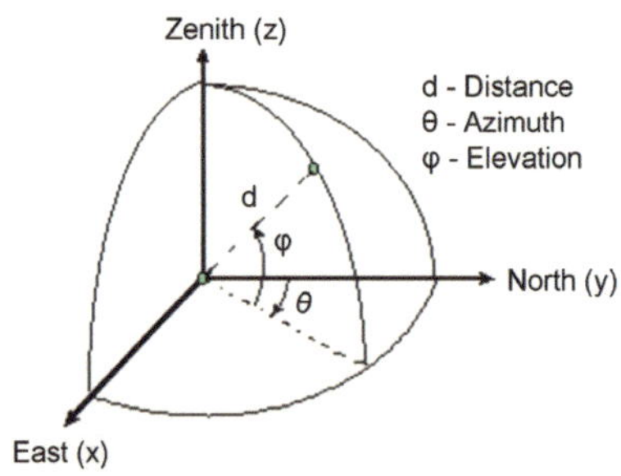

Fig. 2. Commonly used symbols and names in describing spatial hearing coordinates.

One advantage of the polar coordinate system over Cartesian coordinate system is that it can be used in both Euclidean geometry and the spherical, non-Euclidean, geometry that is useful in describing relations between points on a closed surface such as a sphere. In auditory perception studies two spherical systems of coordinates are used. They are referred to as the *single-pole system* and the *two-pole system*. Both are shown in Figure 3.
The head-oriented single-pole system is analogous to the planetary coordinate system of longitudes and latitudes. In the two-pole system, both longitudes and latitudes are represented by series of parallel circles. The single-pole system is widely used in many fields of science. However, in this system the length of an arc between two angles of azimuth depends on elevation. The two-pole system makes the length of the arc between two angles of azimuth the same regardless of elevation. Though less intuitive, this system may be convenient for some types of data presentation (Knudsen, 1982; Makous &

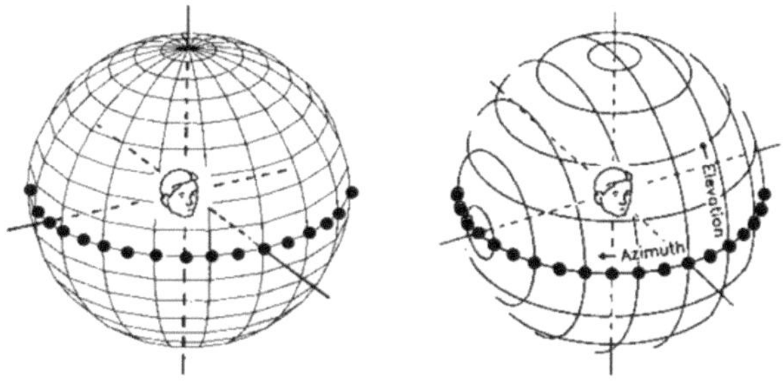

Fig. 3. Single-pole (left) and two-pole (right) spherical coordinate systems. Adapted from Carlile (1996).

Middlebrooks, 1990). Since both these systems share the same concepts of azimuth and elevation, it is essential that the selection of the specific spherical coordinate system always be explicit (Leong & Carlile, 1998).

It should also be noted that there are two conventions for numerically labeling angular degrees that are used in scientific literature: the *360° scheme* and the *±180° scheme*. There are also two possibilities for selecting the direction of positive angular change: clockwise (e.g., Tonning, 1970) or counterclockwise (e.g., Pedersen & Jorgensen, 2005).

The use of two notational schemes is primarily a nuisance that necessitates data conversion in order to compare or combine data sets labeled with different schemes. However, converting angles that are expressed differently in the two schemes from one scheme to the other is just a matter of either adding or subtracting 360°.

In the case of localization studies, where differences between angles are the primary consideration, the ±180° labeling scheme is overwhelmingly preferred. First, it is much simpler and more intuitive to use positive and negative angles to describe angular difference. Second, and more importantly, the direct summing and averaging of angular values can only be done with angles that are contained within a (numerically) continuous range of 180°, such as ±90°. If the 360° scheme is used, then angles to the left and right of 0° (the reference angle) cannot be directly added and must be converted into vectors and added using vector addition.

Less clear is the selection of the positive and negative directions of angular difference. However, if the ±180° scheme is used, the absolute magnitude of angular values is the same regardless of directionality, which is another reason to prefer the ±180° scheme. Under the 360° scheme, the clockwise measurement of any angle other than 180° will have a different magnitude than that same angle measured counterclockwise, i.e., 30° in the clockwise direction is 330° in the counterclockwise direction.

In mathematics (e.g., geometry) and physics (e.g., astronomy), a displacement in a counterclockwise direction is considered positive, and a displacement in a clockwise direction is considered negative. In geometry, the quadrants of the circle are ordered in a counterclockwise direction, and an angle is considered positive if it extends from the x axis in a counterclockwise direction. In astronomy, all the planets of our solar system, when observed from above the Sun, rotate and revolve around the Sun in a counterclockwise direction (except for the rotation of Venus).

However, despite the scientific basis of the counterclockwise rule, the numbers on clocks and all the circular measuring scales, including the compass, increase in a clockwise direction, effectively making it the positive direction. This convention is shown in Figure 2 and is accepted in this chapter. For locations that differ in elevation, the upward direction from a 0° reference point in front of the listener is normally considered as the positive direction, and the downward direction is considered to be the negative direction.

4. Accuracy and precision of auditory localization

The human judgment of sound source location is a noisy process laden with judgment uncertainty, which leads to localization errors. Auditory localization error (LE) is the difference between the estimated and actual directions toward the sound source in space. This difference can be limited to difference in azimuth or elevation or can include both (e.g., Carlile et al., 1997). The latter can be referred to as compound LE.

Once the localization act is repeated several times, LE becomes a statistical variable. The statistical properties of this variable are generally described by *spherical statistics* due to the spherical/circular nature of angular values ($\theta = \theta + 360°$). However, if the angular judgments are within a ±90° range (as is often the case in localization judgments, after disregarding front-back reversals), the data distribution can be assumed to have a linear character, which greatly simplifies data analysis. Front-back errors should be extracted from the data set and analyzed separately in order to avoid getting inflated localization error (Oldfield & Parker, 1984; Makous & Middlebrooks, 1990; Bergault, 1992; Carlile et al., 1997). Some authors (e.g. Wightman & Kistler, 1989) mirror the perceived reverse locations about the interaural axis prior to data analysis in order to preserve the sample size. However, this approach inflates the power of the resulting conclusions. Only under specific circumstances and with great caution should front-back errors be analyzed together with other errors (Fisher, 1987). The measures of linear statistics commonly used to describe the results of localization studies are discussed in Section 5. The methods of spherical (circular) statistical data analysis are discussed in Section 6.

The linear distribution used to describe localization judgments, and in fact most human judgment phenomena, is the normal distribution, also known as the Gaussian distribution. It is a purely theoretical distribution but it well approximates distributions of human errors, thus its common use in experiments with human subjects. In the case of localization judgments, this distribution reflects the random variability of the localizations while emphasizing the tendency of the localizations to be centered on some direction (ideally the true sound source direction) and to become (symmetrically) less likely the further away we move from that central direction.

The normal distribution has the shape of a bell and is completely described in its ideal form by two parameters: the mean (μ) and the standard deviation (σ). The mean corresponds to the central value around which the distribution extends, and the standard deviation describes the range of variation. In particular, approximately 2/3 of the values (68.2%) will be within one standard deviation from the mean, i.e., within the range $[\mu - \sigma, \mu + \sigma]$. The mathematical formula and graph of the normal distribution are shown in Figure 4.

Based on the above discussion, each set of localization judgments can be described by a specific normal distribution with a specific mean and standard deviation. Ideally, the mean of the distribution should correspond with the true sound source direction. However, any lack of symmetry in listener hearing or in the listening conditions may result in a certain bias in listener responses and cause a misalignment between the perceived location of the sound source and its actual location. Such bias is called constant error (CE).

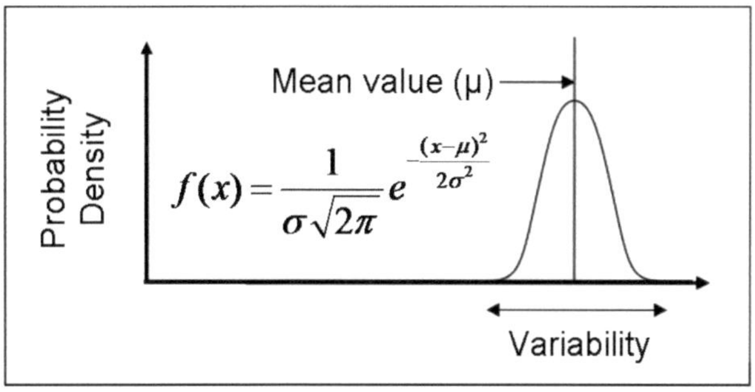

Fig. 4. Normal distribution. Standard deviation (σ) is the range of variability around the mean value ($\mu \pm \sigma$) that accounts for approximately 2/3 of all responses.

Another type of error is introduced by both listener uncertainty/imprecision and random changes in the listening conditions. This error is called random error (RE). Therefore, LE can be considered as being composed of two error components with different underlying causes: constant error (CE) resulting from a bias in the listener and/or environment and random error (RE) resulting from the inherent variability of listener perception and listening conditions. If LE is described by a normal distribution, CE is given by the difference between the true sound source location and the mean of the distribution (x_o) and RE is characterized by the standard deviation (σ) of the distribution.

The concepts of CE and RE can be equated, respectively, with the concepts of precision and accuracy of a given set of measurements. The definitions of both these terms, along with common synonyms (although not always used correctly), are given below:

> Accuracy (constant error, systematic error, validity, bias) is the measure of the degree to which the measured quantity is the same as its actual value.

> Precision (random error, repeatability, reliability, reproducibility) is the measure of the degree to which the same measurement made repeatedly produces the same results.

The relationship between accuracy and precision and the normal distribution from Figure 4 are shown in Figure 5.

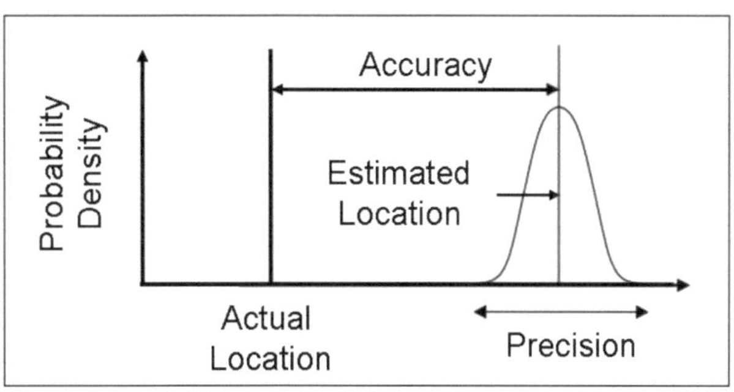

Fig. 5. Concepts of accuracy in precision in localization judgments.

Localization accuracy depends mainly on the symmetry of the auditory system of the listener, the type and behavior of the sound source, and the acoustic conditions of the surrounding space. It also depends on the familiarity of the listener with the listening conditions and on the non-acoustic cues available to the listener. For example, auditory localization accuracy is affected by eye position (Razavi et al., 2007). Some potential bias may also be introduced by the reported human tendency to misperceive the midpoint of the angular distance between two horizontally distinct sound sources. Several authors have reported the midpoint to be located 1° to 2° rightward (Cusak et al., 2001; Dufour et al., 2007; Sosa et al., 2010), although this shift may be modulated by listener handedness. For example, Ocklenburg et al. (2010) observed a rightward shift for left-handed listeners and a leftward shift for right-handed listeners.

Localization precision depends primarily on fluctuations in the listener's attention, the type and number of sound sources, their location in space, and the acoustic conditions of the surrounding space. In addition, both localization accuracy and precision depend to a great degree on the data collection methodology (e.g., direct or indirect pointing, verbal identification, etc). In general, the overall goodness-of-fit of the localization data to the true target location can be expressed in terms of error theory as (Bolshev, 2002) as:

$$p(\theta)=\frac{1}{\sqrt{2}}\times\frac{1}{\sqrt{(CE^2+RE^2)}}\,. \tag{2}$$

5. Linear statistical measures

The two fundamental classes of measures describing probability distributions are measures of central tendency and measures of dispersion. Measures of central tendency, also known as measures of location, characterize the central value of a distribution. Measures of dispersion, also known as measures of spread, characterize how spread out the distribution is around its central value. In general, distributions are described and compared on the basis of a specific measure of central tendency in conjunction with a specific measure of spread.

For the normal distribution, the mean (μ), a measure of central tendency, and the standard deviation (σ), a measure of dispersion, serve to completely describe (parametrize) the distribution. There is, however, no way of directly determining the true, actual values of these parameters for a normal distribution that has been postulated to characterize some population of judgments, measurements, etc. Thus these parameters must be estimated on the basis of a representative sample taken from the population. The sample arithmetic mean (x_o) and the sample standard deviation (SD) are the standard measures used to estimate the mean and standard deviation of the underlying normal distribution.

The sample mean and standard deviation are highly influenced by outliers (extreme values) in the data set. This is especially true for smaller sample sizes. Measures that are less sensitive to the presence of outliers are referred to as robust measures (Huber & Ronketti, 2009). Unfortunately, many robust measures are not very efficient, which means that they require larger sample sizes for reliable estimates. In fact, for normally distributed data (without outliers), the sample mean and standard deviation are the most efficient estimators of the underlying parameters.

A very robust and relatively efficient measure of central tendency is the median (ME). A closely related measure of dispersion is the median absolute deviation (MEAD), which is also very robust but unfortunately also very inefficient. A more efficient measure of dispersion that is however not quite as robust is the mean absolute deviation (MAD). Note that the abbreviation "MAD" is used in other publications to refer to either of these two measures. The formulas for both the standard and robust sample measures discussed above are given below in Table 1. They represent the basic measures used in calculating LE when traditional statistical analysis is performed.

Measure Name	Symbol	Definition/Formula	Comments
Arithmetic Mean	x_o	$x_O = \frac{1}{n}\sum_{i=1}^{n} x_i$	
Standard Deviation	SD	$SD = \sqrt{\frac{1}{n}\sum_{i=1}^{n}(x_i - x_O)^2}$	V (variance) = SD^2.
Median	ME	middle value of responses	
Median Absolute Deviation	MEAD	middle value of the absolute deviations from the median	
Mean Absolute Deviation	MAD	$MAD = \frac{1}{n}\sum_{i=1}^{n} \lvert x_i - x_O \rvert$	

Table 1. Basic measures used to estimate the parameters of a normal distribution.

Strictly speaking, the sample median estimates the population median, which is the midpoint of the distribution, i.e., half the values (from the distribution) are below it and half are above it. The median together with the midpoints of the two halves of the distribution on either side of the median divide the distribution into 4 four parts of equal probability. The three dividing points are called the 1st, 2nd, and 3rd quartiles (Q1, Q2 and Q3), with the 2nd quartile simply being another name for the median. Since the normal distribution is symmetric around its mean, its mean is also its median, and so the sample median can be used to directly estimate the mean of a normal distribution.

The median absolute deviation of a distribution does not coincide with its standard deviation, thus the sample median absolute deviation does not give a direct estimate of the population standard deviation. However, in the case of a normal distribution, the median absolute deviation corresponds to the difference between the 3rd and 2nd quartiles, which is proportional to the standard deviation. Thus for a normal distribution the relationship between the standard deviation and the MEAD is given by (Goldstein & Taleb, 2007):

$$\sigma \approx 1.4826(Q3 - Q2) = 1.4826(MEAD) \tag{3}$$

The SD is the standard measure of RE, while the standard measure of CE is the mean signed error (ME), also called mean bias error, which is equivalent to the difference between the sample mean of the localization data (x_o) and the true location of the sound source. The unsigned, or absolute, counterpart to the ME, the mean unsigned error (MUE) is a measure of total LE as it represents a combination of both the CE and the RE. The MUE was used among others by Makous and Middlebrooks (1990) in analyzing their data. Another error measure that combines the CE and RE is the root mean squared error (RMSE). The relationship between these three measures is given by the following inequality, where n is the sample size (Willmott & Matusuura, 2005).

$$|ME| \leq MUE \leq RMSE \leq \sqrt{n}MUE\,. \tag{4}$$

The RE part of the RMSE is given by the sample standard deviation (SD), but the RE in the MUE does not in general correspond to any otherwise defined measure. However, if each localization estimate is shifted by the ME so as to make the CE equal to zero, the MUE of the data normalized in this way is reduced to the sample mean absolute deviation (MAD). Since the MAD is not affected by linear transformations, the MAD of the normalized data is equal to the MAD of the non-normalized localizations and so represents the RE of the localizations. Thus, the MAD is also a measure of RE. For a normal distribution, the standard deviation is proportional to the mean absolute deviation in the following ratio (Goldstein & Taleb, 2007):

$$\sigma = \sqrt{\frac{\pi}{2}}MAD \approx 1.253(MAD) \tag{5}$$

This means that for sufficiently large sample sizes drawn from a normal distribution, the normalized MUE (=MAD) will be approximately equal to 0.8 times the SD. The effect of sample size on the ratio between sample MAD and sample SD for samples from a normal distribution in shown below in Fig. 6.

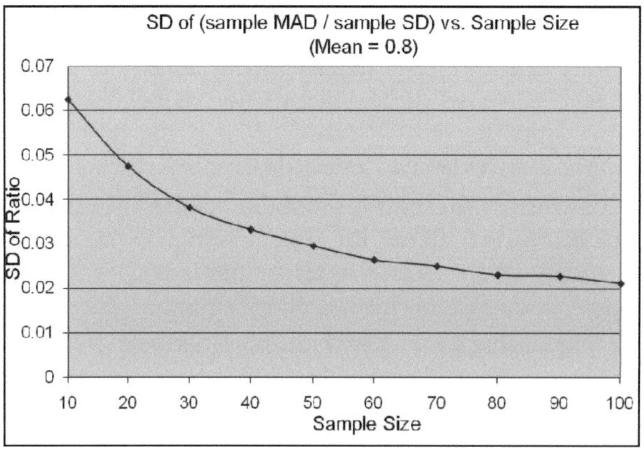

Fig. 6. The standard deviation of the ratios between sample MAD and sample SD for 1000 simulated samples plotted against the size of the sample.

Note that unlike the RMSE, which is equal to the square root of the sum of the squares of the CE (ME) and RE (σ), the MUE is not expressible as a function of CE (ME) and RE (MAD). The formulas for the error measures are given below in Table 2.

The formulas listed in Table 2 and the above discussion apply to normal or similar unimodal distributions. In the case of a multimodal data distribution, these measures are in general not applicable. However, if there are only a few modes that are relatively far apart, then these measures (or similar statistics) can be calculated for each of the modes using appropriate subsets of the data set. This is in particular applicable to the analysis of front-back errors, which tend to define a separate unimodal distribution.

Measure Name	Symbol	Type	Definition/Formula	Comments
Mean Error (Mean Signed Error)	ME	CE	$ME = \frac{1}{n}\sum_{i=1}^{n}(x_i - \eta) = x_o - \eta$	
Mean Absolute Error (Mean Unsigned Error)	MUE	CE & RE	$MUE = \frac{1}{n}\sum_{i=1}^{n}\lvert x_i - \eta\rvert$	$\lvert ME\rvert \le MUE \le \lvert ME\rvert + MAD$
Root-Mean-Squared Error	RMSE	CE & RE	$RMSE = \sqrt{\frac{1}{n}\sum_{i=1}^{n}(x_i - \eta)^2}$	$RMSE^2 = ME^2 + SD^2$
Standard Deviation	SD	RE	$SD = \sqrt{\frac{1}{n}\sum_{i=1}^{n}(x_i - x_0)^2}$	
Mean Absolute Deviation	MAD	RE	$MAD = \frac{1}{n}\sum_{i=1}^{n}\lvert x_i - x_0\rvert$	

Table 2. Basic measures used to calculate localization error (η denotes true location of the sound source).

There is a continuing debate in the literature as to what constitutes a front-back error. Most authors define front-back errors as any estimates that cross the interaural axis (Carlile et al., 1997; Wenzel, 1999). Other criteria include errors crossing the interaural axis by more than 10° (Schonstein, 2008) or 15° (Best et al., 2009) or errors that are within a certain angle after subtracting 180°. An example of the last case is using a ±20° range around the directly opposite angle (position) which corresponds closely to the range of typical listener uncertainty in the frontal direction (e.g., Carlile et al., 1997). The criterion proposed in this chapter is that only estimates exceeding a ±150° error should be considered nominal front-back errors. This criterion is based on a comparative analysis of location estimates made in anechoic and less than optimal listening conditions.

The extraction and separate analysis of front-back errors should not be confused with the process of trimming the data set to remove outliers, even though they have the same effect. Front-back errors are not outliers in the sense that they simply represent extreme errors. They represent a different type of error that has a different underlying cause and as such should be treated differently. Any remaining errors exceeding ±90° may be trimmed (discarded) or winsorized to keep the data set within the ±90° range. Winsorizing is a strategy in which the extreme values are not removed from the sample, but rather are replaced with the maximal remaining values on either side. This strategy has the advantage of not reducing the sample size for statistical data analysis. Both these procedures mitigate the effects of extreme values and are a way of making the resultant sample mean and standard deviation more robust.

The common primacy of the sample arithmetic mean and sample standard deviation for estimating the population parameters is based on the assumption that the underlying distribution is in fact perfectly normal and that the data are a perfect reflection of that distribution. This is frequently not the case with human experiments, which have numerous potential sources for data contamination. In general, this is evidenced by more values farther away from the mean than expected (heavier tails or greater kurtosis) and the presence of extreme values, especially for small data sets. Additionally, the true underlying distribution may deviate slightly in other ways from the assumed normal distribution (Huber & Ronchetti, 2009).

It is generally desired that a small number of inaccurate results should not overly affect the conclusions based on the data. Unfortunately, this is not the case with the sample mean and standard deviation. As mentioned earlier the mean and, in particular, the standard deviation are quite sensitive to outliers (the inaccurate results). Their more robust counterparts discussed in this section are a way of dealing with this problem without having to specifically identify which results constitute the outliers as is done in trimming and winsorizing. Moreover, the greater efficiency of the sample SD over the MAD disappears with only a few inaccurate results in a large sample (Huber & Ronchetti, 2009). Thus, since there is little chance of human experiments generating perfect data and a high chance of the underlying distribution not being perfectly normal, the use of more robust measures for estimating the CE (mean) and RE (standard deviation) may be recommended.

It is also recommended that both components of localization error, CE and RE, always be reported individually. A single compound measure of error such as the RMSE or MUE is not sufficient for understanding the nature of the errors. These compound measures can be useful for describing total LE, but they should be treated with caution. Opinions as to whether RMSE or MUE provides the better characterization of total LE are divided. The overall goodness-of-fit measure given in Eq. 2 clearly uses RMSE as its base. Some authors also consider RMSE as "the most meaningful single number to describe localization performance" (Hartmann, 1983). However, others argue that MUE is a better measure than RMSE. Their criticism of RMSE is based on the fact that RMSE includes MUE but is additionally affected by the square root of the sample size and the distribution of the squared errors which confounds its interpretation (Willmott & Matusuura 2005).

6. Spherical statistics

The traditional statistical methods discussed above were developed for linear infinite distributions. These methods are in general not appropriate for the analysis of data having a spherical or circular nature, such as angles. The analysis of angular (directional) data requires statistical methods that are concerned with probability distributions on the sphere and circle. Only if the entire data set is restricted to a ±90° range can angular data be analyzed as if coming from a linear distribution. In all other cases, the methods of linear statistics are not appropriate, and the data analysis requires the techniques of a branch of statistics called spherical statistics.

Spherical statistics, also called directional statistics, is a set of analytical methods specifically developed for the analysis of probability distributions on spheres. Distributions on circles (two dimensional spheres) are handled by a subfield of spherical statistics called circular statistics. The fundamental reason that spherical statistics is necessary is that if the numerical difference between two angles is greater than 180°, then their linear average will point in the opposite direction from their actual mean direction. For example, the mean direction of 0° and 360° is actually 0°, but the linear average is 180°. Note that the same issue occurs also with the ±180° notational scheme (consider -150° and 150°). Since parametric statistical analysis relies on the summation of data, it is clear that something other than standard addition must serve as the basis for the statistical analysis of angular data. The simple solution comes from considering the angles as vectors of unit length and applying vector addition. The Cartesian coordinates X and Y of the mean vector for a set of vectors corresponding to a set of angles θ about the origin are given by:

$$X = \frac{1}{n}\sum_{i=1}^{n} \sin(\theta_i) \tag{6}$$

and

$$Y = \frac{1}{n}\sum_{i=1}^{n} \cos(\theta_i) \tag{7}$$

The angle θ_o that the mean vector makes with the X-axis is the mean angular direction of all the angles in the data set. Its calculation depends on the quadrant the mean vector is in:

$$\theta_o = \begin{cases} \tan^{-1}(Y/X) & X > 0 \\ \pi + \tan^{-1}(Y/X) & X < 0, Y \geq 0 \\ -\pi + \tan^{-1}(Y/X) & X < 0, Y < 0 \\ \pi/2 & X = 0, Y \geq 0 \\ -\pi/2 & X = 0, Y < 0 \end{cases} \tag{8}$$

The magnitude of the mean vector is called the *mean resultant length* (R):

$$R = \sqrt{X^2 + Y^2}\,. \tag{9}$$

R is a measure of concentration, the opposite of dispersion, and plays an important role in defining the circular standard deviation. Its magnitude varies from 0 to 1 with R = 1 indicating that all the angles in the set point in the same direction. Note that R = 0 not only for a set of angles that are evenly distributed around the circle but also for one in which they are equally divided between two opposite directions. Thus, like the linear measures discussed in the previous section, R is most meaningful for unimodal distributions.

One of the most significant differences between spherical statistics and linear statistics is that due the bounded range over which the distribution is defined, there is no generally valid counterpart to the linear standard deviation in the sense that intervals defined in terms of multiples of the standard deviation represent a constant probability independent of the value of the standard deviation. Clearly, as the circular standard deviation increases, fewer and fewer standard deviations are needed to cover the whole circle.

The circular counterpart to the linear normal distribution is known as the von Mises distribution (Fisher, 1993)

$$f(\theta,\kappa) = \frac{1}{2\pi I_0(\kappa)} e^{\kappa\cos(\theta-\theta_0)}, \tag{10}$$

where θ_o is the mean angle and $I_0(\kappa)$ the modified Bessel function of order 0. The κ parameter of the von Mises function is not a measure of dispersion, like the standard deviation, but, like R, is a measure of concentration. At $\kappa = 0$, the von Mises distribution is equal to the uniform distribution on the circle, while at higher values of κ the distribution becomes more and more concentrated around its mean. As κ continues to increases above 1, the von Mises distribution begins to more and more closely resemble a wrapped normal distribution, which is a linear normal distribution that has been wrapped around the circle

$$f(\theta)=\frac{1}{\sigma\sqrt{2\pi}}\sum_{k=-\infty}^{\infty} e^{-\frac{(\theta-\theta_0+2\pi k)^2}{2\sigma^2}}, \tag{11}$$

where θ_0 and σ are the mean and standard deviation of the linear distribution.

A reasonable approach to defining the circular standard deviation would be to base it on the wrapped normal distribution so that for a wrapped normal distribution it would coincide with the standard deviation of the underlying linear distribution. This can be accomplished due to the fact that for the wrapped normal distribution there is a direct relationship between the mean resultant length, R, and the underlying linear standard deviation

$$R=e^{\frac{-\sigma^2}{2}}. \tag{12}$$

The above equality provides the general definition of the circular standard deviation as:

$$\sigma_C=\sigma=\sqrt{-2\ln(R)}. \tag{13}$$

The sample circular mean direction and sample circular standard deviation can be used to describe any circular data set drawn from a normal circular distribution. However, if the angular data are within ±90°, or within any other numerically continuous 180° range, then linear measures can still be used. Since standard addition applies, the linear mean can be calculated, and it will be equal to the circular mean angle. The linear standard deviation will also be almost identical to the circular standard deviation as long as the results are not overly dispersed. In fact, the relationship between the linear standard deviation and the circular standard deviation is not so much a function of the the range of the data as of its dispersion. For samples drawn from a normal linear distribution, the two sample standard deviations begin to deviate slightly at about $\sigma = 30°$, but even at $\sigma = 60°$ the difference is not too great for larger sample sizes. Results from a set of simulations in which the two sample standard deviations were compared for 500 samples of size 10 and 100 are shown in Fig. 6. The samples were drawn from linear normal distributions with standard deviations randomly selected in the range $1° \le \sigma \le 60°$.

So, for angular data that are assumed to come from a reasonably concentrated normal distribution, as would be expected in most localization studies, the linear standard deviation can be used even if the data spans the full 360°, as long as the mean is calculated as the circular mean angle. This does not mean that localization errors greater than 120° (front-back errors) should not be excluded from the data set for separate analysis.

Once the circular mean has been calculated, the formulas in Table 2 in Section 5 can be used to calculate the circular counterparts to the other linear error measures. The determination of the circular median, and thus the MEAD, is in general a much more involved process. The problem is that there is in general no natural point on the circle from which to start ordering the data set. However, a defining property of the median is that for any data set the average absolute deviation from the median is less than for any other point. Thus, the circular median is defined on this basis. It is the (angle) point on the circle for which the average absolute deviation is minimized, with deviation calculated as the length of the shorter arc between each data point and the reference point. Note that a circular median does not necessarily always exist, as for example, for a data set that is uniformly distributed around the

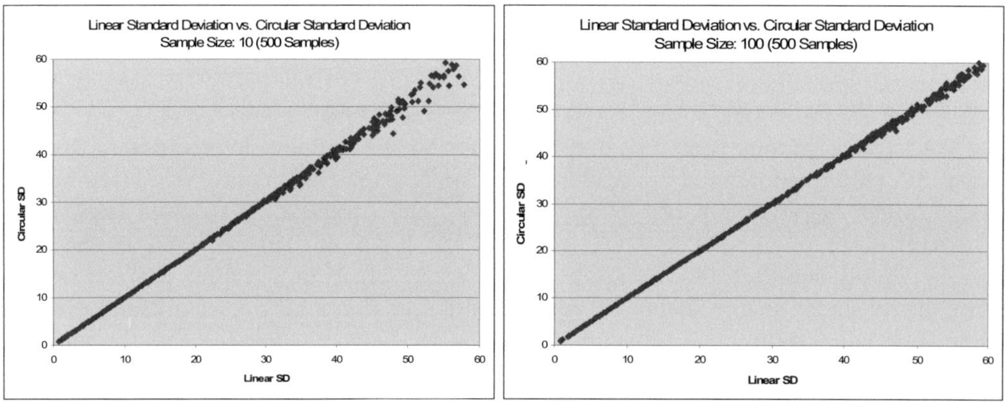

(a) (b)

Fig. 6. Comparison of circular and linear standard deviations for 500 samples of (a) small (n=10) and (b) large (n=100) size.

circle (Mardia, 1972). If however, the range of the data set is less than 360° and has two clear endpoints, then the calculation of the median and MEAD can be done as in the linear case.

Two basic examples of circular statistics significance tests are the nonparametric Rayleigh z test and the Watson two sample U^2 test. The Rayleigh z test is used to determine whether data distributed around a circle are sufficiently random to assume a uniform distribution. The Watson two sample U^2 test can be used to compare two data distributions. Critical values for both tests and for many other circular statistics tests can be found in many advanced statistics books (e.g., Batschelet, 1981; Mardia, 1972; Zar, 1999; Rao and SenGupta, 2001). The special-purpose package Oriana (see http://www.kovcomp.co.uk) provides direct support for circular statistics as do add-ons such as SAS macros (e.g., Kölliker, M. 2005), A MATLAB Toolbox for Circular Statistics (Berens, 2009), and CircStat for S-Plus, R, and Stata (e.g., Rao and SenGupta, 2001).

7. Relative (discrimination) and categorical localization

The LE analysis conducted so far in this text was limited to the absolute identification of sound source locations in space. Two other types of localization judgments are relative judgments of sound source location (location discrimination) and categorical localization.

The basic measure of relative localization acuity is the minimum audible angle (MAA). The MAA, or *localization blur* (Blauert, 1974), is the minimum detectable difference in azimuth (or elevation) between locations of two identical but not simultaneous sound sources (Mills, 1958; 1972; Perrott, 1969). In other words, the MAA is the smallest perceptible difference in the position of a sound source. To measure the MAA, the listener is presented with two successive sounds coming from two different locations in space and is asked to determine whether the second sound came from the left or the right of the first one. The MAA is calculated as half the angle between the minimal positions to left and right of the sound source that result in 75% correct response rates. It depends on both frequency and direction of arrival of the sound wave. For wideband stimuli and low frequency tones, MAA is on the order of 1° to 2° for the frontal position, increases to 8-10° at 90° (Kuhn, 1987), and decreases again to 6-7° at the rear (Mills, 1958; Perrott, 1969; Blauert, 1974). For low frequency tones arriving from the frontal position, the MAA corresponds well with the difference limen (DL) for ITD (~10 μs), and for high frequency tones, it matches well with the difference limen for IID (0.5-1.0 dB), both measured by earphone experiments. The MAA is largest for mid-high frequencies, especially for angles exceeding 40° (Mills, 1958; 1960; 1972). The vertical MAA is about 3-9° for the frontal position (e.g., Perrott & Saberi, 1990; Blauert, 1974).

The MAA has frequently been considered to be the smallest attainable precision (difference limen) in absolute sound source localization in space (e.g., Hartmann, 1983; Hartmann & Rakerd, 1989; Recanzone et al., 1998). However, the precision of absolute localization judgments observed in most studies is generally much poorer than the MAA for the same type of sound stimulus. For example, the average error in absolute localization for a broadband sound source is about 5° for the frontal and about 20° for the lateral position (Hofman & Van Opstal, 1998; Langendijk et al., 2001). Thus, it is possible that the acuity of the MAA, where two sounds are presented in succession, and the precision of absolute localization, where only a single sound is presented, are not well correlated and measure two different human capabilities (Moore et al., 2008). This view is supported by results from animal studies indicating that some types of lesions in the brain affect the precision of absolute localization but not the acuity of the MAA (e.g., Young et al., 1992; May, 2000). In another set of studies, Spitzer and colleagues observed that barn owls exhibited different MAA acuity in anechoic and echoic conditions while displaying similar localization precision across both conditions (Spitzer et al., 2003; Spitzer & Takahasi, 2006). The explanation of these differences may be the difference in the cognitive tasks and the much greater difficulty of the absolute localization task.

Another method of determining LE is to ask listeners to specify the sound source location by selecting from a set of specifically labeled locations. These locations can be indicated by either visible sound sources or special markers on the curtain covering the sound sources (Butler et al., 1990; Abel & Banerjee, 1996). Such approaches restrict the number of possible directions to the predetermined target locations and lead to categorical localization judgments (Perrett & Noble, 1995). The results of categorical localization studies are normally expressed as percents of correct responses rather than angular deviations. The distance between the labeled target locations is the resolution of the localization judgments and describes the localization precision of the study. In addition, if the targets are only distributed across a limited region of the space, this may provide cues resolving potential front-back confusion (Carlile et al., 1997).

Although categorical localization was the predominant localization methodology in older studies, it is still used in many studies today (Abel & Banerjee, 1996; Vause & Grantham, 1999; Van Hosesel & Clark, 1999; Macaulay et al., 2010). Additionally, the Source Azimuth Identification in Noise Test (SAINT) uses categorical judgments with a clock-like array of 12 loudspeakers (Vermiglio et al., 1998) and a standard system for testing the localization ability of cochlear implant users is categorical with 8 loudspeakers distributed in symmetric manner in the horizontal plane in front of the listener with 15.5° of separation (Tyler & Witt, 2004).

In order to directly compare the results of a categorical localization study to an absolute localization study, it is necessary to extract a mean direction and standard deviation from the distribution of responses over the target locations. If the full distribution is known, then by treating each response as an indication of the actual angular positions of the selected target location, the mean and standard deviation can be calculated as usual. If only the percent of correct responses is provided, then as long as the percent correct is over 50%, a normal distribution z-Table (giving probabilities of a result being less than a given z-score) can be used to estimate the standard deviation. If d is the angle of target separation (i.e., the angle between two adjacent loudspeakers), p the percent correct and z the z-score corresponding to $(p+1)/2$, then the standard deviation is given by

$$\sigma = \frac{d}{2z} \tag{14}$$

and the mean by the angular position of the correct target location. This is based on the assumption that the correct responses are normally distributed over the range delimited by the points half way between the correct loudspeaker and the two loudspeakers on either side. This range spans the angle of target separation (d) and thus $d/2$ is the corresponding z-score for the actual distribution. The relationship between the standard z-score and the z-score for a normal distribution N(μ,σ) is given by:

$$z_{N(\mu,\sigma)} = \mu + \sigma \cdot z\,. \tag{15}$$

In this case, the mean, μ, is 0 as the responses are centered around the correct loudspeaker position, so solving for the standard deviation gives Equation 14. As an example, consider an array of loudspeakers separated by 15° and an 85% correct response rate for some individual speaker. The z-score for (1+.85)/2 = .925 is 1.44, so the standard deviation is estimated to be 7.5°/1.44 = 5.2°.

An underlying assumption in the preceding discussion is that the experimental conditions of the categorical judgment task are such that the listener is surrounded by evenly spaced target locations. If this is not the case, then the results for the extreme locations at either end may have been affected by the fact that there are no further locations. In particular this is a problem when the location with the highest percent of responses is not the correct location and the distribution is not symmetric around it. For example, this appears to be the case for the speakers located at ±90° in the 30° loudspeaker arrangement used by Abel & Banerjee (1996).

8. Summary

Judgments of sound source location as well as the resultant localization errors are angular (circular) variables and in general cannot be properly analyzed by the standard statistical methods that assume an underlying (infinite) linear distributions. The appropriate methods of statistical analysis are provided by the field of spherical or circular statistics for three- and two-dimensional angular data, respectively. However, if the directional judgments are relatively well concentrated around a central direction, the differences between the circular and linear measures are minimal, and linear statistics can effectively be used in lieu of circular statistics. The criteria under which the linear analysis of directional data is justified has been a focus of the present discussion. Some basic elements of circular statistics have been also presented to demonstrate the fundamental differences between the two types of data analysis. It has to be stressed that in both cases, it is important to differentiate front-back errors from other gross errors and analyze the front-back errors separately. Gross errors may then be trimmed or winsorized. Both the processing and interpretation of localization data becomes more intuitive and simpler when the ±180° scale is used for data representation instead of the 0-360° scale, although both scales can be successfully used.

In order to meaningfully interpret overall localization error, it is important to individually report both the constant error (accuracy) and random error (precision) of the localization judgments. Error measures like root mean squared error and mean unsigned error represent a specific combination of these two error components and do not on their own provide an adequate characterization of localization error. Overall localization error can be used to characterizes a given set of results but does not give any insight into the underlying causes of the error.

Since the overall purpose of this chapter was to provide information for the effective processing and interpretation of sound localization data, the initial part of the chapter was devoted to differentiating auditory spatial perception from auditory localization and to summarizing the basic terminology used in spatial perception studies and data description. This terminology is not always consistently used in the literature and some standardization would be beneficial. In addition, prior to the discussion of circular data analysis, the most common measures used to describe directional data were compared, and their advantages and limitations indicated. It has been stressed that the standard statistical measures for assessing constant and random error are not robust measures, as they are quite susceptible to being overly influenced by extreme values in the data set. The robust measures discussed in this chapter are intended to provide a starting point for researchers unfamiliar with robust statistics. Given that localization studies, like many experiments involving human judgment, are apt to produce some number of outlying or inaccurate results, it may often be beneficial to utilize robust alternatives to the standard measures. In any case, researchers should be aware of this consideration.

All of the above discussion was related to absolute localization judgments as the most commonly studied form of localization. Therefore, the last section of the chapter deals briefly with location discrimination and categorical localization judgments. The specific focus of this section was to indicate how results from absolute localization and categorical localization studies could be directly compared and what simplifying assumptions are made in carrying out these types of comparisons.

9. References

Abel, S.M. & Banerjee, P.J. (1966). Accuracy versus choice response time in sound localization. *Applied Acoustics,* 49, 405-417.

APA (2007). *APA Concise Dictionary of Psychology.* American Psychology Association, ISBN 1-4338-0391-7, Washington (DC).

Barron, M. & Marshall, A.H. (1981). Spatial impression due to early lateral reflections in concert halls: The derivation of physical measure. *Journal of Sound and Vibration,* 77 (2), 211-232.

Batschelet, E. (1981). *Circular Statistics in Biology.* Academic Press ISBN 978-0120810505, New York (NY).

Batteau, D.W. (1967). The role of the pinna in human localization. *Proceedings of the Royal Society London. Series B: Biological Sciences,* 168, 158-180.

Berens, P. (2009). CircStat: A MATLAB Toolbox for Circular Statistics. *Journal of Statistical Software,* 31 (10), 1-21.

Bergault, D.R. (1992). Perceptual effects of synthetic reverberation on three-dimensional audio systems. *Journal of Audio Engineering Society,* 40 (11), 895-904.

Best, V., Brungart, D., Carlile, S., Jin, C., Macpherson, E., Martin, R.L., McAnally, K.I., Sabin, A.T., & Simpson, B. (2009). A meta-analysis of localization errors made in the anechoic free field, *Proceedings of the International Workshop on the Principles and Applications of Spatial Hearing (IWPASH).* Miyagi (Japan): Tohoku University.

Blauert, J. (1974). *Räumliches Hören*. Sttutgart (Germany): S. Hirzel Verlag (Availabe in English in Blauert, J. Spatial Hearing. Cambridge (MA): MIT, 1997.)

Bloom, P.J. (1977). Determination of monaural sensitivity changes due to the pinna by use of the minimum-audible-field measurements in the lateral vertical plane. *Journal of the Acoustical Society of America* 61, 820-828.

Bolshev, L.N. (2002). Theory of errors. In: M. Hazewinkiel (Ed.), *Encyclopaedia of Mathematics*. Springer Verlag, ISBN 1-4020-0609-8, New York (NY).

Butler, R.A. & Belendiuk, K. (1977). Spectral cues utilized in the localization of sound in the median sagittal plane. *Journal of the Acoustical Society of America,* 61, 1264-1269.

Butler, R.A., Humanski, R.A., & Musicant, A.D. (1990). Binaural and monaural localization of sound in two-dimensional space. *Perception,* 19, 241-256.

Carlile, S. (1996). *Virtual Auditory Space: Generation and Application*. R. G. Landes Company, ISBN 978-1-57059-341-3, Austin (TX).

Carlile, S., Leong, P., & Hyams, S. (1997). The nature and distribution of errors in sound localization by human listeners. *Hearing Research,* 114, 179-196.

Cusak, R., Carlyon, R.P., & Robertson, I.H. (2001). Auditory midline and spatial discrimination in patients with unilateral neglect. *Cortex,* 37, 706-709.

Dietz, M., Ewert, S.D., & Hohmann, V. (2010). Auditory model based direction estimation of concurrent speakers from binaural signals. *Speech Communication* (in print).

Dufour, A., Touzalin, P., & Candas, V. (2007). Rightward shift of the auditory subjective straight Ahead in right- and left-handed subjects. *Neuropsychologia* 45, 447-453.

Emanuel, D. & Letowski, T. (2009). *Hearing Science*. Lippincott, Williams, & Wilkins, ISBN 978-0781780476, Baltimore (MD).

Fisher, N.I. (1987). Problems with the current definition of the standard deviation of wind direction. *Journal of Climate and Applied Meteorology,* 26, 1522-1529.

Fisher, N.I. (1993). *Statistical Analysis of Circular Data*. Cambridge University Press, ISBN 978-0521568906, Cambridge (UK).

Goldstein, D.G. & Taleb, N.N. (2007) We don't quite know what we are talking about when we talk about volatility. *Journal of Portfolio Management,* 33 (4), 84-86.

Griesinger, D. (1997). The psychoacoustics of apparent source width, spaciousness, and envelopment in performance spaces. *Acustica,* 83, 721-731.

Griesinger, D. (1999). Objective measures of spaciousness and envelopment, *Proceedings of the 16th AES International Conference on Spatial Sound Reproduction*, pp. 1-15. Rovaniemi (Finland): Audio Engineering Society.

Hartmann, W.M. (1983) Localization of sound in rooms. *Journal of the Acoustical Society of America,* 74, 1380-1391.

Hartmann, W. M. & Rakerd, B. (1989). On the minimum audible angle - A decision theory approach. *Journal of the Acoustical Society of America,* 85, 2031-2041.

Henning, G.B. (1974). Detectability of the interaural delay in high-frequency complex waveforms. *Journal of the Acoustical Society of America,* 55, 84-90.

Henning, G.B. (1980). Some observations on the lateralization of complex waveforms. *Journal of the Acoustical Society of America,* 68, 446-454.

Hofman, P.M. & Van Opstal, A.J. (1998). Spectro-temporal factors in two-dimensional human sound localization. *Journal of the Acoustical Society of America,* 103, 2634-2648.

Houghton Mifflin (2007). *The American Heritage Medical Directory*. Orlando (FL): Houghton Mifflin Company.

Huber, P.J. & Ronchetti, E. (2009), *Robust Statistics* (2nd Ed.). John Wiley & Sons, ISBN: 978-0-470-12990-6, Hoboken (NJ).

Illusion. (2010). In: *Encyclopedia Britannica*. Retrieved 16 September 2010 from Encyclopedia Britannica Online: http://search.eb.com/eb/article-46670 (Accessed 15 Sept 2010).

Iwaya, Y., Suzuki, Y., & Kimura, D. (2003). Effects of head movement on front-back error in sound localization. *Acoustical Science and Technology*, 24 (5), 322-324.

Jin, C., Corderoy, A., Carlile, SD., & van Schaik, A. (2004). Contrasting monaural and interaural spectral cues for human sound localization. *Journal of the Acoustical Society of America*, 115, 3124-3141.

Knudsen, E.I. (1982). Auditory and visual maps of space in the optic tectum of the owl. *Journal of Neuroscience*, 2 (9), 1177-1194.

Kölliker, M. (2005). Circular statistics Macros in SAS. Freely available online at http://www.evolution.unibas.ch/koelliker/misc.htm (Accessed 15 Sept 2010).

Kuhn, G.F. (1987). Physical acoustics and measurements pertaining to directional hearing. In: W.A. Yost & G. Gourevitch (eds.), *Directional Hearing*, pp. 3-25. Springer Verlag, ISBN 978-0387964935, New York (NY).

Langendijk, E., Kistler, D.,J., & Wightman, F.L. (2001). Sound localization in the presence of one or two distractors. *Journal of the Acoustical Society of America*, 109, 2123-2134.

Langendijk, E. & Bronkhorst, A.W. (2002). Contribution of spectral cues to human sound localization. *Journal of the Acoustical Society of America*, 112, 1583-1596.

Leong, P. & Carlile, S. (1998). Methods for spherical data analysis and visualization. *Journal of Neuroscience Methods*, 80, 191-200.

Lopez-Poveda, E.A., & Meddis, R. (1996). A physical model of sound diffraction and reflections in the human concha. *Journal of the Acoustical Society of America*, 100, 3248-3259.

Macaulay, E.J., Hartman, W.M., & Rakerd, B. (2010). The acoustical bright spot and mislocalization of tones by human listeners. *Journal of the Acoustical Society of America*, 127, 1440-1449.

Makous, J. & Middlebrooks, J.C. (1990). Two-dimensional sound localization by human listeners. *Journal of the Acoustical Society of America*, 92, 2188-2200.

Mardia, K.V. (1972). *Statistics of Directional Data*. Academic Press, ISBN 978-0124711501, New York (NY).

May, B.J. (2000). Role of the dorsal cochlear nucleus in sound localization behavior in cats. *Hearing Research*, 148, 74-87.

McFadden, D.M. & Pasanen, E. (1976). Lateralization of high frequencies based on interaural time differences. *Journal of the Acoustical Society of America*, 59, 634-639.

Mills, A.W. (1958). On the minimum audible angle. *Journal of the Acoustical Society of America*, 30, 237-246.

Mills, A.W. (1960). Lateralization of high-frequency tones. *Journal of the Acoustical Society of America*, 32, 132-134.

Mills, A.W. (1972). Auditory localization. In: J. Tobias (Ed.), *Foundations of Modern Auditory Theory, vol 2* (pp. 301-345). New York (NY): Academic Press.

Moore, B.C.J. (1989). *An Introduction to the Psychology of Hearing* (4th Ed.). Academic Press, ISBN 0-12-505624-9, San Diego (CA).

Moore, J.M., Tollin, D.J., & Yin, T. (2008). Can measures of sound localization acuity be related to the precision of absolute location estimates? *Hearing Research*, 238, 94-109.

Morfey, C.L. (2001). *Dictionary of Acoustics*. Academic Press, ISBN 0-12-506940-5, San Diego (CA).

Morimoto, M. (2002). The relation between spatial impression and precedence effect, *Proceedings of the 8th International Conference on Auditory Display (ICAD2002)*. Kyoto (Japan): ATR

Musicant, A.D. and Butler, R.A. (1984). The influence of pinnae-based spectral cues on sound localization. *Journal of the Acoustical Society of America,* 75, 1195-1200.

Ocklenburg, S., Hirnstein, M., Hausmann, M., & Lewald, J. (2010). Auditory space perception by left and right-handers. *Brain and Cognition,* 72(2), 210-7.

Oldfield, S.R. & Parker, S.P.A. (1984). Acuity of sound localization: A topography of auditory space I. Normal hearing conditions. *Perception,* 13, 581-600.

Pedersen, J.A. & Jorgensen, T. (2005). Localization performance of real and virtual sound sources, *Proceedings of the NATO RTO-MP-HFM-123 New Directions for Improving Audio Effectiveness Conference,* pp. 29-1 to 29-30. Neuilly-sui-Seine (France): NATO.

Perrett, S. & Noble, W. (1995). Available response choices affect localization of sound. *Perception and Psychophysics,* 57, 150-158.

Perrett, S. & Noble, W. (1997). The effect of head rotation on vertical plane sound localization. *Journal of the Acoustical Society of America,* 102, 2325-2332.

Perrott, D.R. (1969). Role of signal onset in sound localization. *Journal of the Acoustical Society of America,* 45, 436-445.

Perrott, D.R. & Saberi, K. (1990). Minimum audible angle thresholds for sources varying in both elevation and azimuth. *Journal of the Acoustical Society of America,* 87, 1728-1731. *Acoustical Society of America* 56, 944-951.

Pierce, A.H. (1901). *Studies in Auditory and Visual Space Perception.* Longmans, Green, and Co, ISBN 1-152-19101-2, New York (NY).

Rao Jammalamadaka, S. & SenGupta, A. (2001). *Topics in Circular Statistics.* World Scientific Publishing, ISBN 9810237782, River Edge (NJ).

Razavi, B., O'Neill, W.E., & Paige, G.D. (2007). Auditory spatial perception dynamically realigns with changing eye position. *Journal of Neurophysiology,* 27 (38), 10249-10258

Recanzone, G.H., Makhamra, S., & Guard, D.C. (1998). Comparison of absolute and relative sound localization ability in humans. *Journal of the Acoustical Society of America,* 103, 1085-1097.

Rogers, M.E. & Butler, R.A. (1992). The linkage between stimulus frequency and covert peak areas as it relates to monaural localization. *Perception and Psychophysics,* 52, 536-546.

Schonstein, D., Ferre, L., & Katz, F.G. (2009). Comparison of headphones and equalization for virtual auditory source localization, *Proceedings of the Acoustics'08 Conference.* Paris (France): European Acoustics Association.

Sosa, Y., Teder-Sälejärvi, W.A., & McCourt, M.E. (2010). Biases in spatial attention in vision and audition. *Brain and Cognition,* 73, 229-235.

Spitzer, M.W., Bala, A., Takahashi, T.T. (2003). Auditory spatial discrimination by barn awls in simulated echoic environment. *Journal of the Acoustical Society of America,* 113, 1631-1645.

Spizer, M.W. & Takahashi, T.T. (2006). Sound localization by barn awls in a simulated echoic environment. *Journal of Neurophysiology,* 95, 3571-3584.

Steinhauser, A. (1879). The theory of binaural audition. A contribution to the theory of sound. *Philosophical Magazine (Series 5),* 7, 181-197.

Strutt, J.W. (Lord Rayleigh). (1876). Our perception of the direction of a source of sound. *Nature,* 7, 32-33.

Strutt, J.W. (Lord Rayleigh). (1907). On our perception of sound direction. *Philosophical Magazine (Series 5),* 13, 214-232.

Tonning, F.M. (1970). Directional audiometry. I. Directional white-noise audiometry. *Acta Otolaryngologica,* 72, 352-357.

Tyler, R.S, & Witt, S. (2004). Cochlear implants in adults: Candidacy. In: R.D. Kent (ed.), *The MIT Encyclopedia of Communication Disorders*, pp. 450-454. Cambridge (MA): MIT Press.

Van Hosesel, R.M. & Clark, G.M. (1999). Speech results with a bilateral multi-channel cochlear implant subject for spatially separated signal and noise. *Australian Journal of Audiology*, 21, 23-28.

Van Wanrooij, M.M. & Van Opstal, A.J. (2004). Contribution of head shadow and pinna cues to chronic monaural sound localization. *Journal of Neuroscience*, 24 (17), 4163-4171.

Vause, N. & Grantham, D.W. (1999). Effects of earplugs and protective headgear on auditory localization ability in the horizontal plane. *Journal of the Human Factors and Ergonomics Society*, 41 (2), 282-294.

Vermiglio, A., Nilsson, M., Soli, S., & Freed, D. (1998). Development of virtual test of sound localization: the Source Azimuth Identification in Noise Test (SAINT), Poster presented at the American Academy of Audiology Convention. Los Angeles (CA): AAA.

Wallach, H. (1939). On sound localization. *Journal of the Acoustical Society of America*, 10, 270-274.

Wallach, H. (1940). The role of head movements and the vestibular and visual cues in sound localization. *Journal of Experimental Psychology*, 27, 339-368.

Watkins, A.J. (1978). Psychoacoustical aspects of synthesized vertical locale cues. *Journal of the Acoustical Society of America*, 63, 1152-1165.

Wenzel, E.M. (1999). Effect of increasing system latency on localization of virtual sounds, Proceedings of the 16th AES International Conference on Spatial Sound Reproduction, pp. 1-9. Rovaniemi (Finland): Audio Engineering Society.

White, G.D. (1987). *The Audio Dictionary*. University of Washington Press, ISBN 0-295965274, Seattle (WA).

Wightman, F.L. & Kistler, D.J. (1989). Headphone simulation of free field listening. II: Psychophysical validation. *Journal of the Acoustical Society of America*, 85, 868–878.

Willmott, C.J. & Matsuura, K. (2005). Advantages of the mean absolute error (MAE) over the root mean square error (RMSE) in assessing average model performance. *Climate Research*, 30, 79–82.

Wilson, H.A. & Myers, C. (1908). The influence of binaural phase differences on the localization of sounds. *British Journal of Psychology*, 2, 363-385.

Yost, W.A. & Gourevitch, G. (1987). *Directional Hearing*. Springer, ISBN 978-0387964935, New York (NY).

Yost, W.A. & Hafter, E.R. (1987). Lateralization. In: W.A. Yost & G. Gourevitch (eds.), *Directional Hearing*, pp. 49-84. Springer, ISBN 978-0387964935, New York (NY).

Yost, W.A., Popper, A.N., & Fay, R.R. (2008). *Auditory Perception of Sound Sources*. Springer, ISBN 978-0-387-71304-5, New York (NY).

Young, P.T. (1931). The role of head movements in auditory localization. *Journal of Experimental Psychology*, 14, 95-124.

Young, E.D., Spirou, G.A., Rice, J.J., & Voigt, H.F. (1992). Neural organization and response to complex stimuli in the dorsal cochlear nucleus. *Philosophical Transactions of the Royal Society London B: Biological Sciences*, 336, 407-413.

Zahorik, P., Brungart, D.S., & Bronkhorst, A.W. (2005). Auditory distance perception in humans: A summary of past and present research. *Acta Acustica*, 91, 409-420.

Zar, J. H. (1999). *Biostatistical Analysis* (4th ed.). Prentice Hall, ISBN 9780131008465, Upper Saddle River (NJ).

5

HRTF Sound Localization

Martin Rothbucher, David Kronmüller, Marko Durkovic, Tim Habigt and Klaus Diepold
Institute for Data Processing, Technische Universität München
Germany

1. Introduction

In order to improve interactions between the human (operator) and the robot (teleoperator) in human centered robotic systems, e.g. Telepresence Systems as seen in Figure 1, it is important to equip the robotic platform with multimodal human-like sensing, e.g. vision, haptic and audition.

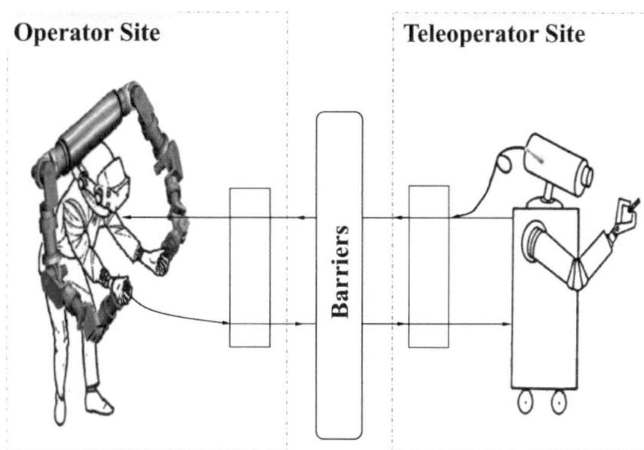

Fig. 1. Schematic view of the telepresence scenario.

Recently, robotic binaural hearing approaches based on Head-Related Transfer Functions (HRTFs) have become a promising technique to enable sound localization on mobile robotic platforms. Robotic platforms would benefit from this human like sound localization approach because of its noise-tolerance and the ability to localize sounds in a three-dimensional environment with only two microphones.

As seen in Figure 2, HRTFs describe spectral changes of sound waves when they enter the ear canal, due to diffraction and reflection of the human body, i.e. the head, shoulders, torso and ears. In far field applications, they can be considered as functions of two spatial variables (elevation and azimuth) and frequency. HRTFs can be regarded as direction dependent filters, as diffraction and reflexion properties of the human body are different for each direction. Since the geometric features of the body differ from person to person, HRTFs are unique for each individual (Blauert, 1997).

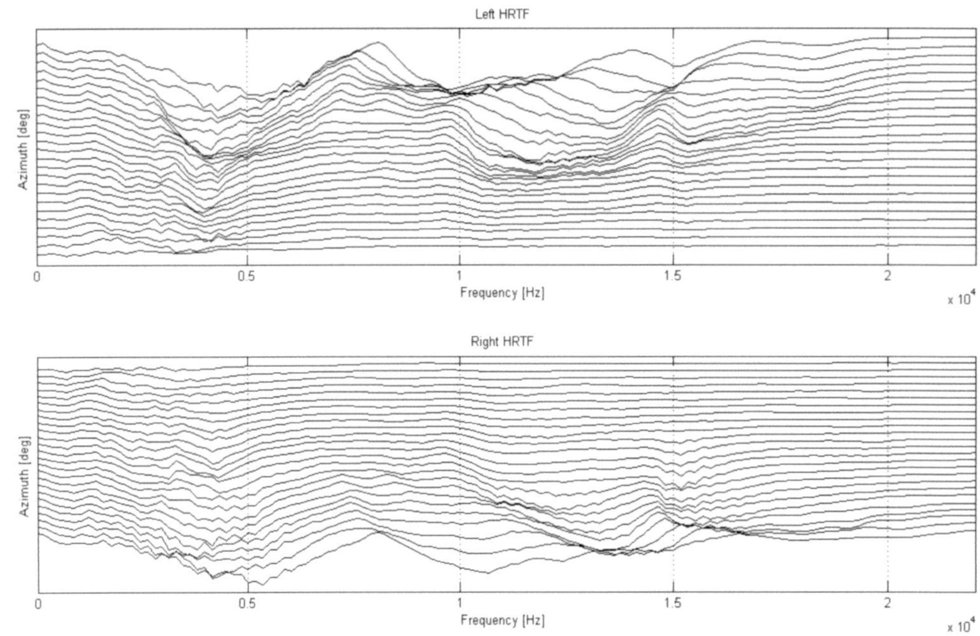

Fig. 2. HRTFs over varying azimuth and constant elevation

The problem of HRTF-based sound localization on mobile robotic platforms can be separated into three main parts, namely the HRTF-based localization algorithms, the HRTF data reduction and the application of predictors that improve the localization performance.
For robotic HRTF-based localization, an incoming sound signal is reflected, diffracted and scattered by the robot's torso, shoulders, head and pinnae, dependent on the direction of the sound source. Thus both left and right perceived signals have been altered through the robot's HRTF, which the robot has learned to associate with a specific direction. We have investigated several HRTF-based sound localization algorithms, which are compared in the first section.
Due to its high dimensionality, it is inefficient to utilize the robot's original HRTFs. Therefore, the second section will provide a comparison of HRTF reduction techniques. Once the HRTF dataset has been reduced and restored, it serves as the basis for localization.
HRTF localization is computational very expensive, therefore, it is advantageous to reduce the search region for sound sources to a region of interest (ROI). Given a HRTF dataset, it is necessary to check the presence of each HRTF in the perceived signal individually. Simply applying a brute force search will localize the sound source but may be inefficient. To improve upon this, a search region may be defined, determines which HRTF-subset is to be searched and in what order to evaluate the HRTFs.
The evaluation of the respective approaches is made by conducting comprehensive numerical experiments.

2. HRTF Localization Algorithms

In this section, we briefly describe four HRTF-based sound localization algorithms, namely the Matched Filtering Approach, the Source Cancellation Approach, the Reference Signal Approach and the Cross Convolution Approach. These algorithms return the position of the sound source using the recorded ear signals and a stored HRTF database. As illustrated in Figure 3, the unknown signal S emitted from a source is filtered by the corresponding left and right HRTFs, denoted by H_{L,i_0} and H_{R,i_0}, before being captured by a humanoid robot, i.e., the left and right microphone recordings X_L and X_R are constructed as

$$\begin{aligned} X_L &= H_{L,i_0} \cdot S, \\ X_R &= H_{R,i_0} \cdot S. \end{aligned} \quad (1)$$

The key idea of the HRTF-based localization algorithms is to identify a pair of HRTFs corresponding to the emitting position of the source, such that correlation between left and right microphone observations is maximized.

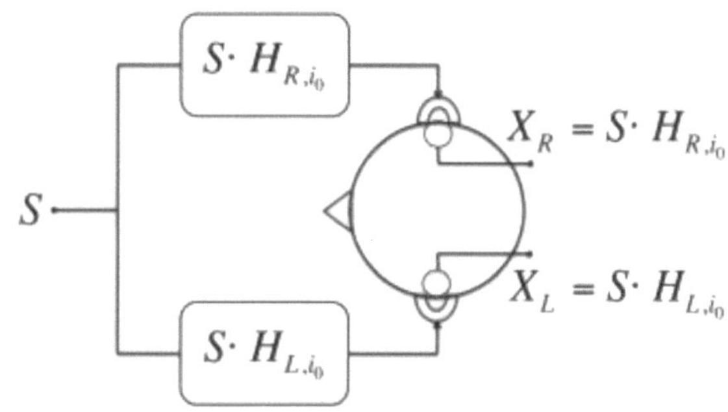

Fig. 3. Single-Source HRTF Model

2.1 Matched Filtering Approach

The Matched Filtering Approach seeks to reverse the H_{R,i_0} and H_{L,i_0}-filtering of the unknown sound source S as illustrated in Figure 3. A schematic view of the Matched Filtering Approach is given in Figure 4.

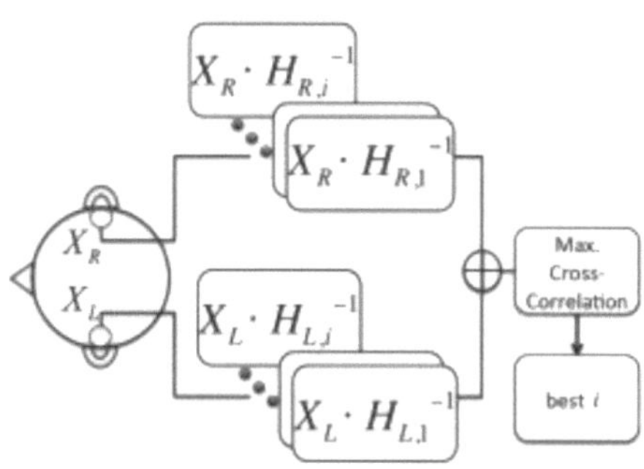

Fig. 4. Schematic view of the Matched Filtering Approach

The localization algorithm is based on the fact that filtering X_L and X_R with the inverse of the correct emitting HRTFs yields identical signals $\tilde{S}_{R,i}$ and $\tilde{S}_{L,i}$, i.e. the original mono sound signal S in an ideal case:

$$\begin{aligned}\tilde{S}_{L,i} =& H_{L,i}^{-1} \cdot X_L \\ =& H_{R,i}^{-1} \cdot X_R \\ =& \tilde{S}_{R,i} \Longleftrightarrow i = i_0.\end{aligned} \tag{2}$$

In real case, the sound source can be localized by maximizing the cross-correlation between $\tilde{S}_{R,i}$ and $\tilde{S}_{L,i}$,

$$arg \max_i \left\{ (\tilde{S}_{R,i}) \oplus (\tilde{S}_{L,i}) \right\}, \tag{3}$$

where i is the index of HRTFs in the database and $\oplus$ denotes a cross-correlation operation. Unfortunately the inversion of HRTFs can be problematic due to instability. This is mainly due to the linear-phase component of HRTFs responsible for encoding ITDs. Hence a stable approximation must be made of the instable version, retaining all direction-dependent information. One method is to use outer-inner factorization, converting an unstable inverse into an anti-causal and bounded inverse (Keyrouz et al., 2006).

2.2 Source Cancellation Algorithm

The Source Cancellation Algorithm is an extension of the Matched Filtering Approach. Equivalently to cross-correlating all pairs $X_L \cdot H_{L,i}^{-1}$ and $X_R \cdot H_{R,i}^{-1}$, the problem can be restated as a cross-correlation between all pairs $\frac{X_L}{X_R}$ and $\frac{H_{L,i}}{H_{R,i}}$. The improvement is that the ratio of HRTFs does not need to be inverted and can be precomputed and stored in memory (Keyrouz & Diepold, 2006; Usman et al., 2008).

$$arg \max_i \left\{ \left(\frac{X_L}{X_R}\right) \oplus \left(\frac{H_{L,i}}{H_{R,i}}\right) \right\} \tag{4}$$

2.3 Reference Signal Approach

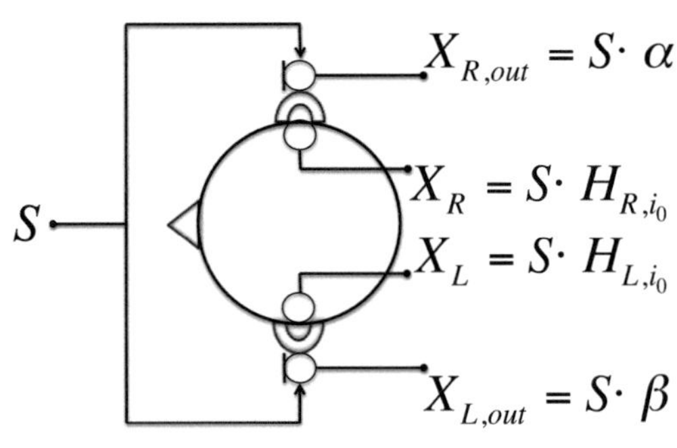

Fig. 5. Schematic view of the Reference Signal Approach setup

This approach uses four microphones as shown in Figure 5: two for the HRTF-filtered signals (X_L and X_R) and two outside the ear canal for original sound signals ($X_{L,out}$ and $X_{R,out}$). The previous algorithms used two microphones, each receiving the HRTF-filtered mono sound signals. The four signals now captured are:

$$X_L = S \cdot H_L \tag{5}$$

$$X_R = S \cdot H_R \tag{6}$$

$$X_{L,out} = S \cdot \alpha \tag{7}$$

$$X_{R,out} = S \cdot \beta \tag{8}$$

α and β represent time delay and attenuation elements that occur due to the heads shadowing. From these signals three ratios are calculated. $\frac{X_L}{X_{L,out}}$ and $\frac{X_R}{X_{R,out}}$ are the left and right HRTFs respectively and $\frac{X_L}{X_R}$ is the ratio between the left and right HRTFs. The three ratios are then cross correlated with the respective reference HRTFs (HRTF ratios in case of $\frac{X_L}{X_R}$). The cross-correlation coefficients are summed, and the HRTF pair yielding the maximum sum

$$arg \max_i \left\{ \left(\frac{X_L}{X_{L,out}} \oplus H_{L,i} \right) + \left(\frac{X_L}{X_R} \oplus \frac{H_{L,i}}{H_{R,i}} \right) + \left(\frac{X_R}{X_{R,out}} \oplus H_{R,i} \right) \right\} \tag{9}$$

defines the incident direction (Keyrouz & Abou Saleh, 2007). The advantage of this system is that HRTFs can be directly calculated yet retain the original undistorted sound signals $X_{L,out}$ and $X_{R,out}$. Thus the direction-dependent filter can alter the incident spectra without regard to the contained information, possibly allowing for better localization. However, the need for four microphones diverges from the concept of binaural localization, exhibiting more hardware and consequently higher costs.

2.4 Convolution Based Approach

To avoid the instability problem, this approach is to exploit the associative property of convolution operator (Usman et al., 2008). Figure 6 illustrates the single-source cross-convolution localization approach. Namely, left and right observations $\tilde{S}_{R,i}$ and $\tilde{S}_{L,i}$ are filtered with a pair of contralateral HRTFs. The filtered observations turn to be identical at the correct source position for the ideal case:

$$\begin{aligned} \tilde{S}_{L,i} =& H_{R,i} \cdot X_L \\ =& H_{R,i} \cdot H_{L,i_0} \cdot S \\ =& H_{L,i} \cdot H_{R,i_0} \cdot S \\ =& H_{L,i} \cdot X_R \\ =& \tilde{S}_{R,i} \Longleftrightarrow i = i_0. \end{aligned} \tag{10}$$

Similar to the matched filtering approach, the source can be localized in real case by solving the following problem:

$$arg \max_i \left\{ (\tilde{S}_{R,i}) \oplus (\tilde{S}_{L,i}) \right\}. \tag{11}$$

2.5 Numerical Comparison

In this section, the previously described localization algorithms are compared by numerical simulations. We use the CIPIC database (Algazi et al., 2001) for our HRTF-based localization experiments. The spatial resolution of the database is 1250 sampling points ($N_e = 50$ in elevation and $N_a = 25$ in azimut) and the length is 200 samples.

In each experiment, generic and real-world test signals are virtually synthesized to the 1250 directions of the database, using the corresponding HRTF. The algorithms are then used to localized the signals and a localization success rate is computed. Noise robustness of the algorithm is investigated by different signal-to-noise ratios (SNRs) of the test signals. It should be noted that testing of the localization performance is rigorous, meaning, that we

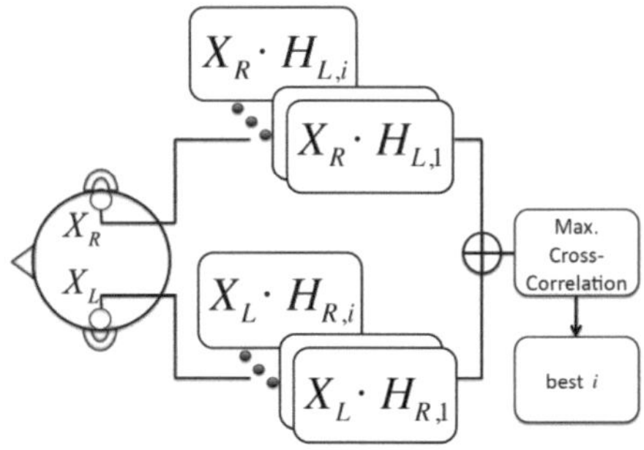

Fig. 6. Schematic view of the cross-convolution approach

do not apply any preprocessing to avoid e.g. instability of HRTF inversion. The localization algorithms are implemented as described above.

Figure 7 shows the achieved localization results of the simulation. The Convolution Based Algorithm, where no HRTF-inversion has to be computed, outperforms the other algorithms in terms of noise robustness and localization success. Furthermore, the best localization results are achieved with white Gaussian noise sources as these ideally cover the entire frequency spectrum. A more realistic sound source is music. It can be seen in Figure 7(d), that the localization performance is slightly degraded compared to the white Gaussian sound sources. The reason for this is that music generally does not inhabit the entire frequency spectrum equally. Speech signals are even more sparse than music resulting in localization success rates worse than for music signals.

Due to the results of the numerical comparison of the different HRTF-based localization algorithms, only the Convolution Based Approach will be utilized to evaluate HRTF data reduction techniques in Section 3 and predictors in Section 4.

3. HRTF Data reduction techniques

In general, as illustrated in Figure 8, each HRTF dataset can be represented as a three-way array $\mathcal{H} \in \mathbb{R}^{N_a \times N_e \times N_t}$.

The dimensions N_a and N_e are the spatial resolutions of azimuth and elevation, respectively, and N_t the time sample size. By a Matlab-like notation, in this section we denote $\mathcal{H}(i,j,k) \in \mathbb{R}$ the (i,j,k)-th entry of $\mathcal{H}$, $\mathcal{H}(l,m,:) \in \mathbb{R}^{N_t}$ the vector with a fixed pair of (l,m) of $\mathcal{H}$ and $\mathcal{H}(l,:,:) \in \mathbb{R}^{N_e \times N_t}$ the l-th slide (matrix) of $\mathcal{H}$ along the azimuth direction.

3.1 Principal Component Analysis (PCA)

Principal Component Analysis expresses high-dimensional data in a lower dimension, thus removing information yet retaining the critical features. PCA uses statistics to extract the adequately named principal components from a signal (in essence being the information that defines the target signal).

The dimensionality reduction of HRIRs by using PCA is described as follows. First of all, we construct the matrix

$$H := [\mathrm{vec}(\mathcal{H}(:,:,1))^\top, \ldots, \mathrm{vec}(\mathcal{H}(:,:,N_t))^\top] \in \mathbb{R}^{N_t \times (N_a \cdot N_e)}, \quad (12)$$

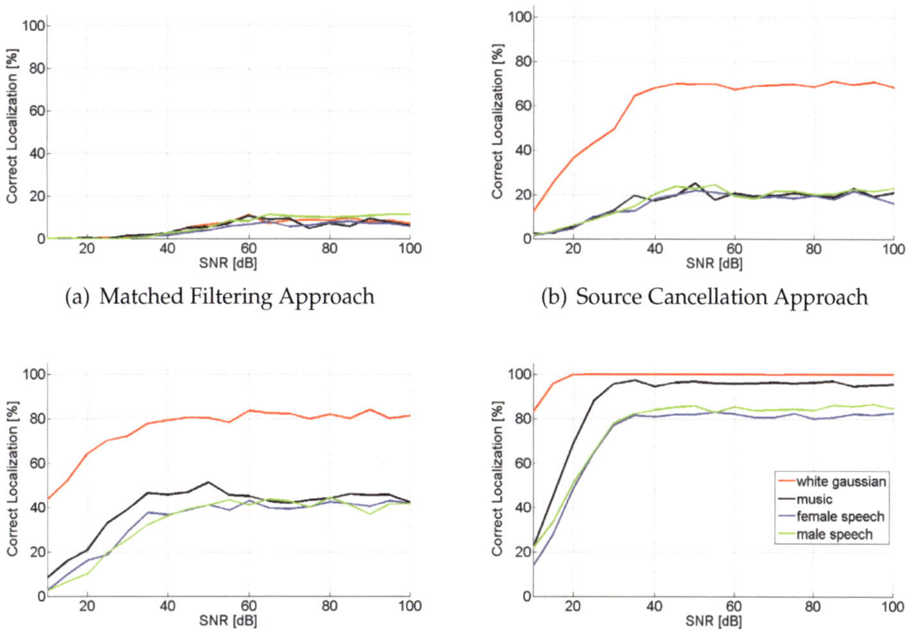

(a) Matched Filtering Approach
(b) Source Cancellation Approach
(c) Reference Signal Approach
(d) Convolution Based Approach

Fig. 7. Comparison of HRTF-based sound localization algorithms.

where the operator vec$(\cdot)$ puts a matrix into a vector form. Let $H = [h_1, \ldots, h_{N_t}]$. The mean value of columns of H is then computed by

$$\mu = \frac{1}{N_t} \sum_{i=1}^{N_t} h_i. \tag{13}$$

After centering each row of H, i.e. computing $\widehat{H} = [\widehat{h}_1, \ldots, \widehat{h}_{N_t}] \in \mathbb{R}^{N_t \times (N_a \cdot N_e)}$ where $\widehat{h}_i = h_i - \mu$ for $i = 1, \ldots, N_t$, the covariance matrix of $\widehat{H}$ is computed as follows

$$C := \frac{1}{N_t} \widehat{H} \widehat{H}^\top. \tag{14}$$

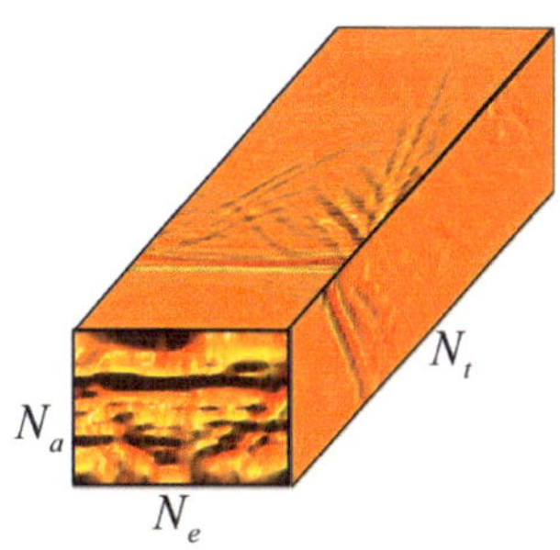

Fig. 8. HRIR dataset represented as a three-way array

Now we compute the eigenvalue decomposition of C and select q eigenvectors $\{x_1, \ldots, x_q\}$ corresponding to the q largest eigenvalues. Then by denoting $X = [x_1, \ldots, x_q] \in \mathbb{R}^{N_t \times q}$, the HRIR dataset can be reduced by the following

$$\widetilde{H} = X^\top \widehat{H} \in \mathbb{R}^{q \times (N_a \cdot N_e)}. \tag{15}$$

Note, that the storage space for the reduced HRIR dataset depends on the value of q. Finally to reconstruct the HRIR dataset one need to compute

$$H_r = X\widetilde{H} + \mu \in \mathbb{R}^{N_t \times (N_a \cdot N_e)}. \tag{16}$$

We refer to (Jolliffe, 2002) for further discussions on PCA.

3.2 Tensor-SVD of three-way array

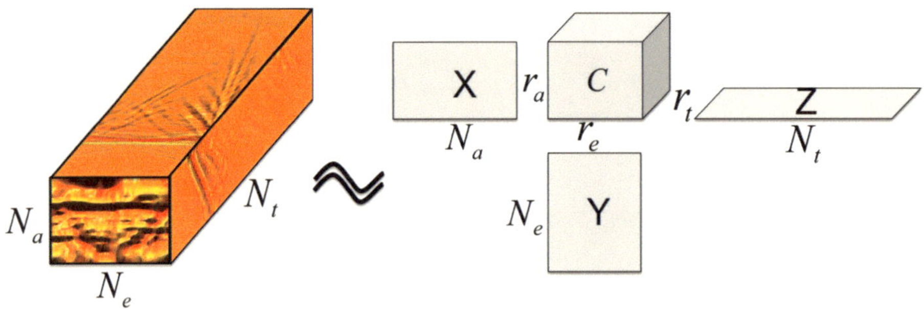

Fig. 9. Schematic view of the Tensor-SVD.

Unlike the PCA algorithm vectorizing the HRIR dataset, Tensor-SVD keeps the structure of the original 3D dataset intact. As shown in Figure 9, given a HRIR dataset $\mathcal{H} \in \mathbb{R}^{N_a \times N_e \times N_t}$, Tensor-SVD computes its best multilinear $rank - (r_a, r_e, r_t)$ approximation $\widehat{\mathcal{H}} \in \mathbb{R}^{N_a \times N_e \times N_t}$, where $N_a > r_a$, $N_e > r_e$ and $N_t > r_t$, by solving the following minimization problem

$$\min_{\widehat{\mathcal{H}} \in \mathbb{R}^{N_a \times N_e \times N_t}} \left\| \mathcal{H} - \widehat{\mathcal{H}} \right\|_{\mathrm{F}}, \tag{17}$$

where $\|\cdot\|_{\mathrm{F}}$ denotes the Frobenius norm of tensors. The $rank - (r_a, r_e, r_t)$ tensor $\widehat{\mathcal{H}}$ can be decomposed as a *trilinear* multiplication of a $rank - (r_a, r_e, r_t)$ core tensor $\mathcal{C} \in \mathbb{R}^{r_a \times r_e \times r_t}$ with three full-rank matrices $X \in \mathbb{R}^{N_a \times r_a}$, $Y \in \mathbb{R}^{N_e \times r_e}$ and $Z \in \mathbb{R}^{N_t \times r_t}$, which is defined by

$$\widehat{\mathcal{H}} = (X, Y, Z) \cdot \mathcal{C} \tag{18}$$

where the (i, j, k)-th entry of $\widehat{\mathcal{H}}$ is computed by

$$\widehat{\mathcal{H}}(i, j, k) = \sum_{\alpha=1}^{r_a} \sum_{\beta=1}^{r_e} \sum_{\gamma=1}^{r_t} x_{i\alpha} y_{j\beta} z_{k\gamma} \mathcal{C}(\alpha, \beta, \gamma). \tag{19}$$

Thus without loss of generality, the minimization problem as defined in (17) is equivalent to the following

$$\begin{aligned} &\min_{X,Y,Z,\mathcal{C}} \|\mathcal{H} - (X, Y, Z) \cdot \mathcal{C}\|_{\mathrm{F}}, \\ &\text{s.t.} \quad X^\top X = I_{r_a}, Y^\top Y = I_{r_e} \text{ and } Z^\top Z = I_{r_t}. \end{aligned} \tag{20}$$

We refer to (Savas & Lim, 2008) for Tensor-SVD algorithms and further discussions.

3.3 Generalized Low Rank Approximations of Matrices

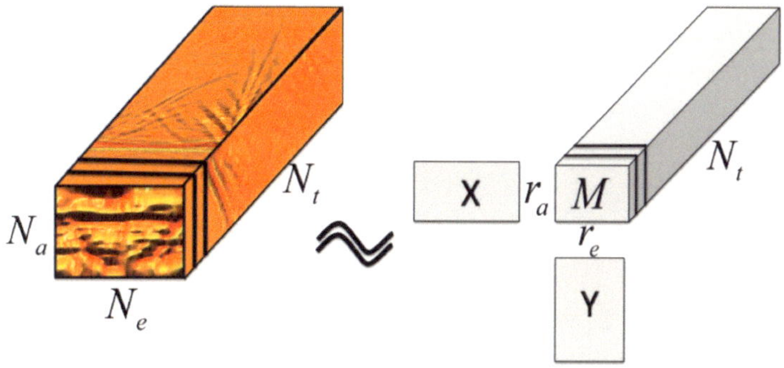

Fig. 10. Schematic view of the Generalized Low Rank Approximations of Matrices

Similar to Tensor-SVD, GLRAM methods, shown in Figure 10 do not require destruction of a 3D tensor. Instead of compressing along all three directions as Tensor-SVD, GLRAM methods work with two pre-selected directions of a 3D data array.
Given a HRIR dataset $\mathcal{H} \in \mathbb{R}^{N_a \times N_e \times N_t}$, we assume to compress $\mathcal{H}$ in the first two directions. Then the task of GLRAM is to approximate slides (matrices) $\mathcal{H}(:,:,i)$, for $i = 1, \ldots, N_t$, of $\mathcal{H}$ along the third direction by a set of low rank matrices $\{XM_iY^\top\} \subset \mathbb{R}^{N_a \times N_e}$, for $i = 1, \ldots, N_t$, where the matrices $X \in \mathbb{R}^{N_a \times r_a}$ and $Y \in \mathbb{R}^{N_e \times r_e}$ are of full rank, and the set of matrices $\{M_i\} \subset \mathbb{R}^{r_a \times r_e}$ with $N_a > r_a$ and $N_e > r_e$. This can be formulated as the following optimization problem

$$\min_{X,Y,\{M_i\}_{i=1}^{N_t}} \sum_{i=1}^{N_t} \left\| (\mathcal{H}(:,:,i) - XM_iY^\top) \right\|_F, \quad \text{s.t.} \quad X^\top X = I_{r_a} \text{ and } Y^\top Y = I_{r_e}. \tag{21}$$

Here, by abuse of notations, $\|\cdot\|_F$ denotes the Frobenius norm of matrices. Let us construct a 3D array $\mathcal{M} \in \mathbb{R}^{r_a \times r_e \times N_t}$ by assigning $\mathcal{M}(:,:,i) = M_i$ for $i = 1, \ldots, N_t$. The minimization problem as defined in (21) can be reformulated in a Tensor-SVD style, i.e.

$$\min_{X,Y,\mathcal{M}} \|\mathcal{H} - (X, Y, I_{N_t}) \cdot \mathcal{M}\|_F, \quad \text{s.t.} \quad X^\top X = I_{r_a} \text{ and } Y^\top Y = I_{r_e}. \tag{22}$$

We refer to (Ye, 2005) for more details on GLRAM algorithms.
GLRAM methods work on two pre-selected directions out of three. There are then in total three different combinations of directions to implement GLRAM on an HRIR dataset. Performance of GLRAM in different directions might vary significantly. This issue will be investigated and discussed in section 3.5.

3.4 Diffuse Field Equalization (DFE)

A technique that provides good compression performance is diffuse field equalization. The technique reduces the number of samples per HRIR, yet retains the original characteristics.
We define the matrix H containing the HRTFs as

$$H := [\text{vec}(\mathcal{H}(:,:,1)), \ldots, \text{vec}(\mathcal{H}(:,:,N_t))] \in R^{(N_a \cdot N_e) \times N_t}, \tag{23}$$

where the operator $vec(\cdot)$ puts a matrix into a vector form. Let $H = [h_1, \ldots, h_{(N_a \cdot N_e)}]$.
DFE removes the time delay at the beginning of each HRTF and then calculates the average power spectrum from all HRTFs, which then is deconvolved from each HRTF, thus removing direction-independent information. The average power $\widetilde{h}$ is computed by

$$\widetilde{h} = \mathcal{F}^{-1}\{\frac{1}{(N_a \cdot N_e)} \sum_{i=1}^{(N_a \cdot N_e)} |\mathcal{F}\{h_i\}|^2\}, \tag{24}$$

where $\mathcal{F}\{\cdot\}$ denotes the fourier transform. Then, $\widetilde{h}$ is shifted circularily by half the kernel length:

$$\widetilde{h}_1 = [\widetilde{h}(\frac{N_t}{2} + 1 \ldots N_t) \ \widetilde{h}(1 \ldots \frac{N_t}{2})]. \tag{25}$$

The filter kernel $\widetilde{h}_1$ is inverted and minimum phase reconstruction is applied, yielding $\widetilde{h}_1^{-1}$. The diffused field equalized dataset is retrieved by

$$h_{\mathrm{DFE}} = [(h_1 * \widetilde{h}_1^{-1}), \ldots, (h_{(N_a \cdot N_e)} * \widetilde{h}_1^{-1})]. \tag{26}$$

After retrieving the dataset h_{DFE} the time delay samples at the beginning of each HRIR can be removed. To achieve higher compression of the dataset, also samples at the end of each HRTFs, which do not contain crucial direction dependent information, can be removed. For further information on DFE see (Moeller, 1992).

3.5 Numerical Comparison

In this section, PCA, GLRAM, Tensor-SVD and Diffused Field Equalization are applied to a HRTF-based sound localization problem, in order to evaluate performance of these methods for data reduction. In each experiment, left and right ear KEMAR HRTF are reduced with one of the introduced reduction methods. A test signal, which is white noise is virtually synthesized using the corresponding original HRTF. The convolution based sound localization algorithm as described in Section 2.4, is fed with the restored databases and used to localize the signals. Finally, the localization success rate is computed.
As already mentioned, GLRAM works on two preselected directions out of three. Therefore, we conduct localization experiments for a subset of directions (35 randomly chosen locations) to detect a combination of well working parameters for GLRAM. After finding a suitable combination of the variables, localization experiments for all 1250 directions are conducted. Firstly, the dataset is reduced for the first two directions, i.e. elevation and azimuth. The contour plot given in Figure 11(a) shows the localization success rate for a fixed pair of values (N_{r_a}, N_{r_e}). Similar results with respect to the pairs (N_{r_a}, N_{r_t}) and (N_{r_e}, N_{r_t}) are ploted in Figure 11(b) and Figure 11(c), respectively. Clearly, applying GLRAM on the pair of (N_{r_e}, N_{r_t}) outperforms the other two combinations.
The application of GLRAM in the directions of elevation and time performs best, therefore, we compare this optimal GLRAM with the standard PCA and Tensor-SVD. As mentioned in section 3.3, GLRAM is a simple form of Tensor-SVD with leaving one direction out. Thus, we investigate the effect of additionally reducing the third direction, whereas the dimensions in elevation and time are fixed to the parameters of the optimal GLRAM. Figure 13 shows that additionally decreasing the dimension in azimuth leads to a huge loss of localization accuracy.
After determining the optimal parameters for GLRAM, the simulations are conducted for all 1250 directions of the CIPIC dataset. Figure 12 shows the localization success rate in dependency of the compression rate for GLRAM and PCA. It can be seen that an optimized GLRAM outperforms the standard PCA in terms of compression.

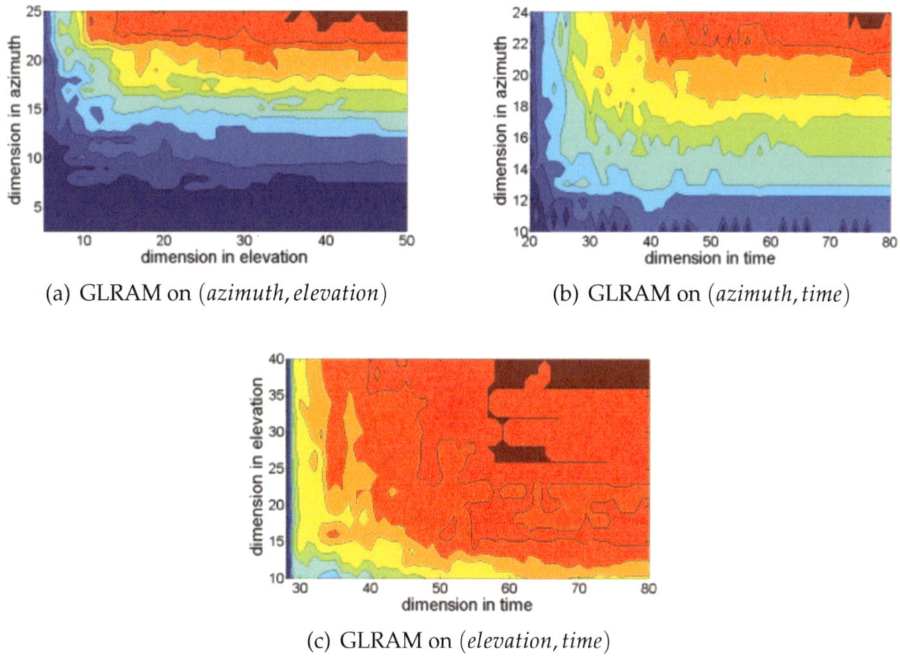

Fig. 11. Contour plots of localization success rate of using GLRAM in different settings.

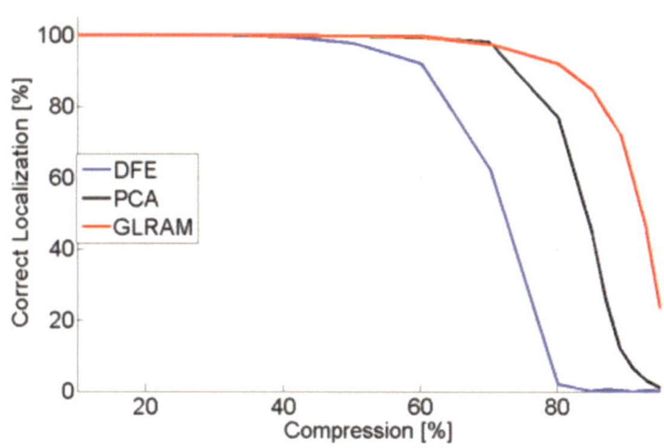

Fig. 12. Comparison between DFE, PCA and GLRAM

4. Predictors for HRTF sound localization

To reduce the computational costs of HRTF-based sound localization, especially for moving sound sources, it is advantageous to determine a region of interest (ROI) as illustrated in Figure 15. A ROI constricts the 3D search space around the robotic platform leading to a reduced set of eligible HRTFs.

Various tracking models have been implemented in microphone sound localization. Primarily they predict the path of a sound source as it is traveling and thus acquiring faster and more accurate non-ambiguous localization results (Belcher et al., 2003; Ward et al., 2003). Most of these filters are updated periodically in scans. In this section, three predictors, namely Time

Delay of Arrival, Kalman filter and Particle filter, are briefly introduced to determine a ROI to reduce the set of eligible HRTFs to be processed to localize moving sound sources.

4.1 Time Delay of Arrival

The time delay between the two signals $x_i[n]$ and $x_j[n]$ is found when the cross-correlation value $R_{ij}(\tau)$ is maximal. Given that τ has been determined, the time delay is calculated by

$$\Delta T = \frac{\tau}{f_s}, \tag{27}$$

where f_s is the sampling rate. Knowing the geometry (distance between the robot's ears) of the microphones and the delays between microphone pairs, a number of locations for the sound source can be disregarded (Brandstein & Ward, 2001; Kwok et al., 2005; Potamitis et al., 2004; Valin et al., 2003). Then, an HRTF-based localization algorithm only evaluates the remaining possible locations of the source.

4.2 Kalman Filter

The Kalman filter is a frequently used predictor (usage for microphone array localization described in (Belcher et al., 2003)). The discrete version exhibits two main states: time update (prediction) and measurement update (correction). The Kalman filter predicts the state of x_k at time k given the linear stochastic difference equation

$$\mathbf{x}_k = A\mathbf{x}_{k-1} + Bu_{k-1} + w_{k-1} \tag{28}$$

and measurement

$$z_k = H\mathbf{x}_k + v_k. \tag{29}$$

Matrices A, B and H provide relation from discrete time $k-1$ to k for their respective variables x (the state) and u (optional control input). w and v add noise to the model. A set of time and measurement update equations are used to predict the next state (Kalman, 1960). The state vector is defined by current location coordinates x and y and the velocity components v_x and v_y (Potamitis et al., 2004; Usman et al., 2008). Note that here the predictor is applied to two dimensional space.

$$\mathbf{x} = [x, v_x, y, v_y]^T \tag{30}$$

An unreliable location estimate during initialization of the the Kalman filter may be a source of error. To improve upon this, particle filters have been implemented in (Chen & Rui, 2004).

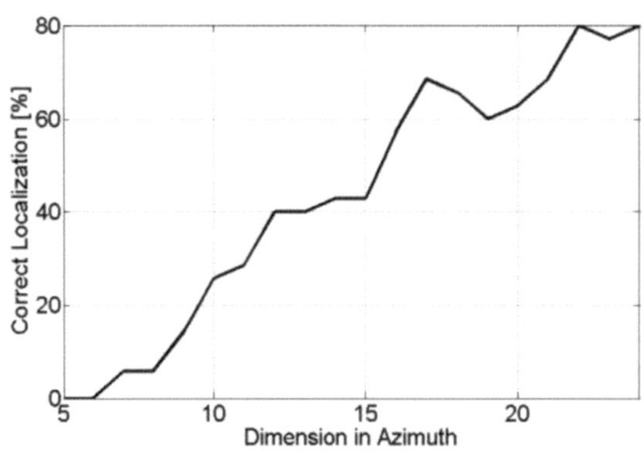

Fig. 13. Localization success rate by Tensor-SVD

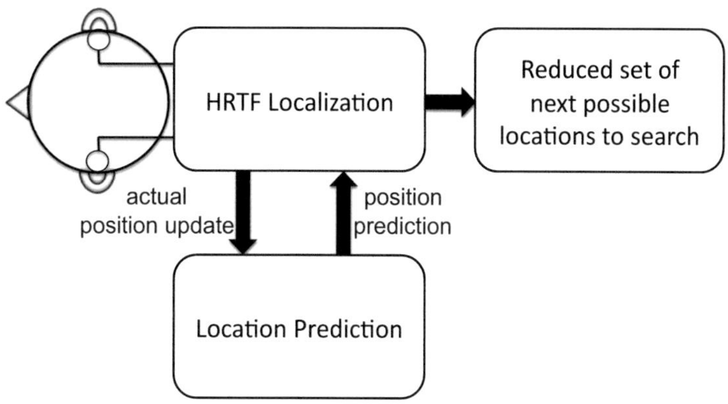

Fig. 14. Schematic view of the application of predictors in HRTF-based localization.

4.3 Particle Filter

The particle filter bases itself on the idea of randomly generating samples from a distribution and assigning weights to each to define their reliability. The particles and their associated weights define an averaged center which is the predicted value for the next step. Each weight w_k^i is associated to a particle x^i in iteration k. A set of N particles is initially drawn from a distribution $q(x_i|x_{k-1}^i, z_k)$ with z_k being the current observed value. For each particle the weight is calculated by

$$w_k^i = w_{k-1}^i \frac{p(z_k|x_k^i)p(x_k^i|x_{k-1}^i)}{q(x_k^i|x_{0:k-1}^i, z_{1:k})}. \tag{31}$$

Once all weights are calculated, their sum is normalized. To determine the predicted value, the weighted average of the particles is taken:

$$\bar{x} = \frac{1}{N}\sum_{i=1}^{N} w_k^i \cdot x_i \tag{32}$$

Over time it may occur that very few particles possess most of the weight. This case requires resampling to protect from particle degeneration. The variance of the weights is used as a measure to check for this case and if required, the set of weights is exchanged with a better approximation (Gordon et al., 1993).

Many particle filter variations exist, such as the Monte Carlo approximations and Sampling Importance Resampling. However a particle filter may find only a local optimum and thus never reaching the global optimum. Evolutionary estimation is proposed in (Kwok et al., 2005) to overcome such problems. Initially a set of potential speaker locations are estimated and then a heuristic search is performed. The speaker locations are called chromosomes and can only move within a defined region. After the initialization, the Time Delay of Arrival (TDOA) is evaluated for each potential location as well as each microphone. The difference v_i between expected and actual TDOAs is used to define a fitness function for each chromosome i together with error variance σ_τ^2:

$$\omega_i = e^{-0.5\frac{v_i^2}{\sigma_\tau^2}} \tag{33}$$

ω_i is then scaled such that $\sum_{i=1}^{n} \omega_i = 1 \rightarrow \widetilde{\omega}_i$ The new estimate of source location is given by

$$s_x = \sum_{i=1}^{n} \widetilde{\omega}_i s_{xi}. \tag{34}$$

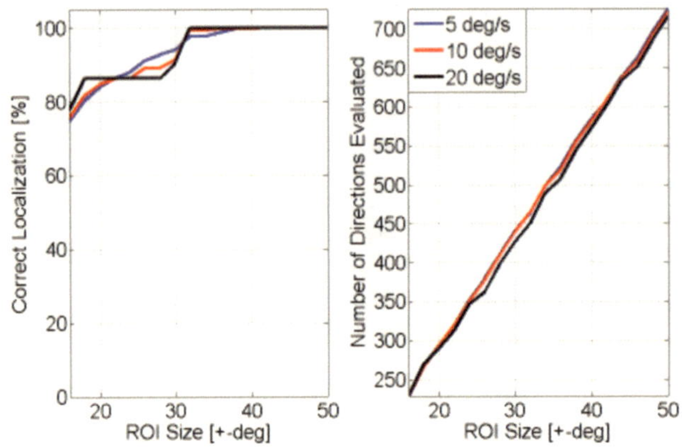

(a) Time Delay of Arrival

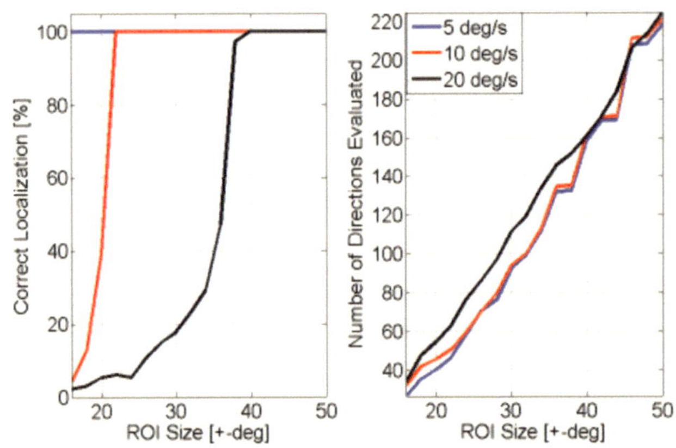

(b) Particle Filter

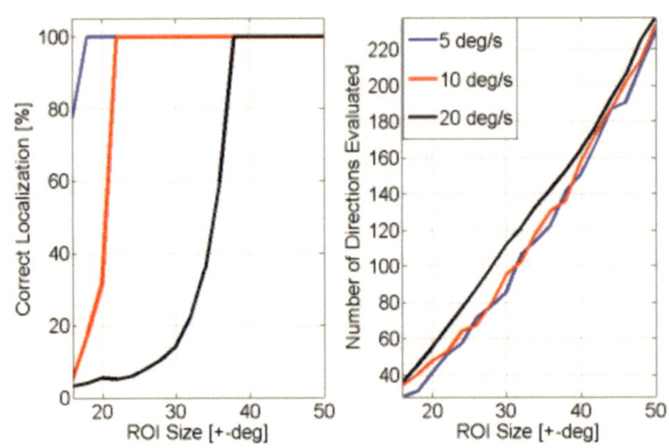

(c) Kalman Filter

Fig. 15. Comparison of predictors for HRTF Sound Localization.

Chromosomes are then selected according to a linearly spaced pointer spanning the fitness magnitude scale, with higher fitness chromosomes being selected more often. The latter chromosomes receive less mutation as compared to weaker chromosomes depending on r_g, the zero mean Gaussian random number variance, and d_m, the distance for mutation (Kwok et al., 2005).

$$s_{xi+1} = s_{xi} + r_g d_m \tag{35}$$

4.4 Numerical comparison

This section gives a performance overview of the applied predictors in a HRTF-based sound localization scenario. We simulate moving sound by virtually synthesizing a sound source, which is white noise, using different pairs of HRTFs. This way, a random path of 500 different source positions is generated, simulating a moving sound source. Then, Time Delay of Arrival, the Kalman filter and the Particle filter seek to reduce the search region for the HRTF-based sound localization to a region of interest. The Convolution Based Algorithm is utilized to localize the moving sound source. The experiments were conducted three times with different speed of the sound source.
Figure 15 summarizes the results of applying predictors to HRTF-based sound localization. The left plots show the localization success rates in dependency of the size of the region of interest. In the right plots the number of directions that have to be evaluated within the localization algorithms are shown. The bigger the region of interest, the more HRTF-pairs have to be utilized to maximize the cross correlation (11) resulting in a higher processing time. On the other hand, the smaller the region of interest, the higher the danger of excluding the HRTF pair that is maximizing the cross correlation (11), leading to false localization results. Our simulation results show that the number of HRTFs to be evaluated for the Convolution Based Algorithm can be significantly reduced to speed up HRTF-based localization for moving sources. Time Delay of Arrival is reducing the search region to 500 directions while reaching hundred percent correct localization of the path, meaning all 500 source positions are detected correctly for the different speeds of the sources. Particle- and Kalman filter are able to reduce the search region to 130 directions in case of sound sources with a speed of $20 deg/s$. For slower sources, only 60 directions need to be taken into account.

Acknowledgements

This work was fully supported by the German Research Foundation (DFG) within the collaborative research center SFB-453 "High Fidelity Telepresence and Teleaction".

5. References

Algazi, V. R., Duda, R. O., Thompson, D. M. & Avendano, C. (2001). The CIPIC HRTF database, *IEEE ASSP Workshop on Applications of Signal Processing to Audio and Acoustics*, New Paltz, NY, pp. 21–24.

Belcher, D., Grimm, M. & Kroschel, K. (2003). Speaker tracking with a microphone array using a kalman filter, *Advances in Radio Science* 1: 113–117.

Blauert, J. (1997). An introduction to binaural technology, *Binaural and Spatial Hearing*, R. Gilkey, T. Anderson, Eds., Lawrence Erlbaum, Hilldale, NJ, USA, pp. 593–609.

Brandstein, M. & Ward, D. (2001). *Microphone arrays - signal processing techniques and applications*, Springer.

Chen, Y. & Rui, Y. (2004). Real-time speaker tracking using particle filter sensor fusion, *Proceedings of the IEEE* 92(3): 485–494.

Gordon, N., Salmond, D. & Smith, A. (1993). Novel approach to nonlinear/non-Gaussian Bayesian state estimation, *Radar and Signal Processing, IEE Proceedings F* 140(2): 107 –113.

Jolliffe, I. T. (2002). *Principal Component Analysis*, second edn, Springer.

Kalman, R. (1960). A new approach to linear filtering and prediction problems, *Transactions of the ASME - Journal of Basic Engineering* 82(Series D): 35–45.

Keyrouz, F. & Abou Saleh, A. (2007). Intelligent sound source localization based on head-related transfer functions, *IEEE International Conference on Intelligent Computer Communication and Processing*, pp. 97–104.

Keyrouz, F. & Diepold, K. (2006). An enhanced binaural 3D sound localization algorithm, *2006 IEEE International Symposium on Signal Processing and Information Technology*, pp. 662–665.

Keyrouz, F., Diepold, K. & Dewilde, P. (2006). Robust 3D Robotic Sound Localization Using State-Space HRTF Inversion, *IEEE International Conference on Robotics and Biomimetics, 2006. ROBIO'06*, pp. 245–250.

Kwok, N., Buchholz, J., Fang, G. & Gal, J. (2005). Sound source localization: microphone array design and evolutionary estimation, *IEEE International Conference on Industrial Technology*, pp. 281–286.

Moeller, H. (1992). Fundamentals of binaural technology, *Applied Acoustics* 36(3-4): 171–218.

Potamitis, I., Chen, H. & Tremoulis, G. (2004). Tracking of multiple moving speakers with multiple microphone arrays, *IEEE Transactions on Speech and Audio Processing* 12(5): 520–529.

Savas, B. & Lim, L. (2008). Best multilinear rank approximation of tensors with quasi-Newton methods on Grassmannians, *Technical Report LITH-MAT-R-2008-01-SE*, Department of Mathematics, Linkpings University.

Usman, M., Keyrouz, F. & Diepold, K. (2008). Real time humanoid sound source localization and tracking in a highly reverberant environment, *Proceedings of 9th International Conference on Signal Processing*, Beijing, China, pp. 2661–2664.

Valin, J., Michaud, F., Rouat, J. & Letourneau, D. (2003). Robust sound source localization using a microphone array on a mobile robot, *IEEE/RSJ International Conference on Intelligent Robots and Systems*, Vol. 2.

Ward, D. B., Lehmann, E. A. & Williamson, R. C. (2003). Particle filtering algorithms for tracking an acoustic source in a reverberant environment, *IEEE Transactions on Speech and Audio Processing* 11(6): 826–836.

Ye, J. (2005). Generalized low rank approximations of matrices, *Machine Learning* 61(1-3): 167–191.

6

Effect of Space on Auditory Temporal Processing with a Single-Stimulus Method

Martin Roy, Tsuyoshi Kuroda and Simon Grondin
Université Laval, Québec
Canada

1. Introduction

The exact nature of the relation between space and time is certainly one of the most fundamental issues in physics (Buccheri, Saniga, & Stuckey, 2003), but it is also an intriguing question for experimental psychologists (Casasanto, Fotakopoulou, & Boroditsky, 2010). A function of perception is to form mental representations indicating what object exists, where it is located, and how it acts, i.e., how the object moves in space with the lapse of time. Space and time are integrated in the perceptual system to cause the perception of motion and speed, and such integration is required to determine the performance of the motor system (e.g., hand movement; see Lee, 2000). How space and time exert mutual influence is a question that was addressed many years ago (Abe, 1935; Helson, 1930), notably by J. Piaget, who studied the ontogenesis of the relations between time, distance and speed (Piaget, 1955).

Time perception has been often explained with the "internal-clock hypothesis," which is notable in discussing the perceptual relation between space and time. An internal clock is usually assumed to be a pacemaker-counter device, with the first module emitting pulses accumulated by the second one (Grondin, 2001, 2010). The amount of accumulation decides the perceived time duration. The performance level is varied, however, when some variation of nontemporal factors are introduced in experiments. This variability, in a duration discrimination task for instance, can be observed by varying the time intervals' structure (filled or empty; Grondin, 1993), or by varying the sensory modality to be stimulated. Space is a nontemporal factor, which is susceptible to vary the performance level of the internal clock.

There are two illusions concerning the perceptual relation between space and time, which have been studied since the early 20th century (see Jones & Huang, 1982; Sarrazin, Giraudo, & Pittenger, 2007; ten Hoopen, Miyauchi, & Nakajima, 2008). The *tau effect* takes place typically in the successive presentation of three signals, say, X, Y, and Z, with Y somewhere between X and Z (Helson, 1930; Helson & King, 1931; Henry, McAuley, & Zaleha, 2009). They are delivered from different sources spaced at equal intervals, resulting in two equal intervals in space, X-Y and Y-Z. These intervals are perceived as unequal in their distance, however, if the signals are presented at unequal intervals in time; if the time interval defined by X and Y is shorter (longer) than the time interval defined by Y and Z, the spatial distance between X and Y is perceived as shorter (longer) than the spatial distance between Y and Z. In other words, the spatial-interval ratio is perceived as if it were similar to the time-interval ratio. Such interaction between space and time can be caused in the opposite direction with the same signal configuration, i.e., the time-interval ratio is perceived as if it were similar to the spatial-interval ratio. This opposite-direction effect was named the *kappa effect* (Cohen, Hansel, & Sylvester, 1953; Price-Williams, 1954).

The *kappa effect* in the auditory mode was investigated in the present study. The *kappa effect* has been tested more often in the visual mode (Cohen, Hansel, & Sylvester, 1953, 1955; Collyer, 1977; Miyatani, 1984-1985; Sarrazin, Giraudo, Pailhous, & Bootsma, 2004), and even in the tactile mode (Goldreich, 2007; Suto, 1952, 1955, 1957). There are researches testing the *kappa effect* in the auditory mode, but most of them focused on the effects of frequency distance (difference), instead of spatial distance, on the perception of time duration (Cohen, Hansel, & Sylvester, 1954; Henry & McAuley, 2009; Jones & Huang, 1982; Shigeno, 1986; Yoblick & Salvendy, 1970). In a typical case, three successive signals were different in their frequency, causing two intervals in time and in frequency, and the time-interval ratio was perceived as if it had been similar to the frequency-interval ratio. There is little evidence for the occurrence of the *kappa effect* in the perception of space and time in the auditory mode (Sarrazin, Giraudo, Pittenger, 2007; see Ouellet, 2003).

The *kappa effect* indicates that time duration increases perceptually in proportion to spatial distance between two signals, but this effect has been demonstrated with three successive signals, where two intervals are bounded on each other. Few researches have examined whether or not the similar effect can take place when a single interval is presented.

It is important in this context to indicate that there are cases where time duration is perceived as shorter when spatial distance is increased in the visual mode (Guay & Grondin, 2001). This result was observed in an experiment employing a single-stimulus method, where a categorization judgment was conducted after the presentation of one interval. The interval was defined by two signals delivered from different sources, which were selected from three sources (above, middle and below) located in front of participants on the same vertical plane. All location pairs were presented in random order within each block. The interval was more often perceived as shorter when it was marked by the above and below sources, in comparison with intervals marked by the above and middle sources or the middle and below sources.

The purpose of the present study was to verify if space exerts influence on time perception (1) when intervals to be measured perceptually are marked by sounds delivered from sources having different distances between them, and (2) when these intervals are presented according to a single-stimulus method.

2. Method

Participants

Twelve 19- to 26-year-old volunteer students at Université Laval (six females and six males) with no hearing problems participated in this experiment. They were paid CAN $20 for their participation.

Apparatus and stimuli

A time interval was defined by two sound stimuli of 20 ms. The stimuli were 1-kHz sinusoidal sounds generated by IBM PC running E-Prime software (version 1.1.4.1 - SP3). The computer was equipped with an SB Audigy 2 sound card, and the stimuli were delivered by Logitech Z-640 loudspeakers. Participants pressed "1" or "3" on the computer keyboard to indicate that the interval was short or long, respectively.

Procedure

The single-stimulus method was employed (Allan, 1979; Morgan, Watamaniuk, & McKee, 2000), i.e., each trial consisted of presenting one interval. The duration of the time intervals was controlled as follows: Eight values of time-interval duration were distributed around a mid-point value which is called the base duration. Four values below the base duration were called the "short" duration, and four values above the base duration were called the "long" duration. There were two base-duration conditions, 125 and 250 ms. In the former case, the "short" intervals lasted 104, 110, 116 and 122 ms, and the "long" intervals 128, 134, 140 and 146 ms. In the latter case, the "short" intervals lasted 208, 220, 232 and 244 ms, and the "long" intervals 256, 268, 280 and 292 ms.

The participants were asked to judge whether the presented interval belonged to the "short" or to the "long" category. A 1.5-s feedback signal was presented immediately after the response on the computer screen and indicated whether the response was correct or not.

There were two conditions of spatial distance between the auditory sources (loudspeakers), 1.1 m and 3.3 m (see Figure 1), and there were two conditions of the direction of stimulus presentation, right to left and left to right.

Each participant completed eight sessions, four for 125-ms base duration and four for 250-ms base duration. Six participants completed the 125-ms sessions before the 250-ms sessions, and six completed the 250-ms sessions before the 125-ms sessions. Four sessions in each base duration corresponded to four spatial conditions (2 distances x 2 directions), and they were carried out in random order. Each session had six blocks of 64 trials where the eight intervals were presented eight times in random order, and thus 48 responses were obtained in each interval in each spatial condition.

Data analysis

The two direction conditions were collapsed, resulting in four conditions in the data analysis (2 distance and 2 base-duration conditions). For each participant and for each condition, an 8-point psychometric function was traced, plotting the time interval duration on the *x-axis* and the "long" response proportion on the *y-axis*. Each point on the psychometric function was based on 96 presentations.

The *pseudo-logistic model* (Killeen, Fetterman, & Bizo, 1997) was employed to calculate psychometric functions that were fitted to the resulting curves. Two indices of performance were estimated from each psychometric function, one for sensitivity and one for perceived duration. As an indicator of temporal sensitivity, one standard deviation (SD) on each psychometric function was employed. Using one SD (or variance) is a common procedure to express temporal sensitivity (Grondin, 2008; Grondin, Roussel, Gamache, Roy, & Ouellet, 2005; Killeen & Weiss, 1987).

The other parameter was the temporal bisection point (BP). In the context of the *kappa effect*, this dependent variable is the most important. The BP can be defined as the x value corresponding to the 0.50 proportion on the y-axis. The observed shift of the BP for different conditions can be interpreted as an indication of differences in perceived duration. If an interval is perceived as longer, the "long" response takes place more frequently, which causes the downward shift of the BP. If an interval is perceived as shorter, the "long" response takes place less frequently, which causes the upward shift of the BP.

3. Results

Figure 2 reports the grouped psychometric function for each of the four experimental conditions: 2 Distances x 2 Base Durations. In order to allow direct comparisons between the base duration conditions, two dependent variables were calculated from the above parameters. One is the *Constant Error*, which is the BP minus the base duration. The other is the *Coefficient of Variation*, which is the SD divided by the BP.

Figure 3 shows the results for the *Constant Error*. Essentially, it reveals higher values in the 3.3-m condition than in the 1.1-m condition. A 2 x 2 ANOVA with repeated measures revealed that both the distance effect, $F(1,11) = 10.55$, $p < .01$, $\eta_p^2 = .49$ and the base duration effect, $F(1,11) = 4.91$, $p < .05$, $\eta_p^2 = .31$, were significant. The interaction effect was also significant, $F(1,11) = 8.36$, $p < .05$, $\eta_p^2 = .43$.

Figure 4 shows the results for the *Coefficient of Variation*. The 2 x 2 ANOVA with repeated measures revealed that both the distance effect, $F(1,11) = 7.23$, $p < .05$, $\eta_p^2 = .40$, and the base duration effect, $F(1,11) = 19.80$, $p < .001$, $\eta_p^2 = .64$, were significant. The interaction effect was not significant, $F(1,11) = .60$, $p = .45$, $\eta_p^2 = .05$.

4. Discussion

The results of the present experiment clearly indicate that increasing the distance between sound sources marking time intervals leads to a decrease of the perceived duration (a higher constant error). These results are inconsistent with what is usually reported when referring to the *kappa effect* but consistent with results obtained in the visual mode with a single-stimulus method (Guay & Grondin, 2001). Other results linking space and time in the auditory mode revealed no such effect of distance between sound sources when sequences of four sounds from four sources were used (Ouellet, 2003). The present experiment also revealed that increasing distance between marker's sources results in a higher coefficient of variation.

The present results can be explained on the basis of the internal-clock hypothesis, where the accumulation process is controlled by an attentional mechanism, with more attention to time resulting in a higher accumulation of pulses (Grondin & Macar, 1992; Grondin & Plourde, 2007; Macar, Grondin, & Casini, 1994). When the two stimuli were farther away from each other in space, more attentional resources were allocated to their location perception (Mondor & Zattore, 1995; Rhodes, 1987; Roussel, Grondin, & Killeen, 2009), which caused the decrease of the resources allocated to the time perception. There were less accumulated pulses in the counter of the internal clock, and thus the time duration was perceived as shorter. This explanation is also consistent with the results obtained with the coefficient of variation. Allocating more resources to the spatial perception caused more variance (more categorization errors - higher coefficient of variation) in the observers' judgment.

Finally, the results also revealed higher coefficients of variation in the 125-ms base duration than in the 250-ms base duration. This finding is consistent with a generalized form of Weber's law applied to time perception in which sensory noise (nontemporal noise due to attention disturbance) causes more damage to performance with briefer intervals.

5. Acknowledgements

This research was made possible by a grant awarded to SG by the Natural Sciences and Engineering Research Council of Canada (NSERC). We would like to thank Marie-Claude Simard and Karine Drouin for their help with data collection. Correspondence should be addressed to Simon Grondin, École de psychologie, 2325 rue des Bibliothèques, Université Laval, Québec, Qc, Canada, G1V 0A6 (E-mail: simon.grondin@psy.ulaval.ca)

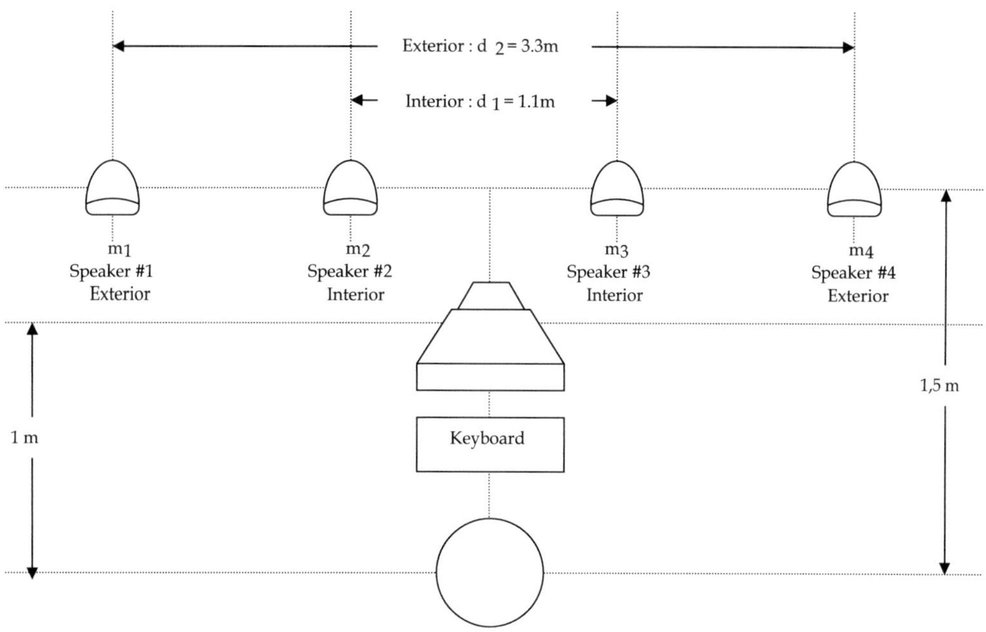

Fig. 1. Experimental set-up

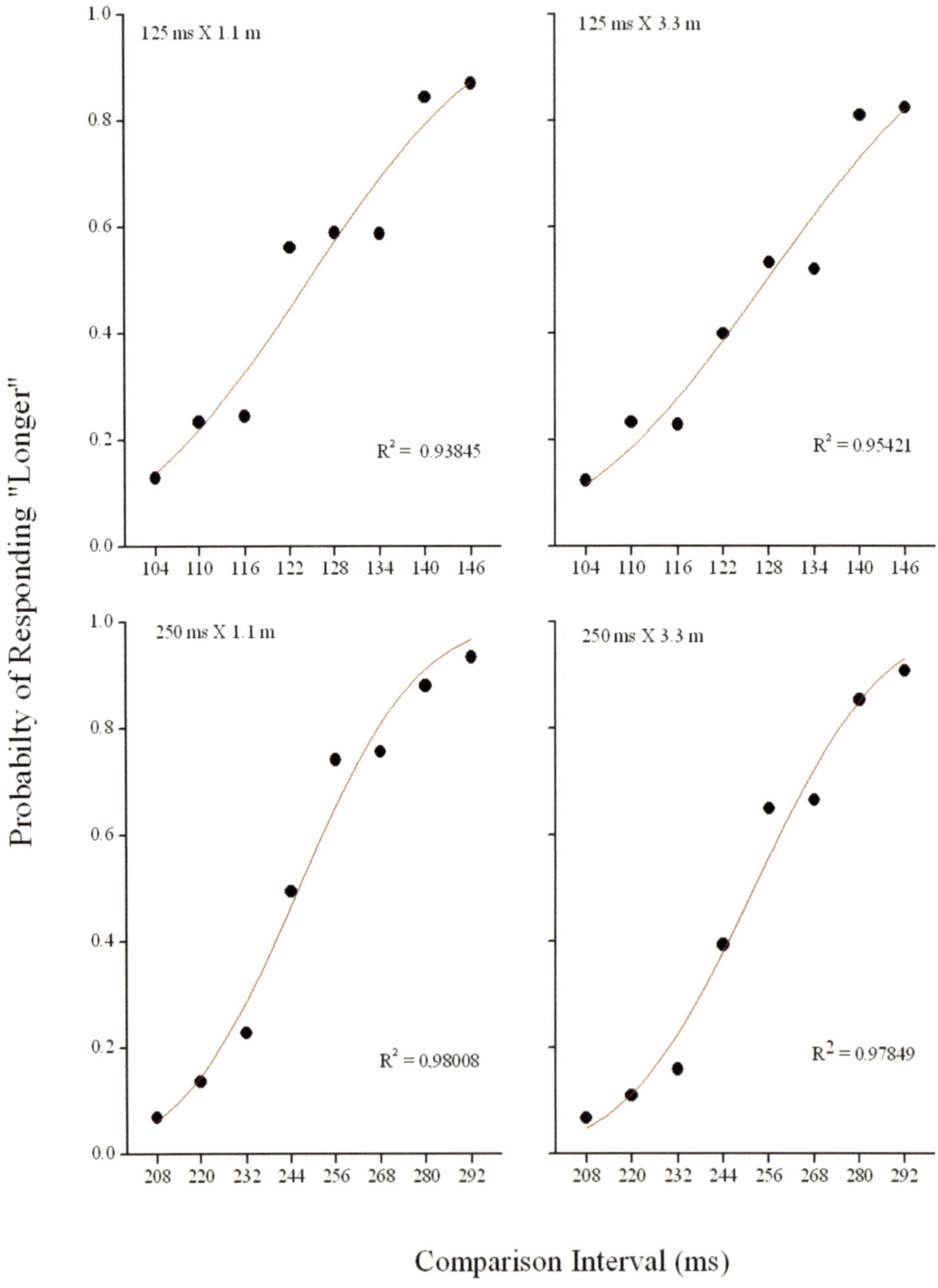

Fig. 2. Psychometric function (grouped data) in each experimental condition.

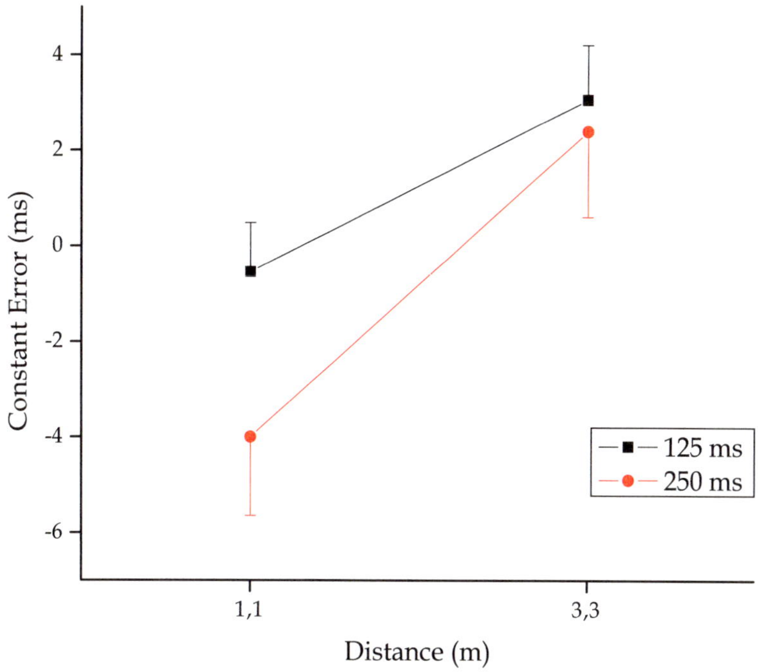

Fig. 3. Mean *Constant Error* as a function of distance between auditory markers. (Bars are standard errors)

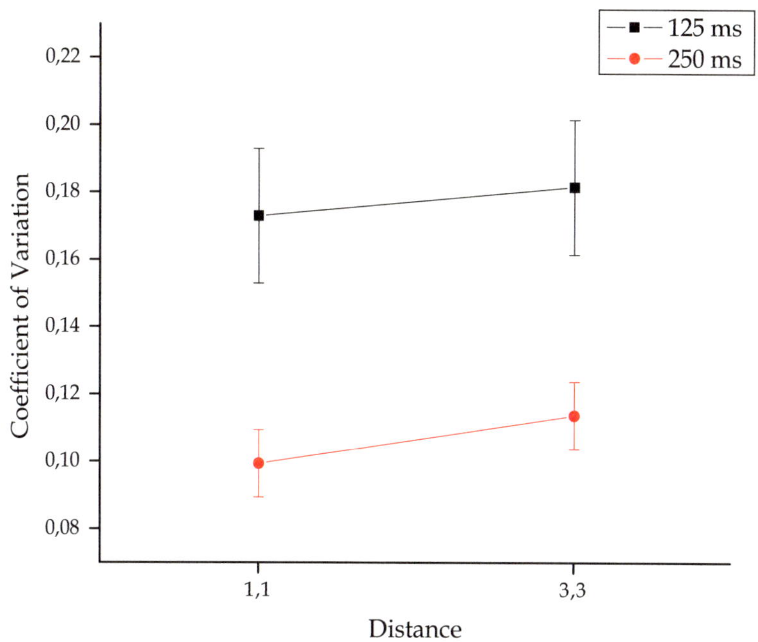

Fig. 4. Mean *Coefficient of Variation* as a function of distance between auditory markers. (Bars are standard errors)

6. References

Abe, S. (1935). Experimental study on the co-relation between time and space. *Tohoku Psychologica Folia, 3,* 53–68.

Allan, L. G. (1979). The perception of time. *Perception & Psychophysics, 26,* 340-354.

Buccheri, R., Saniga, M., & Stuckey, W. M. (Eds.). (2003). *The Nature of Time: Geometry, Physics and Perception*. Dordrecht, Netherlands: Kluwer Academic Publishers.

Casasanto, D., Fotakopoulou, O., & Boroditsky, L. (2010). Space and time in the child's mind: Evidence for a cross-dimensional asymmetry. *Cognitive Science, 34,* 387-405.

Cohen, J., Hansel, C. E. M., & Sylvester, J. D. (1953). A new phenomenon in time judgment. *Nature, 172,* 901.

Cohen, J., Hansel, C. E. M., & Sylvester, J. D. (1954). Interdependence of temporal and auditory judgments. *Nature, 174,* 642-644.

Cohen, J., Hansel, C. E. M., & Sylvester, J. D. (1955). Interdependence in judgments of space, time and movement. *Acta Psychologica, 11,* 360-372.

Collyer, C. E. (1977). Discrimination of spatial and temporal intervals defined by three light flashes : Effects of spacing on temporal judgments and of timing on spatial judgments. *Perception & Psychophysics, 21,* 357-364.

Goldreich, D. (2007). A Bayesian perceptual model replicates the cutaneous rabbit and other tactile spatiotemporal illusions. *PLoS ONE, 2,* e333.

Grondin, S. (1993). Duration discrimination of empty and filled intervals marked by auditory and visual signals. *Perception & Psychophysics, 54,* 383-394.

Grondin, S. (2001). From physical time to the first and second moments of psychological time. *Psychological Bulletin, 127,* 22-44.

Grondin, S. (2008). Methods for studying psychological time. In S. Grondin (Ed.). *Psychology of time* (pp. 51-74). Bingley, UK: Emerald Group Publishing.

Grondin, S. (2010). Timing and time perception: A review of recent behavioral and neuroscience findings and theoretical directions. *Attention, Perception, & Psychophysics, 72,* 561-582.

Grondin, S., & Macar, F. (1992). Dividing attention between temporal and nontemporal tasks: A performance operating characteristic -POC- analysis. In F. Macar, V. Pouthas, & W. J. Friedman (Eds.), *Time, Action, Cognition: Towards Bridging the Gap.* (pp. 119-128). Dordrecht, Netherlands: Kluwer Academic Publishers.

Grondin, S., & Plourde, M. (2007). Discrimination of time intervals presented in sequences: Spatial effects with multiple auditory sources. *Human Movement Science, 26,* 702-716.

Grondin, S., Roussel, M.-È., Gamache, P.-L., Roy, M., & Ouellet, B. (2005). The structure of sensory events and the accuracy of time judgments. *Perception, 34,* 45-58.

Guay, I., & Grondin, S. (2001). Influence on time interval categorization of distance between markers located on a vertical plane. In E. Sommerfeld, R. Kompass, & T. Lachman (Eds.), *Proceedings of the 17th Annual Meeting of the International Society for Psychophysics* (pp. 391-396). Berlin, Germany: Pabst Science Publishers.

Helson, H. (1930). The tau effect: An example of psychological relativity. *Science, 71,* 536–537.

Helson, H., & King, S. M. (1931). The tau effect: An example of psychological relativity. *Journal of Experimental Psychology, 14,* 202–217.

Henry, M. J. & McAuley, J. D. (2009). Evaluation of an imputed pitch velocity model of the auditory kappa effect. *Journal of Experimental Psychology: Human Perception and Performance, 35*, 551-564.

Henry, M. J., McAuley, J. D., & Zaleha, M. (2009). Evaluation of an imputed pitch velocity model of the auditory tau effect. *Attention, Perception & Psychophysics, 71*, 1399-1413.

Jones, B., & Huang, Y. L. (1982). Space-time dependencies in psychophysical judgment of extent and duration: Algebraic models of the tau and kappa effect. *Psychological Bulletin, 91*, 128-142.

Killeen, P. R., Fetterman, J. G., & Bizo, L. A. (1997). Time's cause. In C. M. Bradshaw & E. Szabadi (Eds). *Time and Behavior: Psychological and Neurobehavioral Analyses* (pp. 79-131). Amsterdam, Netherlands: North-Holland/Elsevier Science.

Killeen, P. R., & Weiss, N. A. (1987). Optimal timing and the Weber function. *Psychological Review, 94*, 455-468.

Lee, D. (2000). Learning of spatial and temporal patterns in sequential hand movements. *Cognitive Brain Research, 9*, 35-39.

Macar, F., Grondin, S., & Casini, L. (1994). Controlled attention sharing influences time estimation. *Memory & Cognition, 22*, 673-686.

Mondor, T. A., & Zattore, R. J. (1995). Shifting and focusing auditory spatial attention. *Journal of Experimental Psychology: Human Perception and Performance, 21*, 387-409.

Morgan, M. J., Watamaniuk, S. N. J., & McKee, S. P. (2000). The use of an implicit standard for measuring discrimination thresholds. *Vision Research, 40*, 2341-2349.

Miyatani (1984-1985). The time and distance judgments at different levels of discriminability of temporal and spatial information. *Hiroshima Forum for Psychology, 10*, 45-55.

Ouellet, B. (2003). *L'influence de la distance entre des marqueurs statiques sur la discrimination d'intervalles temporels : À la recherche de l'effet kappa classique en modalité auditive.* Unpublished master's dissertation, Université Laval, Québec, Canada.

Piaget, J. (1955). The development of time concepts in the child. In P. H. Hoch, J. Zubin (Eds.), *Psychopathology of Childhood* (pp. 34-44). New York, USA: Grune & Stratton.

Price-Williams, D. R. (1954). The kappa effect. *Nature, 173*, 363-364.

Rhodes, G . (1987). Auditory attention and the representation of spatial information. *Perception & Psychophysics, 42*, 1-14.

Roussel, M.-È., Grondin, S., & Killeen, P. (2009). Spatial effects on temporal categorization. *Perception, 38*, 748-762.

Sarrazin, J.-C., Giraudo, M.-D., Pailhous, J., & Bootsma, R. J. (2004). Dynamics of balancing space and time in memory: Tau and kappa effects revisited. *Journal of Experimental Psychology: Human Perception and Performance, 30*, 411-430.

Sarrazin, J.-C., Giraudo, M.-D., & Pittenger, J. B. (2007). Tau and kappa effects in physical space: The case of audition. *Psychological Research, 71*, 201-218.

Shigeno, S. (1986). The auditory tau and kappa effects for speech and nonspeech stimuli. *Perception & Psychophysics, 40*, 9-19.

Suto, Y. (1952). The effect of space on time estimation (S-effect) in tactual space. I. *Japanese Journal of Psychology, 22*, 45-57.

Suto, Y. (1955). The effect of space on time estimation (S-effect) in tactual space. II : The role of vision in the S effect upon skin. *Japanese Journal of Psychology, 26*, 94-99.

Suto, Y. (1957). Role of apparent distance in time perception. *Research Reports of Tokyo Electrical Engineering College, 5*, 73-82.

ten Hoopen, G., Miyauchi, R., & Nakajima, Y. (2008). Time-based illusions in the auditory mode. In S. Grondin (Ed.). *Psychology of Time* (pp. 139-188). Bingley, UK: Emerald Group Publishing.

Yoblick, D. A., & Salvendy, G. (1970). Influence of frequency on the estimation of time for auditory, visual, and tactile modalities: The kappa effect. *Journal of Experimental Psychology, 86,* 157-164.

Part 2

Sound Localization Systems

7

Sound Source Localization Method Using Region Selection

Yong-Eun Kim[1], Dong-Hyun Su[2], Chang-Ha Jeon[2], Jae-Kyung Lee[2], Kyung-Ju Cho[3] and Jin-Gyun Chung[2]
[1]Korea Automotive Technology Institute in Chonan,
[2]Chonbuk National University in Jeonju,
[3]Korea Association Aids to Navigation in Seoul,
Korea

1. Introduction

There are many applications that would be aided by the determination of the physical position and orientation of users. Some of the applications include service robots, video conference, intelligent living environments, security systems and speech separation for hands-free communication devices (Coen, 1998; Wax & Kailath, 1983; Mungamuru & Aarabi, 2004; Sasaki et al., 2006; Lv & Zhang 2008). As an example, without the information on the spatial location of users in a given environment, it would not be possible for a service robot to react naturally to the needs of the user.

To localize a user, sound source localization techniques are widely used (Nakadai et al., 2000; Brandstein & Ward, 2001; Cheng & Wakefield, 2001; Sasaki et al., 2006). Sound localization is the process of determining the spatial location of a sound source based on multiple observations of the received sound signals. Current sound localization techniques are generally based upon the idea of computing the time difference of arrival (TDOA) information with microphone arrays (Knnapp & Cater, 1976; Brandstein & Silverman, 1997).

An efficient method to obtain TDOA information between two signals is to compute the cross-correlation of the two signals. The computed correlation values give the point at which the two signals from separate microphones are at their maximum correlation. When only two isotropic (i.e., not directional as in the mammalian ear) microphones are used, the system experiences front-back confusion effect: the system has difficulty in determining whether the sound is originating from in front of or behind the system. A simple and efficient method to overcome this problem is to incorporate more microphones (Huang et al., 1999).

Various weighting functions or pre-filters such as Roth, SCOT, PHAT, Eckart filter and HT can be used to increase the performance of time difference estimation (Knnapp & Cater, 1976). However, the performance improvement is achieved with the penalty of large power consumption and hardware overhead, which may not be suitable for the implementation of portable systems such as service robots.

In this chapter, we propose an efficient sound source localization method under the assumption that three isotropic microphones are used to avoid the front-back confusion effect. By the proposed approach, the region from 0° to 180° is divided into three regions and only one of the three regions is selected for the sound source localization. Thus considerable amount of computation time and hardware cost can be reduced. In addition, the estimation accuracy is improved due to the proper choice of the selected region.

2. Sound localization using TDOA

If a signal emanated from a remote sound source is monitored at two spatially separated sensors in the presence of noise, the two monitored signals can be mathematically modeled as

$$\begin{aligned} x_1(t) &= s_1(t) + n_1(t), \\ x_2(t) &= \alpha s_1(t-D) + n_2(t), \end{aligned} \tag{1}$$

where α and D denote the relative attenuation and the time delay of $x_2(t)$ with respect to $x_1(t)$, respectively. It is assumed that signal $s_1(t)$ and noise $n_i(t)$ are uncorrelated and jointly stationary random processes. A common method to determine the time delay D is to compute the cross correlation

$$R_{x_1x_2}(\tau) = E[x_1(t)x_2(t-\tau)], \tag{2}$$

where E denotes expectation operator. The time argument at which $R_{x_1x_2}(\tau)$ achieves a maximum is the desired delay estimate.

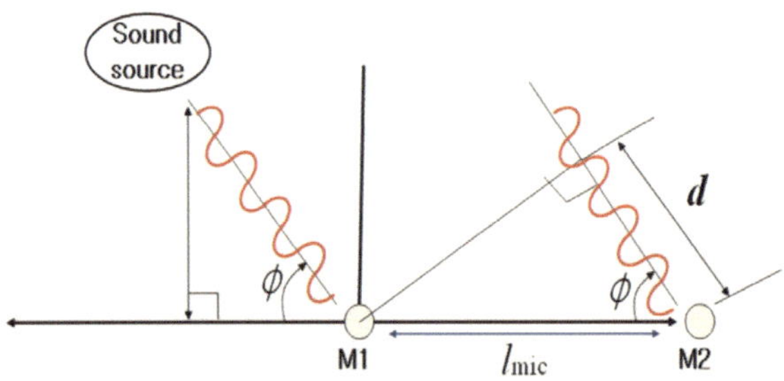

Fig. 1. Sound source localization using two microphones

Fig. 1 shows the sound localization test environments using two microphones. We assume that the sound waves arrive in parallel to each microphone as shown in Fig. 1. Then, the time delay D can be expressed as

$$D = \frac{d}{v_{sound}} = \frac{l_{mic}\cos\phi}{v_{sound}}, \tag{4}$$

where v_{sound} denotes the sound velocity of 343m/s. Thus, the angle of the sound source is computed as

$$\phi = \cos^{-1}\frac{D\,v_{sound}}{l_{mic}} = \cos^{-1}\frac{d}{l_{mic}}. \tag{5}$$

If the sound wave is sampled at the rate of f_s, and the sampled signal is delayed by n_d samples, the distance d can be computed as

$$d = \frac{v_{sound} n_d}{f_s}. \tag{6}$$

In Fig. 1, since d is a side of a right-angled triangle, we have

$$d < l_{mic}. \tag{7}$$

Thus, when $d = l_{mic}$ in (6), the number of maximum delayed samples $n_{d,\max}$ is obtained as

$$n_{d,\max} = \frac{f_s l_{mic}}{v_{sound}}. \tag{8}$$

3. Proposed sound source localization method

3.1 Region selection for sound localization

The desired angle in (5) is obtained using the inverse cosine function. Fig. 2 shows the inverse cosine graph as a function of d. Since the inverse cosine function is nonlinear, Δd (estimation error in d) has different effect on the estimated angle depending on the sound source location. Fig. 3 shows the estimation error (in degree) of sound source location as a function of Δd. As can be seen from Fig. 3, Δd has smaller effect for the sources located from 60° to 120°. As an example, when the source is located at 90° with the estimation error Δd = 0.01, the mapped angle is 89.427°. However, if the source is located at 0° with the estimation error Δd = 0.01, the mapped angle is 8.11°. Thus, for the same estimation error Δd, the effect for the source located at 0° is 14 times larger than that of the source at 90°. To efficiently implement the inverse cosine function, we consider the region from 60° to 120° as approximately linear as shown in Fig. 2.

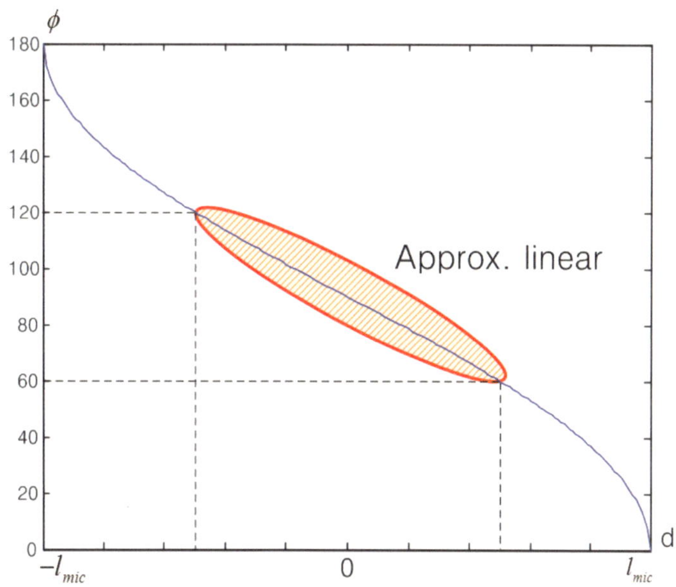

Fig. 2. Inverse cosine graph as a function of d

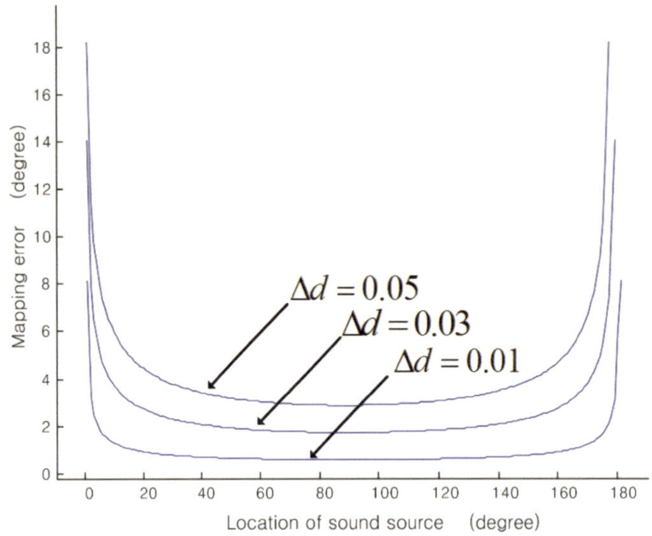

Fig. 3. Estimation error of sound source location as a function of Δd

Fig. 4 shows the front-back confusion effect: the system has difficulty in determining whether the sound is originating from in front of (sound source A) or behind (sound source B) the system. A simple and efficient method to overcome this problem is to incorporate more microphones. In Fig. 5, three microphones are used to avoid the front-back confusion effect, where L, R and B mean the microphones located at the left, right and back sides, respectively. In this chapter, to apply the cross-correlation operation in (2), for each arrow between the microphones in Fig. 5, the signal received at the tail part and the head part are designated as $x_1(t)$ and $x_2(t)$, respectively.

In conventional approaches, correlation functions are calculated between each microphone pair and mapped to angles as shown in Fig. 6-(a), (b) and (c). Notice that, due to the front-back confusion effect, each microphone pair provides two equivalent maximum values. Fig. 6-(d) is obtained by adding the three curves. In Fig. 6-(d), the angle corresponding to the maximum magnitude is the desired sound source location.

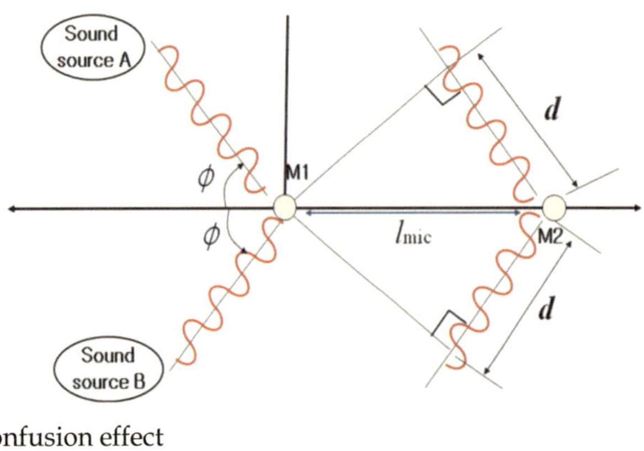

Fig. 4. Front-back confusion effect

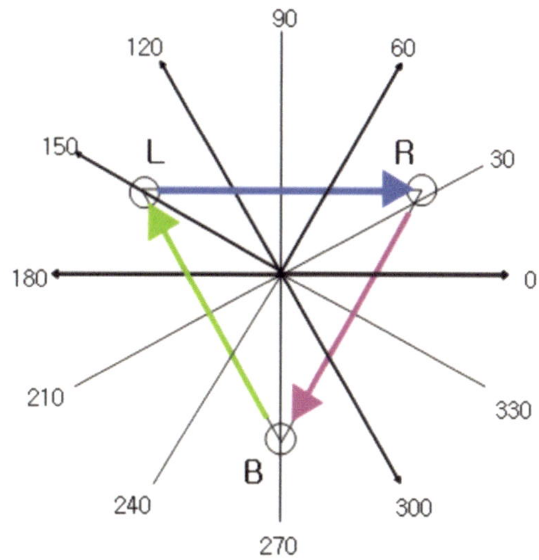

Fig. 5. Sound source localization using three microphones

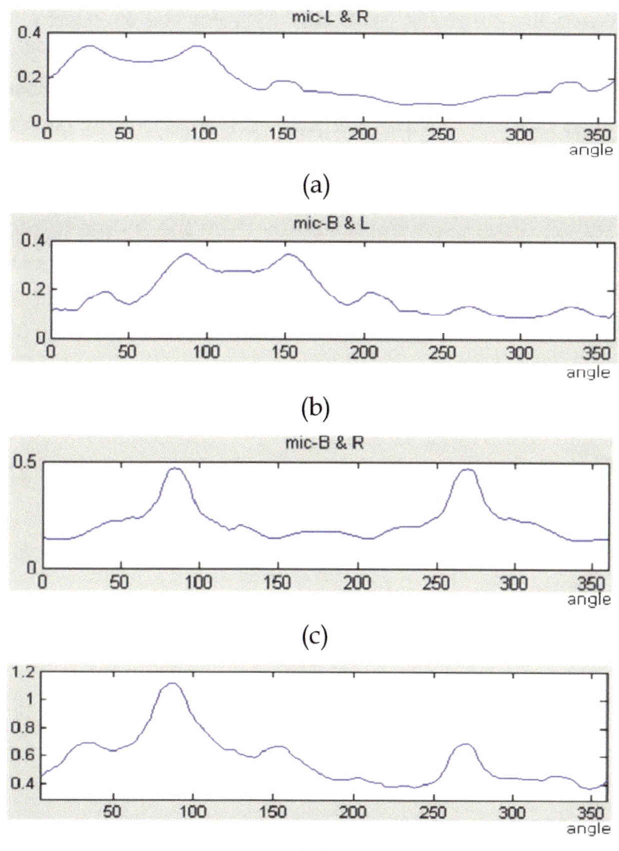

Fig. 6. Angles obtained from microphone pairs: (a) L-R, (b) B-L, (c) R-B, and (d) (L-R)+(B-L)+(R-B)

Source location(angle)	Proper microphone pair
60°~120°, 240°~300°	R-L
120°~180°, 300°~360°	B-R
180°~240°, 0°~60°	L-B

Table 1. Selection of proper microphone pair for six different source locations.

Due to the nonlinear characteristic of the inverse cosine function, the accuracy of each estimation result is different depending on the source location. Notice that in Fig. 5, wherever the source is located, exactly one microphone pair has the sound source within its approximately linear region (60°~120° or 240°~300° for the microphone pair). As an example, if a sound source is located at 30° in Fig. 5, the location is within the approximately linear region for L-B pair. Table 1 summarizes the choice of proper microphone pairs for six different source locations.

The proper selection of microphone pairs can be achieved by comparing the time index τ_{max} values (or, the number of shifted samples) in (2) at which the maximum correlation values are obtained. Fig. 7 shows the comparison of the correlation values obtained from three microphone pairs when the source is located at 90°. For the smallest estimation error, we select the microphone pair whose τ_{max} value is closest to 0. Notice that the correlation curve in the center (by the microphone pair R-L) has the τ_{max} value which is closest to 0.

In fact, for the smallest estimation error, we just need to select the correlation curve in the center. As an example, assume that a sound source is located at 90° in Fig. 5. Then, for the microphone pair R-L, the two signals arrived at the microphones R and L have little difference in their arrival times since the distances from the source to each microphone are almost the same. Thus, the cross correlation has its maximum around $\tau = 0$. However, for L-B pair, the microphone L is closer to the source than the microphone B. Since the received signals at microphones B and L are designated as $x_1(t)$ and $x_2(t)$, respectively, the cross

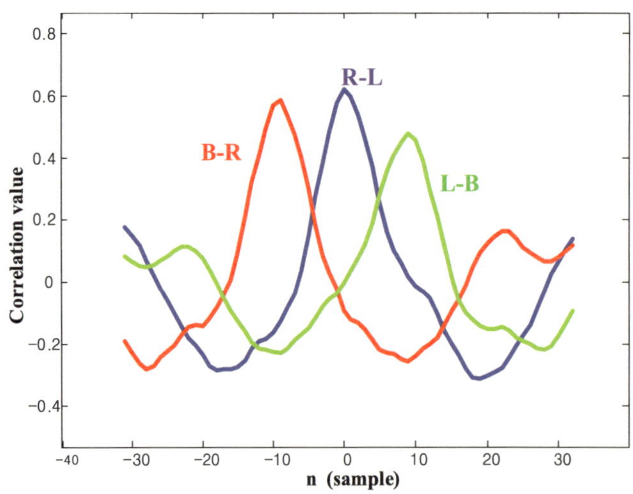

Fig. 7. Comparison of the correlation values obtained from three microphone pairs for the source located at 90°

correlation in (2) gets its maximum when $x_2(t)$ is shifted to the right ($\tau > 0$). The opposite is true for the microphone pair B-R as can be seen from Fig. 7.
Table 2 shows that proper microphone pairs can be simply selected by comparing maximum correlation positions (or, τ_{max} values from each microphone pair).

Maximum correlation positions	Proper Mic.	Front / Back
τ_{max} (BR)≤ τ_{max} (RL) ≤ τ_{max} (LB)	R-L	Front
τ_{max} (BR)≤ τ_{max} (LB) ≤ τ_{max} (RL)	L-B	Front
τ_{max} (RL)≤ τ_{max} (BR) ≤ τ_{max} (LB)	B-R	Front
τ_{max} (LB)≤ τ_{max} (RL) ≤ τ_{max} (BR)	R-L	Back
τ_{max} (RL)≤ τ_{max} (LB) ≤ τ_{max} (BR)	L-B	Back
τ_{max} (LB)≤ τ_{max} (BR) ≤ τ_{max} (RL)	B-R	Back

Table 2. Selecetion of proper microphone pair

If the sampled signals of $x_1(t)$ and $x_2(t)$ are denoted by two vectors X_1 and X_2, the length of the cross-correlated signal R_{X1X2} is determined as

$$n(R_{X1X2}) = n(X_1) + n(X_2) - 1, \tag{9}$$

where $n(X)$ means the length of vector X. In other words, to obtain the cross-correlation result, vector shift and inner product operations need to be performed by $n(R_{X1X2})$ times.
It is interesting to notice that, once the distance between the microphones and the sampling rate are determined, the maximum time delay between two received signals is bounded by $n_{d,max}$ in (8). Thus, instead of performing vector shift and inner product operations by $n(R_{X1X2})$ times as in the conventional approaches, it is sufficient to perform the operations by only $n_{d,max}$ times. Specifically, we perform the correlation operation from $n = -n_{d,max}/2$ to $n = n_{d,max}/2$ (for sampled signals, $\tau = n / f_s$, integer n). In the simulation shown in Fig. 7, $n(X_1) = n(X_2) = 256$ and $n_{d,max} = 64$. Thus, the number of operations for cross-correlation is reduced from 511 to 65 by the proposed method, which means the computation time for cross-correlation can be reduced by 87%.

3.2 Simplification of angle mapping using linear equation

Conventional angle mapping circuits require a look-up table for inverse cosine function. Also, an interpolation circuit is needed to obtain a better resolution with reduced look-up table. However, since the proposed region selection approach uses only the approximately linear part of the inverse cosine function, the use of look-up table and interpolation circuit can be avoided. Instead, the approximately linear region is approximated by the following equation:

$$y = ax + b, \tag{10}$$

where

$$a = \frac{-60}{(\cos\pi/3 - \cos 2\pi/3)\times l_{mic}},$$
$$b = 120 + \frac{60\cos 2\pi/3}{(\cos\pi/3 - \cos 2\pi/3)}. \quad (11)$$

When the distance between the two microphones is given, the coefficients a and b in (10) can be pre-calculated. Thus, angle mapping can be performed using only one multiplication and one addition for a given value of d.

Fig. 8 shows the block diagrams of the conventional sound source localization systems and the proposed system.

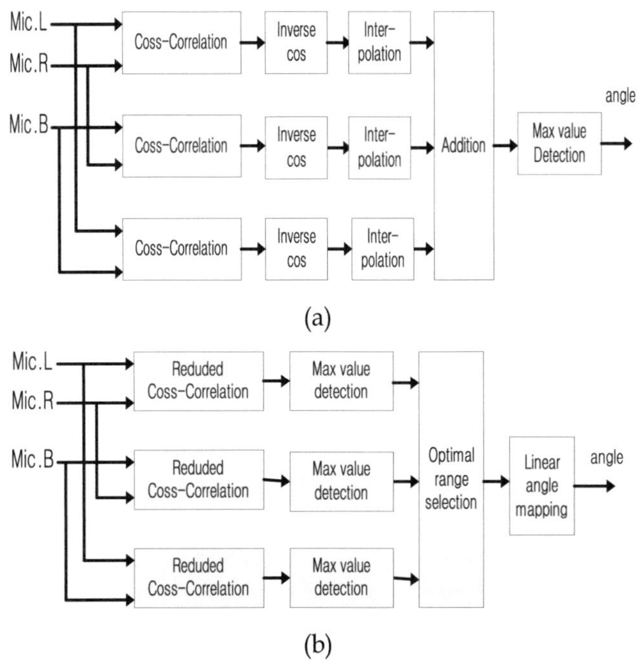

Fig. 8. Block diagrams of conventional and proposed methods: (a) conventional method, and (b) proposed method.

4. Simulation results

Fig. 9 shows the sound source localization system test environments. The distance between the microphones is 18.5cm. The sound signals received using three microphones are sampled at 16 KHz and the sampled signals are sent to the sound localization system implemented using Altera stratix II FPGA. Then, the estimation result is transmitted to a host PC through two FlexRay communication systems. The test results are shown in Table 3. Notice that the average error of the proposed method is only 31% of that of the conventional method. To further reduce the estimation error, we need to increase the sampling rate and the distance between the microphones.

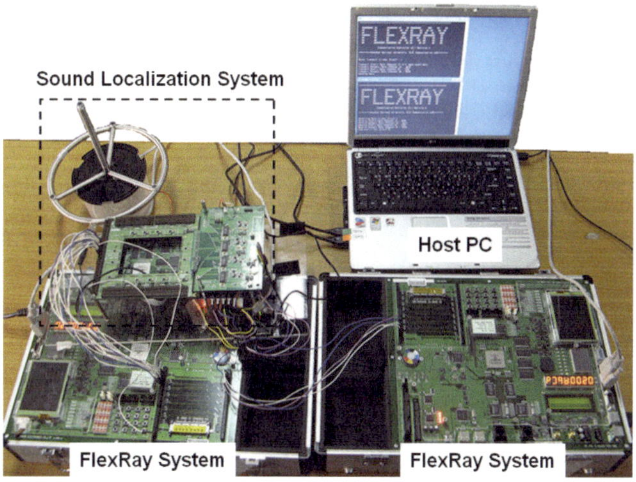

Fig. 9. Sound localization system test environments

Distance	0°	30°	60°	90°
1m	0°	27°	56°	88°
2m	0°	27°	59°	85°
3m	0°	27°	59°	88°
4m	2.5°	34°	57°	95°
5m	4.1°	37°	67°	82°
Maximum absolute error	4.1°	7°	7°	8°
average error	1.32°	4°	3.2°	4.4°

(a)

Distance	0°	30°	60°	90°
1m	0°	32.7°	60°	87.2°
2m	0°	32°	59°	85°
3m	0°	32.7°	60°	87.2°
4m	1	28°	62°	86°
5m	2	33°	61°	92°
Maximum absolute error	2°	3°	2°	4°
average error	0.6°	2.48°	0.8°	3.32°

(b)

Table 3. Simulation results: (a) conventional method, and (b) proposed method

5. Conclusion

Compared with conventional sound source localization methods, proposed method achieves more accurate estimation results with reduced hardware overhead due to the new region selection approach. By the proposed approach, the region from 0° to 180° is divided into three regions and only one of the three regions is selected such that the selected region corresponds to the linear part of the inverse cosine function. By the proposed approach, the

computation time for cross correlation is reduced by 87%, compared with the conventional approach. By simulations, it is shown that the estimation error by the proposed method is only 31% of that of the conventional approach.
The proposed sound source localization system can be applied to the implementation of portable service robot systems since the proposed system requires small area and low power consumption compared with conventional methods. The proposed method can be combined with generalized correlation method with some modifications.

6. Acknowledgment

This research was financially supported by the Ministry of Education, Science Technology (MEST) and National Research Foundation of Korea (NRF) through the Human Resource Training Project for Regional Innovation.

7. References

Brandstein M. S. & Silverman H. (1997). A practical methodology for speech source localization with microphone arrays. *Comput. Speech Lang.*, Vo.11, No.2, pp. 91-126, ISSN 0885-2308

Brandstein M. & Ward D. B. (2001). Robust *Microphone Arrays: Signal Processing Techniques and Applications*, New York: Springer, ISBN 978-3540419532

Cheng I. & Wakefield G. H. (2001). Introduction to head-related transfer functions (HRTFs): representations of HRTFs in time, frequency, and space. *J. Audio Eng. Soc.*, Vol. 49, No.4, (April, 2001), pp. 231-248, ISSN 1549-4950

Coen M. (1998). Design principles for intelligent environments, *Proceedings of the 15th National Conference on Artificial Intelligence*, pp. 547-554

Huang J.; Supaongprapa T.; Terakura I.; Wang F.; Ohnishi N. & Sugie N. (1999) A model-based sound localization system and its application to robot navigation. *Robot. Auton. Syst.*, Vol.27, No.4, (June,1999), pp. 199-209, ISSN 0921-8890

Knnapp C. H. & Cater G. C. (1976). The generalized correlation method for estimation of time delay. *IEEE Trans. Acoust. Speech Signal Process.*, Vol.24, No.4, (August 1976), pp.320-327, ISSN 0096-3518

Lv X. & Zhang M. (2008). Sound source localization based on robot hearing and vision, *Proceedings of ICCSIT 2008 International Conference of Computer Science and Information Technology*, pp. 942-946, ISBN 978-0-7695-3308-7, Singapore, August 29-September 2 2008

Mungamuru, B. & Aarabi, P. (2004). Enhanced sound localization. *IEEE Trans. Syst. Man Cybern. Part B- Cybern.*, Vol.34, No.3, (June, 2004), pp. 1526-1540, ISSN 1083-4419

Nakadai K.; Lourens T.; Okuno H. G. & Kitano H. (2000). Active audition for humanoid, *Proceedings of the 17th National Conference on Artificial Intelligence* and *12th Conference on Innovative Applications of Artificial Intelligence*, pp. 832-839

Sasaki Y.; Kagami S. & Mizoguchi H. (2006). Multiple sound source mapping for a mobile robot by self-motion triangulation, *Proceedings of the 2006 IEEE/RSJ International Conference on Intelligent Robots and Systems*, pp. 380-385, ISBN 1-4244-0250-X, Beijing, China, October, 2006

Wax M. & Kailath T. (1983). Optimum localization of multiple sources by passive arrays. *IEEE Trans. Acoust. Speech Signal Process.*, Vol.31, No.6, (October,1983). pp. 1210-1217, ISSN 0096-3518

8

Robust Audio Localization for Mobile Robots in Industrial Environments

Manuel Manzanares, Yolanda Bolea and Antoni Grau
Technical University of Catalonia, UPC, Barcelona
Spain

1. Introduction

For autonomous navigation in workspace, a mobile robot has to be able to know its position in this space in a precise way that means that the robot must be able to self-localize to move and perform successfully the different entrusted tasks. At present, one of the most used systems in open spaces is the GPS navigation system; however, in indoor spaces (factories, buildings, hospitals, warehouses…) GPS signals are not operative because their intensity is too weak. The absence of GPS navigation systems in these environments has stimulated the development of new local positioning systems with their particular problems. Such systems have required in many cases the installation of beacons that operate like satellites (similar to GPS), the use of landmarks or even the use of other auxiliary systems to determine the robot's position.

The problem of mobile robot localization is a part of a more global problem because in autonomous navigation when a robot is exploring an unknown environment, it usually needs to obtain some important information: a map of the environment and the robot's location in the map. Since mapping and localization are related to each other, these two problems are usually considered as a single problem called simultaneous localization and mapping (SLAM). The problem of Simultaneous Localization and Map Building is a significant open problem in mobile robotics which is difficult because of the following paradox: to localize itself the robot needs the map of the environment, and, for building a map the robot location must be known precisely.

Mobile robots use different kinds of sensors to determine their position: for instance it is very common the use of odometric or inertial sensors, however it is remarkable to consider that in wheel slippage, sensor drifts a noise causing error accumulation, thus leading to erroneous estimates. Another kind of external sensors used in robotics in order to solve localization are for instance CCD cameras, infrared sensor, ultra sonic sensor, mechanical wave and laser. Other sensors recently applied are the instruments sensible to the magnetic field known as the electronic compass (Navarro & Benet, 2009). Mobile robotics are interested on those able to measure the Earths magnetic field and express it through an electrical signal. One type of electronic compass is based on magneto-resistive transducers, whose electrical resistance varies with the changes on the applied magnetic field. This type of sensors presents sensitivities below 0.1 milligauss, with response times below 1 sec, allowing its reliable use in vehicles moving at high speeds (Caruso, 2000). In SLAM some applications with electronic compass have been developed working simultaneously with other sensors such as artificial vision (Kim et al., 2006) and ultrasonic sensors (Kim et al., 2007).

In mobile robotics, due to the use of different sensors at the same time to provide localization information the problem of data fusion rises and many algorithms have been implemented. Multisensor fusion algorithms can be broadly classified as follows: estimation methods, classification methods, inference methods, and artificial intelligence methods (Luo et al., 2002); in the latter are remarkable neural networks, fuzzy and genetic algorithms (Begum et al., 2006); (Brunskill & Roy, 2005). Related with the provided sensors information processing in SLAM context, many works can be found, for instance in (Di Marco et al., 2000), where estimation of the position of the robot and the selected landmarks are derived in terms of uncertainty regions, under the hypothesis that the errors affecting all sensor measurements are unknown but bounded, or in (Begum et al., 2006) where an algorithm processes sensor data incrementally and therefore, has the capability to work online.
Therefore a comprehensive collection of researches have been reported on SLAM, most of which stem from the pioneer work of (Smith et al. 1990). This early work provides a Kalman Filter (KF) based statistical framework for solving SLAM. The KF based SLAM algorithms require feature extraction and identification from sensor data, for estimating the pose and the parameters. In the situation that the system noise and measurement obey a Gaussian amplitude distribution, KF uses the state recursive equation that is with the noise estimates the optimal attitude of mobile robots. But there would be generated errors of localization, if the noise does not obey the distribution. KF is also able to the merge low graded multisensor data models. Particle filter is the next probabilistic technique that has earned popularity in SLAM literature. The hybrid SLAM algorithm proposed in (Thrun, 2001) uses particle filter for posterior estimation over a robot's poses and is capable to map large cyclic environments. Another method of fusion broadly used is Extended Kalman Filter (EKF); the EKF can be used where the model is nonlinear, but it can be suitably linearized around a stable operating point.
Several systems have been researched to overcome the localization limitation. For example, the Cricket Indoor Location (Priyantha, 2000) which relies on active beacons placed in the environment. These beacons transmit simultaneously two signals (a RF and an ultrasound wave). Passive listeners mounted, for example, on mobile robots can, by knowing the difference in propagation speed of the RF and ultrasound signals, estimate their own position in the environment. GSM and WLAN technologies can also be used for localization. Using triangulation methods and measuring several signal parameters such as the signal's angle and time of arrival, it becomes possible to estimate the position of a mobile transmitter/receiver in the environment (Sayed et al., 2005). In (Christo et al., 2009), a specific architecture is suggested for the use of multiples iGPS Web Services for mobile robots localization.
Most of the mobile robot's localization systems are based on robot vision, and robot vision is also a hot spot in the research of robotics. Camera which is the most popular visual sensor is widely used for the localization of mobile robots just now. However some difficulties occur because of the limitation of camera's visual field and the dependence on light condition. If the target is not in the visual field of camera or the lighting condition is poor, the visual localization system of the mobile robot cannot work effectively. Nowadays, the role of acoustic perception in autonomous robots, intelligent buildings and industrial environments is increasingly important and in the literature there are different works (Yang et al., 2007); (Mumolo et al., 2003); (Csyzewski, 2003).
Comparing to the study on visual perception, the study on auditory is still in its infancy stage. The human auditory system is a complex and organic information processing system,

it can feel the intensity of sound and space orientation information. Compared with vision, audition has several unique properties. Audition is omni-directional. The sound waves have strong diffraction ability; audition also is less affected by obstacles. Therefore, the audio ability possessed by robot can make up the restrictions of other sensors such as limited view or the non-translucent obstacles. Nevertheless, audio signal processing presents some particular problems such as the effect of reverberations and noise signals, complex boundary conditions and near-field effect, among others, and therefore the use of audio sensors together with other sensors is common to determine the position and also for autonomous navigation of a mobile robot, leading to a problem of data fusion. There are many applications that would be aided by the determination of the physical position and orientation of users. As an example, without the information on the spatial location of users in a given environment, it would not be possible for a service robot to react naturally to the needs of the user. To localize a user, sound source localization techniques are widely used. Such techniques can also help a robot to self-localize in its working area. Therefore, the sound source localization (one or more sources) has been studied by many researchers (Ying & Runze, 2007); (Sasaki et al., 2006); (Kim et al., 2009). Sound localization can be defined as the process of determining the spatial location of a sound source based on multiple observations of the received sound signals. Current sound localization techniques are generally based upon the idea of computing the time difference of arrival (TDOA) information with microphone arrays (Brandstein & Silverman, 1997); (Knapp & Carter, 1976), or interaural time difference (ITD) (Nakashima & Mukai, 2005). The ITD is the difference in the arrival time of a sound source between two ears, a representative application can be found in (Kim & Choi, 2009) with a binaural sound localization system using sparse coding based ITD (SITD) and self-organizing map (SOM). The sparse coding is used for decomposing given sounds into three components: time, frequency and magnitude, and the azimuth angle are estimated through the SOM. Other works in this field use structured sound sources (Yi & Chu-na, 2010) or the processing of different audio features (Rodemann et al., 2009), among other techniques.

The works that authors present in this Chapter are developed with audio signals generated with electric machines that will be used to mobile robots localization in industrial environments. A common problem encountered in industrial environments is that the electric machine sounds are often corrupted by non-stationary and non-Gaussian interferences such as speech signals, environmental noise, background noise, etc. Consequently, pure machine sounds may be difficult to identify using conventional frequency domain analysis techniques as Fourier transform (Mori et al., 1996), and statistical techniques such as Independent Component Analysis (ICA) (Roberts & Everson, 2001).

The wavelet transform has attracted increasing attention in recent years for its ability in signal features extraction (Bolea et al., 2003); (Mallat & Zhang, 1993), and noise elimination (Donoho, 1999). While in many mechanical dynamic signals, such as the acoustical signals of an engine, Donoho's method seems rather ineffective, the reason for their inefficiency is that the feature of the mechanical signals is not considered. Therefore, when the idea of Donoho's method and the sound feature are combined, and a de-noising method based on Morlet wavelet is added, this methodology becomes very effective when applied to an engine sound detection (Lin, 2001). In (Grau et al., 2007), the authors propose a new approach in order to identify different industrial machine sounds, which can be affected by non-stationary noise sources.

It is also important to consider that non-speech audio signals have the property of non-stationary signals in the same way that many real signals encountered in speech processing, image processing, ECG analysis, communications, control and seismology. To represent the behaviour of a stationary process is common the use of models (AR, ARX, ARMA, ARMAX, OE, etc.) obtained from the experimental identification (Ljung, 1987). The coefficient estimation can be done with different criteria: LSE, MLE, among others. But in the case of non-stationary signals the classical identification theory and its results are not suitable.

Many authors have proposed different approaches to modelling this kind of non-stationary signals, that can be classified: i) assuming that a non stationary process is locally stationary in a finite time interval so that various recursive estimation techniques (RLS, PLR, RIV, etc.) can be applied (Ljung, 1987); ii) a state space modelling and a Kalman filtering; iii) expanding each time-varying parameter coefficients onto a set of basis sequences (Charbonnier et al., 1987); and iv) nonparametric approaches for non-stationary spectrum estimation such a local evolving spectrum, STFT and WVD are also developed to characterize non-stationary signals (Kayhan et al., 1994).

To overcome the drawbacks of the identification algorithms, wavelets could be also considered for time varying model identification. The distinct feature of a wavelet is its multiresolution characteristic that is very suitable for non-stationary signal processing (Tsatsanis & Giannakis, 1993).

The work to be presented in this Chapter will investigate different approaches based on the study of audio signals with the purpose of obtaining the robot location (in x-y plane) using as sound sources industrial machines. For their own nature, these typical industrial machines produce a stationary signal in a certain time interval. These resultant stationary waves depend on the resonant frequencies in the plant (depending on the plant geometry and dimensions) and also on the different absorption coefficients of the wall materials and other objects present in the environment.

A first approach that authors will investigate is based on the recognition of patterns in the acquired audio signal by the robot in different locations (Bolea et al., 2008). These patterns will be found through a process of feature extraction of the signal in the identification process. To establish the signal models the wavelet transform will be used, specifically the Daubechies wavelet, because it captures very well the characteristics and information of the non-speech audio signals. This set of wavelets has been extensively used because its coefficients capture the maximum amount of the signal energy.

A MAX model (Moving Averaging Exogenous) represents the sampled signals in different points of the space domain because the signals are correlated. We use the closest signal to the audio source as signal input for the model. Only the model coefficients need to be stored to compare and to discriminate the different audio signals. This would not happen if the signals were represented by an AR model because the coefficients depend on the signal itself and, with a different signal in every point in the space domain, these coefficients would not be significant enough to discriminate the audio signals. When the model identification is obtained by wavelets transform, the coefficients that do not give information enough for the model are ignored.

The eigenvalues of the covariance matrix are analyzed and we reject those coefficients that do not have discriminatory power. For the estimation of each signal the approximation signal and its significant details are used following the next process: i) model structure selection; ii) model parameters calibration with an estimation model (the LSE method can be

used for its simplicity and, furthermore a good identified model coefficients convergence is assured); iii) validation of the model.
Another approach that will also be investigated is based on the determination of the transfer function of a room, denoted RTF (Room Transfer Function), this model is an LPV (Linear Parameters Varying) because the parameters of the model vary along the robot's navigation (Manzanares et al., 2009).
In an industrial plant, there are different study models in order to establish the transmission characteristics of a sound between a stationary audio source and a microphone in closed environments: i) the beam theory applied to the propagation of the direct audio waves and reflected audio waves in the room (Kinsler et al., 1995); ii) the development of a lumped parameters model similar to the model used to explain the propagation of the electromagnetic waves in the transmission lines (Kinsler et al., 1995) and the study of the solutions given by the wave equation (Kuttruff, 1979). Other authors propose an RTF function that carries out to industrial plant applied sound model (Haneda et al., 1992); (Haneda et al., 1999); (Gustaffson et al., 2000). In these works the complexity to achieve the RTFs is evident as well as the need of a high number of parameters to model the complete acoustic response for a specific frequency range, moreover to consider a real environment presents an added difficulty.
In this research we study how to obtain a real plant RTF. Due that this RTF will be used by a mobile robot to navigate in an industrial plant, we have simplified the methodology and our goal is to determinate the x-y coordinates of the robot. In such a case, the obtained RTF will not present a complete acoustic response, but will be powerful enough to determine the robot's position.

2. Method based on the recognition of patterns of the audio signal

This method is based on the recognition of patterns in the acquired audio signal by the robot in different locations, to establish the signals models the Daubechies wavelets will be used. A MAX model (Moving Averaging Exogenous) represents the sampled signals in different points of the space domain, and for the estimation of each signal the approximation signal and its significant details are used following the process steps mentioned previously: i) model structure selection; ii) model parameters calibration with an estimation model; iii) validation of the model.
Let us consider the following TV-MAX model and be S_i = y(n),

$$y(n) = \sum_{k=0}^{q} b(n;k)u(n-k) + \sum_{k=0}^{r} c(n;k)e(n-k) \tag{1}$$

where $y(n)$ is the system output, $u(n)$ is the observable input, which is assumed as the closest signal to the audio source, and $e(n)$ is a noise signal. The second term is necessary whenever the measurement noise is colored and needs further modeling. The coefficients for the different models will be used as the feature vector, which can be defined as X_S, where

$$X_S = (b_1, b_2, \ldots^{q+1)} \ldots, c_1, c_2, \ldots^{r+1)}) \tag{2}$$

where q+1 and r+1 are the amount of b and c coefficients respectively. From every input signal a new feature vector is obtained representing a new point in the (q+r+2)-dimensional

feature space, *fs*. For feature selection, it is not necessary to apply any statistical test to verify that each component of the vector has enough discriminatory power because this step has been already done in the wavelet transform preprocessing.
This feature space will be used to classify the different audio signals entering the system. Some labeled samples with their precise position in the space domain are needed. In this chapter a specific experiment is shown. When an unlabeled sample enters the feature space, the minimum distance to a labeled sample is computed and this measure of distance will be used to estimate the distance to the same sample in the space domain. For this reason a transformation function f_T is needed which converts the distance in the feature space in the distance in the space domain, note that the distance is a scalar value, independently of the dimension of the space where it has been computed.
The Euclidean distance is used, and the distance between to samples S_i and S_j in the feature space is defined as

$$d_{fs}\left(S_i,S_j\right)=\sqrt{\sum_{k=0}^{q}\left(b_{kS_i}-b_{kS_j}\right)^2+\sum_{k=0}^{r}\left(c_{kS_i}-c_{kS_j}\right)^2} \quad (3)$$

where bkS_i and ckS_i are the b and c coefficients, respectively, of the wavelet transform for the S_i signal. It is not necessary to normalize the coefficients before the distance calculation because they are already normalized intrinsically by the wavelet transformation.
Because there exist the same relative distances between signals with different models, and with the knowledge that the greater the distortion the farther the signal is from the audio source, we choose those correspondences (d_{xy}, d_{fs}) between the samples that are closest to the audio source equidistant in the d_{xy} axis. These points will serve to estimate a curve of n-order, that is, the transformation function f_T. An initial approximation for this function is a polynomial of 4th order and there are several solutions for a unique distance in the feature space, that is, it yields different distances in the x-y space domain.

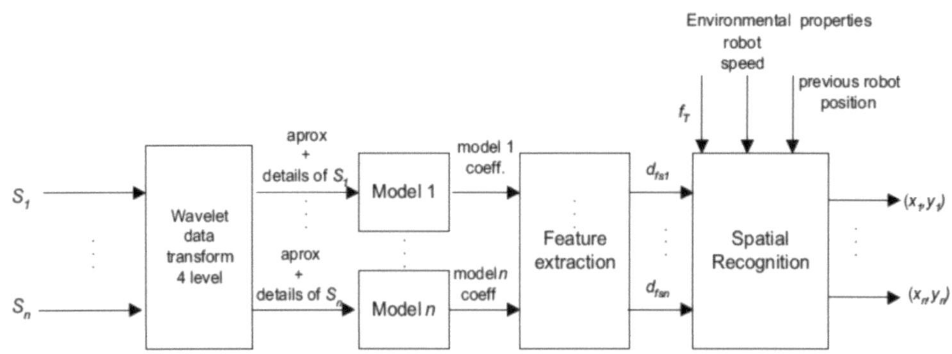

Fig. 1. Localization system in space domain from non-speech audio signals.

We solve this drawback adding a new variable: previous position of the robot. If we have an approximate position of the robot, its speed and the computation time between feature extraction samples, we will have a coarse approximation of the new robot position, coarse enough to discriminate among the solutions of the 4th-order polynomial. In the experiments section a waveform for the f_T function can be seen, and it follows the model from the sound derivative partial equation proposed in (Kinsler et al., 1995) and (Kuttruff, 1979).

In Figure 1 the localization system can be shown, including the wavelet transformation block, the modeling blocks, the feature space and the spatial recognition block which has as input the environment of the robot and the function f_T.

2.1 Sound source angle detection

As stated in the Introduction section, in order to locate sound sources several works have been developed using a microphone array. Because we work with a unique source of sound, and in order to simplify the number of sensors, we propose a system that detects the direction in which the maximum sound intensity is received and, in this way, emulating the response of a microphone array located in the perimeter of a circular platform. To achieve this effect we propose a turning platform with two opposed microphones. The robot computes the angle respect the platform origin (0°) and the magnetic north of its compass. Figure 2 depicts the blocks diagram of the electronic circuit to acquire the sound signals. The signal is decoupled and amplified in a first stage in order to obtain a suitable range of work for the following stages. Then, the maximum of the mean values of the rectified sampled audio signal determines the position of the turning platform.

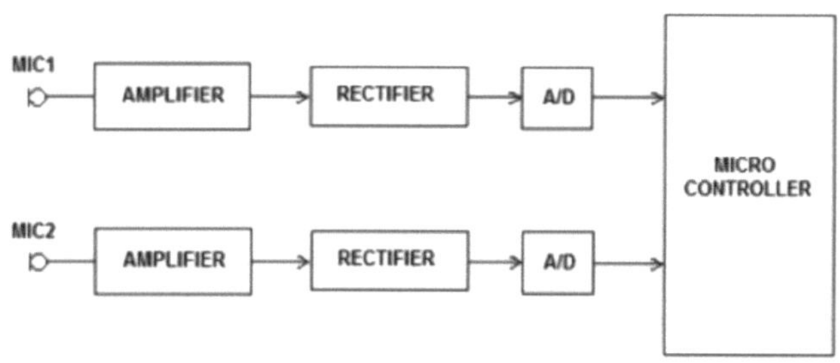

Fig. 2. Angle detection block diagram.

There are two modes of operation: looking for local values or global values. To find the maximum value the platform must turn 180° (because there are two microphones), this mode warranties that the maximum value is determined but the operation time is longer than using the local value detection, in which the determination is done when the system detects the first maximum. In most of the experiments this latter operation mode is enough.

2.2 Spatial recognition

This distance computation between the unlabelled audio sample and labeled ones is repeated for the two closest samples to the unlabelled one. Applying then the transformation function f_T two distances in the x-y domain are obtained. These distances indicate where the unlabelled sample is located. Now, with a simple process of geometry, the position of the unlabelled sample can be estimated but with a certain ambiguity, see Figure 3. In (Bolea et al., 2003) we used the intersection of three circles, which theoretically gives a unique solution, but in practice these three circles never intersect in a point but in an area that induces to an approximation, and thus, to an error (uncertainty) in the localization point.
The intersection of two circles (as shown in Figure 3) leads to a two-point solution. In the correct discrimination of these points the angle between the robot and the sound source is computed.

Since the robot computes the angle between itself and the sound source, the problem is to identify the correct point of the circles intersection. Figure 4 shows the situation. I_1 and I_2 are the intersection points. For each point the angle respect the sound source is computed (α_1 and α_2), because the exact source position is known (x_s, y_s).

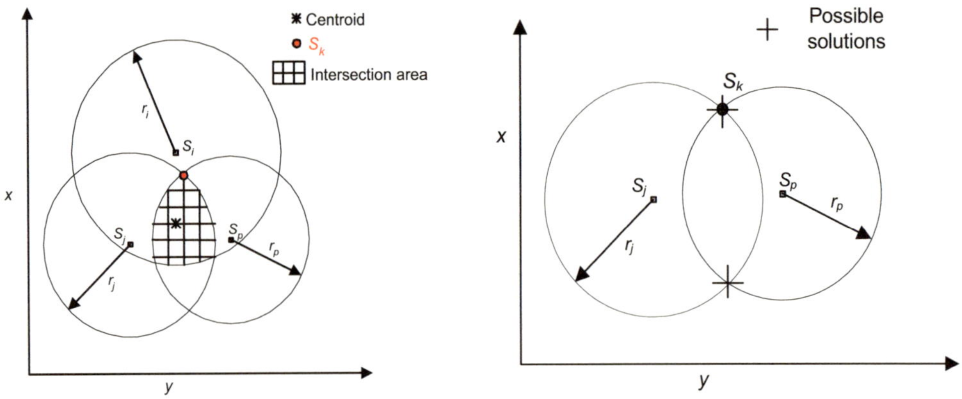

Fig. 3. Geometric process of two (right) or three (left) circles intersection to find the position of unlabeled sample S_k.

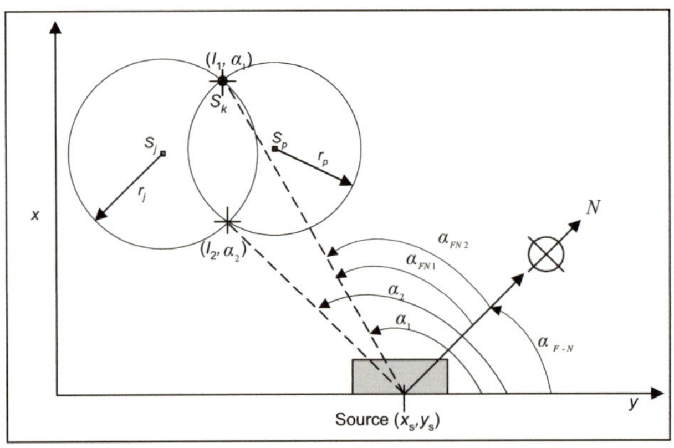

Fig. 4. Angles computation between ambiguous robot localization and sound source.

Angles α_1 and α_2 correspond to:

$$\alpha_1 = arctg\frac{y_{I_1} - y_s}{x_{I_1} - x_s}, \qquad \alpha_2 = arctg\left(\frac{y_{I_2} - y_s}{x_{I_2} - x_s}\right) \tag{4}$$

These angles must be corrected respect the north in order to have the same offset than the angle computed aboard the robot:

$$\alpha_{FN1} = \alpha_1 - \alpha_{F\text{-}N}; \quad \alpha_{FN2} = \alpha_2 - \alpha_{F\text{-}N} \tag{5}$$

being $\alpha_{F\text{-}N}$ the angle between the room reference and the magnetic north (previously calibrated).

Now, to compute the correct intersection point is only necessary to find the angle which is closer to the angle computed on the robot with the sensor.

3. Method based on the LPV model with audio features

In this second approach we study how to obtain a real plant RTF. Due that this RTF will be used by a mobile robot to navigate in an industrial plant, we have simplified the methodology and our goal is to determinate the x-y coordinates of the robot. In such a case, the obtained RTF will not present a complete acoustic response, but will be powerful enough to determine the robot's position. The work investigates the feasibility of using sound features in the space domain for robot localization (in x-y plane) as well as robot's orientation detection.

3.1 Sound model in a closed room

The acoustical response of a closed room (with rectangular shape), where the dependence with the pressure in a point respect to the defined (x,y,z) position is represented by the following wave equation:

$$L_x \frac{\partial^2 p}{\partial x^2} + L_y \frac{\partial^2 p}{\partial y^2} + L_z \frac{\partial^2 p}{\partial z^2} + k^2 p = 0 \tag{6}$$

L_x, L_y and L_z denote the dimensions of the length, width and height of the room with ideally rigid walls where the waves are reflected without loss, Eq. (6) is rewritten as:

$$p(x,y,z) = p_1(x)p_2(y)p_3(z) \tag{7}$$

when the evolution of the pressure according to the time is not taken into account.
Then Eq. (7) is replaced in Eq. (6), and three differential equations can be derived and it is the same for the boundary condition. For example, p1 must satisfy the equation:

$$\frac{d^2 p_1}{dx^2} + k_x^2 p_1 = 0 \tag{8}$$

With boundary conditions in $x = 0$ and $x = L_x$:

$$\frac{dp_1}{dx} = 0$$

k_x, k_y and k_z constants are related by the following expression:

$$k_x^2 + k_y^2 + k_z^2 = k^2 \tag{9}$$

Equation (8) has as general solution:

$$p_1(x) = A_1 \cos(k_x x) + B_1 \sin(k_x x) \tag{10}$$

Through Eq. (8) and limiting this solution to the boundary conditions, constants in Eq. (10) take the following values:

$$k_x = \frac{n_x \pi}{L_x};\ k_y = \frac{n_y \pi}{L_y} \text{ and } k_z = \frac{n_z \pi}{L_z}$$

being n_x, n_y and n_z positive integers. Replacing these values in Eq. (10) the wave equation eigenvalues are obtained:

$$k_{n_x n_y n_z} = \pi \left[\left(\frac{n_x}{L_x} \right)^2 + \left(\frac{n_y}{L_y} \right)^2 + \left(\frac{n_z}{L_z} \right)^2 \right]^{1/2} \tag{11}$$

The eigenfunctions or normal modes associated with these eigenvalues are expressed by:

$$p_{n_x n_y n_z}(x, y, z) = C_1.\cos\left(\frac{n_x \pi x}{L_x} \right).\cos\left(\frac{n_y \pi y}{L_y} \right).\cos\left(\frac{n_z \pi z}{L_z} \right).e^{j\omega t}$$
$$e^{jwt} = \cos(wt) - j\sin(wt) \tag{12}$$

being C_1 an arbitrary constant and introducing the variation of pressure in function of the time by the factor e^{jwt}. This expression represents a three dimensional stationary wave space in the room. Eigenfrequencies corresponding to Eq. (11) eigenvalues can be expressed by:

$$f_{n_x n_y n_z} = \frac{c}{2\pi} k_{n_x n_y n_z}$$
$$f_{n_x n_y n_z} = \sqrt{f_{nx}{}^2 + f_{ny}{}^2 + f_{nz}{}^2} \tag{13}$$
$$f_{n_x n_y n_z} = \sqrt{\left(\frac{n_x c}{2L_x} \right)^2 + \left(\frac{n_y c}{2L_y} \right)^2 + \left(\frac{n_z c}{2L_z} \right)^2}$$

where c is the sound speed. Therefore, the acoustic response of any close room presents resonance frequencies (eigenfrequencies) where the response of a sound source emitted in the room at these frequencies is the highest. The eigenfrequencies depend on the geometry of the room and also depend on the materials reflection coefficients, among other factors. Microphones obtain the environmental sound and they are located at a constant height (z_1) respect the floor, and thus the factor:

$$\cos\left(\frac{n_z \pi z_1}{L_z} \right) \tag{14}$$

is constant and therefore, if temporal dependency pressure respect the time is not considered, Eq. (12) is:

$$p_{n_x n_y n_z}(x, y) = C_2.\cos\left(\frac{n_x \pi x}{L_x} \right).\cos\left(\frac{n_y \pi y}{L_y} \right) \tag{15}$$

In our experiments, L_x = 10.54m, L_y = 5.05m and L_z = 4m, considering a sound speed propagation of 345m/s. When Eq. (15) is applied in the experiments rooms, for mode (1, 1,

2), this equation indicates the acoustic pressure in the rooms depending on the x-y robot's position, and this is:

$$p_{n_x n_y n_z}(x,y) = C_2.\cos\left(\frac{\pi x}{10,54}\right).\cos\left(\frac{\pi y}{5,05}\right) \tag{16}$$

With these ideal conditions and for an ideal value for constant $C_2 = 2$, the theoretic acoustic response in the rooms for this absolute value of pressure, and for this propagation mode, can be seen in Figure 5.

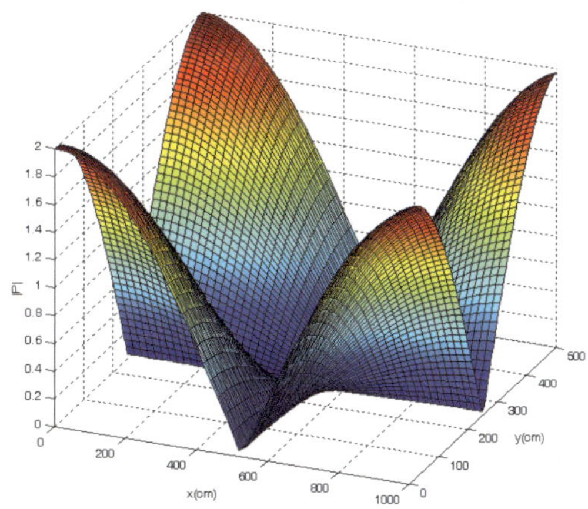

Fig. 5. Room response for propagation mode (1,1,2).

The shape of Figure 5 would be obtained for a sound source that is excited only this propagation mode, really the acoustic response will be more complex as we increase the propagation modes excited by the sound source.

3.2 Transfer function in a closed room

In (Gustaffson et al., 2000) a model based in the sum of second order transfer functions is proposed; these functions have been built between a sound source located in a position d_s emitting an audio signal with a specific acoustic pressure P_s and a microphone located in d_m which receives a signal of pressure P_m; each function represents the system response in front to a propagation mode.

The first contribution of this work is to introduce an initial variation to this model considering that the sound source has a fixed location, and then this model can be expressed as:

$$\frac{P_m(d_m,s)}{P_s(s)} = \sum_{n=1}^{M} \frac{K[d_m]s}{s^2 + 2\xi_n\omega_n s + \omega_n^2} \tag{17}$$

Because our objective is not to obtain a complete model of the acoustic response of the industrial plant, it will not be necessary to consider all the propagation modes in the room and we will try to simplify the problem for this specific application without the need to work with models of higher order.

To implement this experiment the first step is to select the frequency of interest by a previous analysis of the audio signal frequency spectrum emitted by the considered sound source (an industrial machine). Those frequency components with a significant acoustic power will be considered with the only requirement that they are close to one of the resonant frequencies of the environment. The way to select those frequencies will be through a band-pass digital filter centered in the frequency of interest. Right now, the term M in the sum of our model will have the value N, being this new value the propagation modes resulting from the filtering process.

The spectra of the sound sources used in our experiments show an important component close to the frequency of 100Hz for the climatic chamber, and a component of 50Hz for the PCB insulator, see Figure 10 (right) and Figure 11 (right).

For a concrete propagation mode, the variation that a stationary audio signal receives at different robot's position can be modeled, this signal can be smoothed by the variation of the absorption coefficient of the different materials that make up the objects in the room; those parameters are named $K[d_m]$ and $\xi[d_m]$, and Eq. (17) results:

$$H(s,d_m) = \frac{P_m(d_m,s)}{P_s(s)} = \sum_{n=1}^{N} \frac{K[d_m]s}{s^2 + 2\xi_n[d_m]\omega_n s + \omega_n^2} \tag{18}$$

where the gain (K), smooth coefficient (ξ_n) and the natural frequency (ω_n) of the transfer function room system depend on the room characteristics: d_m, n_x, n_y, L_x, and L_y, yielding an LPV indoor model.

Using Eq. (17) the module of the closed room in a specific transmission mode ω_{n1} is:

$$|H(j\omega_{n1},d_{m1})| = \frac{K}{2\xi_{n1}\omega_{n1}} \tag{19}$$

The room response in the propagation mode ω_{n1} (z_1 is a constant), assuming that the audio source only emits a frequency ω_{n1} for a specific coordinate (x,y) of the room, is:

$$|H| = \left|\frac{P_m}{P_s}\right|_{n_x,n_y} = \left|C\cos\left(\frac{n_x\pi x}{L_x}\right)\cos\left(\frac{n_y\pi y}{L_y}\right)\right| \tag{20}$$

with $f_{n_1} = \sqrt{f_{nx}^2 + f_{ny}^2}$, $\omega_{n_1} = 2\pi f_{n1}$.

Equaling Eq. (19) and (20), it results:

$$\xi_{n1} = \frac{k}{2\omega_{n1}\left|\cos\left(\frac{n_x\pi x}{L_x}\right)\cos\left(\frac{n_y\pi y}{L_y}\right)\right|} \tag{21}$$

If the filter is non-ideal then more than one transmission mode could be considered and therefore the following expression is obtained:

$$\sum_{l=1}^{m} \frac{K_{nl}}{2\xi_{nl}\omega_{nl}} = \sum_{l=1}^{m} C\left|\cos\left(\frac{n_{xl}\pi x}{L_x}\right)\cos\left(\frac{n_{yl}\pi y}{L_y}\right)\right| \tag{22}$$

The best results in the identification process in order to determine the robot's position have been obtained, for each considered propagation mode, keeping $K[d_m]$ coefficient constant and observing the different variations in the acquired audio signal in the smoothing coefficient $\xi[d_m]$.
If the zeros of the system are forced to be constant in the identification process for different robot's locations, and we admit that the emitted signal power by the sound sources is also constant and the audio signal power acquired with the microphones varies along the robot's position, then the pole positions in the s plane, for the considered propagation mode, will vary in the different robot's positions and their values will be:

$$s_{1n}[d_m] = -\xi_n[d_m]\omega_n + \omega_n\sqrt{(\xi_n[d_m])^2 - 1} \tag{23}$$

$$s_{2n}[d_m] = -\xi_n[d_m]\omega_n - \omega_n\sqrt{(\xi_n[d_m])^2 - 1} \tag{24}$$

It is worth noting that this model of reduced order gives good results in order to determine the robot's position and, although it does not provide a complete physical description of the evolution of the different parameters in the acoustic response for the different robot's positions, we can admit that according to the physical model given by the wave equation in Eq. (16), the modules of the proposed transfer functions will vary following a sinusoidal pattern and the pole position in the s plane will show those variation in the same fashion.

4. Experiments and discussions

4.1 Method based on the recognition of patterns of the audio signal

In the first proposed method based on the recognition of patterns of the audio signal, in order to prepare a setting as real as possible, we have used a workshop with a CNC milling machine as non-speech audio source. The room has a dimension of 7 meters by 10 meters and we obtain 9 labeled samples (from S_1 to S_9), acquired at regular positions, covering the entire representative workshop surface. With the dimensions of the room, these 9 samples are enough because there is not a significant variance when oversampling.
In Figure 6 the arrangement of the labelled samples can be observed. The robot enters the room, describes a predefined trajectory and gets off. In its trajectory the robot picks four unlabeled samples (audio signals) that will be used as data test for our algorithms (S_{10}, S_{11}, S_{12} and S_{13}). The sample frequency is 8 kHz following the same criteria as (Bielińska, 2002) in order to choose the sampling frequency because its similarity to speech signals.
First, in order to obtain the 9 models coefficients corresponding to the 9 labeled non-stationary audio signals, these signals are decomposed by the wavelet transform in 4 levels, with one approximation signal and 4 detail signals, Figure 7. For the whole samples, the relevance of every signal is analyzed. We observe the more significant decomposition to formulate the prediction model, that is, those details containing the more energy of the signal. With the approximation ($A4_i$) and the detail signal of 4th level ($D4_i$) is enough to represent the original signal, because the mean and deviation for the $D3_i$, $D2_i$ and $D1_i$ detail signals are two orders of magnitude below $A4_i$ and $D4_i$. Figure 7 (bottom left) shows the difference between the original signal and the estimated signal with $A4_i$ and $D4_i$. Practically there is no error when overlapped. In this experiment we have chosen the Daubechies 45 wavelets transform because it yields good results in identification (Tsatsanis & Giannakis, 1993), after testing different Daubechies wavelets.

After an initial step for selecting the model structure, it is determined that the order of the model has to be 20 (10 for the $A4_i$ and 10 for $D4_i$ coefficients), and an MAX model has been selected, for the reasons explained above. When those 9 models are calibrated, they are validated with the error criteria of FPE (Function Prediction Error) and MSE (Mean Square Error), yielding values about 10e(-6) and 5% respectively using 5000 data for identification and 1000 for validation. Besides, for the whole estimated models the residuals autocorrelation and cross-correlation between the inputs and residuals are uncorrelated, indicating the goodness of the models.

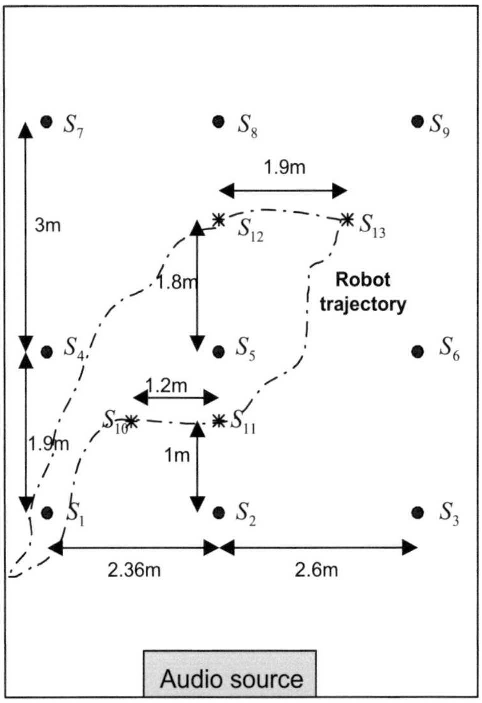

Fig. 6. Robot environment: labeled audio signals and actual robot trajectory with unlabelled signals (S_{10}, S_{11}, S_{12}, S_{13}).

These coefficients form the feature space, where the relative distances among all the samples are calculated and related in the way explained in section 2 in order to obtain the transform function f_T. With these relations, the curve appearing in Figure 8 is obtained, under the minimum square error criteria, approximated by a 4th-order polynomial with the following expression:

$$f_T = d_{fs} = 9.65e(10)d_{xy}^{4} + 1.61e(5)d_{xy}^{3} - 8.49e(2)d_{xy}^{2} + 144.9d_{xy} + 107.84$$

which is related with the solution of the sound equation in (Kinsler et al., 1995); (Kuttruff, 1979) with a physical meaning.
With the transform function f_T we proceed to find the two minimum distances in the feature space to each unlabelled sample respect the labeled ones, that is, for audio signals S_{10}, S_{11}, S_{12} and S_{13}, respect to S_1, ..., S_9.

We obtain four solutions for each signal because each distance in the feature space crosses four times the f_T curve. In order to discard the false solutions we use the previous position information of the robot, that is the $(x_i,y_i)_{prev}$ point. We also know the robot speed (v = 15cm/sec) and the computation time between each new position given by the system, which is close to 3 sec. If we consider the movement of the robot at constant speed, the new position will be $(x_i,y_i)_{prev} \pm (450,450)$mm.

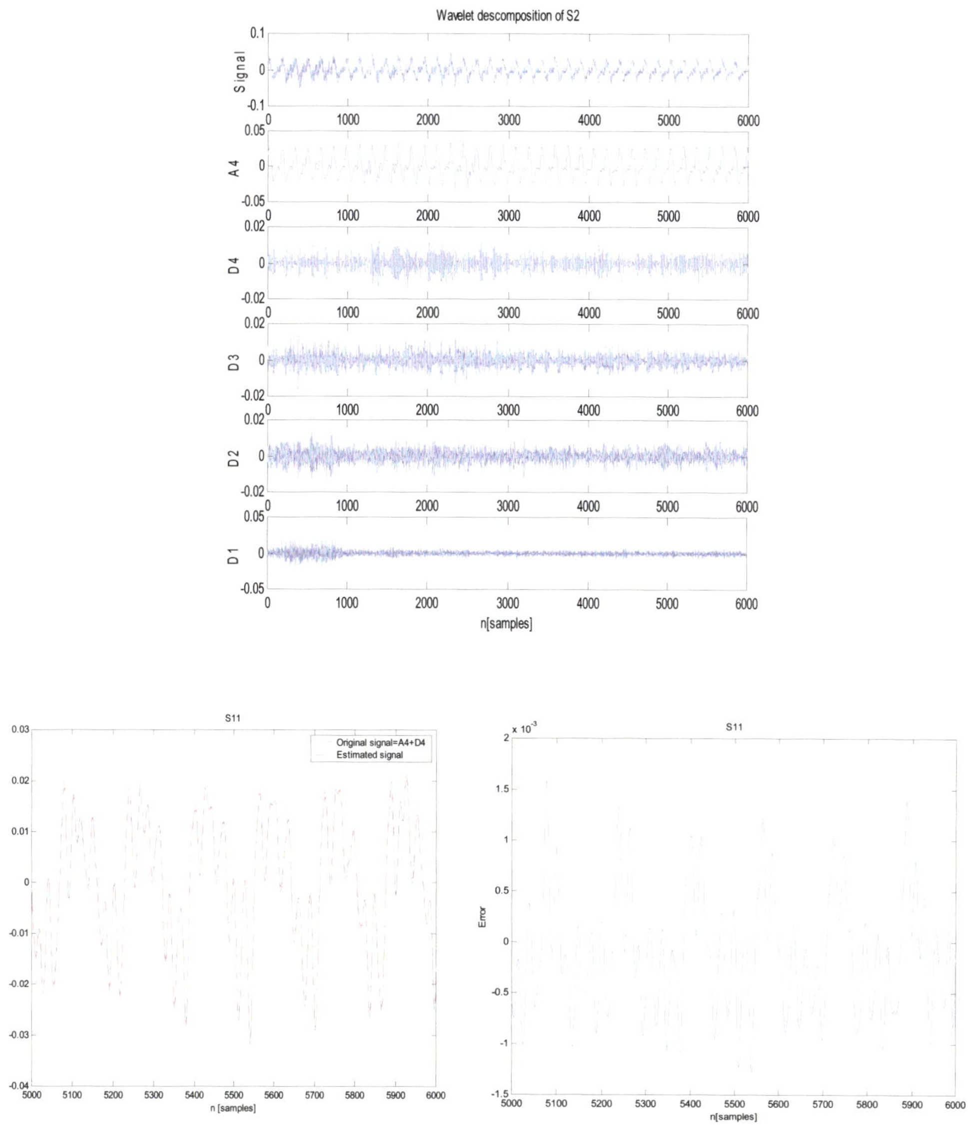

Fig. 7. (Up) Multilevel wavelet decomposition of a non-speech signal (S_2) by an approximation signal and four signal details; (down) comparison between (left) original signal (A4+D4) and the estimated signal and (right) its error for S_{11}.

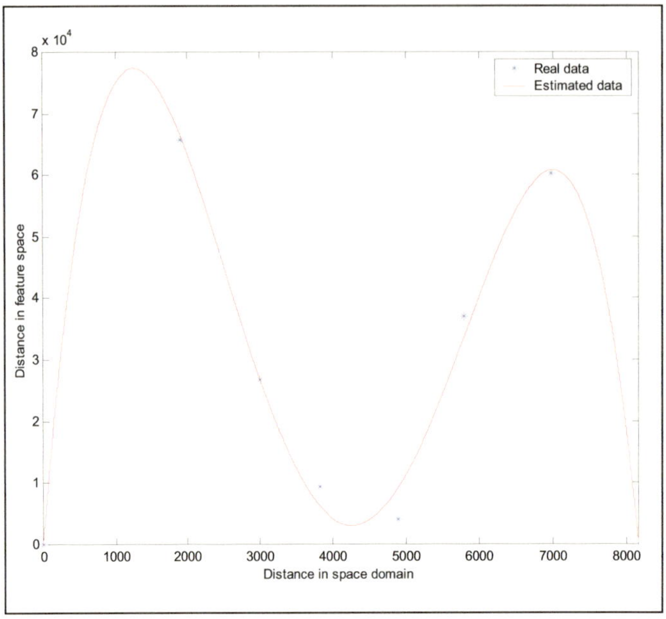

Fig. 8. Transform function f_T.

With this information we choose the solution that best fits with the crossing circles solution and the possible robot movement. In order to solve the ambiguity of the two intersection points, the angle that the robot has computed (γ) is compared with the angles (α_{FN1}, α_{FN2}) analytically computed between the two intersection points and the sound source (corrected respect the magnetic north). The solution is the angle α_{FNi} closer to γ. The uncertainty of this location is bounded by d*sin(ε), being ε the difference between the actual angle of the robot respect the sound source and the max{α_{FNi}, γ}, and d is the actual distance of the robot to the sound source. In our experiments, we have verified that ε is limited to 1.9° for d=1m and 1° for d=2.5m (between 3.3 and 4.3 cm of absolute error in localization).

4.2 Method based on the LPV model

In the second proposed method based on the LPV model, the methodology applied to determine the robot's position is the following:

1. The robot acquires an audio signal in its current position and performs an identification process taking as input signal the filtered sound source signal and as output signal the acquired and filtered signal. The parameters corresponding to the obtained poles in this identification process will be the features components for further steps.
2. The Euclidean distances in the feature space are calculated between the current position and the different labeled samples.
3. The two first samples are chosen and the distance between them and the robot's position are then calculated. Through a transformation function f_T, in the same way that the previous approach, the distance in the feature domain is converted to a distance in the space domain. These two distances in the space domain give two possible positions by the crossing circles of distances.
4. To discriminate between both possible solutions, the angle between each one and the platform containing the microphone array (which contains a compass) are calculated,

and the closest one to the platform angle will be chosen as discriminatory variable to select the current robot's position.

5. Steps 3 and 4 are repeated with the remaining labeled samples, and the solution is chosen among the closest angle to the robot's platform.

The acoustic response of the environment is very directional, and this fact leads to consider some uncertainty in the determination of the transformation function which relates the distance in the feature space and the domain space.

The robot, in order to determine its location, will perform the identification process between the emitted sound signal by the sound sources and the acquired signal by the microphone. As it can be seen in Figure 9, the robot follows the trajectory indicated by the arrows. In the map sound sources are indicated (climate chamber and PCB insulator). Two experiments are carried out using both sound sources separately. There are two kind of audio samples: R_1, R_2, R_3, R_4, R_5, R_6 and R_7 which are used in the recognition step whereas M_1, M_2, M_3, M_4 and M_5 are labeled samples used in the learning step.

The acquired signal in the climatic chamber will be used in the identification process. This signal is time-continuous and, initially, non-stationary; but because the signal is generated by revolving electrical machines it has some degree of stationariety when a high number of samples is used, in this case, 50,000 samples (1.13 seconds).

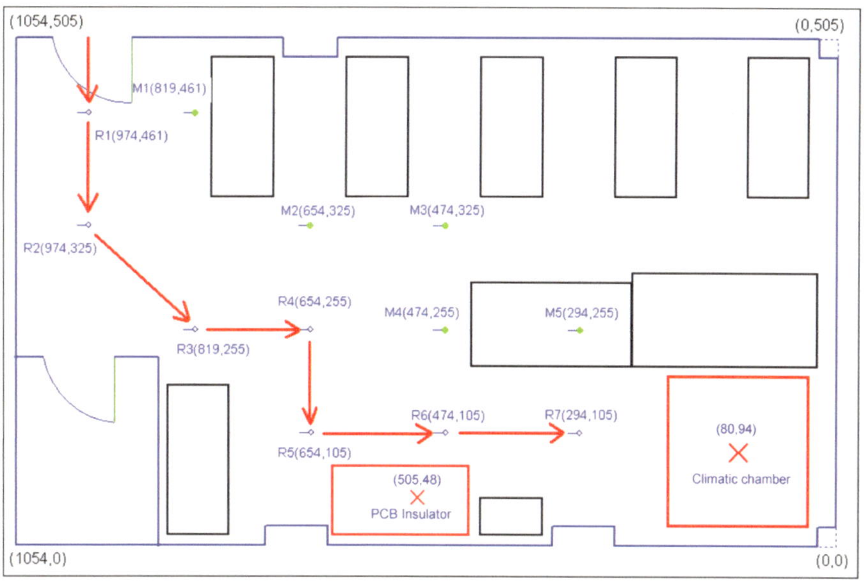

Fig. 9. Robot environment: labeled audio signals and actual robot trajectory with unlabeled signals (R_1, R_2, R_3, R_4, R_5, R_6 and R_7).

The fundamental frequency is located at 100Hz, see Figure 10, and there are also some significant harmonics above and below it. In order to simplify the identification process only the fundamental frequency at 100Hz will be taken into account.

In this approach the sampling frequency is 44,100Hz. Other lower frequencies could be used instead, avoiding working with a high number of samples, but this frequency has been chosen because in a near future a voice recognition system will be implemented aboard the robot and it will be shared with this audio localization system.

The emitted signal for the PCB insulator machine and its spectrum can be seen in Figure 11. To facilitate the plant identification process centering its response in the 100Hz component, the input and output signals will be filtered and, consequently, the input-output relationship in linear systems is an ARX model.

To do that, a band-pass filter is applied to the acquired sound signals by the robot, specifically a 6th-order digital Cauer filter. Figure 12 shows the results of the filter for the input signal in, for instance, robot position R_4 in the climatic chamber (experiment 1).

After an initial step for selecting the model structure, an ARX has been selected, for the reasons explained above of stationery (Charbonnier et al., 1987), with $n_a = 10$, $n_b = 4$ and a delay of 2 for the case of the climatic chamber (experiment 1), and $n_a = 10$, $n_b = 2$ and a delay of 4 in the case of PCB insulator (experiment 2). When those 5 models are calibrated, they are validated with the error criteria of FPE (Function Prediction Error) and MSE (Mean Square Error), yielding values about 10e(-10) and 3% respectively using 5000 data for identification and 3000 for validation. Besides, for the whole estimated models the residuals

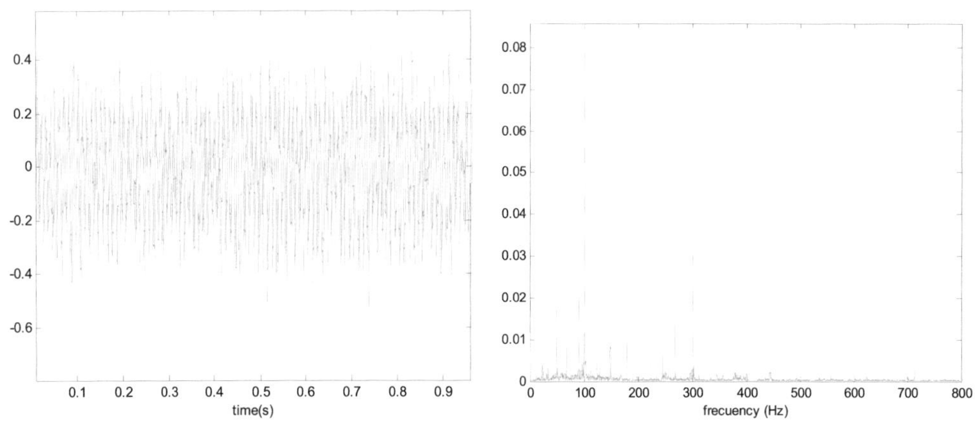

Fig. 10. Source signal (climate chamber) and its frequency spectrum.

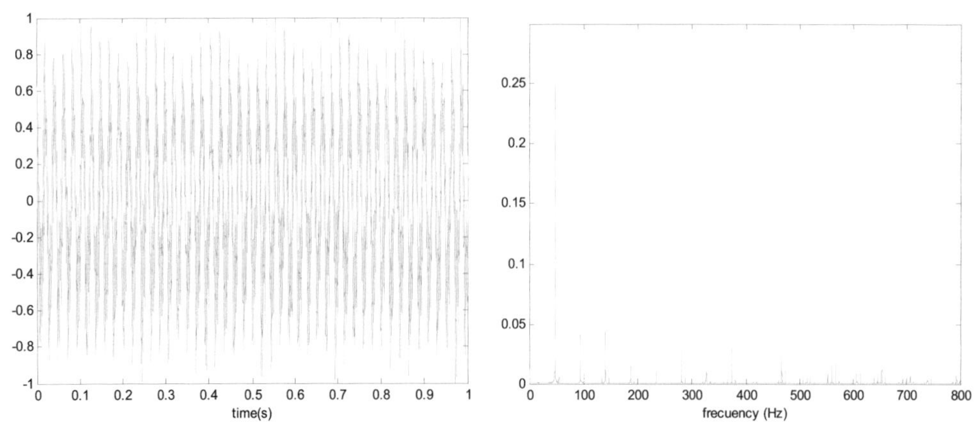

Fig. 11. Source signal (PCB insulator) and its frequency spectrum.

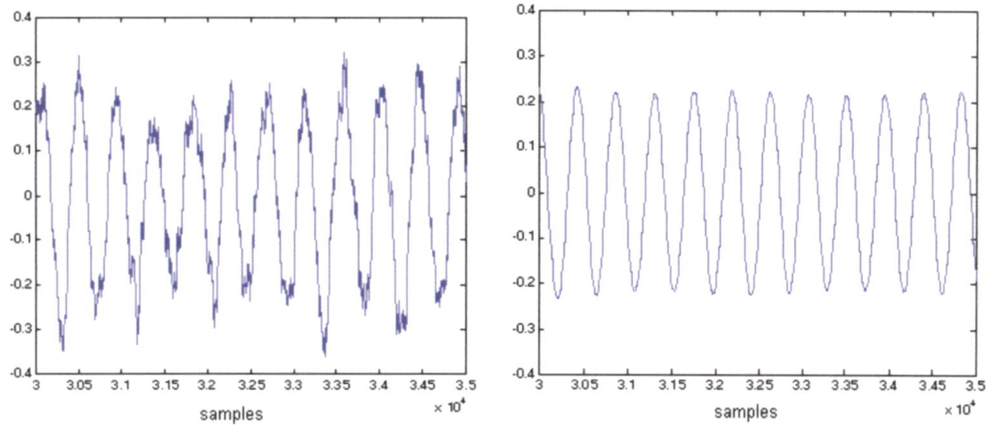

Fig. 12. R4 sound signal (left) and its filtered signal (right).

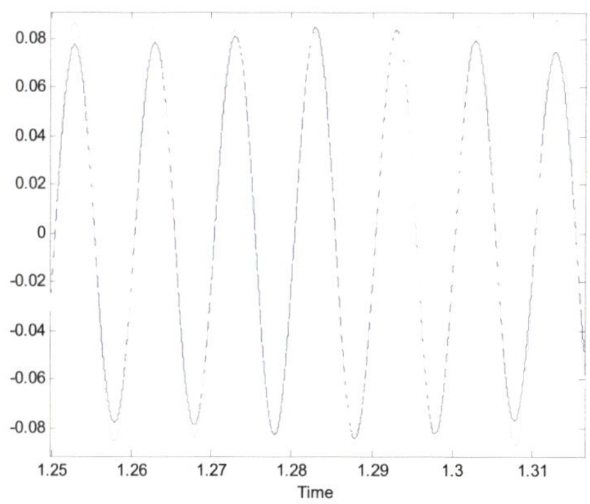

Fig. 13. Original M_5 signal and its estimation.

autocorrelation and cross-correlation between the inputs and residuals are uncorrelated, indicating the goodness of the models.

For instance, for labeled M_5 sample the signal and its estimation can be seen in Figure 13 in the first experiment, validating the model.

When observing the diagram of poles and zeros for the different transfer function models in the identification process for the labeled signals, there exists no difference between the zero positions, and, in the other hand, there is a significant variation in pole positions, due mainly to obstacles presence, reverberations among other effects, see Figure 14. Therefore, we will focus in poles to determinate the points in the feature space.

In experiment 1, in order to determine the transformation function, for every point in the feature space, the distances between them and the source signal are calculated, and these distances are plotted together with their corresponding distances in the space domain.

With these values, after an interpolation process, the transform function f_T is computed. In order to estimate the robot localization, we use other information such as the robot speed (in this case 15cm/sec), the computation time between each new position (3 sec). This fact is a source of uncertainty that adds in average ± 45 cm in the robot's position.

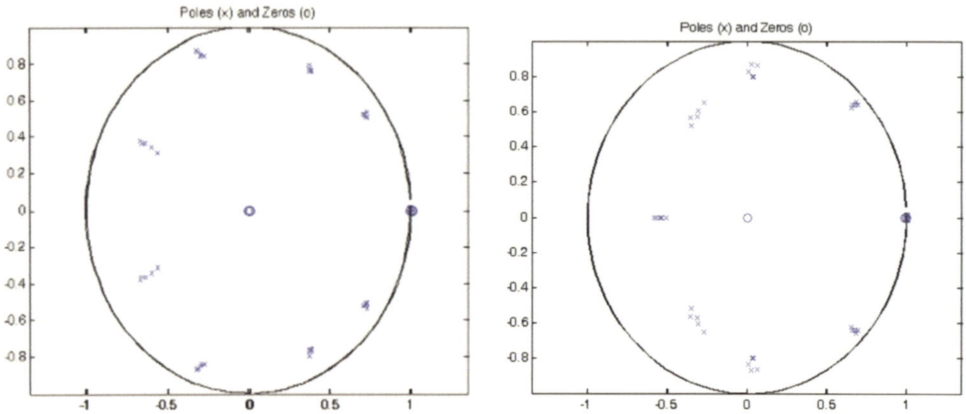

Fig. 14. Poles and zeros positions in experiment 1 (left) and 2 (right).

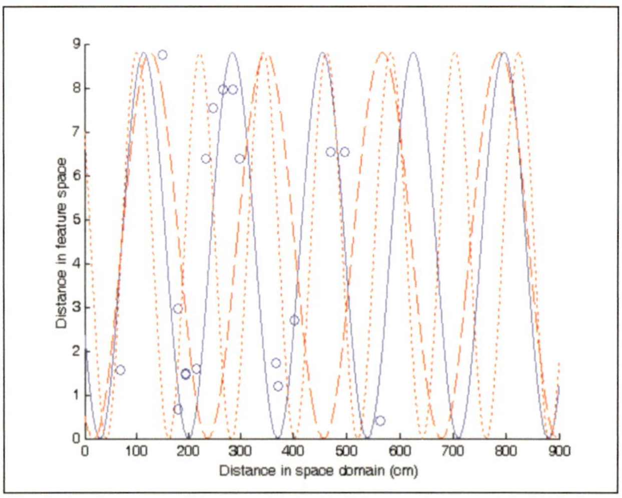

Fig. 15. Nominal transformation function and the limits of the interval for the uncertainty in experiment 1.

In experiment 1, when the climatic chamber is used as sound source the obtained transformation function is:

$$y = 4{,}4 + 4{,}4.\sin\left(\frac{2\pi x}{170} - \frac{80\pi}{170}\right)$$

Now, if an uncertainty interval is supposed (± 50 cm) the transformation function that covers this variability in the robot's position can be expressed (for both experiments) as:

$$y = A + A.\sin\left(\frac{2\pi x}{170 \pm 50} - \frac{\phi}{170 \pm 50}\right)$$

In Figure 15, the nominal transformation function and the limits for the uncertainty interval transformation functions can be seen.

There exists another uncertainty of about ±7.5 degrees in the angle determination due to the rotary platform in the robot that contains the microphones. Finally, to determine the current robot's position the solution that provides the closest angle to the robot's platform will be chosen. The results of our experiments yield an average error in the X axis of -1.242% and in the Y axis of 0.454% in experiment 1 and 0.335% in the X axis and -0.18% in the Y axis, providing estimated x-y positions good enough and robust.

5. Conclusion

With the approaches presented in this Chapter we have achieved some interesting results that encourage the authors to keep on walking in this research field. The room feature extraction is carried out by identification of the sound signals. Besides to reinforce the localization, avoiding ambiguity and reducing uncertainty and incorporating robustness, a sensorial system is used aboard the robot to compute the angle between itself and the sound source. The obtained feature space is related with the space domain through a general approach with acoustical meaning. The validation of this novel approach is tested in different environments obtaining good results. The results keep on being very good when the uncertainty is incorporated in the transformation function.

6. References

Navarro, D. & Benet, G. (2009). Magnetic Map Building for Mobile Robot Localization Purpose, *14th International Conference on Emerging Technologies and Factory Automation*, Palma de Mallorca, September, 2009.

Caruso, M. (2000). Applications of magnetic sensors for low cost compass systems, *IEEE Position Location and Navigation Symposium*, pp. 177–184, San Diego, CA, USA, March 2000.

Kim, H.-D.; Kim, D.-W. & Sim, K.B. (2006). Simultaneous Localization and Map Building using Vision Camera and Electrical Compass, *SICE-ICASE International Joint Conference*, Korea, October, 2006.

Kim, H.-D.; Seo, S-.W.; Jang, I.-H. & Sim, K.B. (2007). SLAM of Mobile Robot in the indoor Environment with Digital Magnetic Compass and Ultrasonic Sensors, *International Conference on Control, Automation and Systems*, Oct. 17-20, 2007, Seoul, Korea.

Luo, R.C.; Yih, C.-C. & Su, K.L. (2002). Multisensor Fusion and Integration: Approaches, Applications and Future Research Directions, *IEEE Sensors Journal*, Vol. 2, no. 2, April 2002.

Begum, M.; Mann, G.K.I. & Gosine, R. (2006). A Fuzzy-Evolutionary Algorithm for Simultaneous Localization and Mapping of Mobile Robots, *IEEE Congress on Evolutionary Computation*, Canada, July, 2006.

Brunskill, M. & Roy, N. (2005). SLAM using Incremental Probabilistic PCA and Dimensionality Reduction, *Proc. of the IEEE International Conference on Robotics and Automation*, Spain, April, 2005.

Di Marco, M.; Garulli, A.; Lacroix, S. & Vicino, A. (2000). A Set Theoretic Approach to the Simultaneous Localization and Map Building Problem, *Proceedings of the 39th IEEE Conference on Decision and Control*, Sydney, Australia, December, 2000.

Smith, R.; Self, M. & Cheeseman, P. (1990). Estimating uncertain spatial relationships in robotics, *Autonomous Robot Vehicles*, vol. 8, pp. 167–193, 1990.

Thrun, S. (2001). A probabilistic online mapping algorithm for teams of mobile robots, *Journal of Robotics Research*, vol. 20, no. 5, pp. 335-363, 2001.

Begum, M.; Mann, G.K.I. & Gosine, R.G. (2006). An Evolutionary SLAM Algorithm for Mobile Robots, *Proceedings of the IEEE/RSJ International Conference on Intelligent Robots and Systems*, October 9 - 15, 2006, Beijing, China.

Priyantha, N.B. (2000). The Cricket location-support system, *Proc. of the 6th Annual International Conference on Mobile Computing and Networking*, pp. 32-43, 2000.

Sayed, A.H.; Tarighat, A. & Khajehnouri, N. (2005). Network-Based Wireless Location: Challenges faced in developing techniques for accurate wireless location information, *IEEE Signal Processing Magazine*, vol. 22, no. 4, July 2005.

Christo, C.; Carvalho, E.; Silva, M.P. & Cardeira, C. (2009). Autonomous Mobile Robots Localization with Multiples iGPS Web Services, *14th International Conference on Emerging Technologies and Factory Automation*, Palma de Mallorca, September, 2009.

Yang, P.; Sun, H. & Zu, L. (2007). An Acoustic Localization System Using Microphone Array for Mobile Robot, *International Journal of Intelligent Engineering & Systems*, 2007.

Mumolo, E.; Nolich, M. & Vercelli, G. (2003). Algorithms for acoustic localization based on microphone array in service robotics, *Robotics and Autonomous Systems*, vol. 42, pp. 69-88, 2003.

Csyzewski, A. (2003). Automatic identification of sound source position employing neural networks and rough sets, *Pattern Recognition Letters*, vol. 24, pp. 921-933, 2003.

Ying, J. & Runze, Y. (2007). Research Status and Prospect of the Acoustic Localization Techniques, *Audio Engineering*, vol. 31, no. 2, pp. 4-8, 2007.

Sasaki, Y.; Kagami S. & Mizoguchi, H. (2006). Multiple sound source mapping for a mobile robot by selfmotion triangulation, *Proceedings of the IEEE/RSJ International Conference on Intelligent Robots and Systems*, Beijing, China, 2006.

Kim, Y.-E.; Su, D.-H.; Chung, G.-J.; Huang, X. & Lee, C.-D. (2009). Efficient Sound Source Localization Method Using Region Selection, *IEEE International Symposium on Industrial Electronics*, ISIE 2009, Seoul Olympic Parktel, Seoul, Korea July 5-8, 2009.

Brandstein, M.S. & Silverman, H. (1997). A practical methodology for speech source localization with microphone arrays, *Computer Speech and Language*, vol. 11, no. 2, pp. 91-126, 1997.

Knapp, C.H. & Carter, G.C. (1976). The generalized correlation method for estimation of time delay, *IEEE Trans. on Acoustics, Speech and Signal Processing*, vol. Assp-24, no. 4, 1976.

Nakashima, H. & Mukai, T. (2005). 3D Sound source localization system based on learning of binaural hearing, *IEEE International Conference on Systems, Man and Cybernetics*, vol. 4, pp. 3534-3539, 2005.

Kim, H.S. & Choi, J. (2009). Binaural Sound Localization based on Sparse Coding and SOM, *IEEE/RSJ International Conference on Intelligent Robots and Systems*, October 11-15, 2009, St. Louis, USA.

Yi, H. & Chu-na, W. (2010). A New Moving Sound Source Localization Method Based on the Time Difference of Arrival, Proc. of the International Conference on Image Analysis and Signal Processing, pp. 118-122, 9-11 April, 2010, Zhejiang, China.

Rodemann, T.; Joublin, F. & Goerick, C. (2009). Audio Proto Objects for Improved Sound Localization, *IEEE/RSJ International Conference on Intelligent Robots and Systems*, October 11-15, 2009, St. Louis, USA

Mori, K.; Kasashima, N.; Yoshioha, T. & Ueno, Y. (1996). Prediction of Spalling on a Ball Bearing by Applying the Discrete Wavelet Transform to Vibration Signals, *Wear*, vol. 195, no. 1-2, pp. 162-168, 1996.

Roberts, S. & Everson, R. (2001). Independent Component Analysis: Principles and Practice, *Cambridge Univ. Press*, Cambridge, UK, 2001.

Bolea, Y.; Grau, A. & Sanfeliu, A. (2003). Non-speech Sound Feature Extraction based on Model Identification for Robot Navigation, *8th Iberoamerican Congress on Pattern Recognition*, CIARP 2003, Lectures Notes in Computer Science, LNCS 2905, pp. 221-228, Havana, Cuba, November 2003.

Mallat, S. & Zhang, Z. (1993). Matching pursuits with time-frequency dictionaries, *IEEE Trans. on Signal Processing*, vol.45, no.12, pp. 3397-3415, 1993.

Donoho, D.-L. (1999). De-noising by soft-thresholding, *IEEE Trans. on Information Theory*, vol. 33, no. 7, pp. 2183-2191, 1999.

Lin, J. (2001). Feature Extraction of Machine Sound using Wavelet and its Application in Fault Diagnosis, *NTD&E International*, vol. 34, pp. 25-30, 2001.

Grau, A.; Bolea, Y. & Manzanares, M. (2007). Robust Industrial Machine Sounds Identification based on Frequency Spectrum Analysis, *12th Iberoamerican Congress on Pattern Recognition*, CIARP 2007, Lecture notes in Computer Science, LNCS 4756, pp. 71-77, Valparaiso, Chile, November 2007.

Ljung, L. (1987). System identification: Theory for the user. *Prentice-Hall*, 1987.

Charbonnier, R.; Barlaud, M.; Alengrin, G. & Menez, J. (1987). Results on AR-modeling of nonstationary signals, *IEEE Trans. Signal Processing*, vol. 12, no. 2, pp. 143-151.

Kayhan, A.S.; Ei-Jaroudi, A. & Chaparro, L.F. (1994). Evolutionary periodogram for nonstationary signals, *IEEE Trans. Signal Process*, vol. 42, no. 6, pp. 1527-1536.

Tsatsanis, M.K. & Giannakis, G.B. (1993). Time-varying system identification and model validation using wavelets, *IEEE Trans. Signal Process*, vol. 41, no. 12, pp. 3512-3523.

Kinsler, L.; Frey, A.; Coppens, A. & Sanders, J. (1995). Fundamentals of Acoustics, *Limusa Ed.*, Barcelona, 1995.

Kuttruff, H. (1979). Room Acoustics, *Applied Science Publishers Ltd.*, 1979.

Haneda, Y.; Makino, S. & Kaneda, Y. (1992). Modeling of a Room Transfer Function Using Common Acoustical Poles, *IEEE International Conference on Acoustics, Speech, and Signal Processing*, ICASSP-92, vol.2, pp. 213–216, 1992.

Haneda, Y.; Kaneda, Y. & Kitawaki, N. (1999). Common-Acoustical-Pole and Residue Model and Its Application to Spatial Interpolation and Extrapolation of a Room Transfer Function, *IEEE Transactions on Speech and Audio Processing*, vol. 7, no. 6, Nov. 1999.

Gustaffson, T.; Pota, H.R.; Vance, J.; Rao, B.D. & Trivedi, M.M. (2000). Estimation of Acoustical Room Transfer Functions, *Proceedings of the 39th IEEE Conference on Decision and Control*, Sydney, Australia, December 2000.

Bolea, Y.; Manzanares, M. & Grau, A. (2008). Robust robot localization using non-speech sound in industrial environments, *IEEE International Symposium on Industrial Electronics*, ISIE 2008, Cambridge, United Kingdom, 30 June- 2 July 2008.

Manzanares, M.; Guerra, E.; Bolea, Y. & Grau, A. (2009). Robot Localization Method by Acoustical Signal Identification, 14th *IEEE Emerging Tech and Factory Automation*, ETFA'09, Palma de Mallorca, Spain, 2009.

Bielińska, E. (2002). Speaker identification, *Artificial Intelligence Methods*, AI-METH, 2002.

9

Source Localization for Dual Speech Enhancement Technology

Seungil Kim[1], Hyejeong Jeon[2], and Lag-Young Kim[2]
[1]*Speech Innovation Consulting Group*
[2]*Information Technology Lab. LG Electronics*
Korea

1. Introduction

Many researchers have investigated multi-channel speech enhancement techniques which can be used for the pre-processing of the speech recognition system. Numerous microphones can give high performance, but they require additional hardware costs and generate the design problem about microphone position. Therefore speech enhancement technique using two microphones is preferred in mobile phone such as LG KM900, iPhone 4 and Nexus One. For enhancing the speech with two or more microphones, the spatial information from the input signal's incident angle should be used. Therefore, various sound source localization(SSL) methods have been used to estimate the talker's direction-of-arrival(DOA). There are two main approaches to localization (Brandstein, 1995), (Dibase, 2000): the steered-beamformer approach, which includes various kinds of beamformers; and time-difference of arrival (TDOA) approach, which includes a generalized cross-correlation (GCC). The steered-beamformer approach has the capability of enhancing a desired signal that originates from a particular direction. The beamformer can steer its response at a particular angle; it can then find the spatial information required to maximize the beamformer output by scanning over a predefined spatial region. For this purpose, we can use a simple conventional delay-and-sum beamformer or many optimum beamfomers (Naguib, 1996). The TDOA approach uses classical time delay estimation techniques, such as cross-correlation, GCC, adaptive time delay estimation, and the adaptive eigenvalue decomposition algorithm (Chen et al., 2006). The most common time delay estimation method is the GCC, which consists of various types such as the unfiltered type, the maximum likelihood (ML) type, and the phase transform (PHAT) type. The GCC-PHAT is a widely used for TDOA estimation method because it works well in a realistic environment.
The resolution of the DOA estimator is deeply related to the aperture size of the array and the number of microphone. A large aperture size and microphones make an accurate estimation result. Therefore, SSL method using two microphones cannot give the accurate direction-of-arrival (DOA) estimation result. Moreover, the implementation of a TDOA estimator requires a voice activity detector (Araki et al., 2007) or a speech/non speech detector (Lathoud, 2006). However, the TDOA estimation often shows a failed result in spite of these kinds of additional processing. Hence, reliable SSL algorithm is needed for dual channel speech enhancement system.

In this chapter, we will define the reliability measure based on waterbed effect of DOA estimator and then show a method of increasing the accuracy of DOA estimation by using reliability measure(Jeon et al., 2007).

2. Dual Speech Enhancement technology

Dual Speech Enhancement (DSE™) is a trademark of advanced two-channel speech enhancement technology developed by LG Electronics. It has been shown that DSE would be competitive to the other state-of-art speech enhancement technologies. DSE technology can be divided into two sub-technologies according to its function and aim. One is the Dual Speech Enhancement for Talk (DSE.T™) and the other is the Dual Speech Enhancement for Recording (DSE.R™).

DSE.T is a solution for speech communication system. Comfortable call is unfortunately impossible in noisy environments. DSE.T can be a new solution to enhance speech quality in noisy place. DSE.T technology was introduced at CES 2009 in Las Vegas via Woo-hyun Baek (LG CTO) as one of the representative technologies prepared by LG Electronics.

DSE.R makes clear video recording with directionality. In DSE.R, two omni-directional microphones are processed and virtually make them as one directional microphone. One more useful thing in DSE.R is the function of electrical steering. Therefore, the user can select the direction of sound focusing. If user wants to record the voice of person who is pictured, user only needs to select "Producer Mode". If user wants to record the landscape or something else, user will select "Narrator Mode".

DSE technology was applied to commercial LG mobile phone, KM900 Arena as shown in Fig. 1. Four related video clips are available in following link and they will be also presented in the multimedia appendix of this book.

LG DSE.T technology - http://goo.gl/TUEo

LG's dual mic noise reduction demo at CES 2009 - http://goo.gl/QlFx

LG DSE.R technology - http://goo.gl/dbz3

DSE.R Test : LG KM900 Arena Video Recording - http://goo.gl/kwzJ

Fig. 1. LG KM900 Arena Phone

3. Sound source localization for Dual Speech Enhancement technology

The direction of talker can be used for DSE technology. However, two microphones are not enough to get high angular resolution at the DOA estimator. Therefore it is needed to reject some unreliable results and select the good one. For the reliable DOA estimation, we adopt the new scheme "Reliability Measure" which arises from waterbed effect. If the obtained reliability measure is lower than a predefined threshold, the result will be rejected. By using the reliable results only, we can decrease the detection failure of the signal.

3.1 Waterbed effect in DOA estimation

The waterbed effect means that if somewhere the amplification characteristic is pushed down, it goes up somewhere else. This term is usually used in filter response. Stoica and Ninness showed the waterbed effect would appear in spectral estimation (Stoica & Ninness, 2004) (Ninness, 2003). He proved that the power spectral density estimated by a periodogram has a constant average relative variance. The searching method for the DOA estimation is very similar to a spectral estimation such as a periodogram. Therefore the waterbed effect in DOA estimation can be obtained by similar process to spectral estimation (Jeon et al. 2007). Let $\Phi(\omega)$ be the power spectral density of a Gaussian white noise process and $\hat{\Phi}(\omega)$ be the periodogram estimate of $\Phi(\omega)$. We can then show that the variance of $\hat{\Phi}(\omega)$ is proportional to the square of the power spectral density (Hayes, 1996). Thus,

$$\mathrm{var}\left\{\hat{\Phi}(\omega)\right\} = \Phi^2(\omega). \tag{1}$$

The average relative variance of $\hat{\Phi}(\omega)$ has the following form:

$$\begin{aligned}&\text{average relative variance}\left\{\hat{\Phi}(\omega)\right\}\\&= \frac{1}{2\pi}\int_{-\pi}^{\pi}\frac{\mathrm{var}\left\{\hat{\Phi}(\omega)\right\}}{\Phi^2(\omega)}d\omega\\&=1.\end{aligned} \tag{2}$$

This phenomenon has been called the waterbed effect.
The waterbed effect in DOA estimation can be reduced by a similar process.
If $R(\theta)$ is the cross-correlation value of GCC-PHAT, then

$$R(\theta) = \sum_{k=0}^{K-1}\frac{X_1[k]X_2^*[k]}{\left|X_1[k]X_2^*[k]\right|}e^{-j\frac{2\pi k}{K}\frac{d\sin\theta}{c}}, \quad \theta\in[-\pi,\pi]. \tag{3}$$

In addition, a power pattern estimate of $R(\theta)$, $\hat{P}(\theta)$, is expressed by

$$\hat{P}(\theta) = \frac{1}{K}\left|R(\theta)\right|^2 = \frac{1}{K}\sum_{k=0}^{K-1}\sum_{l=0}^{K-1}\frac{X_1[k]X_2^*[k]}{\left|X_1[k]X_2^*[k]\right|}\frac{X_1^*[l]X_2[l]}{\left|X_1^*[l]X_2[l]\right|}e^{-j\frac{2\pi(k-l)}{K}\frac{d\sin\theta}{c}}. \tag{4}$$

Furthermore, the expected value of the power pattern is

$$E\left\{\hat{P}(\theta)\right\} = \frac{1}{K}\sum_{k=0}^{K-1}\sum_{l=0}^{K-1} E\left\{ e^{j\frac{2\pi(k-l)}{K}\frac{d\sin\theta_0}{c}} \right\} e^{-j\frac{2\pi(k-l)}{K}\frac{d\sin\theta}{c}}. \tag{5}$$

Let the input signal be a spatially white noise process, and note that the signal is assumed to be Gaussian white noise in the spectral estimation. For spatially white noise, the expected value of $e^{j\frac{2\pi(k-l)}{K}\frac{d\sin\theta_0}{c}}$ is equal to unity when $k = l$ only; otherwise it is equal to zero. Thus, the expected value of the power pattern in (5) is equal to unity. That is,

$$E\left\{\hat{P}(\theta)\right\} = 1. \tag{6}$$

The second-order moment of the power pattern estimate is

$$E\left\{\hat{P}^2(\theta)\right\} = \frac{1}{K^2}\sum_{k=0}^{K-1}\sum_{l=0}^{K-1}\sum_{m=0}^{K-1}\sum_{n=0}^{K-1} E\left\{ e^{j\frac{2\pi(k-l+m-n)}{K}\frac{d\sin\theta_0}{c}} \right\} e^{-j\frac{2\pi(k-l+m-n)}{K}\frac{d\sin\theta}{c}}, \tag{7}$$

and separated by sum of two parts as follows:

$$\begin{aligned} E\left\{\hat{P}^2(\theta)\right\} &= \frac{1}{K^2}\sum\sum_{k+m=l+n}\sum\sum E\left\{ e^{j\frac{2\pi(k-l+m-n)}{K}\frac{d\sin\theta_0}{c}} \right\} e^{-j\frac{2\pi(k-l+m-n)}{K}\frac{d\sin\theta}{c}} \\ &+ \frac{1}{K^2}\sum\sum_{o.w}\sum\sum E\left\{ e^{j\frac{2\pi(k-l+m-n)}{K}\frac{d\sin\theta_0}{c}} \right\} e^{-j\frac{2\pi(k-l+m-n)}{K}\frac{d\sin\theta}{c}}. \end{aligned} \tag{8}$$

The number that can be satisfy $k + m = l + n$ is $\frac{(K+1)(K+2)(2K+3)}{6} - 3K$. Hence, the equation (8) can be simplified to

$$E\left\{\hat{P}^2(\theta)\right\} = \frac{1}{K^2}\left[\frac{(K+1)(K+2)(2K+3)}{6} - 3K\right]. \tag{9}$$

The variance of $\hat{P}(\theta)$ is

$$\begin{aligned} Var\left\{\hat{P}(\theta)\right\} &= E\left\{\hat{P}^2(\theta)\right\} - E\left\{\hat{P}(\theta)\right\}^2 \\ &= \frac{2K^3 + 3K^2 - 5K + 6}{6K^2} \approx \frac{K}{3}. \end{aligned} \tag{10}$$

By using (6) and (10), we can calculate the average relative variance of $\hat{P}(\theta)$ as follows:

$$\begin{aligned} &\text{average relative variance}\left\{\hat{P}(\theta)\right\} \\ &= \frac{1}{2\pi}\int_{-\pi}^{\pi}\frac{\text{var}\left\{\hat{P}(\theta)\right\}}{P^2(\theta)}d\theta \\ &= \frac{2K^3 + 3K^2 - 5K + 6}{6K^2} \approx \frac{K}{3}. \end{aligned} \tag{11}$$

This equation is the waterbed effect in the DOA estimation.

Figure 2 shows the result of the DOA estimation. The input signal which had the source in the angle of 30° location was used. The result showed that the direction was correctly estimated and the waterbed effect appeared in the angle of -30°. Even though there was no other signals, the result showed that there is the negative value in the angle of -30°.

3.2 Reliability measure

The concept of reliability measure was presented in (Jeon et al., 2007) and (Jeon, 2008). Figure 3 shows the cross-correlation value of the GCC-PHAT when the speech source is present at a direction of 0° and when the speech source is absent.

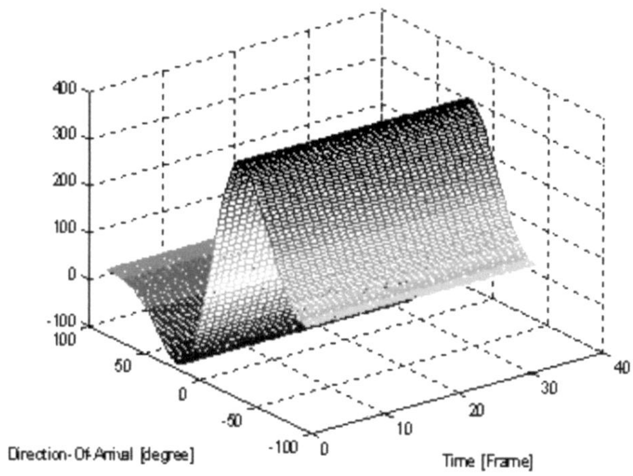

Fig. 2. Waterbed Effect in the DOA Estimation

To test the waterbed effect, we seated the talker in front of the dual microphone receiver. When a dominant source exists, the waterbed effect should cause the mainlobe to be prominent. If there is no directional source, $R(\theta)$ has a flat pattern for all directions. The reliability measure (z), which indicates the prominence of the lobe of $R(\theta)$, is defined as

$$z = f\left(R_{\max} - R_{\min}\right), \tag{12}$$

where $f(x)$ is any monotone-increasing function, $R_{\max}$ is the maximum value of R and $R_{\min}$ is the minimum. We used the formula $f(x) = \left|\frac{x}{K}\right|^2$.

In Fig. 3, the reliability (z) is 0.0177 when speech is absent and the reliability (z) is 0.9878 when speech is present. Because the reliability measure refers to the directivity of the sound source, we only selected the DOA estimation results that had a high reliability value and we clustered those results.

If we assume that a reliable DOA estimation result can be obtained when a dominant directional input exists, we can consider the following two hypotheses of reliability decision problem:

Assuming that reliable DOA estimation result can be obtained when dominant directional input exists, two hypotheses of reliability decision problem are as follow:

H_0 : unreliable DOA estimation result
H_1 : reliable DOA estimation result

And the hypothesis test equation can be defined as

$$z = \frac{1}{K^2}\left(R_{\max} - R_{\min}\right)^2 \underset{d_0}{\overset{d_1}{\gtrless}} \eta, \tag{13}$$

where η is the threshold for the selection of reliable results.

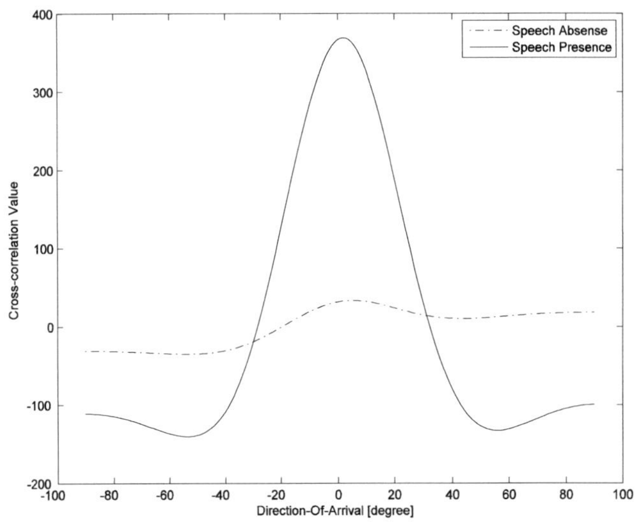

Fig. 3. The cross-correlation value when the speech source is present and when speech source is absent.

3.3 Determination of the threshold

To determine whether the estimate is reliable or not, we need to find the optimum threshold for detection. In (Kim et al., 2008), the optimum threshold was calculated based on maximum likelihood criteria. If we assume that the structure of z is known, reliable source detection can be considered as a simple binary decision problem. To determine which probabilistic model is fit to z, we made observations of z. The recorded data used to calculate the value of z was measured in a quiet conference room. The microphones were 8 cm apart and a single talker was located in front of the microphones. We visually determined that z could be modeled with a Rayleigh pdf as follows:

$$p(z \mid H_0) = \frac{z}{\sigma_0^2}\exp\left(-\frac{z^2}{2\sigma_0^2}\right) \tag{14}$$

$$p(z \mid H_1) = \frac{z}{\sigma_1^2} \exp\left(-\frac{z^2}{2\sigma_1^2} \right). \tag{15}$$

The ML estimation for the unknown parameter $\left(\sigma_0^2, \sigma_1^2\right)$ is given by the maximum value of the log-likelihood function (Schmidt et al., 1996). If we have N_0 items of observation data for z, which is in a decision region Z_0, then

$$\sigma_0^2 = \frac{1}{2N_0} \sum_{i=1}^{N_0} z_i^2 \quad , z_i \in Z_0. \tag{16}$$

Similarly, σ_1^2 can be easily obtained as follows:

$$\sigma_1^2 = \frac{1}{2N_1} \sum_{j=1}^{N_1} z_i^2 \quad , z_j \in Z_1. \tag{17}$$

Figure 4 depicts the observation data distributions fitted with a Rayleigh model. In the quiet conference room, the estimated variances σ_0 and σ_1 are 0.0183 and 0.1997, respectively. If we make use of the likelihood ratio

$$\Lambda(z) = \frac{p(z \mid H_1)}{p(z \mid H_0)}, \tag{18}$$

the decision rule can be represented by

$$\Lambda(z) = \frac{\sigma_0^2}{\sigma_1^2} \exp\left(\frac{\sigma_1^2 - \sigma_0^2}{2\sigma_0^2\sigma_1^2} \cdot z^2 \right) \underset{d_0}{\overset{d_1}{\gtrless}} \lambda. \tag{19}$$

If we take the natural logarithm of both sides of (19), then

$$\ln\left(\frac{\sigma_0^2}{\sigma_1^2} \right) - \left(\frac{\sigma_1^2 - \sigma_0^2}{2\sigma_0^2\sigma_1^2} \cdot z^2 \right) \underset{d_0}{\overset{d_1}{\gtrless}} \ln \lambda. \tag{20}$$

Because the reliability measure, z, always has a positive value in (13),

$$z \underset{d_0}{\overset{d_1}{\gtrless}} \sqrt{\frac{2\sigma_0^2\sigma_1^2}{\sigma_1^2 - \sigma_0^2} \cdot \left\{ \ln \lambda + \ln\left(\frac{\sigma_1^2}{\sigma_0^2} \right) \right\}} = \eta. \tag{21}$$

When $\ln \lambda$ is equal to zero, the threshold of the ML decision rule (Melsa & Cohn, 1978) can be determined by

$$\eta_{ML} = \sqrt{\frac{2\sigma_0^2\sigma_1^2}{\sigma_1^2 - \sigma_0^2} \cdot \ln\left(\frac{\sigma_1^2}{\sigma_0^2} \right)}. \tag{22}$$

If we use $\left(\sigma_0^2, \sigma_1^2\right) = (0.0183, 0.1997)$, which is previously calculated, η_{ML} becomes 0.0567 for Fig. 4.

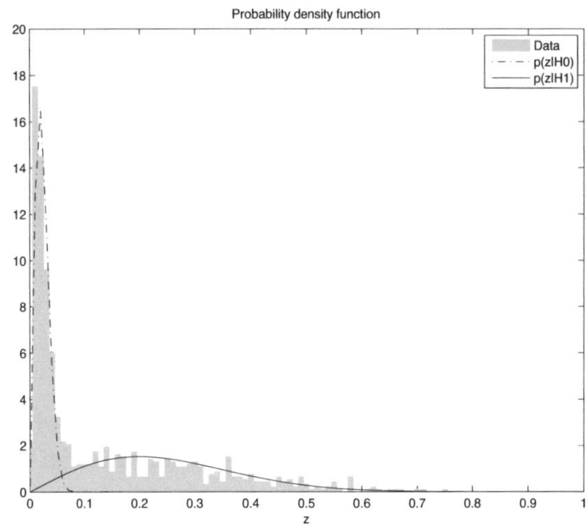

Fig. 4. The cross-correlation value when the speech source is present and when speech source is absent.

4. Performance evaluations

4.1 Simulations

The simulation was performed with the male talker's speech signal. The input speech came from the 30° and the spatially white random noise was mixed to make the SNR of 5dB, 10 dB, 15 dB, and 20 dB. The distance between two microphones was assumed to be 8cm.
The comparison of the estimated DOA is shown in Fig. 5. When the reliability measure and the threshold selection were applied, the average value of the estimated DOA was close to the speech direction. Also, the standard deviation and the RMS error was drastically reduced.

4.2 Experiments

To evaluate the performance of the proposed method, we applied it to the speech data recorded in a quiet conference room. The size of room was 8.5m x 5.5m x2.5m. This conference room, which was suitable for a conference with the several people, generated a normal reverberation effect. The impulse response of the conference room is shown in Fig. 6. The room had various kinds of office furniture such as tables, chairs, a white board standing on the floor, and a projector fixed to the ceiling. The two microphones were placed on the table in the center of the room, and the distance between the microphones was set to 8 cm. Figure 7 shows the experimental setup. The sampling rate of the recorded signal was 8 kHz, and the sample resolution of the signal was 16 bits.
Because the proposed method worked efficiently for the probabilistic model of reliability, we found it useful to eliminate the perturbed results of the estimated DOA in the speech recorded in this room. We compared the results with the normal GCC-PHAT method.

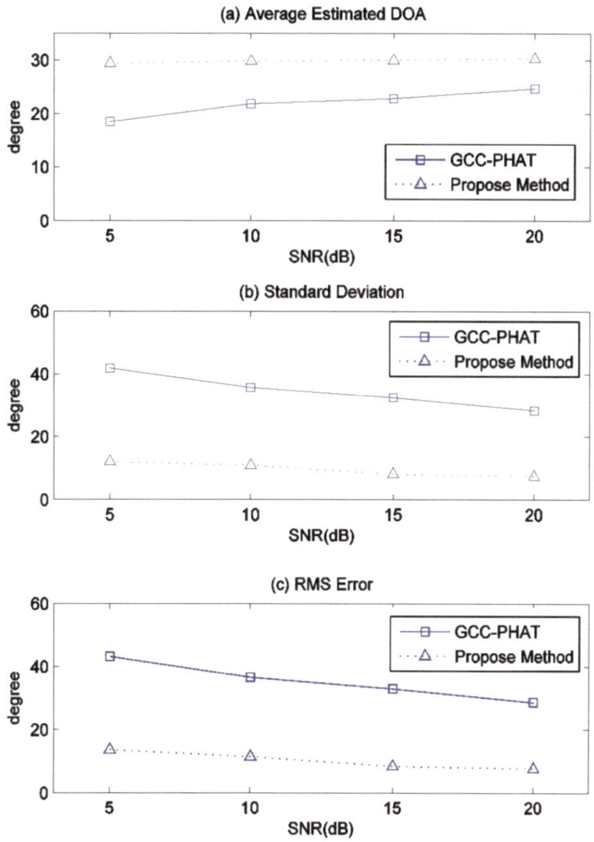

Fig. 5. (a) The average estimated DOA (b) The standard deviation (c) The RMS error when the SNR was 5 dB, 10 dB, and 20 dB

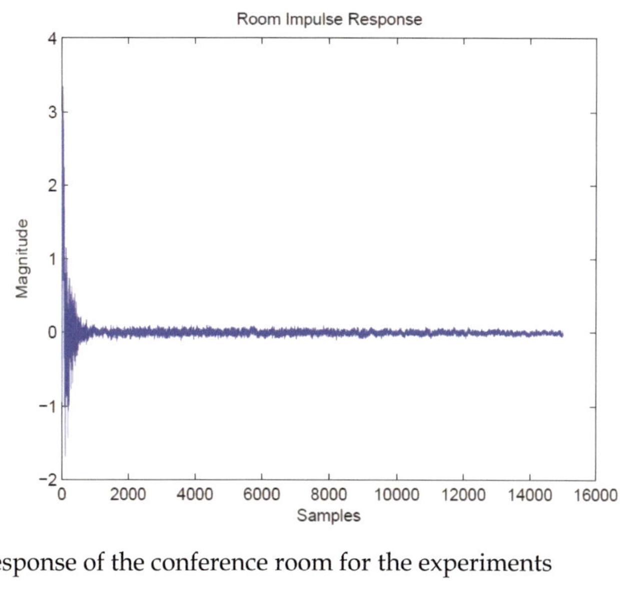

Fig. 6. Impulse response of the conference room for the experiments

4.2.1 Reliability

As shown in Fig. 7 and Fig. 8, we performed the experiment of the DOA estimator for a talker's speech from a direction of 60°. White noise and tone noise resulted from the fan of the projector.

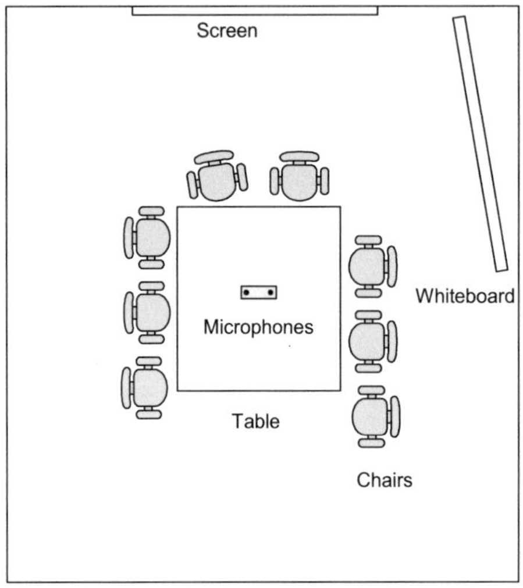

Fig. 7. The Experimental Setup

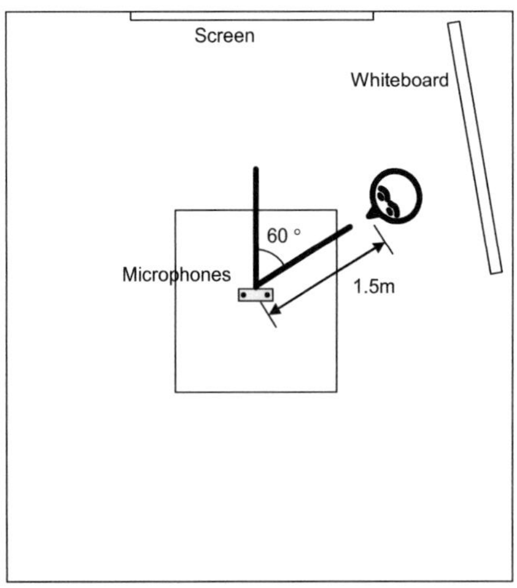

Fig. 8. The Recording Setup for Fixed Talker's Location

Figure 9(a) shows the waveform of the talker's speech. We calculated the direction of the talker's speech on the basis of the GCC-PHAT, and the result is shown in Fig. 9(b). The small circles in the figure indicate the results of the estimated DOA. There are many incorrect results for the estimated DOA, especially in periods when the talker didn't talk. Because of the estimated DOA results for when the talker didn't talk, there was a drastic drop in the performance of the estimated DOA. We calculated the reliability values of the given speech and applied the results to the estimated DOA.

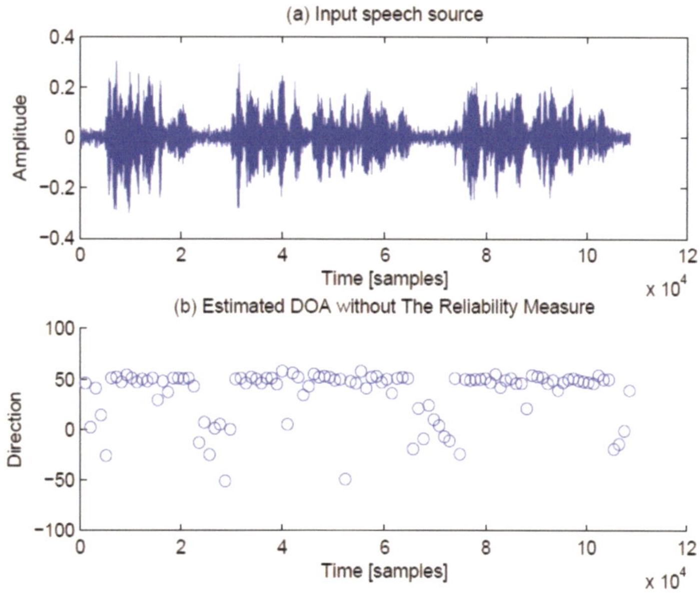

Fig. 9. (a) A waveform of the talker's speech (b) DOA estimation results of GCC-PHAT. It doesn't use the reliability measure.

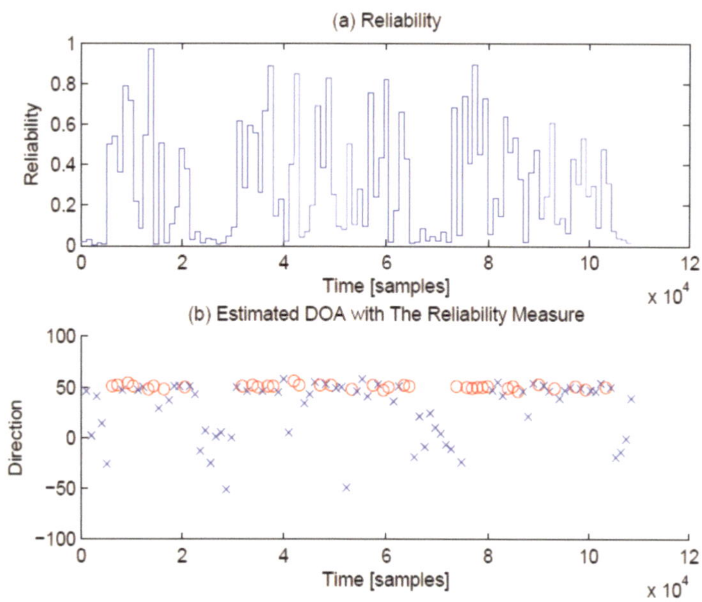

Fig. 10. (a) The calculated reliability for Fig. 9(a). (b) DOA estimation results of GCC-PHAT. It uses the reliability measure and eliminates unreliable estimates.

Figure 10(a) shows the reliability measures of the given speech, and Fig. 10(b) shows the estimated DOA after the removal of any unreliable results. We set the threshold, η, to 0.15. The x-marks indicate the eliminated values; these values were eliminated because the reliability measure revealed that those results were perturbed.

We can trace the talker's direction by using this method. In the experiment, the talker spoke some sentences while walking around the table, and the distance from the talker to the microphones was about 1.5 m. Figure 11 shows the talker's path in the room.

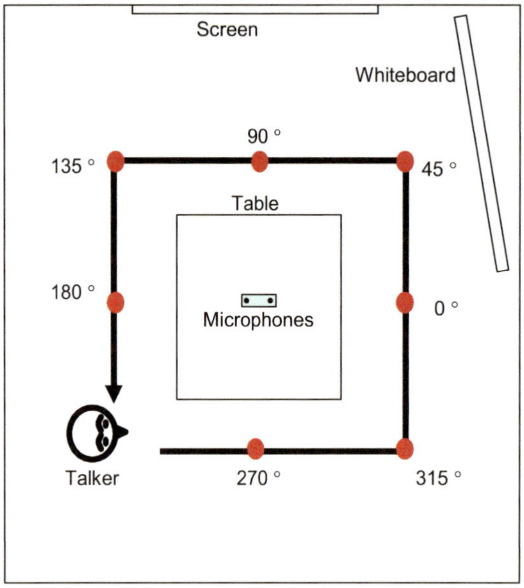

Fig. 11. The Recording Setup for Moving Talker

Figure 12(a) and Fig. 12(b) show the waveform and the estimated DOA based on the GCC-PHAT. The results of the estimated DOA are very disturbed because of the perturbed results. Figure 13(a) shows the calculated reliability values for the speech. By applying the reliability measure, as shown in Fig. 13(b), we can eliminate the perturbed values and produce better results for the estimated DOA. The x-marks represent the eliminated results. By eliminating the perturbed results, we can ensure that the estimated DOA is more accurate and has a smaller variance.

There is a degree of difference between the source direction and the average estimated DOA value. The difference occurs with respect to the height of the talker's mouth. Basically, we calculated the direction of the source from the phase difference of the two input signals. When we set the source direction, we thought the source was located on the same horizontal plane as the microphones. Thus, when the height of the source is not the same as the table, the phase difference cannot be the intended value as shown in Fig. 14. Even though we set the source direction at 90°, the actual source direction was $90^\circ - \theta_h$, where θ_h is

$$\theta_h = \tan^{-1}\left(\frac{h}{d}\right) \tag{23}$$

Because we used the source signal incident from the direction of 60° in Fig. 8, the actual source direction would be 48.5507° by using (23). The same phenomenon also occured in the next experiment; hence, the estimated DOA range was reduced to $(-90^\circ + \theta_h, 90^\circ - \theta_h)$, not $(-90^\circ, 90^\circ)$.

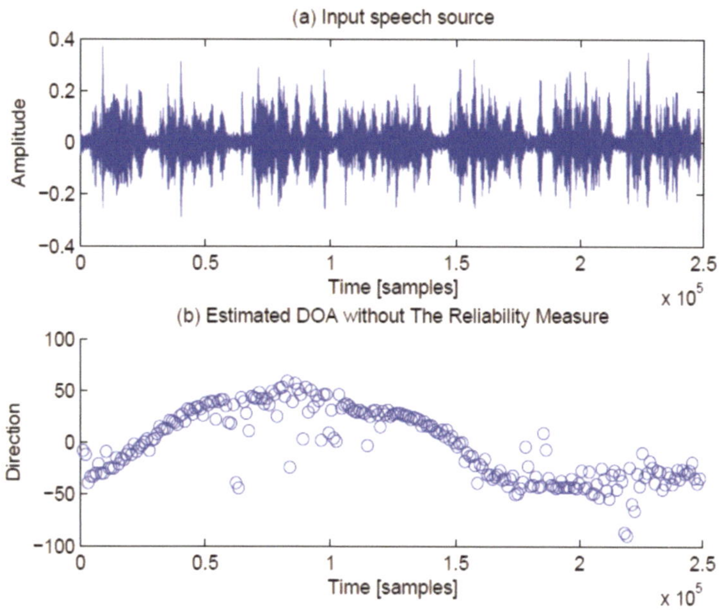

Fig. 12. A waveform of the talker's speech (b) DOA estimation results of GCC-PHAT. It doesn't use the reliability measure.

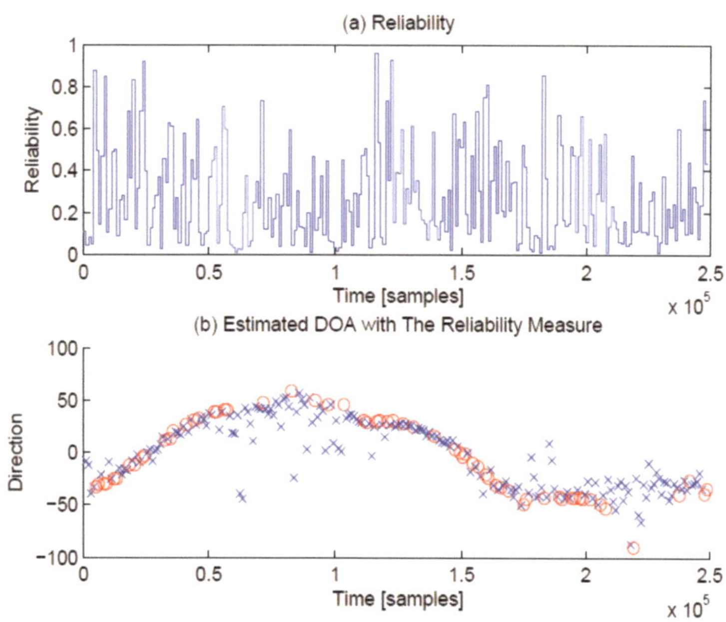

Fig. 13. (a) The calculated reliability for Fig. 11(a). (b) DOA estimation results of GCC-PHAT. It uses the reliability measure and eliminates unreliable estimates.

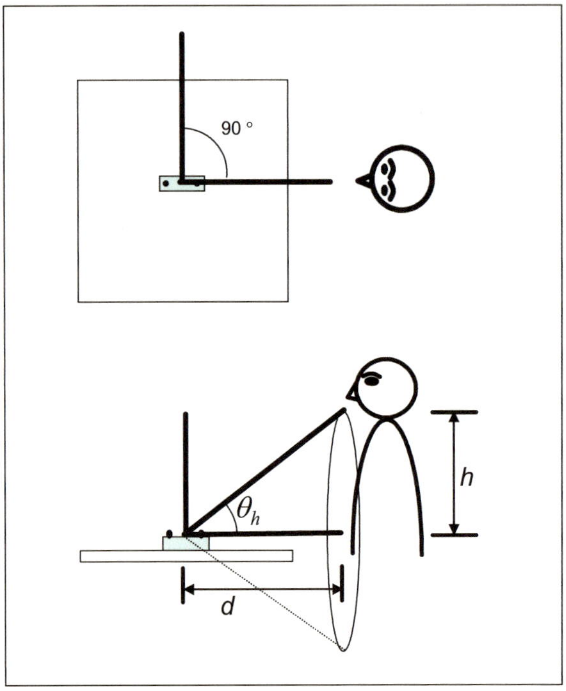

Fig. 14. The Recording Setup for Moving Talker

4.2.2 Speech recognition with DSE technology

The source localization has played an important role in the speech enhancement system. We applied the proposed localization method to the speech recognition system and evaluate its performance in a real car environment (Jeon, 2008).

The measurements were made in a mid-sized car. The input microphones were mounted on a sun visor for speech signal to impinge toward the input device (at the direction of 0°) as shown in Fig. 15. And a single condenser microphone was mounted between the two microphones. It was installed for the comparison with DSE output. The reference microphone was set in front of speaker. We controlled the background noise with the driving speed. In the high and low noise condition, the speed of car was 80-100km/h and 40-60km/h, respectively.

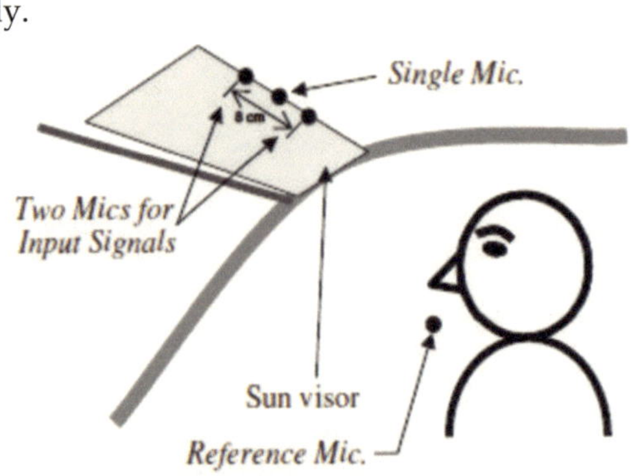

Fig. 15. The experiment setup in a car

For speech recognition test, we used the Hidden Markov Model Toolkit (HTK) 3.4 version as speech recognizer. HTK is a portable toolkit for building and manipulating hidden Markov models. HTK is primarily used for speech recognition research (http://htk.eng.cam.ac.uk/). We used 30 Korean phonemes word set for the experiments. The 30 words were composed of commands which were indispensable to use the telematics system. The speech recognition result is shown in Table 1. The speech recognition rate was decreased according as the background noise was increased.

Noise Type	Speech Recognition Rate
Low (low speed)	73.33
High (high speed)	58.83

Table 1. The speech recognition rate results : No pre-processing

We tested the DSE technology and source localization method using reliability measure. For evaluation, signal-to-noise ratio (SNR) and speech recognition rate were used. The SNR results are shown in table 2. The SNR for the low noise environment was increased from 9.5 to 18.5 and for the high noise from 1.8 to 14.9.
The increased performance of the DSE technology affected to the speech recognition rate. The speech recognition rate is shown in table 3 when the DSE technology was adopted. Without reliability measure, the speech recognition system for the high noise environment didn't give a good result as table 1. However the speech recognition rate was increased from 58.83 to 65.81 for the high noise environment when DSE technology was used.

Method	Low Noise	High Noise
Single Microphone	9.5	1.8
DSE w/o reliability measure	5.2	2.7
DSE with reliability measure	18.5	14.9

Table 2. SNR comparison results

Noise Type	Speech Recognition Rate
Low (low speed)	77.42
High (high speed)	65.81

Table 3. Speech recognition rate results : DSE pre-processing with reliability measure

5. Conclusions

We introduced a method of detecting a reliable DOA estimation result. The reliability measure indicates the prominence of the lobe of the cross-correlation value, which is used to find the DOA. We derived the waterbed effect in the DOA estimation and used this effect to calculate the reliability measure. To detect reliable results, we then used the maximum likelihood decision rule. By using the assumption of the Rayleigh distribution of reliability, we calculated the appropriate threshold and then eliminated the perturbed results of the

DOA estimates. We evaluated the performance of the proposed reliability measure in a fixed talker environment and a moving talker environment. Finally we also verified that DSE technology using this reliable DOA estimator would be useful to speech recognition system in a car environment.

6. References

S. Araki, H. Sawada, and S. Makino (2007). "Blind speech separation in a meeting situation with maximum SNR beamformers," *IEEE International Conference on Acoustics, Speech, and Signal Processing*, vol. I, p. 41-44.

M. Brandstein (1995). A Framework for Speech Source Localization Using Sensor Arrays, *Ph. D Thesis,* Brown University.

J. Chen, J. Benesty, and Y. Huang (2006). "Time delay estimation in room acoustic environments: An overview," *EURASIP Journal on Applied Signal Processing,* Vol. 2006, pp. 1-19.

J. Dibase (2000). A High-Accuracy, Low-Latency Technique for Talker Localization in reverberant Environments Using Microphone Arrays, *Ph. D Thesis,* Brown University.

M. Hayes (1996). Statistical Digital Signal Processing and Modeling, *John Wiley & Sons.*

H. Jeon, S. Kim, L. Kim, H. Yeon, and H. Youn (2007). "Reliability Measure for Sound Source Localization," IEICE Electronics Express, Vol.5, No.6, pp.192-197.

H. Jeon (2008). Two-Channel Sound Source Localization Method for Speech Enhancement System, *Ph. D Thesis,* Korea Advanced Institute of Science and Technology.

G. Lathoud (2006). Spatio-Temporal Analysis of Spontaneous Speech with Microphone Arrays, *Ph. D Thesis,* Ecole Polytechnique Fédérale de Lausanne.

J. Melsa, and D. Cohn (1978). Decision and Estimation Theory, *McGraw-Hill.*

A. Naguib (1996). Adaptive Antennas for CDMA Wireless Networks, *Ph. D Thesis,* Stanford University.

B. Ninness (2003). "The asymptotic CRLB for the spectrum of ARMA processes," IEEE Transactions on Signal Processing, Vol. 51, No. 6, pp. 1520-1531.

F. Schmitt, M. Mignotte, C. Collet, and P. Thourel (1996). "Estimation of noise parameters on SONAR images", *in SPIE International Society for Optical Engineering - Technical Conference on Application of Digital Image Processing XIX - SPIE'96* , Vol. 2823, pp. 1-12, Denver, USA.

P. Stoica, J. Li, and B. Ninness (2004). "The Waterbed Effect in Spectral Estimation," *IEEE Signal Processing Magazine,* Vol. 21, pp. 88-100.

10

Underwater Acoustic Source Localization and Sounds Classification in Distributed Measurement Networks

Octavian Adrian Postolache[1,2], José Miguel Pereira[1,2]
and Pedro Silva Girão[1]
[1]*Instituto de Telecomunicações,*
[2]*ESTSetúbal/IPS (LabIM)*
Portugal

1. Introduction

Underwater sound signals classification, localization and tracking of sound sources, are challenging tasks due to the multi-path nature of sound propagation, the mutual effects that exist between different sound signals and the large number of non-linear effects that reduces substantially the signal to noise ratio (SNR) of sound signals. In the region under observation, the Sado estuary, dolphins' sounds and anthropogenic noises are those that are mainly present. Referring to the dolphins' sounds, they can be classified in different types: narrow-band-frequency-modulated continuous tonal sounds, referred to as whistles, broadband sonar clicks and broadband burst pulse sounds.

The system used to acquire the underwater sound signals is based on a set of hydrophones. The hydrophones are usually associated with pre-amplifying blocks followed by data acquisition systems with data logging and advanced signal processing capabilities for sound recognition, underwater sound source localization and motion tracking. For the particular case of dolphin's sound recognition, dolphin localization and tracking, different practical approaches are reported in the literature that combine time-frequency representation and intelligent signal processing based on neural networks (Au et al., 2000; Wright, 2002; Carter, 1981).

This paper presents a distributed virtual system that includes a sound acquisition component expressed by 3 hydrophones array, a sound generation device, expressed by a sound projector, and two acquisition, data logging, data processing and data communication units, expressed by a laptop PC, a personal digital assistant (PDA) and a multifunction acquisition board. A water quality multiparameter measurement unit and two GPS devices are also included in the measurement system.

Several filtering blocks were designed and incorporated in the measurement system to improve the SNR ratio of the captured sound signals and a special attention was dedicated to present two techniques, one to locate sound signals' sources, based on triangulation, and other to identify and classify different signal types by using a wavelet packet based technique.

2. Main principles of acoustics' propagation

Sound is a mechanical oscillating pressure that causes particles of matter to vibrate as they transfer their energy from one to the next. These vibrations produce relatively small changes in pressure that are propagated through a material medium. Compared with the atmospheric pressure, those pressure variations are very small but can still be detected if their amplitudes are above the hearing threshold of the receiver that is about a few tenths of micro Pascal.

Sound is characterized by its amplitude (i.e., relative pressure level), intensity (the power of the wave transmitted in a particular direction in watts per square meter), frequency and propagation speed.

This section includes a short review of the basic sound propagation modes, namely, planar and spherical modes, and a few remarks about underwater sound propagation.

2.1 Plane sound waves

Considering an homogeneous medium and static conditions, i.e. a constant sound pressure over time, a stimulation force applied in YoZ plane, originates a plane sound wave traveling in the positive x direction whose pressure value, according to Hooke's law, is given by,

$$p(x) = -Y \cdot \varepsilon \tag{1}$$

where p represents the differential pressure caused the sound wave, Y represents the elastic modulus of the medium and ε represents the relative value of its mechanical deformation caused by sound pressure.

For time-varying conditions, there will be a differential pressure across an elementary volume, with a unitary transversal area and an elementary length dx, given by,

$$dp = \frac{\partial p(x,t)}{\partial x} \cdot dx \tag{2}$$

Using Newton's second law and the relationships (1) and (2), it is possible to obtain the relation between time pressure variation and the particle speed caused by the sound pressure,

$$\frac{\partial p}{\partial x} = -\rho \cdot \frac{\partial u(x,t)}{\partial t} \tag{3}$$

where ρ represents the density of the medium and u(x,t) represents the particle speed at a given point (x) and a given time instant (t).

Considering expressions (1), (2) and (3), it is possible to obtain the differential equation of sound plane waves that is expressed by,

$$\frac{\partial^2 p}{\partial t^2} = \frac{Y}{\rho} \cdot \frac{\partial^2 p}{\partial x^2} \tag{4}$$

where Y represents the elastic modulus of the medium and ρ represents its density.

2.2 Spherical sound waves

This approximation still considers a homogeneous and lossless propagation medium but, in this case, it is assumed that the sound intensity decreases with the square value of the

distance from sound source ($1/r^2$), that means, the sound pressure is inversely proportional to that distance ($1/r$).
In this case, for static conditions, the spatial pressure variation is given by (Burdic, 1991),

$$\nabla p = \frac{\partial p}{\partial x} \cdot \hat{u}_x + \frac{\partial p}{\partial y} \cdot \hat{u}_y \frac{\partial p}{\partial z} \cdot \hat{u}_z \tag{5}$$

where $\hat{u}_x$, $\hat{u}_y$ and $\hat{u}_z$ represent the Cartesian unit vectors and ∇ represents the gradient operator.
Using spherical polar coordinates, the sound pressure (p) dependents only on the distance between a generic point in the space (r, θ, ϕ) and the sound source coordinates that is located in the origin of the coordinates' system. In this case, for time variable conditions, the incremental variation of pressure is given by,

$$\frac{1}{r} \cdot \frac{\partial^2 (r \cdot p)}{\partial t^2} = \frac{\partial^2 p}{\partial t^2} \tag{6}$$

where r represents the radial distance between a generic point and the sound source.
Concerning sound intensity, for spherical waves in homogeneous and lossless mediums, its value decreases with the square value of the distance (r) since the total acoustic power remains constant across spherical surfaces.
It is important to underline that this approximation is still valid for mediums with low power losses as long as the distance from the sound source is higher than ten times the sound wavelength ($r>10 \cdot \lambda$).

2.3 Definition of some sound parameters

There are a very large number of sound parameters. However, according the aim of the present chapter, only a few parameters and definitions will be reviewed, namely, the concepts of sound impedance, transmission and reflection coefficients and sound intensity.
The transmission of sound waves, through two different mediums, is determined by the sound impedance of each medium. The acoustic impedance of a medium represents the ratio between the sound pressure (p) and the particle velocity (u) and is given by,

$$Z_m = \rho \cdot c \tag{7}$$

where, as previously, ρ represents the density of the medium and c represents the propagation speed of the acoustic wave that is, by its turn, equal to the product of the acoustic wavelength by its frequency ($c=\lambda \cdot f$).
Sound propagation across two different mediums depends on the sound impedance of each one, namely, on the transmission and reflection coefficients. For the normal component of the acoustic wave, relatively to the separation plane of the mediums, the sound reflection and transmission coefficients are defined by,

$$\Gamma_R = \frac{Z_{m1} - Z_{m2}}{Z_{m1} + Z_{m2}} \qquad \Gamma_T = \frac{2 \cdot Z_{m2}}{Z_{m1} + Z_{m2}} \tag{8}$$

where Γ_R and Γ_T represent the refection and transmission coefficients, and, Z_{m1} and Z_{m2}, represent the acoustic impedance of medium 1 and 2, respectively.
For spherical waves, the acoustic intensity that represents the power of sound signals is defined by,

$$I = \frac{1}{r^2} \cdot \frac{\left(p^2\right)_{av}}{\rho \cdot c} \tag{9}$$

where $(p^2)_{av}$ is the mean square value of the acoustic pressure for r=1 m and the others variables have the meaning previously defined. The total acoustic power at a distance r, from the sound source, is obtained by multiplying the previous result by the area of a sphere with radius equal r. The results that is obtained is given by,

$$P = 4\pi \cdot \frac{\left(p^2\right)_{av}}{2\rho \cdot c} \tag{10}$$

This constant value of sound intensity was expected since it is assumed a sound propagation in a homogenous lossless propagation medium.
In which concerns the sound pressure level, it is important to underline that this parameter represents, not acoustic energy per time unit, but acoustic strength per unit area. The sound pressure level (SPL) is defined by,

$$SPL = 20 \cdot \log_{10} (p/p_{ref}) \tag{11}$$

where the reference power (p_{ref}) is equal to 1 μPa for sound propagation in water or others liquids. Similarly, the logarithmic expression of sound intensity level (SIL) and sound power level (SL) are defined by,

$$\begin{aligned} I &= 10 \cdot \log_{10} (I/I_{ref})\, dB(SIL) \\ S_{WL} &= 10 \cdot \log_{10}(W / W_{ref}) \end{aligned} \tag{12}$$

where the reference values of intensity and power are given by $I_{ref}=10^{-12}$ W/m^2 and $W_{ref}=10^{-12}$ W, respectively.

2.4 A few remarks about underwater sound propagation

It should be noted that the speed of sound in water, particularly seawater, is not the same for all frequencies, but varies with aspects of the local marine environment such as density, temperature and salinity. Due mainly to the greater "stiffness" of seawater relative to air, sound travels approximately with a velocity (c) about 1500 m/s in seawater while in air it travels with a velocity about 340 m/s. In a simplified way it is possible to say that underwater sound propagation velocity is mainly affected by water temperature (T), depth (D) and salinity (S). A simple and empirical relationship that can be used to determine the sound velocity in salt water is given by (Hodges, 2010),

$$\begin{aligned} & c(T,S,DP) \cong A_1 + A_2 \cdot T + A_3 \cdot T^2 + A_4 \cdot T^3 + (B_1 - B_2 \cdot T) \cdot (S - C_1) + D_1 \cdot D \\ & [A_1, A_2, A_3, A_4] \cong [1449, 4.6, -0.055,\ 0.0003] \\ & [B_1, B_2, C_1, D_1] \cong [1.39, 0.012, 35, 0.017] \end{aligned} \tag{13}$$

where temperature is expressed in °C, salinity in expressed in parts per thousand and depth in m.
The sensitivity of sound velocity depends mainly on water temperature. However, the variation of temperature in low depth waters, that sometimes is lower than 2 m in river estuaries, is very small and salinity is the main parameter that affects sound velocity in estuarine salt waters. Moreover, salinity is estuarine zones depends strongly on tides and each sound monitoring measuring node must include at least a conductivity/salinity transducer to compensate underwater sound propagation velocity from its dependence on salinity (Mackenzi, 1981). As a summary it must be underlined that underwater sound transmission is a very complex issue, besides the effects previously referred, the ocean surface and bottom reflects, refracts and scatters the sound in a random fashion causing interference and attenuation that exhibit variations over time. Moreover, there are a large number of non-linear effects, namely temperature and salinity gradients, that causes complex time-variable and non-linear effects.

3. Spectral characterization of acoustic signals

Several MATLAB scripts were developed to identify and to classify acoustic signals. Using a given dolphin sound signal as a reference signal, different time to frequency conversion methods (TFCM) were applied to test the main characteristics of each one.

3.1 Dolphin sounds

In which concerns dolphin sounds (Evans, 1973; Podos et al., 2002), there are different types with different spectral characteristics. Between these different sound types we can refer whistles, clicks, bursts, pops and mews, between others.
Dolphin whistles, also called signature sounds, appear to be an identification sound since they are unique for each dolphin. The frequency range of these sounds is mainly contained in the interval between 200 Hz and 20 kHz (Reynolds et al., 1999). Clicks sounds are though to be used exclusively for echolocation (Evans, 1973). These sounds contains mainly high frequency spectral components and they require data acquisition systems with high analog to digital conversion rates. The frequency range for echolocation clicks includes the interval between 200 Hz and 150 kHz (Reynolds et al., 1999). Usually, low frequency clicks are used for long distance targets and high frequency clicks are used for short distance targets. When dolphins are closer to an object, they increase the frequency used for echolocation to obtain a more detailed information about the object characteristics, like shape, speed, moving direction, and object density, between others. For long distance objects low frequency acoustic signals are used because their attenuation is lower than the attenuation that is obtained with high frequency acoustic signals. By its turn, burst pulse sounds that include, mainly, pops, mews, chirps and barks, seem to be used when dolphins are angry or upset. These signals are frequency modulated and their frequency range includes the interval between 15 kHz and 150 kHz.

3.2 Time to frequency conversion methods

As previously referred, in order to compare the performance of different TFCM, that can be used to identify and classify dolphin sounds a dolphin whistle sound will be considered as reference. In which concerns signals' amplitudes, it makes only sense, for classification purposes, to used normalized amplitudes. Sound signals' amplitudes depend on many factors, namely on the distance between sound sources and the measurement system, being this distance variable for moving objects, for example dolphins and ships. A data acquisition sample rate equal to 44.1 kS/s was used to digitize sound signals and the acquisition period was equal to 1 s. Figure 1 represents the time variation of the whistle sound signal under analysis.

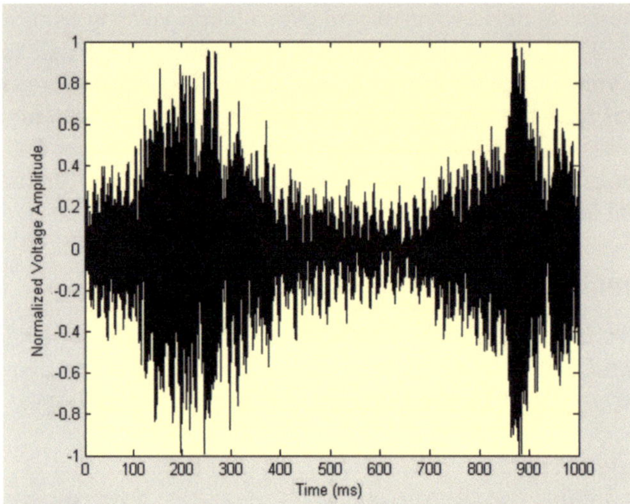

Fig. 1. Time variation of the dolphin whistle sound signal under analysis

Fourier time to frequency conversion method

The first TFCM that will be considered is the Fourier transform method (Körner, 1996). The complex version of this time to frequency operator is defined by,

$$X(f) = \int_{-\infty}^{+\infty} x(t) \cdot e^{-j2\pi \cdot f \cdot t} dt \tag{14}$$

where x(t) and X(f) represent the signal and its Fourier transform, respectively.

The results that are obtained with this FTCM don't give any information about the frequency contents of the signal over time. However, some information about the signal bandwidth and its spectral energy distribution can be accessed. Figure 2 represents the power spectral density (PSD) of the sound signal represented in figure 1. As it is clearly shown, the PSD of the signal exhibits two peaks, one around 2.8 kHz and the other, with higher amplitude, is a spectral component whose frequency is approximately equal to 50 Hz. This spectral component is caused by the mains power supply and can be strongly attenuated, almost removed, by hardware or digital filtering.

It is important to underline that this FTCM is not suitable for non-stationary signals, like the ones generated by dolphins.

Short time Fourier transform method

Short time Fourier transform (STFT) is a TFCM that can be used to access the variation of the spectral components of a non-stationary signal over time. This TFCM is defined by,

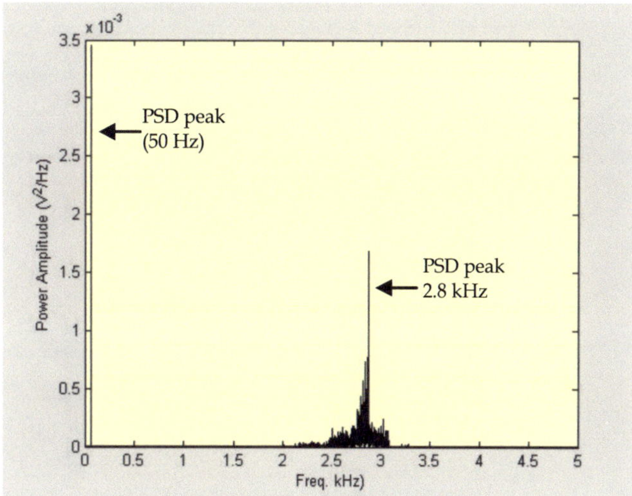

Fig. 2. Power spectral density of the dolphin whistle sound signal

$$X(t,f) = \int_{-\infty}^{+\infty} x(t) \cdot w(t-\tau) \cdot e^{-j2\pi \cdot f \cdot t} dt \quad t \in \Re \tag{15}$$

where x(t) and X(t,f) presents the signal and its STFT, respectively, and w(t) represents the time window function that is used the evaluate the STFT. With this TFCM it is possible to obtain the variation of the frequency contents of the signal over time. Figure 3 represents the spectrogram of the whistle sound signal when the STFT method is used. The spectogram considers a window length of 1024 samples, an overlap length of 128 samples and a number of points that are used for FFT evaluation, in each time window, equal to 1024.
However, the STFT of a given signal depends significantly on the parameters that are used for its evaluation. Confirming this statement, figure 4 represents the spectrogram of the whistle signal obtained with a different window length, in this case equal to 64 samples, an overlap length equal to 16 samples and a number of points used for FFT evaluation, in each time interval, equal to 64. In this case, it is clearly shown that different time and frequency resolutions are obtained. The STFT parameters previously referred, namely, time window length, number of overlapping points, and the number of points used for FFT evaluation in each time window, together with the time window function, affect the time and frequency resolution that are obtained. Essentially, if a large time window is used, spectral resolution is improved but time resolution gets worst. This is the main drawback of the STFT method, there is a compromise between time and frequency resolutions. It is possible to demonstrate (Allen & Rabiner, 1997; Flandrin, 1984) that the constraint between time and frequency resolutions is given by,

$$\Delta f \geq \frac{1}{4\pi \cdot \Delta t} \tag{16}$$

where Δf and Δt represent the frequency and time resolutions, respectively.

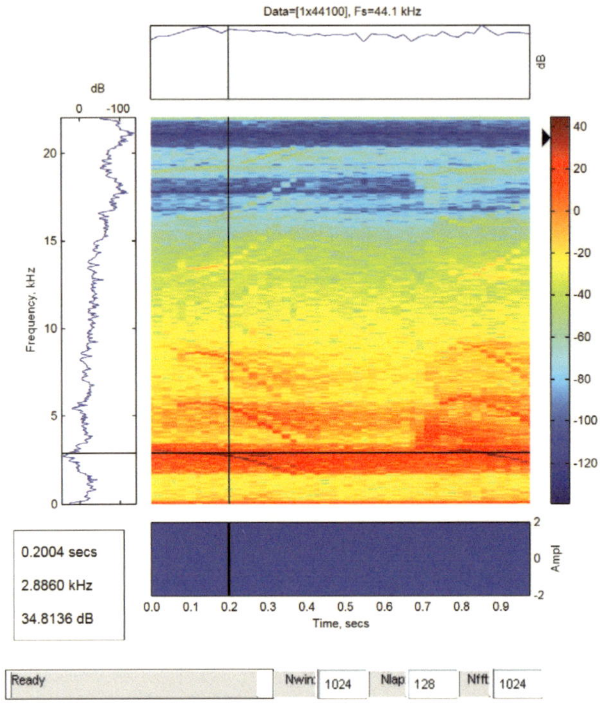

Fig. 3. Spectogram of the whistle sound signal (window length equal to1024 samples, overlap length equal to 128 samples and a number of points used for FFT evaluation equal to 1024)

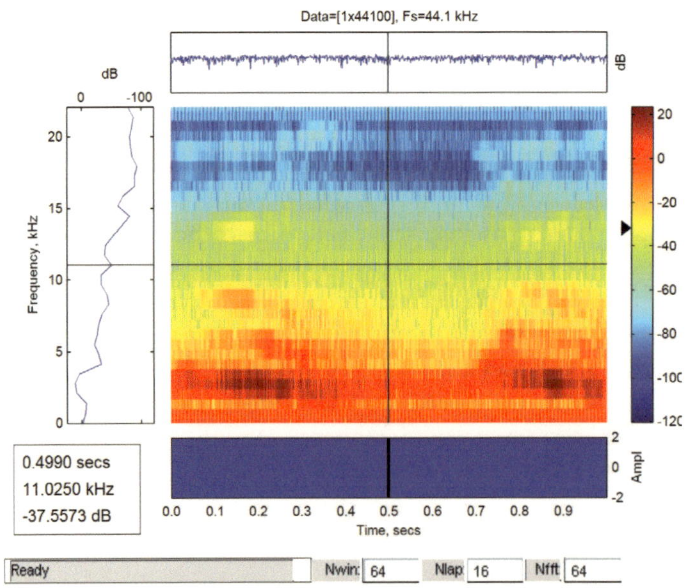

Fig. 4. Spectogram of the whistle sound signal (window length equal to 64 samples, overlap length equal to 16 samples and a number of points used for FFT evaluation equal to 64)

Time to frequency conversion methods based on time-frequency distributions

When the signal exhibits slow variations in time, and there is no hard requirements of time and frequency resolutions, the STFT, previously described, gives acceptable results. Otherwise, time-frequency distributions can be used to obtain a better spectral power characterization of the signal over time (Claasen & Mecklenbrauker, 1980; Choi & Williams, 1989). A well know case of these methods is the Choi-Williams time to frequency transform that is defined by,

$$X(t,f) = \int_{-\infty}^{+\infty} e^{-j2\pi 2\pi} \int_{-\infty}^{+\infty} \sqrt{\sigma/4\pi \cdot \tau^2} \cdot e^{-\frac{\sigma(\mu-t)^2}{4\tau^2}} \cdot x(\mu+\tau/2) \cdot x^*(\mu-\tau/2) \cdot d\mu \cdot d\tau \quad (17)$$

where $x(\mu+\tau/2)$ represents the signal amplitude for a generic time t equal to $\mu+\tau/2$ and the exponential term is the distribution kernel function that depends on the value of σ coefficient. The Wigner-Ville distribution (WVD) time to frequency transform is a particular case of the Choi-Williams TFCM that is obtained when $\sigma \to \infty$, and its time to frequency transform operator is defined by,

$$X(t,f) = \int_{-\infty}^{+\infty} e^{-j2\pi 2\pi} \cdot x(\mu+\tau/2) \cdot x^*(\mu-\tau/2) d\tau \quad (18)$$

These TFCM could give better results in which concerns the evaluation of the main spectral components of non-stationary signals. They can minimize the spectral interference between adjacent frequency components as long as the distributions kernel function parameters' are properly selected. These TFCM provide a joint function of time and frequency that describes the energy density of the signal simultaneously in time and frequency. However, Choi-Williams and WVD TCFM based on time-frequency distributions depends on non-linear quadratic terms that introduce cross-terms in the time-frequency plane. It is even possible to obtain non-sense results, namely, negative values of the energy of the signal in some regions of the time-frequency plane. Figure 5 represents the spectrogram of the whistle sound signal calculated using the Choi-Williams distribution. The graphical representation considers a time window of 1 s, a unitary default Kernel coefficient (σ=1), a time smoothing window (Lg) equal to 17, a smoothing width (Lh) equal to 43 and a representation threshold equal to 5 %.

Wavelets time to scale conversion method

Conversely to others TFCM that are based on Fourier transforms, in this case, the signal is decomposed is multiple components that are obtained by using different scales and time shifts of a base function, usually known as the mother wavelet function. The time to scale wavelet operator is defined by,

$$X(\tau(\alpha) = \int_{-\infty}^{+\infty} x(t) \cdot \alpha^{0.5} \cdot \psi\left(\alpha\overset{\sim 1}{(t-\tau)}\right) dt \quad (19)$$

where ψ is the mother wavelet function, α and τ are the wavelet scaling and time shift coefficients, respectively.

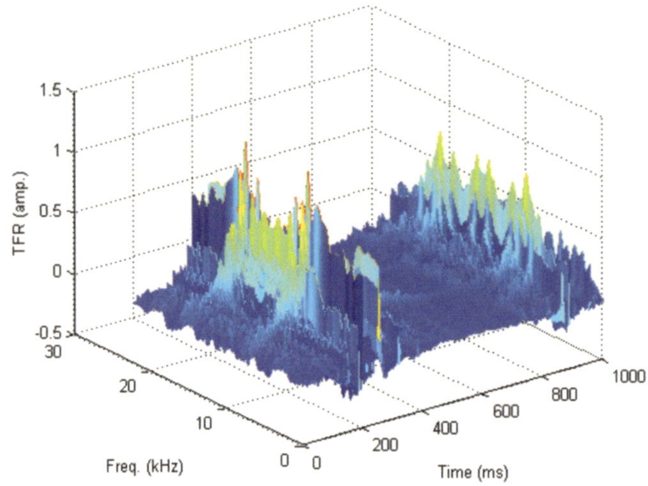

Fig. 5. Spectrogram of the whistle sound signal using the Choi-Williams distribution (time window=1 s, a unitary default Kernel coefficient, time smoothing window=17, a smoothing width=43)

It is important to underline that the frequency contents of the signal is not directly obtained from its wavelet transform (WT). However, as the scale of the mother wavelet gets lower, a lower number of signal's samples are contained in each scaled mother wavelet, and there the WT gives an increased knowledge of the high frequency components of the signal.

In this case, there is no compromise between time and frequency resolutions. Moreover, wavelets are particularly interesting to detect signals' trends, breakdowns and sharp peaks variations, and also to perform signals' compressing and de-noising with minimal distortion.

Figure 6 represents the scalogram of the whistle sound signal when a Morlet mother wavelet with a bandwidth parameter equal to 10 is used (Cristi, 2004; Donoho & Johnstone, 1994). The contour plot uses time and frequency linear scales and a logarithmic scale, with a dynamic range equal to 60 dB, to represent scalogram values. The scalogram was evaluated with 132 scales values, 90 scales between 1 and 45.5 with 0,5 units' increments, an 42 scales between 46 and 128 with 2 units' increments.

The scalogram shows clearly that the main frequency components of the whistle sound signal are centered on the amplitude peaks of the signal, confirming the results previously obtained with the Fourier based TFCM.

3.3 Anthropogenic sound signals

In which concerns underwater sound analysis it is important to analyze anthropogenic sound signals because they can disturb deeply the sounds generated by dolphins' sounds.

Anthropogenic noises are ubiquitous, they exist everywhere there is human activities. The powerful anthropogenic power sources come from sonars, ships and seismic survey pulses. Particularly in estuarine zones, noises from ships, ferries, winches and motorbikes, interfere with marine life in many ways (Holt et al., 2009).

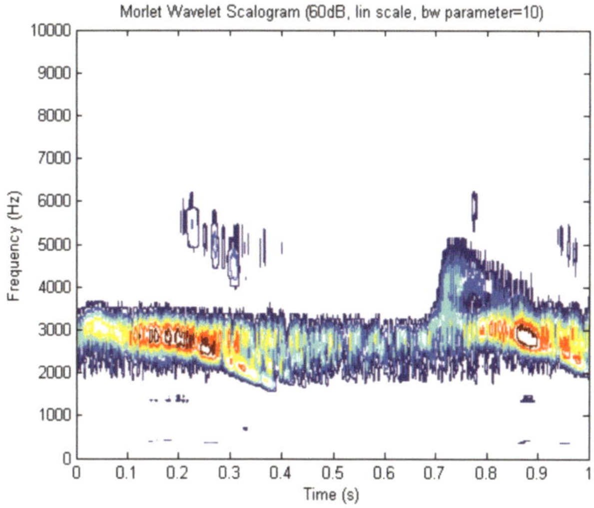

Fig. 6. Scalogram of the whistle sound signal when a Morlet mother wavelet with a bandwidth parameter equal to 10 is used

Since the communication between dolphins is based on underwater sounds, anthropogenic noises can originate an increase of dolphin sounds' amplitudes, durations and repetition rates. These negative effects happen, particularly, whenever anthropogenic noises frequencies overlap the frequency bandwidth of the acoustic signals used by dolphins. It is generally accepted that anthropogenic noises can affect dolphins' survival, reproduction and also divert them from their original habitat (NRC, 2003; Oherveger & Goller, 2001).

Assuming equal amplitudes of dolphin and anthropogenic sounds, it is important to know their spectral components. Two examples of the time variations and scalograms of anthropogenic sounds signals will be presented. Figures 7 and 8 represent the time variations and the scalograms of a ship-harbor and submarine sonar sound signals, respectively. As it is clearly shown, both signals contain spectral components that overlap the frequency bandwidth of dolphin sound signals, thus, affecting dolphins' communication and sound signals' analysis.

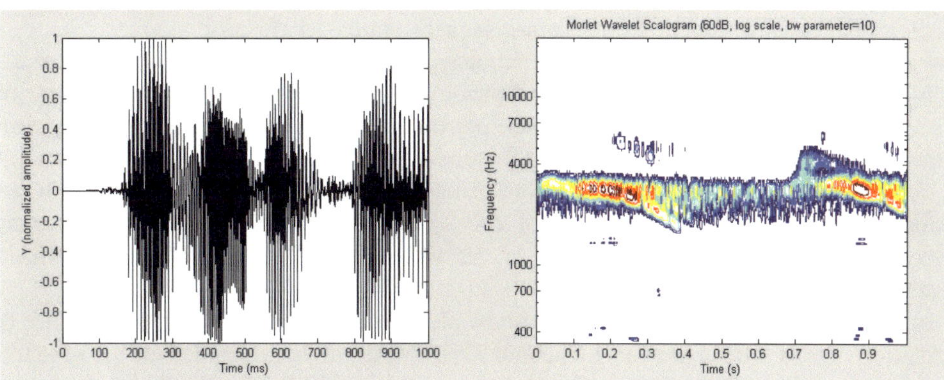

(a) (b)

Fig. 7. Ship-harbour signal: (a) time variation and (b) scalogram

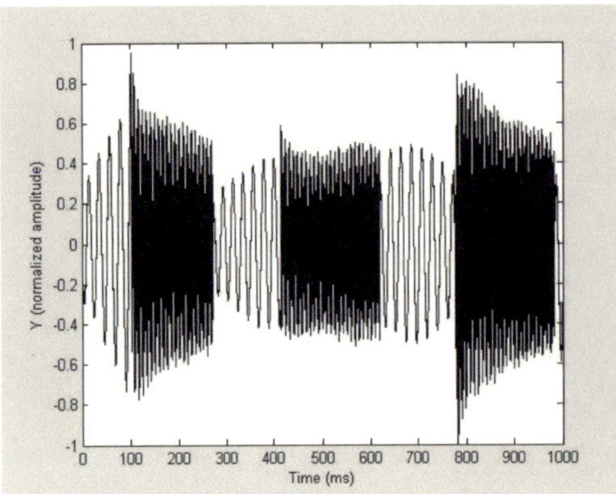

(a)

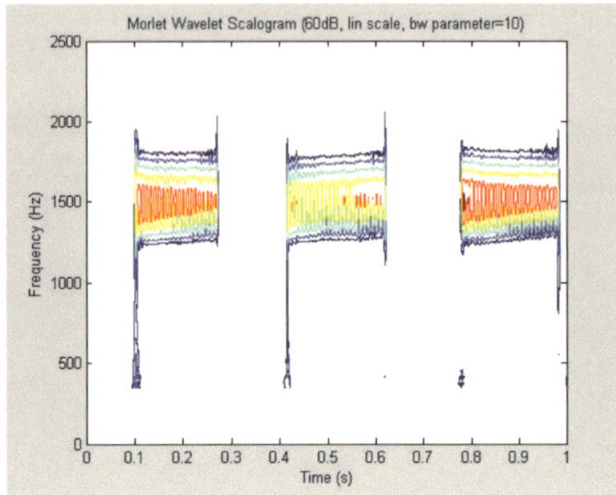

(b)

Fig. 8. Submarine sonar signal: (a) time variation and (b) scalogram

4. Measurement system

The measurement system includes several measurement units than can, by its turn, be integrated in a distributed measurement network, with wireless communication capabilities (Postoloche et al., 2006). Each measurement unit, whose description will be considered in the present section, includes the acoustic devices that establish the interface between the electrical devices and the underwater medium, a water quality measurement unit that is used for environmental assessment purposes, and the signal conditioning, data acquisition and data processing units.

4.1 Hardware

Figure 9 represents the intelligent distributed virtual measurement system that was implemented for underwater sound monitoring and sound source localization. The system includes two main units: a base unit, where the acoustic signals are detected and digitized, and a remote unit that generates testing underwater acoustic signals used to validate the implemented algorithms for time delay measurement (Carter, 1981; Chan & Ho, 1994), acoustic signal classification and underwater acoustic source localization.

A set of three hydrophones (Sensor Technology, model SS03) are mounted on a 20m structure with 6 buoys that assure a linear distribution of the hydrophones. The number and the linear distribution of the hydrophones permit to implement a hyperbolic algorithm (Mattos & Grant, 2004; Glegg et al., 2001) for underwater acoustic source localization, and also to perform underwater sound monitoring tasks including sound detection and classification. The main characteristics of the hydrophones includes a frequency range between 200 Hz and 20 kHz, a sensitivity of -169 dB relatively to 1 V/μPa and a maximum operating depth of 100 m.

The azimuth angle (φ) obtained from hydrophone array structure, together with the information obtained from the GPS1 (Garmin 75GPSMAP) device, installed on the base unit, and the information obtained from the fluxgate compass (SIMRAD RFC35NS) device, are used to calculate the absolute position of the remote underwater acoustic source. After the estimation of the underwater acoustic source localization, a comparison with the values given by the GPS2 is carried out to validate the performance of the algorithms that are used for sound source localization.

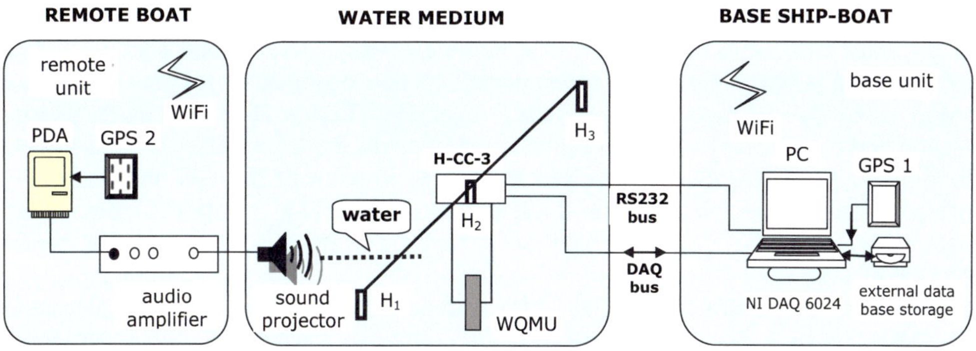

Fig. 9. The architecture of the distributed virtual system for underwater acoustic signal acquisition, underwater sound source localization and sound analysis (H1,H2,H3- hydrophones, H-CC-3- channels' conditioning circuits, WQMU- water quality measurement unit, NI DAQCard-6024 - multifunction DAQCard, GPS1 and GPS2- remote and base GPS units).

As it is presented in figure 9, a three-channel hydrophones' conditioning circuit (H-CC) provides the analog voltage signals associated with the captured sounds. These signals are acquired using three analog channels' inputs (ACH0, ACH1 and ACH2) of the DAQCard using a data acquisition rate equal to 44.1S/s. The azimuth angle information, expressed by $V \cdot \sin\varphi$ and $V \cdot \cos\varphi$ voltages delivered by the electronic compass, is acquired using ACH3 and ACH4 channels of the DAQCard.

The water quality parameters, temperature and salinity, are acquired using a multiparameter Quanta Hydrolab unit (Eco Environmental Inc.) that is controlled by the laptop PC trough a RS232 connection. During system's testing phase, acoustic signals generation is triggered through a Wi-Fi communication link that exists between the PC and the PDA, or by a start-up table that is stored in the PDA and in the PC. Thus, at pre-defined time instants, a specific sound signal is generated by the sound projector (Lubell LL9816) and it is acquired by the hydrophones. The acquisition time delays are then evaluated and localization algorithms, based on the time difference of arrivals (TDOA), are used to locate sound sources. The main characteristics of the sound projector includes a frequency range (±3 dB) between 200 Hz and 20 kHz, a maximum SPL of 180 dB/μPa/m at a frequency equal to 1 kHz and a maximum cable voltage-to-current ratio equal to 20 V_{rms}/3 A.

Temperature and salinity measurements, obtained for the WQMU (Postolache et al, 2002; Postolache et al., 2006; Postolache et al., 2010) are used to compensate sound source localization errors caused by underwater sound velocity variations (13).

4.2 Software

System's software includes two mains parts. One is related with dolphin sounds classification and the other is related with the GIS (Postolache et al., 2007). Both software parts are integrated in a common application that simultaneously identify sound sources and locate them in the geographical area under assessment. In this way, it is possible to locate and pursue the trajectory of moving sound sources, particularly dolphins in a river estuary.

4.2.1 Dolphin sounds classification based on wavelets packets

This software part performs basically the following tasks: hydrophone channel voltage acquisition and processing, fluxgate compass voltage data acquisition and processing, noise filtering, using wavelet threshold denoising (Mallat, 1999; Guo et al., 2000), digital filtering, detection and classification of sound signals. Additional software routines were developed to perform data logging of the acquired signals, to implement the GIS and to perform geographic coordinates' analysis based on historical data. The laptop PC software was developed in LabVIEW (National Instruments) that, by its turn, includes some embedded MATLAB scripts.
The generation of the acoustic signals, at the remote unit, is controlled by the distributed LabVIEW software (laptop PC software and PDA software). The laptop software component triggers the sound generation by sending to the PDA a command using the TCP/IP client-server communication's protocol. The sound type (e.g. dolphin's whistle) and its time duration are defined using a specific command code.
In which concerns the underwater acoustic analysis, the hydrophones' data is processed in order to extract the information about the type of underwater sound source by using a wavelet packet (WP) signal decomposition and a set of neural network processing blocks.
Features' extraction of sound signals is performed using the root mean square (RMS) values of the coefficients that are obtained after WP decomposition (Chang & Wang, 2003). Based on the WP decomposition it is possible to obtain a reduced set of features parameters that characterize the main type of underwater sounds detected in the monitored area.
It is important to underline that conversely to the traditional wavelet decomposition method, where only the approximation portion of the original signal is split into successive approximation and details, the proposed WP decomposition method extends the capabilities of the traditional wavelet decomposition method by decomposing the detail part as well. The complete decomposition tree for a three level WP decomposition is represented in Fig. 10.

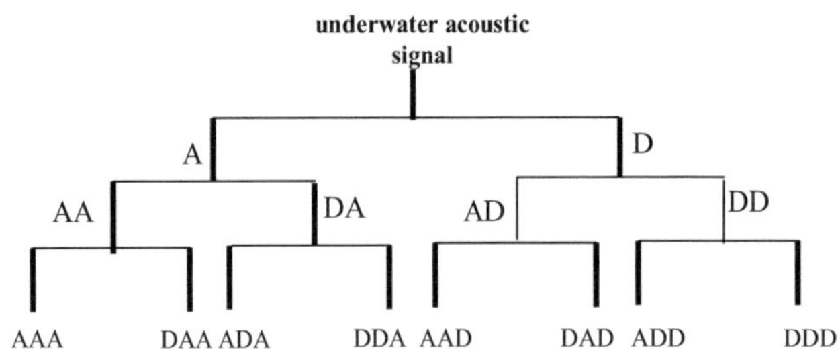

Fig. 10. Decomposition tree for a three level WP decomposition (D-details associated with the high-pass decimation filter, A-approximations associated with the low pass decimation filter)

4.2.2 Geographic Information System: their application to locate sound sources

This software part implements the GIS and provides a flexible solution to locate and pursue moving sound sources. The main components, included in this software part, are the hyperbolic bearing angle and range algorithms, both related with the estimation of sound sources' localizations.

In order to transform the relative position coordinates, determined by the system of hydrophones (Hurell & Duck, 2000), into absolute position coordinates, it is necessary to transform the GPS data, obtained from Garmin GPSMAP76, into a cartographic representation system. The mapping scale used to represent the geographical data is equal to 1/25000. This scale value was selected taking into account the accuracy of the GPS device that was used for testing purposes. The conversion from relative to absolute coordinates is performed in three steps: Molodensky (Clynch, 2006) three-dimensional transformation, Gauss-Krüger (Grafarend & Ardalan, 1993) projection and, finally, absolute positioning calculation. In the last step, a polar to Cartesian coordinates' conversion is performed considering the water surface as a reference plane (XoY), and defining the direction of X-axis by using the data provided by the electronic compass.

Figure 11 is a pictorial representation of the geometrical parameters that are used to locate underwater acoustic sources.

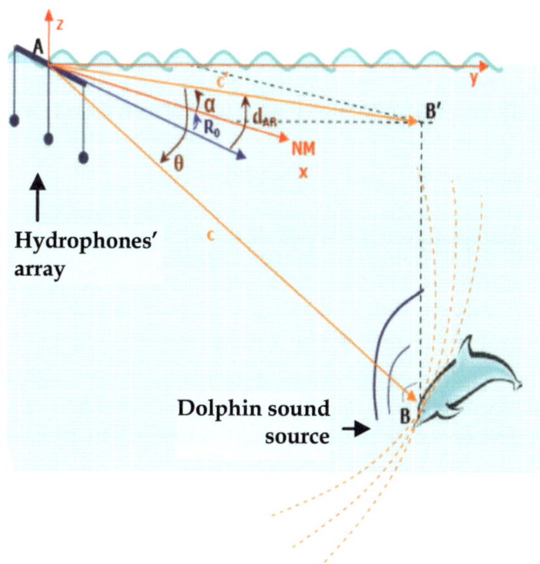

Fig. 11. Geometrical parameters that are used to locate underwater acoustic sources

The main software tasks performed by the measurement system are represented in figure 12.

Finally, it is also important to refer that the underwater sound source localization is calculated and displayed on the user interface together with some water quality parameters, namely, temperature, turbidity and conductivity, that are provided by the WQMU.

Future software developments can also provide improved localization accuracy by profiling the coverage area into regions where multiple measurements results of reference sound sources are stored in an allocation database. The best match between localization measurement data and the data that is stored in the localization database is determined and then interpolation can be used to improve sound source localization accuracy.

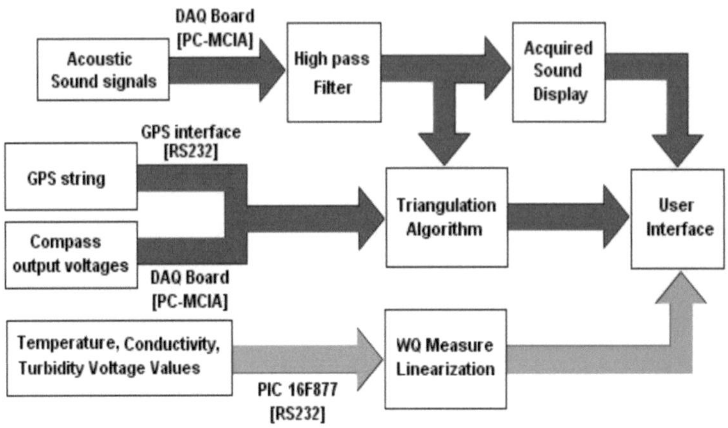

Fig. 12. Measurement system software block diagram

5. Experimental results

To evaluate de performance of the proposed measurements system two experimental results will be presented. The first one is related with the capabilities provided by the sound source localization algorithms, and second one is related with the capabilities of wavelet based techniques to detect and classify dolphin sounds.

5.1 Sound source localization

Several laboratory experiments were done to test the different measuring units including the WQMU. Field tests, similar to the ones previously performed in laboratory, were performed in Sado estuary. During field test of the measurement system, dolphins were sighted but none produced a clear sound signal that could be acquired or traced to the source. In order to fill this gap, a number of experiments took place involving pre-recorded dolphin sounds. For sound reproduction an underwater sound projector was used, which allowed the testing of the sound source localization algorithms. The sound projector, installed in a second boat (remote boat), away from the base ship-boat, was moved way from the hydrophones' structure and several pre-recorded sounds were played. This methodology was used to test the performance of the hydrophones' array structure and the TDOA algorithm to measure the localization of the sound source for different values of distances and azimuth angles. Using the time delay values between the sounds captured by the hydrophones and the data obtained from the electronic compass (zero heading), sounds can be traced to their sources.
Table I represents the localization errors that are expected to estimate the localization of the sound source as a function of the angle resolution, that can be defined by the electronic compass, and by the distance between the hydrophones' and the position of the sound source. The data contained in the table considers that the distance from the hydrophones to sound source is always lower than 500 m.
As it can be verified, in order to obtain the desired precision, characteristic of a 1/25000 scale representation, the resolution in the zero heading acquisition angle, for distances lower than 500 m, cannot be greater then ±½ degree. Experimental results that were performed using GPS_1 and GPS_2 units gave an absolute error lower 10 m. This value is in accordance

with the SimRad RFC35NS electronic compass characteristics whose datasheet specifies an accuracy better than 1° and repeatability equal to ±0.5°.

		Distance (m)					
		10	50	100	150	300	500
Angle(°)	0,25	0,04	0,22	0,44	0,65	1,31	2,18
	0,5	0,09	0,44	0,87	1,31	2,62	4,36
	1	0,17	0,87	1,75	2,62	5,24	8,73
	2,5	0,44	2,18	4,36	6,54	13,09	21,81
	5	0,87	4,36	8,72	13,07	26,15	43,58
	10	1,74	8,68	17,36	26,05	52,09	86,82

Table 1. Localization errors as a function of the zero heading angle accuracy and of the distance between the hydrophones and the sound source

5.2 Wavelet based classification of dolphin sounds

The WP based method that was used to identify and classify sound signals enables a large flexibility to choice the best combination of the WP features that are used for detection and classification purposes.

During the design of the features extraction algorithm, different levels of decomposition, varying between 2 and five, and different sounds' periods, varying between 30 ms and 1000 ms, were used. Referring the wavelet packet decomposition, used for underwater sound features' extraction, a practical approach concerning the choice of the best level of decomposition and mother wavelet function was carried out. Thus, the capabilities of Daubechies, Symlets and Coiflets functions as mother wavelets were tested.

For the studied cases, the RMS of the WP coefficients for different bands of interests were evaluated. As an example, figure 13 represents the features' values obtained with a three level decomposition tree and a db1 mother wavelet, when different sound types are

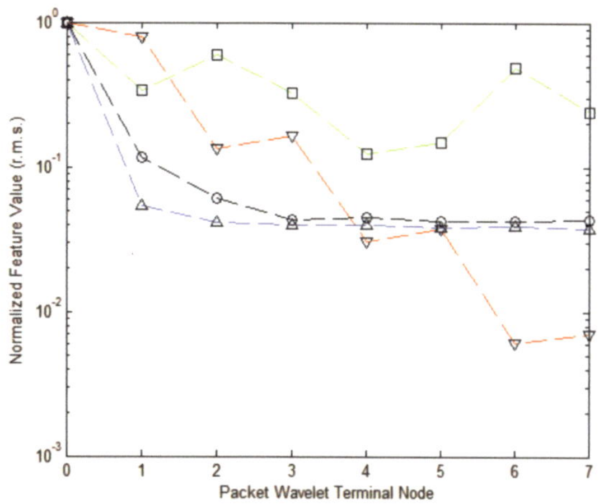

Fig. 13. Wavelet based feature extraction using a three level decomposition tree (green line- dolphin chirp, red line- dolphin whistle, blue line- motorbike, black line- ping sonar).

considered, namely, a dolphin chirp, a dolphin whistle, and two anthropogenic noises, in this case, a water motorbike and a ping sonar sound. As it can be easily verified, the features' values are significantly different for a third level WP decomposition.
Figure 14 represents the WP coefficients for a dolphin whistle when it is used a six level WP decomposition tree. In this case, the number of terminal nodes is higher, 64 instead of 8, but the features' variation profile over packet wavelet terminal nodes have a similar pattern and the sound classification performance is better. As expected, there is always a compromise between the data processing load and the sounds' classification performance.

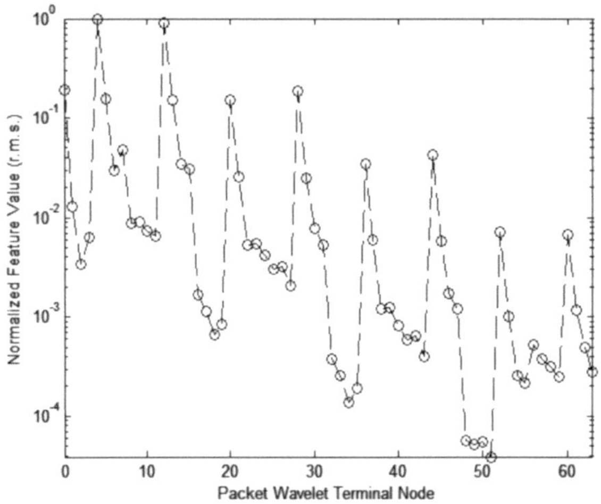

Fig. 14. Wavelet based feature extraction of a dolphin whistle using a six level decomposition tree

Neural Network Classifier

The calculated features for different types of sound signals, real or artificially generated by the sound projector located in the remote unit, are used to train a neural network sound classifier (NN-SC) characterized by a Multilayer Perceptron architecture with 8 neurons in the input layer, a set of 10 to 20 neurons in the hidden layer and one or more neurons (n_{out}) in the output layer (Haykin, 1994). Each input neuron collects one of the 8 features obtained from the WP decomposition. During the training phase of NN-SC the target vector elements are defined according the different sound types used for training purposes. For a similar sound type, for example dolphin whistles, the target vectors' values are within a pre-defined interval. It is important to underline that all values that are used in NN-SC are normalized to its maximum amplitude in order to improve sound identification performance of the neural network.
The number of output neurons (n_{out}) of the NN-SC depends on the number of different signal types to identify. Thus, in the simplified case of the detection of dolphins sounds, independently of their types, the NN-SC uses a single neuron in the output layer with two separated features' range intervals that corresponds to "dolphin sound" and "no dolphin sound" detected, respectively. In order to identify different sound sources and types, it is required more than two features' ranges. This is the case when it is required to classify different sound types, like dolphins bursts, whistles and clicks, or other anthropogenic

sounds, like water motorbikes, ship sound or other underwater noise sounds. Figure 15 represents the NN-SC features' range amplitudes that are used for sound classification.

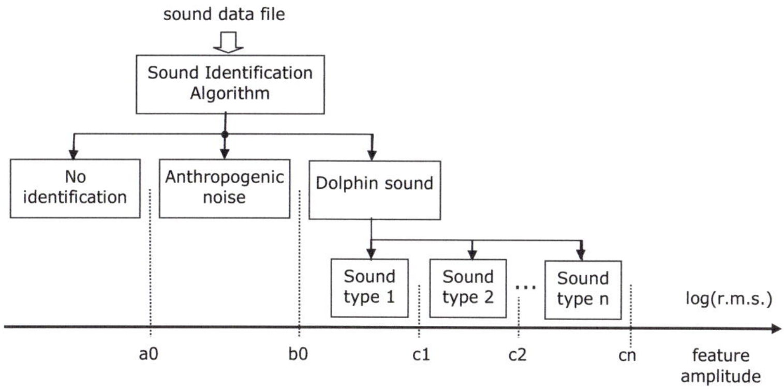

Fig. 15. NN-SC features' amplitudes that are used for sound classification

To test the performance of the sound classification algorithm the following features' amplitudes were considered: between 0.1 and 0.3 for anthropogenic noise sounds; between 0.5 and 0.7 for dolphin whistles and between 0.9 and 1.1 for dolphin chirps. When features' amplitudes are outside the previous intervals there is no sound identification. This happens if the NN-SC gives an erroneous output or if the training set is reduced in which concerns the number of different sound types to be identified.

Figure 16 represents NN-SC normalized output values that are obtained for a third level wavelet packet decomposition, a training and validation sets with 16 elements (sound signals), each one, a root-mean-square training parameter goal equal to 10^{-5}, a number of hidden layer neurons equal to eight and when the Levenberg-Marquardt minimization algorithm is used to evaluate the weights and bias of each ANN neuron.

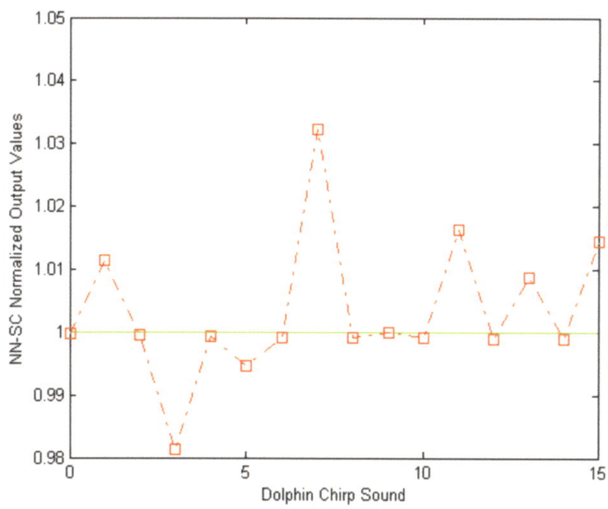

Fig. 16. NN-SC normalized output values that were obtain using a third level wavelet packet decomposition ANN classifiers

The results that were obtained present a maximum relative error almost equal to 3.23 % and the standard deviation of the errors values is almost equal to 1.14 %. Since, in this example, all dolphin sounds' features are in the range between 0.9 and 1.1, we conclude that there is no classification error. Different tests were performed with others sound types and it was verified a very good performance, higher than 95 % of right classifications, as long as there is no NN-SC features' identification ranges overlapping.

6. Conclusions

This chapter includes a review of sound propagation principles, the presentation of different TFCM that can be used to represent frequency contents of non-stationary signals and, finally, the presentation of a measurement system to acquire and process measurement data.
In which concerns sound propagation principles, a particular attention was dedicated to the characterization of plane and spherical sound propagation modes and to the definition of power related acoustic parameters. Some details about underwater sound propagation are presented, particularly the ones that affect sound propagation speed. Variations of sound propagation speed in estuarine waters, where salinity can exhibit large variations, must be accounted to minimize measurement errors of sound sources' localizations.
Referring to TFCM a particular attention was dedicated to short time to frequency transforms and wavelet characterization of underwater sounds. Several examples of the application of these methods to characterize dolphin sounds and anthropogenic noises were presented.
Several field tests of the measurement system were performed to evaluate its performance for sound signals' detection and classification, and to test its capability to locate underwater sound sources. To validate triangulation algorithms an underwater sound projector and an array of hydrophones were used to obtain a large set of measurement data. Using the GPS coordinates of the sound projector, located in a remote boat, and the GPS coordinates of the base ship-boat, where the hydrophones' array and data acquisition units are located, validation of relative and absolute localization of sound sources were performed for distances lower than 500 m and for a frequency range between 200 Hz and 20 kHz. The measurement system also includes data logging and GIS capabilities. The first capability is important to evaluate changes over time in dolphin's habitats and the second one to locate and pursue the trajectory of moving sound sources, particularly dolphins in a river estuary.
In which concerns the detection and classification of underwater acoustic sounds, a wavelet packet technique, based on a third level decomposition and on a RMS features' extraction of the terminal nodes' coefficients, followed by an ANN classification method, is proposed. The classification results that were obtained present a maximum relative error almost equal to 3.23 % and a standard deviation, of the error values, almost equal to 1.14 %.
Further tests are required to evaluate the sound detection and classification algorithms when different sound sources interfere mutually, particularly when dolphin sounds are mixed with anthropogenic noises.

7. References

Allen, J. & Rabiner, L. (1977). "A Unified Approach to Short-Time Fourier Analysis and Synthesis" Proc. IEEE, Vol. 65, No. 11, pp. 1558-64, 1977

Au, W.; Popper, A.N & Fay, R.F. (2000). "Hearing by Whales and Dolphins", New York: Springer-Verlag

Burdic, W.S. (1991). "Underwater Acoustic System Analysis", 2nd edition, Prentice Hall, Inc., Peninsula Publishing, California, U.S.A., 1991

Carter (1981). "Time Delay Estimation for Passive Signal Processing", IEEE Transactions on Acoustics, Speech and Signal Processing, Vol. ASSP-29, No.3, pp. 463-470, 1981

Chang, S.H. & Wang, F. (2003). "Underwater Sound Detection based on Hilbert Transform Pair of Wavelet Bases", Proceedings OCEANS'2003, pp.1680-1684, San Diego, USA, 2003

Choi, H. & Williams, W.J. (1989). "Improved Time-Frequency Representation of Multicomponent Signals Using Exponential Kernels," IEEE Trans. ASSP, Vol. 37, No. 6, pp. 862-871, June 1989

Claasen, T. & Mecklenbrauker, W. (1980). "The Wigner Distribution - A Tool for Time-Frequency Signal Analysis" 3 parts Philips J. Res., Vol. 35, No. 3, 4/5, 6, pp. 217-250, 276-300, 372-389, 1980

Clynch, J.R. (2006). "Datums - Map Coordinate Reference Frames
Part 2 - Datum Transformations", Feb. 2006 (available at
http://www.gmat.unsw.edu.au/snap/gps/clynch_pdfs/Datum_ii.pdf).

Cristi, R. (2004). "Modern Digital Signal Processing", Thompson Learning Inc., Brooks/Cole, 2004

Donoho, D.L & Johnstone, I.M. (1994). "Ideal Spatial Adaptation by Wavelet Shrinkage", Biometrika, Vol 81, pp. 425-455, 1994.

Evans, W.E. (1973). "Echolocation by marine dauphines and one species of fresh-water dolphin", J. Acoust. Soc. Am. 54:191-199, 1973

Eco Environmental, "Hydrolab Quanta" (available at http://www.ecoenvironmental.com.au/eco/water/hydrolab_quantag.htm)

Flandrin, P. (1984). "Some Features of Time-Frequency Representations of Multi-Component Signals" IEEE Int. Conf. on Acoust. Speech and Signal Proc., pp. 41.B.4.1-41.B.4.4, San Diego (CA), 1984

Glegg, S.; Olivieri, M.; Coulson, R. & Smith, S.(2001). "A Passive Sonar System Based on an Autonomous Underwater Vehicle", IEEE Journal of Oceanic Engineering, Vol. 26, No. 4, pp. 700-710, October 2001.

Grafarend, E. & Ardalan, A. (1993). "World Geodetic Datum 2000", Journal of Geodesy 73, pp. 611-623, 1993

Guo, D.; Zhu, W.; Gao, Z. & Zhang, J. (2000). "A study of Wavelet Thresholding Denoising", Proceedings of IEEE ICSP2000, pp. 329-332, 2000.

Haykin (1994), "Neural Networks", Prentice Hall, NewJersey, USA, 1999.

Chan, Y. & Ho, k. (1994). "A Simple and Efficient Estimator for Hyperbolic Location", IEEE Transactions on Signal Processing, Vol.42, No. 8, pp. 1905-1915, Aug. 1994

Hodges, R.P. (2010). "Underwater Acoustics: analysis, design and performance of SONAR", John Wiley & Sons, Ltd, 2010

Oherveger, K. & Goller, F. (2001). "The Matabolic Cost of Bird Song roduction", Journal Exp. Biol., (204), pp. 3379-3388, 2001.

Holt, M.M.; Noren, D.P; Veirs, V.;Emmons, C.K. & Veirs, S. (2009). "Speaking Up: Killer Whales, Increase their Call Capability in Response to Vessel Noise", Journal of Acoustics Society of America, 125(1), 2009

Hurell, A.& Duck, F. (2000). "A two-dimensional hydrophone array using piezoelectric PVDF", IEEE Transactions on Ultrasonics, Ferroelectrics and Frequency Control, Vol. 47, Issue 6, pp.1345-1353, Nov. 2000

Körner, T.W. (1996). "Fourier Analysis",Cambridge, Cambridge University Press, United Kingdom, 1996

Mallat, S. (1999). "A Wavelet Tour of Signal Processing", Elsevier, 1999

Mattos, L. & Grant, E. (2004). "Passive Sonar Applications: Target Tracking and Navigation of an Autonomous Robot", Proceedings of IEEE International Conference on Robotics and Automation, pp.4265-4270, New Orleans, 2004

Mackenzi, K.V. (1981). "Discussion of Sea Water Sound-Speed Determination", Journal of the Acoustic Society of America, (70), pp. 801-806, 1981

National Instruments (2005). "LabVIEW Advanced Signal Processing Toolbox", Nat. Instr. Press, 2005.

NRC (National Research Council) (2003). "Ocean Noise and Marine Animals", National Academies Press, Washington DC, 2003

Podos, J; Silva, V.F. & Rossi-Santos, M. (2002). "Vocalizations of Amazon River Dolphins, Inia geoffrensis: Insights into the Evolutionary Origins of Delphinid Whistles", Ethology, Blackwell Verlag Berlin, 108, pp. 601-612, 2002

Postolache, O.; Pereira, J.M.D. & Girão, P.S. (2002). "An Intelligent Turbidity and Temperature Sensing Unit for Water Quality Assessment", IEEE Canadian Conference on Electrical & Computer Engineering, CCECE 2002, pp. 494-499, Manitoba, Canada, May 2002

Postolache, O.; Girão, P.S.; Pereira, M.D. & Figueiredo, M. (2006). "Distributed Virtual System for Dolphins' Sound Acquisition and Time-Frequency Analysis", IMEKO XVIII World Congress, Rio de Janeiro, Brasil, Sept.

Postolache, O.; Girão, P.S. & Pereira, J.M.D. (2007). "Intelligent Distributed Virtual System for Underwater Acoustic Source Localization and Sounds Classification", Intelligent Data Acquisition and Advanced Computing Systems: Technology and Applications (IDAACS'2007), pp. 132-135, Dortmund, Germany, September 2007

Postolache, O.; Girão, P. & Pereira, M. (2010). "Smart Sensors and Intelligent Signal Processing in Water Quality Monitoring Context", 4th International Conference on Sensing Technology (ICST'2010), Lecce, Itália, June 2010

Reynolds, J.E.; Odell D.H. & Rommel, A. (1999). "Biology of Marine Mammals", edited by John E. Reynolds III and Sentiel A. Rommel, Melbourne University Press, Australia, 1999

Wright, D. (2001). "Undersea with Geographical Information Systems", ESRI Press, USA

11

Using Virtual Acoustic Space to Investigate Sound Localisation

Laura Hausmann and Hermann Wagner
RWTH Aachen, Institute of Biology II
Germany

1. Introduction

It is an important task for the future to further close the gap between basic and applied science, in other words to make our understanding of the basic principles of auditory processing available for applications in medicine or information technology. Current examples are hearing aids (Dietz et al., 2009) or sound-localising robots (Calmes et al., 2007). This effort will be helped by better quantitative data resulting from more and more sophisticated experimental approaches.

Despite new methodologies and techniques, the complex human auditory system is only accessible in a restricted way to many experimental approaches. This gap is closed by animal model systems that allow a more focused analysis of single aspects of auditory processing than human studies. The most commonly used animals in auditory research are birds (barn owls, chicken) and mammals (monkeys, cats, bats, ferrets, guinea pigs, rats and gerbils).

When these animals are tested with various auditory stimuli in behavioural experiments, the accuracy (distance of a measured value to the true value) and precision (repeatability of a given measured value) of the animal's behavioural response allows to draw conclusions on the difficulty with which the animal can use the stimulus to locate sound sources. An example is the measurement of minimum audible angles (MAA) to reveal the resolution threshold of the auditory system for the horizontal displacement of a sound source (Bala et al., 2007). Similarly, one can exploit the head-turn amplitude of humans or animals in response to narrowband or broadband sounds as a measure for the relevance of specific frequency bands, as well as binaural and monaural cues or perception thresholds (e.g. May & Huang, 1995; Poganiatz et al., 2001; Populin, 2006).

The barn owl (*Tyto alba*) is an auditory specialist, depending to a large extent on listening while localising potential prey. In the course of evolution, the barn owl has developed several morphological and neuronal adaptations, which may be regarded as more optimal solutions to problems than the structures and circuits found in generalists.

The owl has a characteristic facial ruff, which amplifies sound and is directionally sensitive for frequencies above 4 kHz (Coles & Guppy, 1988). Additionally, the left and right ear openings and flaps are asymmetrically with the left ear lying slightly higher than the right one. This asymmetry creates a steep gradient of interaural level differences (ILDs) in the owl's frontal field (Campenhausen & Wagner, 2006). These adaptations to sound localisation are one of the reasons why barn owl hearing was established as an important model system during the last decades.

This chapter will focus on the application of a powerful technique for the investigation of sound processing, the virtual auditory space technique. Its basics, relevance and applications will be discussed for human listeners as well as in barn owls, supplemented by a comparison with other species.
Sound localisation is based on extraction of physical cues of the sound reaching the eardrums. Such physical cues are the monaural spectral properties of the sound as well as differences between the sounds reaching the left and rights ears, leading to binaural cues. These cues vary systematically with sound source position relative to an animal's head.
A sound originates from a source and travels through the air until it reaches the eardrums of a listener. Several distortions (reflection, attenuation) are imposed on the sound along its path. Sound parameters may be measured at or close to the eardrum. The comparison of the measured sound at the eardrum with the sound emitted by the source allows for a determination of the distortion and is unique for each individual. The resulting transfer functions are called the head-related transfer functions (HRTFs) referring to the major influence of head shape in the process of distortion. HRTFs carry information about the location of a sound source. Note that the term HRTF refers to the frequency domain, whereas one speaks of the head-related impulse response (HRIR) when the signal is represented in the time domain. Both signals may be transformed from one domain to the other by means of a Fourier transformation (Blauert, 1997).
In monaural spectra the large decreases in amplitude, termed notches, carry information about sound source direction due to their systematic directional variations. Animals and humans use this information during sound localisation, in particular when resolving front-back confusions (Gardner & Gardner, 1973, Hebrank & Wright, 1974). The comparison of the HRTFs measured at the two ears yields two major binaural parameters: Interaural time difference (ITD) and interaural level difference (ILD). The ITD depends on the angle of incidence as well as on the distance between the two ears. This cue may be further divided into envelope and carrier ITDs.
Envelope ITDs occur specifically at the onset and end of a sound and are then called onset ITDs, whereas ITDs derived from the carrier occur in the ongoing sound and are, therefore, called ongoing ITDs. ITD is constant along a circle describing a surface of a cone, termed "cone of confusion" because for sound sources along this cone surface, the identical ITDs do not allow unambiguous localisation of narrowband stimuli (cf. Blauert, 1997). This leads to ambiguities with respect to front and back, and, therefore, this cone is also known as cone of confusion. ILDs arise from the frequency and position depended attenuation of sound by the pinna, the head and the body that typically differs between the two ears.

2. Investigation of sound localisation - current approaches and problems

The simplest approach to find out more about the relevance of the sound parameters is to replay natural sounds from a loudspeaker and measure the subject's reaction to the sounds. These experiments are typically carried out in rooms having walls that strongly suppress sound reflections. If the distance between source and listener is large enough, we have a free-field situation, and the approach is called free-field stimulation.
Free-field sounds have a major disadvantage: the physical cues to sound location cannot be varied independently, because a specific ITD resulting from a given spatial displacement of the sound source also involves a change in the ILD and the monaural spectra. This renders it difficult or even impossible to derive the contribution of single cues to sound localisation.

On the other hand, free-field sounds contain all relevant cues a subject may use in behaviour. Although free-field stimulation allows for an investigation of how relevant specific sound characteristics are, such as the frequency spectrum, the limits of this technique are obvious. Since this chapter focusses on the virtual space technique, we will not review the results from the numerous studies dealing with free-field stimulation.

One way to overcome the problems inherent in free-field stimulation is the dichotic stimulation via headphones, allowing the independent manipulation of ITDs or ILDs in the stimulus. Dichotic stimulation was used to prove that humans use ITDs for azimuthal sound localisation for frequencies up to 1.5 kHz and ILDs for frequencies above 5 kHz (reviewed in Blauert, 1997). The upper frequency limit for ITD extraction seems to be determined by the ability of neurons to encode the phase of the signal's carrier frequency, which in turn is necessary to compare phase differences between both ears.

The lower border for ILD extraction, likewise, seems to be related to the observation that the head of an animal only creates sufficiently large ILDs above a certain frequency. These conclusions are supported by data from animals such as the cat, the ferret, monkey and the barn owl (Koeppl, 1997; Koka et al., 2008; Moiseff and Konishi 1981; Parsons et al., 2009; Spezio et al., 2000; Tollin & Koka, 2009). The use of both ITDs and ILDs in azimuthal sound localisation is known as duplex theory (Blauert, 1997; Macpherson & Middlebrooks, 2002; Rayleigh, 1907).

In the barn owl, the filtering properties of the facial ruff together with the asymmetrical arrangement of the ear openings and the preaural flaps in the vertical plane cause ILDs to vary along an axis inclined to the horizontal plane. This allows the barn owl to use ILDs for elevational sound localisation (Moiseff 1989, Campenhausen & Wagner 2006; Keller et al. 1998). In contrast, mammals use ILDs for high-frequency horizontal localisation (reviewed in Blauert, 1997).

The ability of the owl's auditory neurons to lock to the signal's phase within almost the whole hearing range (Köppl, 1997) - again in contrast to most mammals - together with the use of ILDs for elevational localisation is one of the reasons that make the barn owl interesting for auditory research, despite the mentioned differences to mammals.

With earphone stimulation, binaural cues can be manipulated independently. For example, the systematic variation of either ITDs or ILDs while keeping the other cue constant is nowadays a commonly used technique to characterise neuronal tuning to sound or to investigate the impact of the cues on sound localisation ability (reviewed in Butts and Goldman, 2006). Another example is the specific variation of ongoing ITDs, but not onset ITDs (Moiseff & Konishi, 1981; von Kriegstein et al., 2008) or a systematic variation of the degree of interaural correlation in binaurally presented noises (Egnor, 2001).

Although dichotic stimulation helped to make progress in our understanding of sound localisation, one disadvantage of this method is that human listeners perceive sources as lying inside the head (Hartmann & Wittenberg, 1996; Wightman & Kistler, 1989b) rather than in outside space. Consequently, when only ITDs or ILDs are introduced, but no spectral cues, the sound is "lateralised" towards a direction corresponding to the amplitude of ITD or ILD, respectively. For human listeners, this may yield a horizontal displacement of the sound image sufficient for many applications. However, both vertical localisation and distance estimation are severly hampered, if possible at all. In contrast, a free-field sound source or an appropriately simulated sound is really "localised".

This means that dichotic stimuli do not contain all physical cues of free-field sounds. A method to overcome the problems of dichotic stimulation as described so far, while preserving its advantages, is the creation of a virtual auditory space (VAS), the method and implementation of which is the topic of this chapter. The work of Wightman and Kistler (1989a,b) and others showed that free-field sources could be simulated adequately by filtering a sound with the personal head-related transfer functions (HRTFs). Bronkhorst (1995) reported that performance degraded when subjects were stimulated with very high frequency virtual sounds. This observation reflects the large difficulties in generating veridical virtual stimuli at high-frequencies.

3. The virtual space technique

While dichotic stimulation does not lead to externalisation of sound sources, the use of HRTF-filtered stimuli in a virtual auditory space does (Hartmann & Wittenberg, 1996; Plenge, 1974; Wightman & Kistler, 1989b). For that reason, numerous attempts have been made to develop virtual auditory worlds for humans or animals. The main goal there is that virtual sound sources in VAS should unambiguously reflect all free-field sound characteristics.

A second goal, especially in human research, is to create virtual auditory worlds that are universally applicable across all listeners. This requires a trade-off between the realistic simulation of free-field characteristics and computational power, that is, one wants to discard nonessential cues while preserving all relevant cues. For that purpose, knowledge is required on which cues are the relevant cues for sound localisation and which are not.

The method involved in creating VAS originated in the 1950ies when systematic experiments using artificial head manikins were undertaken (reviewed by Paul, 2009). However, it is only the computational power developed within the last two decades that allows for elaborate calculations and manipulations of virtual auditory stimuli.

Measuring the HRTFs is usually done by inserting small microphones into the ear canals of the subject, as sketched in Figure 1. The sound impinging on the eardrum is measured. Sound is replayed from a free-field loudspeaker (see Fig. 1a). The loudspeaker signal should contain all relevant frequencies within the hearing range of the subject. It has been shown that measurement at or close to the eardrum is adequate, because the measured signal contains the important information (Wightman & Kistler, 1989a,b).

When the signal arrives at the eardrum, it has been filtered by the outer ear, the head and the body of the subject. The amplitude and phase spectra at the eardrum represent the HRTF for the given ear and the respective position. The monaural spectrum of a specific sound at a given position may be obtained by filtering the sound with the respective HRTF. ITDs and ILDs occurring at a given position are derived by comparing the respective measured HRTFs at the two ears.

The procedure of replaying a free-field sound and recording the resulting impulse response at the subject's eardrum is usually carried out for representative spatial locations, i.e., the free-field speaker is positioned at a constant distance at different locations, for example by moving it along a circular hoop (Fig. 1a). In this way, the desired spatial positions in both azimuthal and elevational planes may be sampled (Fig. 1b+c).

One may use click stimuli (Dirac pulses, see Blauert, 1997) as free-field sounds. In this way, a broad range of frequencies can be presented in a very short time. However, such stimuli do not contain much energy and as a consequence have to be repeated many times (typically 1000) in order to increase the energy provided to the listener (reviewed in Blauert, 1997; see also Poganiatz & Wagner, 2001; Poganiatz et al., 2001).

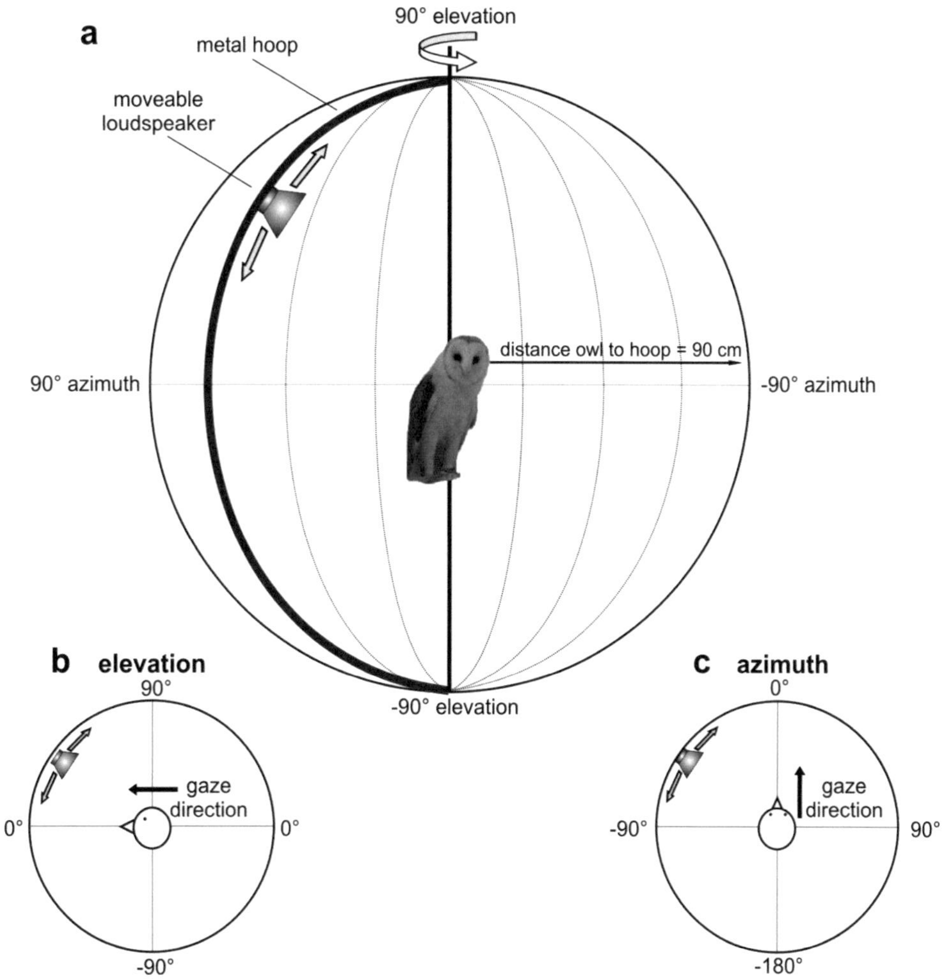

After Hausmann et al. (2010).

Fig. 1. **Schematic of a setup for HRTF measurements.** A) During HRTF measurements, the anesthetised owl is fixated with the help of a cloth jacket in the center of a metal hoop. A loudspeaker can be moved upwards or downwards along the hoop, allowing variation of the vertical stimulus angle as shown in panel B). The hoop can be rotated about its vertical axis, which allowed positioning of the hoop at various azimuthal values, with 0° being directly in front of the owl as shown in panel C).

Other stimuli are so-called sweep signals, which run from low to high frequencies or vice versa in a given time interval. For example, logarithmically rising sweeps have successfully been used for HRTF recordings in the owl (Campenhausen & Wagner, 2006; Hausmann et al., 2010). Such sweep signals have the advantage that a small number of repetitions of sound emissions suffices to yield reproducible measurements, while containing energy in all desired frequencies within the subject's hearing range.

The quality of HRTF recordings measured with both types of stimuli is comparable, as demonstrated by similar shape of HRTFs and localisation performance in the owl for HRTF-filtered stimuli recorded either during application of click noise (Poganiatz & Wagner, 2001; Poganiatz et al., 2001) or of sweeps (Campenhausen & Wagner, 2006; Hausmann et al., 2009; Hausmann et al., 2010). Short click stimuli have also commonly been used for HRTF measurements in other animals and humans, leading to localisation performance comparable to free-field stimulation (Delgutte et al., 1999; Musicant et al., 1990; Tollin & Yin, 2002; Wightman & Kistler, 1989b).

The impulse responses recorded at the subject's eardrum are influenced by the individual transfer functions not only of the subject itself, but also of the equipment used for the recordings such as the microphones, loudspeaker and hardware components. In order to provide an accurate picture of the transfer characteristics, all impulse responses recorded with the subject (specific for each azimuthal (α) and elevational (ε) position) have to be corrected for the transfer characteristics of the system components (T_{sys}).

The correction can be easily done by transforming each impulse response into the frequency domain via Fast Fourier transformation (FFT), and then divide each subject-specific FFT ($H_{\alpha\varepsilon}$) by the reference measurement recorded for the system components including the microphone, but without the subject (T_{sys}) following equation 1.

$$H_{\alpha\varepsilon} = \frac{H_{\alpha\varepsilon} \cdot T_{sys}}{T_{sys}} \qquad (1)$$

In both behavioural and electrophysiological experiments, HRTF-filtered stimuli open a wide range of possible manipulations to analyse single characteristics of sound processing and allow for prediction of localisation behaviour based on HRTF characteristics (humans: Getzmann & Lewald, 2010; Hebrank & Wright, 1974; cat: Brugge et al., 1996; May & Huang, 1996; guinea pig: Sterbing et al., 2003; owl: Hausmann et al., 2009; Poganiatz et al., 2001; Witten et al., 2010).

Virtual auditory worlds can be created for all animals whose HRTFs are measured. An advantage of using the barn owl rather than many mammalian species is that the owl performs saccadic head-turns towards a sound source when sitting on a perch (Knudsen et al., 1979), while the eyes or pinnae can barely be moved (Steinbach, 1972). In contrast, many mammals may move their eyes and pinnae. This allows for example cats or monkeys to locate sound sources even with restrained head to a certain extent (Dent et al., 2009; Populin, 2006; Populin & Yin, 1998). The owl's saccadic head-turn response allows to use the owls' head-turn angle as a measure for the perceived sound source location (Knudsen & Konishi, 1978).

The next section will review how HRTF-filtered stimuli have been implemented in the barn owl as a model system to tackle specific issues of sound localisation which are also relevant for human sound localisation.

4. Virtual auditory space and its applications in an auditory specialist

One of the first applications of VAS for the barn owl was the work of Poganiatz and coworkers (2001). The authors conducted a behavioural study in which individualised HRTFs of barn owls were manipulated in that the broadband ITD was artificially set to a specific value, irrespective of their natural ITD. This artificial ITD was either -100 µs, corresponding to a position of approximately -40° of azimuth (by definition left of the animal) based on a change of 2.5 μs per degree (Campenhausen & Wagner, 2006), or to +100 μs, corresponding to +40° of azimuth (by definition right of the animal).

All other cues such as the ILD and monaural spectra were preserved. That is, the stimuli were ambiguous in that the ITD might point towards a different hemisphere than did all the remaining cues. The authors of the study predicted that the owl should turn its head towards the position encoded by the ITD if the ITD was the relevant cue for azimuthal sound localisation. Similarly, the owl should turn towards the position encoded by ILD and monaural spectra if these cues were relevant for azimuthal localisation.

When these manipulated stimuli were replayed via headphones to the owls, the animals always turned their heads towards the position that was encoded by the ITD and not by the remaining cues. From these findings, Poganiatz et al. (2001) concluded that the owls used exclusively the ITD to determine stimulus azimuth. As we will show below, this may hold for a large range of auditory space. However, the resolution of spatially ambiguous ITDs in the frontal and rear hemispheres requires further cues.

The same approach of manipulating virtual stimuli was used for investigating the role of ILD for elevational sound localisation by setting the broadband ILD in HRTF-filtered stimuli to a fixed value (Poganiatz & Wagner, 2001). Such experiments showed that barn owls' elevational head-turn angles depend partly, but not exclusively on ILDs. The role of ILDs and other cues for elevational localisation will be tackled in more detail below.

Thus, these earlier studies did not resolve the cues needed to resolve front-back confusion or localisation of phantom sources that occur at positions that can be predicted from a narrowband sound's period duration and ITD. Both phenomena are commonly known problems in humans especially for localisation in the median plane (Gardner & Gardner 1973; Hill et al. 2000; Wenzel et al. 1993; Zahorik et al. 2006). Furthermore, it is still unclear which cues, apart from broadband ILD, contribute to elevational sound localisation. The owl's ability to locate sound source elevation is essentially based on its ear asymmetry and facial ruff. Going a step further and utilise the morphological specialisations of the barn owl, a possible application for humans might thus be to mimic an owl's facial ruff to achieve better localisation performance in humans.

We extended the use of HRTF-filtered stimuli to answer some of the above raised questions. The method introduced by Campenhausen & Wagner (2006) allowed us to measure the influence of the barn owl's facial ruff for a closer analysis of the role of external filtering as well as of the interplay of the owl's asymmetrically placed ears with the characteristically heart-shaped ruff.

Using VAS enabled us to analyse the contribution of the facial ruff and the asymmetrically placed ear openings independently from each other, an important aspect if one wants to implement the owls' specialisations for engineering of sound localisation devices.

Virtual ruff removal (Hausmann et al., 2009) was realised by recording HRTFs for anesthetized barn owls a) with intact ruff of the animal that was tested in behavioural experiments later on (individualised HRTFs), b) for a reference animal with intact ruff (reference owl, normal non-individualised HRTFs) and c) for the same reference animal after successive removal of all feathers of the facial disk, leaving only the rear body feathers intact (see also Campenhausen & Wagner, 2006), named "ruffcut" condition.

The advantage of simulating ruff removal rather than actually removing the ruff of the behaving owls consisted in a better reproducibility of stimulus conditions over the course of the experiments, as the feathers regrow after removal and thus stimulus conditions change. Furthermore, responses to the stimuli were comparable between subjects since the stimulus conditions were equal for all three owls included in the study. And third, virtual ruff removal is a more animal-friendly approach than real removal of the feathers, since the behaving owls are not hampered in their usual localisation behaviour, as would be the case if one actually removed their facial ruff.

The measurements yielded three sets of HRTFs. In behavioural experiments, broadband noise (1-12 kHz) was filtered with these HRTFs to simulate ruff removal for three owls. The former two stimulus conditions with intact ruff were required for comparison of the normal localisation performance with that in response to simulated ruff removal. In parallel, the changes of binaural cues were analysed. Virtual ruff removal resulted not only in a reduction of the ITD range in the periphery (Fig. 2B), but also in a corresponding reduction of azimuthal head-turn angles for ruffcut versus normal stimuli (Fig. 2A). That is, the virtual ruff removal induced a change in the localisation behaviour that could be correlated with the accompanying changes of the ITD as the relevant cue for azimuthal localisation (see also Poganiatz et al., 2001).

Virtual ruff removal influenced behaviour in two major ways. First, it caused the ILDs in the frontal field to become smaller, and the ILDs did no longer vary with elevation, as is the case in HRTFs recorded with intact ruff (see also Keller et al., 1998). Correspondingly, the owls lost their ability to determine stimulus elevation.

Second, while owls having a normal ruff could discriminate stimuli coming from the rear from those coming from the front even if the stimuli had the same ITD (Hausmann et al., 2009), this ability to distinguish between front and back in HRTFs having the same ITD was lost after virtual ruff removal. This finding implies that the ITD is indeed the only relevant cue for azimuthal localisation in the frontal field, as suggested by Poganiatz et al. (2001), but stimulus positions with equal ITD in the front and in the rear, respectively, may not be discriminated based on the ITD alone.

Hence, the ruff provides cues other than ITD to resolve position along the cone of confusion (see Blauert, 1997). Potential candidates for the cues provided by the ruff for front-back disambiguation are ILDs and monaural spectral cues, both of which are altered after ruff removal (Campenhausen & Wagner, 2006; Hausmann et al., 2009).

The role of ILDs and spectral cues can be investigated by keeping the ILD in virtual auditory stimuli constant, while the ITD and spectral cues vary with location according to their natural amplitude. Such an approach was pursued in an earlier study by Poganiatz & Wagner (2001), where ILDs in virtual acoustic stimuli were set to a fixed value of either -6 dB (left ear louder) or +6 dB in the frequency range from 4 to 10 kHz. In response to those manipulated stimuli, the owls responded with a positive elevational head-turn to the +6 dB stimuli and with a negative head-turn to the -6 dB stimuli. When the stimulus ILD was set to +6 dB, the owls' head-turn was directed to a relatively constant elevational position. In response to stimuli whose ILD was set to -6 dB, however, the elevational head-turn amplitude was constant at positive stimulus azimuth but increased with incrementally negative stimulus angle, or vice versa. The localisation behaviour depended on the stimulus position, meaning that the elevational localisation was not exclusively defined by the mean broadband ILD.

The study of Poganiatz & Wagner (2001) argued against monaural spectral cues as this additional cue, since the spectra had been preserved according to the natural shape. However, the owls' elevational head-turn angles did not follow a simple and clear relationship, which renders conclusions on the contribution of single binaural and monaural cues difficult. The question remained of whether owls needed broadband ILDs to determine the elevation of virtual sound sources, or whether ILDs in single frequency bands could be used as well.

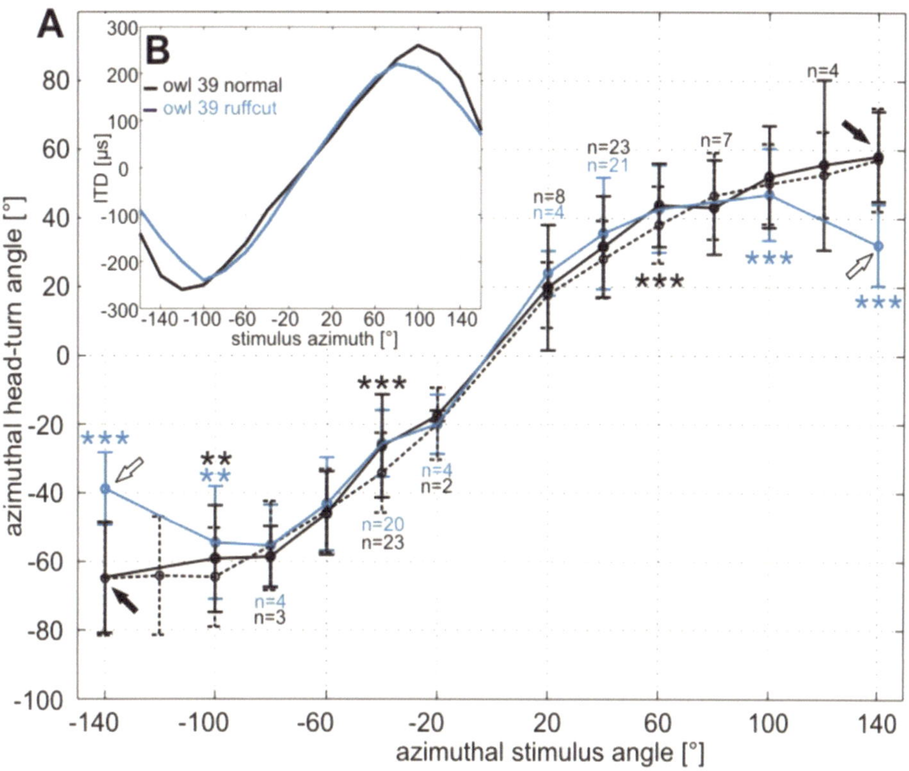

After Hausmann et al. (2009)

Fig. 2. **ITDs and azimuthal head-turn angle under normal and ruffcut conditions.** A) The azimuthal head-turn angles of owls in response to azimuthal stimulation (x-axis) with individualised HRTFs (dotted, data of two owls), non-individualised HRTFs of a reference animal (normal, black, three owls) and to the stimuli from the reference owl after ruff removal (ruffcut, blue, three owls). Arrows mark ±140° stimulus position in the periphery, where azimuthal head-turn angle decreased for stimulation with simulated ruff removal, in contrast to stimulation with intact ruff (individualised and reference owl normal) where they approach a plateau at about ±60°. Significant differences between stimulus conditions are marked with asterisks depending on the significance level (**p<0.01, ***p<0.001) in black (individualised versus reference owl normal) respectively in blue (reference owl normal versus ruffcut). Each data point includes at least 96 trials, unless indicated otherwise by the number of trials (n). B) The ITD in µs contained in the HRTFs at 0° elevation is plotted against stimulus azimuth in degree for the reference owl normal (black) and ruffcut (blue). Note the sinusoidal course of the ITD and the smaller ITD range after ruff removal. ITDs decrease at peripheral azimuths for both intact and removed ruff.

Due to the complex variations of ILDs with both elevation and azimuth in the barn owl, the influence of specific cues on elevational localisation is difficult to investigate. Furthermore, as we have just seen, elevational localisation is influenced by cues other than the ILD, which stands in contrast to the exclusive dependence of azimuthal head-turn angle on ITDs at least in the frontal field (but see Hausmann et al. 2009 for azimuthal localisation in the rear).

Since ILDs are strongly frequency-dependent, the next step we took was the stimulation of barn owls with narrowband stimuli to investigate elevational localisation, so to narrow down the range of relevant frequencies used for elevational localisation. Again, the virtual space technique allowed for a manipulation of stimuli in which ILD cues are preserved for each narrow frequency band, while spectral cues are sparse.

This stimulus configuration may answer the question of whether owls can make use of narrowband spectral cues. If they do, their localisation behaviour should resemble that for non-manipulated stimuli of the same frequency. On the other hand, if monaural narrowband spectra cannot be used, the owls' localisation behaviour for stimuli with virtually removed ILD should differ from that to stimuli containing the naturally occurring ILD. We tested barn owls in the proposed stimulus setup.

We first created narrowband noises. The ILD in such stimuli was then set to a fixed value of zero dB ILD, similar to the approach of Poganiatz & Wagner (2001), without changing the remaining localisation cues. In response to those stimuli, barn owls exhibited elevational head-turn angles that varied with stimulus elevation, indicating that narrowband ILD was sufficient to discriminate sound source elevation.

In addition, the owls were able to resolve azimuthal coding ambiguities, so-called phantom sources, when the virtual stimuli contained ILDs, but not when the ILD was set to zero. This finding implied that owls may use narrowband ILDs to determine the hemisphere a sound originates from, or in other word, to resolve coding ambiguities. The formation of phantom sources will be reviewed in more detail in the following.

5. Coding ambiguities

Coding ambiguities arise if one parameter occurs more than once in auditory space. Coding ambiguities lead to the formation of phantom sources. Many animals perceive phantom sound sources (Lee et al. 2009; Mazer, 1998; Saberi et al., 1998, 1999; Tollin et al. 2003). The main parameter for azimuthal localisation in the frontal hemisphere is the ITD. In the use of ITD, ambiguities occur for narrowband and tonal stimuli when the period duration of the center frequency or tone is shorter than the time that the sound needs to travel around the head of the listener.

For narrowband and tonal stimuli, ITD is equivalent to the interaural phase difference. The sound's phase at one ear can either be matched with the preceding (leading) phase or with the lagging phase at the other ear. Both comparisons may yield valid azimuthal sound source positions if the ITD corresponding to the interaural phase difference of the stimulus falls within the ITD range the animal can experience. For example, a 5 kHz tone has a period duration of 200 μs. In the owl, stimulation from -40° azimuth (i.e., 40° displaced to the left side of the owl's midsagittal plane) corresponds to about -100 μs ITD, based on a change of about 2.5 μs per degree (Campenhausen & Wagner, 2006). In this case, the 5 kHz tone is leading at the owl's left ear by 100 μs, which would result in calculation of the correct sound source azimuth.

However, it is also possible to match the lagging phase at the left ear with the next leading phase at the right ear, resulting in a phantom source at +40° azimuth in the right

hemisphere. A study by Saberi et al. (1998) showed that in case of ambiguous sound images, the owls either turned their heads towards the more frontal sound source, be it a real or a phantom source, or else they turned towards the more peripheral sound source.
With increasing stimulus bandwidth, the neuronal tuning curves for the single frequencies are still cyclic and, therefore, ambiguous as we have just seen. However, there is always one peak at the real ITD, while the position of the phase multiples (side peaks) is shifted according to the period duration, which varies with frequency (Wagner et al., 1987).
Integration, or summation, across a wider band of frequencies thus yields a large peak at the true ITD and smaller side peaks. Hence, for wideband sounds, integration across frequencies reduces ITD coding ambiguities via side-peak suppression in broadband neurons (Mazer, 1998; Saberi et al., 1999; Takahashi & Konishi 1986; Wagner et al., 1987). Sidepeak suppression reduces the neuronal responses to the phantom sources (corresponding to the phase equivalents of the real ITD) compared to the response to the real ITD. Mazer (1998) and Saberi et al. (1999) showed in electrophysiological and behavioural experiments that a bandwidth of 3 kHz was sufficient to reduce phase ambiguities and to unambiguously determine the real ITD.
Thus, in many cases, a single cue does not allow to determine the veridical spatial position unambiguously. This was also shown by electrophysiological recordings of the spatial receptive fields for variations in ILD, but constant ITD (Euston & Takahashi, 2002). In this stimulus configuration, ILDs exhibited broad regions where the ILD amplitude was equal, thus ambiguous.
Aross-frequency integration also reduces such ILD ambiguities, which are based on the response properties of single cells for example in the external nucleus of the inferior colliculus (ICX). Such neurons respond to a narrowband stimulus having a given ITD but varying ILDs with an increased firing rate at wide spatial regions. That is, this neuron's response does not code for a single spatial position, but for a variety of positions which cannot be distinguished based on the neuronal firing rate alone. Only the combination of a specific ITD with a specific ILD results in unambiguous coding of spatial positions and results in the usual narrowly restricted spatial receptive fields (Euston & Takahashi, 2002; Knudsen & Konishi, 1978; Mazer, 1998). In the case of the owl, the natural combinations of ITD and ILD that lead to sharply tuned spatial receptive fields are created by the characteristic filtering properties of the ruff (Knudsen & Konishi, 1978).
To summarise the preceding sections, the ruff plays a major role for the resolution of coding ambiguities. However, it is only the interaction of the ruff with the asymmetrically placed ear openings and flaps that creates the unique directional sensitivity of the owl's auditory system (Campenhausen & Wagner, 2006; Hausmann et al., 2009). This finding should be taken into account if one wants to mimic the owl's facial ruff in engineering science
It is interesting that humans can learn to listen and localise sound sources quite accurately when provided with artificial owl ears (Van Wanrooij et al., 2010). The human subjects in that study wore ear moulds that were scaled to the size of the listener, during an uninterrupted period of several weeks. The ear moulds were formed to introduce asymmetries just as observed in the barn owl. The ability of the subjects to localise sound sources in both azimuth and elevation was tested repeatedly to measure the learning plasticity in response to the unusual hearing experience. At the beginning of the experiments, localisation accuracy in both planes was severely hampered. After few weeks, not only azimuthal localisation performance was close to normal again, but also elevational localisation of broadband sounds, and only these. That is, the hearing performance

apparently underlies a certain plasticity, meaning that a listener can learn to locate sounds accurately even with unfamiliar cues, which opens interesting fields of application.

Similar plasticity was observed in ferrets whose ears were plugged, who learned to localize azimuthal sound sources accurately again after several weeks of training (Mrsic-Flogel et al. 2001).

These experiments underline that auditory representations in the brain are not restricted to individual species, but rather that humans or animals can learn new relationships between a specific combination of localisation cues and a specific spatial position. Despite this plasticity, in everyday applications, it may not seem feasible when listeners need a long period of time to learn a new relationship. However, when familiarity to sound spectra is established via training, localisation performance is improved, a fact that is amongst others exploited for cochlear implant users (Loebach & Pisoni 2009).

Now what are the implications of the above revised findings for the creation of auditory worlds for humans?

First, it is crucial to preserve low-frequency ITDs in virtual stimuli, since these are not only required, but also seem to be dominant for azimuthal localisation (reviewed in Blauert, 1997 for humans; owl: Witten et al., 2010).

Second, ILD cues are necessary in the high-frequency range for accurate elevational localisation in many animal species including humans (e.g. Blauert, 1997; Gardner & Gardner, 1973; Huang & May, 1996; Tollin et al., 2002; Wightman & Kistler, 1989b). In the low-frequency range, the small attenuation by the head results in only small ILDs that hardly vary with elevation (human: Gardner & Gardner 1973; Shaw 1997; cat: May & Huang 1996; monkey: Spezio et al., 2000; owl: Campenhausen & Wagner, 2006; Keller et al., 1998; Hausmann et al., 2010), which makes ILDs a less useful cue for low-frequency sound localisation. However, a study by Algazi et al. (2000) claims that human listeners could determine stimulus elevation surprisingly accurate even when the stimulus contained only frequencies below 3 kHz, although the listeners' performance was degraded compared to a baseline condition with wideband noise. These two cues allow for relatively accurate determination of sound source position in the horizontal plane in humans (see Blauert 1997).

However, ITD and ILD variations alone may as well be introduced to dichotic stimuli presented via headphones, without the requirement of measuring the complex individual transfer functions. That is, as long as pure lateralisation (Plenge 1974; Wightman & Kistler 1989a,b) outside the median plane suffices to fulfil a given task, it should be easier to introduce according ITDs and ILDs to the stimuli. However, for a sophisticated simulation of free-field environments, as well as for unambiguous allocation of spatial positions to the frontal and rear hemispheres, one should use HRTF-filtered stimuli. This holds the more as ILD cues seem to be required for natural sounding of virtual stimuli in human listeners (Usher & Martens, 2007).

Since an inherent feature of HRTFs is the fact that they are individually different, the question arises of whether HRTF-filtered stimuli are feasible for general application, that is, if they can in some way be generalised across listeners to prevent the necessity of measuring HRTFs for each potential listener individually. The latter would be critical anyway because for numerous applications, the future user of the virtual auditory space is unknown in advance. The issue of the extent to which HRTFs can be used for stimulation of different subjects without loosing informational content will be tackled in the following section.

6. Localisation with non-individualized HRTFs – does everybody hear differently?

Meanwhile, there are many studies that attempt to generate sets of "universal" HRTFs, which create the impression of free-field sound sources across all (human) listeners. Such HRTFs eliminate the inter-individually different characteristics which are not crucial for accurate localisation while preserving all relevant characteristics. Even though the listener's performance should not be impaired by the presence of naturally occurring, but unnecessary cues in virtual stimuli, discarding those cues may be advantageous. The preservation of the cues that are indispensable for sound localisation, while eliminating the cues which are not crucial, minimises the effort and time required for computing stimuli.

Across-listener generalised HRTFs intend to prevent the need for measuring the HRTFs of each individual separately, and thereby simplify the creation of VAS for numerous fields of application. At the same time, it is important to prevent artifacts such as front-back confusions, one of the reasons which justify the extended research in the field of HRTFs and virtual auditory spaces.

Whenever HRTF-filtered stimuli are employed, the problem arises of how inter-individually different refractional properties of the head or pinna or differences in head diameter affect localisation performance in response to virtual stimulation. It would be of no use to possess sophisticated virtual auditory worlds, if these were not reliably perceived as being externalised, or else if the virtual space did not unambiguously simulate the intended free-field sound source. A global application of, for example, virtual auditory displays can only be achieved when VASs are really listener-independent to a sufficient extent.

Hence, great efforts have been made to develop universally applicable sets of HRTFs across all listeners, but discarding cues that are not required. An even more important aspect, of course, is to resolve any ambiguities that occur with virtual stimuli but not with natural stimuli. HRTF-filtered stimuli have been used to investigate whether the use of individualised versus non-individualised HRTFs influenced localisation behaviour in various species (e.g. humans: Hofman & Van Opstal, 1998; Hu et al., 2008; Hwang et al., 2008; Wenzel et al., 1993; owl: Hausmann et al., 2009; ferret: King et al., 2001; Mrsic-Flogel et al., 2001). It was shown that one of the main problems when using non-individualised HRTFs for stimulation was that the listeners committed front-back or back-front reversals, that is, they localised stimuli coming from the frontal hemisphere in the rear hemisphere or vice versa.

For many mammalian species, it was shown that in particular, notches in the high-frequency monaural spectra are relevant for sound localisation in the vertical plane (Carlile, 1990; Carlile et al., 1999; Koka & Tollin, 2008; Musicant et al., 1990; Tollin & Yin, 2003), and may help, together with ILD cues, to resolve front-back or back-front reversals as discussed in Hausmann et al. (2009). Whether this effect indeed occurs in the barn owl has yet to be proved.

In what concerns customisation of human HRTF-filtered signals, Middlebrooks (1999) proposed in his study how frequency-scaling of peaks and notches in directional transfer functions of human listeners allows generalisation of non-individualised HRTFs while preserving localisation characteristics. Such an approach may render extensive measurements for each individual unnecessary. Likewise, customisation of median-plane HRTFs is possible if the principal-component basis functions with largest inter-subject variations are tuned by one subject while the other functions are calculated as the mean for all subjects in a database (Hwang et al., 2008). Since localisation accuracy is preserved even when HRTFs for human listeners account for only 30% of individual differences (Jin et al., 2003), slight customisation of measured HRTFs already yielded large improvements in localisation ability.

When individualised HRTF-filtered stimuli are used, the percepts in virtual auditory displays are identical to free-field percepts when the spatial resolution of HRTF-measurements is 6° or less (Langendijk & Bronkhorst, 2000). For 10 to 15° resolution, the percepts are still comparable (Langendijk & Bronkhorst, 2000), which implies that the spatial resolution for HRTF-measurements should not fall below 10°. This issue is of extreme importance in dynamic virtual auditory environments, because here it is required that transitions (switching) between HRTFs needed for the simulation of motion are inaudible to the listener. In other words, the listener should experience a smoothly moving sound image without disturbing clicks or jumps when the HRTF position is changed. Hoffman & Møller (2008) determined the minimum audible angles for spectral switching (MASS) to be 4-48° depending on the direction, and for temporal switching (minimum audible time switching MATS) to be 5-10 μs. That is, this resolution should not be under-run when switching between adjacent HRTF either temporally or spectrally. Interpolation of measured HRTFs is especially important if listeners are moving in the auditory world, to prevent leaps or gaps in the auditory percept. This interpolation has to be done carefully in order to preserve the natural auditory percept (Nishimura et al., 2009).

Standard sets of HRTFs are available on internet databases (e.g. on www.ais.riec.tohoku.ac.jp/lab/db-hrtf/). The availability of standard HRTFs recorded with artificial heads (reviewed in Paul, 2009) and of information and technology provided by head-acoustics companies allows scientists and private persons to benefit from sophisticated virtual auditory environments. Especially in what concerns users of cochlear implants, knowledge on the impact of individual HRTF features such as spectral holes (Garadat et al., 2008) on speech intelligibility has helped to improve hearing performance in those patients. Last but not least, much effort has been made to enhance the perceived "spaciousness" of virtual sounds for example to improve the impression of free-field sounds while listening to music (see Blauert, 1997).

7. Advantage, disadvantages and future prospects of virtual space techniques

There are still many challenges for the calculation of VASs. For instance, HRTFs have to be measured and interpolated very thoroughly for the various spatial positions in order to preserve the distributions of physical cues that occur in natural free-field sounds. This is to some extent easier for the largely frequency-independent ITDs, whereas slight mispositioning of the recording microphones can induce larger errors to the measured ILDs and spectral cues especially in the high-frequency range, which then may lead to mislocalisation of sound source elevation (Bronkhorst, 1995).

When measuring HRTFs, it is also important to carefully control the position of the recording microphone relative to the eardrum, since the transfer characteristics of the ear canal can vary throughout its length (Keller et al., 1998; Spezio et al., 2000; Wightman & Kistler, 1989a).

Another aspect is that the computational efforts for the complex and time-consuming creation of virtual stimuli may be reduced by reversing the positions of microphones and sound source during HRTF measurements. The common approach, which has also been described in the present chapter, is placement of the microphone into the ear canal and subsequent application of sound from outside. In this case, the position of the sound source is varied systematically across representative spatial positions, in order to reflect the amplitude of each physical cue after filtering by the outer ear and ear canal.

However, it is also possible to take the reverse approach, that is, placing the sound source into the ear canal and record the signal that is arriving at a microphone after filtering by the ear canal and outer ear (e.g. Zotkin et al., 2006). The microphones that record the output signals are then positioned at the exact spatial locations where usually the loudspeaker would be. The latter approach has a huge advantage compared to the conventional way, because it saves an immense amount of time. Rather than placing the sound source sequentially to various locations in space, waiting until the signal has been replayed, reposition the sound source and repeat the measurement for another position, one single application of the sound suffices as long as an array of microphones is installed at each spatial location one wants to record an impulse response for. The time consuming conventional approach, however, has the advantage that only a single recording device is required. Furthermore, in the conventional approach, the loudspeaker is not as limited in size as is an in-ear loudspeaker. It may be difficult to build an in-ear loudspeaker with satisfying low-frequency sound emission.

Another possibility to save time when recording impulse responses is to use a microphone moving along a circle, which allows recording of impulse responses for each angle along the horizontal plane in less than one second (Ajdler et al., 2007). Also in this technique, the sound emitter is placed in the ear and the receiver microphone is placed outside the subject's ear canal.

Thus, depending on the purpose of an HRTF measurement, an experimenter has several choices and may simply decide which approach is more useful for his or her requirements.

Another important, but often neglected aspect of sound localisation that still awaits closer investigation is the role of auditory distance estimation. Kim et al. (2010) recently presented HRTFs for the rabbit, which show variances in HRTF characteristics for varying sound source distances. Overestimation of sound sources in the near field occur as commonly as underestimation of source distance in the far field (e.g. Loomis et al. 1998; Zahorik 2002), which again seems to be a phenomenon that is not due to headphone listening, but a common feature of sound localisation.

Loomis & Soule (1996) showed that distance cues are reproducible with virtual acoustic stimuli. The human listeners in their study experienced virtual sounds in considerable distance of several meters, even though the perceived distances were still subject to misjudgements. However, since the latter problem occurs also in free-field sounds (overestimation of near targets and underestimation of far targets), further efforts need to be spent to unravel distance perception in humans.

That is, it is possible to simulate auditory distance with stimuli provided via headphones. Noteworthy, virtual auditory stimuli may be scaled so that they simulate a specific distance, even if a corresponding free-field sound would be under- or overestimated, respectively. This is a considerable advantage of the virtual auditory space technique, because naturally occuring perceptional "errors" may be overcome by in- or decreasing the amplitude of virtual auditory stimuli according to the respective requirements. Fontana and coworkers (2002) developed a method to simulate the acoustics inside a tube in order to successfully provide distance cues in a virtual environment. It is also possible to calibrate distance

estimation using psychophysical rating methods, so to get a valid measure for distance cues (Martens, 2001).
How good distance cues, among which intensity, spectrum and direct-to-reverberant energy are especially important, are preserved with current HRTF recording techniques, i.e., how good they coincide with the natural distance cues, is still to be evaluated more closely.
In sum, the virtual space technique offers a wide range of powerful applications, not only for the general investigation of sound localisation properties, but also for its implementation in daily life. Once the cues that contribute to specific aspects of sound localisation are known, not only established techniques such as hearing aids may be improved, for example for the reduction of background noise or for better separation of several concurring sound sources, but the VAS also allows to introduce manipulations to sound stimuli that would naturally not occur. The latter possibility may be useful to create auditory illusions for various applications. Among these are auditory displays for navigational tasks for example during flight (Bronkhorst et al., 1996) or travel aids for both healthy and blind people (Loomis et al., 1998; Walker & Lindsay, 2006), as well as communicational applications such as telephone conferencing (see Martens, 2001).
However, it is indispensable to further evaluate if recording of HRTFs and creation of VASs indeed reflect all relevant aspects of sound localisation cues, in order to prevent unwanted artifacts that might confound the perceived spatial position.
Although a major goal of basic research has to be the long-time implementation of the gained knowledge for applications in humans, the extended use of animal models for the auditory system can yield valuable data on basic auditory processes, as was shown throughout this chapter.

8. References

Ajdler, T.; Sbaiz, L. & Vetterli, M. (2007). Dynamic measurement of room impulse responses using a moving microphone. *J Acoust Soc Am* 122, 1636-1645

Bala, A.D. ; Spitzer, M.W. & Takahashi, T.T. (2007). Auditory spatial acuity approximates the resolving power of space-specific neurons. *PLoS One,* 2, e675

Blauert, J. (1997). Spatial Hearing. The Psychophysics of Human Sound Localization, MIT Press, ISBN 3-7776-0738-X, Cambridge, Massachussetts.

Bronkhorst, A.W. ; Veltman, J.A. & Van Vreda, L. (1996). Application of a Three-Dimensional Auditory Display in a Flight Task. *Human Factors* 38, 23-33

Butts, D.A. & Goldman, M.S. (2006). Tuning Curves, Neuronal Variability, and Sensory Coding. *PLoS Biol* 4, e92

Calmes, L.; Lakemeyer, G. & Wagner, H. (2007). Azimuthal sound localization using coincidence of timing across frequency on a robotic platform. *J Acoust Soc Am,* 121, 2034-2048

Campenhausen, M. & Wagner, H. (2006). Influence of the facial ruff on the sound-receiving characteristics of the barn owl's ears. *J Comp Physiol A,* 192, 1073-1082

Carlile, S. (1990). The auditory periphery of the ferret. II: The spectral transformations of the external ear and their implications for sound localization. *J Acoust Soc Am* 88, 2195-2204

Carlile, S.; Leong, P. & Hyams, S. (1997). The nature and distribution of errors in sound localization by human listeners. *Hear Res* 114, 179-196

Carlile, S.; Delaney, S. & Corderoy, A. (1999). The localisation of spectrally restricted sounds by human listeners. *Hear Res* 128, 175-189

Coles, R.B. & Guppy, A. (1988). Directional hearing in the barn owl (Tyto alba). *J Comp Physiol A*, 163, 117-133

Delgutte, B. ; Joris P.X. ; Litovsky, R.Y. & Yin, T.C.T. (1999). Receptive fields and binaural interactions for virtual-space stimuli in the cat inferior colliculus. *J Neurophysiol* 81, 2833-2851

Dent, M.L.; Tollin, D.J. & Yin, T.C.T. (2009). Influence of Sound Source Location on the Behavior and Physiology of the Precedence Effect in Cats. *J Neurophysiol* 102, 724-734

Dietz, M. ; Ewert, S.D. & Hohmann, V. (2009). Lateralization of stimuli with independent fine-structure and envelope-based temporal disparities. *J Acoust Soc Am*, 125, 1622-1635

Drager, U. & Hubel, D. (1975). Physiology of visual cells in mouse superior colliculus and correlation with somatosensory and auditory input. *Nature* 253, 203-204

DuLac, S. & Knudsen, E.I. (1990). Neural maps of head movement vector and speed in the optic tectum of the barn owl. *J Neurophysiol*, 63, 131-146

Fontana, F. ; Rocchesso, D. & Ottaviani, L. (2002). A Structural Approach to Distance Rendering in Personal Auditory Displays. In : *Proceedings of the 4th IEEE International Conference on Multimodal Interfaces*, ISBN : 0-7695-1834-6, p. 33.

Garadat, S.N.; Litovsky, R.Y.; Yu, G. & Zeng, F.-G. (2009). Effects of simulated spectral holes on speech intelligibility and spatial release from masking under binaural and monaural listening. *J Acoust Soc Am* 127,2,977-989

Gardner, M.B. & Gardner, R.S. (1973). Problem of Localization in the Mediean Plane, Effect of Pinna Caity Occlusion. *J Acoust Soc Am* 53, 400-408

Harris, L. ; Blakemore, C. & Donaghy, M. (1980). Integration of visual and auditory space in the mammalian superior colliculus. *Nature* 5786, 56-59

Hartline P. ; Vimal, R. ; King, A. ; Kurylo, D. & Northmore, D. (1995). Effects of eye position on auditory localization and neural representation of space in superior colliculus of cats. *Exp Brain Res* 104, 402-408

Hartmann, W. & Wittenberg, A. (1996). On the externalization of sound images. *J Acoust Soc Am*, 99, 3678-3688

Hausmann, L.; von Campenhausen, M. ; Endler, F. ; Singheiser, M. & Wagner, H. (2009). Improvements of Sound Localization Abilities by the Facial Ruff of the Barn Owl (Tyto alba) as Demonstrated by Virtual Ruff Removal. *PLoS One*, 4, e7721

Hausmann, L.; von Campenhausen, M. & Wagner, H. (2010). Properties of low-frequency head-related transfer functions in the barn owl (tyto alba). *J Comp Physiol A*, epub ahead of print

Hebrank, J. & Wright, D. (1974). Are Two Ears Necessary for Localization of Sound Sources in the Median Plane? *J Acoust Soc Am* 56, 935-938

Hill, P. ; Nelson, P. ; Kirkeby, O. & Hamada, H. (2000). Resolution of front-back confusion in virtual acoustic imaging systems. *J Acoust Soc Am* 108, 2901-2910

Hoffman, P.F. & Møller, H. (2008). Audibility of Direct Switching Between Head-Related Transfer Functions. *Acta Acustica united with Acustica* 94, 955-964

Hofman, P.M.; Van Riswick, J.G.A. & Van Opstal, A.J. (1998). Relearning sound localization with new ears. *Nature Neuroscience* 1, 417-421

Hu, H.; Zhou, L.; Ma, H. & Wu, Z. (2007). HRTF personalization based on artificial neural network in individual virtual auditory space. *Applied Acoustics* 69, 163-172

Hwang, S. ; Park, Y. & Park, Y. (2008). Modeling and Customization of Head-Related Impulse Responses Based on General Basis Functions in Time Domain. *Acta Acustica united with Acustica*, 94, 965-980

Jin, C. ; Leong, P. ; Leung, J. ; Corderoy, A. & Carlile, S. (2000). Enabling individualized virtual auditory space using morphological measurements. Proceedings of the First IEEE Pacific-Rim Conference on Multimedia, pp. 235-238

Keller, C. ; Hartung, K. & Takahashi, T. (1998). Head-related transfer functions of the barn owl : measurement and neural responses. *Hear Res* 118, 13-34

King, A. & Calvert, G. (2001). Multisensory integration : perceptual grouping by eye and ear. *Curr Biol* 11, R322-R325

King, A. ; Kacelnik, O. ; Mrsic-Flogel, T. ; Schnupp, J. ; Parsons, C. & Moore, D. (2001). How plastic is spatial hearing? *Audiol Neurootol* 6, 182-186

Krämer, T. (2008). Attempts to build an artificial facial ruff mimicking the barn owl (Tyto alba). *Diploma thesis*, RWTH Aachen, Aachen

Knudsen, E.I. & Konishi, M. (1979). Mechanisms of sound localisation in the barn owl (Tyto alba). *J Comp Physiol A*, 133, 13-21

Knudsen, E.I. ; Blasdel, G.G. & Konishi, M. (1979). Sound localization by the barn owl (Tyto alba) measured with the search coil technique. *J Comp Physiol A* 133, 1-11

Knudsen, E.I. (1981). The Hearing of the Barn Owl. *Scientific American*, 245, 113-125

Koeppl, C. (1997). Phase locking to high frequencie in the auditory nerve and cochlear nucleus magnocellularis of the barn owl, Tyto alba. *J Neurosci* 17, 3312-3321

Koka, K. & Tollin, D. (2008). The acoustical cues to sound location in the rat : measurements of directional transfer functions. *J Acoust Soc Am* 123, 4297-4309

Lee, N. ; Elias, D.O. & Mason, A.C. (2009). A precedence effect resolves phantom sound source illusions in the parasitoid fly *Ormia ochracea*. *Proc Natl Acad Sci USA*, 106(15), 6357-6362

Loebach, J.L. & Pisoni, D. (2008). Perceptual learning of spectrally degraded speech and environmental sounds. *J Acoust Soc Am* 123, 2, 1126-1139

Loomis, J.M. & Soule, J.I. (1996). Virtual acoustic displays for real and virtual environments. In: *Proceedings of the Society for Information Display 1996 International Symposium*, pp. 965-968. San Jose, CA : Society for Information Display.

Loomis, J.M. ; Klatzky, R.L., Philbeck, J.W. & Golledge, R.G. (1998). Assessing auditory distance perception using perceptually directed action. *Perception & Psychophysics* 60, 6, 966-980

Loomis, J.M. ; Golledge, R.G. & Klatzky, R.L. (1998). Navigation System for the Blind : Auditory Display Modes and Guidance. *Presence*, 7, 193-203

Makous, J. & Middlebrooks, J.C. (1990). Two-dimensional sound localization by human listeners. *J Acoust Soc Am* 87, 2188-2200

Martens, W.L. (2001). Psychophysical calibration for controlling the range of a virtual sound source: multidemensional complexity in spatial auditory display. Proceedings of the 2001 International Conference on Auditory Display, Espoo, Finland, July 29-August 1.

May, B.J. & Huang, A.Y. (1995). Sound orientation behavior in cats. I. Localization of broadband noise. *J Acoust Soc Am* 100, 2, 1059-1069

Mazer, J.A. (1998). How the owl resolves auditory coding ambiguity. *Proc Natl Acad Sci USA*, 95, 10932-10937
Meredith, M. & Stein, B. (1986). Visual, auditory, and somatosensory convergence on cells in superior colliculus results in multisensory integration. *J Neurophysiol* 56, 640-662
Middlebrooks, J. & Knudsen, E.I. (1984). A neural code for auditory space in the cat's superior colliculus. *J Neurosci* 4, 2621-2634
Moiseff, A. & Konishi, M. (1981). Neuronal and behavioral sensitivity to binaural time differences in the owl. *J Neurosci* 1, 1, 40-48
Mrsic-Flogel, T.; King, A.; Jenison, R. & Schnupp, J. (2001). Listening through different ears alters spatial response fields in ferret primary auditory cortex. *J Neurophysiol* 86, 1043-1046
Musicant, A.; Chan, J. & Hind, J. (1990). Direction-dependent spectral properties of cat external ear: New data and cross-species comparisons. *J Acoust Soc Am* 87, 757-781
Nishimura, R.; Kato, H. & Inoue, N. (2009). Interpolation of head-related transfer functions by spatial linear prediction. IEEE 1901-1904
Parsons, C.H.; Lanyon, R.G.; Schnupp, J.W.H. & King, A.J. (1999). Effects of Altering Spectral Cues in Infancy on Horizontal and Vertical Sound Localization by Adult Ferrets. *J Neurophysiol* 82, 2294-2309
Paul, S. (2009). Binaural Recording Technology: A Historical Review and Possible Future Developments. *Acta Acustica united with Acustica* 95, 767-788
Plenge, G. (1974). On the differences between localization and lateralization. *J Acoust Soc Am* 56, 944-951
Poganiatz, I. & Wagner, H. (2001). Sound-localization experiments with barn owls in virtual space : influence of broadband interaural level difference on head-turning behavior. *J Comp Physiol A*, 187, 225-233
Poganiatz, I.; Nelken, I. & Wagner, H. (2001). Sound-localization experiments with barn owls in virtual space : influence of interaural time difference on head-turning behavior. *JARO* 2, 1-21
Populin, L.C. (2006). Monkey Sound Localization: Head-Restrained versus Head-Unrestrained Orienting. *J Neurosci* 26, 38, 9820-9832
Populin, L.C. & Yin, T.C.T. (1998). Pinna movements of the cat during sound localization. *J Neurosci* 18, 4233-4243
Rayleigh, Lord (1907). On our perception of sound direction. *Philos Mag* 13, 214-232
Saberi, K.; Farahbod, H. & Konishi, M. (1998). How do owls localize interaurally phase-ambiguous signals? *PNAS* 95, 6465-6468
Saberi, K. ; Takahashi, Y. ; Farahbod, H. & Konishi, M. (1999). Neural bases of an auditory illusion and its elemination in owls. *Nat Neurosci* 2, 656-659
Searle, C.L.; Braida, L.D. ; Cuddy, D.R. & Davis, M.F. (1975). Binaural pinna disparity : another auditory localization cue. *J Acoust Soc Am* 57, 2, 448-455
Spezio, M.L.; Keller, C.H.; Marrocco, R.T. & Takahashi, T.T. (2000). Head-related transfer functions of the Rhesus monkey. *Hear Res* 144, 73-88
Steinbach, M. (1972). Eye movements of the owl. *Vision Research* 13, 889-891
Takahashi, T.T. & Konishi, M. (1986). Selectivity for interaural time difference in the owl's midbrain. *J Neurosci* 6, 3413-3422

Tollin, D.J. & Koka, K. (2009). Postnatal development of sound pressure transformation by the head and pinnae of the cat: Monaural characteristics. *J Acoust Soc Am* 125, 2, 980-994

Tollin, D.J. & Yin, T.C.T. (2002). The Coding of Spatial Location by Single Units in the Lateral Superior Olive of the Cat. I. Spatial Receptive Fields in Azimuth. *J Neuroscience*, 22, 4, 1454-1467

Tollin, D.J. & Yin, T.C.T. (2003). Spectral cues explain illusory elevation effects with stereo sounds in cats. *J Neurophysiol* 90, 525-530

Usher, J. & Martens, W.L. (2007). Naturalness of speech sounds presented using personalized versus non-personalized HRTFs. *Proceedings of the 13th International Conference on Auditory Display*, Montréal, Canada, June 26-29.

Van Wanrooij, M.M., Van Der Willigen, R.F. & Van Opstal, A.J. (2010). Learning Sound Localization with the Barn-owl's Ears. *Abstracts to FENS 2010*, Poster number 169.25, Amsterdam, Netherlands, July 2010

Wagner, H. ; Takahashi, T. & Konishi, M. (1987). Representation of interaural time difference in the central nucleus of the barn owl's inferior colliculus. *J Neurosci* 7, 3105-3116

Walker, B.N. & Lindsay, J. (2006). Navigation Performance With a Virtual Auditory Display : Effects of Beacon Sound, Capture Radius, and Practice. *Human Factors* 48, 2, 265-278

Wenzel, E. ; Arruda, M. Kistler, D. & Wightman, F. (1993). Localization using nonindividualized head-related transfer functions. *J Acoust Soc Am* 94, 111-123

Wightman, F.L. & Kistler, D.J. (1989a). Headphone simulation of free field listening. I : Stimulus synthesis. *J Acoust Soc Am* 85, 2, 858-867

Wightman, F.L. & Kistler, D.J. (1989b). Headphone simulation of free field listening. II : Psychophysical validation. *J Acoust Soc Am* 85, 2, 868-878

Zahorik, P. (2002). Assessing auditory distance perception using virtual acoustics. *J Acoust Soc Am* 111, 4, 1832-1846

Zahorik, P.; Bangayan, P.; Sundareswaran, V.; Wang, K. & Tam, C. (2006). Perceptual recalibration in human sound localization: learning to remediate front-back reversals. *J Acoust Soc Am* 120, 343-359

Zotkin, D.N.; Duraiswami, R.; Grassi, E. & Gumerov, N.A. (2006). Fast head-related transfer function mesurement via reciprocity. *J Acoust Soc Am* 120, 4, 2202-2215

12

Sound Waves Generated Due to the Absorption of a Pulsed Electron Beam

A. Pushkarev, J. Isakova, G. Kholodnaya and R. Sazonov
Tomsk Polytechnic University
Russia

1. Introduction

Over the past 30–40 years, a large amount of research has been devoted to gas-phase chemical processes in low-temperature plasmas. When the low-temperature plasma is formed by a pulsed electron beam, there is a significant reduction, compared to many other methods of formation, in the power consumption for conversion of gas-phase compounds. Analysis of experimental studies devoted to the decomposition of impurities of various compounds (NO, NO_2, SO_2, CO, CS_2, etc.) in air by a pulsed electron beam showed (Pushkarev et al., 2006) that the the energy of the electron beam required to decompose one gas molecule is lower than its dissociation energy. This is due to the fact that under the action of the beam, favourable conditions for the occurrence of chain processes are formed. At low temperatures, when the initiation of a thermal reaction does not occur, under the influence of the plasma there are active centres—free radicals, ions or excited molecules, which can start a chain reaction. This chain reaction will take place at a temperature 150–200 degrees lower than a normal thermal process, but with the same speed. The impact of the plasma facilitates the most energy intensive stage, which is the thermal initiation of the reaction. A sufficient length of the chain reaction makes it possible to reduce the total energy consumption for the chemical process. The main source of energy in this case is the initial thermal energy or the energy of the exothermic chemical reactions of the chain process (e. g., oxidation or polymerization). It is important to note that when conducting a chemical process at a temperature below the equilibrium, one may synthesize compounds which are unstable at higher temperatures or for which the selectivity of the synthesis is low at higher temperatures. For efficient monitoring of the chemical processes, optical techniques are used (emission and absorption spectroscopy, Rayleigh scattering, etc.), chromatography and mass spectrometry (Zhivotov et al., 1985) which all require sophisticated equipment and optical access to the reaction zone.

When the energy of a pulsed excitation source (spark discharge, pulsed microwave discharge, pulsed high-current electron beam, etc.) is dissipated in a closed plasma reactor, then, as a result of the radiation-acoustic effect (Lyamshev, 1996), acoustic oscillations are formed due to the heterogeneity of the excitation (and, thereafter, heating) of the reagent gas. The measurement of sound waves does not require the use of sophisticated equipment, which give a lot of information about the processes occurring in the plasma reactor (Pushkarev et al., 2002; Remnev et al., 2001; Remnev et al., 2003a).

2. Experimental installation

This paper presents the results of the study of sound waves generated in gas mixtures when the energy dissipation of a pulsed high-current electron beam in a closed plasma reactor occurs. The scheme of measurements is shown in Fig. 1.

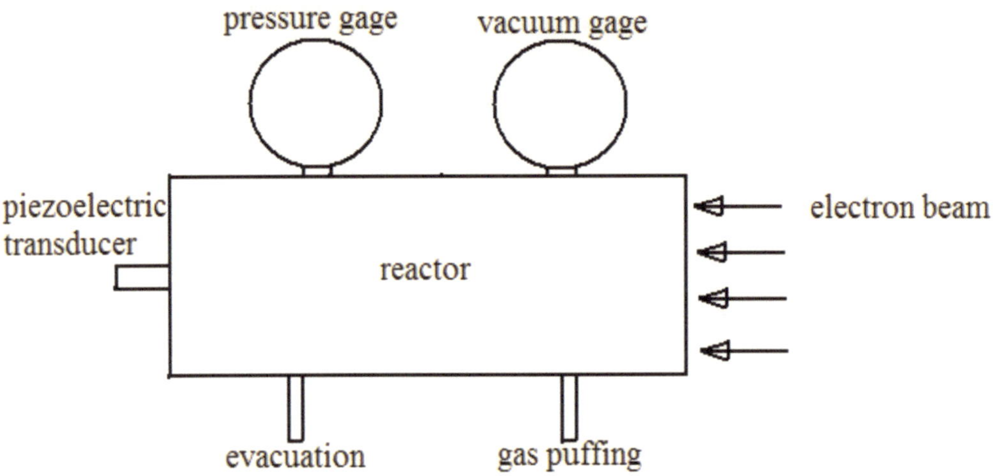

Fig. 1. Experimental scheme

The signal from a piezoelectric transducer was recorded using an oscilloscope Tektronix 3052B (500 MHz, $5 \cdot 10^9$ measurements/s). The source of the high-current electron beam is the accelerator TEA-500 (Remnev et al., 2004a, 2004b). Fig. 2 shows an external view of the TEA-500 accelerator.

In Fig. 3, typical oscilloscope traces of voltage and total electron beam current are shown.

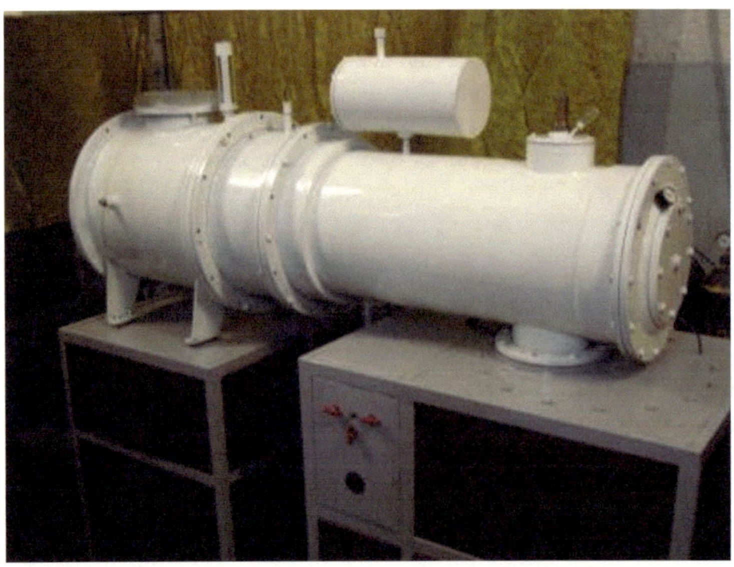

Fig. 2. The TEA-500 accelerator

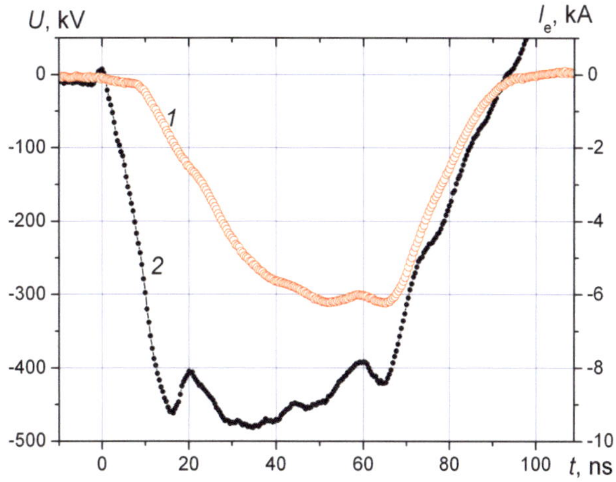

Fig. 3. Oscilloscope traces of electron current (1) and accelerating voltage (2)

These graphs are averaged for 10 pulses with a frequency of 1 impulse/s after operating the cathode for 10–20 pulses. The parameters of the electron beam are given in Table 1.

Electron energy	450–500 keV
Ejected electron current	up to 12 kA
Half-height pulse duration	80 ns
Pulse repetition rate	up to 5 pulses/s
Pulse energy	up to 200 J

Table 1. Parameters of the high-current pulsed electron beam

In Fig. 4 the spatial distribution of the energy density of the electron beam formed by the diode with a cathode made from a carbon fibre is illustrated.

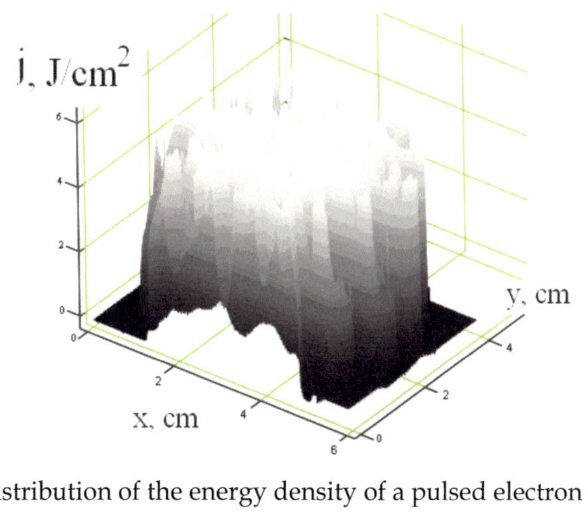

Fig. 4. The spatial distribution of the energy density of a pulsed electron beam

Most of the experiments were carried out with the reactor, comprised of a cylinder of quartz glass with an inner diameter of 14.5 cm and a volume of 6 litres. It is constructed in a tubular form; the electron injection begins from the titanium foil at the end of the tube. At the output flange of the plasma reactor there are a number of tubes used to connect a vacuum gauge and a manometer, a piezoelectric transducer, for an initial injection of the reagent mixture and for the evacuation of the reactor before a gas pumping. Other reactors, with a diameter of 6 cm and a length of 11.5 cm, with a diameter 9 cm and a length of 30 cm, were used as well. Fig. 5 shows a photograph of the plasma chemical reactor.

Fig. 5. Plasma chemical reactor with a volume of 6 litres

The sound waves were recorded by a piezoelectric transducer. Throughout the study, gas mixtures of argon, nitrogen, oxygen, methane, silicon tetrachloride and tungsten hexafluoride were used. When measuring pressure in the reactor using the piezoelectric transducer, we recorded the standing sound waves. An electrical signal coming from the piezoelectric transducer does not require any additional amplification. A typical oscilloscope trace of the signal is shown in Fig. 6. The reactor length is 39 cm and its inner diameter is 14.5 cm.

Test measurements were performed on an inert gas (Ar, 1 atm) to avoid any contribution of chemical transformations under the influence of the electron beam at a change in frequency of sound waves. For further signal processing it was necessary to transform it into digital form. In Fig. 7, a spectrum obtained by Fourier transformation of the signal shown in Fig. 6 is presented.

In our experimental conditions, the precision of measurement of the frequency is ±1.5 Hz.

3. Investigations of the frequency of the sound waves

In a closed reactor with rigid walls, after the dissipation of a pulsed electron beam, whose frequency for an ideal gas is equal to (Isakovich, 1973):

$$f_n = \frac{n}{2 \cdot l}\sqrt{\frac{\gamma RT}{\mu}} \quad , \qquad (1)$$

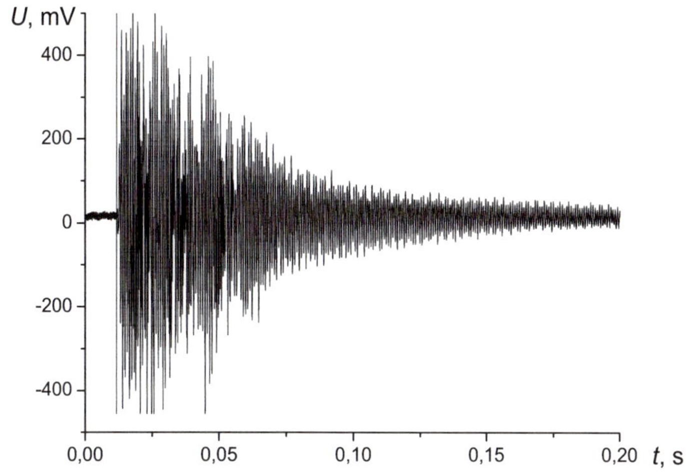

Fig. 6. Signal from the piezoelectric transducer

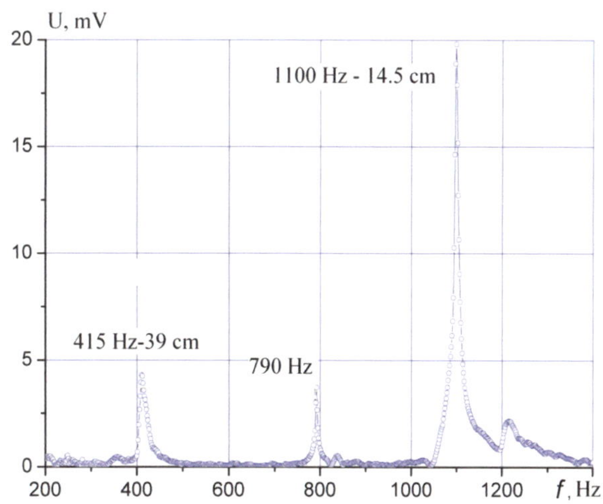

Fig. 7. The frequency spectrum of the signal from piezoelectric transducer. 415 Hz corresponds to the longitudinal sound waves, 1100 Hz is for transverse sound waves

where n is the harmonic number (n = 1, 2, ...), l is the length of the reactor, γ is the adiabatic exponent, R is the universal gas constant, and T and μ are, respectively, the temperature and molar mass of the gas in the reactor.

In the experiments we recorded the sound vibrations that correspond to the formation of standing waves along the reactor and across. For this study, the low-frequency component of the sound waves corresponding to the fundamental frequency (n = 1) waves propagating along the reactor was chosen.

The dependence of the frequency of the sound waves in the plasma reactor on the parameter $(\gamma/\mu)^{0.5}$ for different single-component gases is shown in Fig. 8 for the reactor with a length of 11.5 and 30 cm. The figure shows that in the explored range of frequencies the sound vibrations are well described by the relation for the ideal gases.

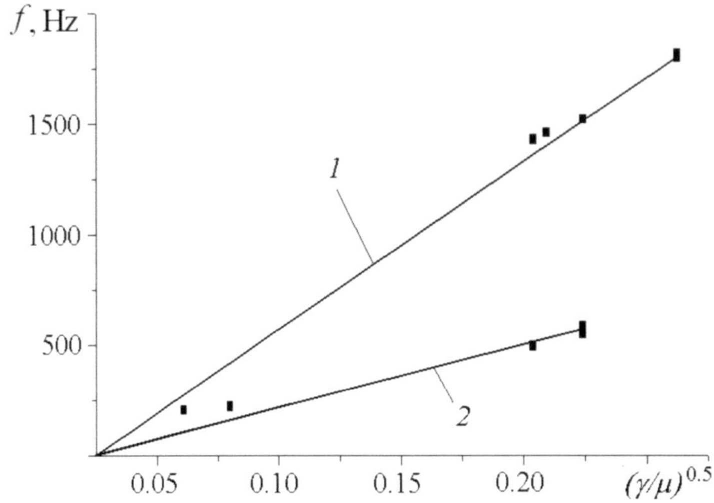

Fig. 8. The dependence of the frequency of the sound vibrations in the reactor on the ratio of the adiabatic exponent to the molar mass of single-component gases. Dots correspond to experimental data, lines are the calculations by (Eq.1) at l = 11.5 cm (1) and 30 cm (2).

In real plasma chemical reactions, multicomponent gas mixtures are used and the reaction products also contain a mixture of gases. When calculating the frequency of acoustic oscillations a weighting coefficient of each component of the gas mixture should be taken into account and the calculation should be performed using the following formula (Yaworski and Detlaf, 1968):

$$f_{sound} = \frac{\sqrt{RT}}{2l\sqrt{m_0}}\sqrt{\sum_i \frac{\gamma_i m_i}{\mu_i}}, \qquad (2)$$

where m_0 is the total mass of all components of the gas mixture; and m_i, γ_i, μ_i are, respectively, the mass, adiabatic exponent and molar mass of the i-th component.
Given that the mass of the i-th component is equal to

$$m_i = 1.66 \cdot 10^{-27} \mu_i N_i = K\mu_i \frac{P_i V}{P_0},$$

where N_i is the number of molecules of i-th component, P_i is its partial pressure, V is the reactor volume, P_0=760 Torr, and K is a constant.
Then (2) can be written in a more convenient way:

$$f_{sound} = \frac{\sqrt{RT}\sqrt{\sum_i \gamma_i P_i}}{2l\sqrt{\sum_i \mu_i P_i}}. \qquad (3)$$

Fig. 9 shows the dependence of the frequency of sound vibrations, resulting in a plasma chemical reactor when an electron beam is injected into two- and three-component mixtures, on the parameter φ defined by

$$\varphi = \frac{\sqrt{\sum_i \gamma_i P_i}}{\sqrt{\sum_i \mu_i P_i}}.$$

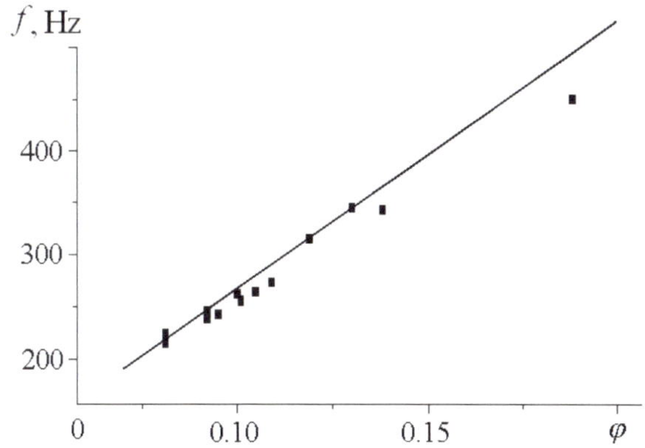

Fig. 9. The dependence of the frequency of sound oscillations in the plasma chemical reactor with a length of 30 cm on the parameter φ for gas mixtures. The points correspond to the experimental values, the lines are calculated from (3).

The frequency measurements of sound vibrations which arise in the plasma chemical reactor from the injection of pulsed electron beams into two- and three-component mixtures, showed that the calculation using (3) leads to a divergence between the calculated and experimental values of under 10%, and at frequencies below 400 Hz, less than 5%.
From (2) and (3) it is observed that the frequency of the sound waves depends on the gas temperature in the reactor, so the temperature should also be monitored. Let us determine the measurement accuracy which is necessary to measure the temperature so that the measurement error of the conversion level does not exceed the error due to the limitations in the accuracy of the frequency measurement. For transverse sound waves in argon (γ = 1.4, μ = 40, l = 0.145 m), (2) gives us that $f_{sound} = 64.2 \cdot (T)^{0.5}$.
This dependence is shown in Fig. 10.

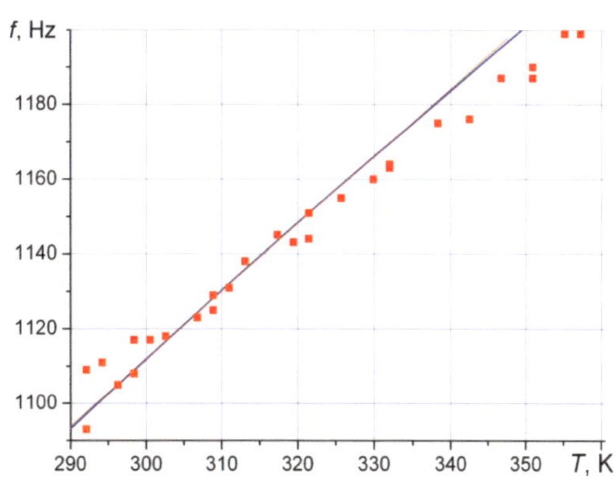

Fig. 10. The dependence of the frequency of transverse sound waves on the gas temperature. The points correspond to the experimental values, the lines - calculation by (2).

The calculated dependence of the frequency of transverse waves on temperature for the range 300–350 K is approximated by the formula f_{calc} = 570+1.8T. It follows that if the accuracy of measuring the frequency of the sound waves is 1.5 Hz it is necessary to control the gas temperature with an accuracy of 0.8 degrees. When measuring the spectrum of sound waves in the reactor, which has different temperatures over its volume, the profile of the spectrum is expanding. But it does not interfere with determining the central frequency for a given harmonic.

4. Investigation of the energy of sound waves

In a closed plasma chemical reactor when an electron beam is injected, standing waves are generated whose shape in our case is close to being harmonic. Then the energy of these sound waves is described by (Isakovich, 1973):

$$E = 0.25\beta \Delta P_s V \tag{4}$$

where β is the medium compressibility, ΔP_s is the sound wave amplitude, and V is the reactor volume.
At low compression rates ($\Delta P_s \ll 1$) and if the momentum conservation law is implemented (under damping), the medium compressibility can be calculated by the formula (Isakovich, 1973):

$$\beta = \left(\rho C_s^2\right)^{-1}, \tag{5}$$

where ρ is the density of the gas and C_s is the velocity of sound in the gas.
Fig. 11 shows the change in pressure in the reactor after the injection of the beam (Pushkarev et al., 2001).

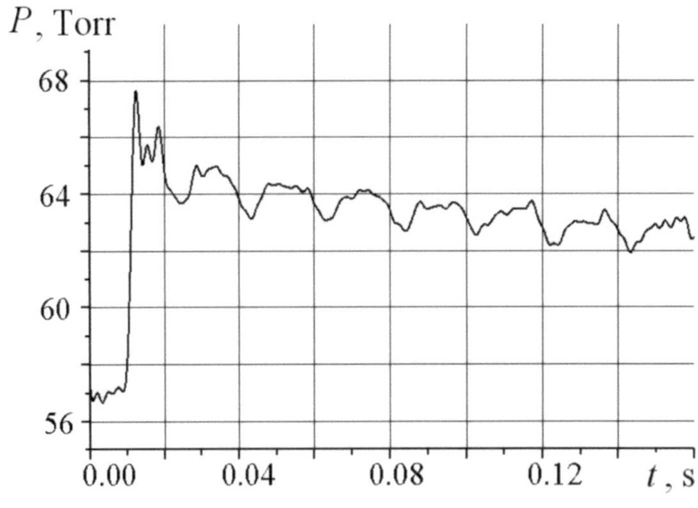

Fig. 11. The change of pressure in the reactor filled with a mixture of hydrogen and oxygen, after the injection of a pulsed electron beam in an absence of combustion.

It is evident that a decrease in pressure in the reactor (due to cooling of the gas), after a sharp increase is sufficiently slow and hence the speed of response of the pressure sensor is adequate for recording a complete change of pressure in the reactor. The dependence of the energy of the sound vibrations in the reactor on the electron beam energy, which is absorbed in the gas, is shown in Fig. 12.

The electron beam energy absorbed by the gas was calculated as the product of the heat capacity of the gas and the mass and the change in the gas temperature. The change of the gas temperature in a closed reactor was determined from the equation of the ideal gas state regarding a change in pressure. Pressure changes were recorded with the help of a fast pressure sensor SDET-22T-B2. The energy of the sound waves was about 0.2% of the electron beam energy absorbed in the gas.

For nitrogen and argon in a wide pressure range (and thus of energy input of the beam into the gas), a good correlation between the energy of sound vibrations and the energy input of the beam into a gas was obtained, which allows evaluation of the energy input of the beam into a gas using the amplitude of the sound waves.

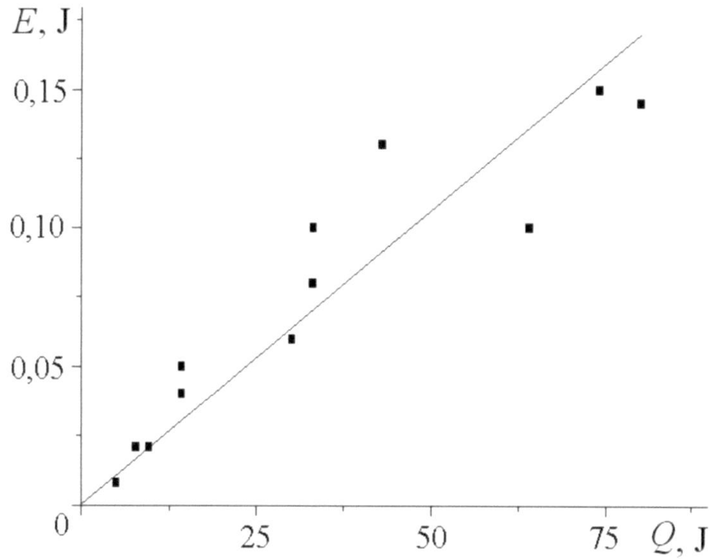

Fig. 12. Dependence of the energy of the sound vibrations in the reactor on the energy of the electron beam absorbed in the gas. The points correspond to the experimental values, the line is an approximation by a polynomial of the first order

5. Investigation of the acoustic attenuation

The presence of gas particles whose size is much larger than the gas molecules causes an increase in the attenuation of sound waves propagating in a gas. You can observe such an effect when watching the attenuation of sound in fog. Sound propagation in the suspensions of microparticles in the gas was studied in (Molevich and Nenashev, 2000) when propagating in open space. To control the process of formation of particles in a volume of the reactor to measure an attenuation of acoustic vibrations is needed, but it is necessary to estimate the sound attenuation in the reactor in the case of absence of microparticles, or aerosol.

Since the waveform of sound oscillations generated in the reactor during the injection of high-current electron beam is close to being harmonic, then a change of energy of the sound waves due to absorption is as follows (Isakovich, 1973):

$$E(t) = E_0 e^{-\alpha t},$$

where α is the time coefficient of absorption.

When the sound waves are propagating in a tube closed from both sides, the absorption coefficient is (Isakovich, 1973):

$$a = a_1 + a_2 + a_3 + a_4$$

where α_1 is the sound absorption coefficient when propagating in an unbounded gas, α_2 is the sound absorption coefficient for reflection from the side walls of the pipe when propagation along the pipe, α_3 is the absorption coefficient when reflection is on the ends of the pipe, and α_4 is the absorption coefficient due to friction on the pipe wall.

5.1 The absorption coefficient of sound when propagating in an unbounded gas

The absorption coefficient of sound wave energy in a gas due to thermal conduction and the shear viscosity of a gas can be calculated by the Stokes–Kirchhoff formula (Isakovich, 1973):

$$\alpha_1 = \frac{(2\pi f_{sound})^2}{2\rho C_{sound}^2}\left[\frac{4}{3}\eta + \chi\left(\frac{1}{C_v} - \frac{1}{C_p}\right)\right], \tag{6}$$

where η is the shear-viscosity coefficient (g/cm·s), χ is the heat conductivity coefficient (cal/cm·s·deg), and C_v and C_p are, respectively, the heat capacity of the gas when the volume is constant, and when the pressure is constant (cal/g·deg).

5.2 The absorption coefficient when reflected from the side walls of the pipe

For a low-frequency sound wave which is propagating through a circular pipe, provided that $\lambda > 1.7d$ (where λ is the wavelength and d is the pipe diameter) the wave front is flat and the damping coefficient of the energy of sound wave when it is propagating along the pipe with ideal thermally conductive walls can be calculated by Kirchhoff's formula (Konstantinov, 1974):

$$\alpha_2 = \frac{1}{r_0}\sqrt{\frac{\pi f_s}{\rho}}\left[(\gamma - 1)\sqrt{\frac{\chi}{\gamma C_p}} + \sqrt{\eta}\right],$$

where r_0 is the pipe radius.

Taking into account that $\rho = \rho_0 P/P_0$, where P is the gas pressure in reactor and ρ_0 is the density of gas at normal conditions, P_0=760 Torr, then:

$$\alpha_2 = \frac{K_1}{r_0}\sqrt{\frac{f_s}{P}} \quad , \; K_1 = \sqrt{\frac{\pi P_0}{\rho_0}}\left[(\gamma-1)\sqrt{\frac{\chi}{\gamma C_p}} + \sqrt{\eta}\right] \quad . \tag{7}$$

5.3 The absorption coefficient of sound when reflected from the ends of the pipe

The energy of sound waves reflected from the wall is

$$E = E_0(1\text{-}\delta),$$

where δ is the coefficient of energy absorption of sound wave for a single reflection and E_0 is the energy of the incident wave.

After n reflections $E_n = E_0(1\text{-}\delta)^{-n}$. After n reflections during the time t a sound wave will pass the distance $L = n{\cdot}l = C_s{\cdot}t$, therefore it can be written

$$n = \frac{C_s\, t}{l}$$

Then the change in the energy of sound waves reflected at the ends of a pipe is

$$E(t) = E_0(1-\delta)^{-\frac{C_s\, t}{l}} \tag{8}$$

If the change in energy of sound waves when reflected at the ends of the pipe is written as

$$E(t) = E_0 e^{-\alpha_3\, t} \tag{9}$$

then from (8) and (9) we obtain

$$\alpha_3 = \frac{C_s \ln(1-\delta)}{l} \tag{10}$$

Under the normal incidence of a flat wave on a metal wall, which is a good heat conductor, the absorption coefficient of sound wave energy is (Molevich and Nenashev, 2000):

$$\delta = 4(\gamma-1)\sqrt{\frac{\pi f_s \chi}{\gamma C_p P}} \quad . \tag{11}$$

But if we consider only a normal incidence of a sound wave onto the ends of the reactor, we would neglect the absorption of sound waves reflected from the side walls of the reactor (i.e., $\alpha_2 = 0$). The absorption of sound at the same time will be forced by the thermal conductivity and viscosity of the gas and by absorption when reflected from the ends of the reactor, as in (6) and (10). As will be shown below, the experimentally measured absorption coefficients of the sound wave energy in the reactor is several times higher than the values which are calculated by (6) and (10). Therefore, when there is a reflection from the ends of the reactor, the dependence of the absorption coefficient on the angle of the incidence should be taken into account and the calculation should be performed by the formula (Molevich Nenashev, 2000):

$$\delta = \sqrt{\frac{f_s}{\gamma P}}\left[0.39(\gamma-1)\sqrt{\frac{\chi}{C_p}} + 0.37\sqrt{\eta}\right], \tag{12}$$

From (10) and (12) we obtain (noting that when $\delta \ll 1$, the quantity $\ln(1-\delta)\approx\delta$):

$$\alpha_3 = \frac{K_2}{l}\sqrt{\frac{f_s}{P}}, \quad \text{where} \quad K_2 = C_s\left[0.39(\gamma-1)\sqrt{\frac{\chi}{\gamma C_p}} + 0.37\sqrt{\frac{\eta}{\gamma}}\right], \tag{13}$$

To take into account the energy losses of the sound wave due to friction on the wall is important in cases where the diameter of the pipe is comparable to the mean free path of gas molecules, i.e. for capillaries. In our case, it can be assumed that $\alpha_4 \approx 0$.

5.4 Calculation of the total absorption coefficient

Then the total absorption coefficient of sound wave energy in a closed reactor can be written as (Pushkarev, 2002):

$$\alpha = \left(\frac{K_1}{r_0} + \frac{K_2}{l}\right)\sqrt{\frac{f_s}{P}}, \tag{14}$$

where K_1 and K_2 are calculated by the (7) and (13), r_0 and l are taken in cm, f_s is in Hz, and P is in Torr. The coefficients K_1 and K_2 for the investigated gases are summarized in Table 2.

gas	K_1	K_2
N_2	25	218
O_2	27	189
Ar	32	254
WF_6	6.5	52
$SiCl_4$	5.9	45.7

Table 2. Calculated damping coefficients

Numerical estimates of the contribution of the different mechanisms of absorption of the sound waves in the reactor show that the influence of volume absorption (due to the thermal conductivity and viscosity of the gas) is insignificant. For the sound waves which are generated in the reactor 30 cm long, filled with nitrogen, at a pressure of 500 Torr: $\alpha_1=1.8\cdot10^{-3}$ s^{-1}, $\alpha_2=5.9$ s^{-1}, $\alpha_3=7.7$ s^{-1}. The total absorption coefficient, which takes into account only the normal incidence of sound-sound wave (i.e., $\alpha_2=0$, $\alpha_3=1.2$ s^{-1}) is much smaller than the experimentally measured coefficient for these conditions (14.7 s^{-1}). It is important to note that when the sound waves are propagating in a closed reactor the main contribution (60%-80%) to the absorption is made by the viscosity of the gas. The magnitude of the second term in (7) and (13) is more than the first term by 3-9 times (for different gases). The contribution of the side walls of the reactor and its ends to the absorption of sound waves is approximately the same for a large reactor. The ratio of energy absorption of sound waves in the reactor for different harmonics is shown in Fig. 13.

An electron beam was injected into the 30 cm long reactor, filled with argon at different pressures. To compare the attenuation coefficients in the different plasma-chemical reactors (lengths 11.5 and 30 cm) and in different gases, the value of the attenuation coefficient was normalized by the coefficient K:

$$K = \left(\frac{K_1}{r_0} + \frac{K_2}{l} \right),$$

Fig. 13 shows that the magnitude of the absorption coefficient is proportional to the square root of frequency of sound waves, in accordance with the (14).

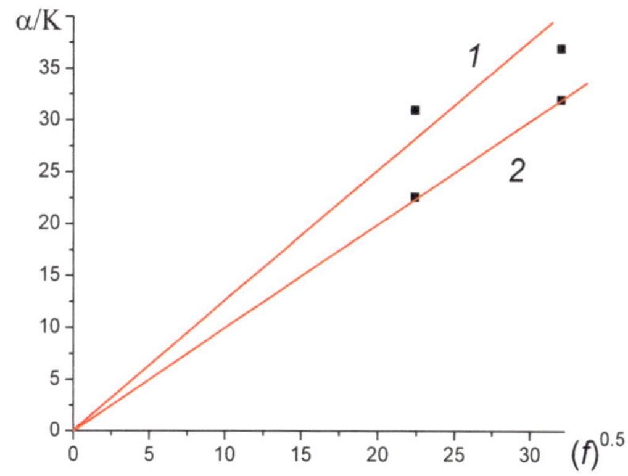

Fig. 13. The dependence of the absorption coefficient of energy of sound waves in the reactor on f_s. The points correspond to the experimental values, lines, to calculation by (14). The gas is argon, 1–400 Torr and 2–500 Torr.

6. Analysis of the conversion level of gas-phase compounds with respect to a change of sound wave frequency

By simple calculations, it can be shown that for a chemical reaction in which both the initial mixture of reagents and the mixture after the reaction are a gas (no phase transition), then the frequency of sound vibrations after the reaction, is equal to that in the initial mixture. A slight change in frequency of standing sound waves can be associated only with changes in the adiabatic exponent. But if a reaction produces solid or liquid products, the frequency of the sound waves will vary. For the pyrolysis reaction of methane:

$$CH_4 = 2H_2 + C$$

The decrease of methane on the value of ΔP will lead to the formation of $P_i = 2\Delta P$ of hydrogen. Let us denote the methane conversion level as $\alpha = \Delta P/P_0$. Then from (3) we obtain:

$$f_s = \frac{\sqrt{RT} \cdot \sqrt{\gamma_1(1-\alpha) + 2\gamma_2\alpha}}{2l \cdot \sqrt{\mu_1(1-\alpha) + 2\mu_2\alpha}}, \qquad (15)$$

where γ_1 and μ_1 are, respectively, the adiabatic index and molar mass of methane, and γ_2 and μ_2 are, respectively, the adiabatic index and molar mass of hydrogen.

For the reactor with an inner diameter of 14.5 cm, the dependence of the level of methane conversion on the frequency of the transverse sound waves is shown in Fig. 14.

When the measurement accuracy of the sound waves frequency is 1.5 Hz and the accuracy of the temperature is 0.8 degrees, the developed method allows controlling the level of conversion of methane to carbon with an accuracy within 0.1% (Pushkarev et al., 2008). A weak damping of sound waves makes possible to measure the oscillation frequency with a fine resolution.

When measuring the frequency of sound waves in a reactor filled with methane, the frequency spectrum obtained is shown in Fig. 15.

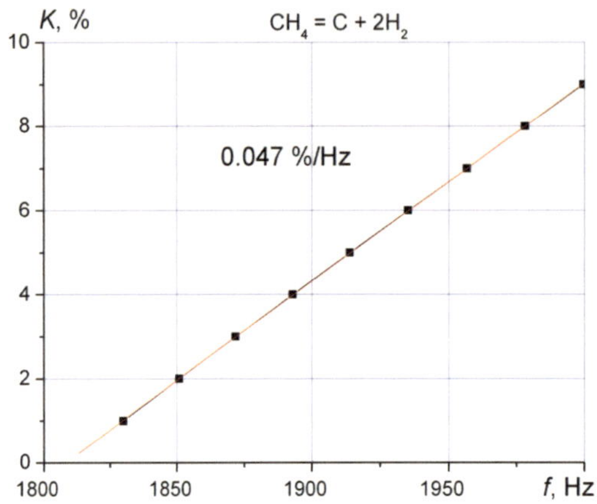

Fig. 14. Dependence of the conversion level of methane in the pyrolysis reaction on the frequency of transverse sound waves.

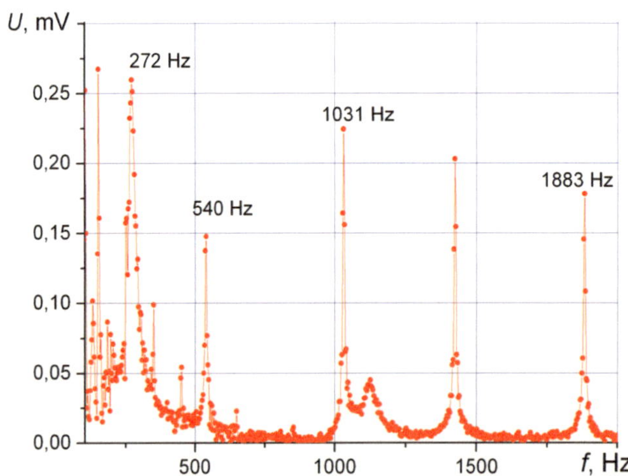

Fig. 15. The spectrum of sound waves in a reactor filled with methane after injection of high-current electron beam.

In Fig. 16, the frequency spectrum in the range of higher harmonics is shown.

In Fig. 17 the dynamics of methane conversion under the influence of a pulsed electron beam is shown.

Adding argon causes an increase in the error in determinating the level of conversion. In the reaction:

$$CH_4 + \beta \cdot Ar = CH_4 \cdot (1\text{-}\alpha) + 2\alpha \cdot H_2 + \beta \cdot Ar + \alpha \cdot C$$

where a is the conversion level of methane and β is the contents of argon in a mixture.

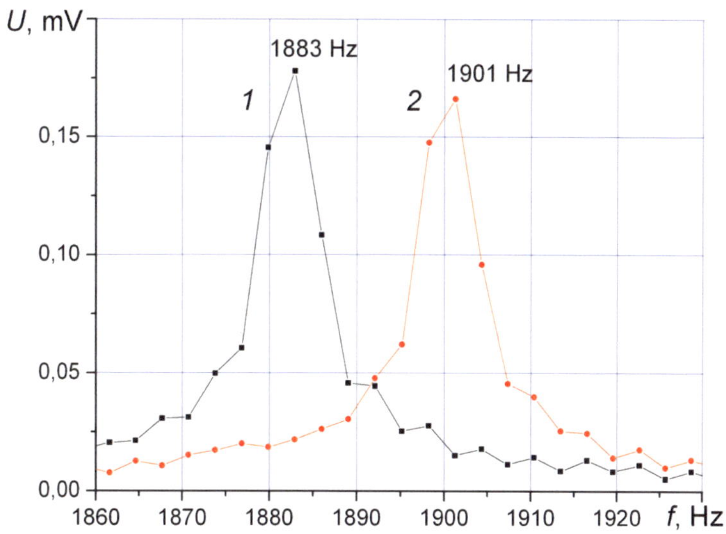

Fig. 16. The spectrum of sound waves in methane: initially (1), and after irradiation by 500 pulses: (2).

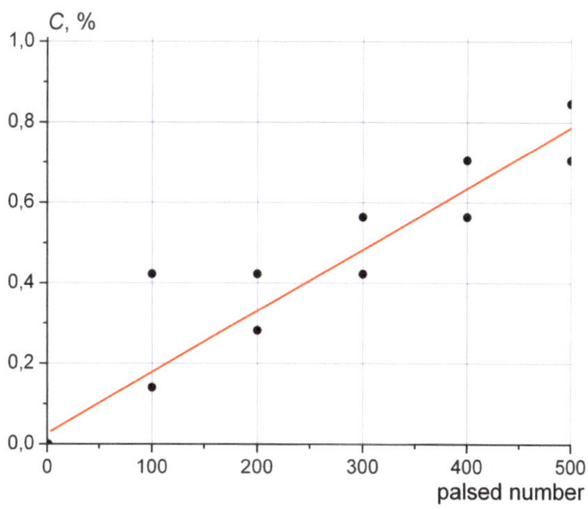

Fig. 17. The dependence of the conversion level of methane pyrolysis on the number of pulses in the electron beam.

In Fig. 18 the dependence of the error in determining the level of conversion of methane (in the pyrolysis reaction) on the percentage content of argon in a mixture of methane and argon is shown. The content of argon was calculated in relation to the volume of methane. 50% argon corresponds to 33% of the total volume of the mixture in the reactor.

The technique developed is designed for operational control of the technology process and can also be used to control the processes of recovery of halogenides of various compounds (Remnev et al., 2004c). Fluoride compounds are widely used in the production of rare earth metals and isotopic enrichment. When restoring halogenides, compounds with a high chemical activity (F_2, HF, HCl, etc.) are formed. Registration of the sound waves does not

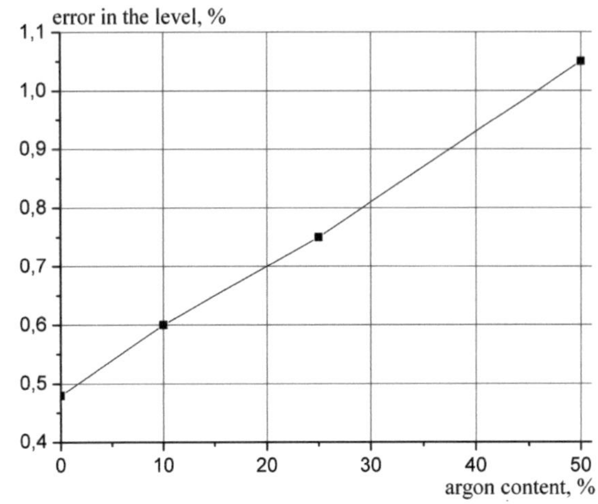

Fig. 18. Dependence of error in determining the level of conversion of methane mixed with argon on the argon content.

require access to the reaction zone, which prevents the destruction of diagnostic equipment. Measurements of the frequency of sound vibrations generated during the dissipation of energy of a pulsed excitation source, make it possible to control the progress of a plasma chemical reaction in the formation of solid products, such as the reaction of $CO_2 \rightarrow C+O_2$; $WF_6 \rightarrow W+3F_2$, etc.

This method gives the average degree of conversion over the reactor volume, which is of particular advantage in avoiding the errors associated with the localization of sampling inherent in other methods. Time measurement and signal processing does not exceed 0.2 s, which allows using this method in systems of automated process control. Measurement of the frequency of acoustic waves is performed with piezoelectric transducer and does not require sophisticated equipment.

The proposed method for controlling the progress of plasma chemical reactions regarding a change in the frequency of sound waves has been used to study the direct recovery of tungsten from tungsten hexafluoride under the influence of high-current electon beams (Vlasov, 2004). A good agreement with the data which were obtained by the weighing of a substrate placed in the reactor was obtained. An acoustic method to control the phase transition was also used for studying the synthesis of nanopowder oxides (Remnev, 2004d).

7. Measurement of energy absorbed by the gas in a closed reactor with exothermic reactions

The measurements show that for nitrogen and argon in a wide range of pressures (and therefore of beam energy input in a gas) a good correlation between the energy of sound vibrations and the electron beam energy input in a gas was obtained. This allows evaluating the energy input of a beam into gas using the amplitude of sound waves. The energy of sound waves is about 0.2% of the electron beam energy absorbed in the gas (Remnev et al., 2003a).

To measure the beam energy input into a gas, calorimetric measurements are commonly used. Here the energy of a pulsed power source is estimated by changes in temperature of a calorimeter. The energy absorbed by the gas is obtained as the difference between the calorimeter readings and the known energy of the pulsed energy source. A disadvantage of this method is that some significant errors in measuring of the source energy which is absorbed by a high pressure take place. When the density of energy absorbed by the gas is more than 10 J/liter and at pressures above 400 Torr, the gas is heated to a temperature which is above that of the heating temperature of the calorimeter. The calorimeter is heated in this case not only by the influence of external energy, but also by the surrounding gas.

The energy of a pulsed electron beam absorbed by the gas in a closed reactor can be measured more precisely by a change in pressure (Moskalev and Sergeev, 1991). The electron beam energy absorbed by the gas is calculated as the product of the specific heat capacity of gas and the mass and the change in the gas temperature. The change of the gas temperature in the reactor is determined from the equation of state for an ideal gas due to a change of pressure in a closed reactor. A disadvantage of this method is that it is impossible to measure the electron beam energy absorbed by the gas in a closed reactor, in the case when an exothermic reaction takes place in the reactor. The change in pressure during the flow of exothermic reactions is due to the heating of the gas caused by absorption of the electron beam energy and heating because of the energy release in exothermic reactions.

In case of exothermic reactions in a closed reactor, the energy of a pulsed power source absorbed by a gas can be also measured by the amplitude of the standing sound waves. We have obtained that when a high-current electron beam with the following parameters: electron energy is 300–350 keV, beam current is 10–12 kA, pulse duration is 50–60 ns, is injected into a mixture of silicon tetrachloride with hydrogen, then a change of pressure in the reactor is 10–15 times bigger than changes due to heating of a gas by absorption of the electron beam energy. The dependence of the energy used for heating the gas in the reactor during the injection of an electron beam, on the gas mass is shown in Fig. 19. Curve 1 corresponds to the vapour-phase $SiCl_4$ or its mixture with argon and hydrogen (in this case m is the partial weight of silicon tetrachloride), curve 2, to that of argon and nitrogen. To prevent $SiCl_4$ condensation on the walls at pressures above 200 Torr (saturation vapour pressure at 30 °C) the reactor was heated to a temperature of 60 °C (the boiling point of $SiCl_4$ at normal conditions is 57 °C).

The dependence of the beam energy input on the gas mass for nitrogen and argon has an usual form: the growth at low pressures (200–600 Torr in the reactor is used for electrons with an energy of 300 keV), when the range of electrons in the beam and in the ionization cascade exceeds the length of the reactor. A part of the beam energy in this case is absorbed by the rear wall of the reactor. When the pressure is above 600 Torr (the mass is more than 0.08 mole) the electron beam is almost completely absorbed by the gas and in curve 2 a plateau is observed. The maximum energy used for gas heating and measured by the pressure jump is 70 J. The energy of the electron beam in a pulse is 90 J.

Calculation of the energy used for heating the silicon tetrachloride and its mixture with other gases during the electron beam injection (curve 1, Fig. 19), based on the indications of the pressure sensor showed that in this case, the release of energy exceeds by an order of magnitude the total energy of the beam and can not be explained only by the absorption of the electron beam in a gas. In addition, the dependence 1 in the investigated range of pressures (and the gas mass) is different than in curve 2 of Fig. 19, which also indicates another source of the heating of $SiCl_4$, beyond the heating by the electron beam. This can be explained by the occurrence in the reactor of the following exothermic reactions:

$$Cl + H_2 = HCl + H + 0.05\ eV$$

$$Cl + Cl + M = Cl_2 + M + 2.5\ eV$$

$$H + Cl_2 = HCl + Cl + 2\ eV$$

$$H + HCl = H_2 + Cl - 0.05\ eV$$

We have found that the frequency of standing sound waves generated in the 30 cm long reactor exceeds 100 Hz for the investigated gases. If the external effects of energy will exceed the period of these sound waves, the conditions for the formation of standing harmonic waves will be violated. Therefore, the duration of a pulsed energy source should be no more than 10^{-2} seconds.

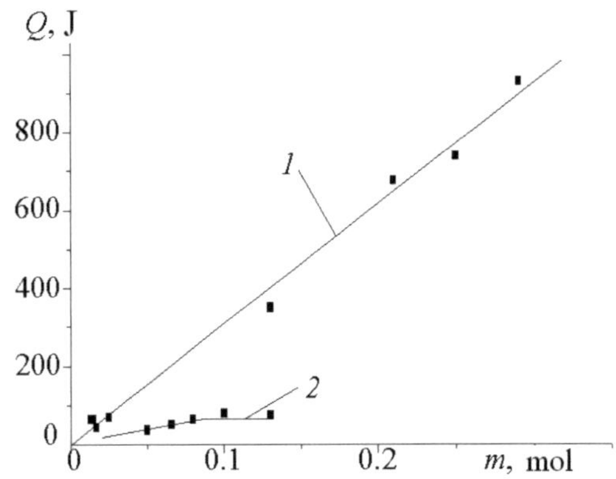

Fig. 19. The dependence of the energy used for heating the gas during the electron beam injection on the gas mass—$SiCl_4$ (1), argon, nitrogen—(2).

8. Registration of the conversion of gas-phase compounds in nanodispersed clusters by means of an increase in the absorption coefficient of sound waves

The presence in a gas of particles whose size is much larger than the gas molecules causes an increase in the attenuation of the sound waves propagating in the gas. The dependence of the absorption coefficient of energy of sound vibrations in the reactor on the pressure for different gases is shown in Fig. 20.

To compare the attenuation coefficients in the different plasma-chemical reactors (lengths 11.5 and 30 cm) and in different gases, the value of the attenuation coefficient was normalized by the coefficient *K*, calculated by the formula:

$$K = \sqrt{f_{sound}}\left(\frac{K_1}{r_0} + \frac{K_2}{l}\right) ,$$

The points in Fig. 20 correspond to the experimental measurements, curve 1 is the calculation by (14). For nitrogen, argon and oxygen, the divergence between the calculated and experimental values of the absorption coefficient does not exceed 30%.

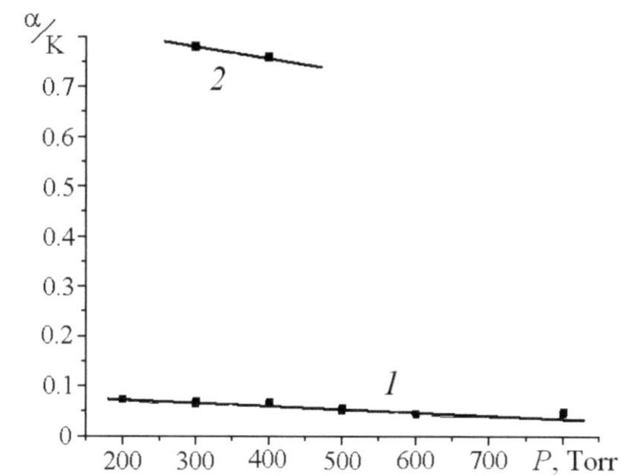

Fig. 20. The dependence of the normalized sound absorption coefficient on the pressure.

For the sound waves generated during the dissipation of a pulsed electron beam in the vapor-phase WF_6 (curve 2 in Fig. 20), the experimentally determined value of the absorption coefficient is much higher than (by a factor of 14–15) that calculated by (14). This can be explained by the formation of clusters in the reactor during the injection of the electron beam. The presence of large particles in the gas increases the absorption of sound waves. When the injection of a high-current electron beam in WF_6 occurs, a direct recovery of tungsten in the form of nanosized particles causes not only an increase in the frequency of the sound waves (Remnev et al., 2004c), but also a significant increase in the energy absorption of sound vibrations.

9. Acoustic diagnostics of a pulsed electron beam

When an electron beam passes through the target, it loses energy. The energy losses create a temperature field in the target. As a result, thermoelastic stresses appear and cause acoustic oscillations, whose shape corresponds to the distribution of electrons in the beam. This method allows of measuring the following parameters of a pulsed electron beam (Remnev et al., 2003b):

- Geometry of the electron beam
- Distribution of the energy over the beam cross-section
- The total energy of the beam
- Electron beam displacement across the plasma reactor

The experimental procedure consists in the following: directly in the beam arrival area a target-rod or a wire was placed. The target was placed perpendicularly towards the motion of the particles. A piezoelectric sensor was fixed at one end of the rod; and the other end has a conical shape to absorb the vibrations propagating in the opposite direction from the sensor. The piezoelectric transducer is clamped between the rod and cone, and the cone is also designed to absorb the vibrations passing through the sensor. The scheme of the detector is shown in Fig. 21.

As a target, a copper wire of rectangular cross-section of 2 mm × 5 mm was used. The wire was placed in a plasma-chemical reactor, being clamped between two rubber vacuum seals. A beam of electrons was directed through the foil of an exit window, the pressure in the reactor was atmospheric. After the beam interacts with the target, the waves propagate along the wire in both directions from the place where the electrons hit, they should be

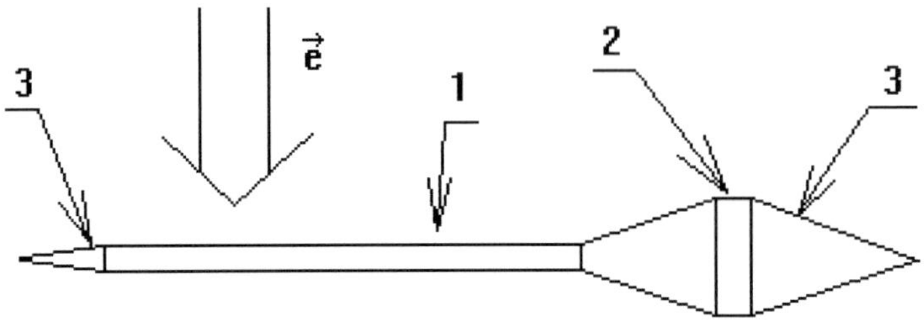

Fig. 21. The scheme of the acoustic diagnostics of the electron beam: 1-target, 2-piezoelectric sensor, 3-extinguishing cones.

diluted over time. This was achieved by increasing the length of the wires coming from opposite sides of the chamber. In order to eliminate the noise caused by an increase in the target potential when it is hit by the electrons, the wire was grounded.

9.1 Determination of the beam geometry

The duration of the first pulse of an acoustic signal, shown in Fig. 22, is determined by the diameter of the electron beam. Knowing the duration and sound velocity in a material of the absorber, the diameter of the beam cross-section can be determined.

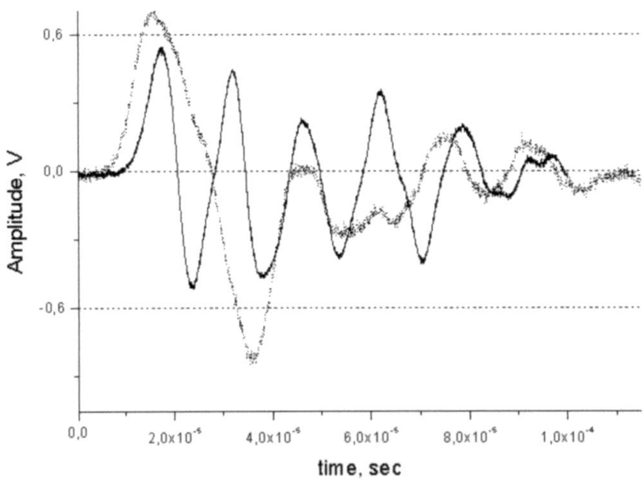

Fig. 22. Oscilloscope trace of the electrical pulses from the piezoelectric transducer when an electron beam is subjected to an absorber

We now determine the geometric dimensions of the beam. The length of the first pulse is 11 microseconds. Knowing the sound speed in the absorber, we obtain:

$$\Delta d = v_{sound} \cdot \Delta t = 3562 \cdot 10.96 \cdot 10^{-6} = 3.9\, cm$$

The diameter of the beam impress on a dosimetric film is 4 cm (see Fig. 4). Fig. 22 also shows an oscilloscope trace of a beam of reduced diameter. The change in the diameter was achieved by screening the absorber so that only the central part of the beam, with a diameter of 20 mm, hits the absorber. Therefore, knowing the duration of the first pulse, the diameter of the electron beam can be determined with a good accuracy.

9.2 Determination of the beam profile

Figure 23 shows the oscilloscope traces obtained at different beam profiles.

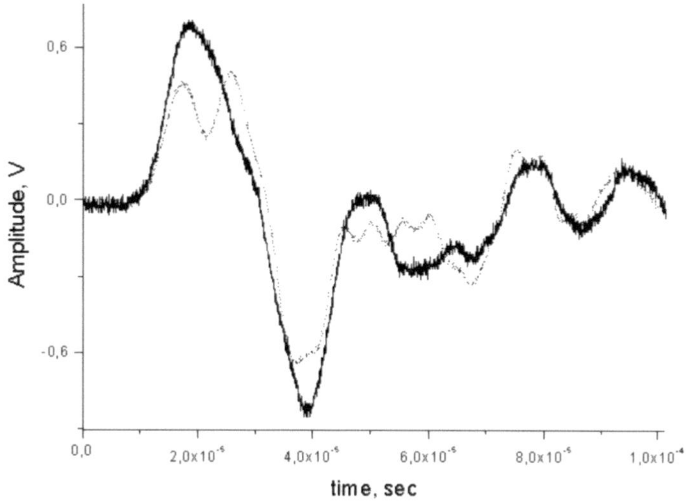

Fig. 23. Oscilloscope traces for different profiles of the beam

A change in the profile was achieved by shielding the central part of the absorber (see Figure 24), thus, the beam can be considered as hollow. The screen size was 10 mm.

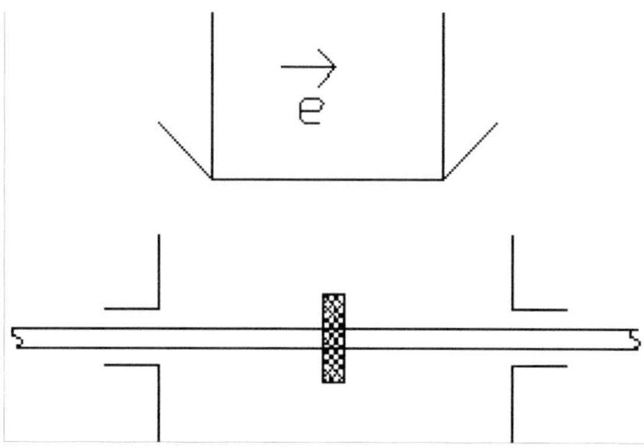

Fig. 24. Screening the center of the beam

A dip in the curve, due to a lack of a central part of the beam, is clearly visible in the graph.

9.3 Determination of the total energy of the beam

The measurements showed that the amplitude of the signal from the piezoelectric transducer is proportional to the total energy of the electron beam in a pulse. Figure 25 shows acoustic pulses for various values of the energy per pulse.

The beam energy was regulated by a grounded grid, which was located between the anode foil and the absorber. The dependence of the integral of the acoustic pulse on the total energy of the beam is shown in Figure 26. The presented dependence has a linear character, which allows a further use of the sensor as a probe of the total beam energy.

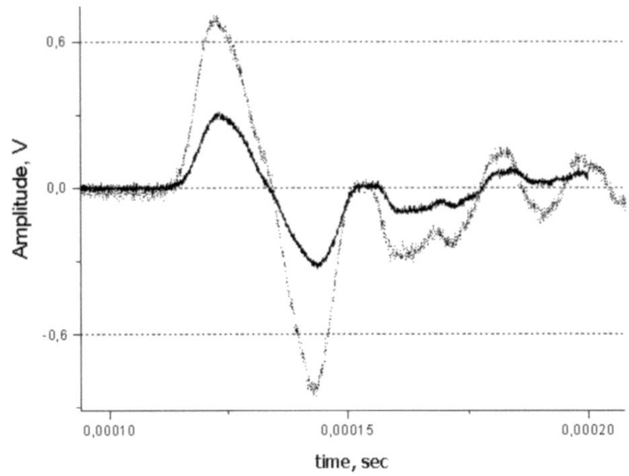

Fig. 25. Oscilloscope traces for different total energies of the beam.

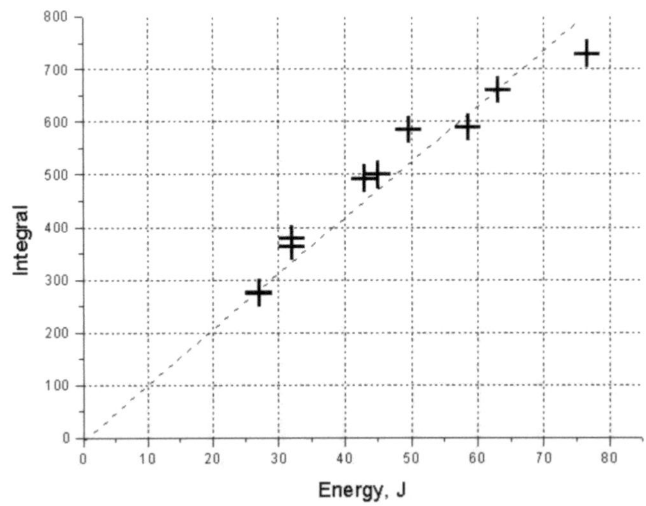

Fig. 26. The dependence of the integral of the acoustic pulse on the total electron beam energy

The advantages of radiation-acoustic diagnostic systems are the high noise immunity of the informative acoustic signal and the possibility of rapid analysis. The system can be used in any manufacturing plants or for research purposes without a significant change in the design of the accelerator chamber. The device is not exposed to the influence of a chemically active media, which makes it promising for the study of plasma chemical reactions initiated by a pulsed electron beam.

10. Conclusion

Our investigation of the sound waves generated in a closed reactor during the absorption of a pulsed electron beam shows that a simple experimental setup, recording the acoustic vibrations can accurately control the plasma process, accompanied by a change in the phase composition of the initial reagent mixture. The formation of clusters in the volume of the reactor causes a change in the frequency of sound waves and a significant increase in the attenuation of the oscillations amplitude. Diagnostics regarding the sound waves in the reactor can be used for operational control of the plasma process. When reducing the diameter of the reactor, the resonance frequency of transverse acoustic waves increases and the accuracy of measurement of methane conversion increases as well.

11. References

Bondar, Y.F., Zavorotny, S.I., Ipatov, A.L., Mkheidze, G.I., Ovchinnikov, A.A., Savin, A.A. (1982) The study of a relativistic electron beam transport of in a dense gas. Plasma Physics, 8(6): pp. 1192–1198.

Isakovich, M.A. (1973) General acoustics. Moscow: Nauka.

Konstantinov, B.P. (1974) A hydrodynamic sound formation and propagation of the sound in the suspension of microparticles in a gas. Moscow: Nauka.

Lyamshev, L.M. (1996) Radiation acoustics. Moscow: Fizmatlit-Nauka.

Molevich N.E., Nenashev, V.E. (2000) Influence of the volume viscosity on the sound propagation in the suspension of microparticles in a gas. Acoustical Physics,4: pp. 520–525.

Moskalev, V.A., Sergeev, G.I. (1991) The measurement of charged particle beams. Moscow: Energoatomizdat.

Pushkarev, A.I., Pushkarev, M.A., Zhukov, L.L., Suslov, A.I. (2001) Measurement of the energy dissipation of the electron beam in a dense gas by a quick-response differential pressure sensor. Physics, 7: pp. 93–97.

Pushkarev A.I., Pushkarev M.A., Remnev, G.E. (2002) Sound waves generated due to the absorption of a pulsed electron beam in gas. Acoustical Physics, 48(2) pp. 220–224.

Pushkarev, A.I., Novoselov, Y.N., Remnev, G.E. (2006) Chain processes in a low-temperature plasma. Novosibirsk: Nauka.

Pushkarev, A.I. and Sazonov, R.V. (2008) Acoustic method of monitoring the conversion of methane into carbon. Acoustical Physics, 54(1): pp. 135–137.

Remnev, G.E., Pushkarev, A.I., Pushkarev, M.A., Krasilnikov, V.A., Guzeeva, T.I. (2001) Monitoring of the changes in chemical composition of gases in a plasma chemical reactor during condensation of the reaction products using the acoustic-wave frequency data. Russian Physics Journal 44(5): pp. 482-485.

Remnev, G.E., Pushkarev, A.I., Pushkarev, M.A. (2003a) Method of monitoring the changes in the phase composition of a gas mixture in a closed reactor Patent of Russia № 2215799 RF, MPK7 S22V 5 /00. / Applied 04.03.2002. Published. 10.11., Bul. № 31.

Remnev, G.E., Pushkarev, A.I., Ezhov, V.V. (2003b) Radiation-acoustic diagnostic of the profile of a pulsed electron beam. Proceedings of the 11th International Scientific School-Seminar "Physics of the pulsed discharges in condensed media". Nikolaev, August 2003, pp. 77–78.

Remnev, G.E., Furman, E.G., Pushkarev, A.I., Kondratiev, N.A., Goncharov, D.V. (2004a) High-current pulsed accelerator with matched transformer: construction and exploitation characteristics. IEEE Transactions on fundamentals and materials, 124(6): pp. 491–495.

Remnev G.E., Furman E.G., Pushkarev A.I., Karpuzov S.B., Kondrat'ev N.A., and Goncharov D.V. (2004b) A High-Current Pulsed Accelerator with a Matching Transformer // Instruments and Experimental Techniques, , v. 47, №3, p. 394–398.

Remnev, G.E., Pushkarev, A.I., Pushkarev, M.A., Krasilnikov, V.A., Guzeeva, T.I. (2004c) Method of a direct reduction of halides Patent of Russia № 2228239. /. Applied 04.02.2002, published 10.05., Bul. № 13.

Remnev, G.E., and Pushkarev, A.I. (2004d) Research of chain plasmochemical synthesis of superdispersed silicon dioxide by pulse electron beam. IEEE Transactions on fundamentals and materials, 124(6): pp. 483–486.

Vlasov, V.A., Pushkarev, A.I., Remnev, G.E. (2004e) Experimental study and mathematical modelling of the recovery processes of fluoride compounds by a pulsed electron beam. Proceedings of Tomsk Polytechnic University, 307(5): pp. 89–93.

Yaworski, B.M., Detlaf, A.A. (1968) Handbook of physics. Moscow: Nauka.

Zhivotov, V.K., Rusanov, V.D., Fridman A.A. (1985) Diagnostics of non-equilibrium chemically active plasma. Moscow: Energoatomizdat.

Part 3

Auditory Interfaces for Enhancing Human Perceptive Abilities

13

Spatial Audio Applied to Research with the Blind

Brian FG Katz[1] and Lorenzo Picinali[2]
[1]Déptartement Communication Homme-Machine, LIMSI-CNRS, Orsay
[2]Department of Media Technology, De Montfort University, Leicester
[1]France
[2]UK

1. Introduction

Spatial audio technology has long been used for studies relating to human perception, primarily in the area of auditory source localisation. The ability to render individual sounds at desired positions or complex spatial audio scenes, without the need to manipulate any physical equipment, has offered researchers many advantages. Recently, the use of spatial audio has expanded beyond the study of such low level processes as localisation, and has been used as a tool to investigate higher-level cognitive functions. This work presents several recent studies where spatial audio technology has been used in order to expand our understanding of spatial cognition, with a specific focus on the abilities of the visually impaired, in both free-field and interior space exploration. These types of works provide for both an improved understanding within cognitive science and for the research and development into improved high resolution renderings with appropriate auditory cues.

2. Spatial auditory perception of the visually impaired

It is a common belief that blind individuals have a heightened auditory sense when compared to the general sighted population. Numerous studies have been carried out in an attempt to examine this claim, focusing on different aspects, from spatial precision to response time, brain activity and neural plasticity. Previous studies have used either real sound sources (speakers) or Virtual Auditory Displays (VAD) with binaural renderings.

As a baseline reference, Starlinger & Niemeyer (1981) conducted a series of audiological threshold tests for 18 blind and 18 sighted individuals. Comparisons for intensity, interaural-time difference, and auditory reflex thresholds found no significant differences. Frequency discrimination thresholds were slightly improved, though still significantly, in blind individuals. Focusing on pitch discrimination, Gougoux et al. (2004) found that early blind individuals were better than sighted control subjects in detecting pitch changes for a pair of sinusoids.

A collection of studies focusing on spatial audition has shown a clear tendency in support of improved capacities in blind individuals. Focusing on central auditory system tasks, Muchnik et al. (1991) compared localization, temporal resolution, and speech extraction in noise. In comparing groups of congenital or early blind, late blind, and sighted controls, each of about 10 individuals, blind subjects outperformed the sighted control group with the blind group localizing source positions within an error span of $\pm 5^\circ$ at a rate of $\approx 76\%$, while the sighted group had a correct response rate of ≈ 52% for the different sources tested in the horizontal plane. Temporal resolution was improved for both blind groups, and speech discrimination levels for noise masked speech were 6% higher when compared to the sighted control group. In contrast, Zwiers et al. (2001a) found that elevation localization, tested in the frontal square area of $\pm 35^\circ$ in azimuth and elevation, deteriorated more for blind subjects in low signal-to-noise conditions as compared to sighted control, although this was tested only for a small subject population. Doucet et al. (2005) found that blind individuals had better azimuthal localization performance in the presence of spectral degradations, implying that blind individuals are better at processing spectral cues. While not specifically studying spatial audition, Röder & Rösler (2003) found that blind subjects exhibited better memory for sounds, also indicating improved processing of spectral information.

Lessard et al. (1998) found comparable frontal lateralization between sighted and blind subjects while in the case of monaural localization (spectral discrimination) an improvement was evident for blind subjects. Examination of localization precision as a function of source location by Röder et al. (1999) found comparable results in the frontal direction and improved localization for more lateral sounds in blind subjects.

Ashmead et al. (1998), in addition to providing a thorough bibliographic study on spatial audition studies, presented a study comparing Minimum Audible Angle (MAA) for azimuthal changes around positions at 0° and 45° and elevation changes at 0°, and Minimum Audible Distance (MAD) thresholds. Test protocol used a 2-down 1-up staircase method. Subjects included early (22 subjects) and late (13 subjects) blind children, sighted children (18 subjects), and a control group of sighted adults (17 subjects). Results for MAA@0° and MAD showed that blind children outperformed or performed comparably to sighted adults, both being better than sighted children. MAA@45° results were comparable for all groups. A reaching-sound-sources task showed comparable results in azimuth and elevation, but lower absolute and distance errors for blind children, underestimating ≈ 2.5 cm, with sighted subject underestimating ≈ 9.5 cm. Similarly, Voss et al. (2004) examined MAA and MAD thresholds for frontal and eccentric source positions using a same/difference task for early blind, late blind, and sighted control subjects. Results showed that early blind individuals had lower MAA thresholds for lateral positions, and that blind subjects in general had improved overall MAD and MAA for rear positions. Considering frontal horizontal sources, Dufour & Gérard (2000) showed that improved auditory localization performance extends to near-sighted individuals as well.

With improved techniques and understanding of localization task performance, some studies have questioned the actual protocol of a pointing task between sighted and blind subjects. Lewald (2002a) found poorer performance in vertical localization but raised the issue of the pointing task and the definition of the point of reference to extrapolate the indicated position. Zwiers et al. (2001b) went further, finding equal performance when taking into account a modified point of reference between sighted and blind subjects, being head or shoulder based. The need to examine more closely perceptual or cognitive differences between sighted and blind individuals has highlighted a variety of insights. Lewald (2002b) examined localization errors in the context of eccentric head positions. While errors for all groups were of similar magnitude, sighted subjects tended to undershoot, while blind subjects overshoot source position estimations. As a result, the following conclusion was offered: "...in contrast to the widespread opinion of compensation of visual loss by a general sharpening of audition, compensatory plasticity in the blind may specifically be related to enhanced processing of proprioceptive and vestibular information with the auditory spatial input." An equally interesting result was found by Ohuchi et al. (2006) in testing angular and distance localization for azimuthally located sources with and without head movement. Overall, blind subjects outperformed sighted control for all positions. For distance estimations, in addition to being more accurate, errors by blind subjects tended to be overestimations, while sighted control subject errors were underestimations, in accordance with numerous other studies. These studies indicate that one must take a second look at many of the accepted conclusions of auditory perception, especially spatial auditory perception, when considering the blind, who do not necessarily have the same error typologies due to different learning sensory conditions. A number of studies, such as Weeks et al. (2000), have focused on neural plasticity, or changes in brain functioning, evaluated for auditory tasks between blind and sighted subjects. Results by both Elbert et al. (2002) and Poirier et al. (2006) have shown increased activity in typically visual areas of the brain for blind subjects.

While localization, spectral analysis, and other basic tasks are of significant importance in understanding basic auditory perception and differences that may exist in performance ability between sighted and blind individuals, these performance differences are inherently limited by the capacity of the auditory system. Rather, it is in the *exploitation* of this acoustic and auditory information, requiring higher level cognitive processing, where blind individuals are able to excel relative to the sighted population. Navigational tasks are one instance where this seems to be clear. Strelow & Brabyn (1982) performed an experiment where subjects were to walk a constant distance from a simple straight barrier, being a wall or series of poles at 2 m intervals (diameter 15 cm or 5 cm), without any physical contact to the barrier. Footfall noise and finger snaps were the only information. With 8 blind and 14 blindfolded sighted control subjects, blind subjects clearly outperformed sighed subjects, some of whom claimed the task to be impossible. The results showed that blindfolded subjects performed overall as well in the wall condition as blind subject in the two pole conditions. Morrongiello et al. (1995) tested spatial navigation with blind and sighted children (ages 4.5 to 9 years). Within a carpeted room (3.7 m $\times$ 4.9 m), four tactile landmarks were placed at the center of each wall. Subjects, blind or blindfolded, were guided around the room to the different landmarks in order to build a spatial cognitive map. The same paths were used for all subjects, and not all connecting paths were presented. This learning stage was performed with or without an auditory landmark condition, a single metronome placed at the starting position. Subjects were then asked to move from a given landmark to another, with both known and novel paths being tested. Different trajectory parameters were evaluated. Results for sighted subjects indicated improvements with age and with the presence of the auditory landmark. Considering only the novel paths, all groups benefited from the auditory landmark. Analyzing the final distance error, sighted children outperformed blind in both conditions with blind subjects in the auditory landmark condition performing comparably to blindfolded subjects without auditory landmark. It is noted that due to the protocol used, it was not possible to separate auditory landmark and learning effect.

3. Virtual interactive environments for the blind: Academic context

Substantial amounts of work attest to the capacity of the blind and visually impaired to navigate in complex environments without relying on visual inputs (e.g., Byrne & Salter (1983); Loomis et al. (1993); Millar (1994); Tinti et al. (2006)). A typical experiment consists of having blind participants learn a new environment by walking around it, with guidance from the experimenter. How the participants perform mental operations on their internal representations of the environment is then assessed. For example, participants are invited to estimate distances and directions from one location to another (Byrne & Salter (1983)). Results from these experiments seem to attest that blind individuals perform better in terms of directional and distance estimation if the location of the experiment is familiar (e.g. at home) rather than unfamiliar.

Beyond the intrinsic value of the outputs of the research programs reported here, more information still needs to be collected on the conditions in which blind people use the acoustic information available to them in an environment to build a consistent, valid representation of it. It is generally recognized that the quality of such mental representations is predictive of the quality of the locomotor performance that will take place in the actual environment. Is it the case that a learning procedure based upon the systematic exploitation of acoustic cues prepares a visually impaired person to move safely in a new and intricate environment? It then needs to be noted that blind people, who have to learn a new environment in which they will have to navigate, use typically special procedures. For instance, when a blind person gets a new job in a new company, it is common for him/her to begin by visiting the building late in the evening: the objective is to acquire some knowledge of the spatial configuration and of the basic features of the acoustical environment (including reverberation effects, sound of their steps on various floor surfaces, etc.). Later on, the person will get acquainted with the daily sounds attached to every part of the environment.

The following sections present a series of three studies which have been undertaken in order to better understand behaviours in non-visual complex auditory environments where spatial cognition plays a major role. A variety of virtual auditory environments and experimental platforms have been developed and put to the service of cognitive science studies in this domain, with special attention to issues with the visually impaired. These studies help both in improving the understanding of spatial cognitive processing as well as highlighting the current possibilities and limitations of different 3D audio technologies in providing sufficient spatial auditory information to subjects.

The first study employs a full-scale immersive virtual audio environment for the investigation of spatial cognition and localisation. Similar in concept to Morrongiello et al. (1995), this study provides for a more complex scene, and more complex interactions for study. As not all experiments can be performed using a full-scale immersive environment, the second study investigates the need for head-tracking by proposing a novel blind active virtual exploration task. The third and final study investigates spatial cognition through architectural exploration by comparing spatial and architectural understanding in real and virtual environments by blind individuals.

4. Study I. Mental imagery and the acquisition of spatial knowledge without vision: A study of blind and sighted people in an immersive audio virtual environment

Visual imagery can be defined as the representation of perceptual information in the absence of visual input (Kaski (2002)). In order to assess whether visual experience is a pre-requisite for image formation, many studies have focused on the analysis of visual imagery in congenitally blind participants. However, only few studies have described how visual experience affects the metric properties of the mental representations of space (Kaski (2002); Denis & Zimmer (1992)).

This section presents a study that was the product of a joint effort of different research groups in different areas for the investigation of a cognitive issue through the development and implementation of a general purpose Virtual Reality (VR) or Virtual Auditory Display (VAD) environment. The aim of this research project was the investigation of certain mechanisms involved in spatial cognition, with a particular interest in determining how the verbal description or the active exploration of an environment affects the elaboration of mental spatial representations. Furthermore, the role of vision was investigated by assessing whether participants without vision (congenitally or early blind, late blind, and blindfolded sighted individuals) could benefit from these two learning modalities, with the goal of improving the understanding of the effect of visual deprivation on the capacity to mentally represent spatial configurations. Details of this study, the system architecture and the analysis of the results, can be found in Afonso et al. (2005a);Afonso et al. (2005b);Afonso et al. (2005c);Afonso et al. (2010).

4.1 Mental imagery task using a tactile/haptic scene (background experiment)

The development of the VAD experiment followed the results of an initial study performed concerning the evaluation of mental imagery using a tactile or haptic interface. Six imaginary objects were located on the perimeter of a physical disk (diameter 50 cm) placed upright in front of the participants. The locations of these objects were learned by the participants exploiting two different modalities. The first one was a verbal description of the configuration itself, while the second one involved the experimenter placing the hand of the participant at the appropriate positions. After having acquired knowledge about the configuration of the objects through one of the two modalities, the participants were asked to create a mental representation of a given spatial configuration, and then to compare distances between the objects situated on the virtual disk.

The results showed that independent of the type of visual deprivation experienced by the participants and of the learning modality, all participants were able to create a mental representation of the configuration that preserved the metric relations between the objects. The precision of the spatial cognitive maps was evaluated using a mental scanning paradigm. The task consisted in mentally imagining a point moving between two objects, subjects responding when the trajectory was completed. A correlation between response times and scanned distances was obtained for all experimental groups and for both modalities. It was noted that blind subjects needed more time than sighted in order to achieve the same level of performance for all conditions.

The examined hypothesis was that congenital blind individuals, who are not expected to generate visual mental images, are nevertheless proficient at using mental simulation of trajectories. Sighted individuals would be expected to perform better, having experience in generating visual mental images. While no difference was found in precision, a significant difference was found in terms of response times between blind and sighted participants. A new hypothesis attempts to explain this difference by examining the details of the task (allocentric vs. egocentric) as being the cause, and not other factors. This hypothesis could explain the difference in the processing times needed by blind people in contrast to the sighted, and could explicate the tendency for the response times of blind individuals to be shorter after the haptic exploration of the configuration.

In order to test this hypothesis, a new experimental system was designed in which the task was conceived to be more natural for, even to the advantage of, blind individuals. An egocentric spatial scene, rather than the allocentric scene used in the previously described haptic task, was used. An auditory scene was also chosen.

4.2 An immersive audio interface

A large-scale immersive VAD environment was created in which participants could explore and interact with virtual sound objects located within an environment.

The scene in which the experiment took place consisted of a room (both physical and virtual) in which six virtual sound objects were located. The same spatial layout configuration and test positions were employed as in the previous haptic experiment. Six "domestic" ecological sound recordings were chosen and assigned to the numbered virtual sound sources: (1) running water, (2) telephone ringing, (3) dripping faucet, (4) coffee machine, (5) ticking clock, and (6) washing machine.

A virtual scene was constructed to match the actual experimental room dimensions. Monitoring the experiment by the experimenter was possible through different visual renderings of the virtual scene. The arrangement of the scene consisted of six objects representing the six sound sources located on the perimeter of a circle. A schematic view of the real and simulated environment and of the positions of the six sound sources is shown in Fig. 1. Participants were equipped with a head-tracker device, mounted on a pair of stereophonic headphones, as well as with a handheld tracked pointing device, both of which were also included in the scene graph. Collision detection was employed to monitor if a participant approached the boundaries of the physical room or the limits of the tracking system in order to avoid any physical contact with the walls during the experiment. A spatialized auditory alert, wind noise, was then used to warn the participants of the location of the wall in order to avoid contact.

The balance between direct and reverberant sound energy is useful in the perception of source distance (Kahle (1995)). It has also been observed that the reverberant energy, and especially a diffuse reverberant field, can negatively affect source localization. As this study was primarily concerned with a spatially precise rendering, rather than a realistic room acoustic experience, the reverberant energy was somewhat limited. Omitting the room effect creates an "anechoic" environment, which is not habitual for most people. To create a more realistic environment for which the room effect was included, an artificial reverberation was used with a reverberation time of 2 s. To counteract the negative effect on source localization, the direct to reverberant ratio was defined as 10 dB at 1 m. The design goal was for distance perception and precision localisation to be achieved through dynamic cues and subject displacements.

The audio scene was rendered over headphones using binaural synthesis (Begault (1994)) developed in the *MaxMSP*[1] environment. A modified version of *IRCAM Spat*[2] was also developed which allowed for the individualization of Inter-aural Time Delay (ITD) based on head circumference, independent of the selected Head Related Transfer Function (HRTF). The position and head orientation of the participant was acquired using a six Degrees-of-Freedom (6DoF) electromagnetic tracking system. Continuously integrating the updated external positional information, the relative positions of the sound sources were calculated, and the sound scene was updated and rendered, ensuring a stable sound scene irrespective of subject movements. The height of the sound sources was normalized relative to the subject's head height (15 cm above) in order to avoid excessive sound pressure levels when sources were approached very closely. An example of the experiment showing the different phases, including the subjective point of view binaural audio rendering, can be found on-line[3].

---

[1] http://www.cycling74.com

[2] http://forumnet.ircam.fr/692.html

[3] http://www.limsi.fr/Rapports/RS2005/chm/ps/ps11/ExcerptExpeVR.mov

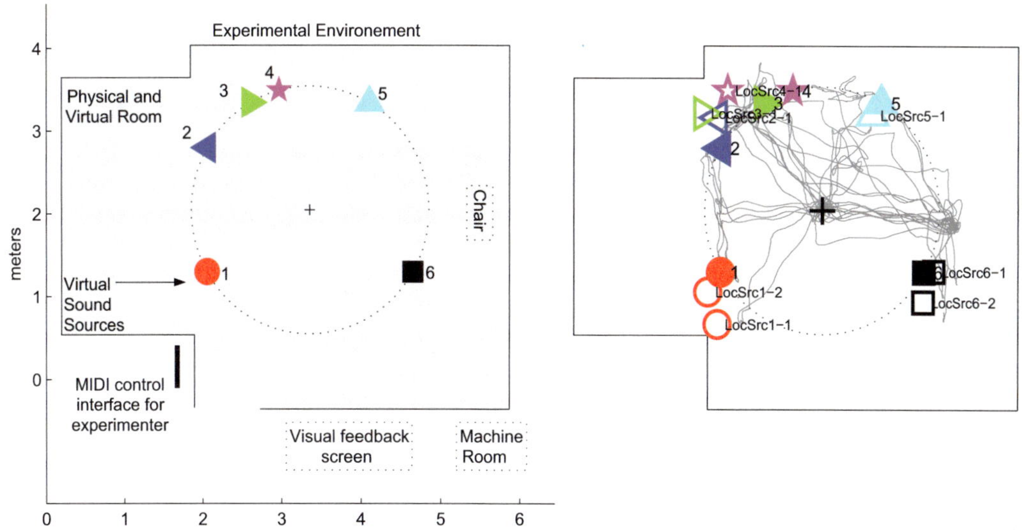

Fig. 1. Schematic view (left) of the real and simulated environment, together with the six sound sources and the reference point chair. Sample visualization (right) of experimental log showing participant trajectory and repositioned source locations (labelled LocSrc*n*-*pass*).

4.3 The task

A total of 54 participants took part in this study. Each one belonged to one of three groups: congenitally or early blind, late blind, and blindfolded sighted. An equal distribution was achieved between the participants of the three groups according to gender, age, and educational and socio-cultural background. These groups were split according to two learning conditions (see Section 4.3.1). Each final group comprised five women and four men, from 25 to 59 years of age.

4.3.1 Learning phase

The learning phase was carried out exploiting one of the two previously tested learning methods: Verbal Description (**VD**) and Active Exploration (**AE**). To begin, each participant was familiarised with the physical room and allowed to explore it for reassurance. They were then placed at the centre of the virtual circle (see Fig. 1) which they were informed had a radius of 1.5 m, and on which the six virtual sound sources were located.

For groups **VD**, the learning phase was passive and purely verbal. The participants were centred in the middle of the virtual circle and informed about the positions of the sound sources by first hearing the sound played in mono (non-spatialized), and then by receiving a verbal description, performed by the experimenter, about its location using conventional clock positions, as are used in aerial navigation, in clockwise order. No verbal descriptions of sound sources were ever used by the experimenter.

For groups **AE**, the learning phase consisted of an active exploration of the spatial configuration. Participants were positioned at the centre of the virtual circle. Upon continuous presentation of each sound source individually (correctly spatialized on the circle), participants had to physically move from the centre to the position of each sound source.

In order to verify that participants correctly learned the spatial configuration, each group was evauated. For groups **AE**, participants returned to the centre of the virtual circle where each sound source was played individually, non-spatialized (mono), in random order, and participants had to point (with the tracked pointer) to the location of the sound sources. The response was judged on the graphical display. The indicated position was valid if the pointer intersected with a sphere (radius = 0.25 m) on the circle (radius = 1.5 m), equating to an angular span of 20 ° centred on the exact position of the sonic object. For groups **VD**, participants had to express verbally where the correct source location was, in hour-coded terms. Errors for both groups were typically of the type linked to confusions between other sources rather than absolute position errors. In the case of any errors, the entire learning procedure was repeated until the responses were correct.

4.3.2 Experimental phase

Following the learning phase, each participant began the experiment standing at the centre of the virtual circle. One sound source was briefly presented, non-spatialized and randomly selected, whose correct position they had to identify. To do this, participants were instructed to place the hand-tracked pointer at the exact position in space where the sound object should be. The height component of the responses was not taken into account in this study. When participants confirmed their positional choice, the sound source was re-activated at the position indicated and remained active (audible) while each subsequent source was added. After positioning the first sound source, participants were led back to the reference chair (see Fig.1). All subsequent sources were presented from this position, rather than from the circle centre. This change of reference point was intentional in order to observe the different strategies used by participants to reconstruct the initial position of sound objects, such as directly walking to the source position or walking first to the circle centre. After placing the final source, all sources were active and the sound scene was complete. This was the first instance in the experiment when the participants could hear the entire scene.
Participants were then returned to the centre of the virtual circle, from where they were allowed to explore the completed scene by moving about the room. Following this, they were repositioned at the centre, with the scene still active. Each sound source was selected, in random order, and participants had the possibility to correct any position they judged incorrect using the same procedure as before.

4.4 Results

Visualization of the experimental phase is possible using the logged information, of which an example is presented in Fig. 1. One can see for several sources two selected positions, equating to the first pass position, and the second pass, refined position.
Evaluation of the experimental phase consisted in measuring the discrepancy between the original spatial configuration and the recreated sound scene. Influence of the learning modality on the preservation of the metric and topological properties of the memorized environment was analyzed in terms of angular, radial, and absolute distance errors as compared with the correct location of the corresponding object.
A summary of these errors is shown in Fig. 2. An ANalysis Of VAriance (ANOVA) was performed on the errors taking into account learning condition and visual condition for each group. Analysis of each error is discussed in the following sections.

4.4.1 Radial error

Radial error is defined as the radial distance error, calculated from the circle centre, between the position of the sound source and the actual position along the circle periphery. For both verbal learning and active exploration, participants generally underestimated the distances

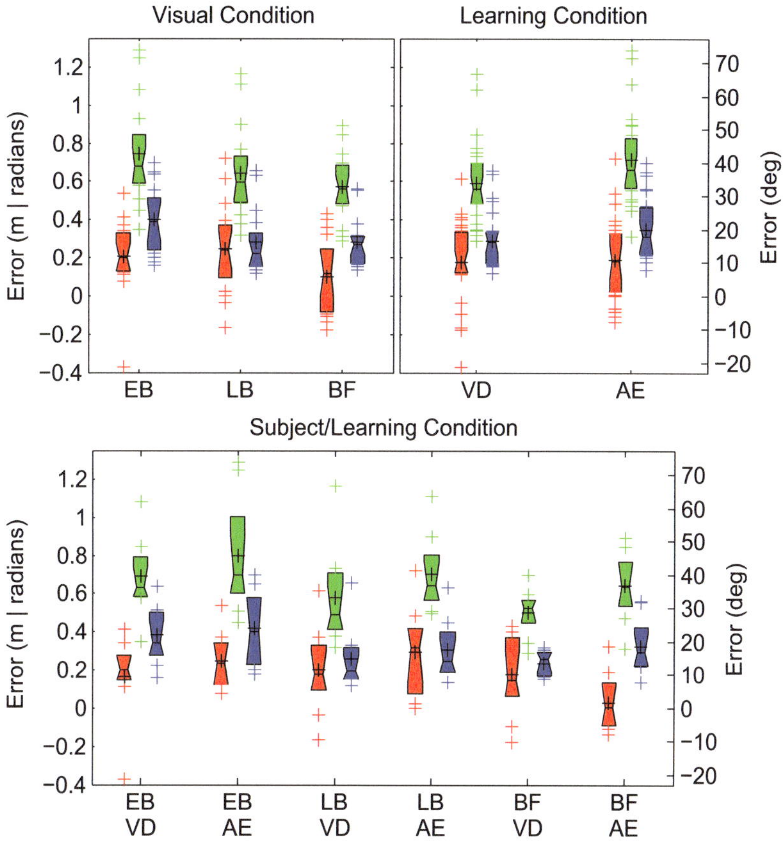

Fig. 2. Overview of the errors collapsed over visual condition (top left), learning condition (top right) and crossed effects (bottom). Radial errors (meters) in red, distance errors (meters) in green, and angular errors (radians left axis, degrees right axis) in blue. Learning conditions are Active Exploration, **AE**, and Verbal Description, **VD**. Visual conditions are Early Blind, **EB**, Late Blind, **LB**, and BlindFolded, **BF**. Black **+** indicate data mean values, notches indicate median values and confidence intervals, and coloured **+** indicate data outliers.

(a positive error) by the same amount (mean = 0.2 m), with similar standard deviation (0.3 m and 0.4 m, respectively). There was no difference among the three groups; each one underestimated the distance with a mean error of 0.2 m for congenitally blind (std = 0.3) and late blind (std = 0.4), and a mean error of 0.1 m for blindfolded (std = 0.3). Interestingly, a significant difference was found for blindfolded participants who learned the spatial configuration from a verbal description, underestimating radial positions (mean = 0.2 m, std = 0.3) when compared with an active exploration (mean = 0.0 m, std = 0.4) [$F(2,48) = 3.32$; $p = 0.045$].

4.4.2 Absolute distance error

Absolute distance error is defined as the distance between the original and selected source positions. Results show a significant effect of learning condition. Active exploration of the virtual environment resulted in better overall estimation of sound source positions (mean = 0.6 m, std = 0.3) as compared to the verbal description method (mean = 0.7 m, std = 0.4)

[F(1,48) = 4.29, p = 0.044]. The data do not reflect any significant difference as a function of visual condition (congenitally blind, mean = 0.7 m, std = 0.4; late blind, mean = 0.6 m, std = 0.3; blindfolded, mean = 0.6 m, std = 0.3).

4.4.3 Angular error

Angular error is defined as the absolute error in degrees, calculated from the position designated by participants in comparison to the circle centre of the reference position of the corresponding sound source. There was no significant difference between learning conditions: verbal description (mean = 17 °, std = 14 °) and active exploration (mean = 20 °, std = 17 °). Congenitally blind participants made significantly larger angular errors (mean = 23 ° , std = 17 °) than late blind (mean = 16 °, std = 15 °) [F(1,32) = 4.52; p = 0.041] and blindfolded sighted participants (mean = 16 °, std = 13 °) [F(1,32) = 6.08; p = 0.019].

4.5 Conclusion

The starting hypothesis was that the learning through active exploration would be an advantage to blind participants when compared to learning via verbal description. If true, this would confirm results of a prior set of experiments which showed a gain in performance of mental manipulations for blind people following this hypothesis (Afonso (2006)). A second hypothesis concerned sighted participants, who were expected to benefit more from a verbal description, being more adapt at generating a visual mental image of the scene, and thus being able to recreate the initial configuration of the scene in a more precise manner.

Considering the scene recreation task, these results suggest that active exploration of an environment enhances absolute positioning of sound sources when compared to verbal description learning. The same improvement appears with respect to radial distance errors, but only for blindfolded participants. Results show that participants underestimated the circle size, independent of the learning modality except for the case of blindfolded participants, with a mean position error close to zero, and that they clearly benefited from learning with perception-action coupling. These results are not in line with previous findings such as Ohuchi et al. (2006) in which blind subjects performed better at distance estimation for real sound sources using only head rotations and verbal position reporting. It clearly appears that an active exploration of the environment improves blindfolded participants' performance, both in terms of absolute position and size of the reconstructed configuration.

It has also been found that subjects blind from birth made significantly more angular positioning errors than late blind or blindfolded groups for both learning conditions. These data are in line with the results of previous studies involving spatial information processing in classic real (non virtual) environments (Loomis et al. (1998)).

5. Study II: A study on head tracking

This study focuses on the role of the Head Movements (HM) a listener uses in order to localize a sound source. Unconscious HM are important for resolving front-to-back ambiguities and for improving localization accuracy (see Wenzel (1998); Wightman & Kistler (1999); Minnaar et al. (2001)). However, previous studies regarding the importance of HM have all been carried out in static situations (participants at a fixed position without any positional displacement). The aim of this experiment is to investigate whether HM are important when individuals are allowed to navigate within the sound scene. In the context of future applications using VAD, it is useful to understand the importance of head-tracking. In this instance, a virtual environment was created employing a joystick for controlling displacement. Elements of this study have been presented by Blum et al. (2006), and additional details can also be found on-line[4].

5.1 Binaural rendering and head tracking

A well-known issue related to the use of non-tracked binaural technology consists in the fact that under normal headphone listening conditions, the sound scene follows HM, such that the scene remains defined in the head-centred reference frame, not in that of the external world, making it unstable relative to HM. In this situation, the individual is unable to benefit from binaural dynamic cues. However, with head orientation tracking, it is possible to update the sound scene relative to the head orientation in real time, correcting this artefact.

In the present experiment, two conditions have been tested: actual orientation head-tracking versus virtual head rotations controlled via joystick. Participants with head-tracking can have pertinent acoustic information from HM as in a natural 'real' situation, whereas participants without head-tracking have to extrapolate cues from other control movements. The hypothesis is that an active exploration task with linear displacements in the VAD is sufficient to resolve localization ambiguities, implying that tracking HM is not always necessary.

5.2 Experimental task

The experiment followed a 'game like' scenario of bomb disposal, and was carried out with sighted blindfolded subjects. Bombs (sound sources simulating a ticking countdown) were located in a virtual open space. Participants had to find them by navigating to their position, using a joystick (displacement control and virtual head rotation relative to the direction, of motion using the twist of the joystick) to move in the VAD. The scene was rendered over headphones (see Section 4.2 for a description of the binaural engine used). For the head-tracked condition, an electromagnetic tracker was employed with a refresh rate of 20 Hz. To provide a realistic auditory environment, artificial reverberation was employed. The size of the virtual scene, and the corresponding acoustics, was chosen to correspond to an actual physical room (the Espace de Projection, *Espro*, at IRCAM) with its variable acoustic characteristics in its more absorbing configuration (reverberation time of 0.4 s). Footstep sounds were included during movement, rendered to aid in the perception of displacement and according to the current velocity.

In the virtual environment, the relation between distances, velocity, and the corresponding acoustic properties was designed so as to fit a real situation. Forward/backward movements of the joystick allowed displacement respectively forward and backward in the VAD. The maximum speed, corresponding to the extreme position, was 5 km/h, which is about the natural human walking speed. With left/right movements, participants controlled body rotation angle, which relates to the direction of displacement. Translation and rotation could be combined with diagonal manipulations. The mapping of lateral joystick position, δx, to changes in navigation orientation angle, α, was based on the relation: $\alpha = (\delta x / x_{max}) 50° \delta t$; where x_{max} is the value corresponding to the maximum lateral position of the joystick, and δt the time step between two updates of δx.[5] For the material used, this equation provides a linear relation between α and δx with a coefficient of 0.001.

The design of the task was centered on the principle that, as with unconscious HM, linear displacements and a stable source position would allow for the resolution of front-back

[4] http://rs2007.limsi.fr/index.php/PS:Page_16

[5] http://www.openscenegraph.org/

confusions. To concentrate on the unconscious aspect, a situation involving two concurrent sources was chosen. While the subject was searching for one bomb, the subsequent target would begin ticking. As such, the conscious effort was focussed on the current target, while the second target's position would become more stable in the mental representation of the scene. This was thought to incite HM for the participant for localizing the new sound while keeping a straight movement toward the current target. As two sources could be active at the same time, two different countdown sounds were used alternatively with equal normalized level.

Each test series included eight targets. The distance between two targets was always 5 m. In order to enforce the speed aspect of the task, a time limit (60 s) was imposed to reach each target (defuse the bomb), after which the bomb exploded. The subsequent target would begin ticking when the subject arrived within a distance of 2 m from the current target. In the event of a failed target, the participant was placed at the position of the failed target and would then resume the task towards the next target. Task completion times and success rates were used to evaluate the effects of the different conditions.

A target was considered found and defused when the participant arrived within a radius of 0.6 m. This 'hit detection radius' of 0.6 m corresponds to an angle of $\pm 6.8\,^{\circ}$ at a distance of 5 m from the source, which is the mean human localization blur in the horizontal plane (Blauert (1996)). As a consequence, if the participant oriented him/herself with this precision when starting to look for a target, this could be reached by going straightforward.

The experiment was composed of six identical trials involving displacement along a succession of eight segments (eight sources to find in each trial). The first trial was considered a training session, and the last segment of each trial was not taken into account as only a single target signal was present for the majority of the search.

In total, $5 \times 6 = 30$ segments per participant were analyzed. The azimuthal angles made by the six considered segments of each trial were balanced between right/left and back/front ($-135\,^{\circ}$, $-90\,^{\circ}$, $-45\,^{\circ}$, $45\,^{\circ}$, $90\,^{\circ}$, $135\,^{\circ}$). Finally, to control a possible sequence effect, two different segment orderings were created and randomly chosen for each participant.

5.3 Recorded data

Twenty participants without hearing deficiencies were selected for this study. Each subject was allocated to one of the two head-tracking conditions (with or without). An equal distribution was achieved between the participants of the two groups according to gender, age, and educational and socio-cultural background. Each group comprised five women and five men with a mean age of 34 years (from 22 to 55 years, std = 10).

Result analysis was based on the following information: *hit time* (time to reach target for each segment), *close time* (time to get within 2 m from target, when the subsequent target sound starts), and the *total percentage* of successful hits (bombs defused).

Position and orientation of the participant in the VAD were recorded during the entire experiment, allowing for subsequent analysis of trajectories. At the end of the experiment, participants were asked to draw the trajectory and source positions on a sheet of paper (the starting point and first target were already represented in order to normalize the adopted scale and drawing orientation).

5.4 Results

Large individual differences in hit time performance ($p < 10^5$) were observed. Some participants showed a mean hit time more than twice the quickest ones. Percentage of

successful hits varied from 13% to 100%, and the participants that were quicker in completing the task, obtained a higher percentage of hits. In fact, some participants were practically unable to execute the task while others exhibited no difficulty. Performance measures of mean hit times and total percentage hit were globally correlated with a linear correlation coefficient of -0.67 ($p = 0.0013$).
The influence of the source position sequence (two different orderings were randomly proposed) and of the type of source (two different sounds were used) was tested. No effect was found for these two control variables.
Analysis of hit times and head-tracked condition did not reveal any significant effect. Mean hit times of the two groups were very similar (19.8 s versus 20.4 s). Table 1 shows that participants in the head-tracked condition for HM did not perform better than those in the non-tracked condition.
A significant effect was found for subject age. Four age groups were defined with four participants between 20 and 25 years, six between 25 and 30, six between 30 and 40 and four between 40 and 60. Table 1 shows the performances for each age group. Young participants had shorter hit times and higher percentage of hits if compared with older ones. A significant effect of age ($p < 0.0001$) and a significant gender×age interaction ($p = 0.0007$) were found: older women had more difficulty in executing the task.

	HM Tracking		Age Groups				Videogame Experience	
Condition	No	Yes	20-25	25-30	30-40	40-60	No	Yes
mean Hit Time (s)	19.8	20.4	17.0	20.4	19.7	29.4	22.1	19.0
Standard Deviation (s)	11.1	11.9	10.3	11.9	9.6	15.2	12.2	10.8
% Hit Sources	86%	80%	92%	94%	94%	35%	69%	94%

Table 1. Performance results as a function of tracking, age, and video game experience.

In a questionnaire filled in before the experiment, participants were asked to report whether they had previous experience with video games. Eleven participants reported they had such experience, while the remaining nine participants did not. Table 1 shows that the experienced group had higher performances results. There was a significant effect of this factor on hit times ($p = 0.004$), and the group with video game experience had 94% hits versus only 69% for the other group. Not surprisingly, individuals familiar with video games seemed more comfortable with immersion in the virtual environment and also with joystick manipulation. This can be related to the age group observation since no participant from group [40-60] reported any experience with video games.
A significant learning effect was found ($p = 0.0047$) between trials, as shown in Table 2. This effect was most likely due to a learning effect of navigation within the VAD rather than a memorization of the position of the sources, since participants did not experience any sequence repetition and reported that they treated each target individually. Results of the post navigation trajectory reconstruction task confirm this by the fact that participants were unable to precisely draw the path of a trial on a sheet of paper when they were asked to do so at the end of the experiment. This lack of reconstruction ability is in contrast to the previous experiment (see Section 4), where subjects were able to reconstruct the sound scene after physical navigation. This can be seen as an argument in favour of the importance of the memorization of sensorimotor (locomotor) contingencies for the representation of space.
Through inspection of the different trajectory paths, it was observed that front/back confusions were present for participants in both tracking conditions. In Fig. 3A and Fig. 3B,

Learning effect					
Trial	2	3	4	5	6
Hit Time Mean (sec.)	22.1	22.4	20.2	19.0	17.0
Standard Deviation	12.5	12.6	11.5	9.8	9.9
% of hit sources	78%	82%	85%	81%	91%

Table 2. Performance as a function of trial sequence over all subjects.

two trajectories with such inversion are presented for two participants in the 'head-tracking' condition. Example A shows front/back confusion in the path between sources 3 and 4: the participant reaches source 3, source 4 is to the rear, but the subject moves forward in the opposite direction. After a certain distance is travelled, the localization inversion is realized and the subject correctly rotates again to go back in the correct direction. Fig. 3B shows a similar event between sources 1 and 2. Overall, in comparing the head orientation vector and the movement orientation vector, participants in the head-tracked condition did not appear to use HM to resolve localization ambiguities, focusing on the use of the joystick, keeping the head straight and concentrating only on frontal targets. It is apparent that rotations were typically made with the joystick at each source to decide the correct direction.

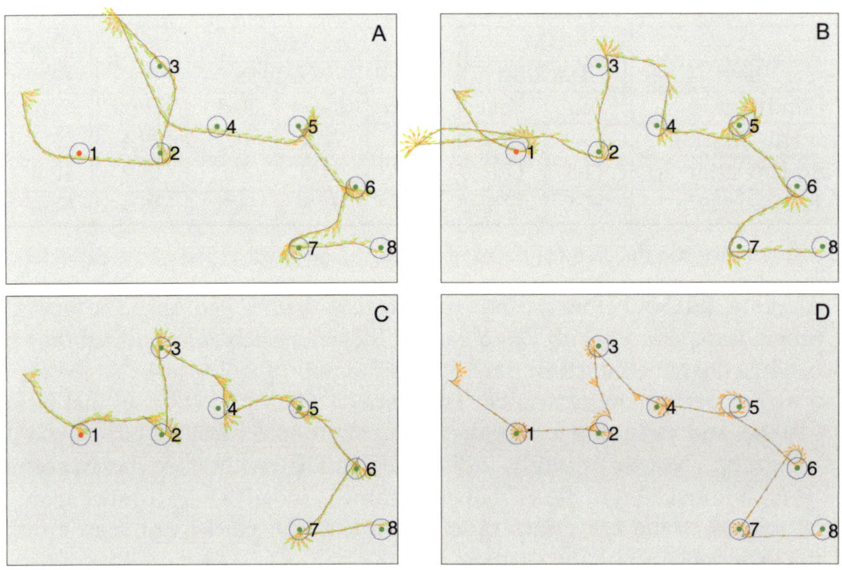

Fig. 3. Examples of trajectories of different participants with (ABC) and without (D) head-tracking. Arrows indicate movement orientation (orange) and the head orientation (green). A-B: examples of front/back confusion. C-D: typical navigation strategies with (C) and without (D) head-tracking condition.

5.4.1 Discussion and perspectives

The inclusion of head-tracking was not found to be necessary for the task proposed in this experiment. Movements of the joystick and virtual displacement were considered sufficient for the participants to succeed in the task. However, the use of a joystick elicits some questions pertaining to subject experience with video games and to the effect on task performance, as well as to the apparent lack of use of HM even when available.

Participants seem to have transferred vestibular modality toward the use of the joystick. This is supported by the typical navigation strategy observable in the participants' trajectories where rotations were made with the joystick (Fig. 3C and Fig. 3D). It is not yet clear how this finding can be extended to other tasks which require a more complex understand of the sound scene. As the subjects were not able to recount the positions of the different targets or their trajectories, it is possible that HM are still required for more complex spatially related tasks.

6. Study III. Creating a Virtual reality system for Visually impaired persons

This research results from collaboration between researchers in psychology and in acoustics on the issue of spatial cognition in interior spaces. Navigation within a closed environment requires analysis of a variety of acoustic cues, a task that is well developed in many visually impaired individuals, and for which sighted individuals rely almost entirely on visual information. Focusing on the needs of the blind, creation of cognitive maps for spaces, such as home or office buildings, can be a long process, for which the individual may repeat various paths numerous times. While this action is typically performed by the individual on-site, it is of some interest to investigate at which point this task can be performed off-site, at the individual's discretion. In short, is it possible for an individual to learn an architectural environment without being physically present? If so, such a system could prove beneficial for navigation preparation in new and unknown environments.
A comparison of three types of learning has been performed: *in situ* real displacement, passive playback of a recorded navigation (with and without HM tracking), and active navigation in a virtual architecture. For all conditions, only acoustic cues are employed.

6.1 Localisation versus spatial perception

Sound source localisation in an anechoic environment is a special and quite unnatural situation. It is more typical to hear sound sources with some amount of reflections, even in outdoor environments, or with a high density of reflections in reverberant spaces. These additional acoustic path returns from the same source can cause certain impairments, such as source localisation confusion and degradation of intelligibility. At the same time, these additional acoustic signals can provide information regarding the dimensions, material properties, as well as cues improving sound source localisation.
In order to be able to localize a sound source in a reverberant environment, the human hearing system gives the most weight to the first signal that reaches the ear, i.e. the signal that comes directly from the sound source. It does not consider the localisation of the other signals resulting from reflections on walls, ceiling, floor, etc. that arrive 20-40 ms after the first signal (these values can change depending on the typology of the signal, see Moore (2003), pp. 253-256). This effect is known as the *Precedence Effect* (Wallach et at. (1949)), and it allows for the localisation of a sound source even in situations when the reflections of the sound are actually louder than the direct signal. There are of course situations where errors occur, if the reflected sound is sufficiently louder and later than the direct sound. Other situations can also be created where false localisation occurs, such as with the Franssen effect (Hartmann & Rakerd (1989)), but those are not the subject of this work. The later arriving signals, while not being useful for localization, are used to interpret the environment.
The ability to directionally analyse the early reflection components of a sound are not thought to be common in sighted individuals for the simple reason that the information gathered from this analysis is often not needed. In fact, as already outlined in Section 3, information about the

spatial configuration of a given environment is mainly gathered though sight, and not through hearing. For this reason, a sighted individual will find information about the direction of the reflected signal components redundant, while a blind individual will need this information in order to gather knowledge about the spatial configuration of an environment. Elements in support of this will be given in Section 6.4 and 6.4.3, observing for example how blind individuals make use of self-generated noise, such as finger snaps, in order to determine the position of an object (wall, door, table, etc.) by listening to the reflections of the acoustic signals.

It is clear that most standard interactive VR systems (e.g. gaming applications) are visually-oriented. While some engines take into account source localisation of the direct sound, reverberation is most often simplified and the spatial aspects neglected. Basic reverberation algorithms are not designed to provide such geometric information. Room acoustic auralization systems though should provide such level of spatial detail (see Vorländer, (2008)). This study proposes to compare the late acoustic cues provided by a real architecture with those furnished both by recordings and by using a numerical room simulation, as interpreted by visual impaired individuals. This is seen as the first step in responding to the need of developing interactive VR systems specifically created and calibrated for blind individuals, a need that represents the principal aim of the research project discussed in the following sections.

6.2 Architectural space

In contrast to the previous studies, this one focuses primarily on the understanding of an architectural space, and not of the sound sources in the space. As a typical example, this study focuses on several (four) corridor spaces in a laboratory building. These spaces are not exceptionally complicated, containing a various assortment of doors, side branches, ceiling material variations, stairwells, and static noise sources. An example of one of the spaces used in this study is shown in Fig. 4. In order to provide reference points for certain validations, some additional sound sources were added. These simulated sources were simple audio loops played back over positioned loudspeakers.

6.3 Comparison of real navigation to recorded walkthrough

Synthesized architectural environments, through the use of numerical modelling, are necessarily limited in their correspondence to a real environment. In contrast, it can be hypothesized that a spatially correct recording performed in an actual space should be able to capture and allow for the reproduction of the actual acoustic cues, without the need to necessarily define or prescribe said cues.

In order to verify this hypothesis, two exploration conditions were tested within the four experimental corridors: real navigation and recorded walkthrough playback. In order to take into account the possible importance of HM, two recording methods were compared. The first, binaural recording, employs a pair of tiny microphones placed at the entrance of the ear canals. This recording method captures the fine detail of the HRTF but is limited in that the head orientation is encoded within the recording. The second method, Ambisonic recording, employs a spatial 3-dimensional recording. This recording, upon playback, can be rotated and as such can take into variations in head orientation during playback.

For the real navigation condition, a blind individual was equipped with in-ear binaural microphones (open meatus in order not to obstruct natural hearing) in order to monitor and be able to analyse any acoustic events. The individual then advanced along the corridor from

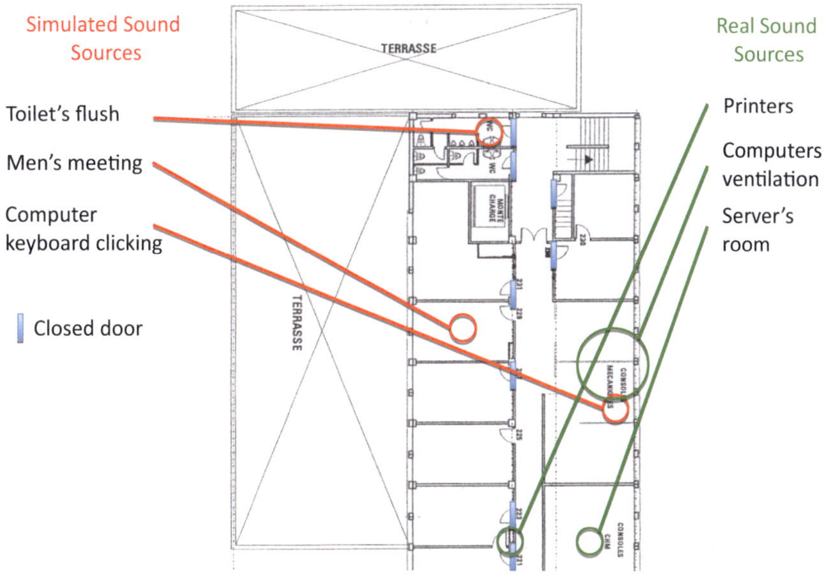

Fig. 4. Plan and positions of real and artificially simulated sound sources for environment 1.

one end to the other, and returned. No other navigation aides were used (cane, guide dog, etc.), but any movements or sounds were allowed. Contact with the environment was to be avoided, and the individual remarkably avoided any collisions. This navigation was tracked using a CCTV camcorder system with visual markers placed throughout the space for later calibration.

In order to have recordings for the playback conditions, an operator equipped with both binaural (in-ear *DPA 4060*) and B-Format (Gerzon (1972)) (*Soundfield ST250*) recording systems precisely repeated the path of the real navigation condition above. Efforts were made to maintain the same speed, and head movements, as well as any self-generated noises. This process was repeated for the four different environments.

6.3.1 Playback rendering system

In the Ambisonic playback condition the B-Format recording was then rendered over binaural headphones employing the approach of *virtual speakers*. This conversion from Ambisonic to stereo binaural signal was realized through the development and implementation of a customized software platform using *MaxMSP* and a head orientation tracking device (*XSens MTi*). The 3D sound-field recorded (B-Format signal) was modified in real-time performing rotations in the Ambisonics domain as a function of participant's head orientation. The rotated signal was then decoded on a virtual loudspeakers system with the sources placed on the vertices of a dodecahedron, at 1 m distance around the centre. These twelve decoded signals were then rendered as individual binaural sources via twelve instances of a binaural spatialization algorithm, which converts a monophonic signal to a stereophonic binaural signal (Fig. 5). The twelve binauralized virtual loudspeaker signals are then summed and rendered to the subject.

The binaural spatialization algorithm used was based on the convolution between the signal to be spatialized and a HRIR (*Head Related Impulse Response*) extracted from the Listen IRCAM database[6]. More information about this approach can be found in McKeag & McGrath. (1996). Full-phase HRIR were employed, rather than minimum-phase simplifications, in order to maintain the highest level of spatial information. A customization of the Interaural Time Differences (ITD), given the head circumference of the tested participant, and an HRTF selection phase were also performed as mentioned in the previously cited studies, so that an optimal binaural conversion could be performed.

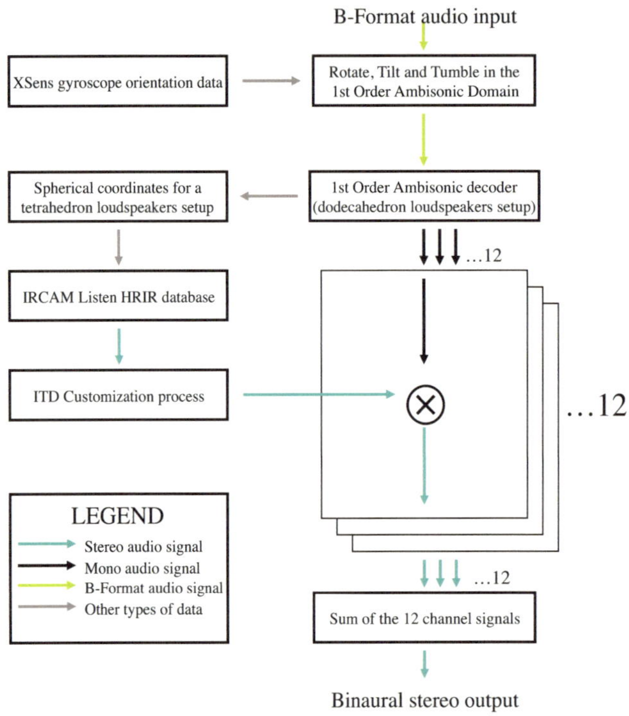

Fig. 5. Schematic representation of the Ambisonic to binaural conversion algorithm.

6.3.2 Protocol and Results: Real versus recorded walkthrough

Two congenitally blind and three late blind participants (two female, three male) took part in this experiment. Each subject was presented with one of the two types of recordings for two of the four environments. Participants were seated during playback.

The learning phase consisted of repeated listings to the playback until the participant felt they understood the environment. When presented with binaural renderings, participants were totally passive, having to remain still. Head orientation in the scene was dictated by the state of the recording. When presented with Ambisonic renderings, they had the possibility to freely perform head rotations, which resulted in real-time modification of the 3D sound environment, ensuring stability of the scene in the world reference frame. Participants were allowed to listen to each recording as many times as desired. As these were playback recordings, performed at a given walking speed, it was not possible to dynamically change the navigation speed or direction. Nothing was asked of the participants in this phase

[6] http://recherche.ircam.fr/equipes/salles/listen/

Two tasks followed the learning phase. Upon a final replay of the playback, participants were invited to provide a verbal description of every sound source or architectural element detected along the path. Following that, participants were invited to reconstruct the spatial structure of the environment using a set of LEGO® blocks. This reconstruction was expected to provide a valid reflection of their mental representation of the environment.
A similar task was demanded to one congenitally blind individual who performed a real navigation within the environments, and was used as a reference.
The verbal descriptions revealed a rather poor understanding of the navigated environments, which was confirmed by the reconstructions. Fig. 7 shows a map of one actual environment and LEGO® reconstruction for different participant conditions. For the real navigation condition, the overall structure and a number of details are correctly represented. The reconstruction shown for the binaural playback condition reflects strong distortions as well as misinterpretations, as assessed by the verbal description. The reconstruction shown following for the Ambisonic playback condition reflects similar poor and misleading mental representation.
Due to the very poor results for this test, indicating the difficulty of the task, the experiment was stopped before all participants completed the exercise. Overall, results showed that listening to passive binaural playback or Ambisonic playback with interactive HM did not allow blind people to build a veridical mental representation of the virtually navigated environment. Participants' comments about the binaural recordings pointed to the difficulties related to the absence of information about displacement and head orientation. Ambisonic playback, while offering head-rotation correction, still resulted in poor performance, worse in some cases relative to binaural recordings, because of the poorer localization accuracy provided by this particular recording technique. Neither condition was capable of providing useful or correct information about displacement in the scene. The most interesting result was that none of the participants understood that recordings were made in a straight corridor with openings on the two sides.
As a final control experiment, after the completion of the reconstruction task, participants were invited to actually explore one of the corridors. They confirmed that they could perceive exactly what they heard during playback, but that it was the sense of *their own displacement* that made them able to describe correctly the structure of the navigated environment. This corroborates findings of previous studies for which the gathering of spatial information is significant for blind individuals when learnt with their own displacements (see Section 4). Further analysis of the reconstruction task can be found in Section 6.4.1.

6.4 Comparison of real and virtual navigation

The results of the preliminary phase of the project outlined how the simulation of navigation through the simple reproduction of signals recorded during a real navigation could not be considered an adequate and sufficiently precise method for the creation of a mental image of a given environment. The missing element seemed to be found in the lack of interactivity and free movement within the simulated environment. For this reason, a second experiment was developed, with the objective of delivering information about the spatial configuration of a closed environment and the positions of sound sources within the environment itself, exploiting interactive virtual acoustic models.
Two of the four closed environments from the initial experience were retained, for which 3D architectural acoustic models were created using the CATT-Acoustics software[7]. Within each

[7] http://www.catt.se

of these acoustic models, in addition to the architectural elements, the different sound sources from the real situation (both real and artificial) were included in order to be able to carry out a distance comparison task (see Section 6.4.1). A third, more geometrically simple model was created for a training phase in order for subjects to become familiar with the interface and protocol. The geometrical model of one experimental space is shown in Fig. 6.

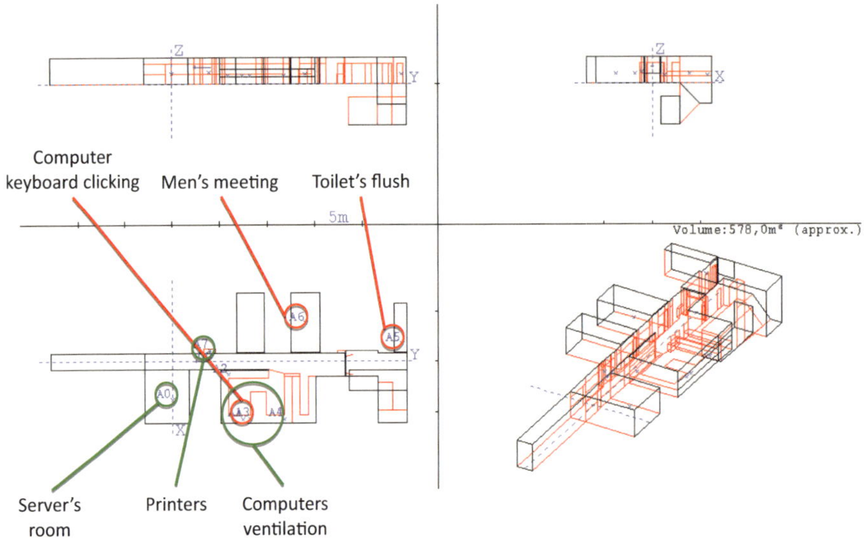

Fig. 6. Geometrical acoustic model of the first space including positions of real (green) and artificially simulated (red) sources.

After observations in the real navigation stage that blind individuals made extensive use of self-produced noises, such as finger snaps and footsteps, in order to determine the position of an object (wall, door, table, etc.) by listening to the reflections of the acoustic signals (see also Section 6.1), a simulation of these noises was included. With the various elements taken into account, a large number of spatial impulse responses were required for the virtual active navigation rendering. A 2nd order Ambisonic rendering engine was used (as opposed to the prerecorded walkthough which used 1st order Ambisonic) to improve spatial precision while still allowing for dynamic head rotation.

Due to the large number of concurrent sources and to the size of 2nd order impulse responses (IR), a real-time accurate rendering was not feasible. Therefore, another approach was elaborated. As a first step, navigation was limited to one dimension only. Due to the fact that both environments were corridors, the user was given the possibility to move along the centreline. Receiver positions were defined at equally spaced positions along this line, at head height, as well as source positions at ground level (for footfall noise) and waist height (finger snap noise). In order to provide real-time navigation of such complicated simulated environments, it was decided to pre-calculate the 2nd order Ambisonic signals for each position of the listener, and then to pan between the different signals during the real-time navigation, rather than performing all the convolutions in real-time, converting finally the Ambisonic signals to binaural using the same approach described in Section 6.3, modified to account for 2nd order Ambisonic.

In the experimental condition, participants were provided with a joystick as a navigation device and a pair of headphones equipped with the head-tracking device (as in Section 6.3).

The footfall noise was automatically rendered in accordance with displacements in the virtual environment. The mobile self-generated finger snap was played each time the listener pressed a button on the joystick.

6.4.1 Protocol: Real versus virtual navigation

The experiment consisted in comparing two modes of navigation along two different corridors, with the possibility offered to the participants to go back and forth along the path at their will. Along the corridor, a number of sources were placed at specific locations, corresponding to those in the real navigation condition. In the real condition, two congenitally blind and three late blind individuals (three females, two males) participated for two corridors. In the virtual condition, three congenitally blind and two late blind individuals (three females, two males) explored the same two corridors.
The assessment of the spatial knowledge acquired in the two learning conditions involved two evaluations, namely a reconstruction of the environment using LEGO® blocks (as in Section 6.3.2) and a test concerning the mental comparison of distances. For the first navigated corridor, the two tasks were executed in one order (block reconstruction followed by distance comparison), while for the second learned corridor the order was reversed.

6.4.2 Block reconstruction

Several measures were made on the resulting block reconstructions: number of sound sources mentioned, number of open doors and staircases identified, number of perceived changes of the nature of the ground, etc. Beyond some distinctive characteristics of the different reconstructions (e.g. representation of wide or narrower corridor), no particular differences could be found between real and virtual navigation conditions; both were remarkably accurate as regards the relative positions of the sound sources (see example in Fig. 7). Door openings into rooms containing a sound source were well identified, while more difficulty was found for openings with no sound source present. Participants were also capable of distinctively perceiving the various surface material changes along the corridors.
An objective evaluation on how similar the different reconstructions are from the actual map of the navigated environment was carried out using bidimensional regression analysis (Nakaya (1997)). After some normalisation, the positions of the numerous reference points, both architectural elements and sound sources (93 coordinates in total) were compared with the corresponding point in the reconstructions, with a mean number of points of 46 ± 12 over all subjects. The bidimensional regression analysis results in a correlation index between the true map and the reconstructed map. Table 3 shows the correlation values of the different reconstructions for real and virtual navigation conditions, together with the correlations for the limited reconstructions done after the binaural and Ambisonic playback conditions, for the first tested environment. Results for the real and virtual navigation conditions are comparable, and both are greater than those of the limited playback conditions. This confirms the fact that playing back 3D audio signals, with and without head-tracking facilities, is not sufficient in order to allow the creation of a mental representation of a given environment due mainly to the lack of displacement information. On the other hand, via real and virtual navigation this displacement information is present, and the amelioration of the quality of the mental reconstruction is confirmed by the similar values in terms of map correlation. Furthermore, correlation values corresponding to the virtual navigation are slightly higher than those for real navigation, confirming the accuracy of the mental reconstruction in the first condition compared with the second.

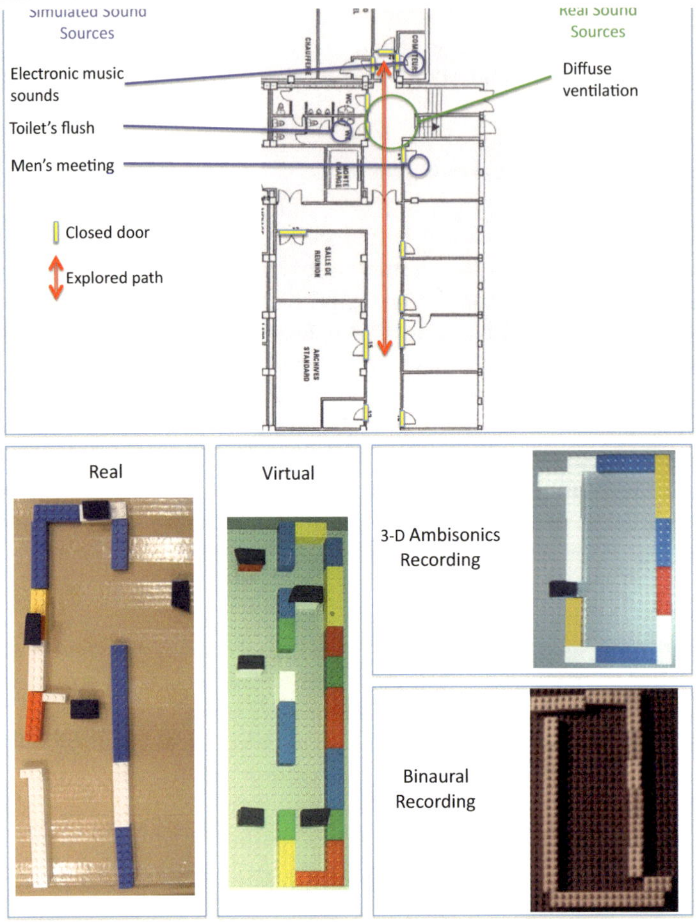

Fig. 7. Examples of LEGO® reconstructions following real navigation, virtual navigation, and binaural and Ambisonic playback.

Correlation of the LEGO® reconstruction				
Condition	Real	Virtual	Ambisonic Rec	Binaural Rec
Correlation Index mean	0.81	0.83	0.71	0.72
Standard Deviation	0.04	0.15	-	-

Table 3. Correlation and standard deviation for bidimensional regression analysis of reconstructions for architectural environment 1. (Std is not available for playback conditions as they contain only 1 entry each.)

6.4.3 Distance comparison

Mental comparison of distances has been typically used in studies intended to capture the topological veridicity of represented complex environments. The major finding from such studies is that when people have to decide which of two known distances is the longer, the frequency of correct responses is lower and the latency of responses is longer for smaller differences. The so-called *symbolic distance effect* is taken as reflecting the analog character of

the mental representations and the capacity of preserving the metrics of the actual distances (Denis (2008); Denis & Cocude (1992); Nordzij & Postma (2005)).
In addition to the starting and the arrival points, three sound sources existed along each path within the two navigated environments (1st: keyboard, men's voices and toilet flush; 2nd: women's voices, electronic sound, and toilet flush). All distances pairs, having one common item for each path (e.g., keyboard-men's voices / keyboard-toilet), have been considered. Distances were classified into three categories: small, medium, and large. Participants were presented with each pair of distances orally, and had to then indicate which was the longer of the two.
Analysis of the results focused on the frequency of correct responses. Table 4 shows the frequency of correct responses for the participants for both real and virtual navigation conditions.

Distance comparison						
Environment	Real			Virtual		
Distance type	Small	Medium	Large	Small	Medium	Large
% correct answers	92.8%	97.6%	100%	83.6%	98.8%	100%
Standard Deviation	2.95	3.29	0	11.28	2.68	0

Table 4. Percent of correct responses for distance comparisons as a function of navigation condition.

Results show that even with a high level of performance for the real navigation condition, there is a confirmation of the symbolic distance effect. The probability of making a correct decision when two distances are mentally compared increased with the size of the difference. A similar trend is seen in the virtual navigation condition. Analysis is difficult as, for both conditions, results are near perfect for medium distances and perfect for large distances. The similarity of results for the two conditions is notable. Both physical displacement (real navigation) and active virtual navigation with a joystick in a virtual architectural acoustic environment allowed blind individuals to create mental representations which preserved the topological and metric properties of the original environment.
Some interesting points were reported by the participants in the virtual navigation condition. They reported that for sound sources that were located at the left or at the right of the corridor, they perceived both the direct signal coming from the sound source and the reflected signal coming from the opposite direction (reflection off the wall), making it possible to locate both the source on one side and the reflecting object (in this case a wall) on the other. The finger snap sound (auditory feedback) was considered extremely useful for understanding some spatial configurations. Both these factors can be considered as extremely important results in light of what has been described in Section 6.1, corroborating the hypothesis that the developed application could indeed offer a realistic and well defined acoustical virtual reality simulation of a given environment, precise enough so that information about the spatial configuration of the total environment, not just source positions, can be gathered by visually impaired users solely through auditory exploration.

7. Acknowledgements

Studies were supported in part by a grant from the European Union (STREP Wayfinding, Contract 12959) and internal research grants from the LIMSI-CNRS (Action Initiative). Experiments conducted were approved by the Ethics Committee of the National Centre for

Scientific Research (*Comité Opérationnel pour l'Ethique en Sciences de la Vie*). The authors are grateful to Michel Denis, for supervision of the cognitive aspects of these studies, and to Christian Jacquemin, for his invaluable contributions to the design of the virtual environments. Finally, the authors would like to thank Amandine Afonso and Alan Blum, collaborators and co-authors of the various studies presented.

8. References

Afonso, A., Katz, B.F.G., Blum, A. & Denis, M. (2005a). Spatial knowledge without vision in an auditory VR environment, *Proc. of the 14th Meeting of the European Society for Cognitive Psychology*, August 31 - September 3, Leiden, the Netherlands.

Afonso, A., Katz, B.F.G., Blum, A., Jacquemin, C. & Denis, M. (2005b). A study of spatial cognition in an immersive virtual audio environment: Comparing blind and blindfolded individuals, *Proc. of the 11th Meeting of the International Conference on Auditory Display*, 6-9 July, Limerick, Ireland.

Afonso, A., Katz, B.F.G., Blum, A. & Denis, M. (2005c). Mental imagery and the acquisition of spatial knowledge without vision: A study of blind and sighted people in an immersive audio virtual environment, *Proc. of the 10th European Workshop on Imagery and Cognition*, 28-30 June, St Andrews, Scotland.

Afonso, A. (2006). *Propriétés analogiques des représentations mentales de l'espace: Etude comparative de personnes voyantes et non-voyantes*, Doctoral dissertation, Université Paris Sud, Orsay, France.

Afonso, A., Blum, A, Katz, B.F.G., Tarroux, P., Borst, G. & Denis, M. (2010). Structural properties of spatial representations in blind people: Scanning images constructed from haptic exploration or from locomotion in a 3-D audio virtual environment, *Memory & Cognition*, Vol. 38, No. 1, pp. 591-604.

Ashmead D.H., Wall R.S., Ebinger K.A., Eaton, S.B., Snook-Hill M-M, & Yang X. (1998). Spatial hearing in children with visual disabilities, *Perception*, Vol. 27, pp. 105-122.

Begault, D. R. (1994). *3-D sound for virtual reality and multimedia*, Cambridge, MA: Academic Press.

Blauert, J. (1996). *Spatial Hearing, the Psychophysic of Human Sound Localization*, Cambridge, Massachusetts, USA: The MIT Press Cambridge.

Blum, A., Denis, M., Katz, B.F.G. (2006). Navigation in the absence of vision : How to find one's way in a 3D audio virtual environment?, *Proc. of the International Conference on Spatial Cognition*, September 12-15, Rome & Perugia, Italy.

Byrne, R. W., & Salter, E. (1983). Distances and directions in the cognitive maps of the blind. *Canadian Journal of Psychology*, Vol. 37, pp. 293-299.

Denis, M. & Zimmer, H. D. (1992). Analog properties of cognitive maps constructed from verbal descriptions, *Psychological Research*, Vol. 54, pp. 286-298.

Denis, M. & Cocude, M. (1992). Structural properties of visual images constructed from poorly or well-structured verbal descriptions, *Memory and Cognition*, Vol. 20, pp. 497-506.

Denis, M. (2008). Assessing the symbolic distance effect in mental images constructed from verbal descriptions: A study of individual differences in the mental comparison of distances, *Acta Psychologica*, Vol. 127, pp. 197-210.

Doucet, M.E., Guillemot, J.P., Lassonde, M., GagnÃl'é J.P., Leclerc, C., & Lepore, F. (2005). Blind subjects process auditory spectral cues more efficiently than sighted individuals, *Experimental Brain Research*, Vol. 160, No. 2, pp. 194-202.

Dufour, A. & Gérard, Y. (2000). Improved auditory spatial sensitivity in nearsighted subjects, *Cognitive Brain Research*, Vol. 10, pp. 159-165.

Elbert, T., Sterr, A., Rockstroh, B., Pantev, C., Muller, M.M. & Taub, E. (2002). Expansion of the Tonotopic Area in the Auditory Cortex of the Blind, *The Journal of Neuroscience*, Vol. 22, pp. 9941-9944.

Gerzon, M. A. (1972). Periphony With-Height Sound Reproduction, *Proc. of the 2nd Convention of the Central Europe Section of the Audio Engineering Society*, Munich, Germany.

Gougoux, F., Lepore, F., Lassonde, M., Voss, P., Zatorre, R.J., & Belin, P. (2004). Neuropsychology: Pitch discrimination in the early blind, *Nature*, Vol. 430, p. 309.

Hartmann, W.M. & Rakerd, B. (1989). Localization of sound in rooms IV: The Franssen effect, *J. Acoust. Soc. Am.*, Vol. 86, pp. 1366-1373.

Kahle, E. (1995) *Validation d'un modéle objectif de la perception de la qualité acoustique dans un ensemble de salles de concerts et d'opéras.* Doctoral dissertation, Université du Maine, Le Mans.

Kasky, D. (2002). Revision: Is visual perception a requisite for visual imagery?, *Perception*, Vol. 31, pp. 717-731.

Lessard, N., Paree, Lepore, F. & Lassonde, M. (1998). Early-blind human subjects localize sound sources better than sighted subjects, *Nature*, Vol. 395, pp. 278-280.

Lewald, J. (2002a). Vertical sound localization in blind humans, *Neuropsychologia*, Vol. 40, No. 12, pp. 1868 âĂŞ 1872.

Lewald J. (2002b). Opposing effects of head position on sound localization in blind and sighted human subjects, *European Journal of Neuroscience*, Vol. 15, pp. 1219-1224.

Loomis, J. M., Klatzky, R. L., Golledge, R. G., Cicinelli, J. G., Pellegrino, J. W. & Fry, P. A. (1993). Nonvisual navigation by blind and sighted: Assessment of path integration ability, *Journal of Experimental Psychology: General*, Vol. 122, pp. 73-91.

Loomis, J. M., Golledge, R. G. & Klatzky, R. L. (1998). Navigation system for the blind: Auditory display modes and guidance, *Presence: Teleoperators and Virtual Environments*, 7, 193-203.

McKeag, A. & McGrath, D. S. (1996). Sound Field Format to Binaural Decoder with Head Tracking, *Proc. of the 101th Audio Engineering Convention*, Los Angeles, CA.

Millar, S. (1994). *Understanding and representing space: Theory and evidence from studies with blind and sighted children*, Oxford, UK: Clarendon Press.

Minnaar, P., Olesen, S.K., Christensen, F., Møller, H. (2001). The importance of head movements for binaural room synthesis, *Proc. of the 7th Meeting of the International Conference on Auditory Display*, Espoo, Finland.

Moore, Brian C. J. (2003). *An Introduction to the Psychology of Hearing*, Fifth Edition, London, UK: Academic Press.

Morrongiello, B.A., Timney, B., Humphrey, G.K., Anderson, S., & Skory, C. (1995). Spatial knowledge in blind and sighted children, *J Exp Child Psychology*, Vol. 59, pp. 211-233.

Muchnik C, Efrati M, Nemeth E, Malin M, & Hildesheimer M. (1991). Central auditory skills in blind and sighted subjects, *Scandinavian Audiology*, Vol. 20, pp. 19-23.

Nakaya,T. (1997): Statistical inferences in bidimensional regression models. *Geographical Analysis*, Vol. 29, pp. 169-186.

Nordzij, M. L. & Postma, A. (2005). Categorical and metric distance information in mental representations derived from route and survey descriptions, *Cognition*, Vol. 100, pp. 321-342.

Ohuchi, M., Iwaya, Y., Suzuki, Y., & Munekata, T. (2006). A comparative study of sound localization acuity of congenital blind and sighted people, *Acoust. Sci. & Tech*, Vol. 27, pp. 290-293.

Poirier, C., Collignon, O., Scheiber, C., & De Volder, A. (2006). Auditory motion processing in early blind subjects, *Cognitive Processing*, Vol. 5, No. 4, pp. 254-256.

Röder, B., Teder-Salejarvi, W., Sterr, A., Rösler, F., Hillyard, S.A., & Neville, H.J. (1999). Improved auditory spatial tuning in blind humans, *Nature*, Vol. 400, pp. 162-166.

Röder B, Rösler F. (2003). Memory for environmental sounds in sighted, congenitally blind and late blind adults: Evidence for cross-modal compensation, *Int J Psychophysiol*, Vol. 50, pp. 27-39.

Starlinger I. & Niemeyer, W. (1981). Do the Blind Hear Better? Investigations on Audiotory Processing in Congenital or Early Acquired Blindness, *Audiology*, Vol. 20, pp. 503-509.

Strelow, E.R. & Brabyn, J.A. (1982). Locomotion of the blind controlled by natural sound cues, *Perception*, Vol. 11, pp. 635-640.

Tinti, C., Adenzato, M., Tamietto, M. & Cornoldi, C. (2006). Visual experience is not necessary for efficient survey spatial cognition: Evidence from blindness, *Quarterly Journal of Experimental Psychology*, Vol. 59, pp. 1306-1328.

Vorländer, M. (2008). *Auralization: Fundamentals of Acoustics, Modelling, Simulation, Algorithms and Acoustic Virtual Reality*, Aachen, Germany: Springer-Verlag. ISBN: 978-3-540-48829-3

Voss, P., Lassonde, M., Gougoux, F., Fortin, M., Guillemot, J-P., Lepore, F. (2004). Early- and Late-Onset Blind Individuals Show Supra-Normal Auditory Abilities in Far-Space, *Current Biology*, Vol 14, pp. 1734-1738.

Wallach, H., Newman, E. B., and Rosenzweig, M. R. (1949). The precedence effect in sound localization, *Journal of Experimental Psychology*, Vol. 27, pp. 339-368.

Weeks, R., Horwitz, B., Aziz-Sultan, A., Tian, B., Wessinger, C. M., Cohen, L.G., Hallett, M., & Rauschecker, J.P. (2000). A Positron Emission Tomographic Study of Auditory Localization in the Congenitally Blind, *J. Neuroscience*, Vol. 20, pp. 2664-2672.

Wenzel, E. M. (1998). The impact of system latency on dynamic performance in virtual acoustics environments, *Proc. of the 16th ICA and 135th ASA International Conference*, Seattle, WA, pp. 2405-2406.

Wightman, F. L. & Kistler, D. J. (1999). Resolution of front-back ambiguity in spatial hearing by listener and source movement, *Journal of the Acoustical Society of America*, Vol. 105, pp. 2842-2853.

Zwiers M.P., Van Opstal A.J., & Cruysberg J.R. (2001a). A spatial hearing deficit in early-blind humans, *The Journal of Neuroscience*, Vol. 21.

Zwiers, M.P., van Opstal, A.l., & Cruysberg J.R.M. (2001b). Two-dimensional sound-localization behavior of early-blind humans., *Experimental Brain Research*, Vol. 140, pp. 206-222.

14

Sonification of 3D Scenes in an Electronic Travel Aid for the Blind

Michal Bujacz, Michal Pec, Piotr Skulimowski,
Pawel Strumillo and Andrzej Materka
Institute of Electronics, Technical University of Lodz
Poland

1. Introduction

Sight, hearing and touch are the sensory modalities that play a dominating role in spatial perception in humans, i.e. the ability to recognize the geometrical structure of the surrounding environment, awareness of self-location in surrounding space and determining in terms of depth and directions the location of nearby objects. Information streams from these senses are continuously integrated and processed in the brain, so that a cognitive representation of the 3D environment can be accurately built whether stationary or in movement. Each of the three senses uses different cues for exploring the environment and features a different perception range (Hall, 1966). Touch provides information on the so called near space (termed also haptic space), whereas vision and hearing are capable of yielding percepts representing objects or events in the so called far space.

Spatial orientation in terms of locating scene elements is the key capability allowing humans to interact with the surrounding environment, e.g. reaching objects, avoiding obstacles, wayfinding (Gollage, 1999) and determining own location with respect to the environment.

An important aspect of locating objects in 3D space is the integration of percepts coming from different senses. Understanding distance to objects (depth perception) has been possible by concurrent binocular seeing and touching experience of near space objects (Millar, 1994). For locating and recognition of far space objects, vision and hearing cooperate in order to determine distance, bearings and the type of objects. The field of view of vision is limited to the space in front of the observer whereas hearing is ominidirectional and sound sources can be located even if occluded by other objects.

Correct reproduction of sensory stimuli is important in virtual reality systems in which 3D vision based technologies are predominantly employed for creating immersive artificial environments. Many applications can greatly benefit from building acoustic 3D spaces (e.g. operators of complex control panels, in-field communication of combating soldiers or firemen). If such spaces are appropriately synthesized, perception capacity and immersion in the environment can be considerably enhanced (Castro, 2006). It has been also evidenced that if spatial instead of monophonic sounds are applied, the reaction time to acoustic stimuli becomes shorter and the listener is less prone to fatigue (Moore, 2004). Because of the enriched acoustic experience such devices offer (e.g. spaciousness and interactivity) they are frequently termed auditory display systems. Recently, such systems gain also in importance

in electronic travel aids (ETA) for the blind. The sensory substitution concept is employed in these devices for auditory navigation of the user (Hersh, 2008), (Strumillo, 2006).
The presented studies focus on the problem of 3D space representation by means of auditory cues for the purposes of aiding the mobility of visually impaired persons. The task is simplified by extensive image processing and scene segmentation calculations, allowing sonification to focus on translating geometrical features and spatial location of major 3D scene elements (e.g. potential obstacles or walls). The number of simultaneously generated sound streams can be limited to the perceptive capacity of a human. Seminal trials were conducted in which different space sonification scenarios were tested with the participation of both sighted and visually impaired volunteers.
The chapter is organized as follows. In Section 2 an overview of ETAs employing auditory displays is given, along with the introduction of the system developed at the Technical University of Lodz. Section 3 discusses the processing steps employed in the prototype ETA - scene reconstruction, segmentation and sonification. Section 4 goes into more detail about sonification and the developed sound coding scheme. Section 5 presents the HRTF measurement system constructed for the project, as well as the results of localization tests with virtual sound sources in 3D space. Section 6 moves on to simulations of the ETA prototype and tests of the developed sonification scheme in virtual reality.

2. Review of sonic outputs of electronic travel aids

One of the first reported electronic travel aids for the visually impaired was built by a Polish scientist Kazimierz Noiszewski in 1897 (Capp, 2000). Dubbed "the artificial eye" - the device used photosensitive Selenium cells and a buzzer to convert light to sounds of strength proportional to the registered brightness. Since Noiszewki's pioneering work there have been many attempts undertaken to use hearing as a sensory substitute for the lost vision, with significant leaps made in the 1960s using ultrasonic sensors and in the 1990s using modern computer technology
Depending on the amount of conveyed information, modern ETAs can be divided into two main groups:

- obstacle detectors, that use laser or ultrasound sensors, but offer limited information
- environmental imagers, which convert scene images or 3D data into rich but complicated sound patterns.

The disadvantage of the simple obstacle detectors is that they are less useful for understanding space, while the main shortcoming of more complex environmental imaging systems is the requirement of a large degree of focus of the user and prolonged training. All sonic devices must also take care not to overly burden the sense of hearing (Bregman, 1999). Consequently, no single ETA has been widely accepted by the community of the visually impaired; however, a few experienced relative success and will be discussed in this section.
The most basic obstacle detectors are built on the concept of the white cane, which remains the basic mobility aid for the blind and can be regarded as an extension of the sense of touch. Such devices as UltraCane (Hoyle, 2003) or Teletact (Damaschini et al., 2005) use ultrasound or laser sensors correspondingly and simple auditory or vibration output for further extension of the cane's reach. These devices provide extra head level protection that is not offered by a standard white cane. Each of these systems uses some form of energy emitted into the environment. The reflected signals are analyzed and if an obstacle is present in the nearby space a simple alert sounds.

More complex obstacle detectors that use ultrasonic waves to scan the scene and convert the reflected signal into sounds are the Sonic Pathfinder (Heyes, 1984) and the KASPA (Kay's Advanced Spatial Perception Aid) system (Kay, 1974). The head-mounted Sonic Pathfinder uses three sonar beams and generates a simple sound code, comprising of frequencies proportional to distances to obstacles. The KASPA system offers better resolution by making use of frequency-modulated (FM) signals for echo location. The ultrasonic transducers are mounted on a standard white cane. KASPA communicates object distance by pitch, but also enhances the scene perception as the timbre of the sounds depends on the texture of scanned objects.

The flaw of ultrasonic devices is that the emitted beams diverge with distance and their angular precision of locating obstacles deteriorates. On the other hand, the laser based devices can be interfered by strong ambient light.

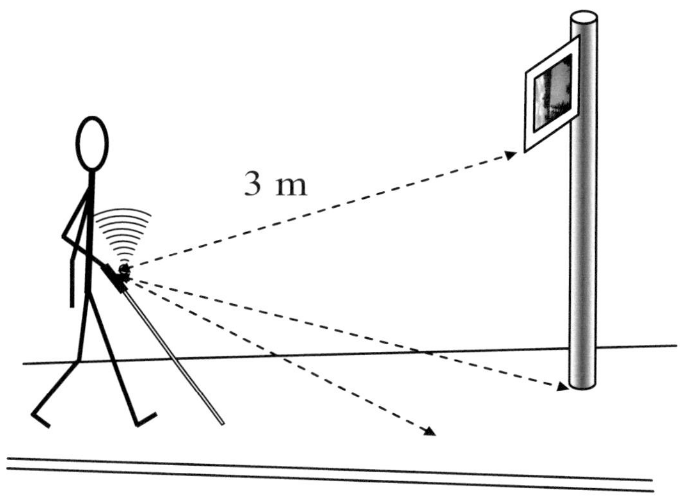

Fig. 1. An example of an electronic obstacle detector in action; laser or ultrasound beams emitted from cane handle are reflected form obstacles and communicated to the user by audio or tactile cues, e.g. providing head level protection that is not offered by a standard white cane

A number of current projects aim at building ETA's that feature functionalities of environment imagers that convert images (e.g. stereoscopic images) of a scene into a scheme of sound codes (auditory icons) or tactile patterns reflecting scene obstacles. Because of the much lower perception capacity of the human auditory (or tactile) system in comparison to visual perception, some form of image analysis or data pre-selection must be applied before the auditory code can be synthesized and presented to the blind user. One such system built in the Nederland, the vOICe, scans the image of a scene in short one-second cycles (Meijer, 1992). Image contents in the columns of pixels that move from left to right are converted into sound patterns. Each column of pixels is used as the spectrum for the synthesized sound, with higher positioned pixels corresponding to higher frequencies and the pixel brightness determining the magnitude of the frequency component. The resulting auditory code is very rich in information, but far from intuitive, thus the users require on average a few months of training before independent navigation trials can be commenced. Note also that the obstacles are not necessarily represented by bright image regions.

Another interesting system was designed and built in the University of La Laguna, Spain (Gonzales-Mora, 1999), under the name of Espacio Acustico Virtual (EAV). This ETA combines stereovision and Head Related Transfer Function (HRTF) technologies. Miniature cameras are mounted into glasses that are attached to headphones of special construction. Stereovision enables 3D reconstruction of a scene that is represented by a collection of equally spaced points. Each such point is characterised by its distance and direction as seen from the ETA. Such a collection of points are sources of acoustic generators which periodically and simultaneously generate short sharp sounds. Locations of points in the environment are determined by a pair of HRTF filters and their distance is reflected by the loudness and phase shift. The shortcoming of the system is the acoustic overload the listener is confronted to. The project has reached the phase of a prototype, but has not been commercialized yet.
The indicated environmental imaging systems, however, do not attempt to pre-process images (or select scene objects) to match the sensory bandwidth mismatch between the human sight and the sense of hearing.

3. 3D scene sonification concept

The ETA system under development at the Technical University of Lodz, Poland, utilizes stereovision and HRTF technologies to provide real-time conversion of video into sound streams. The predominant assumption made in the construction of this system is to limit the amount of auditory information that is presented to a blind user to the most useful minimum. A block diagram of the auditory space representation system is shown in Fig. 2 and the processing steps are described in subsections below. The three main modules of the constructed ETA system perform the following tasks:

- acquisition of stereovision image sequences and their processing for real time 3D scene reconstruction and segmentation,
- coding and synthesis of sound streams associated with selected scene objects,
- filtering of the sound output through individualized HRTFs (Head Related Transfer Functions) for spatial sound illusion via headphones.

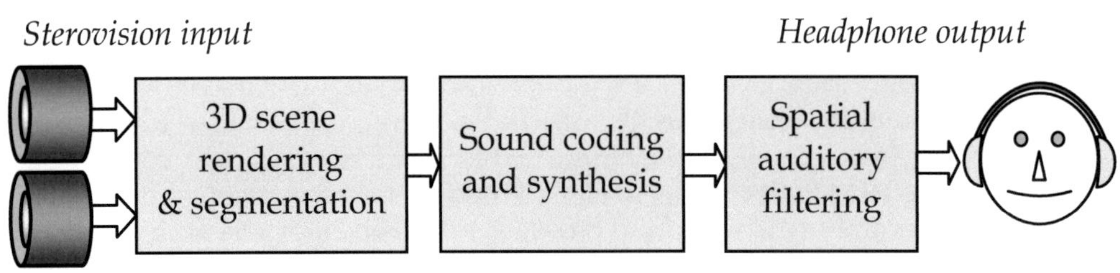

Fig. 2. Schematic of the auditory space representation system

3.1 Scene reconstruction

Scene reconstruction is the determination of 3D coordinates of points belonging to the scene. Usually, a reconstructed scene is presented in the form of a 2.5D image - i.e. a 2D array of depth values.

A number of computer vision solutions were considered for use in the developed ETA, but stereovision was chosen due to its inexpensiveness, passiveness and the experience the authors have had with the technology.
Stereovision reconstruction consists of comparison of images of the same scene as viewed by two cameras from different locations in space. A point in 3D space with coordinates $(X; Y; Z)$ will appear in two different image coordinates when viewed by two cameras $p_L(x_0; y_0)$ and $p_R(x; y)$. If the camera axes are parallel and the images have been rectified and corrected for geometric distortions (Brown, 2003), the disparity can be calculated as the difference between the horizontal coordinates of the point in the left and right image:

$$d(p_L; p_R) = x_0 - x \quad (1)$$

Disparity can directly be used to find the depth of the reconstructed point:

$$Z = Bf/d, \quad (2)$$

where B - distance between cameras' optical axes, f - focal length of cameras.
Disparity is calculated for each point for which a corresponding point can be found in the second camera's image, which is determined using correlation of a small window of pixels (default: 5x5). One of the major problems of stereovision is that the maximum of correlation is often difficult to determine, as smooth surfaces will result in large highly correlated regions. For such regions depth estimation is marked inconclusive. Only points for which the correlation maximum was clearly found are considered correctly reconstructed.
Basic stereovision reconstruction was later improved on by camera ego-motion estimation. Estimating the movement vectors of the cameras allowed to better predict positions of correlating pixel groups, improving framerates, as well as leading to sub-pixel depth accuracy thanks to interpolation of the positions of scene elements. A more detailed description of how the processing of the stereovision data is carried out is given in (Skulimowski, 2007 & 2008).

Fig. 3. Stereovision scene reconstruction - the left and right images and the result of the reconstruction; Grey areas represent pixel groups for which proper depth estimation was not possible

3.2 Scene segmentation

After successful 3D reconstruction, the vision module performs segmentation by implementing original algorithms for building a 3D model of the scene (Skulimowski, 2009). The model consists of planes and other objects interpreted as scene obstacles. This type of model is justified by noting that most man-made environments assume such spatial geometry, e.g. streets and buildings' walls, corridors, halls, rooms etc.

An iterative algorithm detects surfaces basing on the depth map. The map is overlaid with a mesh of triangles, with vertices in points of well determined depth. Each triangle's normal vector and plane equation is calculated, and those with similar coefficients to their neighbours are defined as belonging to a common plane. Iteratively, planes are combined if they have similar normal vectors. After a number of iterations, all remaining points which formed groups too small to qualify as planes are grouped with theirs spatial neighbours and marked as obstacles. Clouds of points classified as obstacles are then analyzed to estimate the obstacle's size, shape and orientation. An example of the segmentation procedure and its output is shown in Fig. 4. Note that the plane representing the floor surface in cancelled out and is not meant for sonification.

Fig. 4. The scene segmentation process: original depth map in pseudo-color (left), detection of planes overlaid on the original image (center), and grouping of objects which remain after removal of segmented surfaces (right)

3.3 Scene sonification

The output of the vision module is further converted into a data stream that encodes the size, distance, and angular direction of the obstacles, as well as the equation coefficients of planes present in the scene. This stream of data is used for controlling the sound synthesis module that generates the auditory streams, each representing a unique scene element. Different sonification schemes are assigned to the selected obstacles and planes.

To create a spatialized perceptive illusion of sound streams attached to scene elements (obstacles and planes), the sound presentation module processes the sound streams using listeners' individual HRTFs. The final output of the system are spatially filtered stereophonic sounds arriving in the listener's headphones.

3.3.1 Development of the sound coding scheme

The term "sound code" is used to describe the method of representing the attributes of a real object with parameters of a virtual audio source, so that the sound carries information useful for a visually impaired traveller. A number of trials where volunteers judged various sound codes (Bujacz, 2006) led to the eventual design of a sonar-like sonification method called "depth scanning".

The first sound coding concept did not utilize scene segmentation and was based on direct depth information. The reconstructed scene was automatically swept with a simulated narrow-beam sensor (Bujacz, 2006) and the sound output was given in the form of MIDI synthesized musical tones corresponding to the distance to the nearest obstacle covered by the beam. A screenshot of the program and the horizontal scanning concept is shown in Fig. 5. The simulated ETA was tested with participation of 10 sighted volunteers. Results showed that the sound coding method was very inefficient; however, it allowed accurate scene recognition after just 3-4 hours of practice. Participants were able to draw the shape of a room observed through the code after 1-3 minutes of observation, as well as slowly navigate simple labyrinths.

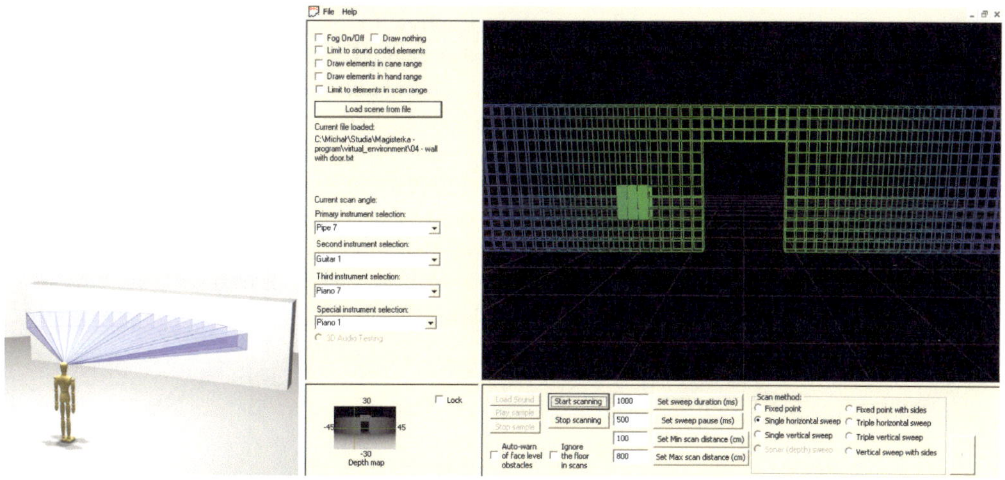

Fig. 5. "Musical range finder" scene sonification concept; the scene is periodically scanned horizontally producing sounds of tone inversely proportional to distance (left); Simulation software (right), note the graphics window highlights the currently sonified scene section

After successful trials of the scene segmentation algorithm, a more efficient sound code was developed to utilize the knowledge of a parameterized scene description (Strumillo, 2006). In addition to MIDI sounds, the simulation programs used formant-based sound synthesis in an attempt to recreate vowel sounds to add an extra dimension to the sound code (Pełczyński, 2006). The formant-based synthesis was also used in further described studies on sound localization.

A wide selection of sound codes was prepared, so that feedback from visually impaired testers could be obtained. The test participants were tasked with identifying the primary sound characteristics they easily and quickly recognized. The majority of testers decided that the ability to quickly interpret information about nearest obstacles and the scene layout was of most importance. Pure musical tones were deemed more pleasant to continuous frequency changes and caused less dissonance during simultaneous playback of multiple sources. Synthesized musical instruments were preferred over artificial sounds, and timbre manipulation through formant filtering was judged too difficult to easily interpret. Methods of informing of multiple scene elements were also tested. Simultaneous playback of all sounds with periods proportional to the distance was deemed overly noisy and difficult to interpret. However, the very instinctive solution of increasing the period of sounds with proximity to an object was noted as having potential for special warnings, such as alerting about very near face-level obstacles. Scanning the scene for obstacles horizontally, vertically or distance-wise was also considered, and the last method proved most useful for travel.

Bregman's theories about the functioning of auditory perception (Bregman, 1990), especially the concept of auditory streams, were of big importance during the sound code design. A sound stream is usually a single sound source in a specific location; however, multiple sounds played in unison or close succession integrate into a single stream. This can be a desired effect, e.g. in music; however, the idea behind the sound code was to achieve an opposite effect. Each scene element was to be perceived as a separate auditory event. For this reason the sound streams were separated temporally (default 0.2 s) and spectrally by pitch and tone.

The final version of the sound code was prepared after the surveys with potential blind users and aimed at being most clear and instinctive to interpret. Distant objects were encoded with more quiet sounds, which were also played back later during the scanning cycle, i.e. object distance was coded with sound amplitude and time delay from the start of a scanning cycle. Larger objects were assigned longer and lower sounds thus encoding object size with tone and duration. HRTF filtering was used to give the sound sources the illusion of originating from scene elements and it will be discussed in more detail in the next section.

In the "depth scanning" sound code, the sources are presented in order of their proximity to the observer. A good way to illustrate this coding concept is to picture a virtual scanning plane that moves through the scene and releases sound sources as it intersects various scene elements. The scene sonification method is illustrated in Fig. 6.

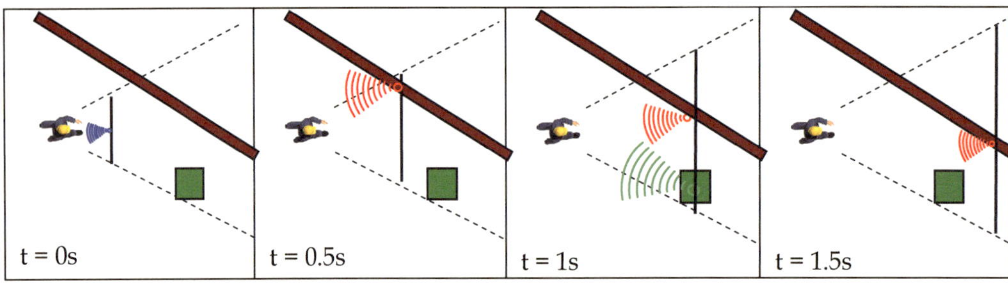

Fig. 6. One cycle of the depth scanning sound code concept. A virtual scanning plane (solid vertical line) moves away from the observer through a scene with one wall (red) and one obstacle (green). The detection range is marked with dashed lines and the sound sources which are released as the scanning surface intersects scene elements are shown with icons whose size is proportional to their loudness. After a cycle completes and a short period of silence (default 0.5s) the scanning starts again.

The number of sound sources presented during a scanning cycle can be limited and is set to 4 nearest elements as default. The scanning can be relatively accurate to a range of 15 m; however, a range of only 5 m was chosen as only the nearest scene elements were deemed important for safe travel. The listener can adjust the speed of the scanning. The default scanning period is 2 s. A number of listeners sped up the scanning to 1.5 s periods; however, most preferred to slow it down to 2.5 - 3 s during training.

Different types of scene elements were assigned sounds of different musical instruments. The prototype only features recognition of two classes of elements - walls and generic obstacles; however, the sound code assumes future expansion of categories to include elements such as moving obstacles, with possible recognition of human silhouettes, and various surface discontinuities, i.e. curbs, drop-offs, stairs and doors.

The sound code utilizes audio files pre-generated with a Microsoft General MIDI synthesizer and modulated with 5% noise (14 dB SNR). They are stored in collections of 5s long wave files of full tones from the diatonic scale (octaves 2 to 4) referred to as banks. The range of pitches in a bank can be selected independently for each scene element type. The default setting for obstacles spans from musical tone G2 (98 Hz) to B4(493 Hz) and for walls from G3 (196 Hz) to G4 (392 Hz). The default instrument for obstacles is a synthesized piano sound (General MIDI Program 1), while the bass end of a calliope synthesizer (General MIDI Program 83) was used for walls. The instruments were chosen during preliminary surveys with blind testers; however, it was the wish of most trial participants to be able to assign their own choice of sounds if possible (Bujacz, 2005).

4. Measurement and verification of HRTFs

In parallel to the ETA project, research on head related transfer functions (HRTFs) was carried out. The HRTFs were measured in an anechoic chamber using special equipment designed for efficiency of data collection (Fig. 7). Data was collected in the full azimuth range (θ = 0° to 360°) with a 5° step and a broad elevation range (φ=-45° to 90°) with a 9° step. Twelve sighted and eight visually impaired volunteers took part in the HRTF measurements. The measurement and processing procedure performed in an anechoic chamber enables recording of the head related impulse responses (HRIRs) to the sounds produced by all speakers, while the listener sits in a revolving chair with the microphones placed at the entrances to his ear canals. The measured HRIRs were converted into a format supported by the NASA SoundLAB environment (Miller, 2001) used for later trails directly or as libraries for custom written software (Pec, 2008).

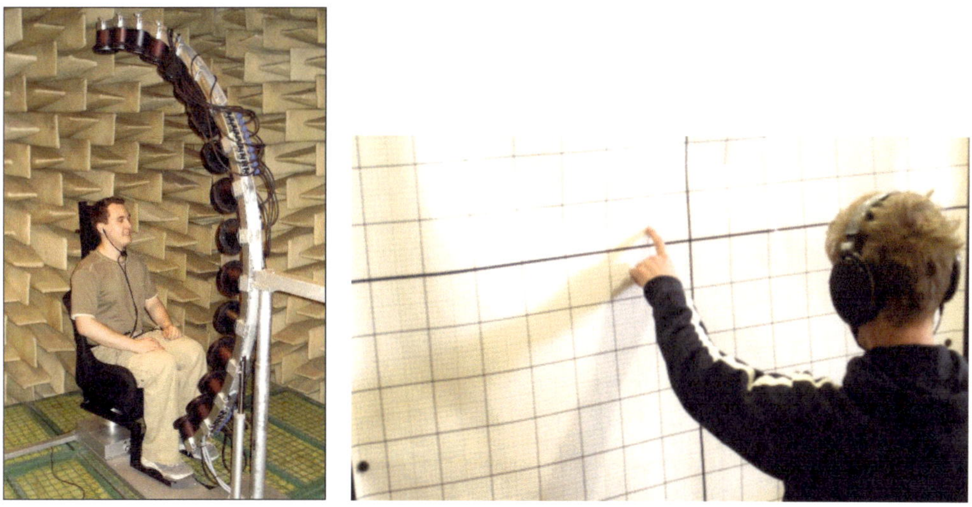

Fig. 7. The HRTF measurement setup in an anechoic chamber (left) and sound localization trials (right)

Most volunteers who took part in the measurements also participated in externalization and localization trials, which served to verify the correctness of data collection and the usefulness of personalized HRTFs. In the first trials, out of 15 volunteers, 12 achieved full sound externalization and 2 gained limited externalization after short training. Different static and moving virtual sources located in the frontal space hemisphere (to avoid front-back confusions) were presented to the volunteers using their personalized and generic HRTFs. Volunteers were to point to the location of perceived sounds on a grid in front of them (Fig. 7). The results of the localization trials with different virtual sound sources for both sighted and blind volunteers are shown in Figs. 8 - 11.

The blind volunteers localized the spatialized sound sources with larger errors (12.5°deg) than sighted individuals (8° on average). The better results for sighted participants are likely due to better trained localization skills thanks to visual feedback training throughout their entire lives (Zahorik 2001). Although the results are not presented in the charts, congenitally blind volunteers made more errors than those who lost sight at a later age. The visually impaired volunteers stressed the need for early-age rehabilitation and training to improve sound localization skills. Despite errors being larger than expected from similar studies (Wersenyi, 2007), the concept of using HRTFs for sonifying the obstacles was proven to be worth considering. Wideband sounds are mandatory for accurate localization, and moving sources are localized with a slightly higher precision than static ones.

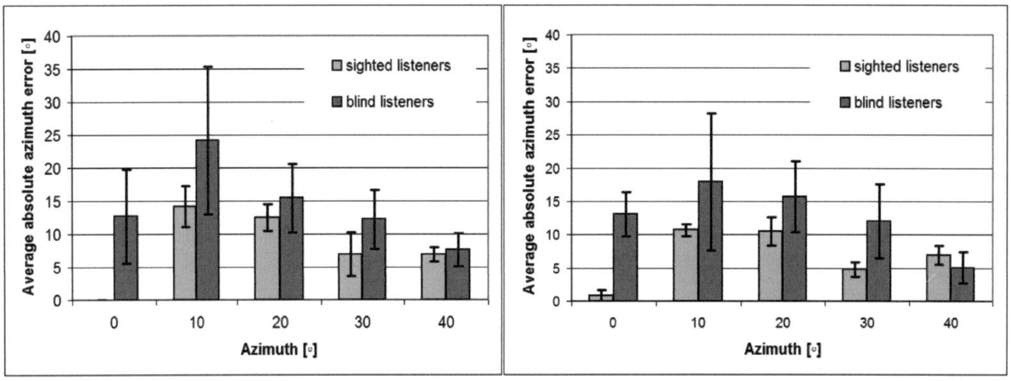

Fig. 8. Comparison of azimuth localization errors for sighted and blind listeners: (left) narrow-band formant synthesized vowel "a" sound, (right) wideband chirp; error bars represent standard deviation among all test participants

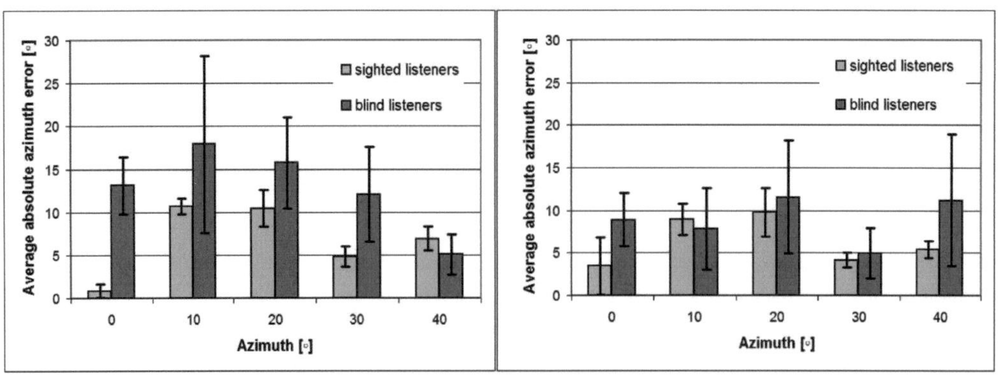

Fig. 9. Comparison of azimuth localization errors for sighted and blind listeners: (left) static wideband sounds (right) oscillating wideband sounds; error bars represent standard deviation among all test participants

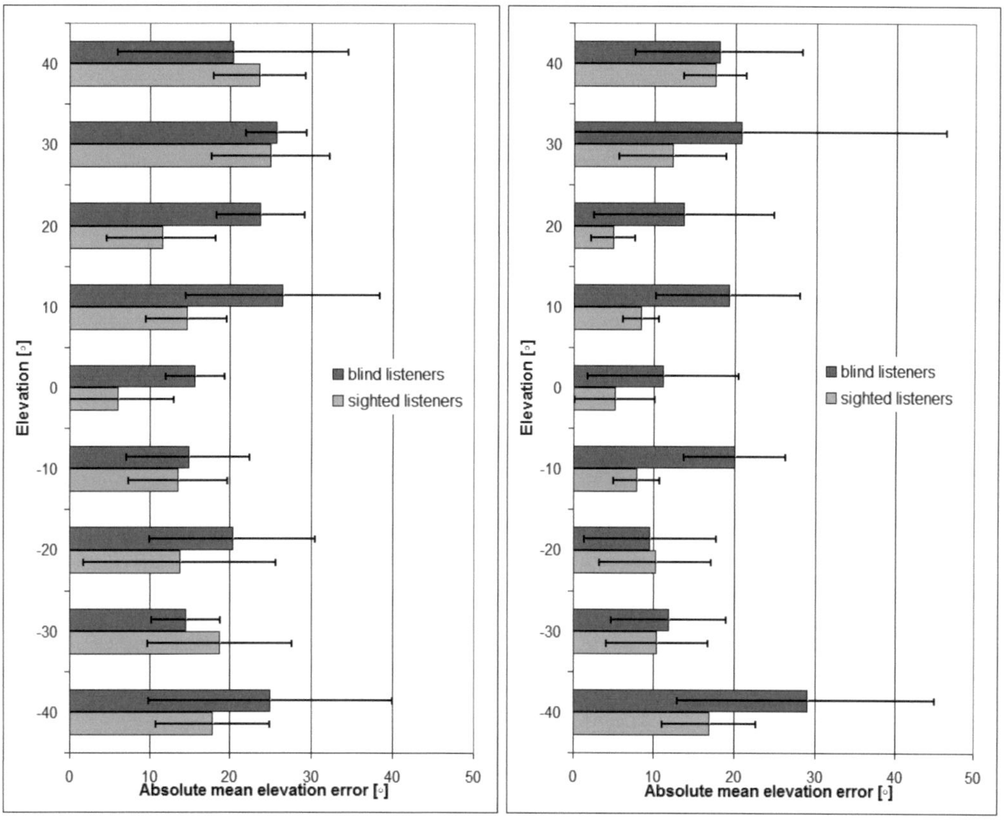

Fig. 10. Comparison of elevation localization errors for sighted and blind listeners: (left) narrow-band formant synthesized vowel "a" sound, (right) wideband chirp; error bars represent standard deviation among all test participants

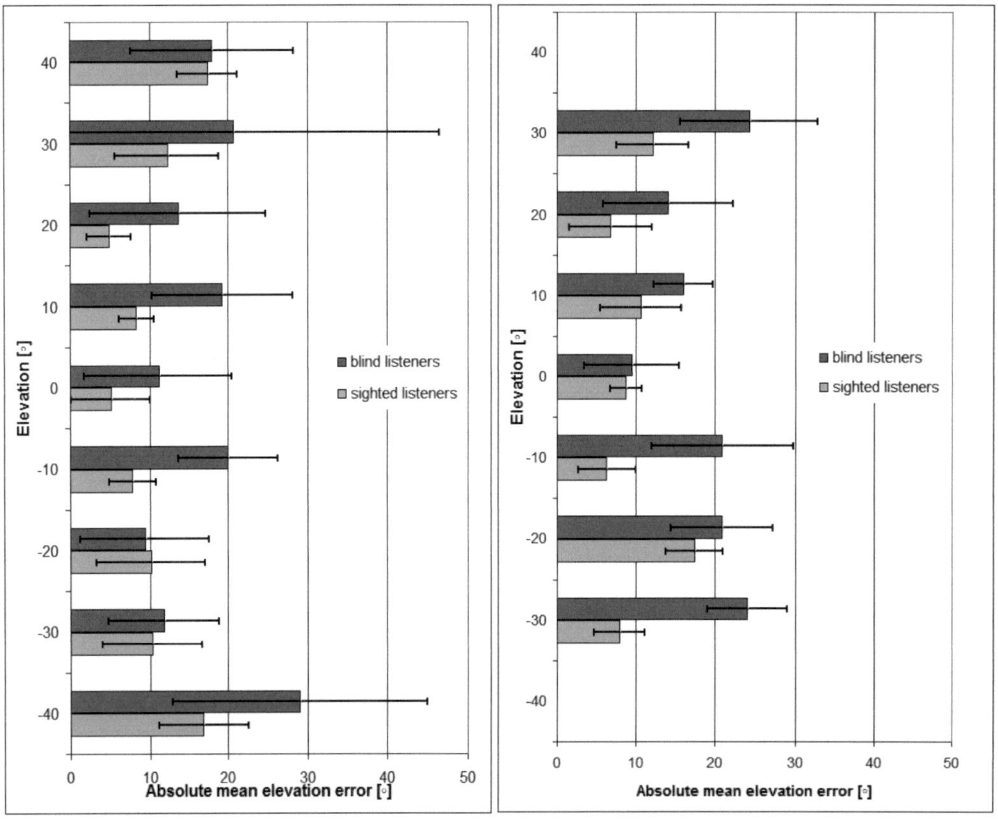

Fig. 11. Comparison of elevation localization errors for sighted and blind listeners: (left) static wideband sounds (right) oscillating wideband sounds; error bars represent standard deviation among all test participants

6. Virtual reality trials

After static trials with virtual sources, we tested the sonification concept in conditions allowing for head movements relative to the sounds, as active perception plays an important role in sound localization (Kato, 2003). A VirtualResearch V6 head mounted display with an InterSense InterTrax 3DOF tracker were used for the trials. Because the tracker only detected rotational movements and previous tests showed that performance in trials requiring translational movements in a VR environment were too dependant on previous experience with 3D environments, such as games (Bujacz, 2005), the VR tests were designed in a way to avoid the necessity of any translational movements.

During the VR trials 10 volunteers, aged 22 to 49, tried to describe simple sonified scenes, such as those presented in Fig. 12 or localize virtual sound sources in discrete positions. Obstacles could be located in three positions along each axis: horizontally - left/center/right, vertically - low/eye-level/high, and depth-wise - near/middle/far. Walls could be positioned in five different ways: parallel to the camera axis on the left or right, at a 45°angle on the left or right, and perpendicularly to the camera axis in front of the observer. Errors of a single position in any direction are referred to as "small", while of two or more positions as "large".

The participants of the trials were all sighted and their personalized HRTF characteristics were collected in the study discussed in Section 5. All participants were trained with the use of the sound code for 15-30 mins by observing scenes visually after hearing them sonified. During the actual trials the first batch of 10 scenes was also considered training and not taken into account for the calculation of final results. The participants continued to learn throughout the trials, as after providing an answer based on the sound code, they were allowed to verify it visually.

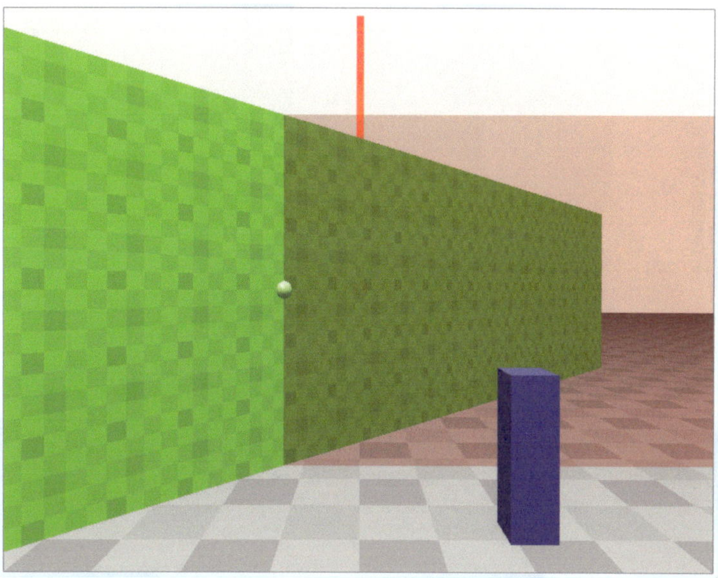

Fig. 12. A sample scene used in the VR sound localization and scene recognition trials; during trials only the floor and zero azimuth marker on the horizon were visible, while participants identified the position of invisible scene elements basing on the sounds heard

The VR trials consisted of three stages, each containing 80 randomized scenes - 40 filtered with personalized HRTFs and 40 with HRTFs of a default individual. In the first stage, the volunteers were tasked with localizing a single object which could appear in one of 27 positions (3x3x3). The sound code was set to fast 1 s cycles and the participants, even though not timed, were asked to make their guesses quickly and instinctively. The second stage consisted of describing simple scenes with a single small obstacle and a wall. This time the sound code was slowed down to 2.5 s cycles. The final third stage consisted of describing scenes with a single wall and two obstacles of different size.

The trial results averaged for the 10 participants, along with standard deviation markings, are shown in Figs. 13 to 16. Personalized HRTFs give a clear advantage in sound localization, especially in terms of large errors; however, small errors are still frequent.

Personalized HRTFs clearly provide better localization accuracy than non-personalized ones; however, errors are still very frequent, especially in locating the vertical position of sources. The main advantage of the personalized HRTFs is visible in the significantly decreased number of large errors, i.e. up-down confusions. A surprising result is the lessened frequency of errors in all other aspects of the sound code, not only those directly related to sound localization. The cause of those improvements might be better externalization or the fact that the sounds seem more natural when filtered through personal HRTFs.

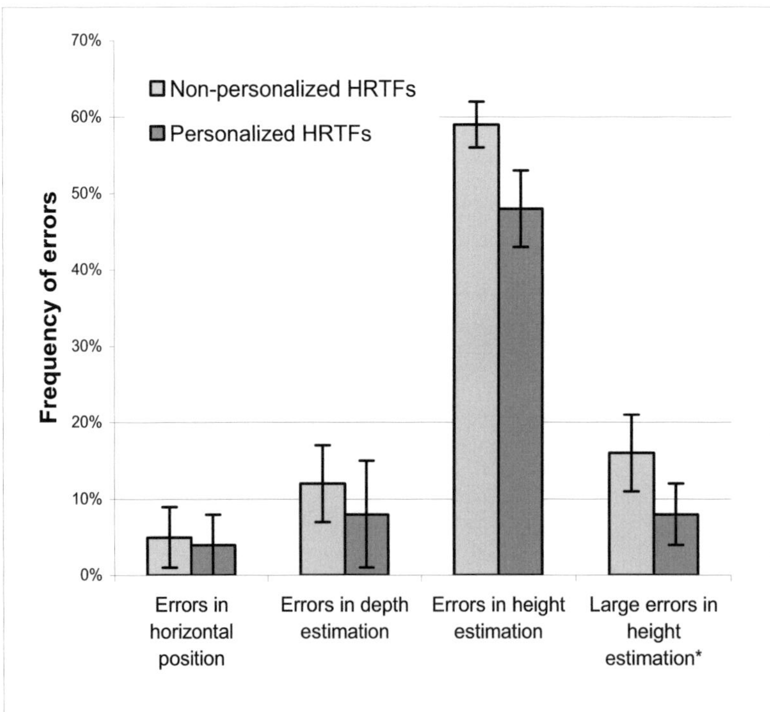

Fig. 13. Percentage of errors in single source localization trials; *) large errors mean a mistake of two discrete positions, which only occurred in cases of up-down confusions

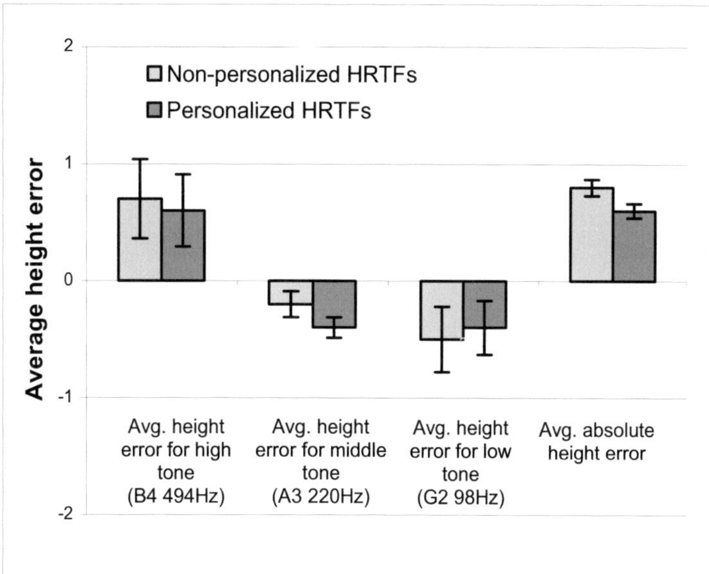

Fig. 14. Directionality of height errors. An error value of +1.0 means a source was localized one level higher than it should be, -1.0 one level lower. Errors were in the range of -2.0 to +2.0, though up-down confusions were not frequent

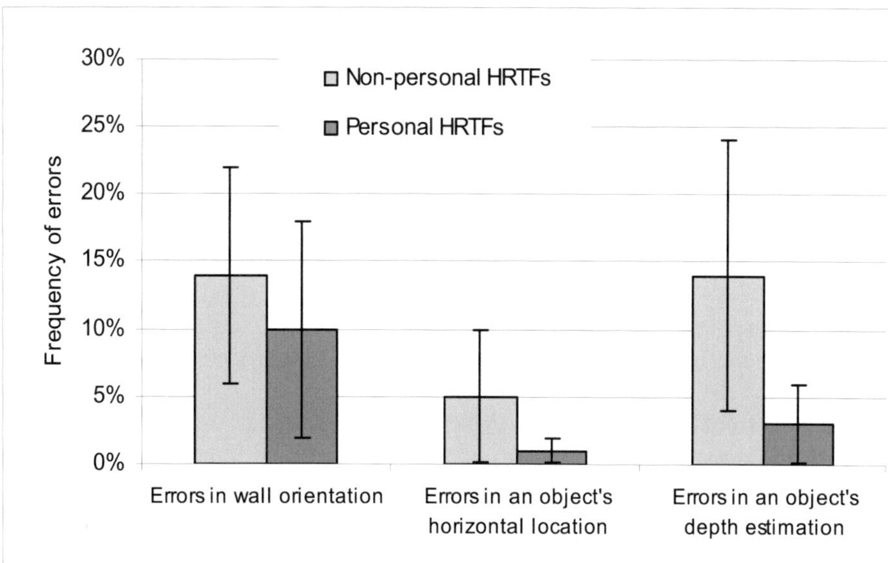

Fig. 15. Percentage of errors during the localization of one wall and one obstacle

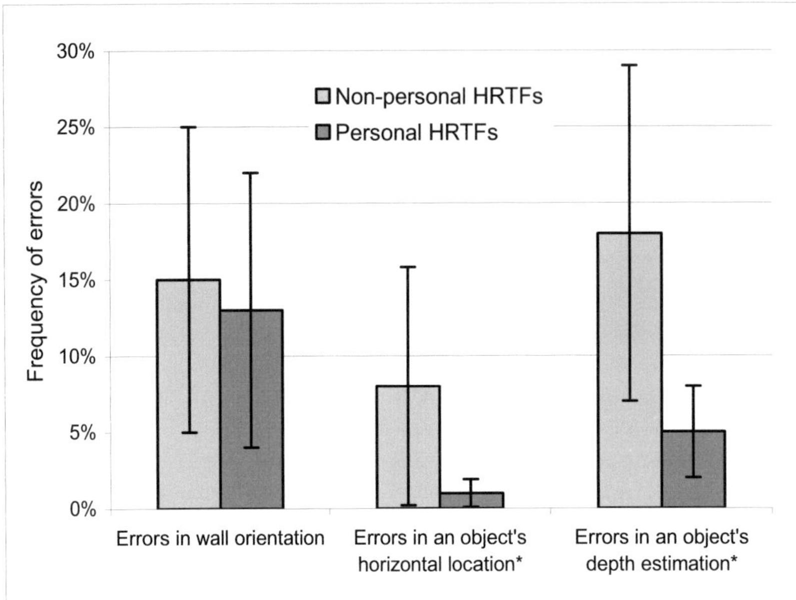

Fig. 16. Percentage of errors during the localization of one wall and two obstacles; the results are averaged for the two obstacles, as errors in decoding the relative position (both horizontal and depth) of two objects never occurred and identification of the larger and smaller object was always correct

Errors in scene recognition were unfortunately frequent. Especially the vertical localization of obstacles leaves a lot to wish for; however, the proposed sound code is effective enough to move on to trials in real scenes once the hardware of the ETA prototype is fully functional.

Results presented in Fig. 14 show a strong psychoacoustic correlation of pitch with perceived height. When in the first stage of the tests closer scene elements were encoded with high pitched sounds, they were nearly always perceived higher than their HRTFs actually placed them and were the main source of the large errors.

7. Conclusions and future work

In this chapter we have outlined main results of the research devoted to sensory substitution systems in which spatial location and geometric features of 3D scene objects are converted into spatialized sound icons in order to offer an auditory scene representation system for the visually impaired.

The conclusions from the presented studies are that the use of personalized HRTFs improves externalization and localization of virtual sources, although the improvement is small compared to non-personal HRTFs. The proposed sound encoding concept is instinctive and easy to learn, as well as efficient at warning of nearby obstacles and orientation in simple scenes. An original scene sonification concept was proposed in which scene obstacles important for user safety are segmented out from the scene images, assigned unique sound icons and selected for auditory display. The cyclic depth scanning method and sequential sonifying of obstacles concur with Bregmans' theory of sound streams, i.e. a

low number of selected sound streams are presented only so that the user can easily track them while in movement.
Further research is needed to judge the usefulness of the prototype when users need to focus on the actual task of walking and navigating in real environments. Real-world trials with a portable prototype and visually impaired participants are in preparation.
Results of the presented work can be of use in virtual reality systems in which immersion in virtual world can be further improved by supporting 3D imaging of objects with 3D auditory sensation of the surrounding acoustic scenes.

8. Acknowledgements

This work has been supported by the Ministry of Science and Higher Education of Poland research grant no. N N516 370536 in years 2009-2010 and grant no. N R02 008310 in years 2010-2013. The third author is a scholarship holder of the project entitled "Innovative education [...]" supported by the European Social Fund.

9. References

Benjamin, J., Malvern, J., (1973), "The new C-5 laser cane for the blind." In: *Proc. Carnahan Conf. on Electronic Prosthetics*, Univ. of Kentucky.

Bourbakis, N. (2008). Sensing Surrounding 3-D Space for Navigation of the Blind, *IEEE Engineering in Medicine and Biology Magazine*, Jan/Febr. 2008, 49–55

Bregman S. (1990). *Auditory Scene Analysis: the Perceptual Organization of Sound,* A Bradford Book, The MIT Press, Cambridge, Massachusetts

Brown, M.Z., Burschka, D. & Hager, G.D. (2003). "Advances in computational stereo", *IEEE Transactions on Pattern Analysis and Machine Intelligence*, Vol. 25, No. 8, 993–1008

Bujacz, M. & Strumillo, P. (2006): Stereophonic Representation of Virtual 3D Scenes – a Simulated Mobilty Aid For the Blind, *XI Symposium AES: New Trends In Audio And Video*, 157–162

Capp, M., Picton, P., (2000) The optophone: an electronic blind aid, *Engineering Science and Education Journal*, June 2000, 137-143

Castro-Toledo, D.; Magal, T.; Morillas, S. & Peris-Fajarnés, G. (2006). 3D Environment Representation through Acoustic Images. Auditory Learning in Multimedia systems, *Current Developments in Technology-Assisted Education*, 735–740

Damaschini, R.; Legras, R.; Leroux, R. & Farcy, R. (2005). Electronic Travel Aid for the Blind people, in *Assistive Technology: from Virtuality to Reality*, Pruski, A. & Knops, H. (Eds.), 251–260

Dobrucki, A., Plaskota, P., Pruchnicki, P., Pec, M., Bujacz, M., Strumillo, P., (2010). Measurement System of Personalized Head Related Transfer Functions and Its Verification by Virtual Source Localization Trials with Visually Impaired and Sighted Individuals, Journal of Audio Engineering Society, vol. 58, no. 9, pp. 724–738.

Gollage, R. G. (Ed.) (1999). *Wayfinding behaviour: cognitive mapping and other spatial processes,* John Hopkins University Press, Baltimore , USA

Gonzalez-Mora, J. L., Rodriguez-Hernandez, A., Rodriguez-Ramos, L. F., Diaz-Saco, L., Sosa, N., (1999). *Engineering Applications of Bio-Inspired Artificial Neural Networks.*

Springer Berlin/Heidelberg, Ch. Development of a new space perception system for blind people, based on the creation of a virtual acoustic space, 321-330.

Hall, E. T. (1966). *The Hidden Dimension*, Doubleday, Garden City, N.Y.

Hersh, M. A. & Johnson, M. A. (Eds.) (2008). *Assistive Technology for Visually Impaired and Blind People*, Springer–Verlag, London Limited

Heyes, D. A., (1984). The sonic pathfinder: A new electronic travel aid. *Journal of Visual Impairment and Blindness* 77, 200-202.

Hoyle, B. S. (2003). The Batcane – mobility aid for the vision impaired and the blind, *IEE Symposium on Assistive Technology*, 18–22

Kato, M., Uematsu, H., Kashino, M., Hirahara, T., (2003) "The effect of head motion on the accuracy of sound localization", *Acoustical Science and Technology*, Vol. 24, No.5, 315-317.

Kay, L., (1964). An ultrasonic sensing probe as a mobility aid for the blind. *Ultrasonics* April-June.

Kay, L., (1974). A sonar aid to enhance spatial perception of the blind : Engineering design and evaluation. *Radio and Electronic Engineer* 44, 605-627.

Meijer, P., (1992). An experimental system for auditory image representations. *IEEE Transactions on Biomedical Engineering* 39, 112-121.

Moore, B. C. J. (2004). *An introduction to the psychology of hearing*, Elsevier Academic Press, London, UK

Millar, S. (1994), *Understanding & representing space*, Clarendon Press, Oxford.

Miller, J., (2001), SLAB: A software-based real-time virtual acoustic environment rendering system. In: *Proceedings of the 2001 International Conference on Auditory Display*, Espoo, Finland.

Pelczynski, P., Strumillo, P., Bujacz, M., Formant-based speech synthesis in auditory presentation of 3D scene elements to the blind, ACOUSTICS High Tatras 2006 - *33rd International Acoustical Conference - EAA Symposium*, Štrbské Pleso, Slovakia, October 4th - 6th, 2006, 346-349.

Skulimowski, P., Bujacz, M., Strumillo, P., (2009). *Image Processing & Communications Challenges.* Academy Publishing House EXIT, Warsaw, Ch. Detection and Parameter Estimation of Objects in a 3D Scene. 308-316

Skulimowski, P. & Strumillo, P. (2007). Obstacle localization in 3D scenes from stereoscopic sequences. *Proc. of the 15th European Signal Processing Conference (EUSIPCO 2007)*, September 3-7, Poznań, Poland, 2095–2099

Skulimowski, P. & Strumillo, P. (2008). Refinement of depth from stereo camera ego-motion parameters, *Electronics Letters*, vol. 44, no. 12, 729–730

Strumillo, P.; Pelczynski, P.; Bujacz, M. & Pec, M. (2006). Space perception by means of acoustic images: an electronic travel aid for the blind, *ACOUSTICS High Tatras 06 - 33rd International Acoustical Conference – EAA Symposium*, Štrbské Pleso, Slovakia, October 4th – 6th, 2006, 296–299

15

Virtual Moving Sound Source Localization through Headphones

Larisa Dunai, Guillermo Peris-Fajarnés,
Teresa Magal-Royo, Beatriz Defez and Victor Santiago Praderas
Universitat Politécnica de València
Spain

1. Introduction

Humans are able to detect, identify and localize the sound source around them, to roughly estimate the direction and distance of the sound source, the static or moving sounds and the presence of an obstacle or a wall [Fay and Popper, 2005]. Sound source localization and the importance of acoustical cues, has been studied during many years [Brungart et al., 1999]. Lord Rayleigh in his "duplex theory" presented the foundations of the modern research on sound localization [Stutt, 1907], introducing the basic mechanisms of localization. Blauert defined the localization as "the law or rule by which the location of an auditory event (e.g., its direction and distance) is related to a specific attribute or attributes of a sound event" [Blauert, 1997].

A great contribution on sound localization plays the acoustical cues, Interaural Time Difference ITD and Interaural Level Diference ILD, torso and pinnae (Brungart et al., 1999), [Bruce, 1959]. [Kim et al., 2001] confirm that the Head Related Transfer Functions (HRTFs) which represent the transfer characteristics of the sound source in a free field to the listener external ear [Blauert, 1997]), are crucial for sound source localization.

An important role in the human life plays the moving sound localization [Al'tman et al., 2005]. In the case of a moving source, changes in the sound properties appear due to the influence of the sound source speed or due to the speed of the used program for sound emission.

Several research have been done on static sound localization using headphones [Wenzel et al., 1993], [Blauert, 1997] but few for moving sound source localization. It is well known that on localization via headphones, the sounds are localized inside the head [Junius et al., 2007], known as "lateralization". Previous studies [Hartmann and Wittenberg, 1996] in their research on sound localization, showed that sound externalization via headphones can be achieved using individual HRTFs, which help listeners to localize the sound out in space [Kulkani et al., 1998], [Versenyi, 2007]. Great results have been achieved with the individual HRTFs, which are artificially generated and measured on a dummy head or taken from another listener. Due to those HRTFs, the convolved sounds are localized as real sounds [Kistler et al., 1996], [Wenzel, 1992].

This chapter presents several experiments on sound source localization. Two experiments are developed using monaural clicks in order to verify the influence of the Inter-click interval on sound localization accuracy.

In the first of these experiments [Dunai et al., 2009] the localization of the position of a single sound and a train of sounds was carried out for different inter-click intervals (ICIs). The initial sound was a monaural delta sound of 5ms processed by HRTFs filter. The ICIs were varying from 10ms to 100ms. The listeners were asked to inform what they listened, the number and the provenience of the listened sound and also if there was any difference between them, evaluating the perceived position of the sound ("Left", "Right" or "Centre"). It was proven that the accurateness in the response improves with the increase of the length of ICI. Moreover, the train of clicks was localized better than the single click due to the longer time to listen and perceive the sound provenience.

In the second study (Dunai et al., 2009), the real object localization based on sensory system and acoustical signals was carried out via a cognitive aid system for blind people (CASBliP). In this research, the blind users were walking along a 14m labyrinth based on four pairs of soft columns should localize the columns and avoid them. The average time of sound externalization and object detection was 3,59 min. The device showed no definitive results due to the acoustical signal speed, which required improvements.

2. Experiment

2.1 Experiment 1. A pair of sounds and a train of sounds source localization

In the Experiment 1, the localization of the static sound source was studied; the saltation perception on the inter-click presence was also analyzed. The experiment is based on monaural click presented at different inter-click intervals (ICI), from 10ms to 100ms. Two types of sounds single click and train of clicks are generated and thereafter tested at different inter-click intervals. At short inter-click intervals, the clicks were perceived as a blur of clicks having a buzzy quality. Moreover, it was proven that the accurateness in the response improves with the increase of the length of ICI.

The present results imply the usefulness of the inter-click interval in estimating the perceptual accuracy. An important benefit of this task is that this enables a careful examination of the sound source perception threshold. This allows detecting, localizing and dividing with a high accuracy the sounds in the environment.

Sound sample

Sound source positions used for stimulus presentation in this experiment were generated for a horizontal frontal plane. A sound of 5ms duration was generated with Above Audition software.

In the first case, the generated sound with duration of 5ms was used as spatial sound and in the second case; the sound was multiplied by six, becoming a train of sound with duration of 30ms.

The sound has been convolved using Head Related Transfer Functions (HRTFs). It is known that the HRTFs are very important for sound localization, because they express the sound pressure at the listener eardrum over the whole frequency range. In the present study, the HRTFs were generated at 80dB at a frequency of 44100 Hz and processed by a computer for the frontal plane, for a distance of 2 m, with azimuth of 64° (32° at the left side of the user and 32° at the right side of the user).

In the experiments the sound were presented randomly in pairs Left-Right and Right-Left, delivered using Matlab version 7.0, on an Acer laptop computer.

Test participants

Ten volunteers, 4 females and 6 males, age range 27-40 years, average 33,5 participate in this experiment. Each subject reported to have normal hearing, they did not reported any hearing deficiencies. All of them were supposed to other acoustical experiments with computer and acoustical mobility devices.

Procedure

The experiment was carried out in a single session. The session consisted of two runs, one for a single sound and one for a train of sound. Each run was based on six sounds. Fig.1 shows the schematic presentation of the sound: a) shows the monaural sound in which, the click comes from (Left) L→R (Right) and R→L, with randomly varying ICIs; b) shows the train of sound, where the presentation procedure is the same as for the single sound, the sound come from L→R and R→L, with randomly varying ICIs. Different interclick intervals (ICI), from 10 ms to 100 ms were used (10ms, 12ms, 25ms, 50ms and 100ms).
Localization test were carried out in a chamber of 4,8m x 2,5m x 12m, where external sounds were present.
Since the experiments described in this chapter were focused on examining the perception in human listeners, it was important to be able to measure spatial capabilities in an accurate and objective way. For the localization test, subject localized auditory sound presented in the headphones, telling the direction of the listened sound. In both cases the experiment begins with various exercises where the subjects are able to hear the sound and train of sound, separately, firstly the left one and afterwards the right one, continuing with the six sounds delivered by the program randomly. Afterwards the subject completed the all six sounds, the new exercises were presented of the combination "Left-Right" and "Right-Left".
For the localization tests, listeners were sitting comfortably in a chair in front of a computer. Before starting the test, the listeners received written and oral instructions and explanations of the procedure. They were asked to pay especial attention and to be concentrated on the experiment.
Before localization experiments, subjects had a training protocol to become familiar with the localization. This protocol included the speech pointing techniques, which requires that the subject verbally informs the evaluator about the perceived localization of a sound. During the experiment, since the subject had not access to the computer screen, the tendency of capturing the sound with the eyes was eliminated.
During the test, the subjects were supposed to listen through the headphones, model HD 201, twelve pairs of sounds; six pairs of single sound and six pairs of trains of sound "Left-Right" and "Right-Left" at different ICIs, from 100 ms to 10 ms in a decreasing succession.
The sounds were delivered in a random position. The sound used in the experiment was the same sound used in the testing procedure. The sound duration was brief enough, so that listener could not make head movements during the sound presentation. Between each two consecutive pair of sound, the decision time (Td) was computed; this was the time needed for evaluating the sound (see Fig. 1).
The subjects were asked what they listened, the number and the provenience of the listened sound and also if there was any difference between them. The subjects where allowed to repeat them, if necessary, after they had evaluated the perceived position for each sound, classifying them as "Left", "Right" or possible "Centre". Once the subject had selected a response, a next pair of sound was presented. Each trial lasted approximately 2 min. The average time per subject for all experiment was around 35 min.
Some distraction cues as: environmental noises, draw away seeing or hearing someone- since the subject remained with opened eyes influenced on the experimental sound source perception and results. Because of this reason, the subjects were allowed to make judgments about the source location independently.

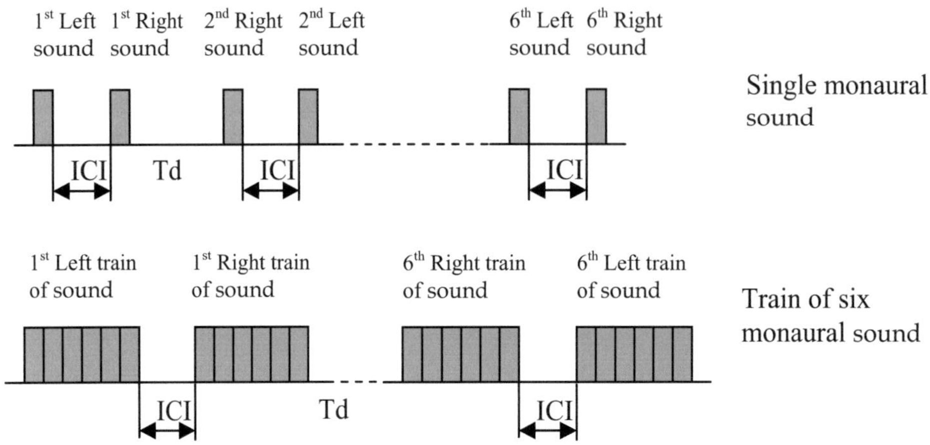

Fig. 1. Schematic presentation of the sound. In both situations the sound is of 5ms. In the first case, the sound has been listened at the different interclick intervals ICI separated by a decision time Td. In the second case, the sound has been substituted by a train of six sound.

The results were collected by the evaluator and introduced manually into a previously prepared table. After the test, localization performances were examined using the analyses described in the following section.

Results

The results from the Experiment 1 were collected for data analysis. Localization performances summary statistics for each subject are listed in Table 1. The graphical user interface was generated by Excel in linear standard model. Subject response was plotted in relation to the Inter-click Interval. The main data for all subjects is presented in Fig. 2 with an error of 5%.

The perception of the single and train of sound and the perceived position of the sound pairs "Left-Right" and "Right-Left" were analyzed. Both factors as well as the interaction with the ICIs were significant.

Fig. 2 shows that the perception of the sound source position decreases when ICIs does. For avoiding errors, the tests results were registered up to an ICI of 10ms. Because ICI was enough short, the sound were perceived as a single entity moving from one ear to another or from one ear to the centre having a buzzing quality.

In the case of the single pair of sound at ICI of 12ms, because the length of the sound and the length of the ICI were too short, the subjects could not distinguish clearly the sound corresponding to the pairs "Left-Right" and "Right-Left".

When comparing the perception of the single sound with the perception of the train of sound Fig. 2 a), a great continuity of the sound position across almost the entire range of ICIs was detected. In other words, the perception of the sound position was stronger for the train of sound. This effect may be a result of the better localization associated with the sound.

$$\sqrt{\frac{\sum(x-\overline{x})^2}{(n-1)}} \quad (1)$$

For ICIs between 25 and 10ms, the subjects perceive the "Right-Left" pair of sounds with a higher precision than that of pairs "Left-Right" for single sound and train of sound.

In other case, for ICIs of 50ms, the perception of the pair of single sound "Right-Left" is higher than the perception of the pair Left-Right. In the case of the train of sound, the perception results are equivalent for both pairs Left-Right and Right-Left.
When trying to explain the sound source perception threshold, we perceive the perception of the saltation illusion. With shorter ICIs, a blur of sound were perceived, in contrast with the individual sound at longer ICIs. As the psychologist Gestalt noted, the perceptual system scrambles for the simplest interpretation of the complex stimuli presented in the real world. Therefore, the studies were based on analyzing and proving that, grouping the sound, the sound source is better perceived and localized.
For longer ICIs, this procedure is not so important, since each sound can be identified and localized. The present results demonstrate the usefulness of the inter-click interval in estimating the perceptual accuracy. A possible benefit of this task is enabling a careful examination of the sound source perception threshold. This allows detecting, localizing and dividing with high accuracy the sounds in the environment.

Sound perception in %			Train of sound perception in %		
interclick ms	Azimuth -30°	azimuth 30°	interclick ms	Azimuth -30°	azimuth 30°
100	100%	100%	100	100%	100%
50	90%	86%	50	100%	100%
25	80%	90%	25	88%	96%
12	83%	95%	12	76%	79%
10	88%	86%	10	75%	86%
8	100%	95%	8	100%	96%
6	100%	95%	6	85%	93%
5	100%	92%	5	100%	95%
1	100%	100%	1	100%	100%

Table 1. Localization performance summary statistics for all subjects (P1-P9) in frontal field. The percentage of the perception experiment is calculated on the basis of the six delivered sounds.

2.1 Experiment 2. The influence of the inter-click interval on moving sound source localization tests

In the Experiment 2, an analysis of moving sound source localization via headphones is presented. Also, the influence of the inter-click interval on this localization is studied. The experimental sound consisted of a short delta sound of 5ms, generated for the horizontal frontal plane, for distances from 0,5m to 5m and azimuth of 32° to both left and right sides, relative to the middle line of the listener head, which were convolved with individual HRTFs. The results indicate that the best accurate localization was achieved for the ICI of 150ms. Comparing the localization accuracy in distance and azimuth, it is deduced that the best results have been achieved for azimuth. The results show that the listeners are able to extract accurately the distance and direction of the moving sound for higher inter-click intervals.

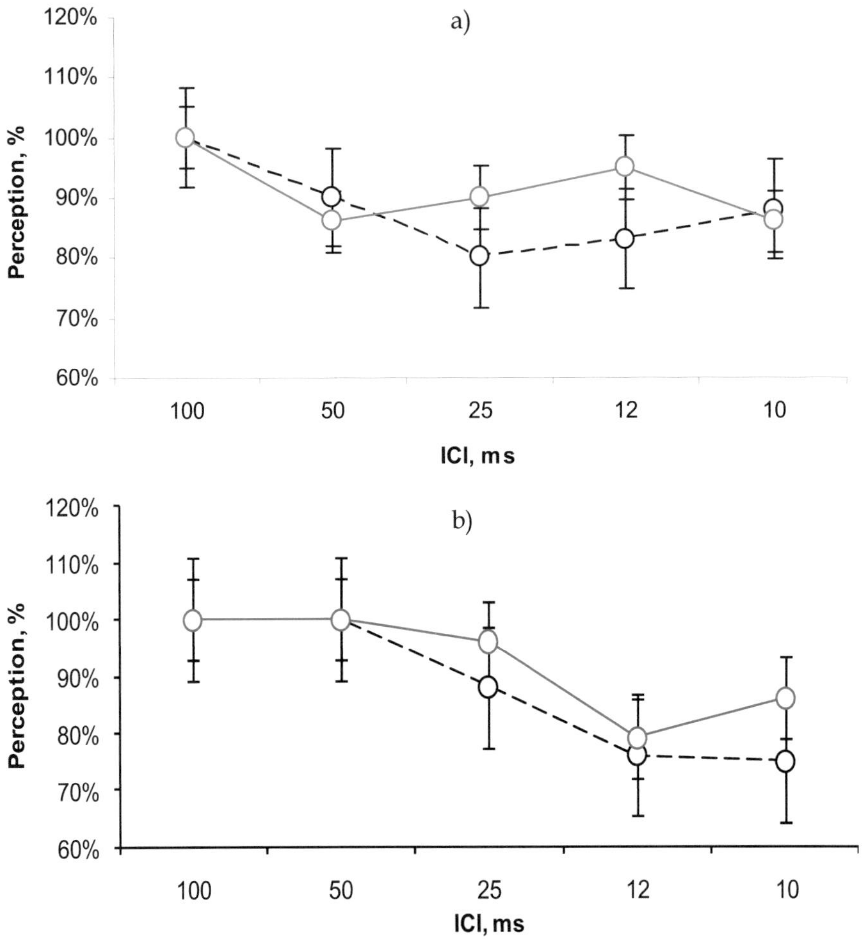

Fig. 2. Mean estimation of the click location: a) shows the sound perception at -30° (left side) and +30° (right side); b) corresponds to the train of sound perception at -30° (left side) and +30° (right side)

Subjects

Nine young subjects students with ages between 25 and 30 years and different gender, all of them had normal vision and hearing abilities, were involved in the experiments. All participants had normal distance estimation and good hearing abilities. They demonstrate a correct perception of the sounds via headphones. A number P1-P9 identified the subjects.
All subjects participated in previous auditory experiments in the laboratory. Each participant received a description of what was expected of him/her and about all procedure. All participants passed the localization training and tests described below.

Stimuli and signal processing

A delta sound (click) of 2048 samples and sampling rate of 44.100 Hz was used. To obtain the spatial sounds, the delta sound was convolved with Head-Related Transfer Function (HRTF) filter measured for each 1° in azimuth (for 32° left and 32° right side of the user) at

each 1cm in distance. The distance range for the acoustical module covers from 0,5m to 5m, an azimuth of 64°, and 64 sounding pixels per image at 2 frames per second.
Recording of Head-Relates Transfer Functions were carried out in an anechoic chamber. The HRTFs measurements system consist on a robotic and acquisition system. The robotic system consists of an automated robotic arm, which includes a loudspeaker, and a rotating chair on an anechoic chamber. A manikin was seated in the chair with a pair of miniature microphones in the ears. In order to measure the transfer function from loudspeaker-microphone as well as for headphone-microphone, the impulse response using Maximum Length Binary Sequence (MLBS) was used. The impulse response was obtained by taking the measured system output circular cross-correlation with the MLBS sequence.
Due to that the HRTF must be measured from the two ears, there is necessary to define the two inputs and output signals. Lets x_1 (n) be the digital register of the sound that must be reproduced by the speakerphone. Lets $y_1(n)$ be the final register recorded by the microphone placed in one of the acoustic channels of the manikin or man, corresponding to the response to $x_1(n)$. Similarly, let $x_2(n)$ be the sound to be reproduced through the headphone and $y_2(n)$ the answer registered by the headphone, respectively for the second ear. The location of the head in the room is assumed to be fixed and is not explicitly included in our explication.
In order to determine $x_1(n)$, it is necessary to generate a $x_2(n)$ such that the $y_2(n)$ is identical to $y_1(n)$. In that way, we achieve that an acoustic stimulus generated from the speakerphone and another generated by the headphones, produce the same results in the auditive channel of the user or manikin. Therefore we obtain the same acoustical and spatial impression.
In order to obtain these stimuli, a digital filter which transforms the $x_1(n)$ into $x_2(n)$ has been developed. In the transformed frequency domain, let be X_1 the representation of the $x_1(n)$ and Y_1 the representation of the y_2 (n).
Then Y_1, which is the registered response of the $x_1(n)$ reproduction, is:

$$Y1 = X1LFM \tag{1}$$

In (1), L represents the grouped transfer function of the speakerphone and all audio reproduction system. F represents the transfer function of the environment situated between the speakerphone and the additive channel (HRTF) and M represents the set of functions composed by the microphone and the whole audio reproduction system.
The response registered by the microphone via headphones, when the $x_2(n)$ is reproduced, can be expressed as follows:

$$Y_2 = X_2HM \tag{2}$$

where H represents the transfer function of the headphone and all reproduction system to the additive channel.
If Y_1=Y_2, isolating X_2 we obtain:

$$X_2 = \frac{X_1LF}{H} \tag{3}$$

Then, for any measurement the digital filter will be defined as follows:

$$T = \frac{LF}{H} \tag{4}$$

Therefore, it will filter the signal $x_1(n)$ and the resulting signal $x_2(n)$ will be reproduced by the headphone; then the signal registered by the microphone, which is placed in the auditive channel must be $y_1(n)$. This signal must be equal to the signal $x_1(n)$, which is reproduced by the speakerphone.
The filter described by (4) describes the speakerphone for a single spatial position for only one ear. For both ears two filters are required for the simulation of each signal source for a determined spatial position.
Assuming that we measure the Y_1 and X_1 transfer functions for different spatial positions for both ears at the same time, the Transfer Function speakerphone-microphone (G_{LM}) is defined as follows:

$$G_{LM} = \frac{Y_1}{X_1} = L \cdot F \cdot M \tag{5}$$

Having the function given by (5) simultaneously for both ears, we measure both transfer functions Y_2 and X_2, on which the transfer functions headphone-microphone G_{HM}, are defined:

$$G_{HM} = \frac{Y_2}{X_2} = H \cdot M \tag{6}$$

The necessary filters for the sound simulation are obtained from the function speakerphone-microphone G_{LM} for each ear, as the reverse of the function headphone-microphone G_{HM} of the same ear (see (4)). So, for both ears:

$$T = \frac{G_{LM}}{G_{HM}} = \frac{L \cdot F \cdot M}{H \cdot M} = \frac{L \cdot F}{H} \tag{7}$$

For both transfer function speakerphone-microphone G_{LM} and headphone-microphone G_{HM}, the measurement technique of the impulse response Maximum Length Binary Responses MLBS was applied with later crossed correlation between the system answer and input of the MLBS.
The impulse response of the system can be obtained through circular crossed correlation between input MLBS of the system and the output answer. This is, if we apply to the system an MLBS, which will called s(n), and measure the output the signal y(n) during the time which MLBS lasts, the impulse response h(n) will be defined as follows:

$$h(n) = \Omega_{sy}(n) = s(n)\Phi y(n) = \frac{1}{L+1}\sum_{k=0}^{L-1} s(k) \cdot y(n+k) \tag{8}$$

where Φ represents the circular or periodic crossed correlation operation, corrupted by the aliasing time, and not a pure impulse response.
In the event that the sequence is enough long, then the resultant aliasing can be rejected. Due to that, the direct implementation of (8) for long sound sequences require high computational time, the equivalent between the correlation and periodic crossed correlation has been used. The obtained information was passed into the frequency domain, where the convolution operation is translated into a vector multiplication.
After this, the results were passed into the frequency domain, where the convolution operation is translated into a vector multiplication.

$$a(n)\Phi b(n) = \frac{1}{L+1} a(-n) * b(n) \tag{9}$$

where the inversion of the first sequence is circular, similar to the convolution. Nevertheless, the computational time results to be enough high, due to that the used Fast Fourier Transform (FFT) have a length of 2^k-1. In order to obtain an increasing performance in time processing the FFT length has to be $(2^k-1)^2$.

Finally, using the Fast Hadamard Transform (FHF), it was possible to reduce the computational time between the two magnitudes. The h(n) is then calculated as follows:

$$h(n) = \frac{1}{(L+1)s[0]} P_2 \left\langle S_2 \left\{ H_{L+1} \left[S_1 \left(P_1 y(n) \right) \right] \right\} \right\rangle \tag{10}$$

In this case to the system has been applied a MLBS s(n) with a length L, after what the result y(n) was registered. The matrix P is the permutation matrix, the matrix S is matrix of rescaling, the H_{L+1} is the matrix Hadamard of degree L+1. After the HRTFs were measured, with the equipment shown in figure 3.13, it was verified if the HRTFs are realistic and externalized. For this purpose, an off-line localization procedure was carried out.

The output signals (the HRTF) are sampled at 22050Hz and a length of 46ms (8192 bit).

The HRTFs were measured for the horizontal frontal plane at the ear level from 0,5 to 5m in distance and in azimuth between 32° left and 32° right with respect to the centre of the listener head (measurements at every 1°). Fig. 4 shows the graphical representation of the sound reproduction.

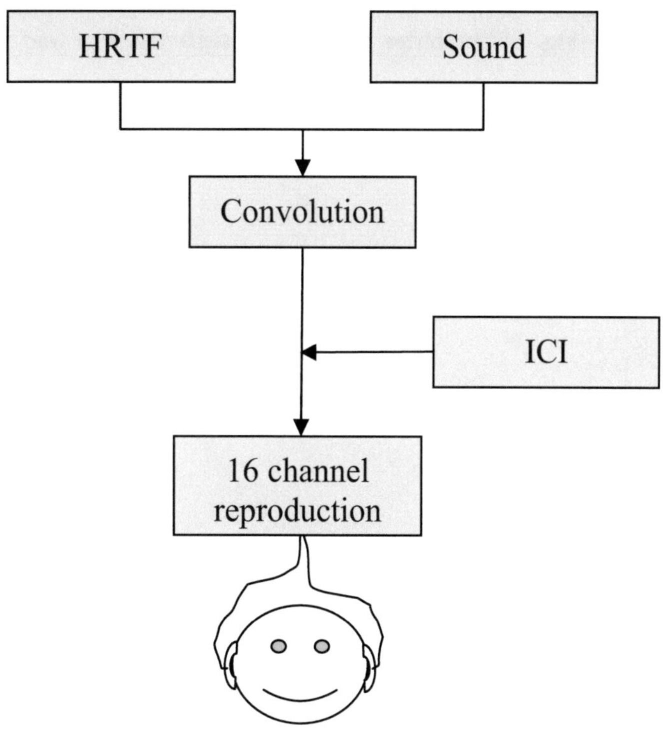

Fig. 4. Method for sound processing and reproduction

Equipment

A Huron system with 80 analogue outputs, eight analogue inputs and eight DSPs 56002, and a computer for off-line sound processing was used for the sound generation and processing. SENNHEISER headphones models HD 201 were used to deliver the acoustical information. MATLAB 7.0 was used as experimental software. The resultant graphical sound trajectory for each experiment was displayed on a separate window and saved for off-line processing. All experiments run on ACER Aspire 5610 computer.

Procedure

The goal of the experiments is to analyze the localization of a moving sound source via headphones and to see how the inter-click interval (ICI) influences the sound localization quality. The comparison between the localization performances enables to evaluate the importance of the inter-click interval parameter for its use in sound localization and acoustical navigation systems.

The movement of the sound source was achieved by switching the convolved sound for a frontal plane at the eyes level at increasing distances from 0,5 to 5m (1 cm increase) and for azimuth between 32° right and 32° left (1° increase) with respect the middle of the head. The sounds were delivered for five inter-click intervals [200ms, 150ms, 100ms, 75ms and 50ms]. Fig. 5 shows one of the trajectories the sound was running. Four different trajectories were created. The delivered trajectory was selected randomly by the computer when the experiment starts.

Before starting the experiment, the training exercises were carried out; the objective and the procedure of the experiment were explained to each individual participant. One sound was delivered for all five ICIs, where the participants were able to see graphically the listened sound trajectory (See Fig. 5). In order to proceed with the test and experiment, the

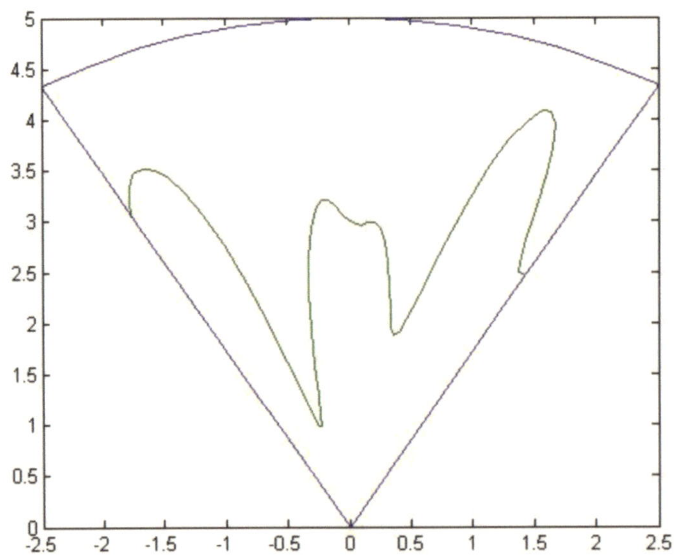

Fig. 5. Sound trajectory example, direction from left to right. The x axis represents the azimuth where the 0 is the centre of the head, which is 0°. The -2.5 is the -32° at left side of the head and 2.5 respectively is 32° at the right side of the head. The y axis represents the distance from 0 to 5m

participants were asked to seat comfortably in the chair in front of a computer. After reading and testing the training exercises, the participants were supposed to carry out the experiment.
A sound at a specific ICI was delivered by the computer via headphones. During the experiment, the participants were free to move. Nevertheless, they were required to move the less possible and to be concentrated on the sound, in order to create a plane of the sound route in the imagination. The test was performed both with open eyes and with closed eyes depending on the participant wishes. In the case of the closed eyes, there was a limitation of effects of the visual inputs. Due to this, the participant achieved a better interpretation of the trajectory image.
The participants were asked to carefully hear the sound and draw the listened trajectory in a paper. They were allowed to repeat the sound if it was necessary. All the participants asked to repeat the sound at least three times. Each participant was supposed to have five trials, one for each ICI. Only one sound trajectory was used per participant for all five ICIs. For all participants, the experiment started with the ICI of 200ms, decreasing it progressively up to 50ms.
After the experiment the participants commented the perceived sound trajectory and they compared the listened sound for each ICI.

Results

The moving sound source localization is an important factor for the navigation task improvement. The main variables analyzed in this paper were the moving sound source localization and the inter-click interval ICI [200, 150, 100, 75, and 50ms]. The study analyzes the interaction between these variables in measurements of distance and azimuth.
Generally, no significant differences on the results were registered between participants. However, great difference was found in the sound localization between higher and lowers inter-click intervals.
The maximum displacement in distance is 1,26m for an ICI of 50ms and the minimum displacement was 0,42m for an ICI of 150ms, the maximum displacement in azimuth was 11,4^{o} for an ICI of 50ms and the minimum 0,71^{o} for an ICI of 150ms.
Average results of sound localization in azimuth and distance as a function of the inter-click interval are shown in Fig. 6. Best results have been achieved for greater ICIs, due to the time needed by the brain to perceive and process the received information. Because the time between two sounds is higher, the sound is perceived as jumping from one position to another from left to right in equal steps. For the ICI of 200ms, the sound was not perceived as a moving sound, but rather as a jumping sound from location to location. However, for the ICIs lower than 100ms the sound was perceived as a moving sound from the left to right, but there was enough difference between the original sound trajectory and the perceived one. The participants had great difficulties to perceive the exact distance and azimuth, because the sound was delivered too fast. Moreover, when the sound trajectory had multiple turning points on a small portion of the space, the participants perceived this portion as one turn-return way. Fig. 7 represents a specific case, corresponding to one of the participants; it shows the moving sound localization at four ICIs. The red colour represents the listened sound trajectory drawn by the participant. The grey colour represents the real sound trajectory drawn by the computer. The x axis represents the azimuth where the 0 value is the centre of the head, the negative values are the values at the left side of the head, whereas the values at the right side of 0 represent the azimuth values at the right side of the human head. The -2.5 represents the 32° at left side of the head and 2.5 the 32° at the right side of the head. The y axis represents the distance from 0 to 5m.

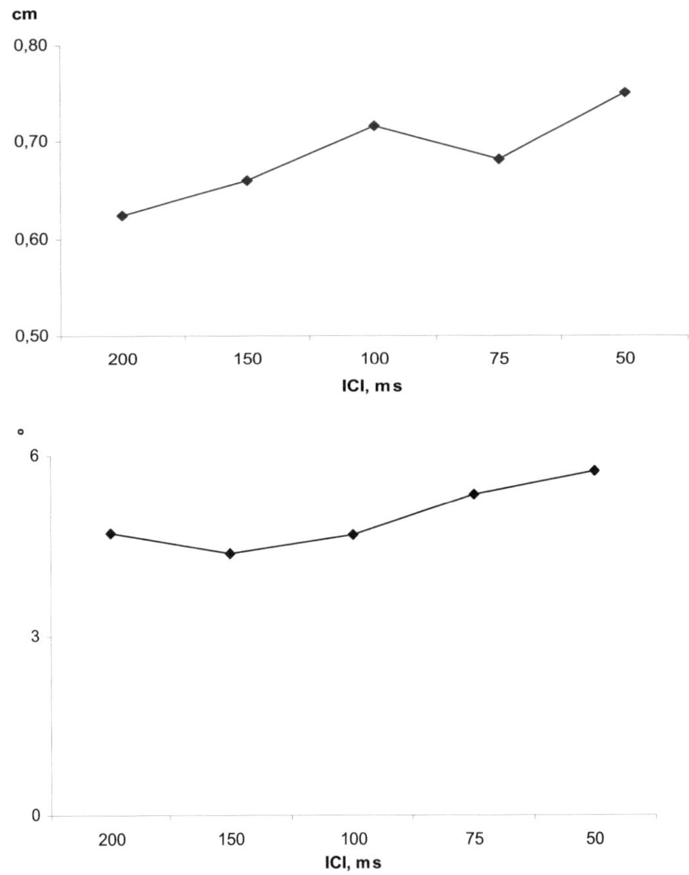

Fig. 6. Average displacements in azimuth and distance for all participants

In some cases, the participants perceived the sound trajectory as an approximate straight line when the inter-click interval was 50ms. Even repeating several times the experiment, the participants were confused regarding the localization of the moving sound. They commented "the sound moves too fast and I feel that it is running from left to right in a straight line". Despite listeners were not able to localize the moving sound source at lower inter-click intervals so well as they were able to localize the moving sound for greater inter-click intervals, they were able to judge about the sound position in azimuth and distance.

Various factors as drawing abilities (how the participants can accurately draw), sound interpretation (how the participants can interpret the heard sounds, by colours, by image etc.), the used hearing methods (with closed or opened eyes), the external noises, etc., influenced the experiment results. Despite all participants were informed about the use of one sound per participant for all ICIs, they draw the trajectories at different distances. This error appears because of the participant drawing ability; it is not so easy to interpret graphically what is listened or the image the brain creates if there is not practice on that. For some of participants, great concentration and relaxation was required, to be able to correctly perceive the sounds.

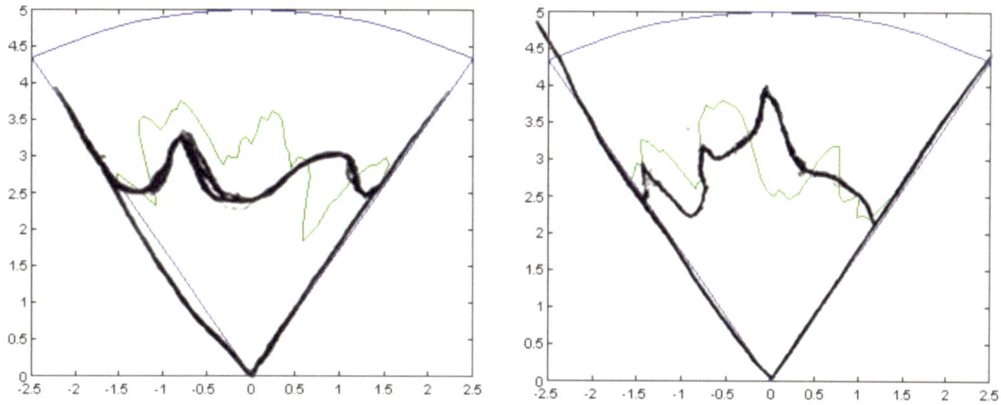

Fig. 7. Sound trajectory for one participant for the ICIs of 50ms and 100ms. The black colour represents the heard sound trajectory drawn by the participant; the green colour represents the real sound trajectory drawn by the computer. The x axes represent the azimuth, in which the 0 value is the centre of the head, the negative value are the values at the left side of the head and the values at the right side of 0 represent the azimuth values at the right side of the human head. -2.5 represents the 32° at left side of the head and 2.5 respectively the 32° at the right side of the head. The y axis represents the distance from 0 to 5m.

Multiple observations on training sound trajectory were given to participants about how to perceive the sound and to be confident of their answer. Two participants were excluded from the main analysis due to the difficulties in localizing the sound. The participants experienced the moving sound localization as a straight line for all inter-click intervals.

3. Conclusion

In the present chapter two sets of experiments are described according to the examined spatial performance involving simple broad-band stimuli. Both experiments measured how well single and train of static and moving sounds are localized in laboratory conditions. These experiments demonstrated that sound source is essential for accurate three-dimensional localization. The approach was to present sounds overlapped in time in order to observe the performance in localization, in order to see how time delay between two sounds (ICI inter-click interval) influences on sound source localization. From the first experiment it was found that better localization performance was achieved for trains of sounds at an ICI of 100ms. If analyzing the localization results at the left and right side of the human head, it must mention that improved results were obtained at the left side for the single click and at the right side for the train of clicks. At short inter-click intervals, the train of clicks was perceived as a blur of clicks. At short inter-click intervals the single clicks was perceived as one click, there were not perceived the difference between the first click and the second one. In this case only the first click was perceived, the second click was perceived as a week eco. Moreover, the sound perception threshold was studied. In the second study the localization of a moving sound source both in distance and azimuth was analyzed. The results demonstrate that the best results were achieved for an inter-click interval ICI of 150ms. When comparing the localization accuracy in distance and azimuth, better results were obtained in azimuth. The maximum error in azimuth is of 11,4° at the ICI of 50ms. The disadvantages of the results at short ICI's are due to that the total time of the sound run is very short, that prevent the user to perceive all the sound coordinates. Regarding the large ICI's, the saltation from one click to another don not allows the user to make the connection between the two clicks. From this motive the user perceive the sounds as diffuse. Spatial cues such as interaural time difference ITD and interaural level difference ILD play an important role in spatial localization due to their attribution on the azimuthal sound localization. They arise due to the separation of the two ears, and provide information about the lateral position of the sound.

4. References

Al`tman Ya.A.; Gurfinkel V.S.; Varyagina O.V.; Levik Yu.S. (2005). The effect of moving sound images on postural responses and the head rotation illusion in humans, Neuroscience and Behavioral Physiology, 35 (1), 103-106

Blauert J. (1997). Spatial Hearing: The Psychophysics of Human Sound Localization, revised edn, The MI Press, Cambridge, MA, USA,

Bruce H., Hirsh D. and I.J., (1959). Auditory Localization of Clicks J. Acoust. Soc.Am., 31(4), 486-492

Brungart D.S Nathaniel I., W. R. Raibiowitz. (1999). Auditory localization of nearby sources II. Localization of a broadband source, J. Acoust. Soc.Am. 106 (4), 1956-1968

[1]Brungart D.S., Rabinowitz W. M. (1999). Auditory localization of nearby sources. Head-related transfer functions, J. Acoust. Soc.Am. 106(3), 1465-1479

Dunai L., Peris F G.., Defez B. G., Ortigosa A.N., Brusola S F. (2009). Perception of the sound source position, Applied Physics Journal, (3), 448-451

[1]Dunai L., Peris F G.., Defez B. G., Ortigosa A.N., (2009). Acoustical Navigation Sysyem for Visual Impaired People, LivingAll European Conference

Dunai L., Peris Fajarnes G., Defez Garcia B., Santiago Praderas V., Dunai I., (2010), The influence of the Inter-Click Interval on moving sound source localization for navigation systems, Applied Physics Journal, (3), 370-375

Hartmann W.M., Wittenberg A., (1996). On the externalization of sound images, J. Acoust. Soc.Am. 99 (6): 3678-3688

Junius D., Riedel H., Kollmeier B., (2007). The influence of externalization and spatial cues on the generation of auditory brainstem responses and middle latency responses, Hearing Research 225, 91-104

Kim H.Y., Suzuki Y, Sh. Takane, Sone T. (2001). Control of auditory distance based on the auditory parallax model, Applied Acoustics 62, 245-270

Kistler D.J., Wightman F.L., (1996). A model of head-related transfer functions based on principal components analysis and minimum-phase reconstruction a, b), J. Acoust. Soc.Am. 91 (3), 1637-1647

Kulkani A., Colburn S.H., (1998). Role of spectral detail in sound-source localization, Nature, 396, 747-749

Strutt J.W. (1907). On our perception of sound direction, Philos. Mag.; Vol 13, 214.232

Versenyi G. (2007). Localization in a Head-Related Transfer Function-based virtual audio sysnthesis using additional high-pass and low-pass filtering of sound sources, Acoust. Science & Technology, 28 (4), 244-250

Wenzel E., Arruda M., Kistler D., Foster S. (1993). Localization using non-individualized head-related transfer functions, J. Acoust. Soc.Am. 94, 111-123

Wenzel E.M., (1992). Localization in virtual acoustic display, Presence Telop Virt. Environ. 1, 80-107

16

Unilateral Versus Bilateral Hearing Aid Fittings

Monique Boymans and Wouter A. Dreschler
Academic Medical Centre, Department of Clinical and Experimental Audiology, Amsterdam The Netherlands

1. Introduction

This study is designed to assess the added value of fitting a second hearing aid: to evaluate the current fitting practices, to assess the effect on spatial hearing, to evaluate this objectively, and to predict a positive effect from diagnostic tests.
The reasons and/or criteria for fitting one or two hearing aids are not always obvious. Many considerations, as localization, seem to play a role both for the hearing-impaired person and for the audiologist. A large asymmetry in hearing loss can be a contra indication for a bilateral fitting, but it is not clear to which limits. The key question in section two is: What are current fitting practices in a large (multi-centre) clinical population and which are the audiometric characteristics of subjects fitted with one or two hearing aids? Section three describes some recent findings in the literature. Section four describes the effects on spatial hearing that can be assessed in the individual patient. Section five addresses the issue whether a successful bilateral fitting can be predicted from apriori tests.

2. What are the current fitting practices

In order to find current fitting practices a large retrospective study (Boymans et al.2006, 2009) was conducted. In this study case history data, audiometric, and rehabilitation data, and subjective fitting results were evaluated in a population of 1000 subjects using modern hearing aids, included from eight Audiological Centers in the Netherlands. All centers are members of the foundation PACT, the Platform for Audiological and Clinical Testing and they are representative for Audiological Centers in the Netherlands. PACT was established as a platform for independent clinical research related to the use of hearing aids. Each center selected 125 consecutively hearing aid fittings and analyzed the clinical files of these subjects.
An extensive questionnaire on long-term outcome measures was conducted. This questionnaire is called the AVETA (the Amsterdam questionnaire for unilateral and bilateral hearing aid fittings). 505 questionnaires were returned from 1000 files/subjects described above after at least two years of hearing aid use. The questionnaire consisted of different components. Besides some general questions parts of existing questionnaires were included like the Hearing Handicap and Disability Inventory (HHDI, van den Brink, 1995), the Amsterdam Inventory of Auditory Disability and Handicap (AIADH, Kramer et al., 1995),

Abreviated Profile of Hearing Aid Benefit (APHAB, Cox et al., 1995), and the International Outcome Inventory for Hearing Aids (IOI-HA, Cox et al., 2000). In addition we asked about the reasons why the patients used one or two hearing aids. The AIADH and APHAB questions were asked for the situation without a hearing aid, with one hearing aid, and with two hearing aids (if applicable). On the basis of 28 questions, 7 categories were composed in which auditory functioning was measured in the different situations: detection of sounds, discrimination or recognition of sounds, speech intelligibility in quiet, speech intelligibility in noise, speech intelligibility in reverberation, directional hearing or localization, and comfort of loud sounds. For each patient and each category the mean scores were calculated only when more than 50% of the questions were available. The total auditory function is the average result of all categories.

The subjective results of the populations with unilateral and bilateral hearing aids were compared with the case-history and audiometric data from the clinical files.

2.1 Percentage bilateral

In our sample of 1000 subjects, 587 Subjects were fitted with two hearing aids (bilaterally). 413 Subjects were fitted with one hearing aid, but in 7 of these subjects a CROS or biCROS fitting was applied. The latter fittings were regarded as unilateral fittings, because the sound presentation was to one ear only (in all of these subjects the hearing loss at the better ear was worse than 30 dB (HL).

2.2 Effects of age

Age appeared not a factor of importance with respect to the distribution of bilateral and unilateral fittings: about 60% of every age decade was fitted bilaterally.

2.3 Effects of hearing loss

Figure 1 shows the absolute numbers of unilateral and bilateral fittings as a function of the average hearing loss at the better ear. For small hearing losses relatively more unilateral fittings than bilateral fittings were found. For larger hearing losses more bilateral fittings were found, ranging from 40% to 69%.

There is a trend that patients with small hearing losses have a preference for unilateral fittings at the poorer ear, while patients with larger hearing losses have a preference for unilateral fittings at the better ear.

2.4 Effects of asymmetry

Figure 2 represents the absolute difference between both ears for the groups with unilateral and bilateral fittings. Most bilaterally fitted patients had a rather symmetric hearing loss (92% had inter aural differences up to 20 dB) but bilateral fittings were also found for asymmetrical losses with interaural differences up to 30-40 dB. The average asymmetry between both ears for unilateral fittings was 22.2 dB (±23.0) and for the bilateral fittings 8.0 dB (±8.7).

In the unilateral fitted group 44 % of the hearing losses was symmetrical (± 10dB), and in 65% of the remaining cases the hearing aid was fitted to the better ear.

There was a trend that a large asymmetry in pure-tone audiogram went along with a large asymmetry in maximum speech discrimination. But there was also a lot of scatter. Sometimes a small asymmetry in pure-tone audiogram went along with a large asymmetry

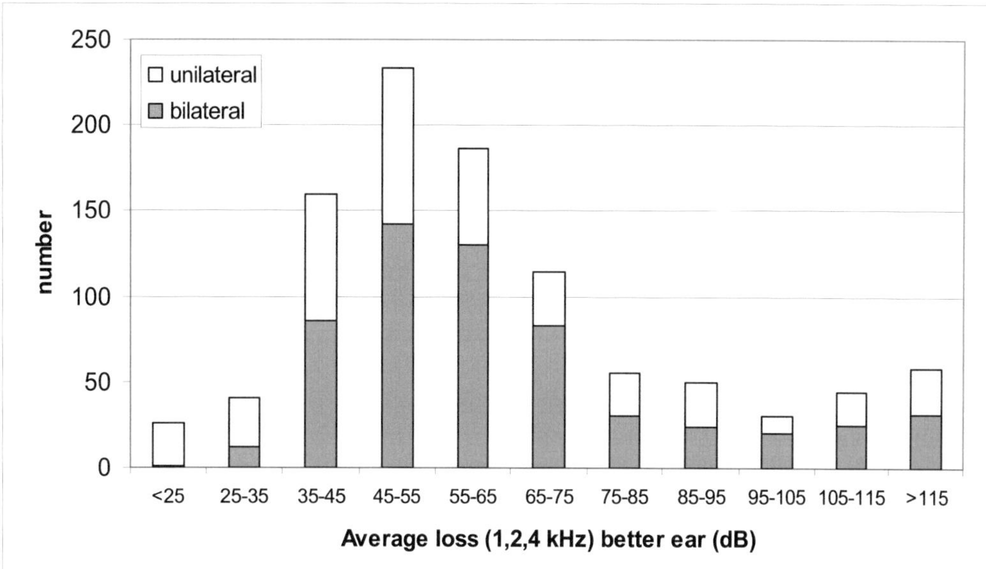

Fig. 1. Cumulative histogram for the numbers of unilateral and bilateral fittings for the total group for different hearing losses at the better ear (average 1,2,4 kHz).

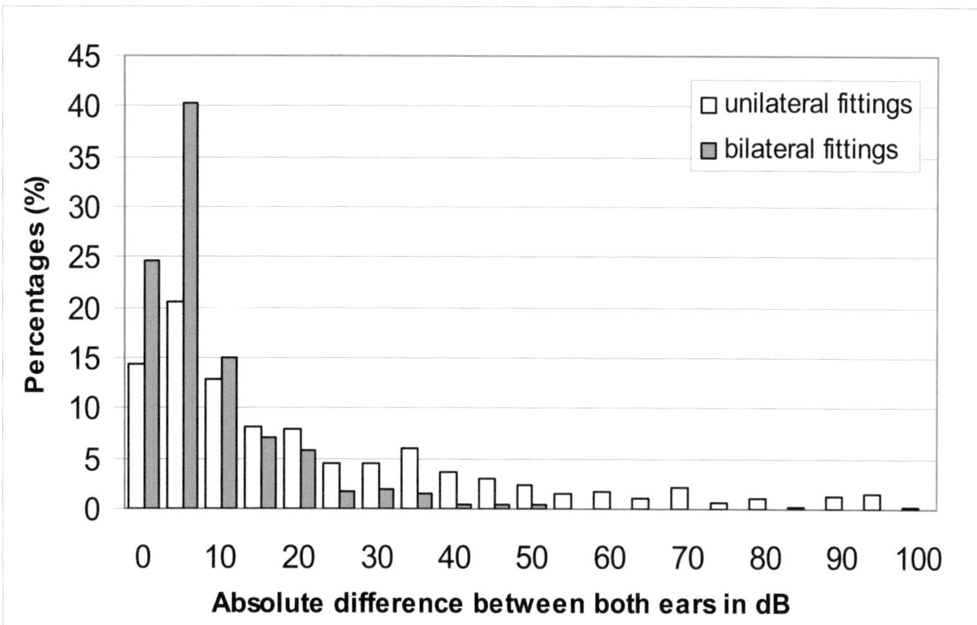

Fig. 2. The absolute difference between the PTA's (1,2,4 kHz) of both ears for the groups with unilateral and bilateral fittings.

in speech discrimination and vice versa. The trend that better-ear fittings were found for larger asymmetries, was predominantly dependent on the asymmetry of the speech discrimination.

In subgroups matched on gender, age, degree of hearing loss and audiometric asymmetry, no significant differences between unilaterally fitted and bilaterally fitted participants were found for hearing aid use, residual handicap, and satisfaction. However significant better results were found for the bilaterally fitted group than for the unilaterally fitted group for localization but also for detection and reverberation when the situation as fitted (bilateral or unilateral) was compared to the unaided situation.

Significantly higher scores for localization and speech in noise were found for the group with "high-end" hearing aids and they showed less residual handicap than the group with more basic hearing aids. However this does not influence the long-term effects of bilateral benefits.

The analysis of the relation between objective parameters from audiometric and case history data and the subjective outcome measures of different subgroups showed that the candidacy for bilateral fittings could not be predicted from age, maximum speech intelligibility, employment, exposure to background noise, and social activities.

2.5 Reasons for choosing bilateral

Part of the questionnaires was devoted to reasons why the patient himself/herself chose for one or two hearing aids. This was partly an open question.

In the group of 210 unilaterally fitted patients 410 times a reason was mentioned to choose for a unilateral fitting. The choice of one hearing aid was frequently based on the residual capacity of the other ear that was still relatively good (70x) or just worse (73x). Also using the telephone with the other ear could be a reason to choose for one hearing aid (43x), or problems with the own voice when fitted bilaterally (39x).

In the group of 295 bilaterally fitted patients 690 times a reason was mentioned to choose for a bilateral fitting. Obviously, the quality of sound was mentioned as the most important reason (150x). Other reasons like better localization, the balance between ears, and listening to both sides occurred in about the same numbers (90x-110x). In only one case it was mentioned that two hearing aids are chosen to stop further deprivation.

3. Some issues on bilateral fitting in literature

3.1 Deprivation

Deprivation effect was frequently described in the literature. When the hearing organ is stimulated insufficiently, speech discrimination ability can deteriorate gradually. Hearing impaired subjects fitted unilaterally and who have bilateral hearing losses may develop a deprivation effect in the unaided ear.

Gelfand et al. (1987) described long-term effects of unilateral, bilateral or no amplification in subjects with bilateral sensorineural hearing losses. They compared audiometric thresholds and speech scores for phonetically balanced (PB) words with results obtained 4-17 years later. Speech recognition scores were not significantly different in both ears for the bilaterally fitted subjects and for the subjects not wearing hearing aids. However, in adults with a unilateral hearing aid fitting, speech recognition performance for the unaided ear was decreased significantly. This might be attributed to the deprivation effect. Silman et al. (1984) also used the deprivation effect as starting point for their research. They investigated

whether deprivation occurs and if it can be found after a long-term follow-up. 44 Adults with bilateral sensorineural hearing losses were fitted unilaterally with hearing aids and 23 with bilateral aids. For all of these subjects data about auditory functioning were obtained prior to the hearing aid evaluation, at the time of the hearing aid evaluation, and 4-5 years after the evaluation. The most important result is that there were significant differences between initial and follow-up speech-recognition scores only for the unaided ears of the unilaterally fitted group. The authors indicate that this is an auditory deprivation effect that was not found in the bilaterally fitted group. Age and hearing sensitivity factors were partial led out. So, these factors could not have influenced the conclusions. A third study is the work of Silman et al. (1993), who investigated both auditory deprivation and acclimatisation. To investigate both aspects, 19 adult subjects were fitted unilaterally, 28 bilaterally and there were 19 matched control subjects. All of them had a bilaterally symmetrical sensorineural hearing impairment. Their speech recognition ability was tested by three different tests (W-22 CID, nonsense syllable test (NST), speech-reception-in-noise (SRT)). They were initially tested six to twelve weeks following the hearing aid fitting. After one year, the follow-up test was performed. The results of the latter test showed a slight improvement in speech perception in the aided ear, in comparison with the initial test, and a larger decrement in the unaided ear. This was visible in the W-22 test as well as in the NST test. The improvements in the aided ear can be regarded as acclimatisation to amplification at the aided ear; the decrements can be ascribed to auditory deprivation in the unaided ear. The difference in magnitude suggests that more time is needed for a significant acclimatisation effect in the aided ears of both the unilaterally and bilaterally aided groups than for an auditory deprivation effect in the unaided ears of the unilaterally aided group.
The occurrence of deprivation is a reason to choose for two hearing aids. Hurley (1999) found that word recognition scores deteriorated in the unaided ear after 5 years of hearing aid use for 25% of the unilaterally fitted subjects. Although there can be some recovery from deprivation, there are also cases known where the auditory deprivation effect is not reversible (Gelfand, 1995). In contrast to other investigators, Jauhiainen (2001) found no indications for the onset of auditory deprivation in unaided ears.

3.2 Horizontal localization

Improved localization is an advantage often mentioned in literature. It means that subjects with two hearing aids are better capable of determining from what direction a sound arrives. Punch et al. (1991) presented objective data of this advantage. Although their research is focused on bilateral fitting strategies, they found that localization with bilateral hearing aids was significantly superior to localization with unilateral hearing aids. Besides this objective advantage, Stephens et al. (1991) found that an improvement of localization was one of the reasons for people to choose for two hearing aids. Dreschler and Boymans (1994) tested localization ability with one and two hearing aids in the same subjects. They found that the localization ability was significantly better with two aids than with one. The results of Byrne et al. (1992) showed that the bilateral advantage was also applicable for subjects with moderate to severe hearing losses. In the experiments of Köbler et al. (2002) the subjects had to repeat sentences and indicate the side where the sentence came from. The results for localization were almost the same for the condition without hearing aids and with two hearing aids. A worse result was found for the condition with only one hearing aid.

In contrast with other studies, Vaughum-Jones et al. (1993) found that the localization ability with two hearing aids was worse than with one hearing aid in some subjects. Their conclusion was that subjects initially should be aided unilaterally and, if necessary, two aids could be considered. Nabelek et al. (1980) investigated the effect of asymmetry in sound pressure levels produced by signals coming from two loudspeakers. By changing the sound pressure level (when the sound level at one side was increased by a certain amount of dB's (ΔL), the sound level at the other side was decreased by the same amount) the position of the sound image in a lateralization experiment varied. In normal-hearing subjects, for sound imagines on the midline, ΔL was zero. In unfitted hearing-impaired subjects with bilateral hearing losses, ΔL was within the normal range. However, in aided balanced (equal gains) and/or unbalanced conditions (10 dB disparity in gains) ΔL for midline images was outside the normal range for some bilaterally fitted subjects. Based on these results, the authors concluded that bilateral hearing aids could give a bias in the symmetry of the presentation levels between both ears.

3.3 Speech perception

Speech intelligibility is one of the most important aspects for the hearing-impaired (if not the most important). Most studies concentrate on the speech perception in noise and in reverberation, because these are the most critical listening situations. The fitting of bilateral hearing aids introduces two sources of improvement: the binaural squelch effect and the removal of head-shadow effect. The squelch effect is the true binaural component and can be described as the difference (in dB) in the critical signal-to-noise ratio (S/N ratio) between monaural and binaural listening. However, the benefits of bilateral fittings for speech intelligibility appear to be related primarily to the compensation of head-shadow effect. When listening with two hearing aids, the difference (in dB) of the critical S/N ratio between near-ear and far-ear listening is about 6-7 dB smaller than for listening with one aid (Markides, 1982). Köbler et al. (2002) used a fixed S/N ratio of + 4dB, and they found a statistically significant advantage of 5% in speech intelligibility when the subjects were fitted bilaterally.

Festen and Plomp (1986) investigated the speech-reception threshold (SRT) in noise with one and with two hearing aids in a group of 24 hearing-aid users. All subjects had a nearly symmetrical hearing loss. The critical S/N ratio measured (the S/N ratio at 50 % speech perception) proved to be hardly better with two hearing aids than with one hearing aid for subjects with moderate hearing losses when speech and noise came from the frontal direction. However, a significant benefit for bilaterally fitted hearing aids was present in subjects with a pure tone average $PTA_{(.5,1,2\ kHz)}$ larger than 60 dB, and when the speech and noise sources were spatially separated. Day et al. (1988) also concluded that subjects with severe hearing losses experience more benefit from two hearing aids than from one. They used a free field audiovisual sentence-in-noise test (FASIN) in a reflection-free room.

Bronkhorst and Plomp (1989) showed that the binaural advantage due to head shadow effect decreased when the hearing loss at high frequencies was more severe. So, the binaural advantage depends on the audiometric configuration of the hearing loss. Also, Bronkhorst and Plomp (1990) found that the binaural advantage due to a spatial separation of speech and noise was smaller for small hearing losses than for large hearing losses. In contrast to this study, Moore et al. (1992) showed a binaural advantage for almost all hearing losses when speech and noise were separated. However, in Moore's test design one ear was blocked for the unilateral situation. This suggests that contribution of the unaided ear was mainly responsible for the fact that the benefit from bilateral fitting depends on the degree of hearing loss.

Hawkins and Yacullo (1984) determined the S/N ratio necessary for a constant performance level of word recognition for normal hearing and for hearing-impaired listeners with bilaterally symmetrical mild-to-moderate sloping sensorineural hearing losses. The showed a bilateral advantage (2-3 dB) and this appeared to be independent of microphone type and reverberation time. In addition, there was a directionality advantage for the conditions with directional microphones compared to the same conditions with omni-directional microphones. These two advantages appeared to be additive (at least at the two shorter reverberation times) because no interaction between the two was found. The results indicated that the optimum performance in noise was achieved when hearing-impaired subjects weared bilateral hearing aids with directional microphones in rooms with short reverberation times.

Nabelek et al. (1981) measured the effects of unilateral and bilateral fittings for 15 subjects with bilateral sensorineural hearing losses in noise and in reverberation. Word recognition scores were significantly higher in bilateral listening modes. The advantage of bilateral listening did not depend strongly on reverberation time or the use of hearing aids. The scores improved by 7 % for a reverberation time of 0.1 s and 3.4 % for a reverberation time of 0.5 s.

Leeuw and Dreschler (1991) found better critical S/N ratios for speech intelligibility in noise (SRT-test) tested by normal-hearing listeners using two BTE hearing aids compared to one BTE hearing aid (mean difference 2.5 dB). This implied a significant advantage of bilateral over unilateral amplification, which proved to be dependent on the type of microphone (omni-directional or directional) and the azimuth of the noise source, except for 0^0. Contrary to the results of Hawkins and Yacullo (1984), the bilateral advantage in speech intelligibility was highest with directional microphones.

Dreschler and Boymans (1994) measured SRTs in noise with a spatial separation between speech and noise in 12 hearing-impaired subjects. The results showed better SRT's for the subjects using bilateral hearing aids. Bilaterally fitted subjects made better use of the spatial separation between speech and noise sources, resulting in 5dB better SRT thresholds. In addition, they applied a dichotic discrimination task, where 3-syllable words and 4-syllable numbers were presented simultaneously from $+45^0$ and -45^0 azimuths. Results only showed a clear bilateral improvement in speech discrimination for the speech material that was presented from the (unilaterally) unaided side. For words and for numbers, this effect was statistically significant.

Not all studies support the findings of improved speech intelligibility. Allen et al. (2000) found a significant evidence of binaural interference for 2 out of 48 elderly subjects ($p<0.05$). Although the small number could easily be explained by normal variability in differences between speech scores, this finding may indicate that for some individuals speech intelligibility scores with two ears could be poorer than with the better ear alone. Bodden (1997) argued that the binaural function of the ears should be restored by hearing aids. When hearing loss deteriorates the binaural function, signal processing should be used as compensation.

Markides (1982) found a difference of 2-3 dB as the bilateral advantage of two hearing aids. His experiments confirm that the effect of the head-shadow compensation are more important than the effect of binaural squelch.

4. What are the effects on spatial hearing objectively and subjectively?

In the literature many advantages of bilateral hearing aid fittings, relative to unilateral fittings, were shown, but it is difficult to obtain hard evidence about the benefits because of methodological limitations. For the correct interpretation one has to keep in mind that blinding was not possible, and that the selection of subjects in these studies partly determined the findings.

In the retrospective study described in section 2 no clear information could be found about the objective outcome measures. Therefore a prospective study was conducted. In that study (Boymans et al 2006, 2008) the same Audiological centers participated. 214 Subjects who were willing to start a trial period with bilateral hearing aids were included, 113 men and 101 women with an average age of 66 years (range: 18-88). For 133 subjects the fitting concerned a first fitting (62%). Most hearing losses were sensorineural hearing losses (79%). The average hearing loss (500 - 4000 Hz) was 47 dB for the right ears as well as for the left ears. After the trial period 200 subjects opted for a bilateral fitting (93%) and 14 subjects (7%) for a unilateral fitting. The small unilateral group was not distinguishable from the bilateral group on base of the asymmetry between both ears.

After the trial period that was long enough to decide between unilateral or bilateral fitting, (trial durations vary individually from 4 weeks to several months), the evaluation tests were conducted in order to evaluate the benefits of a second hearing aid, objectively.

To measure the effect of a second hearing aid a localization test was used as well as a Speech Reception Threshold Test (SRT-test) with spatially separated sound sources. To perform well on the latter test good localization ability was needed. Both measurements were conducted with unilateral and bilateral conditions for all subjects. The ear of the unilaterally fitted hearing aid was based on the preference of the individual subject.

4.1 Localization test

For the localization test 5 loudspeakers were used (-90°, -45°, 0°, +45°, +90°) at 75 cm from the subject. Mixed sounds were randomly presented from different loudspeakers, for example: music, children laughing, dogs barking. All sounds were presented at an average level of 65 dB(A) with adequate roving. Every 0.7 seconds a new sound was generated randomly from the sounds that were not active at that time. The duration of the signals varied between 2.2 and 3.5 seconds. So, during the test three to five signals were presented simultaneously at each moment. There was one target sound: the telephone bell. The subject had to indicate where the target sound came from. The duration between the answer and the next stimuli varied between 4 and 10 seconds. The order of presentations was randomized, in total six measurements for each loudspeaker box. The test was performed with one and with two hearing aids.

In table 1 the results of the unilateral condition (average for the right and the left side) and the bilateral condition are shown. Localization was significantly better for the bilateral condition than for the unilateral condition. Most errors were made within 45°. There was a reduction of errors when a second hearing aid was added, for all degrees of errors ($< 45^0$, 45^0- 90^0, $>90^0$): a bilateral improvement of about 10% for the situation within 45 degrees and 13% for all situations together.

The benefit of a second hearing aid for localization proved to be rather independent of the other data, but showed a small but significant correlation with total auditory functioning (result derived from the AVETA questionnaire ($r = 0.18$; $p < 0.05$)).

	Percentage of errors	
	Unilateral (avg R/L)	Bilateral
< 45 degrees	38,3	28,3
45 - 90 degrees	7,2	5,1
> 90 degrees	2,5	1,3
Total	48,0	34,7

Table 1. The percentage of errors (within 45 degrees, between 45-90 degrees, more than 90 degrees and the total errors) of the localization test, for the unilateral condition (2nd column) and the bilateral condition (3rd column).

4.2 Speech intelligibility with spatially separated sound sources

A test with separated sound sources is a good simulation of daily conversations in real life. Localization plays an important role during this test. Two loudspeakers were positioned in front of the subject. The left loudspeaker was placed at -45 degrees and the right loudspeaker at + 45 degrees at 75 cm from the subject. For both sides a Speech Reception Threshold test was performed. The noise was an interfering (time-reversed) signal of the other gender. In conditions with speech from the left-hand side, the interfering signal came from the right-hand side and vice versa. The sentences (VU98, see Versfeld et.al., 2000) were presented randomly from the left and the right hand side, however, the male and the female speaker did not change from position during one list. So the subjects had to concentrate on where the normal speech came from and had to repeat the sentence. The interfering noise was presented at 65 dB(A).

	Critical S/N ratio in dB	
	Ipsi lateral side	Contra lateral side
Unilateral	-4,6 dB	-1,6 dB
Bilateral	-5,0 dB	-4,9 dB

Table 2. Average critical S/N ratio for the condition with the unilateral hearing aid at the speech side (ipsi-lateral, 2nd column) and the hearing aid at the noise side (contra-lateral, 3rd column) and for the bilateral condition (n=214). Lower values indicate better results.

In table 2 the average critical signal to noise ratios are presented for the ipsi- and contra-lateral condition, lower values represent better results. In this table the results with a female and a male voice have been averaged.

The ipsi-lateral condition (second column) is the condition with the unilaterally fitted hearing aid at the speech side; this is the most favorable condition. In the bilateral condition the second hearing aid is added to the noise side. Nevertheless, an improvement of 0.4 dB is shown. This is a purely binaural effect (binaural squelch effect).

The contra-lateral condition (third column) is the most difficult condition, with the unilaterally fitted hearing aid at the noise side. In the bilateral condition the second hearing aid is added to the speech side. An improvement of 3.3 dB is shown, due to the combined effect of elimination of the head shadow and the effect of binaural squelch. Participants

with more severe hearing losses showed a higher bilateral benefit for the SRT test with spatially separated sources, than did participants with milder hearing losses.
The benefit in speech perception with spatially separated sound sources is related to localization ($r = 0.19$; $p<0.01$).

4.3 Questionnaire

In the prospective study the AVETA questionnaire was used as well. The subjects were asked to complete the questionnaire for the condition without a hearing aid, with one hearing aid and with two hearing aids.
The average results of the group who preferred a unilateral fitting (n=13) showed a significant benefit for one hearing aid compared with the condition without hearing aid, for all categories (detection, discrimination, speech in quiet, speech in noise, localization, and aversiveness of loud sounds) ($p< 0.01$) except for the comfort of loud sounds. For loud sounds the comfort with a hearing aid was significantly lower than without hearing aid ($p<0.001$). There was no significant difference between the unilateral and the bilateral conditions for the group who preferred a unilateral fitting. The group who preferred a bilateral fitting showed significantly better scores with one hearing aid than without a hearing aid for all categories ($p<0.001$) except for the comfort of loud sounds. Again this score decreases with a hearing aid ($p<0.001$). Contrary to the group who preferred one hearing aid, the bilaterally fitted group shows significantly better scores with two hearing aids than with one hearing aid ($p<0.001$), but again the comfort of loud sounds scores significantly worse ($p<0.001$).
In this study the subjects were asked to mention reasons why they preferred one or two hearing aids also. More than one reason was possible. 138 Times a reason was given for the advantage of a unilateral fitting and 649 times for a bilateral fitting. Most mentioned reasons for a unilateral fitting were: own voice was more pleasant with one hearing aid (31%) and the unaided ear was used for the telephone (25%). For the bilateral fittings most mentioned reasons were: intelligibility from both directions (20%), better localization (19%), better sound quality (20%), and a better stereophonic effect/balance (19%).

5. Can we predict the positive effect from diagnostic tests?

To determine whether the bilateral benefit could be predicted from a-priori tests a battery of diagnostic tests (by headphones: Binaural Masking Level Difference, Interaural Time Difference) were applied before the trial period.
Horizontal localization may be assumed to be improved especially in subjects that are able to exploit interaural cues in time and level. Therefore a test on the Interaural Time Difference (ITD) was selected. The differences in arrival time are most effective for low frequencies up to about 1500 Hz (Dillon, 2001). The ability to localize sounds is important, especially in conversations with several people. Binaural cues are also important to exploit phase differences between dichotically presented target signals and masking signals in order to separate both more effectively (binaural squelch). Therefore a test for the Binaural Masking Level Difference (BMLD) was selected to measure the effect of binaural squelch. This is an important aspect of the cocktail party effect (Cherry, 1953; Bronkhorst, 2000) that may yield improved speech intelligibility in conditions with separated (simultaneous) sound sources.

5.1 Interaural Time Difference

The ITD test measures the sensitivity of the binaural system to perceive Interaural Time Differences. The interpretation of the ITD-result is: the smaller the value the better the sensitivity to Interaural Time Differences. In the ITD test every time two brief noise bursts (narrow-band noise of 500 Hz, 125 ms in duration) were presented binaurally. The duration of the temporal gap between the noise bursts was 250 ms. The binaural noise bursts were presented with a short interaural time difference. Because the time difference between both noises in a binaural noise burst (Δt) was very small, it was perceived as one single percept (fusion of the sounds), but the location of the perceived sound image in the head was largely determined by the ear where the noise arrived first (this is called the precedence effect; Gardner, 1968; Moore; 1982; Goverts et al., 2000). In the second binaural noise burst, the order of both noises was reversed. For example, in the first noise burst the noise was presented first at the right ear and Δt later at the left ear. In the second noise burst the noise was presented first at the left ear and then at the right ear. Consequently, the perceptual image of these two noise bursts in this example was as a noise pair moving from the right-hand side of the head to the left-hand side. For Δt is zero the noise bursts would be heard in the middle of the head. Δt was varied adaptively, starting with a temporal shift of 0.3 ms. The subjects were asked to indicate to which side the noises were moving in their heads. A 3-up 1-down procedure was used to determine the ITD.

5.2 Binaural Masking Difference

For the BMLD test, an octave-band noise with a centre frequency of 500 Hz, was presented to both ears. A tone of 500 Hz was also presented binaurally, one measurement with the tone in phase and one measurement with the tone out of phase. The masked thresholds of the tones were determined according to a 3-up 1-down procedure. The Masking Level Difference is calculated by subtracting the in-phase threshold from the out-of-phase threshold. In subjects with normal hearing the threshold of the signal out of phase is considerably lower than for the signal in phase. This means: the more negative the BMLD-value, the better the binaural function (Moore, 1982).

For both adaptive procedures (ITD and BMLD) the thresholds were determined by averaging of eight turning points. The subjects could exercise first until they understood the instruction. Before the ITD and the BMLD test, a matching test at a calculated stimulus level was used, to establish the same loudness of the stimuli in both ears. The stimulus level at the better ear was fixed at 60 dB SPL for average hearing losses up to 40 dB HL (averaged at 500, 1000, 2000, and 4000 Hz). For higher losses the stimulus level was set at the average hearing loss + 20 dB. The stimulus level at the other ear (the poorer ear) was determined by the result of the matching test (the average of three measurements provided that the differences between the test results were smaller than 10 dB. If not, the matching test had to be repeated).

5.3 Results

For the results of the hearing-impaired subjects again both groups were distinguished: the group who preferred one hearing aid (n=14), and the group who preferred two hearing aids (n=200). The median scores and the 25 and 75 percentile scores for the ITD-test and the BMLD-test are presented in Table 3. A lower score means a better result. As a reference also subjects with normal hearing were tested. They showed a better result than the hearing-impaired subjects for the BMLD-test, but there is a considerable overlap between the normal-hearing and the hearing-impaired groups.

	ITD (µsec)		BMLD (dB)	
	Median	*P25 / P75*	*Median*	*P25 / P75*
Unilateral fitting (n=14)	123.6	71.3 / 392.3	-15.5	-18.3 / -11.0
Bilateral fitting (n=200)	158.6	81.4 / 793.5	-14.4	-18.4 / -8.6
Normal hearing (n=10)	40.7	33.2 / 48.5	-19.5	-21.5 /-12.0

Table 3. Results of the unilaterally fitted group, the bilaterally fitted, and the normal-hearing group for the Interaural Time Difference-test. and the Binaural Masking Level Difference-test.

For the ITD-test the differences between the groups are larger, but the trends are similar. There is a clear difference between the hearing-impaired groups and the normal-hearing group. Again, the differences between both hearing-impaired groups are small and there is an overlap between both groups. A few subjects found the test very difficult. The choice for one hearing aid proved to be not related to poor results of the binaural tests.

6. Conclusions

The benefit of a second hearing aid is obvious. This effect is shown objectively with localization tests and speech intelligibility with spatially separated sound sources, but also subjectively with the AVETA questionnaire. Localization proved to be rather independent of the other data, but is related to total auditory functioning

However the benefit of a second hearing aid is hardly to predict with ITD and BMLD tests. The most important factor for predicting different outcome measures (including the bilateral benefit for speech) was the PTA in the better ear. Binaural diagnostic tests by headphones did not contribute significantly to a better prediction of the bilateral benefit.

Our conclusion is that every hearing impaired subject should start with a bilateral fitting to experience the benefits and the drawbacks. It is very useful to evaluate the trial period with good localization tests and with speech intelligibility tests with spatially separated sound sources.

7. References

Allen, R.L., Schwab, B.M., Cranford, J.L., Carpenter, M.D. (2000) Investigation of binaural interference in normal-hearing and hearing-impaired adults. *J.Am.Acad.Audiol.* 11: 494-500.

Bodden M. (1997) Binaural hearing and hearing impairment: Relations, problems, and proposals for solutions. *Seminars in Hearing* 18(4): 375-379.

Boymans, M., Dreschler, W.A. (2006) "Evidence for benefits of bilateral hearing aids." *Phonak proceedings. Hearing care for adults.*

Boymans, M., Goverts, S.T., Kramer, S.E., Festen, J.M., Dreschler, W.A. (2008). A prospective multi-centre study of the benefits of bilateral hearing aids. *Ear & Hearing.* 29(6): 930-41.

Boymans, M., Goverts, S.T., Kramer, S.E., Festen, J.M., Dreschler, W.A. (2009) Candidacy for bilateral hearing aids: a retrospective multicenter study. *Journal of Speech Language & Hearing Research.* 52(1): 130-40. Epub 2008 Jul 29.

Bronkhorst, A.W., & Plomp, R. (1989). Binaural speech intelligibility in noise for hearing impaired listeners. *Journal of the Acoustical Society of America, 86,* 1374-1383.

Bronkhorst, A.W., & Plomp, R. (1990). A clinical test for the assessment of binaural speech perception in noise. *Audiology, 29(5),* 275-285.

Van den Brink, R.H.S. (1995). Attitude and illness behaviour in hearing impaired elderly. Doctoral thesis Groningen University (ISBN 90-9008014-7).

Bronkhorst, A.W. (2000). The Cocktailparty Phenomenon: a review of research on speech intelligibility in multiple-talker conditions. *Acta Acustica 86,* 117-128.

Byrne, D., Noble, W. & LePage, B. (1992). Effects of long-term bilateral and unilateral fitting of different hearing aid types on the ability to locate sounds. *Journal of the American Academy of Audiology, 3,* 369-382.

Cherry, E.C.(1953). Some experiments in the recognition of speech, with one and with two ears. *J. Acoust. Soc. Am. 25*: 975-979.

Cox, R.M., & Alexander, G.C. (1995). The abbreviated profile of hearing aid benefit. *Ear & Hearing, 16, 176-183.*

Cox, R.M., Hyde, M., Gatehouse, S., .Noble, W., Dillon, H., & Bentler, R. et al. (2000). Optimal Outcome Measures, Research priorities and International Cooperation. *Ear & Hearing, 21 (suppl.),* 106S-115S.

Dillon, H. (2001). Binaural and bilateral considerations in hearing aid fitting. Chapter 14 in: *Hearing aids*. Thieme New York ISBN 1-58890-052-5.

Dreschler, W.A., & Boymans, M. (1994). Clinical Evaluation on the advantage of binaural hearing aid fittings. *Audiologische Akustik, 5,* 12-23.

Festen, J.M., Plomp, R. (1986) Speech-reception threshold in noise with one and two hearing aids. *J. of the Acoustical Society of America* 79(2): 465-471.

Gardner, M.B. (1968). Historical background of the Haas and-or precedence effect. *J.Acoust.Soc.Am. Jun;43(6):*1243-8.

Gelfand, S.A., Silman, S., and Ross, L. (1987). Long-term effects of monaural, binaural and no amplification in subjects with bilateral hearing loss. *Scandinavian Audiology* 16(4): 201-207.

Gelfand, S. (1995). A. Long-term recovery and no recovery from the auditory deprivation effect with binaural amplification: six cases. *Journal of the American Academy of Audiology* 6(2): 141-149.

Goverts, S.T., Houtgast, T., van Beek, H.H. (2000). The precedence effect for lateralization at low sensation levels. *Hearing Research Oct; 148(1-2):*88-94.

Hawkins, D.B., Yacullo, W.S. (1984) Signal-to-noise ratio advantage of binaural hearing aids and directional microphones under different levels of reverberation. *JSHD* 49(3): 278-286.

Hurley, R.M. (1999). Onset of auditory deprivation. *J.Am.Acad.Audiol.* 10: 529-534.

Jauhiainen, T. (2001). Progression of sensorineural hearing impairment in aided and unaided ears. *Scand.Audiol.*30:Suppl.52: 28-31.

Köbler, S., & Rosenhall, U. (2002). Horizontal localization and speech intelligibility with bilateral and unilateral hearing aid amplification. *International Journal of Audiology, 41,* 3905-400.

Kramer, S.E., Kapteyn, T.S., Festen, J.M., & Tobi, H. (1995). Factors in subjective hearing disability. *Audiology, 34,* 311-320.

Leeuw, A.R., Dreschler, W.A. (1991) Advantages of directional hearing aid microphones related to room acoustics. *Audiology* 30(6): 330-344.

Markides, A. (1982) The effectiveness of binaural hearing aids. *Scandinavian Audiology* Supplementum. 15: 181-196.

Moore, B.C.J. (1982). *An introduction to the Psychology of hearing*. Academic press, inc. ISBN 0-12-505620- 6: (162-163, 168-176).

Moore, B.C., Johnson, J.S., Clark, T.M., Pluvinage, V. (1992) Evaluation of a dual-channel full dynamic range compression system for people with sensorineural hearing loss. *Ear & Hearing* 13(5): 349-370.

Nabelek, A.K., Letowski, T., and Mason, D. (1980). An influence of binaural hearing aids on positioning of sound images. *Journal of Speech & Hearing Research* 23(3): 670-687.

Nabelek, A.K., Mason, D. (1981) Effect of noise and reverberation on binaural and monaural word identification by subjects with various audiograms. *JSHR* 24(3): 375-383.

Punch, J.L., Jenison, R.L., Allan, J., & Durrant, J.D. (1991). Evaluation of three strategies for fitting hearing aids binaurally. *Ear & Hearing, 12(3),* 205-215.

Silman, S., Gelfand, S.A., and Silverman, C.A. (1984). Late-onset auditory deprivation: effects of monaural versus binaural hearing aids. *Journal of the Acoustical Society of America* 76(5): 1357-1362.

Silman, S., Silverman, C.A., Emmer, M.B., and Gelfand, S.A. (1993). Effects of prolonged lack of amplification on speech-recognition performance: Preliminary findings. *Journal of Rehabilitation Research and Development.* 30(3): 326-332.

Stephens, S.D., Callaghan, D.E., Hogan, S., Meredith, R., Rayment, A., & Davis, A. (1991). Acceptability of binaural hearing aids: a cross-over study. *Journal of the Royal Society of Medicine, 84(5),* 267-269.

Vaughan-Jones, R.H., Padgham, N.D., Christmas, H.E., Irwin, J., & Doig, M.A.. (1993). One aid or two? - More visits please! *Journal of Laryngology & Otology, 107(4),* 329-332.

Versfeld, N.J., Daalder, L., Festen, J.M., Houtgast, T. (2000). Method for the selection of sentence materials for efficient measurement of the speech reception threshold. *J. Acoust. Soc. Am. 107 (3):* 1671-1684.

17

Auditory Guided Arm and Whole Body Movements in Young Infants

Audrey L.H. van der Meer and F.R. (Ruud) van der Weel
Norwegian University of Science and Technology
Norway

1. Introduction

Can infants use auditory information to guide their movements adequately in space, and if so, to what degree? Perceptual development has mostly been considered through the visual system. Similar to vision, audition provides us with spatial information over extended distances. There is generally little research about the use of auditory information for guided movement in the environment, and the similarity between vision and hearing is narrowly attached to a theoretical framework. Effective action is prospective and supposes the pick-up of predictive perceptual information, so as to prepare the body how, when, and where a movement is to be performed. Studies on the use of auditory information for action are rare. This chapter will describe two studies with young infants where it will be shown that the auditory system is equally important as the visual system to the performance of prospective action in the environment. It will be concluded that the auditory system is best conceived as a functional listening system where auditory information is used as a perceptual source for guiding behaviour in the environment.

2. Vision versus hearing

Perception of the environment has mostly been considered through visual information. Similar to vision, audition provides us with spatial information over extended distances. Distance perception by ear under naturalistic conditions is a particularly well-developed human capability (Ashmead et al., 1990; Little et al., 1992; Wightman & Jenison, 1995). Hearing may be even more important than vision in orienting towards distant events. We often hear stimuli before we see them, particularly if they take place behind us or on the other side of opaque objects such as walls.

Auditory information is especially important for guiding behaviour in the environment by lack of visual information, such as with the blind (Millar, 1994). In the absence of vision, auditory localization of events that are behind the listener is thought to be aided by spectral shaping introduced by the ears and head (e.g., Hill et al., 2000), as well as by changes in interaural differences (especially time differences) resulting from head movements (Thurlow et al., 1967; Wallach, 1940; Wightman & Kistler, 1999). There is generally little research about the use of auditory information for guided movement in the environment (Jenison, 1997; Lockman, 1990; Pick, 1990).

According to J.J. Gibson's affordance theory (1979), action is affected by environmental information. Information about the environment can be achieved through different senses (visual, auditory, haptic, etc.). However, adult studies on the use of auditory information for action are rare. For example, Russell and Turvey (1999) posed the question of whether sighted observers with eyes closed could judge correctly whether a wall was wide enough for unimpeded passage based on perception of the distances between a sound-emitting object. Results indicated a limited form of auditory affordance perception: listeners could perceive, with acceptable tolerance, a sound source's azimuth relative to the body's boundaries. The auditory perceptual ability was affected by the source-to-listener distance and visual preview of the spatial layout. Other studies have confirmed the same, reporting variation in the perception of sound location due to source-to-listener distances (Guski, 1990; Loomis et al., 1993) and an improvement of the ability to auditorily control action with previous visual preview of the spatial layout (Warren, 1978).

Research on auditory perception with infants has been concentrated around sound discrimination and auditory localization within specific action systems. It has been shown that infants already from the moment they are born have the ability to turn their heads toward a sound (Muir et al., 1999; Muir & Field, 1979; Wertheimer, 1961). After a while, infants stop to turn after sounds that require large movement of the head (Bower, 1979, 2002; Muir & Nadel, 1998), partly because of changes in the auditory cortex (Clifton et al., 1981), and because of the muscular strength in the neck being too weak in relation to the gravity force. When the infant is 4-5 months old this discrepancy disappears and the infant will turn the head faster and more precise than in the neonatal period (Muir & Clifton, 1985). Head and eye movements can indicate a sound's direction but they cannot inform about distances. Previous research has shown that sighted infants will reach for sounding objects in the absence of visual clues (Ashmead et al., 1987; Clifton, 1992; Clifton et al., 1991; Litovsky & Clifton, 1992; Morrongiello, 1988; Perris & Clifton, 1988). This ability implies a sense of auditory space, a world in which sounding objects are localized in relation to one's body. By 6 months of age, infants are sensitive to changes in the location of sounds as small as 13-19 degrees (Ashmead et al., 1987; Morrongiello, 1988). By 7 months of age, infants have at least a dichotomous discrimination of auditory space, i.e., within and beyond reach (Clifton et al., 1991; Litovsky & Clifton, 1992). This indicates that infants have the ability to differentiate acoustic information and perform adequately in different action systems.

3. Early arm movements

Acting successfully entails perceiving environmental properties in relation to oneself (J.J. Gibson, 1979). Organisms do not perceive objects *per se*, but what these objects afford for action. What any given object affords depends on the size and action possibilities of the perceiver. Affordances are therefore not fixed: they have to be updated during life to accommodate changes in action capabilities and bodily characteristics. This is particularly apparent during infancy, when new skills are constantly appearing and bodily dimensions are changing rapidly (Adolph et al., 1993).

Next, a series of experiments on neonatal arm movements will be described and their possible functional significance for later reaching and grasping will be discussed. Before babies can reach out and successfully grasp objects in the environment, they first have to learn they have an arm and what it can do. It is proposed that arm movements made by young infants during the first four months of life have an important exploratory function, essential for the development of eye-hand coordination. But would neonates also be able to control their arm movements based on sound?

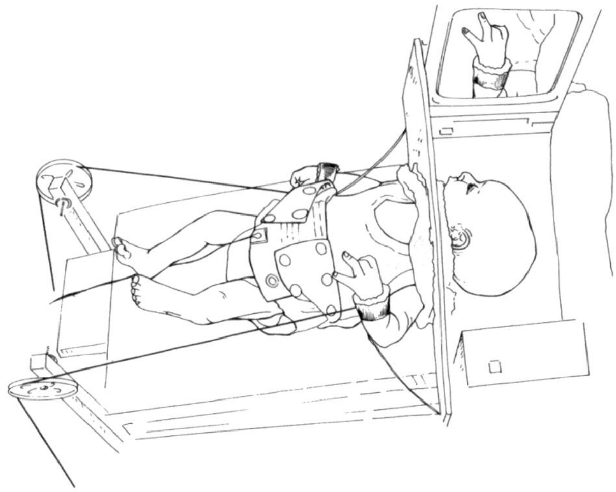

Fig. 1. A newborn baby taking part in the weight-lifting experiment, with sight of the facing hand prevented while the hand on the opposite side of the body is visible in real-time on a small video-monitor.

As a result of the strong influence of the maturation position, newborn babies are still usually considered reflexive organisms, incapable of making voluntary movements. Maturationists consider the existence and disappearance of reflexes as evidence of nervous system growth and development. As maturation of the cortex and the spinal motor nerve pathways proceeds, the cortical centres supposedly inhibit the primitive reflexes. Cortical maturation and the resulting inhibition of reflexive movements are thought to take up most of the four months of a baby's life. Until that time, arm movements made by very young babies are simply dismissed as reflexive, involuntary and purposeless (Van der Meer & Van der Weel, 1995).

Moving a limb or the whole body in a controlled manner requires acting hand-in-hand with gravity and other non-muscular forces, such as the drag of clothing and stiffness of the joints (Bernstein, 1967). As a consequence, movements cannot be represented simply as patterns of efference to the muscles nor in any preprogrammed context-insensitive way. Accurate control requires online regulation of muscular activation based on perceptual information about the dynamics of the limb movement and the external force field, as well as about the movement of the limb relative to objects or surfaces to which it is being guided.

Are neonates capable of such perceptuo-motor control or are their movements to be seen as simply reflexive or due to spontaneous patterned efference to the muscles as is commonly believed? There now is some evidence that newborn babies can move their arms and hands in a purposeful way (Bower et al., 1970; Butterworth & Hopkins, 1988; Von Hofsten, 1982), and we are able to tell that their movements take into account the gravitational and other external forces acting on the limbs (Van der Meer et al., 1995, 1996; Van der Meer, 1997a). However, the question remains whether newborn babies can control their arms based on sound.

3.1 Lifting weights in neonates

To test whether newborn babies take account of external forces in moving their limbs, we recorded spontaneous arm-waving movements while the baby lay supine with its head turned to one side (Van der Meer et al., 1995). Free-hanging weights, attached to each wrist by strings passing over pulleys, pulled on the arms in the direction of the toes. The babies were allowed to see only the arm they were facing, only the opposite arm on a video monitor (see Figure 1), or neither arm because of occluders.

The babies opposed the perturbing force so as to keep an arm up and moving normally, but only when they could see the arm, either directly or on the video monitor. Thus, newborn babies purposely move their hand to the extent that they will counteract external forces applied to their wrists so as to keep the hand in their field of view. In addition, newborns move their arms more when they can see them (Van der Meer et al., 1996).

3.2 Keeping the arm in the limelight

In order to investigate whether newborns are also able to adjust their arm movements to environmental demands in a flexible manner, we investigated whether manipulating where the baby sees the arm has an influence on where the baby holds the arm (Van der Meer, 1997a). Spontaneous arm-waving movements were recorded in the semi-dark while newborns lay supine facing to one side. A narrow beam of light (7 cm in diameter) was shone in one of two positions: high over the baby's nose or lower down over the baby's chest, in such a way that the arm the baby was facing was only visible when the hand encountered the, otherwise, invisible beam of light (see Figure 2).

The babies deliberately changed arm position depending on the position of the light and controlled wrist velocity by slowing down the hand so as to keep it in the light and thus clearly visible. This suggests sophisticated control of position and velocity of the hand rather than excited thrashing of the limbs, the way neonatal arm movements have been described in the past. However, would newborns also be able to control deceleration of the hand in such a precise manner?

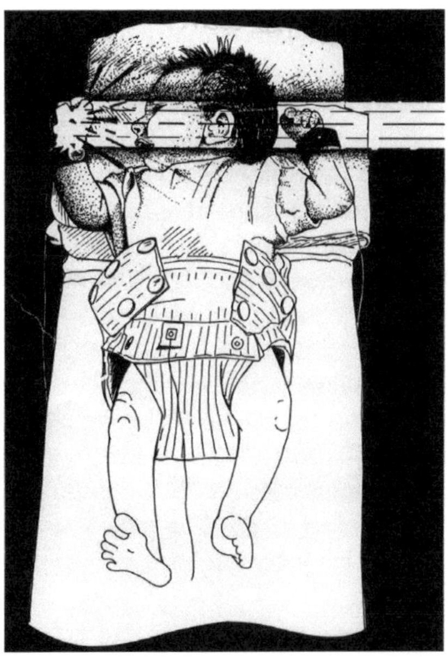

Fig. 2. A 21-day-old baby keeping her arm in the "limelight", a narrow beam of light (7 cm diameter) in an otherwise dark room.

Figure 3 shows a typical position and velocity record of a newborn baby waving its arm. For all instances where the baby's hand entered the light and remained there for 2 seconds or longer, the onset of deceleration (point of peak velocity) of the hand was noted with respect to the position of the light. Surprisingly, in 70 out of all 95 cases (~74%), the babies started to decelerate the arm before entering the light (as in Figure 3), showing evidence of anticipation of, rather than reaction to, the light. On those occasions where the babies appeared not to anticipate the position of the light, more than 70% of these occurred within the first 90 seconds after starting the experiment or changing the position of the light (see Figure 4). Thus, we have shown clear evidence of both learning and memory in newborn babies. By waving their hand through the light in the early stages of the experiment the babies were learning about and remembering the position of the light. This very quickly allowed them to accurately and prospectively control the deceleration of the arm into the light and remain there, while effectively making the arm clearly visible.

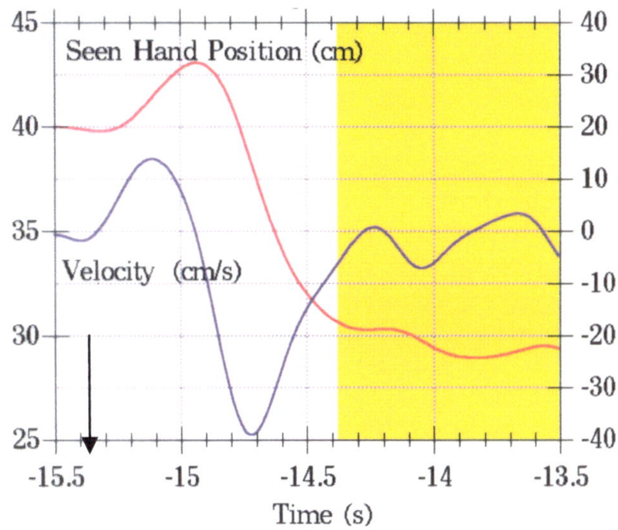

Fig. 3. A typical position and velocity record of a 22-day-old newborn baby waving its arm with light level with chest. The baby could only see the hand clearly in the yellow area, which represents the beam of light. The position trace (red line) indicates that the baby moves its arm 12 cm in the direction of the toes towards the light. The velocity trace (blue line) shows anticipatory deceleration (indicated by the black arrow) starting about 350 ms before the hand enters, and remains in, the shaded visible area. Not that the velocity of the hand in the shaded area is hovering around zero.

3.3 Functional significance of early arm movements

It, thus, seems plausible that the spontaneous arm waving of neonates of the kind measured in our experiments is directed and under precise visual control. Neonates can purposely control the position, velocity and deceleration of their arms so as to keep them clearly visible. Their level of arm control, however, is not yet sufficiently developed that they can reach successfully for toys. Young babies have to do a lot of practising over the first four to five months, after which they can even catch fast-moving toys successfully. What could be the functional significance of neonatal arm movements for later successful reaching and grasping?

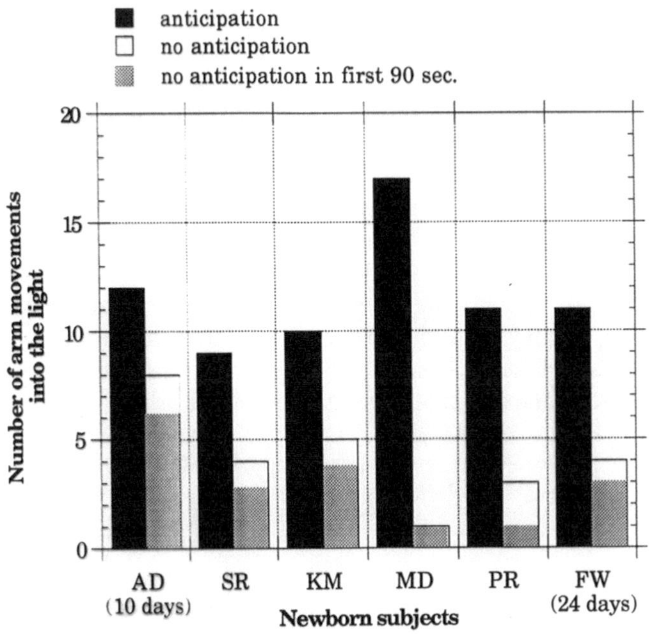

Fig. 4. Total number of cases (n=95) in which the arm was moved into the light and remained there for at least 2 seconds for each newborn subject (increasing in postnatal age from left to right). In a significant number of cases, the babies anticipated the position of the light by starting the deceleration phase of their arm movement before entering the light (black bars). On those 25 occasions where the babies showed no anticipation of the position of the light (white bars), 18 (or 70%) occurred within the first 90 seconds of starting the experiment or changing the position of the light (shaded bars). Note that for 20-day-old MD only one case was recorded where she did not anticipate the position of the light when initiating the deceleration phase of her arm movement into the light, and that this case occurred within one minute into the experiment.

To direct behaviour in the environment successfully, the infant needs to establish a bodily frame of reference for action (Van der Meer & Van der Weel, 1995). Since actions are guided by perceptual information, setting up a frame of reference for action requires establishing informational flow between perceptual input and motor output. It also requires learning about body dimensions and movement possibilities. Thus, while watching their moving arms, newborn babies acquire important information about themselves and the world they move in - information babies need for later successful reaching and grasping beginning at around four to five months of age.

It is widely known that young infants spend many hours looking at their hands (see Figure 5). And so they should, for a vast amount of lessons in practical optics have to be learned in those early weeks before reaching for objects can emerge. First of all, infants have to learn that the hands belong to the self, that they are not simply objects, but that they can be used to touch all sorts of interesting object in the environment. In order to successfully reach out and grasp objects in the environment, infants also have to familiarize themselves with their own body dimensions in units of some body-scaled or, more generally, action-scaled metric (Warren, 1984; Warren & Whang, 1987). In other words, infants have to learn to perceive the shapes and sizes of objects in relation to the arms and hands, as within reach or out of reach, as graspable or not graspable, in terms of their affordances for manipulation.

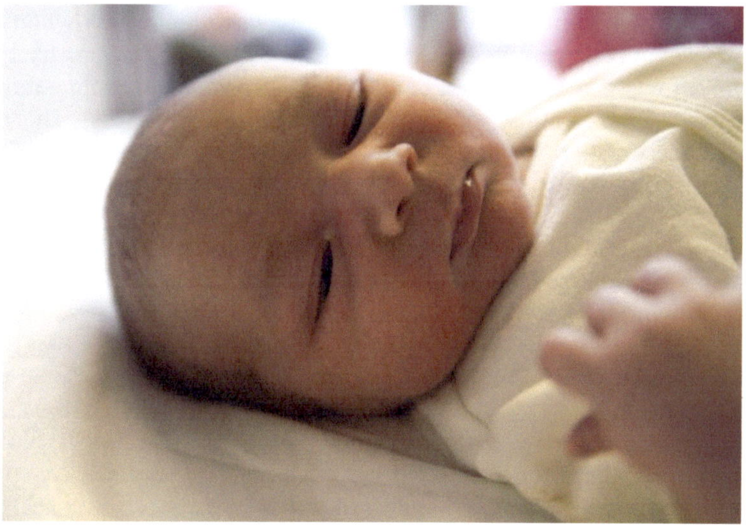

Fig. 5. A newborn baby boy of only a few hours old is studying his hand intensely.

All this relational information has to be incorporated into a bodily frame of reference for action in those early weeks before reaching for objects "emerges". We have all experienced this process of incorporation, namely when learning new perceptuo-motor skills. For instance, tennis rackets, skis, golf clubs and other extensions of the human body such as false teeth and new cars, first have to be incorporated into our habitual frame of reference, before we can use them to their full potential (Tamboer, 1988). At first, we experience those instruments as unmanageable barriers between the environment and ourselves. However, once incorporated into our "bodily" frame of reference, they increase our action possibilities considerably and are almost regarded as our own body parts.

In this context, it is possible to speculate about the role of early arm movements for distance perception in general. Professor Henk Stassen (1994, personal communication) is a mechanical engineer from Delft University of Technology, The Netherlands. He designs artificial arms for babies who are born with two stumps because of genetic disorders or because their mothers has taken the drug thalidomide in the 1960s during pregnancy to prevent miscarriage. He observed that if you fit babies with artificial arms early (around 2 to three months), they do not seem to have any problems avoiding obstacles as soon as they learn to walk. However, if the arms are fitted to late, the babies will have tremendous problems perceiving distance, and they will initially bump into walls and obstacles when they start walking. J.J. Gibson (1979) suggested that we perceive distance in relation to our own nose length. Stassen's observations would suggest that we scale distance according to our arm length, as within reach or out of reach.

During infancy new skills are constantly appearing and bodily dimensions are changing rapidly. In general, the bodily frame of reference has to be updated during life, to accommodate changes in action capabilities and body characteristics. Sudden changes in action capabilities, as after stroke, show this very clearly, as do rapid changes in body size in pregnancy and adolescence. Teenagers, for example, can be notoriously clumsy; they undergo such sudden growth spurts that their bodily frames of reference need to be updated nearly daily.

Successfully reaching out and grasping objects in the environment requires infants to be familiar with their own body dimensions. As infants wave their arms while supine, they learn about their own body and its dimensions through vision. It seems likely that a fast-growing organism will constantly need to calibrate the system controlling movement, and visual proprioceptive information is least susceptible to "growth errors". This being so, our findings could have practical implications for babies with visual deficits and for the early diagnosis of premature babies at risk of brain damage. If early arm movements have an important function for later reaching, then infants with signs of hypoactivity and/or spasticity of the arms should be monitored closely with respect to retardation of developing reaching and possibly other perceptuo-motor skills. In such cases, early intervention should concentrate on helping the baby to explore its arms and hands, both visually (E.J. Gibson, 1988) and non-visually (Fraiberg, 1977). A simple intervention technique that could be used on babies with a visual deficit is the use of brightly coloured, high-contrast mittens, or a string of bells around the baby's wrists. It is a well-known phenomenon that reaching out is the first developmental milestone that blind babies fail to reach on time. The sound of bells always accompanying that particular proprioceptive feeling when the arms move might enable the baby to establish a stable bodily frame of reference for reaching based on auditory exploration of the self.

3.4 Control of early arm movements based on sound

This brings us to the question: Would newborn babies be able to control their arm movements by means of sound? In order to answer this question, newborn babies between three and six weeks of age were placed on their backs with the head in the midline position by means of a vacuum pillow. In this position, both ears were uncovered and available for sound localization (see Figure 6). Miniature loudspeakers of the sort used in telephones were attached to the baby's wrists. The baby's mother was placed in an adjacent, sound proof room where she could see her baby through a window. The mother was instructed to

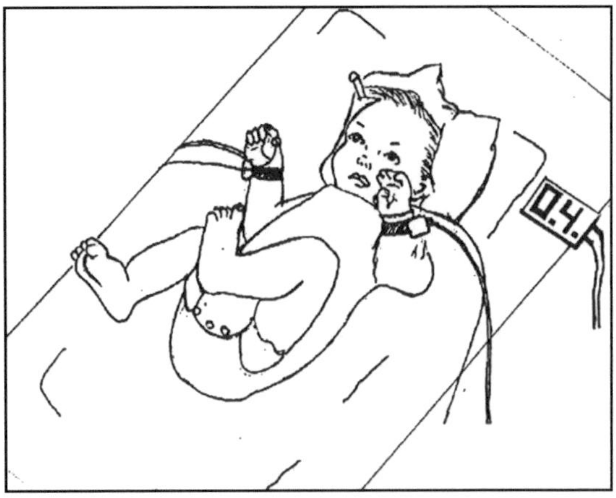

Fig. 6. A four-week-old baby participating in an experiment on auditory localization of the arms. The mother's voice is played softly over one of the small loudspeakers attached to the baby's wrists.

speak or sing to her baby continuously, while the sound of her voice in real time was played softly over one of the loudspeakers attached to the baby's wrist. In order to hear her mother's voice, the baby would have to move the "sounding" wrist close to the ear, and change arms when the mother's voice was played over the other loudspeaker. The results showed that newborn babies were able to control their arms in such a way that the distance of the left and right wrist to the ear was smaller when the mother's voice was played over that wrist than when it was not. Further analyses showed that there were far more reductions than increases in distance between wrist and ear when the sound was on (see Figure 7). However, when the sound was off the number of reductions and increases in distance between wrist and ear was about the same.

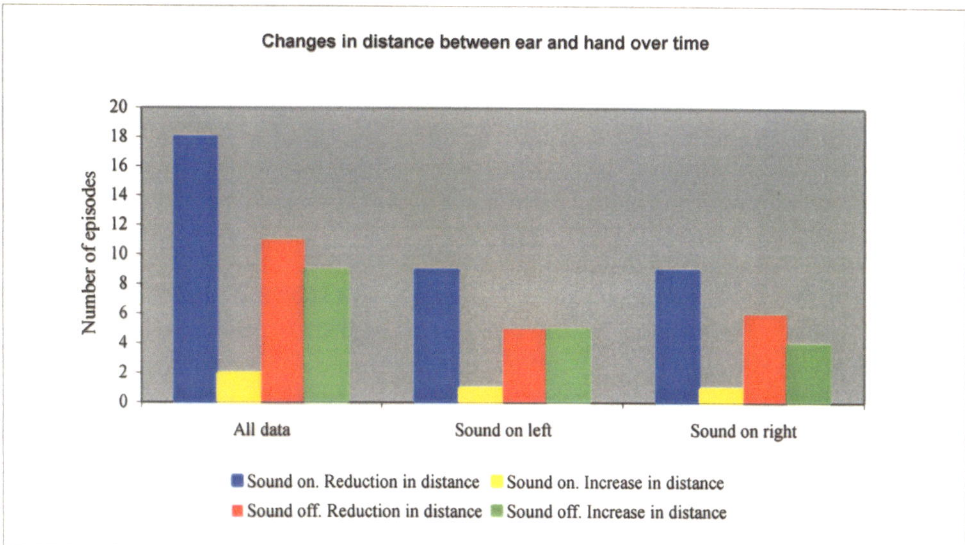

Fig. 7. Changes in distance between ear and hand over time for all data (left), when sound was on the left (middle), and when sound was on the right (right).

Thus, sighted newborn babies can control their arms with help of both sight and sound. This implies that arm movements are not simply reflexive, nor can they be explained away as excited thrashing. Young babies can act intentionally from the start, and they come equipped with perceptual systems that can be used to observe the environmental consequences of their actions. At the same time, an action provides valuable information about oneself. It is this dual process of perceiving oneself and perceiving the consequences of self-produced actions that provides very young infants with knowledge about themselves, in terms of their action capabilities and bodily characteristics. Obviously, during infancy new skills are constantly appearing, and bodily dimensions are changing rapidly. As new action possibilities emerge, infants have to update their perceptions, and vice versa, in a never-ending circular cycle.

4. Getting around with light and sound

Adaptive movement in the environment depends on guidance to a destination, avoidance of obstacles, steering and staying on course, and selecting the most economical route to the goal from several alternatives. Effective action supposes prospective control (Gibson & Pick, 2000; Lee, 1993; Von Hofsten, 1993) so as to prepare the body how, when, and where a movement is to be performed. Vision is unquestionably of prime use in locating environmental resources (Gibson & Schmuckler, 1989). Research on visually guided movements in children mainly involves studies where the child moves to a destination with a partly covered goal while the child has to choose between different routes.

A study of detour behaviour by McKenzie and Bigelow (1986) blocked one path to a goal and left one open to find out whether ambulatory infants could choose the shortest route and also show flexible behaviour when the barrier was moved. At 14 months, all infants changed routes successfully and generally followed shorter and more effective routes. Most studies conclude that experience is of significant importance to adaptive performance in this type of task (Caruso, 1993; Hazen et al., 1978; Lockman, 1984; McKenzie & Bigelow, 1986; Pick, 1993; Pick & Lockman, 1981; Rieser et al., 1982; Rieser & Heiman, 1982). A summary of the studies indicates that besides movement experience, other variables such as exploratory movements (Caruso, 1993; Lockman, 1984), the task complexity and motivation to reach the goal (McKenzie & Bigelow, 1986; Rieser et al., 1982), visualization of the layout and the opportunity to get continuous visual information about the goal (Rieser et al., 1982), postural control (Adolph, 2000; Van der Meer, 1997b), and perception of the meaning of the object and event (Adolph et al., 1993; Bertenthal et al., 1984; Ulrich et al., 1990) supposedly affect infants' abilities of adaptive movement in visual perception tasks.

To what extent are infants able to get around the environment by use of auditory perception in a mobility task? Researchers who have examined aspects of perception and action in infants have found that, in general, functionally appropriate perception of what an object affords emerges as the physical capacity to perform that function or task evolves (Gibson et al., 1987; Ulrich et al., 1990).

4.1 Auditory guided rotation in infants

Rotation on the stomach is one of the first opportunities infants have to respond to an auditory stimulus behind them. This skill requires infants to use their arms and legs to rotate around their own body axis. It emerges when infants are 6-7 months old (Bobath & Bobath, 1975; Illingsworth, 1973), and allows for a new opportunity to interact with the environment. Emergence of rotation skill requires maturation of both skeletal and neuromuscular systems (Thelen et al., 1987), but the ability to interact adaptively with the environment is not just a result of motor skills. Successful actions require both motor skills and perceptual sensitivity, and of course the ability to integrate the two (Adolph et al., 1993; E.J. Gibson, 1988; Gibson & Schmuckler, 1989; Lee, 1993; Schmuckler, 1993, 1996).

One of infants' first opportunities to move in the environment is by use of rotation skill in a prone position. Use of this skill is also the first opportunity for infants to detect what is behind them, and to perform adequate whole-body movements based on auditory perception. Little is known about infants' rotation skill, and the consequences of using this skill in orienting to objects and individuals. Based on affordance research, we investigated whether infants mastering the rotation skill would use auditory perception for rotation along the shortest way to a sound source, relative to their own position in space (Van der Meer et al., 2008).

Twelve healthy, full-term infants between 6 and 9 months, who mastered the rotation skill in both directions, were included in the study. Figure 8 shows the positions of the infant and the mother in the circle, where the infant performed the rotation and the mother gave

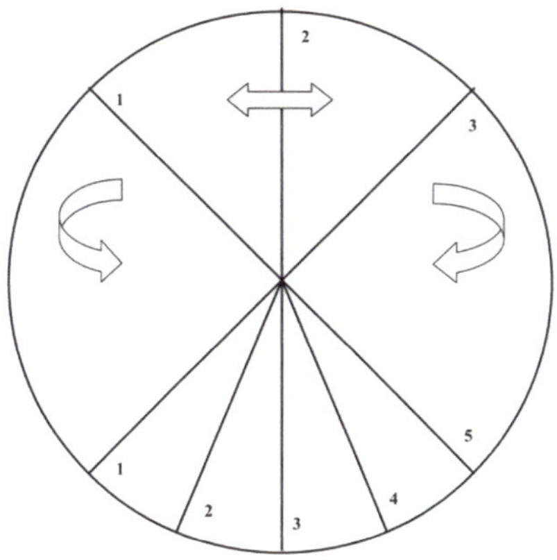

Fig. 8. Illustration of the three different starting positions of the infant (top) and the five different starting positions of the mother (bottom) within the rotation circle. The baby was placed on its stomach with its feet pointing towards the centre of the circle.

continuous auditory stimulation to her baby. To ensure the task remained challenging for the infant, there were three starting positions for the infant and five starting positions for the mother. The coordinate system was constructed with five different angles between the infant's positions and the mother's positions: 90°, 112.5°, 135°, 157.5°, and 180°. Out of a possible 15 combinations, a total of 10 trials were presented in a fixed-random order: four different directional trials where the shortest way would be to rotate to the left and four different directional trials where the shortest way would be to rotate to the right, and two non-directional trials at 180°.

A magnetic tracker system was used to measure the infant's rotations. The system consists of sensors (weighing 25 g each) and a magnetic box which transmits a magnetic field of 3 x 3 x 3 m. The sensors were placed on the infant in the magnetic field (see Figure 9) and their positions (in x, y, and z direction) and angular rotation (azimuth, role, and elevation) were continuously recorded at 100 Hz.

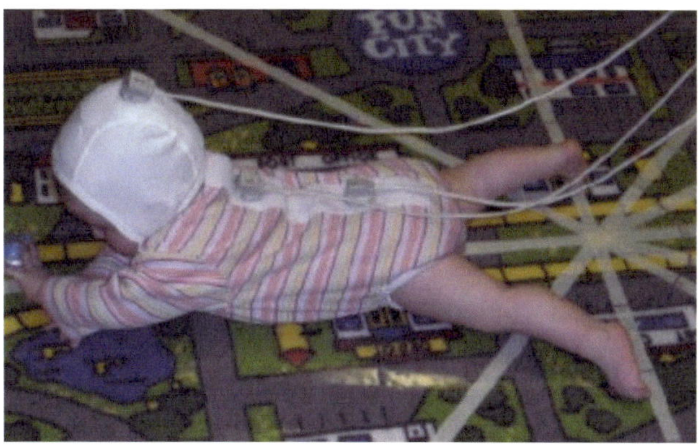

Fig. 9. A 7-month-old infant wearing a special body and hat placed prone in the rotation circle and participating in the experiment. The magnetic trackers to measure the infant's rotation movements were placed on the head, between the shoulder blades, and on the lower back.

Before each trial the experimenter placed the infant in one of the three starting positions in the middle of the rotation circle, with the feet to the centre. The experimenter sat in front of the infant and maintained its attention, while the mother was instructed to position herself quietly and unseen by the infant in one of five positions, as indicated by the experimenter. Her position was 50 cm behind the centre of the circle (behind the infant's feet). As soon as the measuring started, the experimenter stopped interacting with the infant, while the mother gave continuous auditory stimuli with her voice. The mother was instructed to call her baby in a way that came natural to her, and to continue calling until the baby reached her.

In total, 96 directional trials were recorded. The criterion for rotation was that the infant rotated (both with the head and body) in one direction until the mother was visible for the child. Information about the infant's rotation direction was analyzed through video and the kinematic analyses. In each trial, the rotation direction of the infant was encoded as shortest versus longest way in relation to the position of the infant and the position of the mother. Contrary to expectation, infants did not move their heads before rotating, but in general moved their heads and bodies smoothly in one direction as the trial began.

In case of the directional trials, the babies chose the shortest way in 87.5% of the trials (84 out of 96 trials), indicating that infants between 6 and 9 months use auditory information to move along the shortest way to a goal. Four babies consistently chose the shortest way on all their directional trials, five babies made one mistake, two babies made two mistakes, and one baby made three mistakes (out of 8). Infants chose the shortest way in 75.0% for the largest angle to 95.8% for the smallest angle (see Figure 10). Thus, infants are capable of picking the shortest way to rotate to their mothers, even though they make fewer mistakes with the shorter angles than with the larger angles. This suggests that infants experience increased difficulty differentiating more ambiguous auditory information for rotation.

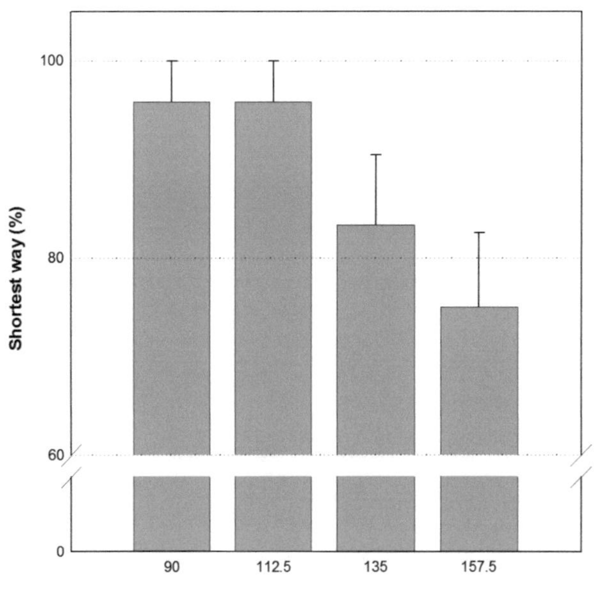

Fig. 10. Average percentages of rotation along the shortest way (including standard error of the mean bars) for the four angle conditions for all twelve participating infants.

To investigate whether infants prospectively adjusted their rotations' angular velocity to the different directional angle conditions, peak angular velocity was calculated for the first couple of pushes that took place within 50% of total rotation time when sight of the mother was unlikely to play a role. Angular velocity was calculated from the azimuth of the marker between the infant's shoulder blades. The azimuth is the direction of the marker referenced to the centre of the rotation circle. The angular velocity is the rate of change of the azimuth. The horizontal and the vertical movements were therefore disregarded in this analysis. As a result, small movements forwards or backwards, but not involving any rotation, showed up as stationary in the data. Figure 11 shows a typical graph of an infant covering an angle of 157.5° towards her mother. An analysis including successful directional trials only showed that the larger the angle between infant and mother, the higher the mean peak angular velocity with which the infants rotated towards her. This finding suggests prospective control of movement, as indicated by a more forceful initial push with the arms and legs in the case of larger angles to be covered.

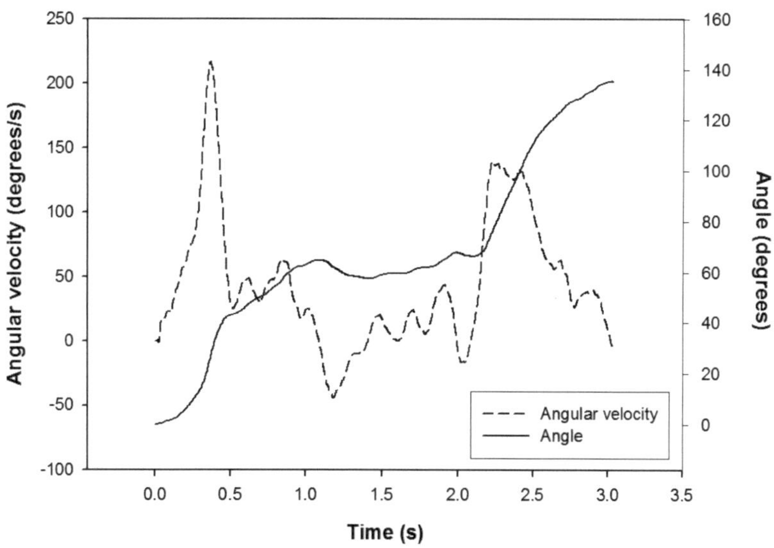

Fig. 11. Illustration of an infant's peak angular velocity (dashed line) during rotation through 157.5° to the left, with a peak angular velocity of 216°/s. Because the angle to the reference point was measured counter clockwise, negative angular velocity indicated clockwise movement. Note that infants typically rotated slightly less than the required angle (here: 140°, solid line, because they would often stop rotating a little short of their mum.

4.2 The role of auditory information in guiding whole body movements in space

By manipulating infants' prone rotations with an auditory stimulus from different angles behind the infant, it was found that young infants can use auditory information to guide their movements adequately in space (Van der Meer et al., 2008). In order to be able to rotate along the shortest way to a goal using auditory perception, infants need to be able to locate and specify the direction of the auditory information, and to perceive the angle between themselves and their mother in terms of their own action capabilities. The findings suggest that 6- to 9-month-old infants are capable of controlling their rotation actions effectively and efficiently. Thus, infants' decisions to rotate in a particular direction are not random, but controlled by means of auditory information specifying the shortest way to their mother.

This study is different from other studies in several respects. Infants in the present study were younger, the task was different, and the main perceptual source of information that was used to guide action was auditory instead of visual. In general, use of auditory perception for action has been a neglected research area in the ecological tradition (but see Russell & Turvey, 1999). The present findings corroborate the results of previous studies that newborns and older infants can differentiate between auditory information from left versus right (e.g., Morrongiello & Rocca, 1987; Muir & Field, 1979; Muir et al., 1999; Perris & Clifton, 1988; Wertheimer, 1961), and that they from the age of about six months can localize auditory information for reaching up to 12-14° precisely (Ashmead et al., 1987; Morrongiello, 1988; Morrongiello et al., 1994).

The findings are also in agreement with studies where the task for the infant was to find its way to mum or an object around obstacles with the help of visual perception (e.g., Caruso, 1993; Hazen et al., 1978; Lockman, 1984; McKenzie & Bigelow, 1986; Pick, 1993; Rieser et al., 1982). It can therefore be concluded that sighted infants can use both visual and auditory information for navigation in the environment. The studies by Rieser et al. (1982) and Lockman (1984) have shown that infants are capable of choosing appropriate routes to a goal using vision around the age of 24 and 14 months, respectively. The degree of difficulty of the task, different motor skills and motivation to reach the goal, as well as different degrees of visual information about the goal can explain the age difference for prospective action in these studies. Van der Meer et al.'s (2008) study, on the other hand, indicates that infants as young as 6-7 months will choose the most efficient way to their mother, based on auditory information and using their rotation skill. A possible reason why this has not been reported earlier is because of the fact that the tasks used to study infants' navigational skills have depended on motor skills that develop later in life, such as crawling and independent walking. The use of the mother's voice can also have contributed to the findings. This is a source of auditory information that is easily recognized by infants (DeCasper & Fifer, 1980), and might have increased the infants' motivation to solve the task.

Contrary to expectation, infants did not noticeably move their heads before deciding which way to turn, nor was there any significant latency before a rotation. Slight head rotations as small as 1 or 2° are considered to be helpful in resolving front-back confusions (Hill et al., 2000), a phenomenon where listeners in the absence of vision indicate that a sound source in the frontal hemifield appears to be in the rear hemifield, or vice versa (Wightman & Kistler, 1999). The infants in the present experiment actually might have used vision to resolve this confusion. For example, for a sound source at 135° the interaural time difference is about the same as for a source at 45°, thus solving the task by means of a cross-model elimination process.

5. Conclusion

The research reported here shows that newborn babies can use auditory information to control their arms in the environment, and that babies before they start crawling at around 9 months can use auditory information to control their whole body movements in space. Our results can contribute to the understanding of the auditory system as a functional listening system where auditory information is used as a perceptual source for guiding behaviour in the environment.

6. References

Adolph. K.E. (2000). Specificity of learning: Why infants fall over a veritable cliff. *Psychological Science*, 11, 290-295, 0033-295X

Adolph, K.E., Eppler, M.A. & Gibson, E.J. (1993). Crawling versus walking infants' perception of affordances for locomotion over sloping surfaces. *Child Development*, 64, 1158-1174, 0009-3920

Ashmead, D.H., Clifton, R.K. & Perris, E.E. (1987). Perception of auditory localization in human infancy. *Developmental Psychology*, 23, 641-647, 0012-1649

Ashmead, D.H., LeRoy, D. & Odom, R.D. (1990). Perception of the relative distances of nearby sound sources. *Perception & Psychophysics*, 47, 326-331, 0031-5117

Bernstein, N.A. (1967). *The Coordination and Regulation of Movements*. Pergamon Press, 0444868135, Oxford.

Bertenthal, B.I., Campos, J.J. & Barrett, K.C. (1984). Self-produced locomotion: An organizer of emotional, cognitive and social development in infancy, In: *Continuities and Discontinuities in Development*, R.N. Emde & R.J. Harmon, (Eds), 175-209, Plenum, 0306415631, New York

Bobath, B. & Bobath, K. (1975). *Motor Development in the Different Types of Cerebral Palsy*, W. Heinemann, 0433033339, London

Bower, T.G.R. (1979). *Human Development*, W.H. Freeman, 0716700581, San Francisco

Bower, T.G.R. (2002). Space and objects, In: *Introduction to Infant Development*, A. Slater & M. Lewis, (Eds), 131-144, Oxford University Press, 0198506465, New York

Bower, T.G.R., Broughton, J.M. & Moore, M.K. (1970). Demonstration of intention in the reaching bahavior of neonate humans. *Nature*, 228, 679-681, 0028-0836

Butterworth, G. & Hopkins, B. (1988). Hand-mouth coordination in the newborn baby. *British Journal of Developmental Psychology*, 6, 303-314, 0261-510X

Caruso, D.A. (1993). Dimensions of quality in infants' exploratory behavior: Relationship to problem-solving activity. *Infant Behavior and Development*, 16, 441-454, 0163-6383

Clifton, R.K. (1992). The development of spatial hearing in human infants, In: *Developmental Psychoacoustics*, L.A. Werner & E.W. Rubel, (Eds), 135-157, American Psychological Association, 9781557981592, Washington, DC

Clifton, R.K., Morrongiello, B.A., Kuling, J.W. & Dowd, J.M (1981). Newborns' orientation towards sound: Possible implications for cortical development. *Child Development*, 52, 833-838, 0009-3920

Clifton, R.K., Perris, E. & Bullinger, A. (1991). Infants' perception of auditory space. *Developmental Psychology*, 27, 187-197, 0012-1649

DeCasper, A.J. & Fifer, W.P. (1980). Of human bonding: Newborns prefer their mothers' voices. *Science*, 208, 1174-1176, 0036-8075

Fraiberg, S. (1977). *Insights from the Blind*. Basic Books, 0465033180, New York

Gibson, E.J. (1988). Exploratory behavior in the development of perceiving, acting and acquiring of knowledge. *Annual Review of Psychology*, 39, 1-41, 0066-4308

Gibson, E.J. & Pick, A.D. (2000). *An Ecological Approach to Perceptual Learning and Development*, Oxford University Press, 0195165497, New York

Gibson, E.J., Riccio, G., Schmuckler, M.A., Stoffregen, T.A. Rosenberg, D. & Taormina, J. (1987). Detection of the traversability of surfaces by crawling and walking infants. *Journal of Experimental Psychology: Human Perception and Performances*, 13, 533-544, 0096-1523

Gibson, E.J. & Schmuckler, M.A. (1989). Going somewhere: An ecological and experimental approach to the development of mobility. *Ecological Psychology*, 1, 3-25, 1040-7413

Gibson, J.J. (1979/1986). *The Ecological Approach to Visual Perception*, Houghton Mifflin, 0898599598, Boston

Guski, R. (1990). Auditory localization: Effects of reflecting surfaces. *Perception*, 19, 819-830, 031-0066

Hazen, N., Lockman, J.J. & Pick, H.L. (1978). The development of children's representations of large-scale environments. *Child Development*, 49, 623-636, 0009-3920

Hill, P.A., Nelson, P.A. & Kirkeby, O. (2000). Resolution of front-back confusion in virtual acoustic imaging systems. *Journal of the Acoustical Society of America*, 108, 2901-2910, 0001-4966

Illingsworth, R.S. (1973). *Basic Developmental Screening: 0-2 years*, Blackwell Scientific, 9780632099306, Oxford

Jenison, R.L. (1997). On acoustic information for motion. *Ecological Psychology*, 9, 131-151, 1040-7413

Lee, D.N. (1990). Getting around with light or sound. In: *The Perception and Control of Self Motion*, R. Warren & A.H. Wertheimer, (Eds), 487-505, Erlbaum, 0805805176, Hillsdale, NJ

Lee, D.N. (1993). Body-environment coupling. In: *The Perceived Self: Ecological and Interpersonal Sources of Self-Knowledge*, U. Neisser, (Ed.), 43-67, Cambridge University Press, 9780521415098, Cambridge

Litovsky, R.Y. & Clifton, R.K. (1992). Use of sound pressure level in auditory distance perception by six-month-old infants and adults. *Journal of the Acoustical Society of America*, 92, 794-802, 0001-4966

Little, A.D., Mershon, D.H. & Cox, P.H. (1992). Spectral content as a cue to perceived auditory distance. *Perception*, 21, 405-416, 031-0066

Lockman, J.J. (1984). The development of detour ability during infancy. *Child Development*, 55, 482-491, 0009-3920

Lockman, J.J. (1990). Perceptuomotor coordination in infancy. In: *Developmental Psychology: Cognitive, Perceptuo-Motor, and Neuropsychological Perspectives*, C.-A. Hauert (Ed.), 85-111, Plenum Press, 0444884270, New York

Loomis, J.M., Klatzky, R.L., Golledge, R.G., Cicinelli, J.G., Pellegrino, J.W. & Fry, R.A. (1993). Nonvisual navigation by blind and sighted: Assessment of path integration ability. *Journal of Experimental Psychology: General*, 122, 73-91, 0096-3445

McKenzie, B.E. & Bigelow, E. (1986). Detour behaviour in young human infants. *British Journal of Developmental Psychology*, 4, 139-148, 0261-510X

Millar, S. (1994). *Understanding and Representing Space: Theory and Evidence from Studies with Blind and Sighted Children*. Clarendon Press, 0198521421, Oxford

Morrongiello, B.A. (1988). Infant's localization of sound along the horizontal axis: Estimates of minimum audible angles. *Developmental Psychology*, 24, 8-13, 0012-1649

Morrongiello, B.A., Fenwich, K.D., Hillier, L. & Chance, G. (1994). Sound localization in newborn human infants. *Developmental Psychobiology*, 27, 519-538, 1098-2302

Morrongiello, B.A. & Rocca, P.T. (1987). Infants' localization of sounds in the horizontal plane: Effects of auditory and visual cues. *Child Development*, 58, 918-927, 0009-3920

Muir, D. & Clifton, R.K. (1985). Infants' orientation to the location of sound sources. In: *The Measurement of Audition and Vision in the First Year of Postnatal Life: A Methodological*

Overview, G. Gottlieb & N.A. Krasnegor (Eds), 171- 194, Ablex, 0893911305, Norwood, NJ

Muir, D. & Field, J. (1979). Newborn infants orient to sound. *Child Development*, 50, 431-436, 0009-3920

Muir, D.W., Humphrey, D.E. & Humphrey, G.K. (1999). Pattern and space perception in young infants. In: *The Blackwell Reader in Developmental Psychology*, A. Slater & D. Muir (Eds), 116-142, Blackwell Science, 0631207198, Boston, MA

Muir D.M. & Nadel, J. (1998). Infant social perception. In: *Perceptual Development: Visual, Auditory, and Speech Perception in Infancy*, A. Slater (Ed.), 247-285). Psychology Press, 086377850X, Hove

Perris, E.E. & Clifton, R.K. (1988). Reaching in the dark toward sound as a measure of auditory localization in infants. *Infant Behavior and Development*, 11, 473-491, 0163-6383

Pick, H.L. (1990). Issues in the development of mobility. In: *Sensory- Motor Organizations and Development in Infancy and Early Childhood*, H. Bloch & B.I. Bertenthal (Eds), 419-439, Kluwer Academic Publishers, 0792308131, Dordrecht

Pick, H.L. (1993). Organization of spatial knowledge in children. In: *Spatial Representation: Problems in Philosophy and Psychology*, N. Eilan, R. McCharthy & B.Brewer (Eds), 31-42, Blackwell, 0631183558, Oxford

Pick, H.L. & Lockman, J.J. (1981). From frames of reference to spatial representations. In: *Spatial Representation and Behavior Across the Life Span: Theory and Application*, L.S. Liben, A.H. Patterson, & W. Newcombe (Eds), 39-61, Academic Press, 0124479804, Orlando, FL

Rieser, J.J., Doxsey, P.A., McCarrell, N.J. & Brooks, P.H. (1982). Wayfinding and toddlers' use of information from an aerial view of a maze. *Developmental Psychology*, 18, 714-720, 0012-1649

Rieser, J.J. & Heiman, M.L. (1982). Spatial self-reference system and shortest-route behavior in toddlers. *Child Development*, 53, 524-533, 0009-3920

Russell, M.K. & Turvey, M. (1999). Auditory perception of unimpeded passage. *Ecological Psychology*, 11, 175-188, 1040-7413

Schmuckler, M.A. (1993). Perception-action coupling in infancy. In: *The Development of Coordination in Infancy*, G.J.P. Savelsbergh (Ed.), 137-173, Elsevier Science Publishers, 0444893288, Amsterdam

Schmuckler, M.A. (1996). Development of visually guided locomotion: Barrier crossing in toddlers. *Ecological Psychology*, 8, 209-236, 1040-7413

Tamboer, J.W.I. (1985). *Mensbeelden achter Bewegingsbeelden*. De Vrieseborch, 9060762126, Haarlem.

Thelen, E., Kelso, J.A.S. & Fogel, A. (1987). Self-organizing systems and infant motor development. *Developmental Review*, 7, 39-65, 0273-2297

Thurlow, W.R., Mangels, J.W. & Runge, P.S. (1967). Head movements during sound localization. *Journal of the Acoustical Society of America*, 42, 489-493, 0001-4966

Ulrich, B.D., Thelen, E. & Niles, D. (1990). Perceptual determinations of action: Stair-climbing choices of infants and toddlers. In: *Advances in Motor Development Research*, J.E. Clark & J. Humphrey (Eds), Vol. 3, 1-15, AMS Publishers, 0120097249, New York

Van der Meer, A.L.H. (1997a). Keeping the arm in the limelight: Advanced visual control of arm movements in neonates. *European Journal of Paediatric Neurology*, 4, 103-108, 1532-2130

Van der Meer, A.L.H. (1997b). Visual guidance of passing under a barrier. *Early Development and Parenting*, 6, 147-157, 1057-3593

Van der Meer, A.L.H., Ramstad, M. & Van der Weel, F.R. (2008). Choosing the shortest way to mum: Auditory guided rotation in 6- to 9-month-old infants. *Infant Behavior and Development*, 31, 207-216, 0163-6383

Van der Meer, A.L.H. & Van der Weel, F.R. (1995). Move yourself, baby! Perceptuo-motor development from a continuous perspective. In: The Self in Infancy: Theory and Research, P. Rochat (Ed.), 257-275, Elsevier Science Publishers, 0444819258, Amsterdam.

Van der Meer, A.L.H., Van der Weel, F.R. & Lee, D.N. (1995). The functional significance of arm movements in neonates. *Science*, 267, 693-695, 0036-8075

Van der Meer, A.L.H., Van der Weel, F.R. & Lee, D.N. (1996). Lifting weights in neonates: Developing visual control of reaching. *Scandinavian Journal of Psychology*, 37, 424-436, 0036-5564

Von Hofsten, C. (1982). Eye-hand coordination in newborns. *Developmental Psychology*, 18, 450- 461, 0012-1649

Von Hofsten, C. (1993). Prospective control: A basic aspect of action development. *Human Development*, 36, 253- 270, 0018-716X

Wallach, H. (1940). The role of head movements and vestibular and visual cues in sound localization. *Journal of Experimental Psychology*, 27, 339-368, 0022-1015

Warren, D.H. (1978). Perception by the blind. In: *Handbook of Perception (Volume X): Perceptual Ecology*, E.C. Carterette & M.P. Frideman (Eds), 65-85, Academic Press, 0121619109, New York

Warren, W.H. (1984). Perceiving affordances: Visual guidance of stair climbing. *Journal of Experimental Psychology: Human Perception and Performance*, 10, 683-703, 0096-1523

Warren, W.H. & Whang, S. (1987). Visual guidance of walking through apertures: Body-scaled information for affordances. *Journal of Experimental Psychology: Human Perception and Performance*, 13, 371-383, 0096-1523

Wertheimer, M. (1961). Psychomotor coordination of auditory and visual space at birth. *Science*, 134, 1692, 0036-8075

Wightman, F.L. & Jenison, R.L. (1995). Auditory spatial layout. In: *Handbook of Perception and Cognition (Vol 5): Perception of Space and Motion*, W. Epstein & S. Rogers (Eds), 365-399, Academic, 0122405307, Boston

Wightman, F.L. & Kistler, D.J. (1999). Resolution of front-back ambiguity in spatial hearing by listener and source movement. *Journal of the Acoustical Society of America*, 105, 2841-2853, 0001-4966

Part 4

Spatial Sounds in Multimedia Systems and Teleconferencing

18

Camera Pointing with Coordinate-Free Localization and Tracking

Evan Ettinger[1] and Yoav Freund[2]
[1]Google Inc., Mountain View, CA
[2]Department of Computer Science and Engineering, UC San Diego, La Jolla, CA
USA

1. Introduction

In this work we consider the problem of using audio localization techniques to locate human speakers and point a pan-tilt-zoom (PTZ) camera in their direction. We study this problem in the context of the *The Automatic Cameraman* (TAC) - an interactive display installation at UC San Diego (Cheamanunkul et al., 2009). A frontal view of TAC is given in Figure 1. TAC is a system which gives the user a hands-free interactive experience through computer vision and audio signal processing technologies. To start the interaction a user must first approach the display and speak. The system then localizes where the speaker is via a microphone array, and directs the camera to point there. In this work we describe exactly this initial part of the system, namely, how to point the camera at sound sources accurately and reliably.
The main novelty of our method is that it does not rely on a-priori knowledge of the position of the microphones and the camera and the orientation of the PTZ camera. Traditional methods for audio localization require specifying these positions and orientations within a coordinate system. We call our method *coordinate-free* as it does not require a-priori specified coordinate system nor does it attempt to construct one. Instead, in this work we take a statistical approach based on machine learning. Our algorithm analyzes the relationships between different measurements and deduces the mapping from microphone delays to pan/tilt angles required to point the camera towards the speaker. The ability to calibrate the system after deployment allows placing the microphones far from each other and with no pre-specified geometry. This, in turn, allows the user to optimize the locations of the microphone according to the acoustics of the particular location.
The application we consider in this work is that of camera pointing, but it is worth noting that our method is not constrained to just this problem alone. Direction of arrival (DOA) estimation is used widely throughout robotics, general sonar applications, beam-forming, and many other domains. Our method applies when knowledge of a precise coordinate system isn't needed, such as pointing a camera at an object, pointing a robot at an object, or simply estimating direction or arrivals relative to a reference point.
The key observation behind audio localization techniques is that spatially separated microphones observe a time-delay between the arrival of a sound source. This is depicted in Figure 2. Estimating these time-delays accurately is a fundamental step in many popular

localization techniques. In the next section, we briefly discuss how to estimate these time-delays which will be a fundamental underpinning of our coordinate-free methodology that follows.

We first describe our technique based on statistical regression to map time-delay information from a frame of audio to a pan-tilt directive for our PTZ camera. This gives a method for estimating from a single frame of audio what direction the sound source is coming from. However, this method analyzes each time frame independently and does not leverage any temporal information, such as the ways speakers move in space.

To address this temporal concern, we introduce a coordinate-free tracking methodology for estimating these time-delays accurately based on a particle filtering approach. We show that a naive implementation of a particle filter does not track these time-delays accurately. Instead, we propose two methods to improve the particle filter for this particular problem. The first is a manifold learning step that learns the low-dimensional structure on which these time-delays live. The second is a new particle filtering framework based on new advances in the online learning community that has several advantages over a traditional approach. We outline the details of these methods and discuss them in more depth in what follows.

The rest of the chapter is organized as follows. In Section 2 we describe the fundamental concepts of the TDOA and the PHAT transform. In Section 3 we discuss traditional coordinate-based methods for localizing a sound source from time-delay estimates. In Section 4 we discuss our coordinate free approach that attempts to learn a regressor that maps time-delay information directly into pan-tilt directives for the PTZ camera. We show that our method lends to an accurate camera pointing method with experiments in Section 5. The system used in these experiments does not take into account noise in the TDOA estimates or information about the way humans move. In Section 6 we present a coordinate-free tracking method which takes this information into account. In Section 7 we describe experiments that demonstrate the improvement in performance that result from incorporating tracking into our system. We conclude the chapter with some final remarks.

2. Time-delay estimation

The basis of sound source localization is that spatially separated pairs of microphones experience a time-delay of arrival from a fixed sound. An illustration of this physical phenomenon in a 2-d setting is shown in Figure 2.

In this work we do not assume any knowledge of microphone or camera positions, however, for the expository discussion in this section it is useful to assume they are known and fixed. Let $m_i \in \mathbf{R}^3$ be the three dimensional Cartesian coordinates for microphone i. For a sound source located at position s and assuming a spherical propagation model, the direct path time delay between microphone i and j can be calculated as

$$\Delta_{ij} = \frac{\|m_i - s\|_2 - \|m_j - s\|_2}{c} \tag{1}$$

where c is the speed of sound in the medium. Δ_{ij} is often called the *time delay of arrival* (TDOA) between microphone i and j. It is worth noting that if f is the sampling rate being used, then the largest the TDOA can be in terms of audio samples is $M = \|m_i - m_j\|_2 f/c$. In other words, Δ_{ij} is always in the range $[-M, M]$ and in practice can only be estimated to the nearest sample. This observation directly reveals the fact that close together microphones cannot have as wide a range of TDOAs as microphones that are spaced further apart. Placing microphones further

Fig. 1. Frontal view of TAC display unit. PTZ camera and four of the seven total microphone are visible.

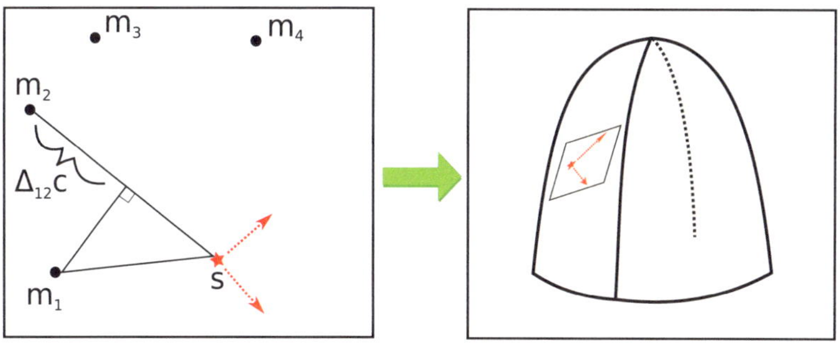

Fig. 2. **Left:** A 2-dimensional world with 4 microphones. Time-delay Δ_{12} is shown between microphones m_1 and m_2. The sound source (red star) is shown with 2 degrees of freedom for movement (red arrows). **Right:** Suppose we restrict our view to the TDOA values Δ_{12}, Δ_{23} and Δ_{34}. The right hand side figure depicts the 2-dimensional manifold created by mapping locations in the 2-dimensional world to these three TDOA variables. The manifold is not affine because of the non-linearities of the geometry. However it is locally affine.Thus the red movement arrows of the figure on the left map to the red arrows of the figure on the right.

apart allows for more variability in the feasible TDOAs, and hence, results in a better ability to discriminate between audio source locations in space.

Given k microphones there are $\binom{k}{2}$ unique pairs of microphones for which Δ_{ij} can be estimated. We let $\vec{\Delta} = (\Delta_{ij})_{i<j} \in \mathbf{R}^{\binom{k}{2}}$ be the vector that contains each of these unique TDOAs for a given audio source location. We will often call $\vec{\Delta}$ the *TDOA vector*.

When given a fixed Δ_{ij} for a pair of microphones, we can deduce from Equation 1 that the set of feasible s positions that could have resulted in the observed Δ_{ij} form one sheet of a 3-d hyperboloid in space (for a 2-d world representation see Figure 3). It follows that for a fixed

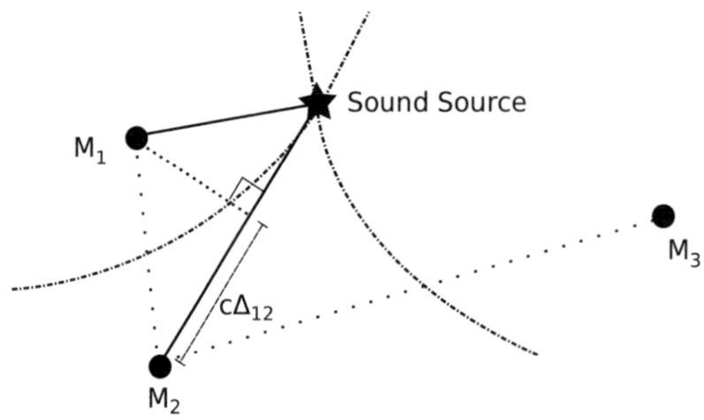

Fig. 3. A 2-d world where 3 microphones are necessary to uniquely determine a sound source's location via multilateration. If given Δ_{12}, Δ_{23} and knowledge of the microphone positions, then one can solve for the intersection of the corresponding hyperbolas for s.

$\vec{\Delta}$, the possible audio source locations that could have generated such a TDOA vector can be determined through finding the intersection among all such hyperboloids. This procedure is known as *multilateration*.

However, in practice we can only estimate each Δ_{ij} from the underlying audio signals. As a result, the estimation procedure faces multiple challenges that easily lead to inaccuracies. First and foremost, sound easily bounces off of many physical materials causing multi-path reflections and reverberations. Secondly, the audio signal is only captured at a finite precision with respect to time since the signal must be digitized with a finite sampling rate. This means we can only estimate TDOAs with a finite precision that depends on the audio sampling rate. These challenges often results in estimation errors in Δ_{ij} and so it is not surprising that in practice the intersection of all the corresponding hyperboloids is empty!

One of the most popular time delay estimation (TDE) techniques, and the method used in this work, is a generalized cross-correlation (GCC) technique that utilizes the phase transform (PHAT), first discussed in the audio localization literature by Knapp and Carter and then further analyzed by many others (Knapp & Carter, 1976; Omologo & Svaizer, 1994; 1996). PHAT is very robust to noise and reverberations compared to other correlation based TDE techniques (J. DiBiase, 2001; Svaizer et al., 1997). Let $X_k(\omega)$ be the Fourier transform of microphone k. The GCC between microphone l and m is

$$R_{lm}(\tau) = \frac{1}{2\pi} \int_{-\infty}^{\infty} \Psi(\omega) X_l(\omega) X_m^*(\omega) e^{j\omega\tau} d\omega \tag{2}$$

where $\Psi(\omega)$ is a weighting function for the GCC and * denotes complex conjugation. The PHAT weighting of the GCC is of the form

$$\Psi(\omega) = \frac{1}{|X_l(\omega) X_m^*(\omega)|} \tag{3}$$

The PHAT weighting has a whitening effect by removing amplitude information in the signals. Compared to standard cross-correlation, PHAT puts all the emphasis on aligning the phase component of the transformed audio signals and none on the amplitudes. Empirically, it has been observed that the result of using the PHAT weighting is often a large spike in the GCC at the true TDOA. Hence the PHAT method for TDOA estimation is to let

$$\Delta_{ij} = \arg\max_{s} R_{ij}(s) \tag{4}$$

The PHAT correlations are typically very pronounced at the estimated TDOA with a small number of significant secondary peaks. It has been observed often that if the true TDOA is not at the largest peak it is often at one of these large secondary peaks (J. DiBiase, 2001). This property has been exploited by many methods, and will be exploited by the particle filtering method that we describe later on.

3. Related work

Sound localization techniques via microphone arrays can be divided into two major paradigms: TDOA two step localization and steered response power (SRP) based. The first technique involves first estimating for a frame of audio the TDOAs between all pairs of microphones and then solving the subsequent geometric multilateration problem. The most popular is a least squares approach to find the 3-d location that is close to all the resulting hyperboloids. One such approach is to simplify the nonlinear least squares problem by linearizing it through either a Taylor expansion (Foy, 1976) or by introducing an extra variable as a function of the source location (Chan & Ho, 1994; Friedlander, 1987; Gillette & Silverman, 2008; Huang et al., 2001; Smith & Abel, 1987; Stoica & Li, 2006). This leads to a closed-form solution to the problem since it becomes a linear least-squares problem, but the resulting variance in the source location estimator is large (Chan & Ho, 1994; Huang et al., 2001). There are many other variations on this approach that could fall in this category as well (Brandstein et al., 1995; Gustafsson & Gunnarsson, 2003; Silverman et al., 2005).

The second category for source localization techniques are all based on maximizing the steered response power (SRP) of a beamformer (J. DiBiase, 2001). For example, a simple instance in this class is to maximize the energy of a delay-and-sum beamformer over a range of steering directions. That is, for each source location x, one first calculates the corresponding TDOA vector, $\Delta(x)$, derived from the array geometry. By delaying the frames of audio by these TDOAs and summing all the signals together, one gets a reconstruction of the original signal. This reconstruction has the most energy when $\Delta(x)$ is correct. Conversely, $\Delta(x)$ can be estimated by maximizing the energy of the reconstructed signal. Probably the most popular of SRP based beamformers is the so called SRP-PHAT beamformer (Do et al., 2007; J. DiBiase, 2001). Here, instead of maximizing the energy of the delay-and-sum reconstruction, one calculates the PHAT correlation, $R_{ij}(\tau)$, for all pairs of microphones and then solves the optimization $\arg\max_x \sum_{i<j} R_{ij}(\Delta_{ij}(x))$.

Both the two step and beamforming based methods require knowledge of a coordinate system wherein microphone positions are known. For small microphone arrays a coordinate system can easily be found by simply measuring the distances between microphones by hand as in (Wang & Chu, 1997). If we want to be able to localize sounds in a large room accurately, then a large microphone array that spreads throughout the room is beneficial. However, measuring accurately by hand the relative distances now becomes much more difficult and positional errors on the order of 1-5cm can seriously degrade beamforming techniques (Sachar et al., 2005).

Since doing such measurements is often too difficult, especially for arrays with many elements, many techniques have been developed to automatically calibrate the positions of the microphone elements (Birchfield & Subramanya, 2005; Hörster et al., 2005; McCowan et al., 2008; Raykar & Duraiswami, 2004; Sachar et al., 2005). These techniques are based on using a carefully designed device that emits a special sound. Delay measurements are made at the array and with the known geometry of the device one can solve for the microphone positions. Typically distances from the device to the microphones, or inter-microphone distances are estimated. For example, if pairwise distances between microphones can be estimated, then multidimensional scaling (MDS) can be used to find the location of the sound source (Birchfield & Subramanya, 2005; Hörster et al., 2005; McCowan et al., 2008; Raykar & Duraiswami, 2004; Sachar et al., 2005).

Note that if we were to use a coordinate based system to estimate the location of the speaker we would need an additional step to map the estimated location to the direction directive for the PTZ camera. To compute this mapping we would need to know the location and orientation of the camera relative to the microphones. Instead we developed a *coordinate free* method which maps the estimated delays directly to pan and tilt commands for the camera. In this way we avoid the need to measure the relative locations of the microphones and the camera.

In order to learn the mapping from delays to pan/tilt (PT) we collect observations consisting of a set of delays between microphones for a fixed source location and the associated PT to center such a source. With this database of samples, we estimate via regression analysis a model for the system. This model allows us to estimate for a fixed $\vec{\Delta}$ what the corresponding PT directive for our camera should be. We describe the methodology and experiments for this method in the next two sections.

4. Coordinate-free localization

In this section we describe the regression models we use for estimating the mapping from $\vec{\Delta}$ to PT. For what follows assume that a training set of size m is given with observations of the form $y_i = (\theta_i, \psi_i)$, for pan and tilt respectively. These observations are paired with an estimated TDOA vector derived from the N microphones, namely $x_i = \vec{\Delta}_i$ with $p = \binom{N}{2}$ coordinates. We organize the training set into matrices $Y \in \mathbf{R}^{N\times 2}$ and $X \in \mathbf{R}^{N\times p}$ where each observation is a row vector. In what follows, we briefly remind the reader of least squares linear regression and a tree based regressor based on principal direction trees (PD-Trees) (Verma et al., 2009).

Least squares linear regression

For each column of Y, denoted Y_i, we fit a separate linear regression model. The linear regression model has the form

$$f(X) = \beta_0 + \sum_{j=1}^{p} X_j \beta_j$$

where X_j is the jth column of X and β is the vector containing the coefficients in the linear model. The least squares (LS) solution to linear regression chooses the model that minimizes the residual sum of squares (RSS)

$$RSS(\beta) = \sum_{i=1}^{N} (y_i - f(x_i))^2$$

When X is full rank the LS solution can be written in closed form as $\beta = (X^T X)^{-1} X^T Y_i$. It is known that if the true model of data generation is linear, then the LS estimator is the minimum variance unbiased estimator of β.

PD-tree

In the experiments described in the next section we will also explore the use of a constant depth PD-Tree with regressors learned in each leaf node. A PD-Tree is a binary partitioning tree that at each node projects the data present in that node onto its principal direction and splits the data into two children nodes based on the median value. We grow a PD-Tree to depth 2 and fit linear least squares regressors in each leaf node. This will act as a piece-wise regression model.

Principal direction trees are chosen since they are known to adapt quickly to low dimensional structure present in data (Verma et al., 2009). We know that our TDOA data, despite being in rather high dimensions has a low dimensional structure since it has underpinnings to a physical location from the generating sound source. Sound sources only have 3 spatial dimensions in which they can vary so as a consequence our TDOAs also have exactly this many degrees of freedom. Although the underlying structure on which these TDOAs is not linear (intersection of hyperboloids), but is locally linear. As we shall see in the next section, a PD-tree of depth three yields a good approximation for most of the area covered by the automatic cameraman.

5. Experiments: Localization bias

In this section we present two experiments. The first one generates a training set and test set with a simple device that helps us collect training examples. The second experiment aims to learn examples over time from people who interact with our display over time. We describe each in further detail in what follows.

5.1 Experiment: Grid dataset

The device used to collect all the data in the experiments to come is shown in Figure 4b. It consists of a simple radio and a green LED attached to a 9V battery with a switch and dimmer all in a plastic encasing. We will call this the *calibration device* from here on. The radio component of the calibration device can be tuned to a nonexistent station that emits noise that is very close to white. This random noise typically has the most consistent TDOA vector estimates using the PHAT technique. A simple color thresholding detector was written to find the LED in the camera's field of view using Max/MSP and Jitter (*Max/MSP website*, n.d.). The result is a real-time control of the PTZ-camera to keep the LED centered in the field of view, and a constant white noise to calculate TDOAs for. The calibration device is used to collect samples of TDOA vectors in unison with where the camera is pointing to center the green LED in its field of view. The camera can be queried as to what the pan and tilt it is currently whenever a TDOA vector is collected. These two pieces of information are recorded together as a complete observation instance.

The result of the training set collection is a dataset of close to 28k observations. We noticed that when a estimate for Δ_{ij} was incorrect, it typically had a very large deviation from what was often consistent. To remove such noisy observations, we performed some simple outlier removal by thresholding the magnitudes of the $\vec{\Delta}$ projections onto the bottom global PCA eigenvectors (orthogonal space) leave approximately 20k observations remaining as our training set. We then did a PCA analysis of just the $\vec{\Delta}$ parts of this training set. Figure 4 shows the percentage of variance explained by the addition of each eigenvector. It's clear that the top two eigenvectors dominate most of the variance explained, and that the 3rd eigenvector seems to have a significant advantage over the remaining ones. The total percent variation captured by the top 3 eigenvectors is nearly 90%. This follows from the fact that there are 3 spatial

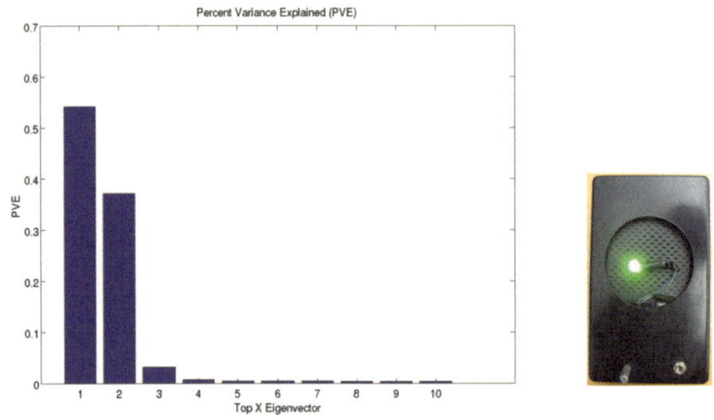

(a) (b)

Fig. 4. Left: Percentage of variance explained by top X eigenvector. The top 3 eigenvectors dominate and the rest are noise. Right: Calibration device used to collect training and grid dataset.

degrees of freedom that were examined during the training data collection period. Moreover, two of these spatial directions had much more spatial variance then the third, ceiling-to-floor, spatial direction. The room is simply much larger in width and breadth than the variance in observation heights, which matched typical heights that human speakers could appear at.

From this training set with outliers removed we have nearly 20k observations with which we learn a simple linear least-squares regression (LS) model and a PD-Tree model of depth 2. We would like to analyze how the bias-variance trade-off of these simple models behaves as function of physical position of the sound source in the lobby. In other words, in what areas do these simple models perform well, and where does the inherent non-linearity of the problem cause large bias?

With these questions in mind we collect a test set of data in a similar fashion to the training set. We place the calibration device at a fixed height (approximately 1m from the floor) and roll it along straight lines using a rolling chair. We repeat this process for each of the 13 lines in the grid depicted in Figure 5b. This results in a variety of observations that cover a representative set of the spatial variability in the room relevant for human speakers. Moreover, using white noise as our sound source will simulate the behavior of our model under conditions where TDE is highly optimized. This gives us insight into isolating the effects of the model assumptions.

Figure 5a depicts the embedding of the TDOA vector components of the entire grid test set onto the top 2 eigenvectors from the PCA learned from the training set. The zoomed in portion depicts lines 9-13 in red and lines 1-6 in blue in the same orientation as the diagram in Figure 5b. The curved nature of each line can be observed from such plots. Even though the spatial location of the sound source is varying along a straight line in space, the corresponding location in the TDOA vector space corresponds to slightly curved trajectories. It is clear that a linear model for spatial location is not going to fully capture all the variation, but nevertheless the grid structure is still very recognizable in even just the top 2 eigenvectors indicating that a linear model is a good approximation in these regions.

Figure 5c compares the predictions from the simple linear LS model to the pan and tilt recorded from the light detector. The dots in black are the predicted pan (or tilt) from the model for each TDOA vector observation. The green line depicts the pan (or tilt) from the light detector. Finally the red line depicts an exponential moving average (EMA) of the model predictions over time. In other words, the EMA prediction, p_t, at time t is calculated with

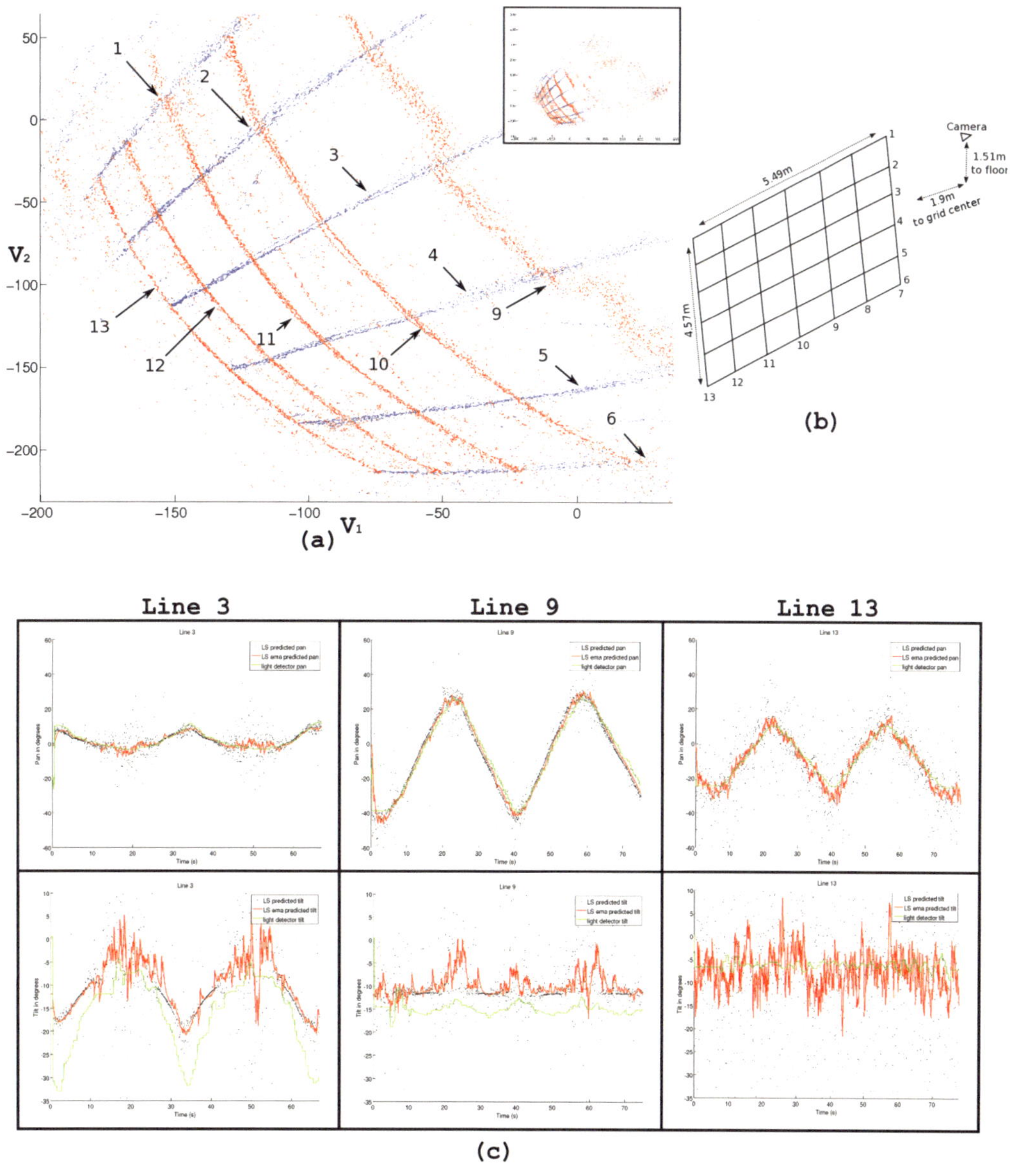

Fig. 5. (a) Embedding of the TDOAs collected from the grid onto top 2 eigenvectors. The entire embedding is shown small in the upper right corner and a zoomed in portion of the same embedding is shown larger. (b) To the right is a diagram of the equispaced grid over which data was collected. (c) Below are 3 selected lines and the LS predicted value for each TDOA collected. Also depicted in red is an exponential moving average of the predictions ($\alpha = 0.10$), and in green where the camera was pointing to center the LED.

Model	Grid Line Number							
	1	3	5	7	9	11	13	avg
LS-pan	4.31	2.77	2.22	5.99	3.56	3.20	3.96	**3.87**
PD-pan	4.22	3.14	3.05	4.14	3.05	2.45	3.88	**3.47**
LS-tilt	5.15	7.57	7.50	3.33	5.63	3.90	4.48	**5.75**
PD-tilt	4.70	4.72	4.65	3.26	4.82	2.95	6.55	**4.55**

Table 1. RMSE (in degrees) of different regression models for each grid line.

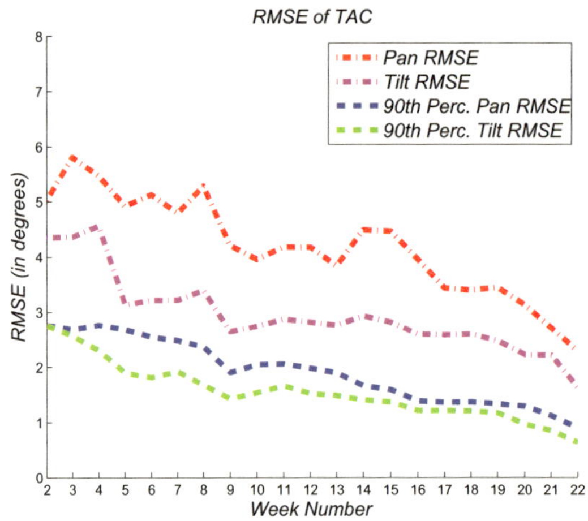

Fig. 6. RMSE for pan and tilt of a PDTree trained each week with new data acquired by TAC.

update $p_t = (1-\alpha)p_{t-1} + \alpha f(\Delta_t)$, where $f(\Delta_t)$ is the prediction of the raw observation at time t. We chose $\alpha = 0.1$. The EMA line should give us a sense of what the true model predictions are by smoothing out the observation noise. In doing so, we can compare the light detector observations to the EMA line and get a sense for the bias in our model.

Table 1 gives the root-mean squared error (RMSE) between the EMA of the model predictions and the observations from the light detector for each of the regression models. The PD-Tree method outperforms a simple linear model. Moreover, the overall averages are very similar to results reported by traditional coordinate based methods, meaning that coordinate-free methods need not sacrifice accuracy (Badali et al., 2009).

5.2 Lifelong learning

We can easily acquire a training set without the aide of a device with help from a face detector. Training examples can be collected whenever a user speaks while their face is centered in the field of view, creating a stable measurement of the form $(\vec{\Delta}, \theta, \phi)$. Many such examples can be collected over time by having the PTZ-camera continually centering the user's face and the user continuing to speak. This is in fact what we do in TAC. Whenever a user is interacting with TAC a log is recorded that records these stable training points. We retrain a PDTree with linear models in the leaves at the end of each week on the entire training set collected up to that point.

We took all the observations TAC has seen over a period of approximately 6 months ($\sim$3000 observations), and split this randomly into a 70/30 training and test set. We then examined how TAC can improve its localization accuracy by retraining a regressor for pan and tilt each week on the data from the training set seen to that point. We averaged root-mean squared

error (RMSE) calculations over 20 such random training/test splits. Figure 6 shows the improvement of this regressor in terms of RMSE. Also shown is the RMSE when the top 10% of squared-residuals are removed from the RMSE calculation.
The improvement is near-linear from week to week. Moreover, many of the errors are near or below one degree in both pan and tilt. This is promising since the locations in the test set are representative of where most users frequent when interacting with TAC. This means we are very accurate (< 1 degree error) in these locations.

6. Coordinate-free TDOA tracking

One deficiency of the methods presented thus far is that the are frame based methods that do not leverage temporal information. For instance, we know that sound sources do not move quickly or disappear and reappear in different locations instantaneously. Therefore, some smoothness assumptions about the variability of TDOAs over time would be beneficial to a general methodology that attempts to localize sound sources using information across many frames.
In TAC we have 7 microphones, causing a 21-D TDOA vector. In what follows we propose a particle filtering methodology for tracking the 21-D TDOA vector over sequential frames of audio. The methodology has three important innovations above the naive median filtering strategy outlined above. The first is that TDOAs from one frame to the next should not vary too much. This assumption should be explicitly integrated into any model of tracking. A second observation is that TDOAs can only occur from a feasible region of the 21-D space in which TDOA vectors lie. We propose a PD-Tree based model of this feasible region. It is well known that particle filters tend to break-down when the object being tracked have many dimensions to their state space. By modeling the feasible region, we alleviate this well known deficiency of particle filters by making the effective dimensionality of the TDOA space much lower.
The last contribution is a new particle weighting and resampling scheme inspired by results in online learning. The resampling scheme is such that we can leverage the PD-Tree model in a novel fashion that allows for averaging over different bandwidths in the tree. We will show in the experiments that this averaging scheme can improve over baseline schemes especially when a sound source enters regions that are not modeled well by a single global linear model. In addition, it is known that the weighting scheme used is much more robust to model mis-specification than traditional particle filters.

6.1 Particle filters

Particle filtering is an approximation technique used to solve the Bayesian filtering problem for state space tracking (Arulampalam et al., 2002). More specifically, assume we have observations y_t and a state space x_t. Often the state space will consist of the position of the object of interest, and sometimes higher moments like velocity or acceleration. The goal of the particle filter is to keep a discrete set of particles that well-approximates the posterior density of the current state given the past observations $p(x_t|y_0,\ldots,y_t)$. In the TDOA tracking problem our observations y_t will be the PHAT correlations for a given frame of audio and the state space x_t will be composed of each of the $D = \binom{k}{2}$ time delays.
The *bootstrap* is one of the most popular particle filtering algorithms (Gordon et al., 1993). Here, a weighting over m particles is chosen to approximate the posterior density. Let $w_t^{(i)}$

be the weight associated with particle i at time t. Then, a single iteration of the algorithm proceeds as follows:

1. **Sample:** draw m particles $x_{t-1}^{(i)}$ from the existing set of particles according to their weights $w_{t-1}^{(i)}$.
2. **Propagate:** Let the particles propagate according to the transition function, $x_t^{(i)} = g(x_{t-1}^{(i)}) + u_t$.
3. **Weight update:** Update weights according to $w_t^{(i)} = w_{t-1}^{(i)} p(y_t|x_t^{(i)})$ and normalize so they sum to one.

The result is a set of particles approximately distributed as the posterior density $p(x_t|y_{1:t})$. This sample set allows for computation of any quantity as a function of the posterior. For example, often we would like to estimate the mean of the posterior distribution which will be our prediction of the current state. This estimate is given by

$$\hat{x}_t = \int x_t p(x_t|y_{1:t}) dx_t \approx \sum_{i=1}^{m} w_t^{(i)} x_t^{(i)} \tag{5}$$

The weights are chosen to approximate the *relative* posterior density for their respective particles.

This popular variant of particle filters has been shown to perform well in the coordinate-based tracking literature (Lehmann & Johansson, 2007). The key decisions for optimizing such a particle filtering algorithm are:

1. **Likelihood:** The choice of likelihood function, $p(y_t|x_t)$, is critical since this will govern how weights are calculated.
2. **Propagation function:** The propagation function $g(\cdot)$ is also essential and needs to be chosen accurately. In coordinate based methods g is chosen to be linear and u_k often to be Gaussian.
3. **Number of particles:** The total number of particles m. The larger m is the more computational load the system must undertake. Optimizing m is of paramount importance for real-time implementations.

More so than the other choices, the likelihood function is by far the most difficult. The true likelihood function for how PHAT observations are generated from a given sound source location seems very difficult to model. Nevertheless, it has been shown that some simple choices for the likelihood function can lead to good tracking performance (Lehmann & Johansson, 2007). In making a choice for the likelihood function, first notice that we must have support over the entire observation space. If we don't meet this requirement, particles that occur with likelihood zero will get weight zero and die immediately. This is not the behavior we would like since particles that were performing well in the past may then suddenly die. Instead, we should want a more graceful way for particles to tend towards zero weight. As a result, often a mixture of a uniform prior over the entire observation space is mixed with the likelihood function to avoid this behavior.

One deficiency of the particle filter is that accurate tracking becomes very difficult when the state space becomes larger than a few positional locations (e.g. 2-D or 3-D locations). In TDOA tracking, the state spaces can potentially be much larger. For example, the seven microphones

Algorithm 1 Generic bootstrap based particle filtering audio tracking algorithm.

Initial Assumptions: At time t-1, we have the set of particles $x_{t-1}^{(i)}$ and weights $w_{t-1}^{(i)}$, $i \in \{1,\ldots,m\}$, being a discrete representation of the posterior $p(x_{t-1}|y_{1:t-1})$.

1: **Dynamics:** Propagate the particles through the transition equation $x_t^{(i)} = g(x_{t-1}^{(i)}, u_t)$.

2: **Weight Update:** Assign each particle a likelihood weight according to $w_t^{(i)} = p(y_t|x_t^{(i)})$. Then, normalize weights so that they sum to 1.

3: **Resample:** Resample m new particles from $\{x_t^{(i)}\}_{i=1}^m$ according to the weight distribution $\{w_t^{(i)}\}_{i=1}^m$. Let these be the new set of particles $\{x_t^{(i)}\}_{i=1}^m$ and assign uniform weight to each.

in TAC give rise to a 21-D TDOA vector space, but with arrays with more microphones the space can be even larger. The difficulty arises in the randomness need in u_k to generate enough variety of particles so that a few are close to a good state representation. One obvious remedy would be to increase the number of particles, but this causes the real-time feasibility of the algorithm to quickly diminish.

To alleviate this problem, when a coordinate system is known, then the state space can be represented as the 3-d position of the audio source. This makes the algorithm feasible with a small number of particles (typically < 100). In our coordinate-free approach, we take a similar dimensionality reduction technique by directly modeling the low dimensional structure on which the TDOAs lie via a PD-Tree. However, before introducing our algorithm we first discuss related work in coordinate based TDOA tracking.

6.2 Related work

Particle filtering methods dominate the audio source tracking literature (Lehmann & Johansson, 2007; Li & Ser, 2010; Pertilä et al., 2008; Talantzis et al., 2009). The seminal work of Ward et. al is the first to popularize the use of particle filtering methods for audio tracking and is still widely regarded as state-of-the art (Ward et al., 2003). Further experiments and slight improvements on this method were presented in Lehmann & Johansson (2007). This method is the focus of what follows realizing that the others mentioned above are all derived from this seminal work.

We reproduce the bootstrap particle filtering method for audio source tracking in Algorithm 1. The predicted state at each step of this algorithm is the weighted mean $\hat{x}_t = \sum_{i=1}^m w_t^{(i)} x_t^{(i)}$. Here the state space is chosen to be 3-d Cartesian coordinates $x_t^{(i)} = [p_x p_y p_z]$ and the dynamics g is chosen to be the identity with spherical Gaussian noise for u_k. The size of the Gaussian noise u_k is a tunable parameter that must coincide with the assumptions about how quickly the objects being tracked can move.

The major choice in the algorithm is how to perform the weight update step, in particular, what choice should be made for the likelihood function $p(y_t|x_t^{(i)})$. The choices for this function can arise either from GCC based methods or steered beamforming based methods. For example, a simple steered beamforming based approach is as follows. For the weight update in Algorithm 1, let $p(y_t|x_t^{(i)}) = F(y_t, \Delta(x_t^{(i)})$ where F calculates the steered response power of the current frame of audio steered towards $x_t^{(i)}$.

More computationally efficient methods for representing the likelihood function were presented in Ward et al. (2003) based on PHAT transforms. The idea for the likelihood here is to define a function that combines how close the current particle is to the largest peaks in the PHAT correlation from each pair of microphones $p \in \{1, \ldots, D\}$. This will be the method we use in the work presented in this chapter. In particular we use the following.

First, to identify the peaks in a given pair's PHAT function we take a simple z-scoring method. Let $[A]_+ = max(0, A)$. Then, for each PHAT correlation R_p let it undergo a z-scoring transform as follows (note from here on we drop the subscript t for ease of notation):

$$Z_p(\tau) = \left[\frac{R_p(\tau) - \mu_p}{\sigma_p} - C\right]_+ \tag{6}$$

where μ_p, σ_p are the mean and standard deviation of R_p over a fixed bounded range of τ, and C is a constant requiring that peaks be at least C standard deviations above the mean. This performs well to find a small, fixed number, of peaks K_p in each R_p.

We now define $p(y|x^{(i)})$ in terms of these peaks:

$$p(y|x^{(i)}) \propto p_0 + \sum_{p=1}^{D} \sum_{l=1}^{K_p} Z_p(\tau_l) \mathbb{N}(\tau_l; \Delta(x^{(i)})_p, \sigma_z^2) \tag{7}$$

where $\Delta(x^{(i)})_p$ the TDOA associated with pair p derived from the 3-D location $x^{(i)}$, $\mathbb{N}(x; \mu, \sigma^2)$ is the density under a normal distribution evaluated at x with mean μ and variance σ^2, and Z_p has K_p non-zero entries each of which are at τ_l. The parameter p_0 is the background likelihood that determines how much likelihood is given to any TDOA regardless of the observation. This parameter is essential for this kind of particle filter so that the likelihood function never evaluates to 0. Otherwise a particle's weight can never abruptly vanish. The variance parameter σ_z^2 controls how much weighting is given relative to how far each state is from the peaks in the corresponding PHAT series. So, a particle will be given high likelihood if the particle's derived TDOA matches well with the largest peaks in the observed PHAT series. Conversely, if the derived TDOA is far from any of the observed peaks it will be given a very low likelihood.

A nice property of this choice of likelihood is that it does not rely solely on the maximum of each PHAT series being accurate (a similar advantage was observed between steered beamformers over the 2-step localization procedure discussed in previous sections). Since often the peaks in the PHAT localization are corrupted due to reverberations or multipath reflections, relying heavily on only these maximum peaks is not robust. The likelihood defined in Equation 7 neither relies too heavily on the accuracy of a single pair of microphones, nor on the largest peak in each pair's PHAT series. Secondary peaks can contribute substantially to the likelihood as well. As we will see, integrating a particle filtering based tracking method into the localizer will lead to a much stabler and robust localization method.

6.3 Normal hedge based particle filter

In this section we introduce the Normal Hedge based particle filter. This particle filter, although very similar to the traditional particle filter introduced above, will have several advantages. First, the resampling scheme will not require particles to be resampled every iteration. In fact, particles will remain "alive" for as long as they perform well. Secondly, the requirements of the algorithm will allow for much more flexibility in specifying a likelihood function. Recall that in Equation 7 we had to define a parameter for the background likelihood p_0, otherwise particles could quickly go to zero weight and die. No such requirement is

needed by the particle filter presented here, moreover, the guarantee that will be given is relative to the defined likelihood function. This means that the resulting Normal Hedge particle filtering algorithm will perform well as long as the likelihood function encourages good tracking performance (i.e. high likelihood scores indicate that the particle matches the observation well).
Before introducing the full Normal Hedge particle filter we first discuss the Normal Hedge online algorithm for predicting from a group of experts' advice, initially presented in Chaudhuri et al. (2009).

Normal Hedge

The Normal Hedge algorithm is a parameter-free online algorithm for hedging over the predictions from a group of N experts (Chaudhuri et al., 2009). One of the barriers to practical implementations of previous online learning algorithms was that they all contained a learning parameter that was very important to tune correctly for good performance. Normal Hedge has no such parameter, yet still has a very strong performance guarantee like that of the previous online algorithms.
The setup for the algorithm is as follows. At each iteration t expert i makes a prediction that has an associated loss $\ell_t^{(i)} \in [0,1]$. The notion of loss in this setting is very general, but in most cases is typically derived as a function of the expert's prediction and the actual observation (e.g. the difference between the prediction and the observation normalized to the $[0,1]$ range). The algorithm maintains a discrete probability distribution over the experts $w_t^{(i)}$. After observing the losses, the learner itself incurs a loss according to the expected loss under this discrete distribution,

$$\ell_t^A = \sum_{i=1}^{N} w_t^{(i)} \ell_t^{(i)} \tag{8}$$

The notion of *regret* is the essential quantity of interest in online learning. The algorithm's *instantaneous regret* is defined as $r_t^{(i)} = \ell_t^A - \ell_t^{(i)}$ and the *cumulative regret* up to time t is defined as

$$R_t^{(i)} = \sum_{\tau=1}^{t} r_\tau^{(i)} \tag{9}$$

Intuitively the cumulative regret measures how well the algorithm is doing relative to a single action chosen to predict at all previous iterations up to t. The goal for an online algorithm is to minimize the cumulative regret of the algorithm relative to any given expert (in particular, the best expert in hindsight).
The Normal Hedge algorithm is given in Algorithm 2. It requires no parameters and the computational needs are also simple. The algorithm must maintain the weights and regrets over each of the N experts and also a line search is needed to solve for c_t in the weight update stage.
The guarantee proved in Chaudhuri et al. (2009) is that the cumulative regret to the best ϵ percentile of experts will be small. In particular at time t the cumulative regret of Normal Hedge to the ϵ percentile expert will be $O(\sqrt{t(1+\ln 1/\epsilon)} + \ln^2 N)$. This is more general than the regret bounds that already existed in the online learning literature which only considered regret to the "best" expert in hindsight. The notion of "ϵ percentile" is a more useful bound in the sense that in many practical situations there are many experts among the N which are almost as good as each other. As a result, guaranteeing performance relative to the

Algorithm 2 Normal Hedge parameter-free online learning algorithm.

Initial Assumptions: At time $t-1$ we're given the cumulative loss of each expert $R_{t-1}^{(i)}$ and the discrete weighting $w_t^{(i)}$. Initially $R_0^{(i)} = 0$ and $w_1^{(i)} = 1/N$ for all i.

1: **Update Losses:** Each action incurs a loss $\ell_t^{(i)}$ and the learner incurs loss $\ell_t^A = \sum_{i=1}^N w_t^{(i)} \ell_t^{(i)}$.

2: **Update Regrets:** Update the cumulative regrets $R_t^{(i)} = R_{t-1}^{(i)} + (\ell_t^A - \ell_t^{(i)})$

3: **Update Weights:** First, find $c_t > 0$ that satisfies $\frac{1}{N}\sum_{i=1}^N \exp\left(\frac{([R_t^{(i)}]_+)^2}{2c_t}\right) = e$. Then, update weight distribution for round $t+1$ by $w_{t+1}^{(i)} = \frac{[R_t^{(i)}]_+}{c_t}\exp\left(\frac{([R_t^{(i)}]_+)^2}{2c_t}\right)$. Normalize the weights so they sum to one.

absolute best is often too strong. Moreover, the bound given in Chaudhuri et al. (2009) is still competitive with other known results when considering the "best" expert case by setting $\epsilon = 1/N$.

NH-pf derivation

Transforming the Normal Hedge algorithm into a particle filtering algorithm is quite natural. We must only transform the terminology "experts" into "particles" and we're most of the way towards a Normal Hedge based particle filtering algorithm. A recent paper was published that was that first to describe how the Normal Hedge algorithm can be used as a particle filter (Chaudhuri et al., 2010).

In the tracking problem we consider an expert to be a predictor of a sequence of hidden states $(x_1, \ldots, x_t)$ up to time t. This sequence of states is a proposed explanation for the sequence of observations $(y_1, \ldots, y_t)$. Instead of a likelihood function $p(y_t|x_t)$ like in particle filters, for the Normal Hedge tracking algorithm we must define a loss function on which to measure each expert's performance. The loss $\ell_t^{(i)}$ for expert i should measure how well an experts sequence of states matches the sequence of observations.

After defining this loss, we nearly have all the components needed to utilize Normal Hedge in the tracking framework. However, there is a computational issue at hand, namely, the exponential explosion in possibilities for state space sequences. Imagine we could run Normal Hedge over this enormous number of experts. Luckily, we'd have one advantage on our side because the Normal Hedge weighting would give many experts weight zero except for a core group that are outperforming the predictions of the algorithm itself. Nevertheless, some approximation is necessary, but this sparsity property will ease the requirements of any approximation. The approach we take will sample from this large set of experts in a very similar fashion to that of the bootstrap particle filter described earlier in this chapter.

Just as the particles in a particle filter are a discrete approximation to the posterior density, we will utilize a set of particles to approximate the induced distribution by Normal Hedge over the set of state sequences. The Normal Hedge algorithm for TDOA tracking is given in Algorithm 3. Notice that a further simplification is taken to the problem by only maintaining the *discounted cumulative regret*

$$R_t^{(i)} = (1-\alpha)R_{t-1}^{(i)} + (\ell_t^A - \ell_t^{(i)}) \tag{10}$$

where the parameter α controls how much memory our tracking algorithm should have in terms of penalizing losses observed in the past. This approximates the need for tracking

Algorithm 3 Normal Hedge based particle filter.

Initial Assumptions: At time t-1, we have the set of particles $x_{t-1}^{(i)}$ and Normal Hedge weights $w_{t-1}^{(i)}, i \in \{1, \ldots, m\}$

1: **Regret Update:** Obtain losses $\ell_t^{(i)}$ for each particle and update *discounted cumulative regrets* $R_t^{(i)}$ (Equation 10).

2: **Resample:** For each particle $x_t^{(i)}$ with $R_t^{(i)} < 0$, resample a new particle in its place

1. Choose a current particle $x_t^{(k)}$ according to $\{w_t^{(i)}\}_{i=1}^m$.
2. Let the new particle $x_t^{(i)} = x_t^{(k)} + u_k$, where u_k is Gaussian noise. **(Coordinate-free only):** Project back onto the TDOA manifold using the PD-tree projection.
3. Assign $R_t^{(i)} = (1-\alpha)R_t^{(k)} + (\ell_t^A - \ell(x_t^{(i)}))$.

3: **Weight Updates:** Update the weights of each particle according to the Normal Hedge procedure (see Algorithm 2). Normalize them to one.

4: **Dynamics:** Propagate the particles through the transition equation $x_t^{(i)} = g(x_{t-1}^{(i)})$.

sequences of states. Since typically we only want to predict the current state, this is an acceptable simplification. The weights can then be calculated according to the steps involved for computing $w_t^{(i)}$ in the Normal Hedge algorithm (see Algorithm 2). What remains as a critical algorithmic choice is how we compute the loss $\ell_t^{(i)}$ for each particle, analogous to the decision of the likelihood function in particle filters.

A nice property of this filter (NH-pf) is that the resampling procedure for particles emerges naturally from the Normal Hedge weighting function. A particle will be given zero weight whenever its cumulative regret has started to perform worse than the algorithm itself, and at this moment it is resampled near a particle that has good historical performance.

Another nice property of NH-pf is that it can still maintain good tracking performance when the loss function has a modeling mismatch with the true observation process (Chaudhuri et al., 2010). Chaudhuri et. al show that if a traditional particle filter has a likelihood function that mismatches the true underlying process, then its performance will break down much quicker than the corresponding NH-pf. This final observation could prove to be advantageous in the TDOA tracking scenario. As stated earlier, choosing a likelihood function for TDOA tracking is somewhat arbitrary since the process for generating a PHAT observation from a given source location is extremely difficult to model. For this tracking problem and many other of practical importance, a model mismatch of this kind is unavoidable.

TDOA tracking with coordinate-free NH-pf

We now describe how we track TDOAs in a coordinate-free fashion. First, we expand the state space to be that of the entire TDOA vector (for TAC a 21-dimensional state space). In addition, we must specify what loss function we will be using to calculate regrets relative to. We will utilize the negative likelihood function described in Equation 7

$$\ell_t^{(i)}(x_t, y_t) = -p(y_t|x_t^{(i)}) \tag{11}$$

As discussed earlier, tracking in this many dimensions becomes difficult, but we also know that our TDOAs lie on a low dimensional manifold. In previous sections we discussed PD-Tree

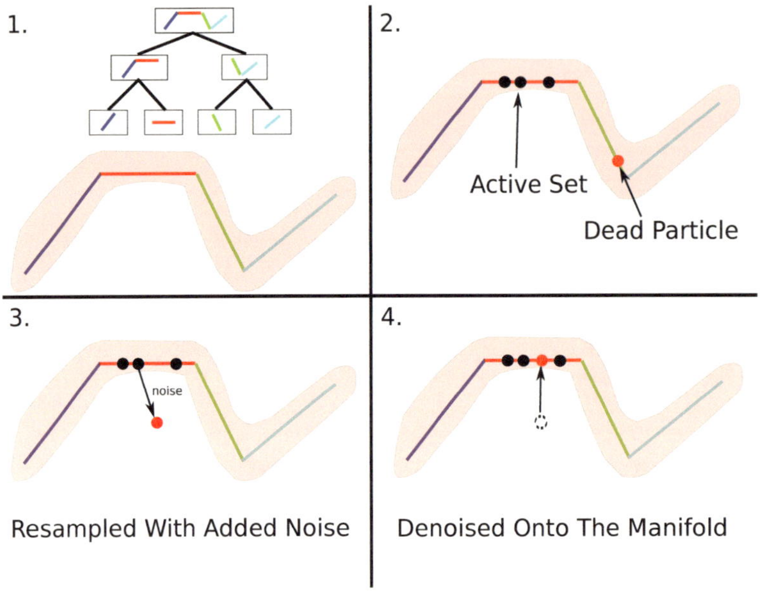

Fig. 7. Depiction of a dead particle resampled and projected back onto the manifold.

based regression models. We again utilize a PD-Tree to alleviate this high dimensional tracking problem, however, at the leaf nodes instead of a regressor we learn a low dimensional affine model via a PCA. During the resampling step for new particles, after noise is added to the newly born particle it is projected back onto the TDOA structure via the appropriate PCA model in the PD-Tree leaf node. This process is depicted in Figure 7.

First, the corresponding leaf node for the newly born particle is found. Then, the particle is denoised by projecting it onto the principal components of local piece of the manifold stored within. The overall effect is a tracking algorithm that is constrained to have a state-space lie on the low dimensional structure captured by the PD-Tree.

Note that there is a bias variance trade-off as you descend the PD-Tree in terms of the principle components models stored in each node. The nodes higher up in the tree have small variance since the bulk of the training data was used to learn the principal directions. However, the bias is also large at these nodes since a linear model is not appropriate at this granularity. As you descend the tree the variance increases, whereas the bias decreases as you approach locally linear regions.

Moreover, the fit from individual nodes may vary across even their own partition region. Consider a given internal node of the PD-Tree. For a certain region of the cell's partition the linear model may be a good fit, whereas in other regions of the cell's partition it may be poor. It is clear from this argument that the best node that fits the true TDOA for a given source location will vary with location both across the tree and possibly in depth as well.

It then makes sense to consider using the entire PD-Tree during the projection step instead of just the leaves at a fixed depth. A natural way to accomplish this emerges from the NH-pf resampling scheme by making a slight alteration to the projection step with the PD-Tree discussed above. After resampling a newly born particle and before projecting it back onto the manifold via the PD-Tree, first pick a depth uniformly at random in the PD-Tree. This will be the depth used for this single newly born particle, and this procedure is repeated for each newly born particle.

This random strategy will have the nice property that it will naively find the correct model depth for the current sound source location over time. Depths that are chosen that are poor models will have particles die soon thereafter, whereas particles that are drawn from depths that perform well will survive. We will examine this procedure in the experiments that follow.

7. Experiments: TDOA tracking

The experiments that follow were conducted from recordings of real speakers talking and moving slowly while facing TAC's microphone array. We describe each individual experiment in detail in what follows.

Setup

To build a PD-tree we first collected a training set of TDOA vectors from our microphone array. We accomplished this by moving a white noise producing sound source around the room near typical locations that sitting or standing people would be interacting with the display. This resulted in approximately 20,000 training TDOA vectors to which we built a PD-tree of depth 2. In each node of the PD-tree we store the mean of the training data and the top $k = 3$ principal directions.

Here are the parameter settings we use for the experiments that follow. We use $m = 50$ particles for each type of particle filter examined. The discounting factor for NH is set to $\alpha = 0.05$.

We made several real audio recordings of a person walking throughout the room facing the array and talking. We describe each experiment in detail in what follows.

Usage of Manifold modeling

This first experiment has a person walking and counting aloud while facing the array. The person's path goes through the center of the room far from each microphone. Since TDOAs evolve more slowly when the sound source is far from each microphone we'd expect this to be well modeled by the root PCA of our PD-tree. We compare using the root PCA versus no projection step at all for both a standard particle filter (PF) and the Normal-Hedge particle filter (NH).

Figure 8 depicts such a comparison. Here we show tracking results from two microphone pairs that are typical of the remaining pairs (i.e. two coordinates of the 21-D TDOA state). In green is shown Z_t^p where its magnitude is represented by the size of the circle marker. The sound source moved in a continuous and slowly moving path so we'd expect each TDOA coordinate to follow a continuous and slowly changing path as well. The trackers with the PCA projection step outperform their counterparts without the projection.

From this single trial run, NH-pca seems to have a slight advantage over PF-pca from time to time, but the two algorithms are competitive in performance. However, a more closer examination shows an advantage to NH-pf. When averaged over 25 independent runs over this audio recording the NH-pf with pca is slightly more accurate and clearly stabler than the standard PF. Figure 9 depicts the RMSE of each tracker averaged over 25 independent trials. The RMSE is calculated coordinate-wise relative to the maximum of the PHAT series for each frame. Since the maximum derived TDOA is often accurate, but sometimes widely inaccurate (especially during periods of silence), we smoothed each RMSE series using an exponential moving average with $\alpha = 0.05$. It is clear from these plots that the pca based methods are outperforming the non-pca ones, and the NH methods have an advantage over the standard PF methods.

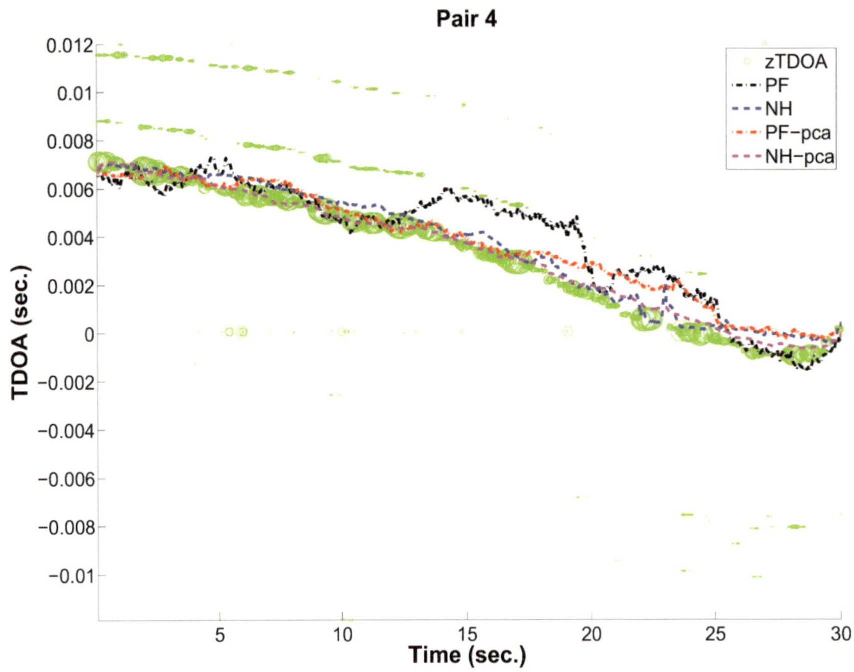

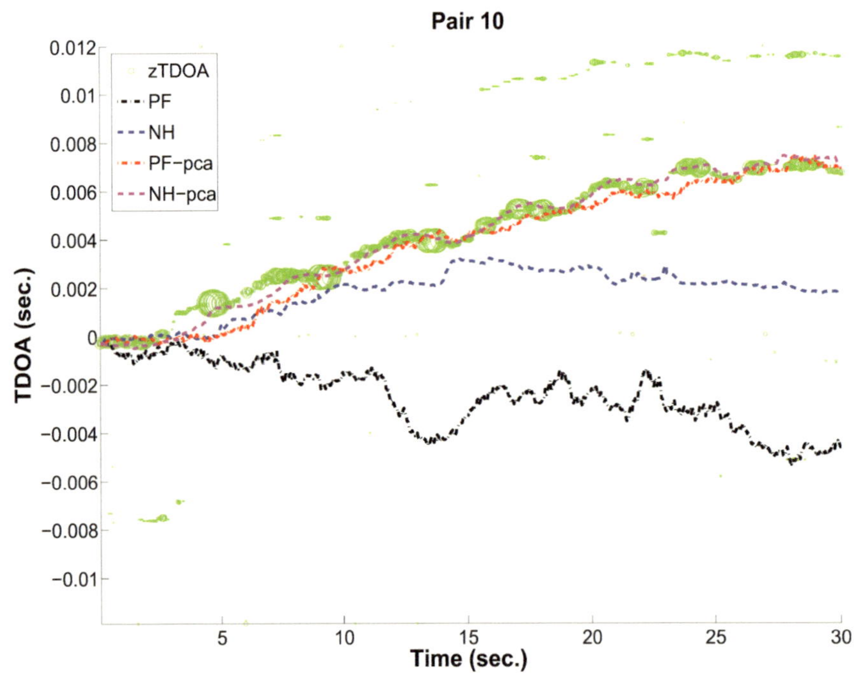

Fig. 8. Performance of NH and PF with and without using a global PCA projection for denoising.

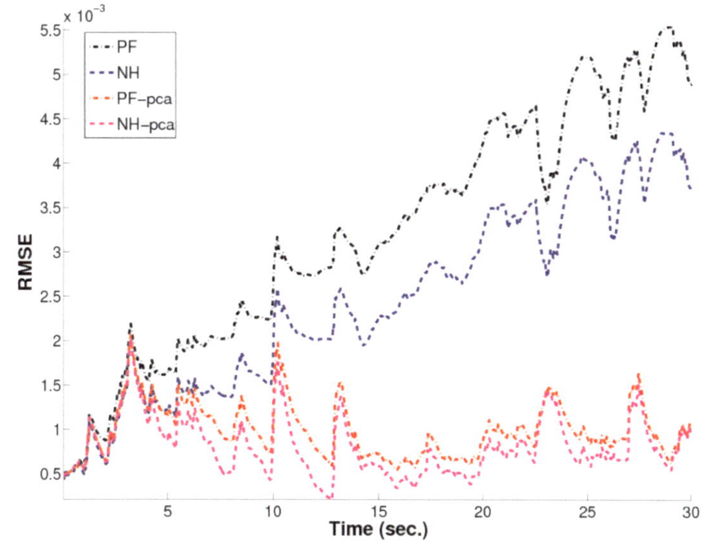

Fig. 9. RMSE over 25 independent runs of each of the trackers.

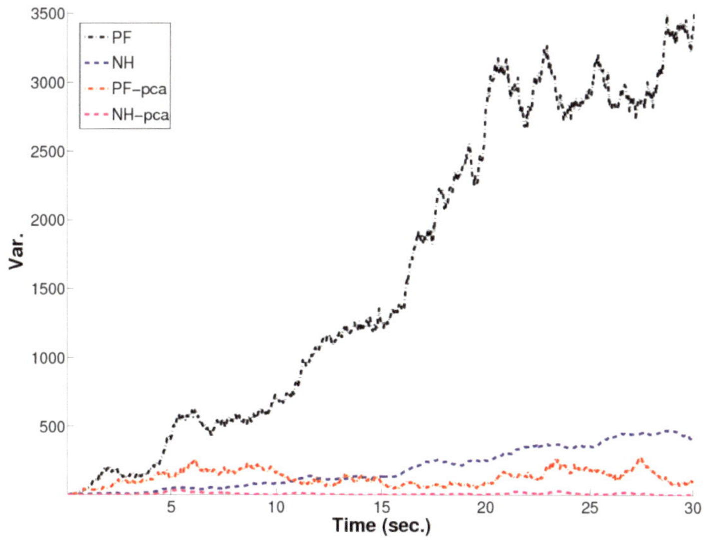

Fig. 10. Variance over 25 independent runs of each of the trackers.

Figure 10 depicts the variance of each method as a function of time. For each time t, first the norm of the state vector was calculated, and then we computed the variance of this norm across the 25 runs. The variance increases with time for all trackers except that of NH-pf with the pca projection. When comparing NH to PF, the NH trackers have much less variance than their PF counterpart. It's clear from these results that the NH-pf with the pca is a very stable and accurate tracker.

Remember that there are only 50 particles to track a state that is 21 dimensional. There are no dynamics involved in our particle filters, so the resampling stage alone has to include enough randomness for the source to be tracked as it moves. When the manifold model is not used the amount of randomness needed is too large for 50 particles to be able to track on all D dimensions. However, when a model of the manifold is used, effective tracking results can be had. Moreover, it should be noted that the NH version uses less randomness since it only resamples when the weight of a particle becomes zero. Despite this, the NH versions are able to have a competitive performance with standard particle filters.

Testing different manifold models

The data collected for the experiment discussed here is exactly the same as the previous experiment except the path the speaker took traveled much closer to some pairs of microphones at certain points in time. When a sound source is moving close to some set of microphones, the TDOAs involved with those microphones will change much more rapidly and non-linearly. With this path we hope to examine the usefulness of deeper nodes in the PD-tree. We will test the PD-Tree using only the nodes at a fixed depth ($d = 0, 1, 2$) and also the randomized scheme for choosing uniformly among the depths (see discussion in previous section for a full description of the randomized strategy).

Since the performance of NH was superior when using the global PCA projection we only examine NH in this experiment. This will allows us to explore the randomized manifold modeling scheme. In a standard particle filter, no benefit is gained by adopting this randomized strategy since all particles are resampled at each iteration. Thus, the random strategy in a standard filter can never allow particles to "gravitate" towards the correct depth. Figure 11 is a similar figure to one found in the previous section. The particle filtering variants examined here use projections at fixed depth zero (NH-0), one (NH-1), and two (NH-2). The random strategy is also examined (NH-rand). It is clear that somewhere between 50s-70s the location of the sound source is modeled poorly by the global PCA at the root and is better modeled by the PCA at level 2. However, it is only for this short duration where this modeling transition takes place. Depth's 0 and 1 performed particularly poorly in this region, while depth 2 has a significant advantage.

However, the best performing tracker was one that utilized the entire tree structure in a random fashion. By allowing particles to birth at a random depth, there is a clear pressure to transition from a depth-0 model to a depth-2 model rather quickly. This can be seen in Figure 12. Here we depict what the depth distribution of each of the 50 particles are at time t for NH-rand. Nearly all the particles during this time period that were sampled from depth-2 are staying alive during this period. This is a rather intuitive result since a particular node's PCA model may only be good for tracking in a small region of the corresponding PD-tree node's partition region. When the sound source exits this area that is modeled well by the node's affine space, some other depth in the tree may become a better model. NH-rand naturally captures such transitions.

8. Discussion

We've given a coordinate-free method for camera pointing via audio localization with a microphone array. We first presented a method of translated time-delays into pan-tilt directives via standard regression and followed that up with an analysis of a TDOA tracking methodology to improve reliability. As a result, our coordinate-free approach allows for

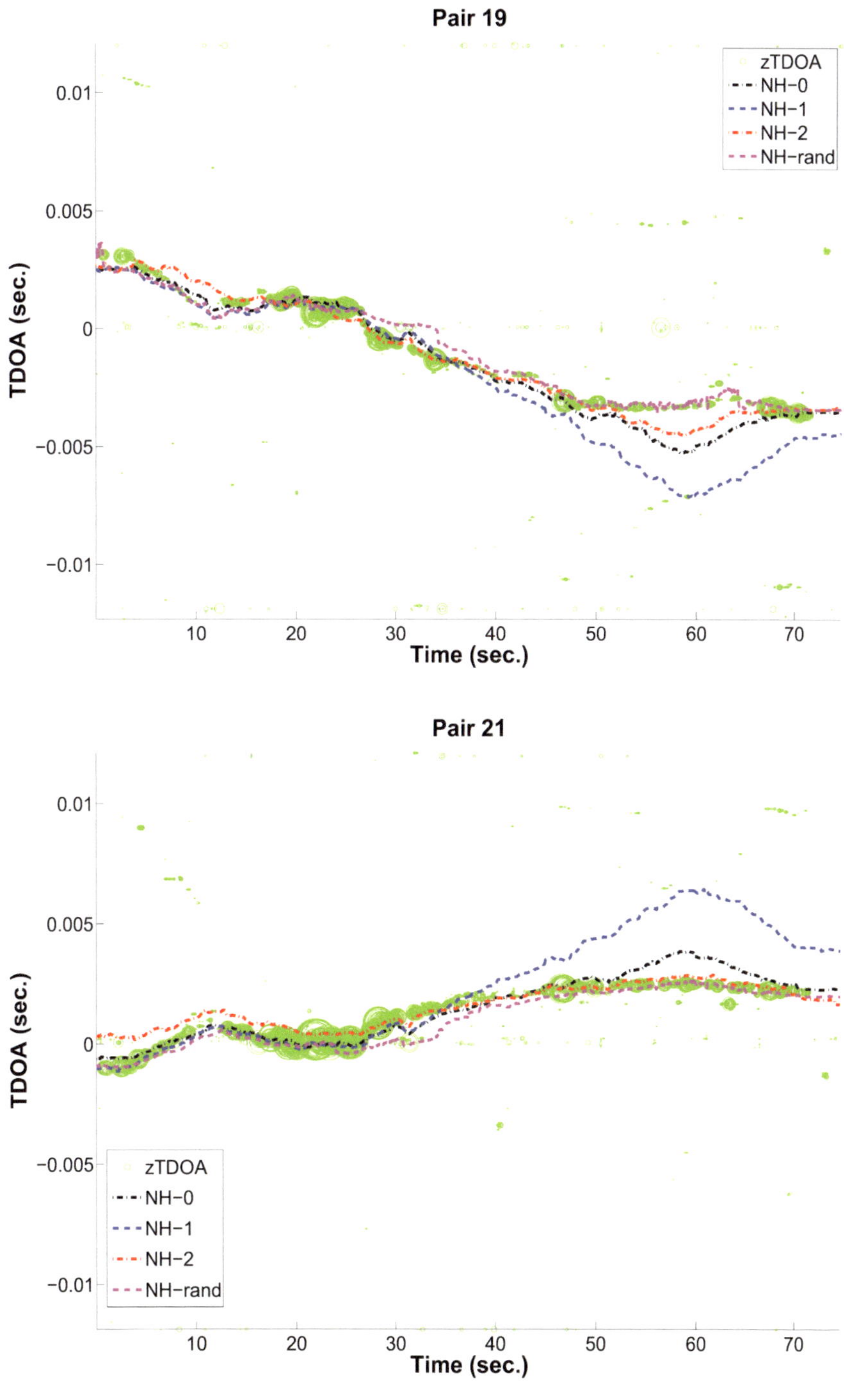

Fig. 11. Using various depths in the PD-tree as part of the projection step.

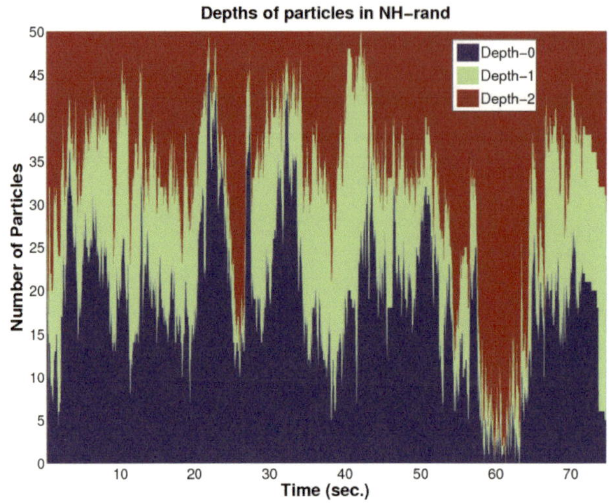

Fig. 12. For NH-rand, the PD-tree depths at time t that the m particles have been sampled from last.

arbitrary placement of sensor elements which can be beneficial for both array geometry considerations and alleviates the need for tedious measurement.

9. References

Arulampalam, M., Maskell, S., Gordon, N. & Clapp, T. (2002). A tutorial on particle filters for online nonlinear/non-Gaussian Bayesian tracking, *IEEE Transactions on signal processing* 50(2): 174–188.

Badali, A., Valin, J.-M., Michaud, F. & Aarabi, P. (2009). Evaluating real-time audio localization algorithms for artificial audition in robotics, *Intelligent Robots and Systems, 2009. IROS 2009. IEEE/RSJ International Conference on*, pp. 2033 –2038.

Birchfield, S. & Subramanya, A. (2005). Microphone array position calibration by basis-point classical multidimensional scaling, *Speech and Audio Processing, IEEE Transactions on* 13(5): 1025–1034.

Brandstein, M., Adcock, J. & Silverman, H. (1995). A closed-form method for finding source locations from microphone-array time-decay estimates, *Acoustics, Speech, and Signal Processing, IEEE International Conference on* 5: 3019–3022.

Chan, Y. & Ho, K. (1994). A simple and efficient estimator for hyperbolic location, *Signal Processing, IEEE Transactions on* 42(8): 1905 –1915.

Chaudhuri, K., Freund, Y. & Hsu, D. (2009). A parameter-free hedging algorithm, *Advances in Neural Information Processing Systems 22*, pp. 297–305.

Chaudhuri, K., Freund, Y. & Hsu, D. (2010). An online learning-based framework for tracking, *UAI 2010, Proceedings of the 26th Conference in Uncertainty in Artificial Intelligence*.

Cheamanunkul, S., Ettinger, E., Jacobsen, M., Lai, P. & Freund, Y. (2009). Detecting, tracking and interacting with people in a public space, *ICMI-MLMI '09: Proceedings of the 2009 International Conference on Multimodal Interfaces*.

Do, H., Silverman, H. & Yu, Y. (2007). A real-time srp-phat source location implementation using stochastic region contraction(src) on a large-aperture microphone array,

Acoustics, Speech and Signal Processing, 2007. ICASSP 2007. IEEE International Conference on, Vol. 1, pp. I–121 –I–124.

Foy, W. (1976). Position-location solutions by taylor-series estimation, *Aerospace and Electronic Systems, IEEE Transactions on* AES-12(2): 187 –194.

Friedlander, B. (1987). A passive localization algorithm and its accurancy analysis, *Oceanic Engineering, IEEE Journal of* 12(1): 234 – 245.

Gillette, M. & Silverman, H. (2008). A linear closed-form algorithm for source localization from time-differences of arrival, *Signal Processing Letters, IEEE* 15.

Gordon, N., Salmond, D. & Smith, A. (1993). Novel approach to nonlinear/non-Gaussian Bayesian state estimation, *IEE proceedings. Part F. Radar and signal processing* 140(2): 107–113.

Gustafsson, F. & Gunnarsson, F. (2003). Positioning using time-difference of arrival measurements, *Acoustics, Speech, and Signal Processing, 2003. Proceedings. (ICASSP '03). 2003 IEEE International Conference on*, Vol. 6.

Hörster, E., Lienhart, R., Kellermann, W. & Bouguet, J.-Y. (2005). Calibration of visual sensors and actuators in distributed computing platforms, *VSSN '05: Proceedings of the third ACM international workshop on Video surveillance & sensor networks*, ACM, New York, NY, USA, pp. 19–28.

Huang, Y., Benesty, J., Elko, G. & Mersereati, R. (2001). Real-time passive source localization: a practical linear-correction least-squares approach, *Speech and Audio Processing, IEEE Transactions on* 9(8): 943 –956.

J. DiBiase, H. Silverman, M. B. (2001). *Robust localization in reverberant rooms. In M. Brandstein and D. Ward Microphone Arrays.*, Springer-Verlag.

Knapp, C. & Carter, G. (1976). The generalized correlation method for estimation of time delay, *Acoustics, Speech and Signal Processing, IEEE Transactions on* 24(4).

Lehmann, E. & Johansson, A. (2007). Particle filter with integrated voice activity detection for acoustic source tracking, *EURASIP J. Appl. Signal Process.* 2007(1): 28–28.

Li, T. & Ser, W. (2010). Three dimensional acoustic source localization and tracking using statistically weighted hybrid particle filtering algorithm, *Signal Process.* 90(5): 1700–1719.

Max/MSP website (n.d.). http://www.cycling74.com.

McCowan, I., Lincoln, M. & Himawan, I. (2008). Microphone array shape calibration in diffuse noise fields, *Audio, Speech, and Language Processing, IEEE Transactions on [see also Speech and Audio Processing, IEEE Transactions on]* 16(3): 666–670.

Omologo, M. & Svaizer, P. (1994). Acoustic event localization using a crosspower-spectrum phase based technique, *Acoustics, Speech, and Signal Processing, 1994. ICASSP-94., 1994 IEEE International Conference on*, Vol. ii, pp. II/273 –II/276 vol.2.

Omologo, M. & Svaizer, P. (1996). Acoustic source location in noisy and reverberant environment using csp analysis, *Acoustics, Speech, and Signal Processing, 1996. ICASSP-96. Conference Proceedings., 1996 IEEE International Conference on*, Vol. 2, pp. 921 –924 vol. 2.

Pertilä, P., Korhonen, T. & Visa, A. (2008). Measurement combination for acoustic source localization in a room environment, *EURASIP J. Audio Speech Music Process.* 2008: 1–14.

Raykar, V. & Duraiswami, R. (2004). Automatic position calibration of multiple microphones, *Acoustics, Speech, and Signal Processing, 2004. Proceedings. (ICASSP '04). IEEE International Conference on* 4: iv–69–iv–72 vol.4.

Sachar, J., Silverman, H. & Patterson, W. (2005). Microphone position and gain calibration for a large-aperture microphone array, *Speech and Audio Processing, IEEE Transactions on* 13(1): 42–52.

Silverman, H., Yu, Y., Sachar, J. & Patterson, W.R., I. (2005). Performance of real-time source-location estimators for a large-aperture microphone array, *Speech and Audio Processing, IEEE Transactions on* 13(4): 593 – 606.

Smith, J. & Abel, J. (1987). Closed-form least-squares source location estimation from range-difference measurements, *Acoustics, Speech and Signal Processing, IEEE Transactions on* 35(12): 1661 – 1669.

Stoica, P. & Li, J. (2006). Lecture notes - source localization from range-difference measurements, *Signal Processing Magazine, IEEE* 23(6): 63 –66.

Svaizer, P., Matassoni, M. & Omologo, M. (1997). Acoustic source location in a three-dimensional space using crosspower spectrum phase, *ICASSP '97: Proceedings of the 1997 IEEE International Conference on Acoustics, Speech, and Signal Processing (ICASSP '97) -Volume 1*, IEEE Computer Society, Washington, DC, USA, p. 231.

Talantzis, F., Pnevmatikakis, A. & Constantinides, A. G. (2009). Audio-visual active speaker tracking in cluttered indoors environments, *Trans. Sys. Man Cyber. Part B* 39(1): 7–15.

Verma, N., Kpotufe, S. & Dasgupta, S. (2009). Which spatial partition trees are adaptive to intrinsic dimension?, *UAI 2009, Proceedings of the 25th Conference in Uncertainty in Artificial Intelligence*.

Wang, H. & Chu, P. (1997). Voice source localization for automatic camera pointing system in videoconferencing, *ICASSP '97: Proceedings of the 1997 IEEE International Conference on Acoustics, Speech, and Signal Processing (ICASSP '97) -Volume 1*, IEEE Computer Society, Washington, DC, USA, p. 187.

Ward, D., Lehmann, E. & Williamson, R. (2003). Particle filtering algorithms for tracking an acoustic source in a reverberant environment, *Speech and Audio Processing, IEEE Transactions on* 11(6): 826 – 836.

19

Sound Image Localization on Flat Display Panels

Gabriel Pablo Nava, Yoshinari Shirai, Kaji Katsuhiko, Masafumi Matsuda, Keiji Hirata and Shigemi Aoyagi
NTT Communication Science Labs
Japan

1. Introduction

In recent years, we have experienced rapid advancements on solid-state luminous displays which have lead to their introduction in many, if not all, multimedia applications. Moreover, former light projectors used in immersive audiovisual spaces have been succeeded by large LCD panels which have overcome requirements such as dark room, occlusion-free sightline between projector and screen, spacious installation, etc. But on the other hand, while the screens used with light projectors allow a relatively free placement of loudspeakers (e.g. behind the screen) to provide realistic sound spatialization, flat LCD displays impose new challenges from the acoustical point of view. One of them is to provide positional correspondence between the video images on the display and their sound image while allowing multiple users to properly localize the sound images on the LCD panels. To achieve this, *auditory displays* have been traditionally implemented with conventional stereo loudspeakers installed at the sides of the video display panel as shown in Fig. 1. However, a fundamental problem with such setups is that correct sound image localization is achieved only by listeners positioned at the symmetrical axis of the stereo array. Listeners standing at the asymmetrical areas tend to perceive a sound image shifted towards their closest loudspeaker. This problem has been extensively studied and attributed to the so-called *precedence effect*: from multiple sound sources radiating similar sound intensities, a listener tends to localize a single sound image close to the nearest source, given that the differences of arrival times of the sounds reaching his ears are between about 1 and 50 ms (Gardner, 1968). As demonstrated by a number of precise experiments (Rakerd, 1986), the precedence effect provides important psychoacoustical cues to the process of sound localization in humans as well as in other species (Litovsky et al., 1999), nevertheless, it represents a fundamental problem in stereo reproduction systems that attempt realistic sound spatialization over a wide area. For this reason, several approaches to expand the area of accurate sound image perception have appeared in the literature. Some of them have coped the problem as to that of developing loudspeakers with radiation characteristics that help to equalize the sound from both channels over a broad area, while others focus on pre-processing the audio signals (either in the analog or digital domain) to create an spatialization effect on the listeners. In the first scenario, Bauer reported a loudspeaker arrangement aiming the optimum trade off between the position of the listener and the balance of acoustic energy distributed from the stereo channels (Bauer, 1960). He concluded that stereo loudspeakers whose radiation pattern follows the cosine law (such as that of the front lobe of an acoustic dipole), together

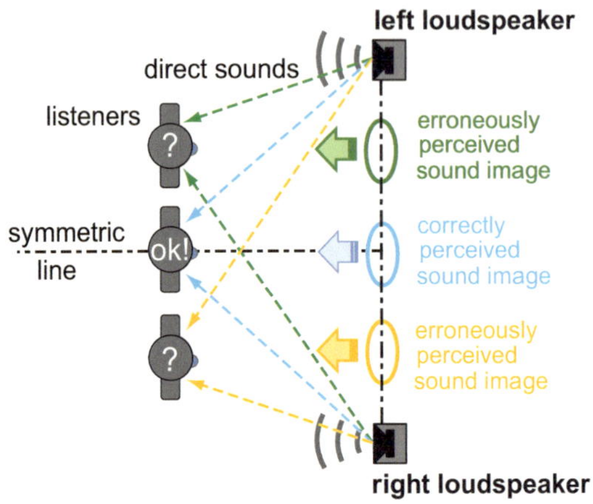

Fig. 1. Traditional stereo loudspeaker setup. Listeners at the asymmetrical areas tend to perceive a sound image shifted towards their closest loudspeaker direction.

with their placement facing in counter fire, can achieve balance of the stereo channels not only at the symmetrical line but also within the (circular) area that encloses the two acoustic dipoles. On the contrary, Kates, based on his model that takes into account the directional characteristics of the human ear (Kates, 1980), debated that the loudspeakers should be more directional (rather than with a spherical radiation) to target an specific audience area. This idea was further supported by Davis who employed arrays of multi-loudspeaker drivers per stereo channel to achieve the desired radiation pattern over the audio band (Davis, 1987). In his implementations, Davis used crossover networks to deliver specific-band signals to each loudspeaker driver of determined frequency response so that the overall interaction of the individual loudspeakers would lead the sought directionality. In the second scenario, but still aiming at the directional radiation issue, Aoki *et. al.* proposed a 6-drivers stereo loudspeaker array (three drivers per channel) capable of maintaining in-head sound image localization within the seating area (a rectangular table), (Aoki & Koizumi, 1987). Each loudspeaker driver of the triplets are excited by either delayed, inverted and/or amplified versions of the original stereo signals. Thus, when the audio signals are radiated, the unwanted acoustic energy at specific areas is suppressed by cancelation of the opposed-phase waves. A more recent approach following similar ideas was introduced by Ródenas *et. al.* in what they called a *position-independent* stereo system (Ródenas et al., 2003). In contrast to the system of Aoki *et. al.*, the position-independent implementation used pairs of loudspeaker drivers in each channel. Ródenas et. al. further proposed psycoacoustic models to achieve optimum directivity of the loudspeakers through their excitation by signals filtered with optimal FIR filters. Another example of binaural reproduction based on Wiener filters was reported by Kim and Wang, (Kim & Wang, 2003). Other approaches, although suited to a single listener, take advantage of the computational power of recent computers to incorporate video tracking systems that provide real-time information of the listener's position (with respect to the loudspeakers) to adjust the audio signals that should be delivered by the stereo loudspeakers in order to render a sound image at the recognized position, (examples: Gardner (1997); Kyriakakis et al. (1998); Merchel & Groth (2009)). Yet, there is another tendency of recent audio reproduction systems which is not limited to stereo channels but rather requires large numbers of collinearly arranged loudspeakers to provide audio spatialization to an specific

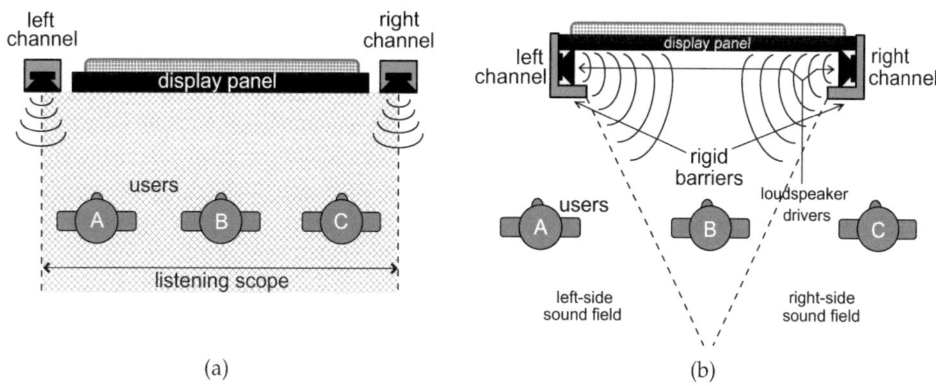

Fig. 2. (a) Conventional stereo loudspeakers with a display panel. (b) The new loudspeaker design with L-like rigid barriers.

audience. Their principle is based on the assumption that any given wave field can be accurately reconstructed by the superposition of elementary spherical waves radiated from a number of individually driven sound sources (loudspeakers). The most representative technique of this approach is the Wave Field Synthesis, or WFS (Berkhout et al., 1993), and its successful performance in applications including joint video and audio/speech has been substantiated by experimental work with setups in which the large array of loudspeakers can be easily hidden behind the thin projection screens (Melchoir et al., 2003; Werner & Boone, 2003). As Werner points out, accurate audio spatialization reproduction enhances the naturalness of the interaction among the participants in a teleconference, leading to a more productive collaborative work (Werner & Boone, 2003). From that perspective, WFS has probed to provide a high quality auditory interface for the target audience. Nevertheless, as modern teleconference-collabotarion systems tend to include solid-state luminous displays large enough (usually forming complete walls) to immerse the users, the loudspeaker placement and the acoustic influence of the flat hard displays becomes a serious problem to straightforwardly implement sound spatialization with existent methods and conventional loudspeakers. At this level, the loudspeaker design and placement seem to play such an important role as that of the audio signal processing involved to render a sound image at a given position. Nonetheless, relatively less research has been reported in this field. It is, therefore, the aim of this chapter to introduce a novel loudspeaker design for flat hard display panels on which localization of sound images is desired. Its simple, yet effective, principle is demonstrated with numerical simulations, objective and subjective experiments, and real-time user-interactive applications.

2. Loudspeakers for flat display panels

2.1 An L-like design to improve sound image localization

Conventional stereo loudspeakers for video displays are typically mounted as independent units placed at the sides of the display panel, facing straight to the audience. In spite of the existent stereo spatialization techniques, the listening area achieved by this arrangement is delimited by the separation of the stereo loudspeakers as illustrated in Fig. 2(a). On the other hand, some acoustic engineers (Bauer, 1960; Kates, 1980), based on experimental work, recommend an angular placement of the loudspeakers facing counter-fire into the target audience. Such loudspeaker angulation results in a an expansion of the overall radiation pattern of the stereo array. Following a similar idea, in the novel design proposed here, the loudspeaker drivers are, in addition, embedded to the interior of L-like rigid structures (referred as *rigid barriers* hereafter) which at the same time are attached to the lateral edges of the display panel. These rigid barriers act as attenuators of sound intensity as a function of the listening direction. In this way, the loudspeaker drivers are, indeed, off the plane of the display surface and pointing in counter-fire into its interior. Fig. 2(b) shows an sketch of the proposed loudspeaker arrangement for flat displays. With such configuration, the design aims to mechanically redistribute the radiated acoustic energy so as to balance the sound strength at the off-symmetrical axis. To achieve that, the design relies on two basic assumptions:

1. the display panel is flat and hard, and
2. monaural sound signals are to be rendered on the panel.

Note that, in principle, compliance of the above statements allows applicability of the design to any display panel of a given size. Nevertheless, the dimensions of the rigid barriers have to be optimized because of the dependency on operative frequency, the size of the panel and the coverage area. Such issue is addressed in further paragraphs.

Let us consider an example where a monaural sound image at the center of the display panel of Fig. 2(b) is desired. Moreover, suppose that the audio signals to drive the left and right channels are identical, i.e. zero phase and equal amplitude. At the symmetrical point B (or along the symmetrical axis), the acoustic signals radiated by the loudspeakers are observed with the same (or nearly the same) time and intensity and through a direct path from both channels. Therefore, as in a conventional stereo loudspeaker setup, a listener at B is able to perceive a central sound image on the display. At the asymmetrical positions A and C, the sound from the closest loudspeaker arrives first through indirect paths (by diffraction on the edges), but its intensity is attenuated by the rigid barrier so as to prevent the masking effect of the sound from the opposite channel (which arrives through a direct path). Note also that, with this configuration, the acoustic energy from one channel is fundamentally radiated to the front of the panel and to the opposite-side field. Thus, the hypothesis is that as the sound from the farthest loudspeaker remains unmasked, the effect of precedence is expected to be reduced as function of the listening position, leading to an improvement of sound image localization at the asymmetrical areas. With the aid of numerical simulations, this hypothetical supposition will be further demonstrated.

2.2 Optimization of the L-like design

Although the simplicity of the L-like design makes it easy to implement, the dimensions of the rigid barriers are not fixed and should be optimized for a particular case. If for example, the sound of an audio-visual application is to be spatialized over an effective listening area defined by the coverage angle θ from the surface of the display panel of size h on Fig. 3(a), then the size δ of the L-like rigid barriers has to be optimized to achieve the desired level of sound intensity radiated over a grid of N (receiving) field points. Therefore, the optimum size $\delta_{mboxopt}$ is the one that minimizes the following objective function within a number β of

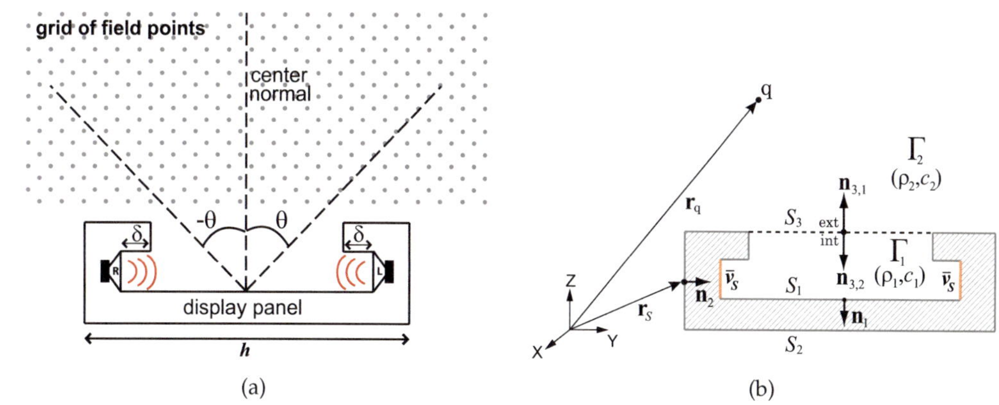

Fig. 3. (a) Optimization of L-like rigid barriers for a target listening area. (b) Theoretical model for the numerical optimization.

discrete frequencies in a given audible frequency band, as

$$\delta_{\text{opt}} = \min_{\delta \in \mathbb{R}^+} \frac{1}{\beta} \sum_{k=1}^{\beta} G(f_k, \delta) \tag{1}$$
$$\text{subject to} \quad 0 < \delta \leq \max \ ,$$

where

$$G(f_k, \delta) = 10 \ \log_{10} \left[\frac{1}{N} \sum_{q=1}^{N} \left(\left| p_q^{\text{des}}(f_k, \theta) \right|^2 - \left| p_q^{\text{sim}}(f_k, \delta) \right|^2 \right| \right] \ . \tag{2}$$

Here, p_q^{des} and p_q^{sim} are the actually desired and the numerically simulated sound pressures at the q-th field point. Alternatively, the sound field p_q^{des} can be further represented as

$$p_q^{\text{des}}(f_k, \theta) = p_q^{\text{des}}(f_k) \cdot W(\theta) \ , \tag{3}$$
$$0 \leq W(\theta) \leq 1 \ ,$$

i.e., the sound pressure p_q weighted by a normalized θ-dependent function $W(\theta)$ to further control the desired directivity. Thus, G measures the average (in dB's) of the error between the (squared) magnitude of p_q^{des} and p_q^{sim}.

The acoustic optimization is usually performed numerically with the aid of computational tools such as finite element (FEM) or boundary element (BEM) methods, being the latter a frequently preferred approach to model sound scattering by vibroacoustic systems because of its computational efficiency over its FEM counterpart. BEM is, indeed, an appropriate framework to model the acoustic display-loudspeaker system of Fig. 3(a). Therefore, the adopted theoretical approach will be briefly developed following boundary element formulations similar to those described in (Ciskowski & Brebbia, 1991; Estorff, 2000; Wu, 2000) and the domain decomposition equations in (Seybert et al., 1990).

Consider the solid body of Fig. 3(b) whose concave shape defines two subdomains, an interior Γ_1 and an exterior Γ_2, filled with homogeneous compressible media of densities ρ_1 and ρ_2, where any sound wave propagates at the speed c_1 and c_2 respectively. The surface of the body is divided into subsegments so that the total surface is $S = S_1 + S_2 + S_3$, i.e. the interior, exterior and an imaginary auxiliary surface. When the acoustic system is perturbed with a harmonic force of angular frequency ω, the sound pressure p_q at any point q in the 3D propagation field, is governed by the Kirchhoff-Helmholtz equation

$$\int_S \left(p_S \frac{\partial \Psi}{\partial \mathbf{n}} - \Psi \frac{\partial p_S}{\partial \mathbf{n}} \right) dS + p_\mathrm{q} = 0 \,, \tag{4}$$

where p_S is the sound pressure at the boundary surface S with normal vector $\mathbf{n}$. The Green's function Ψ is defined as $\Psi = e^{-jkr}/4\pi r$ in which $k = \omega / c$ is the wave number, $r = |\mathbf{r}_S - \mathbf{r}_\mathrm{q}|$ and $j = \sqrt{-1}$. Moreover, if the field point q under consideration falls at any of the domains Γ_1 or Γ_2 of Fig. 3(b), the sound pressure p_q is related to the boundary of the concave body by

- for q in Γ_1:

$$C_\mathrm{q}\, p_\mathrm{q} + \int_{S_1+S_3} \left(\frac{\partial \Psi}{\partial \mathbf{n}}\, p_{S_1+S_3} - \Psi\, \frac{\partial p_{S_1+S_3}}{\partial \mathbf{n}} \right) dS =$$

$$C_\mathrm{q}\, p_\mathrm{q} + \int_{S_1} \left(\frac{\partial \Psi}{\partial \mathbf{n}_1}\, p_{S_1} + j\omega\rho_1\, \Psi\, v_{S_1} \right) dS + \int_{S_3} \left(\frac{\partial \Psi}{\partial \mathbf{n}_{3,1}}\, p_{S_3}^{\mathrm{int}} + j\omega\rho_1\, \Psi\, v_{S_3}^{\mathrm{int}} \right) dS = 0 \tag{5}$$

- for q in Γ_2:

$$C_\mathrm{q}\, p_\mathrm{q} + \int_{S_2+S_3} \left(\frac{\partial \Psi}{\partial \mathbf{n}}\, p_{S_2+S_3} - \Psi\, \frac{\partial p_{S_2+S_3}}{\partial \mathbf{n}} \right) dS =$$

$$C_\mathrm{q}\, p_\mathrm{q} + \int_{S_2} \left(\frac{\partial \Psi}{\partial \mathbf{n}_2}\, p_{S_2} + j\omega\rho_2\, \Psi\, v_{S_2} \right) dS + \int_{S_3} \left(\frac{\partial \Psi}{\partial \mathbf{n}_{3,2}}\, p_{S_3}^{\mathrm{ext}} + j\omega\rho_2\, \psi\, v_{S_3}^{\mathrm{ext}} \right) dS = 0 \tag{6}$$

Note that in the latter equations (5) and (6), the particle velocity equivalent $\partial p / \partial \mathbf{n} = -j\omega\rho\, v$ has been used. Thus, p_{S_i} and v_{S_i} represent the sound pressure and particle velocity on the i-th surface S_i. The parameter C_q depends on the solid angle in which the surface S_i is seen from p_q. For the case when q is on a smooth surface, $C_\mathrm{q} = 1/2$, and when q is in Γ_1 or Γ_2 but not on any S_i, $C_\mathrm{p} = 1$.
To solve equations (5) and (6) numerically, the model of the solid body is meshed with discrete surface elements resulting in a number of L elements for the interior surface $S_1 + S_3$ and M for the exterior $S_2 + S_3$. If the point q is matched to each node of the mesh (collocation method), equations (5) and (6) can be written in a discrete-matrix form

$$\mathbf{A}_{S_1}\, \mathbf{p}_{S_1} + \mathbf{A}_{S_3}^{\mathrm{int}}\, \mathbf{p}_{S_3}^{\mathrm{int}} - \mathbf{B}_{S_1}\, \mathbf{v}_{S_1} - \mathbf{B}_{S_3}^{\mathrm{int}}\, \mathbf{v}_{S_3}^{\mathrm{int}} = 0 \,, \tag{7}$$

and

$$\mathbf{A}_{S_2}\, \mathbf{p}_{S_2} + \mathbf{A}_{S_3}^{\mathrm{ext}}\, \mathbf{p}_{S_3}^{\mathrm{ext}} - \mathbf{B}_{S_2}\, \mathbf{v}_{S_2} - \mathbf{B}_{S_3}^{\mathrm{ext}}\, \mathbf{v}_{S_3}^{\mathrm{ext}} = 0 \,, \tag{8}$$

where the $\mathbf{p}_{S_i}$ and $\mathbf{v}_{S_i}$ are vectors of the sound pressures and normal particle velocities on the elements of the i-th surface. Furthermore, if one collocation point at the centroid of the each element, and constant interpolation is considered, the entries of the matrices $\mathbf{A}_{S_i}$, $\mathbf{B}_{S_i}$, can be computed as

$$a_{l,m} = \begin{cases} \int_{s_m} \frac{\partial \Psi(\mathbf{r}_l, \mathbf{r}_m)}{\partial \mathbf{n}_l} \, ds & \text{for } l \neq m \\ 1/2 & \text{for } l = m \end{cases}$$

$$b_{l,m} = -j\omega\rho_k \int_{s_m} \Psi(\mathbf{r}_l, \mathbf{r}_m) \, ds \,, \tag{9}$$

where s_m is the m-th surface element, the indexes $l = m = \{1,2,\ldots,L \text{ or } M\}$, and $k = \{1,2\}$ depending on which subdomain is being integrated.

When velocity values are prescribed to the elements of the vibrating surfaces of the loudspeaker drivers (see $\bar{v}_S$ in Fig. 3(b)), equations (7) and (8) can be further rewritten as

$$\mathbf{A}_{S_1}\,\mathbf{p}_{S_1} + \mathbf{A}_{S_3}^{\text{int}}\,\mathbf{p}_{S_3}^{\text{int}} - \hat{\mathbf{B}}_{S_1}\,\hat{\mathbf{v}}_{S_1} - \mathbf{B}_{S_3}^{\text{int}}\,\mathbf{v}_{S_3}^{\text{int}} = \bar{\mathbf{B}}_{S_1}\,\bar{\mathbf{v}}_{S_1} \,, \tag{10}$$

$$\mathbf{A}_{S_2}\,\mathbf{p}_{S_2} + \mathbf{A}_{S_3}^{\text{ext}}\,\mathbf{p}_{S_3}^{\text{ext}} - \hat{\mathbf{B}}_{S_2}\,\hat{\mathbf{v}}_{S_2} - \mathbf{B}_{S_3}^{\text{ext}}\,\mathbf{v}_{S_3}^{\text{ext}} = \bar{\mathbf{B}}_{S_2}\,\bar{\mathbf{v}}_{S_2} \,, \tag{11}$$

Thus, in equations (10) and (11), the $\hat{\mathbf{v}}_{S_i}$'s and $\bar{\mathbf{v}}_{S_i}$'s denote the unknown and known particle velocities, and $\hat{\mathbf{B}}_{S_i}$'s and $\bar{\mathbf{B}}_{S_i}$'s their corresponding coefficients.

At the auxiliary interface surface S_3, continuity of the boundary conditions must satisfy

$$\rho_1\, p_{S_3}^{\text{int}} = \rho_2\, p_{S_3}^{\text{ext}} \tag{12}$$

and

$$\frac{\partial p_{S_3}^{\text{int}}}{\partial \mathbf{n}_{3,1}} = -\frac{\partial p_{S_3}^{\text{ext}}}{\partial \mathbf{n}_{3,2}} \,, \quad \text{or} \quad , -j\omega\rho_1\, v_{S_3}^{\text{int}} = j\omega\rho_2\, v_{S_3}^{\text{ext}} \tag{13}$$

Considering that both domains Γ_1 and Γ_2 are filled with the same homogeneous medium (e.g. air), then $\rho_1 = \rho_2$, leading to

$$p_{S_3}^{\text{int}} = p_{S_3}^{\text{ext}} \tag{14}$$

and

$$-\,v_{S_3}^{\text{int}} = v_{S_3}^{\text{ext}} \tag{15}$$

Substituting these interface boundary parameters in equations (10) and (11), and rearranging into a global linear system of equations where the surface sound pressures and particle velocities represent the unknown parameters, yields to

$$\begin{bmatrix} \mathbf{A}_{S_1} & \mathbf{0} & \mathbf{A}_{S_3}^{\text{int}} & -\hat{\mathbf{B}}_{S_1} & \mathbf{0} & -\mathbf{B}_{S_3}^{\text{int}} \\ \mathbf{0} & \mathbf{A}_{S_2} & \mathbf{A}_{S_3}^{\text{ext}} & \mathbf{0} & -\hat{\mathbf{B}}_{S_2} & \mathbf{B}_{S_3}^{\text{ext}} \end{bmatrix} \begin{bmatrix} \mathbf{p}_{S1} \\ \mathbf{p}_{S2} \\ \mathbf{p}_{S3} \\ \hat{\mathbf{v}}_{S1} \\ \hat{\mathbf{v}}_{S2} \\ \mathbf{v}_{S3}^{\text{int}} \end{bmatrix} = \begin{bmatrix} \bar{\mathbf{B}}_{S_1}\,\bar{\mathbf{v}}_{S_1} \\ \bar{\mathbf{B}}_{S_2}\,\bar{\mathbf{v}}_{S_2} \end{bmatrix} . \tag{16}$$

Observe that the matrices $A's$ and $B's$ are known since they depend on the geometry of the model. Thus, once the vibration $\bar{\mathbf{v}}_{S1}$ of the loudspeakers is prescribed, and after equation (16) is solved for the surface parameters, the sound pressure at any point q can be readily computed by direct substitution and integration of equation (5) or (6). Note also that, a

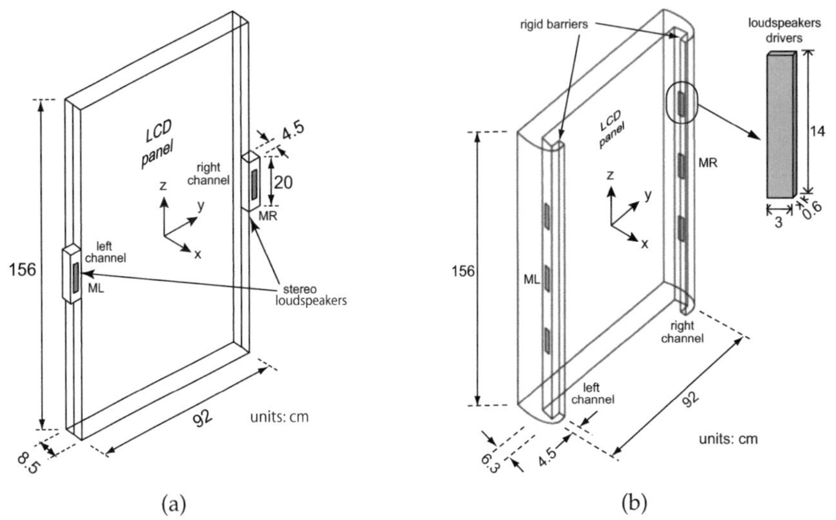

Fig. 4. Conventional stereo loudspeakers (a), and the L-like rigid barrier design (b), installed on 65″ flat display panels.

multidomain approach allows a reduction of computational effort during the optimization process since the coefficients of only one domain (interior) have to be recomputed.

3. Sound field analysis of a display-loudspeaker panel

3.1 Computer simulations

3.1.1 Numerical model

Recent multimedia applications often involve large displays to present life-size images and sound. Fig. 4(a) shows an example of a vertically aligned 65-inch LCD panel intended to be used in immersive teleconferencing. To this model, conventional (box) loudspeakers have been attached at the lateral sides to provide stereo spatialization. A second display panel is shown in Fig. 4(b), this is a prototype model of the L-shape loudspeakers introduced in the previous section. The structure of both models is considered to be rigid, thus, satisfying the requirements of flatness and hardness of the display surface.

In order to appreciate the sound field generated by each loudspeaker setup, the sound pressure at a grid of field points was computed following the theoretical BEM framework discussed previously. Considering the convention of the coordinate system illustrated in Figs. 4(a) and 4(b), the grid of field points were distributed within -0.5 m $\leq x \leq 2$ m and -1.5 m $\leq y \leq 1.5$ m spaced by 1 cm. For the numerical simulation of the sound fields, the models were meshed with isoparametric triangular elements with a maximum size of 4.2 cm which leaves room for simulations up to 1 kHz assuming a resolution of 8 elements per wavelength. The sound source of the simulated sound field was the left-side loudspeaker (marked as ML in Figs. 4(a) and 4(b)) emitting a tone of 250 Hz, 500 Hz and 1 kHz, respectively for each simulation. The rest of the structure is considered static.

3.1.2 Sound field radiated from the flat panels

The sound fields produced by each model are shown in Fig. 5. The sound pressure level (SPL) in those plots is expressed in dB's, where the amplitude of the sound pressure has been

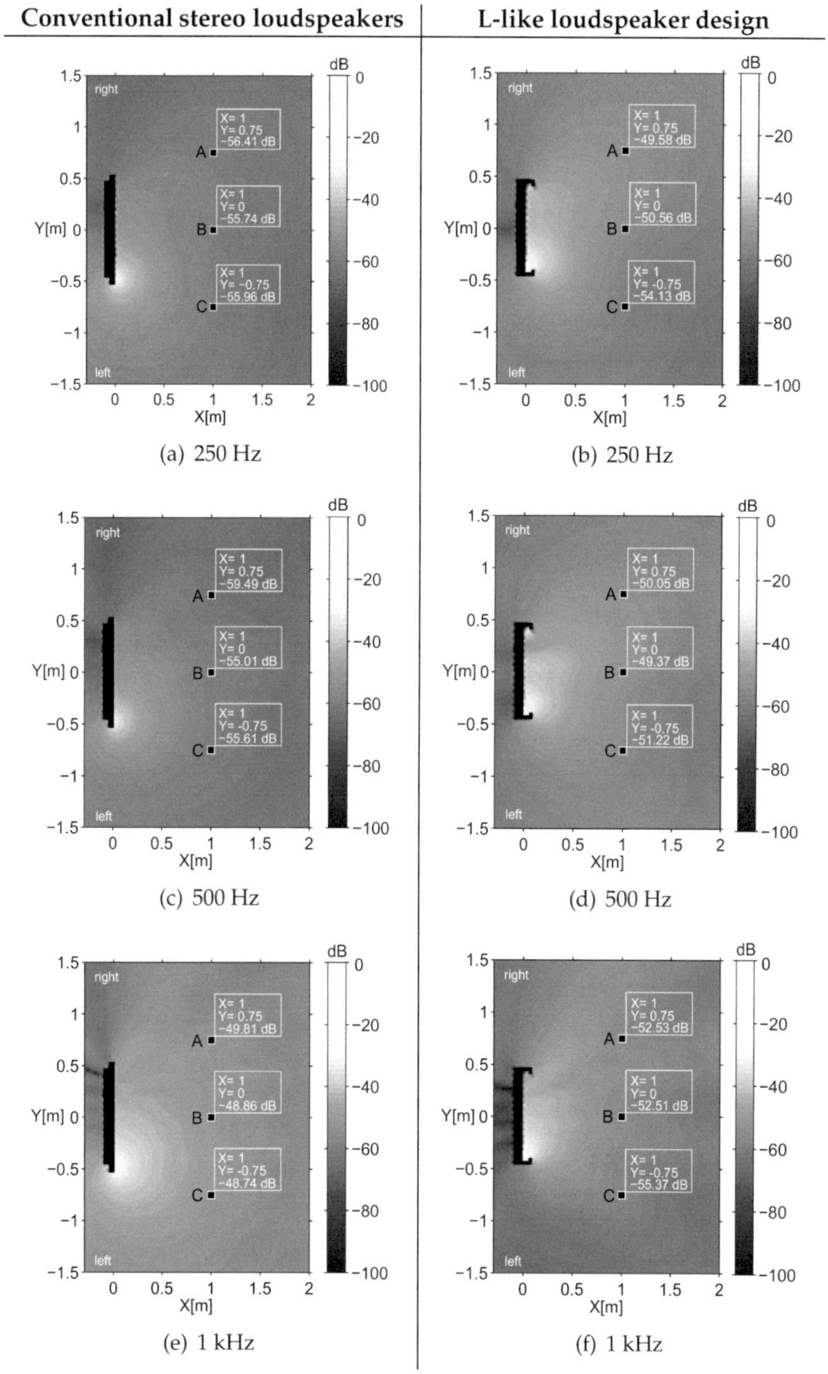

Fig. 5. Sound field generated by a conventional stereo setup (left column), and by the L-like loudspeaker design (right column) attached to a 65-inch display panel. Sound source: left-side loudspeaker (ML) emitting a tone of 250 Hz, 500 Hz, and 1 kHz respectively.

normalized to the sound pressure p_{spk} on the surface of the loudspeaker driver ML. For each analysis frequency, the SPL is given by

$$\text{SPL} = 20\log_{10}\frac{|p_{\text{q}}|}{|p_{spk}|} \tag{17}$$

where a SPL of 0 dB is observed on the surface of ML.
In the plots of Figs. 5(b), 5(d) and 5(f), (the L-like design), the SPL at the points A and B has nearly the same level, while point C accounts for the lowest level since the rigid barriers have effectively attenuated the sound at that area. Contrarily, Figs. 5(a), 5(c) and 5(e), (conventional loudspeakers) show that the highest SPL level is observed at point C (the closest to the sounding loudspeaker), whereas point A gets the lowest. Further note that if the right loudspeaker is sounding instead, symmetric plots are obtained. Let us recall the example where a sound image at the center of the display panel is desired. When both channels radiate the same signal, a listener on point B observes similar arrival times and sound intensities from both sides, leading to a sound image perception on the center of the panel. However, as demonstrated by the simulations, the sound intensities (and presumably, the arrival times) at the asymmetric areas are unequal. In the conventional stereo setup of Fig. 4(a), listeners at points A and C would perceive a sound image shifted towards their closest loudspeaker. But in the loudspeaker design of Fig. 4(b), the sound of the closest loudspeaker has been delayed and attenuated by the mechanical action of the rigid barriers. Thus, the masking effect on the sound from the opposite side is expected to be reduced leading to an improvement of sound image localization at the off-symmetry areas.

3.2 Experimental analysis

3.2.1 Experimental prototype

It is a common practice to perform experimental measurements to confirm the predictions of the numerical model. In this validation stage, a basic (controllable) experimental model is desired rather than a real LCD display which might bias the results. For that purpose, a flat dummy panel made of wood can be useful to play the role of a real display. Similarly, the rigid L-like loudspeakers may be implemented with the same material. An example of an experimental prototype is depicted in Fig. 6(a) which shows a 65-inch experimental dummy panel built with the same dimensions as the model of Fig. 4(b). The loudspeaker drivers employed in this prototype are 6 mm-thick flat coil drivers manufactured by FPS Inc., which can output audio signals of frequency above approximately 150 Hz. This experimental prototype was used to performed measurements of sound pressure inside a semi-anechoic room.

3.2.2 Sound pressure around the panel

The sound field radiated by the flat display panel has been demonstrated with numerical simulations in Fig. 5. In practice, however, measuring the sound pressure in a grid of a large number of points is troublesome. Therefore, the first experiment was limited to observe the amplitude of the sound pressure at a total of 19 points distributed on a radius of 65 cm from the center of the dummy panel, and separated by steps of 10° along the arc $-90^{\text{o}} \leq \theta \leq 90^{\text{o}}$ as depicted in Fig. 6(b), while the left-side loudspeaker ML was emitting a pure tone of 250 Hz, 500 Hz and 1 kHz respectively.

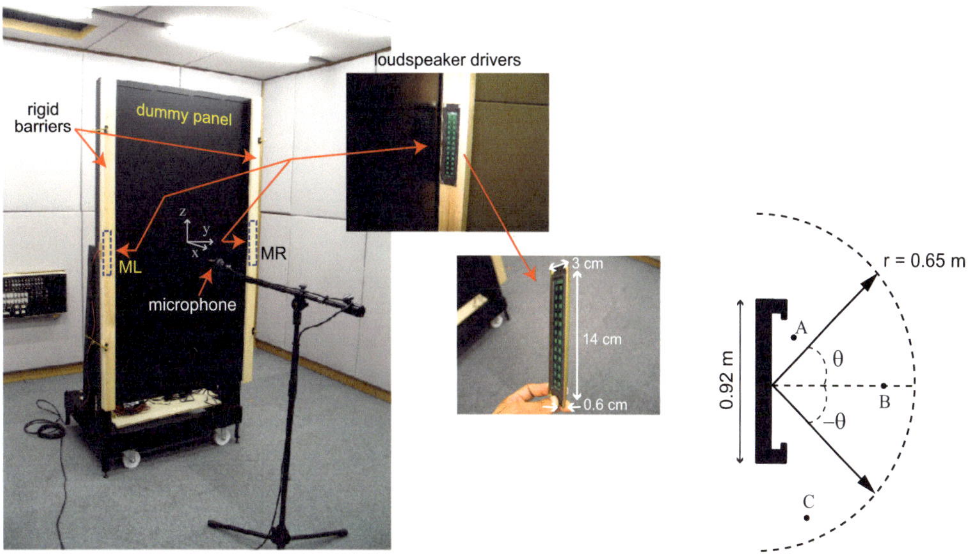

(a) (b)

Fig. 6. (a) Experimental dummy panel made of wood resembling a 65-inch vertically align LCD display. (b) Location of the measurements of the SPL generated by the dummy panel.

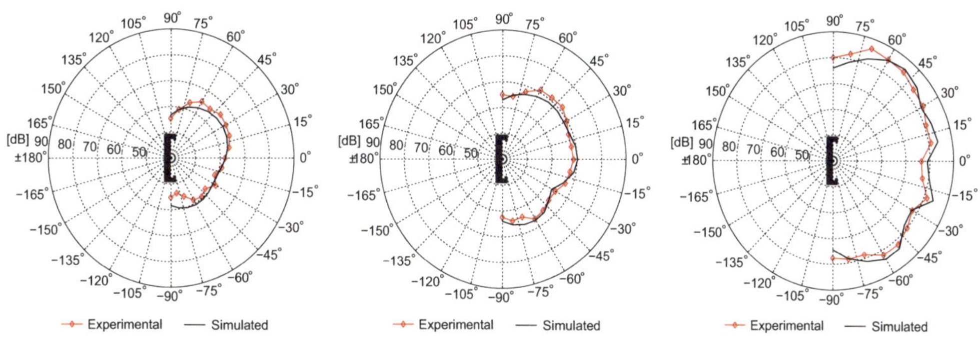

(a) 250 Hz (b) 500 Hz (c) 1 kHz

Fig. 7. Sound pressure at three static points (A, B and C), generated by a 65-inch LCD panel (Sharp LC-65RX) within the frequency band 0.2 - 4 kHz.

The attenuation of sound intensity introduced by the L-like rigid barriers as a function of the listening angle, can be observed on the polar plots of Fig. 7 where the results of the measurements are presented. Note that the predicted and experimental SPL show close agreement and also similarity to the sound fields of Fig. 5 obtained numerically, suggesting that the panel is effectively radiating sound as expected. Also, the dependency of the radiation pattern to the frequency has been made evident by these graphs, reason why this factor is taken into account in the acoustic optimization of the loudspeaker design.

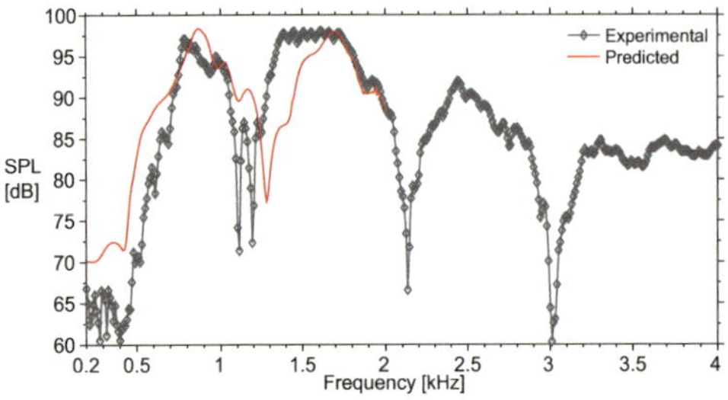

(a) Point A

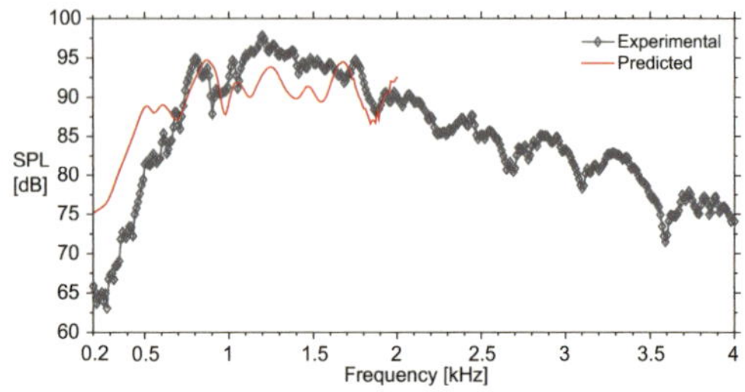

(b) Point B

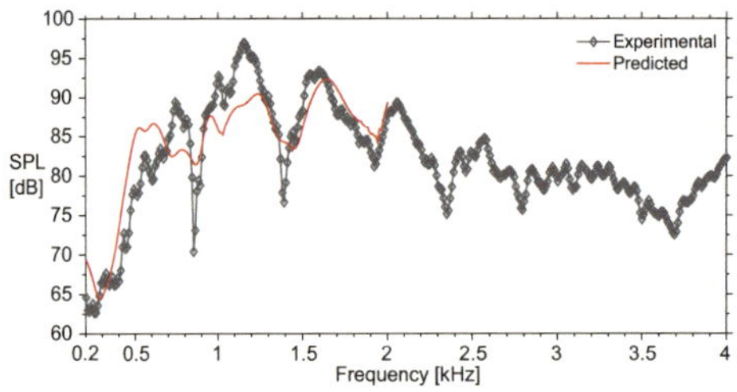

(c) 1 kHz

Fig. 8. Sound pressure level at a radius of 65 cm apart from the center of the dummy panel.

3.2.3 Frequency response in the sound field

A second series of measurements of SPL were performed at three points where, presumably, users in a practical situation are likely to stand. Following the convention of the coordinate system aligned to the center of the panel (see Fig. 6(a)), the chosen test points are $A(0.25, 0.25)$, $B(0.5, 0.0)$ and $C(0.3, -0.6)$ (in meters). At these points, the SPL due to the harmonic vibration of both loudspeakers, ML and MR, was measured within the frequency band 0.2–4 kHz with intervals of 10 Hz. For the case of the predicted data, the analysis was constrained to a maximum frequency of 2 kHz because of computational power limitations. The lower bound of 0.2 kHz is due to the frequency characteristics of the employed loudspeaker drivers.

The frequency response at the test points A, B and C, are shown in Fig. 8. Although there is a degree of mismatch between the predicted and experimental data, both show similar tendencies. It is also worth to note that the panel radiates relatively less acoustic energy at low frequencies (approximately below 800 Hz). This highpass response was originally attributed to the characteristics of the experimental loudspeaker drives, however, observation of a similar effect in the simulated data reveals that the panel, indeed, embodies a highpass behavior. This feature can lead to difficulties in speech perception in some applications such as in teleconferencing, in which case, reenforcement of the low frequency contents may be required.

4. Subjective evaluation of the sound images on the display panel

The perception of the sound images rendered on a display panel has been evaluated by subjective experiments. Thus, the purpose of these experiments was to assess the accuracy of the sound image localization achieved by the L-like loudspeakers, from the judgement of a group of subjects. The test group consisted of 15 participants with normal hearing capabilities whose age ranged between 23 and 56 years old (with mean of 31.5). These subjects were

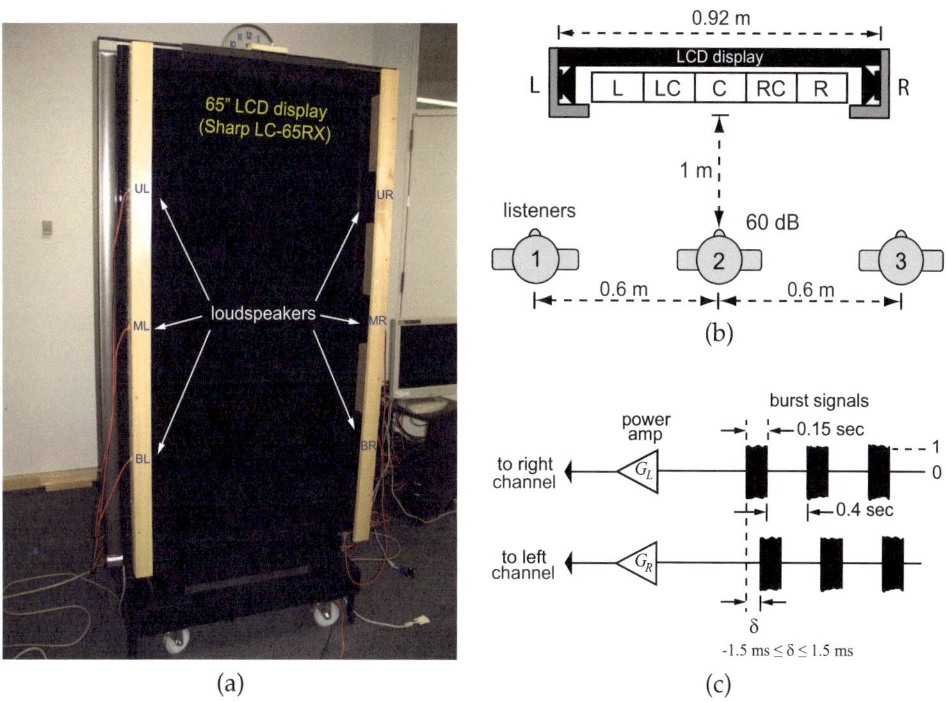

Fig. 9. 65-inch LCD display with the L-like loudspeakers installed (a). Setup for the subjective tests (b). Broadband signals to render a single sound image between ML and MR on the LCD display.

asked to localize the sound images rendered on the surface of a 65-inch LCD display (Sharp LC-65RX) which was used to implement the model of Fig. 9(a).

4.1 Setup for the subjective tests

The 15 subjects were divided into groups of 3 individuals to yield 5 test sessions (one group per session). Each group was asked to seat at one of the positions 1, 2 or 3 which are, one meter away form the display, as indicated in Fig. 9(b). In each session, the participants were presented with 5 sequences of 3 different sound images reproduced (one at a time) arbitrarily at one of the 5 equidistant positions marked as L, LC, C, RC, and R, along the line joining the left (ML) and right (MR) loudspeakers. At the end of each session, 3 sound images have appeared at each position, leading to a total of 15 sound images at the end of the session. After every sequence, the subjects were asked to identify and write down the perceived location of the sound images.

To render a sound image at a given position, the process started with a monaural signal of broadband-noise bursts with amplitude and duration as specified in Fig. 9(c). Therefore, to place a sound image, the gain G of each channel was varied within ($0 \leq G \leq 1$), and the delay δ between the channels was linearly interpolated in the range -1.5 ms $\leq \delta \leq 1.5$ ms. In such a way that, a sound image on the center corresponds to half the gain of the channels and zero delay, producing a sound pressure level of 60 dB (normalized to 20 μP) at the central point (position 2).

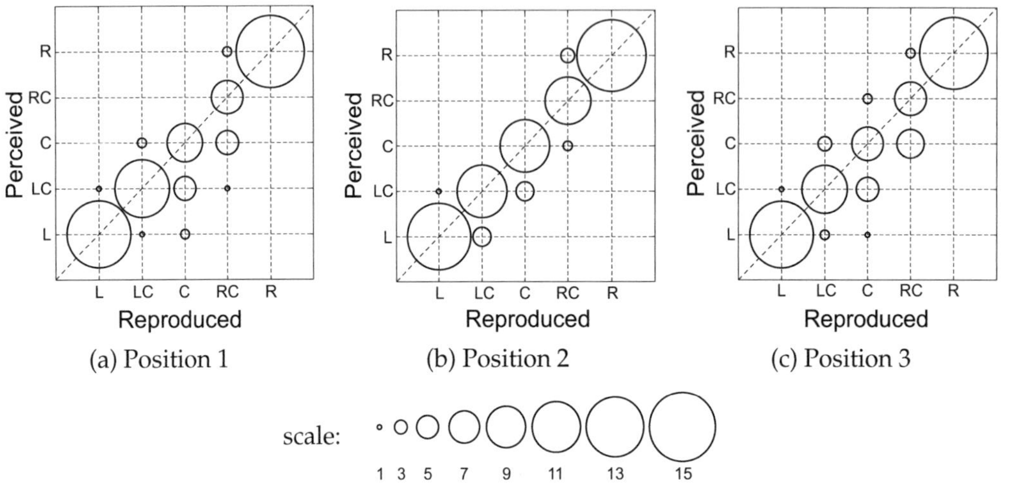

Fig. 10. Results of the subjective experiments.

4.2 Reproduced versus Perceived sound images

The data compiled from the subjective tests is shown in Fig. 10 as plots of *Reproduced* versus *Perceived* sound images. In the ideal case that all the reproduced sound images were perceived at the intended locations, a high correlation is visualized as plots of large circles with no sparsity from the diagonal. Although such ideal results were not obtained, note that the highest correlation between the parameters was achieved at Position 2 (Fig. 10(b)). Such result may be a priori expected since the sound delivered by the panel at that position is similar to that delivered by a standard stereo loudspeaker setup in terms of symmetry. At the lateral Positions 1 and 3, the subjects evaluated the sound images with more confusion which is reflected with some degree of sparsity in the plots of Figs. 10(a) and (c), but yet achieving significant level of correlation. Moreover, it is interesting to note the similarity of the correlation patterns of Figs(a) and (c) which implies that listeners at those positions were able to perceive similar sound images.

5. Example applications: Multichannel auditory displays for large screens

One of the challenges of immersive teleconference systems is to reproduce at the local space, the acoustic (and visual) cues from the remote meeting room allowing the users to maintain the sense of presence and natural interaction among them. For such a purpose, it is important to provide the local users with positional agreement between what they see and what they hear. In other words, it is desired that the speech of a remote speaker is perceived as coming out from (nearby) the image of his/her face on the screen. Aiming such problem, this section introduces two examples of interactive applications that implement multichannel auditory displays using the L-like loudspeakers to provide realistic sound reproduction on large display panels in the context of teleconferencing.

5.1 Single-sound image localization with real-time talker tracking

Fig. 11 of the first application example, presents a multichannel audio system capable of rendering a remote user's voice at the image of his face which is being tracked in real-time by video cameras. At the remote side, the monaural signal of the speech of a speaker (original

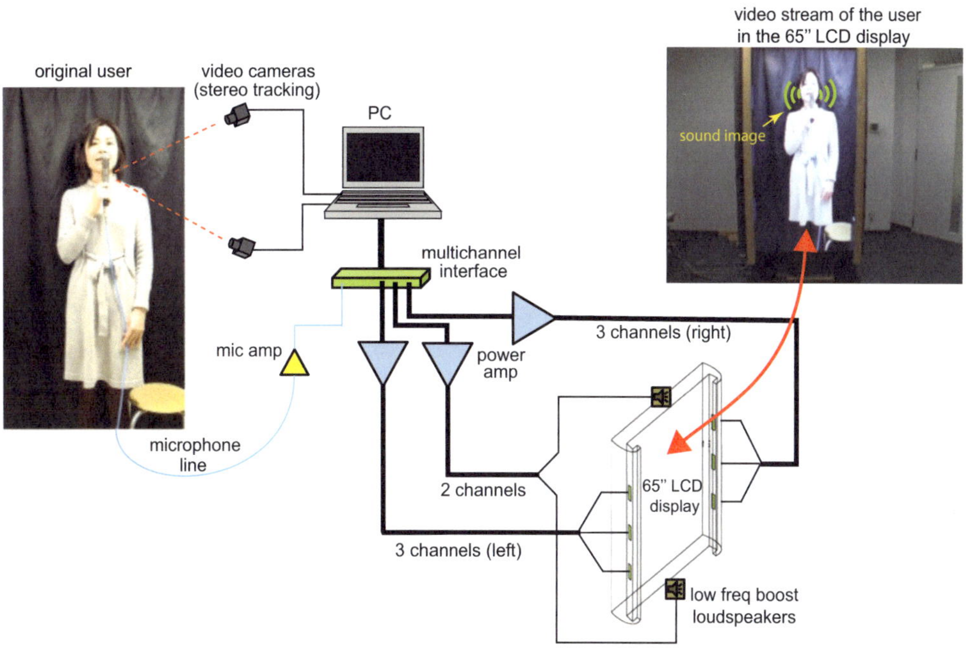

Fig. 11. A multichannel (8 channels) audio system for a 65-inch LCD display, combined with stereo video cameras for real-time talker tracking.

user, Fig.11) is acquired by a microphone on which a visual marker was installed. The position of the marker is constantly estimated and tracked by a set of video cameras. This simple video tracking system assumes that the speaker holds the microphone close to his mouth when speaking, thus, the origin of the sound source can be inferred. Note that for the purpose of demonstration of the auditory display, this basic implementation works, but alternatively it can be replaced by current robust face tracking algorithms to improve the localization accuracy and possibly provide a hands-free interface.

In the local room (top-right picture of Fig. 11), while the video of the remote user is being streamed to a 65-inch LCD screen, the audio is being output through the 6-channel loudspeakers attached to the screen panel. In fact, the 65-inch display used in this real-time interactive application is the prototype model of Fig. 4(a) plus two loudspeakers at the top and bottom to enforce the low frequency contents. Therefore, the signal to drive these booster loudspeakers is obtained by simply lowpass filtering (cut off above 700 Hz) the monaural source signal of the microphone. As for the sound image on the surface of the display, once the position of the sound source (i.e. the face of the speaker) has been acquired by the video cameras, the coordinate information is used to interpolate the sound image (left and right, and up/down), thus, the effect of a moving sound source is simulated by panning the monaural source signal among the six lateral channels in a similar way as described in section 4.1. The final effect is a sound image that moves together with the streaming video of a remote user, providing a realistic sense of presence for a spectator in the local end.

5.2 Sound positioning in a multi-screen teleconference room

The second application example is an implementation of an auditory display to render a remote sound source on the large displays of an immersive teleconference/collaboration room

known as t-Room (see ref. to NTT CS). In its current development stage, various users at different locations can participate simultaneously in a meeting by sharing a *common virtual space* recreated by the local t-Room in which each of them is physically present. Other users can also take part of the meeting by connecting through a mobile device such a note PC. In order to participate in a meeting, a user requires only the same interfaces needed for standard video chat through internet: a web camera, and a head set (microphone and earphones). In Fig. 12 (right lower corner), a remote user is having a discussion from his note PC with attendees of a meeting inside a t-Room unit (left upper corner). Moreover, the graphic interface in the laptop is capable of providing full-body view of the t-Room participants through a 3D representation of the eight t-Room's decagonally aligned displays. Thus, the note PC user is allowed to navigate around the display panels to change his view angle, and with the head set, he can exchange audio information as in a normal full-duplex audio system. Inside t-Room, the local users have visual feedback of the remote user through a video window representing the note PC user's position. Thus, this video window can be moved to the remote user's will, and as the window moves around (and up/down) in the displays, the sound image of his voice also displaces accordingly. In this way, local users who are dispersed within the t-Room space are able to localize the remote user's position not only by visual but also by audible cues.
The reproduction of sound images over the 8 displays is achieved by a 64-channel loudspeaker system (8 channels per display). Each display is equipped with a loudspeaker array similar to that introduced in the previous section: 6 lateral channels plus 2 low frequency booster channels. As in the multichannel audio system with speaker tracking, the sound image of the laptop user is interpolated among the 64 channels by controlling the gain of those channels necessary to render a specific sound images as a function of the video window position. Non-involved channels are switched off at the corresponding moment. For this multichannel auditory display, the position of the speech source (laptop user) is not estimated by video cameras but it is readily known from the laptop's graphic interface used to navigate inside t-Room, i.e., the sound source (face of the user) is assumed to be nearby the center of the video window displayed at the t-Room side.

6. Potential impact of the sound image localization technology for large displays

As display technologies evolve, the future digital environment that surrounds us will be occupied with displays of diverse sizes playing a more ubiquitous role (Intille, 2002; McCarthy et al., 2001). In response to such rapid development, the sound image localization approach introduced in this chapter opens the possibility for a number of applications with different levels of interactivity. Some examples are discussed in what follows.

6.1 Supporting interactivity with positional acoustic cues

Recent ubiquitous computing environments that use multiple displays often output rich video contents. However, because the user's attentive capability is limited by his field of vision, user's attention management has become an issue of research (Vertegaal, 2003). Important information which is displayed on a screen out of the scope of the user's visual attention may just be missed or not realized on time. But on the other hand, since humans are able to accurately localize sound in a 360^0 plane, auditory notifications represent an attractive alternative to deliver information (e.g. (Takao et al., 2002)). Let us consider the specific example of the video interactivity in t-Room. Users have reported discomfort when using

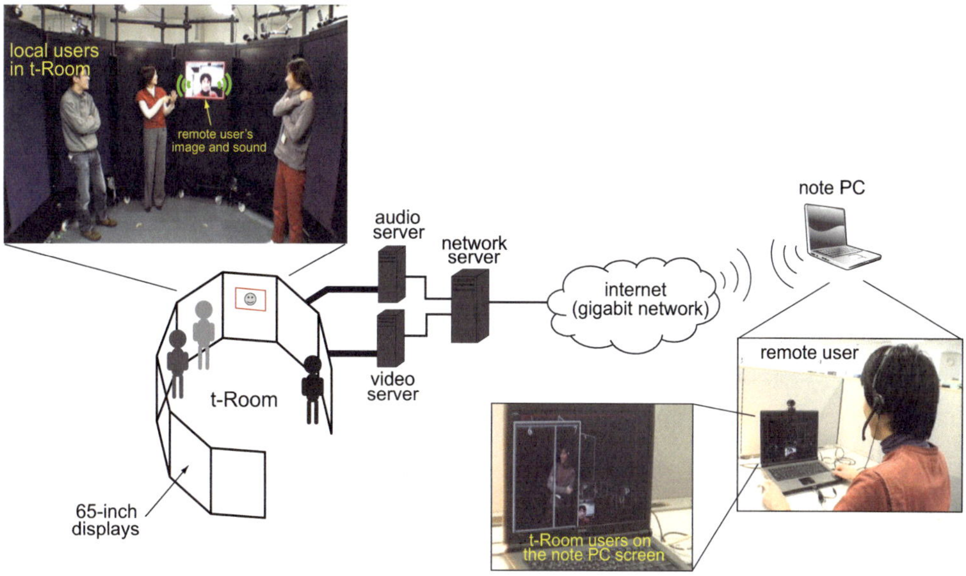

Fig. 12. Immersive teleconference room (t-Room) with a multichannel (64 channels) auditory display to render the sound images of remote participants on the surface of its large LCD displays.

the mouse pointer which is often visually lost among the eight surrounding large screens. This problem is even worsened as users are free to change their relative positions. In this case, with the loudspeaker system introduced in this chapter, it is possible to associate a subtle acoustic image positioned on the mouse pointer to facilitate its localization. Another example of a potential application is in public advertising where public interactive media systems with large displays have been already put in practice (Shinohara et al., 2007). Here, a sound spatialization system with a wide listening area can provide information on the spatial relationship among several advertisements.

6.2 Delivering information with positional sound as a property

In the field of Human Computer Interaction, there is an active research on the user-subconscious interactivity based on the premise that humans have the ability to subconsciously process information which is presented at the background of his attention. This idea has been widely used to build not only ambient video displays but also ambient auditory displays. For example, the whiteboard system of Wisneski et al. (1998) outputs an ambient sound to indicate the usage status of the whiteboard. Combination of musical sounds with the ambient background has been also explored (E. D. Mynatt & Ellis, 1998).

In an ambient display, the source information has to be appropriately mapped into the background in order to create a subtle representation in the ambient (Wisneski et al., 1998). For the case of an auditory ambient, features of the background information have been used to control audio parameters such as sound volume, musical rhythm, pitch and music gender. The controllable parameters can be further extended with a loudspeaker system that in addition allows us to position the sound icons according to information contents (e.g. depending on its relevance, the position and/or characteristics of the sound are changed).

6.3 Supporting position-dependent information

There are situations where it is desired to communicate specific information to a user depending on his position and/or orientation. This occurs usually in places where the users are free to move and approach contextual contents of his interest. For example, at event spaces such as museums, audio headsets are usually available with pre-recorded explanations which are automatically playbacked as the user approaches an exhibition booth. Sophisticated audio earphones with such features have been developed (T. Nishimura, 2004). However, from the auralization point of view, sound localization can be achieved only for the user who wears the headset. If a number of users within a specific listening field is considered, the L-like loudspeaker design offers the possibility to control the desired audible perimeter by optimizing the size of the L-like barriers to the target area and by controlling the radiated sound intensity. Thus, only users within the scope of the information panel listen to the sound images of the corresponding visual contents, while users out of that range remain undisturbed.

7. Conclusions

In this chapter, the issue of sound image localization with stereophonic audio has been addressed making emphasis on sound spatialization for applications which involve large flat displays. It was pointed out that the effect of precedence that occur with conventional stereo loudspeakers setups represents an impairment to achieve accurate localization of sound images over a wide listening area. Furthermore, some of the approaches dealing with this problem were enumerated. The list of the survey was extended with the introduction of a novel loudspeaker design targeting the sound image localization on flat display panels. Compared to existent techniques, the proposed design aims to achieve expansion of the listening area by mechanically altering the radiated sound field through the attachment of L-like rigid barriers and a counter-fire positioning of the loudspeaker drivers. Results from numerical simulations and experimental tests have shown that the insertion of the rigid barriers effectively aids to redirect the sound field to the desired space. The results also exposed the drawbacks of the design, such as the dependency of its radiation pattern with the dimensions of the target display panel and the listening coverage. For such a reason, the dimensions of the L-like barriers have to be optimized for a particular application. The need for low-frequency reenforcement is another issue to take into account in applications where the intelligibility of the audio information (e.g. speech) is degraded. On the other hand, it is worth to remark that the simplicity of the design makes it easy to implement on any flat hard display panel.

To illustrate the use of the proposed loudspeaker design, two applications within the framework of immersive telepresence were presented: one, an audio system for a single 65-inch LCD panel combined with video cameras for real-time talker tracking, and another, a multichannel auditory display for an immersive teleconference system. Finally, the potentiality of the proposed design was highlighted in terms of sound spatialization for human-computer interfaces in various multimedia scenarios.

8. References

Aoki, S. & Koizumi, N. (1987). Expansion of listening area with good localization in audio conferencing, *ICASSP '87*, Dallas TX, USA.

Bauer, B. B. (1960). Broadening the area of stereophonic perception, *J. Audio Eng. Soc.* 8(2): 91–94.

Berkhout, A. J., de Vries, D. & Vogel, P. (1993). Acoustic control by wave field synthesis, *J. Acoustical Soc. of Am.* 93(5): 2764–2778.

Ciskowski, C. & Brebbia, C. (1991). *Boundary Element Methods in Acoustics*, Elsevier, London.

Davis, M. F. (1987). Loudspeaker systems with optimized wide-listening-area imaging, *J. Audio Eng. Soc.* 35(11): 888–896.

E. D. Mynatt, M. Back, R. W. M. B. & Ellis, J. (1998). Designing audio aura, *Proc. of SIGCHI Conf. on Human Factors in Computing Systems*, Los Angeles, US.

Estorff, O. (2000). *Boundary Elements in Acoustics, Advances and Applications*, WIT Press, Southampton.

Gardner, M. B. (1968). Historical background of the haas and/or precedence effect, *J. Acoustical Soc. of Am.* 43(6): 1243–1248.

Gardner, W. G. (1997). *3-D Audio using loudspeakers*, PhD thesis.

Intille, S. (2002). Change blind information display for ubiquitous computing environments, *Proc. of Ubicomp2002*, Göterborg, Sweden, pp. 91–106.

Kates, J. M. (1980). Optimum loudspeaker directional patterns, *J. Audio Eng. Soc.* 28(11): 787–794.

Kim, S.-M. & Wang, S. (2003). A wiener filter approach to the binaural reproduction of stereo sound, *J. Acoustical Soc. of Am.* 114(6): 3179–3188.

Kyriakakis, C., Holman, T., Lim, J.-S., Hong, H. & Neven, H. (1998). Signal processing, acoustics, and psychoacoustics for high quality desktop audio, *J. Visual Com. and Image Represenation* 9(1): 51–61.

Litovsky, R. Y., Colubrn, H. S., Yost, W. A. & Guzman, S. J. (1999). The precedence effect, *J. Acoustical Soc. of Am.* 106(4): 1633–1654.

McCarthy, J., Costa, T. & Liongosari, E. (2001). Unicast, outcast & groupcast: Toward ubiquitous, peripheral displays, *Proc. of Ubicomp2001*, Atlanta, US, pp. 331–345.

Melchoir, F., Brix, S., Sporer, T., Roder, T. & Klehs, B. (2003). Wave field synthesis in combination with 2D video projection, *24th AES Int. Conf. Multichannel Audio, The New Reality*, Alberta, Canada.

Merchel, S. & Groth, S. (2009). Analysis and implementation of a stereophonic play back system for adjusting the "sweet spot" to the listener's position, *126th Conv. of the Audio Eng. Soc.*, Munich, Germany.

NTT CS The future telephone: *t-Room*, NTT Communication Science Labs. http://www.mirainodenwa.com/e_index.html.

Rakerd, B. (1986). Localization of sound in rooms, III: Onset and duration effects, *J. Acoustical Soc. of Am.* 80(6): 1695–1706.

Ródenas, J. A., Aarts, R. M. & Janssen, A. J. E. M. (2003). Derivation of an optimal directivity pattern for sweet spot widening in stereo sound reproduction, *J. Acoustical Soc. of Am.* 113(1): 267–278.

Seybert, A., Cheng, C. & Wu, T. (1990). The resolution of coupled interior/exterior acoustic problems using boundary element method, *J. Acoustical Soc. of Am.* 88(3): 1612–1618.

Shinohara, A., Tomita, J., Kihara, T., Nakajima, S. & Ogawa, K. (2007). A huge screen interactive public media system: mirai-tube, *Proc. of 2th international Conference on Human-Computer interaction: interaction Platforms and Techniques*, Beijin, China, pp. 936–945.

T. Nishimura, Y. Nakamura, H. I. H. N. (2004). System design of event space information support utilizing cobits, *Proc. of Distributed Computing Systems Wrokshops*, Tokyo, Japan, pp. 384–387.

Takao, H., Sakai, K., Osufi, J. & Ishii, H. (2002). Acoustic user interface (aui) for the auditory displays, *Communications of the ACM* 23(1-2): 65–73.

Vertegaal, R. (2003). Attentive user interfaces, *Communications of the ACM* 46(3): 30–33.

Werner, P. J. & Boone, M. M. (2003). Application of wave field synthesis in life-size videoconferencing, *114th Conv. of the Audio Eng. Soc.*, Amsterdam, The Netherlands.

Wisneski, C., Ishii, H. & Dahley, A. (1998). Ambient displays: Turning architectural space into an interface between people and digital information, *Proc. of Int. Workshop on Cooperative Buildings*, Darmstadt, Germany, pp. 22–32.

Wu, T. (2000). *Boundary Element Acoustics, Fundamentals and Computer Codes*, WIT Press, Southampton.

20

Backward Compatible Spatialized Teleconferencing based on Squeezed Recordings

Christian H. Ritz[1], Muawiyath Shujau[1], Xiguang Zheng[1], Bin Cheng[1], Eva Cheng[1,2] and Ian S Burnett[2]
[1]*School of Electrical, Computer and Telecommunications Engineering, University of Wollongong, Wollongong,*
[2]*School of Electrical and Computer Engineering, RMIT University, Melbourne, Australia*

1. Introduction

Commercial teleconferencing systems currently available, although offering sophisticated video stimulus of the remote participants, commonly employ only mono or stereo audio playback for the user. However, in teleconferencing applications where there are multiple participants at multiple sites, spatializing the audio reproduced at each site (using headphones or loudspeakers) to assist listeners to distinguish between participating speakers can significantly improve the meeting experience (Baldis, 2001; Evans et al., 2000; Ward & Elko 1999; Kilgore et al., 2003; Wrigley et al., 2009; James & Hawksford, 2008). An example is Vocal Village (Kilgore et al., 2003), which uses online avatars to co-locate remote participants over the Internet in virtual space with audio spatialized over headphones (Kilgore, *et al.*, 2003). This system adds speaker location cues to monaural speech to create a user manipulable soundfield that matches the avatar's position in the virtual space. Giving participants the freedom to manipulate the acoustic location of other participants in the rendered sound scene that they experience has been shown to provide for improved multitasking performance (Wrigley et al., 2009).
A system for multiparty teleconferencing requires firstly a stage for recording speech from multiple participants at each site. These signals then need to be compressed to allow for efficient transmission of the spatial speech. One approach is to utilise close-talking microphones to record each participant (e.g. lapel microphones), and then encode each speech signal separately prior to transmission (James & Hawksford, 2008). Alternatively, for increased flexibility, a microphone array located at a central point on, say, a meeting table can be used to generate a multichannel recording of the meeting speech A microphone array approach is adopted in this work and allows for processing of the recordings to identify relative spatial locations of the sources as well as multichannel speech enhancement techniques to improve the quality of recordings in noisy environments. For efficient transmission of the recorded signals, the approach also requires a multichannel compression technique suitable to spatially recorded speech signals.

A recent approach for multichannel audio compression is MPEG Surround (Breebaart et al., 2005). While this approach provides for efficient compression, it's target application is loudspeaker signals such as 5.1 channel surround audio rather than microphone array recordings. More recently, Directional Audio Coding (DirAC) was proposed for both compression of loudspeaker signals as well as microphone array recordings (Pulkki, 2007) and in (Ahonen et al., 2007), an application of DirAC to spatial teleconferencing was proposed. In this chapter, an alternative approach based on the authors' Spatially Squeezed Surround Audio Coding (S^3AC) framework (Cheng et al., 2007) will be presented. In previous work, it has been shown that the S^3AC approach can be successfully applied to the compression of multichannel loudspeaker signals (Cheng et al., 2007) and has some specific advantages over existing approaches such as Binaural Cue Coding (BCC) (Faller et al., 2003), Parametric Stereo (Breebaart et al., 2005) and the MPEG Surround standard (Breebaart, et al., 2005). These include the accurate preservation of spatial location information whilst not requiring the transmission of additional side information representing the location of the spatial sound sources. In this chapter, it will be shown how the S^3AC approach can be applied to microphone array recordings for use within the proposed teleconferencing system. This extends the previous work investigating the application of S^3AC to B-format recordings as used in Ambisonics spatial audio (Cheng et al., 2008b) as well as the previously application of S^3AC to spatialized teleconferencing (Cheng et al., 2008a).

For recording, there are a variety of different microphone arrays that can be used such as simple uniform linear or circular arrays or more complex spherical arrays, where accurate recording of the entire soundfield is possible. In this chapter, the focus is on relatively simple microphone arrays with small numbers of microphone capsules: these are likely to provide the most practical solutions for spatial teleconferencing in the near future. In the authors' previously proposed spatial teleconferencing system (Cheng et al., 2008a), a simple four element circular array was investigated. Recently, the authors have investigated the Acoustic Vector Sensor (AVS) as an alternative for recording spatial sound (Shujau et al., 2009). An AVS has a number of advantages over existing microphone array types including their compact size (occupying a volume of approximately 1 cm^3) whilst still being able to accurately record sound sources and their location. In this chapter, the S^3AC will be used to process and encode the signals captured from an AVS.

Fig. 1 illustrates the conceptual framework of the multi-party teleconferencing system with N geographically distributed sites concurrently participating in the teleconference. At each site, a microphone array (in this work an AVS) is used to record all participants and the resulting signals are then processed to estimate the spatial location of each speech source (participant) relative to the array and to enhance the recorded signals that may be degraded by unwanted noise present in the meeting room (e.g. babble noise, environmental noise). The resulting signals are then analysed to derive a downmix signal using the S^3AC representing the spatial meeting speech. The downmix signal is an encoding of the individual speech signals as well as information representing their original location at the participants' site. The downmix could be a stereo signal or a mono signal. For a stereo (two channel) downmix, spatial location information for each source is encoded as a function of the amplitude ratios of the two channels; this requires no separate transmission of spatial location information. For a mono (single channel) downmix, separate information representing the spatial location of the sound sources is transmitted as side information. In either approach, the downmix signal is further compressed in a backwards compatible approach using standard audio coders such as the Advanced Audio Coder (AAC) (Bosi & Goldberg, 2002). Since the application of this chapter is spatial teleconferencing, downmix compression is achieved using the extended Adaptive Multi-Rate Wide Band (AMR-WB+) coder (Makinen, 2005). This coder is chosen as it is one of the best performing standard coders at low bit rates for both speech and audio (Makinen, 2005) and is particularly suited to S^3AC. In Fig. 1, each site must unambiguously spatialise N-1 remote sites and utilizes a standard 5.1 playback system, however, the system is not restricted to this and alternative playback scenarios could be used (e.g. spatialization via headphones using Head Related Transfer Functions (HRTFs) (Cheng et al., 2001).

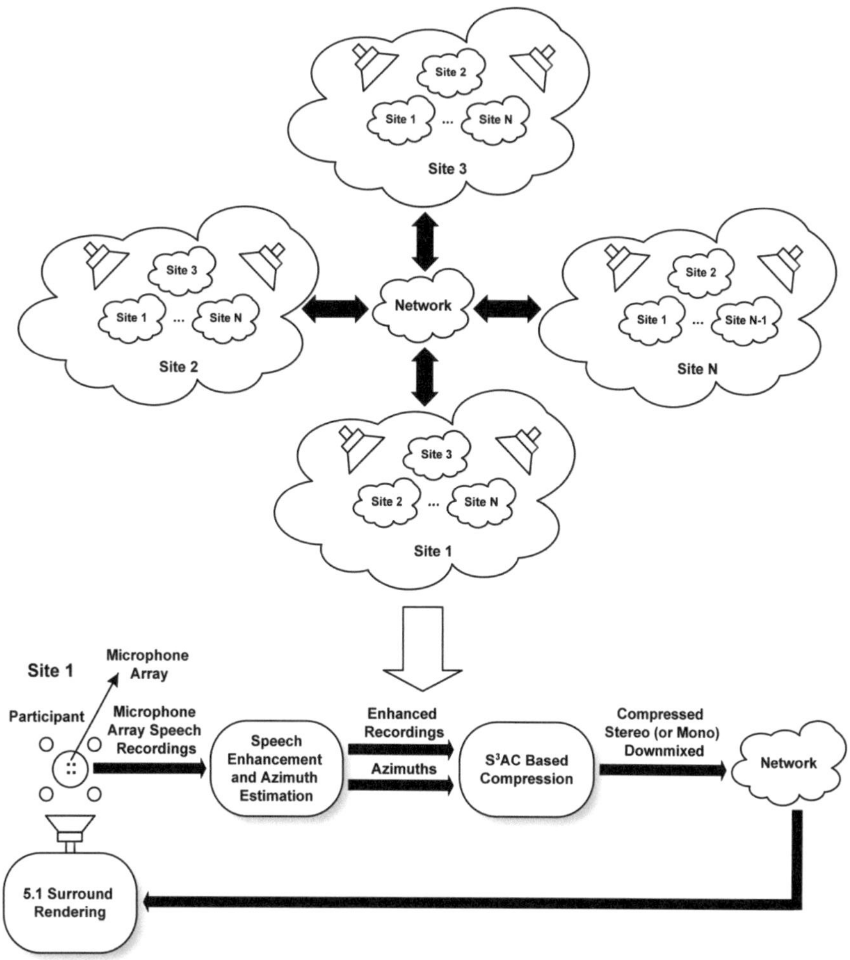

Fig. 1. Conceptual Framework of the Spatial Teleconferencing System. Illustrated are multiple sites each participating in a teleconference as well as a system overview of the S^3AC-based recording and coding system used at each site.

A fundamental principle of S^3AC is the estimation of the location of sound sources and this requires estimation of the location of sources corresponding to each speaker. In (Cheng et al., 2008a), the speaker azimuths were estimated using using the Steered Response Power with PHAse Transform (SRP-PHAT) algorithm (DiBiase et al., 2001). This technique is suited to spaced microphone arrays such as the circular array presented in Fig. 1 and relies on Time-Delay Estimation (TDE) applied to microphone pairs in the array. In the current system, the AVS is a co-incident microphone array and hence methods based on TDE such as SRP-PHAT are not directly applicable. Hence in this work, source location information will be found by performing Directional of Arrival (DOA) estimation using the Multiple Signal Classification (MUSIC) method as proposed in (Shujau et al., 2009).

In this chapter two multichannel speech enhancement techniques are investigated and compared: a technique based on the Minimum Variance Distortionless Response (MVDR) beamformer (Benesty et al., 2008); and an enhancement technique based on sound source separation using Independent Component Analysis (ICA) (Hyvärinen et al., 2001). In contrast to existing work, these enhancement techniques are applied to the coincident AVS microphone array and results will extend those previously described in (Shujau et al., 2010).

The structure of this chapter is as follows: Section 2 will describe the application of S^3AC to the proposed teleconferencing system while Section 3 will describe the recording and source location estimation based on the AVS; Section 4 will describe the experimental methodology adopted and present objective and subjective results for sound source location estimation, speech enhancement and overall speech quality based on Perceptual Evaluation of Speech Quality (PESQ) (ITU-R P.862, 2001) measures; Conclusions will be presented in Section 4.

2. Spatial teleconferencing based on S3AC

In this section, an overview of the S^3AC based spatial teleconferencing system will first be presented followed by a detailed description of the transcoding and decoding stages of the system.

2.1 Overview of the system

Fig. 2 describes the high level architecture of the proposed spatial teleconferencing system based on S^3AC. Each site records one or more sound sources using a microphone array and these recordings are analysed to derive individual sources and information representing their spatial location using the source localisation approaches illustrated in Fig. 1 and described in more detail in Section 3. In this work, spatial location is determined only as the azimuth of the source in the horizontal plane relative to the array. In Fig. 2 sources and their corresponding azimuth are indicated as Speaker 1 + Azimuth to Speaker N + Azimuth.

The resulting signals from one or more sites are input to the S^3AC transcoder that processes the signals using the techniques to be described in Section 2.2 to produce a downmix signal that encodes the original soundfield information. The downmix signal can either be a stereo signal (labeled as S^3AC-SD in Fig. 2), where information about the source location is encoded as a function of the amplitude ratio of the two signals (see Section 2.2) or a mono-signal (labeled as S^3AC-SD in Fig. 2), where side-information is used to encode the source location information. In the implementation described in this work, the downmix is compressed using the AMR-WB+ coder, as illustrated in Fig. 2. This AMR-WB+ coder was chosen to provide backwards compatibility with a state-of-the-art standardised coder that has been shown to provide superior performance for speech and mixtures of speech and other audio at low bit rates (6 kbps up to 36 kbps), which is the target of this work.

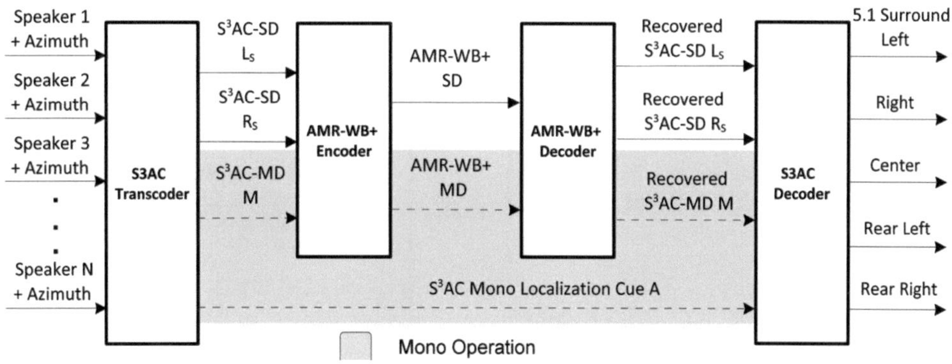

Fig. 2. High Level Architecture of the S³AC based teleconferencing system. S³AC-SD refers to the Stereo Downmix mode while S³AC-MD refers to the optional Mono Downmix mode. Speaker 1 to Speaker N refers to the recorded signals from one or more sites.

At the decoder, following decoding by the speech codec, the received downmix signals are analysed using the S³AC decoder described in Section 2.3 to determine the encoded source signals and information representing their spatial location. It should be noted that the spatial information represents the original location of each speaker relative to a central point at the recording site. The final stage is rendering of a spatial soundfield representing the teleconference, which is achieved using a standard 5.1 Surround Sound loudspeaker system (although alternative spatialization techniques may also be used due to the coding framework representing sound sources and their locations, which provides for alternative spatial rendering).

2.2 S³AC transcoder

An illustration of the S³AC transcoder is shown in Fig. 3 and consists of three main stages: Time-Frequency Transformation, Spatial Squeezing and Inverse Time-Frequency Transformation. Input to the S³AC transcoder are the speaker signals and corresponding azimuths of Fig. 2. Here, $s_{i,j}(n)$ and $\theta_{i,j}(n)$ are defined asthe speech source j and corresponding azimuth at site i, where $i=1$ to N and $j = 1$ to M_j and where N is the number of sites and M_j is the number of participants (unique speech sources) at each site.

In Fig. 3, this notation is used to indicate for site 1, signals representing the recorded sources and their corresponding azimuths. These signals are converted to the Fourier domain using a short time Fourier transform to produce the frequency domain signals $S_{i,j}(n,k)$, where n represents the time frame and k represents discrete frequency. Here, similar to the existing principle of S³AC, a separate azimuth is determined for each time-frequency component using the direction of arrival estimation approaches described in Section 3. While the azimuth is not expected to vary widely with frequency when a single participant is speaking, there will be variation when multiple participants are speaking concurrently; hence azimuths are denoted $\theta_{i,j}(n,k)$. This indicates that at each time and frequency there could be one or more speakers active at one or more sites.

The second stage of the S³AC transcoder is spatial squeezing, which assigns a new azimuth for the sound source in a squeezed soundfield. Conceptually, this involves a mapping of the source azimuth derived for the original 360° soundfield of the recording site to a new azimuth within a smaller region of a virtual 360° soundfield that represents all sources from all sites. This process can be described as:

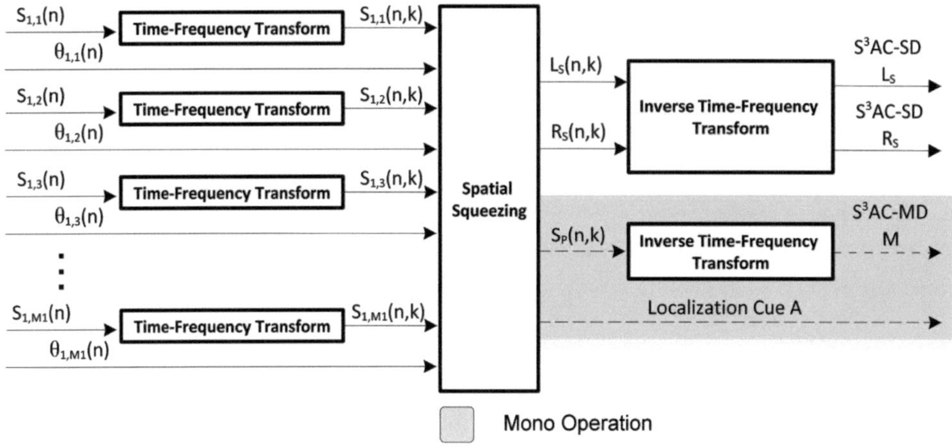

Fig. 3 S³AC Transcoder showing the encoding of multiple spatial speech signals and their azimuths as a time domain stereo (or optional mono) downmix signal.

$$\theta_{i,j}^{s}(n,k) = f\left(\theta_{i,j}(n,k)\right) \tag{1}$$

where f is a mapping function, which can be thought of as a quantization of the original azimuth to the squeezed azimuth. Examples of mapping functions for spatial audio compression are described in (Cheng et al., 2006). Here, a uniform quantization approach is adopted, whereby each azimuth is mapped to a squeezed azimuth equal to one of a possible $360/N$ quantized azimuths; conceptually, this divides the virtual soundfield into N equal regions, each representing one of the N remote sites. Following azimuth mapping, a downmix signal is created using one of two possible. Firstly, a stereo downmix can be created using the approach described by:

$$\begin{aligned} L_S(n,k) &= \frac{S_p(n,k)\cdot\left(\tan\varphi_d + \tan\theta_{p,s}(n,k)\right)}{\sqrt{2\tan^2\varphi_d + 2\tan^2\theta_{p,s}(n,k)}} \\ R_S(n,k) &= \frac{S_p(n,k)\cdot\left(\tan\varphi_d - \tan\theta_{p,s}(n,k)\right)}{\sqrt{2\tan^2\varphi_d + 2\tan^2\theta_{p,s}(n,k)}} \end{aligned} \tag{2}$$

where the left and right channel of the stereo signals, L_S and R_S, have an angular separation of $2\varphi_d$ and this approach encodes the azimuth as the ratio of the downmix signals and hence requires no separate representation (or transmission) of spatial information. In(2), $S_p(n,k)$ represents the primary spatial sound source corresponding to the active speech at a given time at frequency over all participants and sites. This is determined as the source with the highest magnitude using (3).

$$S_p(n,k) = \max_{i,j}\left(\left|S_{i,j}(n,k)\right|\right) \tag{3}$$

For non-concurrently speaking participants, this will correspond to the speech of the only person speaking. In the alternative mono-downmix approach (see Fig. 3), the downmix is simply equal to the primary sound sources, $S_P(n,k)$. This approach requires separate representation (and transmission) of the azimuth information. For either downmix approach, the resulting signal is passed through an inverse time-frequency transform to create a time-domain downmix for each frame. This is the final stage of Fig. 3. The output of the transcoder is then fed to the AMR-WB+ encoder block of Fig. 3 prior to transmission..

2.3 S^{3}AC decoder

The S^{3}AC decoder block of Fig. 2 is illustrated in more detail in Fig. 4. Following speech decoding, the resulting received downmix signals are converted to the frequency domain using the same transform as applied in the S^{3}AC transcoder. These signals are then fed to the spatial repanning stage of Fig. 4. In the stereo-downmix mode, spatial repanning applies inverse tangent panning to the decoded stereo signals $\hat{R}_s(n,k)$ and $\hat{L}_s(n,k)$ to derive the squeezed azimuth of the time-frequency virtual source, $\hat{\theta}_{p,s}(n,k)$, using (4):

$$\hat{\theta}_{p,s}(n,k) = \arctan\left(\frac{\hat{L}_s(n,k) - \hat{R}_s(n,k)}{\hat{L}_s(n,k) + \hat{R}_s(n,k)} \cdot \tan\varphi_d\right) \tag{4}$$

The original azimuth $\hat{\theta}_{i,j}^s(n,k)$ of this virtual source is then recovered using:

$$\hat{\theta}_{i,j}^s(n,k) = f^{-1}\left(\hat{\theta}_{p,s}(n,k)\right) \tag{5}$$

In Equation (4), f^{-1} represents the inverse azimuth mapping function used in Equation (1). Following decoding of the original azimuth of the primary source, an estimate of the primary source $\hat{S}_p(n,k)$ is obtained using Equation (2) and the estimated primary source azimuths and decoded downmix signals.

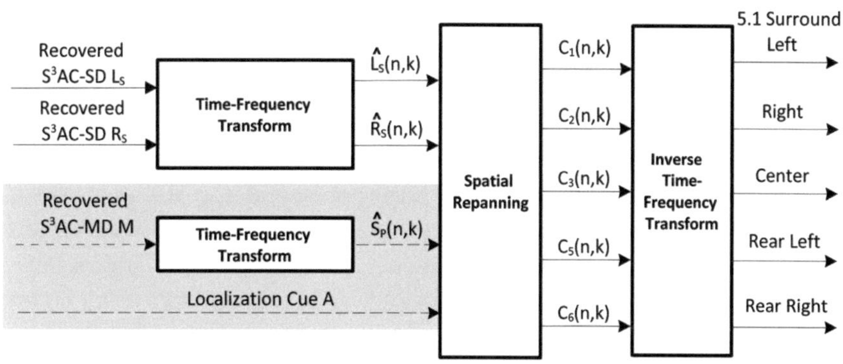

Fig. 4. S^{3}AC Decoder illustrating the processing of time domain signals recovered by the AMR-WB+ decoder to produce time-domain loudspeaker signals for reproduction of the spatial teleconference audio at each site.

The final rendering stage of the spatial re-panning is dependent on the desired playback system at each site. Illustrated in Fig. 4 is the scenario whereby reproduction at each site is achieved using a standard 5.1 channel Surround Sound loudspeaker system and utilizing all channels other than the low frequency effect channel. In this scenario, the estimated primary sources are amplitude panned to the desired location using two channels of the 5 channel system. This can be achieved by re-applying equation (2) using the azimuthal separation of the chosen two channels in the playback system and the estimated primary source azimuth. The output of this stage is a set of frequency-domain loudspeaker channel signals and the final step is to apply an inverse time-frequency transform to obtain the time-domain loudspeaker signals. Other reproduction techniques are also possible (e.g. binaural reproduction using HRTF processing (Cheng, 2008b). Due to the preservation of the original spatial location of each participant at each site, rendering could include accurate spatialization for virtual recreation of remote participants (e.g. for correct positioning of speech signals to correspond with the videoed participants). Alternatively, positioning could be achieved interactively at each site such as described in (Kilgore et al., 2003). In this chapter the primary focus is to ensure the perceptual quality resulting from decoding of each of the received spatial speech signals and hence further discussion on spatial rendering is not included.

3. An AVS for spatial teleconferencing

3.1 Overview of the AVS

An AVS consists of three orthogonally mounted acoustic particle velocity sensors and one omni-directional acoustic pressure microphone, allowing the measurement of scalar acoustic pressure and all three components of acoustic particle velocity (Hawkes & Nehorai, 1996; Lockwood & Jones, 2006). A picture of the AVS used in this work is shown in Fig. 5. Compared to linear microphone arrays, AVS's are significantly more compact (typically occupying a volume of 1 cm^3) (Hawkes & Nehorai, 1996; Lockwood & Jones, 2006; Shujau et al., 2009) and can be used to record audio signals in both the azimuth and elevation plane. Fig. 2 presents a picture of the AVS developed in (Shujau et al., 2009). The acoustic pressure and the 2D (x and y) velocity components of the AVS can be expressed in vector form as:

$$\mathbf{s}(n) = [o(n), x(n), y(n)]^T \tag{6}$$

In (6), $\mathbf{s}$(n), is the vector of recorded samples, where $o(n)$ represents the acoustic pressure component measured by the omni-directional microphone and $x(n)$ and $y(n)$ represent the outputs from two gradient sensors that estimate the acoustic particle velocity in the x and y direction, relative to the microphone position. For the gradient microphones, the relationship between the acoustic pressure and the particle velocity is given by Equation (7) (Shujau et al., 2009):

$$[x(n), y(n)] = g(p(n) - p(n - \Delta n))u \tag{7}$$

Equation (7) assumes a single primary source, where g represents a function of the acoustic pressure difference and:

$$\mathbf{u} = \left[\cos\theta_{i,j} \quad \sin\theta_{i,j}\right]^T \tag{8}$$

is the source bearing vector with $\theta_{i,j}$ representing the azimuth of the single source relative to the microphone array (Shujau et al., 2009).

3.2 Direction of arrival estimation of speech sources using the AVS

Directional information from an AVS can be extracted by examining the relationship between the 3 microphone channels. Accurate Direction of Arrival (DOA) estimates are dependent upon placement of the microphones, the structure that holds the microphones and the polar patterns generated by each microphone. A design that results in highly accurate DOA estimation using the Multiple Signal Classification (MUSIC) method of Schmidt (Schmidt, 1979) was presented in (Shujau et al., 2009) and is adopted here.

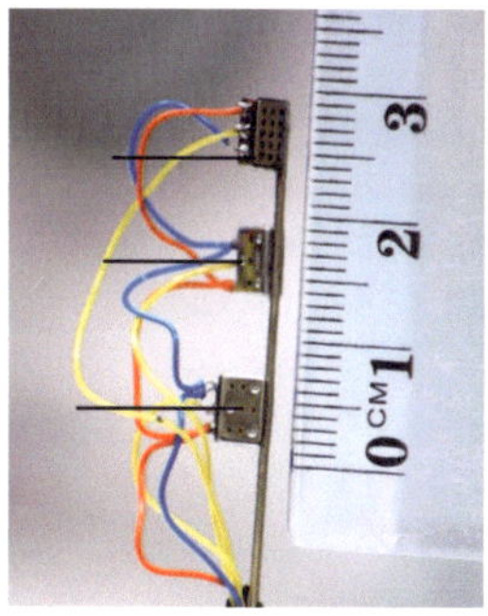

Fig. 5. The Acoustic Vector Sensor (AVS) used for recording of the spatial teleconference at each site.

The MUSIC algorithm allows for the estimation of the source DOA using the eigenvalues and eigenvectors of the covariance matrix formed from the recorded signals (Manolakis et al. 2005; Schmidt, 1979). The covariance matrix formed from the recorded signals is described in Equation (9), where L represents the number of samples used to find the covariance matrix (in this work, L corresponds to a single frame of 20 ms duration).

$$R(n) = \frac{1}{L}\sum_{n=1}^{L} \mathrm{Re}\{s(n)s^*(n)\} \tag{9}$$

The MUSIC algorithm is then used to estimate the azimuth of source j at site i, $\theta_{i,j}$, using Equation (10).

$$\theta_{i,j} = \min_\theta \left[P(\theta) = \frac{1}{\sum \left| \mathbf{V}^H \mathbf{h}(\theta) \right|^2} \right] \tag{10}$$

where $\mathbf{V}$ is the smallest eigenvector of the covariance matrix R from (9) and $\mathbf{h}(\theta)$ is the steering vector for the AVS and $\theta \in (-\pi, \pi)$. Assuming sources are only in the 2D plane, relative to the microphone array, , the steering vector can be described as a function of the azimuth as (Manolakis et al. 2005; Schmidt, 1979):

$$\mathbf{h}(\theta) = \left[\cos(\theta) \quad \sin(\theta) \quad 1 \right] \tag{11}$$

which is formed from the x and y components of Equation (6) and where 1 represents the omni-directional microphone.

3.3 Enhancement of AVS recordings

Speech enhancement for the AVS is achieved using two methods. The first method uses Independent Component Analysis (ICA) (Hyvärinen et al., 2001) while the second method uses the Minimum Variance Distortionless Response (MVDR) beamformer (Benesty et al., 2008).

3.3.1 Enhancement via ICA

The traditional ICA model applied to a multichannel speech recording assumes that microphone frequency responses for each channel are the same and that the mixing matrix is a result only of the acoustic transfer function. However, for the AVS, the microphones have directional polar responses and an approach for ICA for the AVS was previously described in (Shujau et al., 2010). This work applies ICA to recordings of the acoustic pressure gradients. In ICA, the aim is to separate a set of mixed signals into signal representing one or more independent sources. Here, the case for two source signals and 3 microphones is first considered. The recorded signals can be modeled using the mixing model:

$$\hat{\mathbf{s}}(n) = \sum_{j=0}^{2} \mathbf{A_k} \mathbf{s}_j(n) \tag{12}$$

In Equation (12), $\hat{\mathbf{s}}(n)$ represents a model of the recorded signals $\mathbf{s}(n)$ of equation (6), and $\mathbf{s}_j(n) = [s_1(\mathrm{n}), s_2(\mathrm{n})]^T$ represents the vector of source signal samples and $\mathbf{A_k}$ represents the convolutive mixing matrices, each of size 3 × 2. In the case where there is only one speaker in the presence of diffuse noise, the output components following ICA will be the primary speech source as well as residual noise signals. Here, for anechoic recordings, ICA was implemented using the well known FastICA implementation (Hyvärinen et al., 2001) while reverberant recordings were processing using a convolutive FastICA algorithm (Douglas et al., 2005).

3.3.2 Ehancement via MVDR

The MVDR Beamformer is the most widely used beamformer for microphone arrays. The expected outcome of any beamformer for speech is to combine the sensor signals in such a way that the desired speech signal is preserved or enhanced while the interfering signals are reduced without introducing any distortion. In this work a frequency domain MVDR beamformer is implemented. The MVDR beamformer is formed by choosing the coefficients of the filter $\mathbf{w}$ such that output power $E[Z^2] = \mathbf{w}^T \mathbf{R}(n,k)\mathbf{w}$ is minimized without introducing any distortion to the source signal (Benesty et al., 2008) where $\mathbf{R}$ is the covariance matrix of Equation (9) in the frequency domain. For each 20 ms frame, an FFT of 1024 samples is found using a Hamming window with an overlap of 50 %. The frequency domain samples are represent by the components of a vector $\mathbf{S}(n,k) = [x(n,k)\ y(n,k)\ o(n,k)]$ where n is the sample number and k is the frequency bin. The F = 32 most recent frames are buffered and the covariance matrix $\mathbf{R}(n,k)$ of the vector $\mathbf{S}(n,k)$ is found as (Lockwood et al., 2004):

$$\mathbf{R}(n,k) = \begin{bmatrix} \frac{c}{F}\sum_{l=0}^{F-1} x(m_l,k)^* x(m_l,k) & \frac{1}{F}\sum_{l=0}^{F} x(m_l,k)^* y(m_l,k) & \frac{1}{F}\sum_{l=0}^{F} x(m_l,k)^* o(m_l,k) \\ \frac{1}{F}\sum_{l=0}^{F} y(m_l,k)^* y(m_l,k) & \frac{c}{F}\sum_{l=0}^{F} y(m_l,k)^* y(m_l,k) & \frac{1}{F}\sum_{l=0}^{F} y(m_l,k)^* o(m_l,k) \\ \frac{1}{F}\sum_{l=0}^{F} o(m_l,k)^* x(m_l,k) & \frac{1}{F}\sum_{l=0}^{F} o(m_l,k)^* y(m_l,k) & \frac{c}{F}\sum_{l=0}^{F} o(m_l,k)^* o(m_l,k) \end{bmatrix} \quad (13)$$

Where m_l=n-lL and c = 1.03 which is regularization constant to help avoid matrix singularity and * represents complex conjugate. The covariance matrix is updated every 16 frames. The MVDR filter is expressed as (Lockwood et al., 2004):

$$\mathbf{w_k} = \frac{\mathbf{R}_{ky}^{-1}\mathbf{h}}{\mathbf{h}^T\mathbf{R}_{ky}^{-1}\mathbf{h}} \quad (14)$$

where **h** is the steering vector of Equation (10) and and the optimization constraints for each frequency band are described as:

$$\min_{\mathbf{w}} \mathbf{w}^T\mathbf{R}_{ky}\mathbf{w} \qquad \text{subject to} \qquad \mathbf{w}^T\mathbf{h} = 1 \quad (15)$$

The output of the beamformer for each frequency band k is given by (Lockwood et al., 2004):

$$Z_k = \mathbf{w_k^H Y_k} \quad (16)$$

The time domain output is obtained by determining the inverse FFT and performing overlap add reconstruction.

4. Experimental evaluation

In this section, results are presented evaluating the source localisation performance, enhancement performance and overall speech quality using the AVS and S³AC based spatial teleconferencing system.

4.1 Experimental evaluations

An experimental rig was created, where the AVS was mounted on a custom built rotating platform to allow positioning of the microphones relative to the source. Sound sources were produced by self powered speakers (Genelec 8020A) located at 1 m from the array. For source localization experiments, a series of monotone signals each 2 seconds long and of equal energy were recorded in an anechoic room with frequencies ranging from 100 Hz to 10 kHz. The recordings were made with the microphone rotated in 5° intervals corresponding to sources located at azimuths ranging from 0° to 90°, hence covering a full quadrant in the x-y plane. Recordings were also made using speech sources in both anechoic and reverberant conditions (with RT_{60} of 30ms), using 12 speech sentences (six male and six female) from the IEEE speech corpus (IEEE Subcommittee, 1969). Each sentence is 10 s long with 1s of silence at the start and end. Five 10 s segment noise sources are utilised: babble; recordings of factory floor; background noise from a moving vehicle; white; and pink noise, which were taken from an existing database (Institute for Perception-TNO, 1990). Diffuse noise was simulated using 4 loudspeakers located at equal distances on a circle surrounding the array. Recordings were made of a single target speech source in the presence of diffuse noise as well as one or two speech interferers as the noise source. The recordings were sampled at a rate of 48 kHz and two different Signal-to-Noise Ratio (SNR) levels of 0 dB and 20 dB. For source localization experiments, recordings were also made with a Uniform Linear Array (ULA) with a similar number of capsules to the AVS and a SoundField ST-250 microphone (SoundField) and the MUSIC algorithm was also used for the DOA estimation for these recordings. The SoundField microphone was chosen as it provides a direct comparison with an existing co-located microphone array.

4.2 Sound source location estimation

To investigate the performance of the AVS for estimating the source location, a series of experiments were conducted. In these experiments, recordings using the AVS were processed using the MUSIC algorithm to estimate source directions. The source localization error was measured using the Average Angular Error (AAE), defined as the average error over all frames tested between the true and estimated DOA. For monotone sources in anechoic environments, as shown in Fig. 6, DOA estimates obtained from the AVS were found to have an average error of less than 2° for a range of source frequencies, compared with average errors of more than 4.5° for the ULA. In addition to the monotone sources, experiments were carried with speech sources recorded in the presence of diffuse noise. For these experiments, the reverberant recordings were considered rather than the easier scenario of DOA estimation in anechoic conditions of Fig. 6. Further, results were compared with the SoundField microphone rather than the inferior ULA. The results from these experiments (Fig. 7) show that the average error produced by the AVS for localising a speech source in diffuse noise for reverberant conditions is approximately 1.6^0 compared to that of the SoundField Microphone which is 5^0.

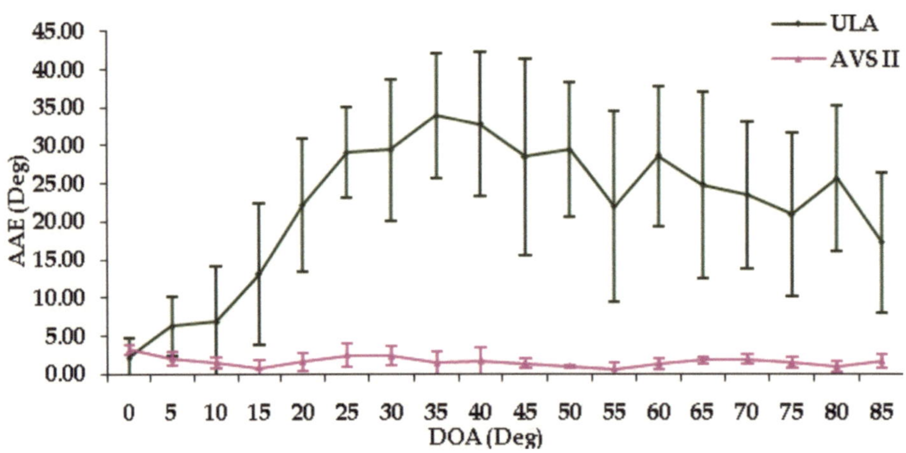

Fig. 6. Average Angular Error (AAE) for theDOA estimated for a series of tone sources with frequencies ranging from 1-10 kHz using both the AVS and the ULA. Recordings were made in anechoic conditions.

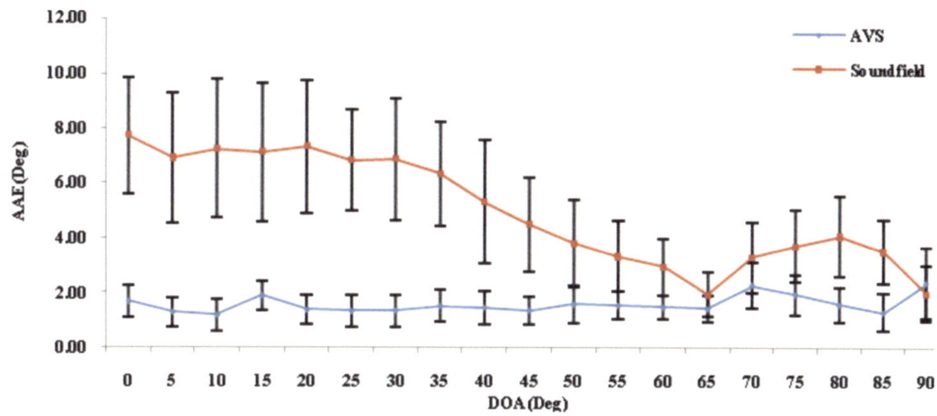

Fig. 7. Average Angular Error (AAE) for theDOA estimated for for recordings of speech from teh AVS and the SoundField microphones. Reocrdings were made in reverberant conditions.

4.3 Evaluation of speech enhancement

The results presented in this section are for two multi channel speech enhancement algorithms for the AVS, namely the ICA and MVDR beamformer. Both enhancement algorithms were used to process the recorded speech databases described in Section 4.1. The outputs from the enhancement algorithms are low pass filtered and down sampled to 16 kHz and then evaluated using the ITU-PESQ software (ITU-R P.862, 2001). When using PESQ, each output from ICA was compared with the original clean source signal to give a Mean Opinion Score for Listening Quality (MOSLQO) (Ma et al., 2009); the highest MOSLQO corresponds to the target source. A difference MOSLQO is generated by subtracting the MOSLQO of an omni-directional recording of the mixed sources (used as the reference) from the highest MOSLQO of the ICA outputs (Ma et al., 2009).

Results in Fig. 8 are for a speech source with both speech and diffuse noise as the interferer in anechoic conditions. The results show that on average when the recordings are enhanced with ICA there is an average improvement in MOSLQO of approximately 0.9 for diffuse noise as interferer and approximately 1.7 for speech as the interferer. In contrast, results for

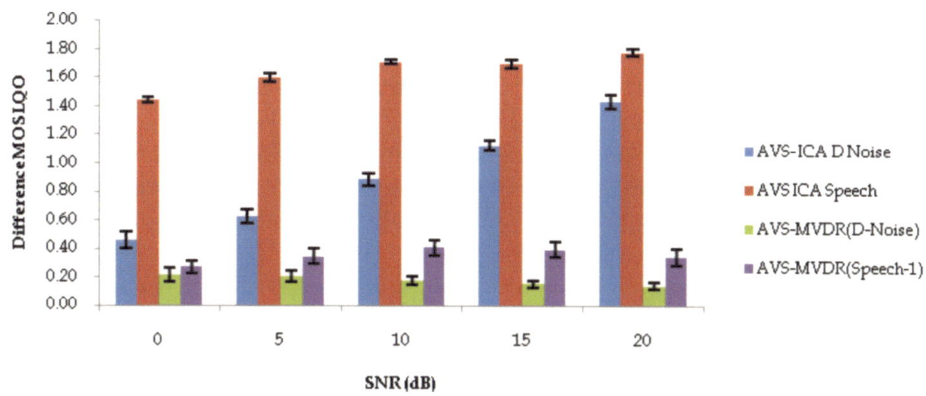

Fig. 8. Difference MOS Vs SNR in Anechoic Conditions

the MVDR based enhancement approach show a MOSLQO improvement of 0.2 and 0.4, respectively, for the diffuse noise and speech interferers. Hence, ICA shows a MOSLQO improvement of 0.7 and 1.3 over MVDR for diffuse noise and speech interferers, respectively.

Results for the reverberant case are shown in Fig. 9, where the difference MOSLQO for ICA is 0.7 for speech as the interferer and 0.5 for diffuse noise as interferer. In contrast, the MVDR enhancer results in an improved MOSLQO for both speech and noise interferers of 0.1. These results show that the ICA based enhancer is superior to the MVDR based enhancer in both anechoic and reverberant environments and for both diffuse and speech noise sources.

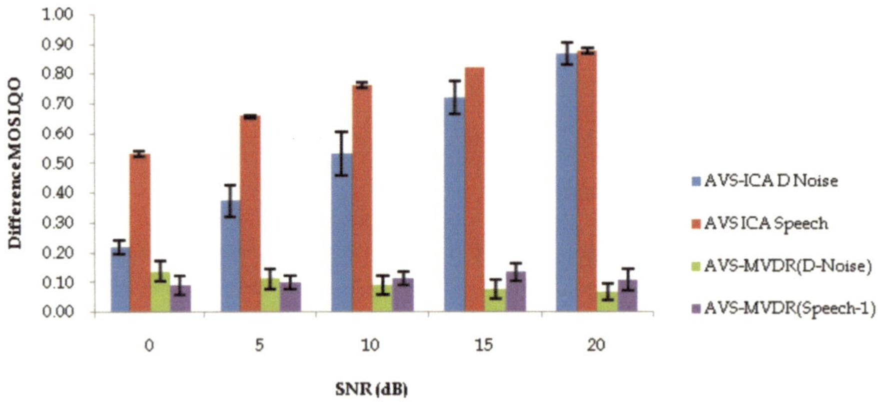

Fig. 9. Difference MOS Vs SNR in reverberant conditions.

4.4 Estimation of overall speech quality

To investigate the performance within the proposed spatial teleconferencing system, the recorded database of Section 4.1 was encoded through the proposed teleconferencing system including AMR-WB+ encoding of the downmix signals. The PESQ measure was used to analyse the resulting quality of the decoded signals that are the output of the proposed system. For the PESQ measures, the original clean sources were used as the reference. The AMR-WB+ coder was operated at each of the possible 31 bit rates ranging from 6 kbps to 36 kbps in increments of 1 kbps.

The first set of results for clean speech, where speech sources did not overlap in time, are shown in Fig. 10. The purpose of this test is to verify that the S^3AC coding framework does not introduce significant distortion additional to that introduced by the downmix compression. The results of Fig. 10 confirm that this is the case, with a gradual increase in PESQ as the bit rate of the AMR-WB+ coder increases. These results agree with existing results for the AMR-WB+ codec (Makin et al. 2005).

The second set of results shown in Fig. 11 is for recordings in the presence of diffuse noise with an SNR of 0 dB. The results in Fig. 11 (a) are for anechoic recordings where PESQ results have been averaged across those obtained for each of the five noise types and error bars represent 95 % statistical confidence intervals. Curves with blue asterisks represent results for PESQ of the decoded outputs from AMR-WB+ compared with the original clean

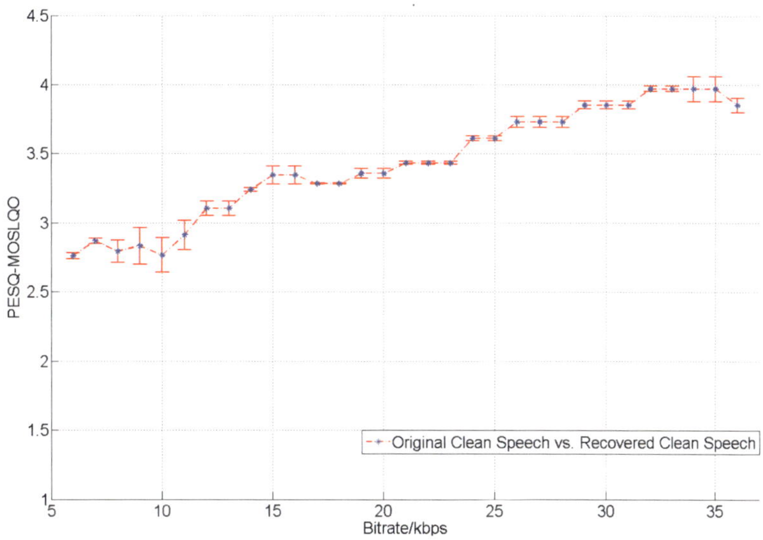

Fig. 10. PESQ results for original clean speech coded using the proposed spatial teleconferencing system including downmix compression via AMR-WB+.

speech. These results show a slight increase in PESQ over the bit rates tested with the average PESQ over all bit rates approximately 1.8. The green dashed line represents the upper limit for when no AMR-WB+ coding is applied to the enhanced recordings and it can be seen that the PESQ is statistically equivalent to the PESQ obtained after AMR-WB+ coding at the highest bit rates. It is proposed that this result is due to the high noise level present and hence further degradation caused by the speech coder does not dramatically reduce the resulting PESQ. The red dashed curve represents the results obtained when coding the non-enhanced recordings (taken as the omni-directional microphone output of the AVS) with the AMR-WB+ coder. As can be seen, the average PESQ is approximately 1.2, which is 0.6 less than results obtained when applying enhancement prior to encoding. The results for echoic recordings display similar trends to those for anechoic recordings, with PESQ results on average 0.4 higher for enhanced recordings compared with those for non-enhanced signals. On average, the PESQ results for enhanced recordings are 0.3 lower for echoic recordings compared to anechoic recordings.

Figure 12 shows results for anechoic and echoic recordings in the presence of noise where the SNR is 20 dB. In anechoic conditions (Fig. 12 (a)), the PESQ results are on average 2.6 for the enhanced signals decoded by AMR-WB+, which is an approximate 1.4 higher MOS prediction than for non-enhanced recordings. In echoic conditions (Fig. 12 (b)), average PESQ scores are approximately 2.2, which are is an approximate 0.9 increase in estimated MOS compared to non-enhanced signals. For bit rates of 14 kbps and above, results are statistically similar to those obtained when no speech coding is applied to the enhanced recordings. Compared with results for an SNR of 0 dB, the PESQ results for an SNR of 20 dB are approximately 0.8 higher. This result is to be expected due to the reduced level of noise present in the 20 dB SNR condition.

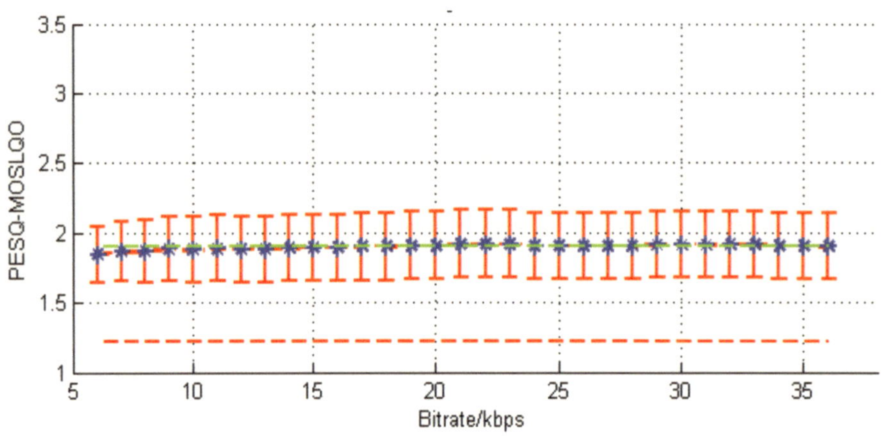

(a)

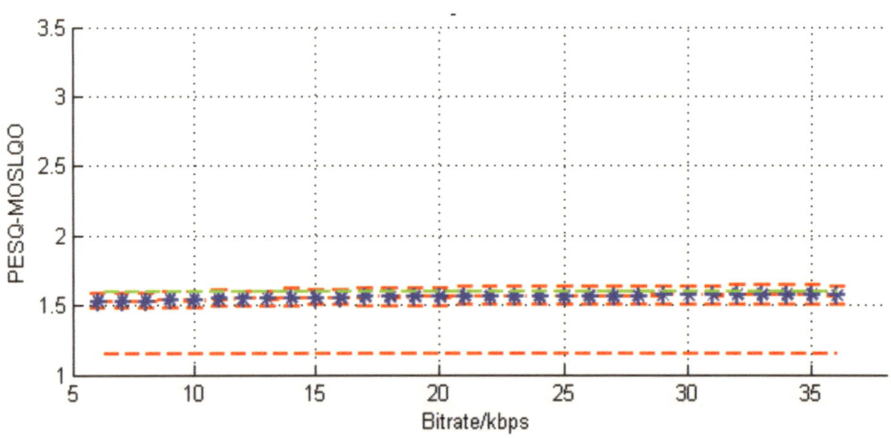

(b)

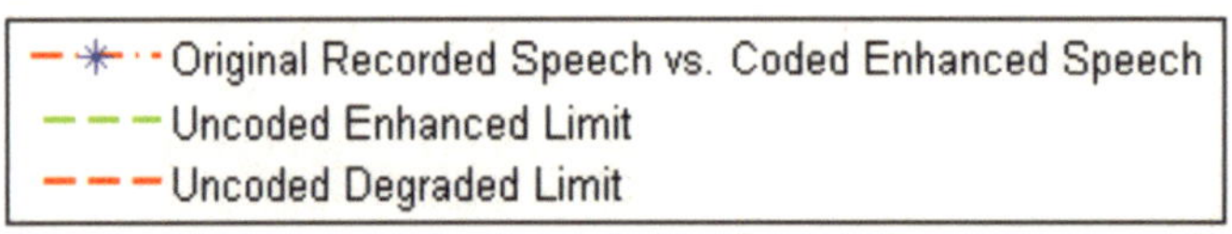

Fig. 11. Average results for recordings in diffuse noise across all noise sources for an SNR of 0 dB (a) Anechoic recordings. (b) Echoic recordings. Blue asterisk curves: results for PESQ of the decoded outputs from AMR-WB+ compared with the original clean speech. Green dashed curves: results when no AMR-WB+ coding is applied to the enhanced recordings. Red dashed curves: results obtained when coding the non-enhanced recordings.

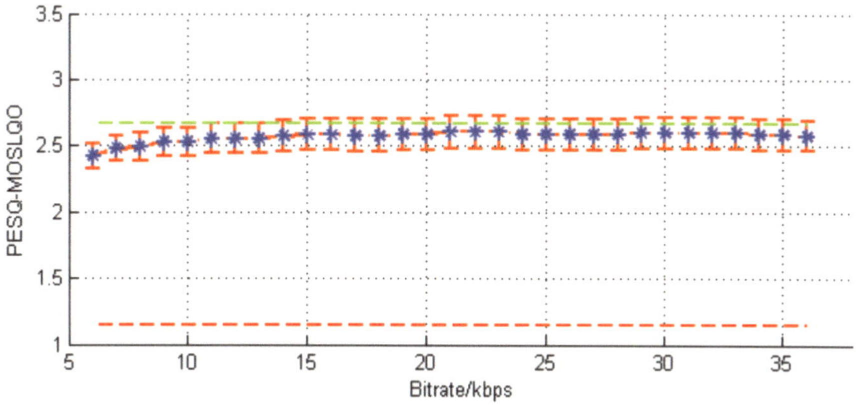

(a)

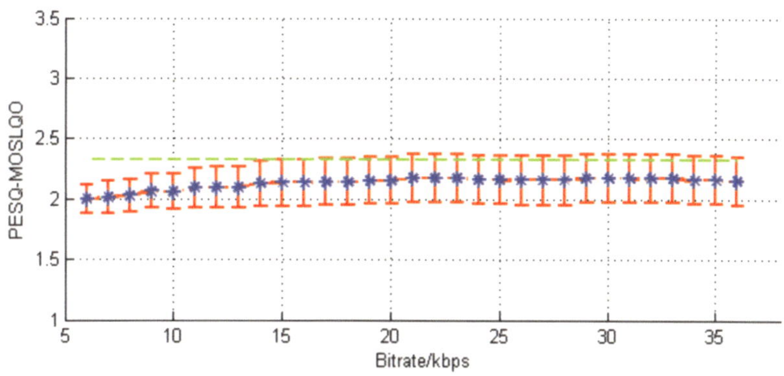

(b)

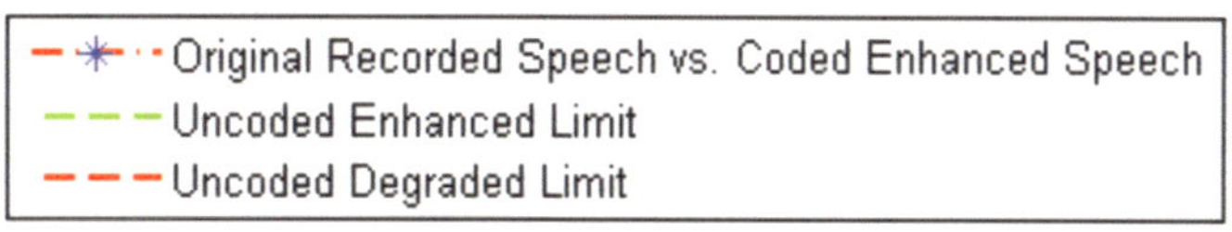

Fig. 12. Average results for recordings in diffuse noise across all noise sources for an SNR of 20 dB. (a) Anechoic recordings. (b) Echoic recordings. Blue asterisk curves: results for PESQ of the decoded outputs from AMR-WB+ compared with the original clean speech. Green dashed curves: results when no AMR-WB+ coding is applied to the enhanced recordings. Red dashed curves: results obtained when coding the non-enhanced recordings.

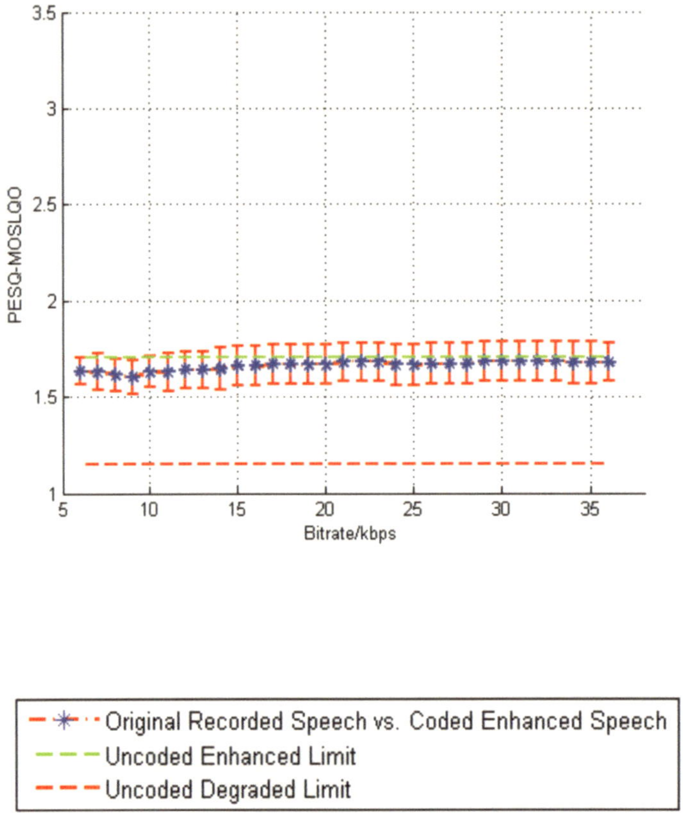

Fig. 13. PESQ results for recordings of two simultaneous speakers. Blue asterisk curves: results for PESQ of the decoded outputs from AMR-WB+ compared with the original clean speech. Green dashed curves: results when no AMR-WB+ coding is applied to the enhanced recordings. Red dashed curves: results obtained when coding the non-enhanced recordings.

The final set of results in Fig. 13 is for recordings of two simultaneous speech sources of equal power (SNR of 0 dB) separated by an angle of 45° and at a distance of 1 m from the microphone array. Here, one speech signal is treated as the desired source and the other as the interfering noise source. Results are for the PESQ of the output of the ICA enhancement that has been compressed and decoded using AMR-WB+, similar to the diffuse noise experiments. This shows that the enhancement results in an approximate 0.5 increase in estimated MOS for all bit rates tested. The PESQ results are also statistically similar to those obtained without AMR-WB+ coding of the ICA output.

5. Conclusion

This chapter has presented an approach to efficient compression for spatialized teleconferencing based on the concept of spatial squeezing of microphone array recordings of speech. Recordings were made using a collocated microphone array known as an AVS. Through encoding estimates of individual speech sources and information representing their spatial location, the proposed framework provides a flexible approach to the spatial rendering of the teleconference at each participants site. Results were presented confirming the accurate prediction of spatial sound sources though processing of the AVS recordings using the MUSIC algorithm. The approach results in a stereo or mono downmix signal representing the entire teleconference, which can then be efficiently compressed and transmitted using existing standard speech coders such as AMR-WB+. Hence, this provides for backward compatibility with existing speech coding and transmission systems.
Results were also presented demonstrating the performance of multichannel speech enhancement using a sound source separation inspired approach based on ICA. Predictions of subjective quality using PESQ showed that ICA-based enhancement results in a significant improvement in the predicted MOS compared to those obtained using the existing MVDR-based speech enhancer designed for microphone arrays. Results were also presented illustrating the performance in terms of PESQ when encoding the signals obtained from ICA enhancement using the proposed spatial teleconferencing system. These results show that the proposed approach does not introduce significant degradation in PESQ when compared with the PESQ obtained without encoding through the spatial teleconferencing compression system. Future work will focus on determining solutions to further enhancing the recorded signals when two or more participants are speaking simultaneously as well as improved methods for speech enhancement in low SNR conditions.

6. References

Ahonen, J.; Pulkki, V.; Lokki, T. (2007). "Teleconference Application and BFormat Microphone Array for Directional Audio Coding," Proc. AES 30th Int. Conf: Intelligent Audio Environments, March 2007.

Baldis, J. J. (2001). "Effects of spatial audio on memory, comprehension, and preference during desktop conferences," Proc. ACM SIGCHI Conference on Human factors in Computing Systems, pp.166-173, Washington, USA, March 2001.

Benesty, J.; Chen J.; Huang, Y. (2008). "Microphone Array Signal; Processing," Springer, Berlin.

Bosi, M.; Goldberg, R.E. (2002). Introduction to Digital Audio Coding and Standards, Springer, ISBN:1-4020-7357-7.

Breebaart, J., et al. (2005a). "Parametric Coding of Stereo Audio", EURASIP Jour. Applied Signal Proc., 1305-1322, Sep. 2005.

Breebaart, J. et al. (2005b). "MPEG Spatial Audio Coding/MPEG Surround: Overview and Current Status", in Proc. 119th AES Convention, New York, USΛ, Oct., 2005.

Cheng, C.I.,; Wakefield, G.H. (2001). Introduction to Head-Related Transfer Functions (HRTFs): Representations of HRTFs in Time, Frequency and Space, J. Audio Eng. Soc., 49(4):231-249, Apr. 2001.

Cheng, B.; Ritz, C.; Burnett, I. (2006). "Squeezing the Auditory Space: A New Approach to Multi Channel Audio Coding", Advances in Multimedia Information Processing - PCM2006, *Proceedings of the 7th Pacific-Rim Conference on Multimedia (PCM2006)*, Hangzhou, China, Nov. 2-4, 2006, Lecture Notes in Computer Science 4261, pp. 572 - 581, Springer-Verlag, 2006.

Cheng, B.; Ritz, C.; Burnett, I. (2007). "Principles and Analysis of the Squeezing Approach to Low Bit Rate Spatial Audio", *Proc. 2007 IEEE International Conf. on Acoustic, Speech and Signal Processing (ICASSP2007)*, Volume 1, Pages 13-16, Apr. 2007.

Cheng, Eva; Cheng, Bin; Ritz, Christian; Burnett, Ian S. (2008a). "Spatialized Teleconferencing: Recording and 'Squeezed' Rendering of Multiple Distributed Sites", *Proc. Australasian Telecommunication Networks and Applications Conference*, Pages 441 - 416, Dec. 2008.

Cheng, B. Ritz C. H.; Burnett, I. S. (2008b). "Binaural reproduction of spatially squeezed surround audio," *Proc. ICSP 2008: 9th International Conference on Signal Processing*, (Beijing, China), 2008, pp. 506-509.

Cheng, B.; Ritz, C.; Burnett, I. (2008c). "A Spatial Squeezing Approach to Ambisonics Audio Compression", Proc. 2008 IEEE International Conf. on Acoustic, Speech and Signal Processing (ICASSP 2008), Las Vegas, USA, Mar. 2008.

Cheng, B.; Ritz, C.; Burnett, I. (2009). "Spatial Audio Coding by Squeezing: Analysis and Application to Compressing Multiple Soundfields", Proc. EUSIPCO 2009, Glasgow, Scotland, p. 909-913, 24-28 August 2009.

DiBiase, J. H.; Silverman, H. F.; Brandstein, M. S. (2001). "Robust localization in reverberant rooms,", Microphone Arrays: Techniques and Applications, M. Brandstein and D. Ward, Eds., Berlin: Springer- Verlag, 2001, pp. 157–180.

Douglas, S. C.; Sawada, H.; Makino, S. (2005). "A spatio-temporal fastICA algorithm for separating convolutive mixtures," IEEE ICASSP05, Vol.5, pp 165-168, March 2005.

Evans, M. J.; Tew A. I.; J.; Angus, A. S. (2000). "Perceived performance of loudspeaker-spatialized speech for teleconferencing," Journal of the Audio Engineering Society, vol. 48, no9, pp. 771-785.

Faller, C.; Baumgarte, F. (2005). "Binaural Cue Coding - Part II: Schemes and Applications", IEEE Trans. on Speech and Audio Proc., vol.11, No.6, Nov., 2003.

Hawkes, M.; Nehorai, A. (1996). "Bearing Estimates With Acoustic Vector Sensor Arrays", American Institute of Physics Con. Proc. Vol: 386, pp 345-358, April 1996.

Hyvärinen, A.; Karhunen, J.; Oja, E. (2001). "Independent Component Analysis," John Wiley & Sons Inc, New York.

IEEE Subcommittee (1969). IEEE Recommended Practice for Speech Quality Measurements", IEEE Trans. Audio and Electro-acoustics, AU-17(3), 225-246.

ITU-R P.862 (2001). ITU-R Recommendation P.862, "Perceptual evaluation of speech quality (PESQ): An objective method for end-to-end speech quality assessment of narrow-band telephone networks and speech codecs".

Institute for Perception-TNO (1990). Noise Data, The Signal Processing Information Base (SPIB), Soesterberg, The Netherlands. Available online: http://spib.rice.edu/spib/select_noise.html

James, B. St.; Hawksford, M. O. J. (2008). "Corpuscular Streaming and Parametric Modification Paradigm for Spatial Audio Teleconferencing", Journal of the Audio Engineering Society, Volume 56 Issue 10 pp. 823-843, October 2008.

Kilgore, R.; Chignell, M.; Smith, P. (2003). "Spatialized audioconferencing: what are the benefits?", Proc. 2003 IBM Conference of the Centre for Advanced Studies on Collaborative Research, pp. 135-144, Ontario, Canada.

Lockwood, M. E.; Jones, D. L.; Bilger, C.; Lansing, C. R.; O'Brien, W. D. Jr.; Wheeler, B. C.; Feng, A. S. (2004). "Performance of time- and frequency-domain binaural beamformers based on recorded signals from real rooms," J. Acoust. Soc. Am. 115 (1), January 2004, pp. 379-391.

Lockwood, M. L.; Jones, D. L. (2006). "Beamformer Performance With Acoustic Vector Sensor In Air", J. Acoust. Soc. Am, 119, 608-619, January 2006.

Ma, J.; Hu, Y.; Loizou, P. C. (2009). "Objective measures for predicting speech intelligibility in noisy conditions based on new band-importance functions ", J. Acoust. Soc. Am., pp-3387-3405, May 2009.

Makinen, J.; Bessette, B.; Bruhn, S.; Ojala, P.; Salami, R.; Taleb, A. (2005). "AMR-WB+: a new audio coding standard for 3rd generation mobile audio services", Proceedings IEEE International Conference on Acoustics, Speech, and Signal Processing, 2005. (ICASSP '05), Vol. 2, pp. 1109 -1112.

Manolakis, D. G.; Ingle, G. K.; Kogon, S. M. (2005)., "Statistical and Adaptive Signal Processing: Spectral Estimation, Signal Modeling, Adaptive Filtering and Array Processing" Boston: Artech House, INC.

Pulkki, V. (2007). "Spatial sound reproduction with directional audio coding," J. Audio Eng. Soc., vol. 55, no. 6, pp. 503–516, June 2007.

Schmidt, R. O. (1979).,"Multiple Emitter and Signal Parameter Estimation," Proceedings,RADC Spectral Estimation Workshop, 243-258, October 1979.

Shujau, M.; Ritz, C.H.; Burnett, I.S. (2009). "Designing Acoustic Vector Sensors for localization of sound sources in air", Proc. EUSIPCO 2009, Glasgow, Scotland , pp 849-853, 24-28 August 2009.

Shujau, M., Ritz, C.H., Burnett, I.S. (2010). "Source Separation using Acoustic Vector Sensors", Proc. IEEE 2010 International Conference on Acoustics, Speech and Signal Processing (ICASSP'2010), Dallas, Texas, March 14-19.

SoundField. User Manual for ST 250, Sound field reserch Ltd, West Yorkshire, England, Issue 1.5.

Ward, D. B.; Elko, G. W. (1999). "Robust and adaptive spatialized audio for desktop conferencing," Journal of the Acoustical Society of America, vol. 105, no. 2, p. 1099, Feb. 1999.

Wrigley, Stuart N.; Tucker, Simon; Brown, Guy J.; Whittaker, Steve (2009). "Audio spatialization strategies for multitasking during teleconferences", Proceedings of the Interspeech 2009, Pages 2935- 2938, September, 2009.

Part 5

Applications in Biomedical and Diagnostic Studies

21

Neurophysiological Correlate of Binaural Auditory Filter Bandwidth and Localization Performance Studied by Auditory Evoked Fields

Yoshiharu Soeta and Seiji Nakagawa
National Institute of Advanced Industrial Science and Technology (AIST)
Japan

1. Introduction

Binaural hearing is specifically useful for our ability to separate a speech from a background noise and localize sounds. Binaural hearing performances are influenced by binaural auditory filter, inteaural time delay (ITD), interaural correlation (IAC), and so on. Some psychological experiments have clarified binaural auditor filter bandwidths (Kollmeier & Holube, 1989; Holube et al., 1998) and performance of sound localization related to ITD and IAC (Mills, 1958; Jeffress et al., 1962). However, little is known about the neural correlates, which makes an important contribution to our understanding of the auditory system. Therefore, we tried to estimate binaural auditory filter bandwidth and localization performance by the response in human auditory cortex.

Frequency selectivity has an important role in many aspects of auditory perception. For example, one sound may be obscured or rendered inaudible in the presence of other sounds. Frequency selectivity represents the ability of the auditory system to separate out or resolve the frequency components of a complex sound and can be characterized by the auditory filter bandwidths. Auditor filter bandwidths have been used to identify a fundamental perceptual unit that defines the frequency resolution of the auditory system - the critical bandwidth (CBW). The critical band (CB) concept has been used to explain a wide range of perceptual phenomena involving complex sounds.

Physiological correlates of the CBW have been described in several studies examining the auditory evoked potential (AEP) or auditory evoked field (AEF) in humans. Zerlin (1986) reported an abrupt increase in the amplitude of wave V of the brainstem AEP responses when the bandwidth of a two-tone complex approximated the CBW. Burrows & Barry (1990) reported that the amplitude of Na of the AEP rapidly increased when the frequency separation of a two-tone complex increased beyond the CBW. Soeta et al. (2005) and Soeta & Nakagawa (2006a) found that the amplitude of the N1m of AEFs increased with increasing the bandwidth of a bandpass noise or the frequency separation of a two-tone complex increased beyond the CBW. These studies have focused on physiological correlates of the monaural auditory filter in human auditory cortex; however, relatively little is known about the physiological correlates of the binaural auditory filter in the human auditory cortex. In natural listening environments, both the monaural and binaural auditory filters contribute to the performance of the auditory system in separating desired a speech from an undesired background noise (Kollmeier & Holube, 1989). Therefore, the physiological correlates of the binaural auditory filter in human auditory cortex merit investigation.

Performance of sound localization is also important in natural listening environments. There are two possible cues as to the sound localization: an ITD and an interaural level difference (ILD). Consider a sinusoidal sound source located to one side of the head in the horizontal plane with an azimuth of 45° and an elevation of 0°. The sound reaching the farther ear is delayed in time and is less intense than that reaching the nearer ear. Owing to the physical nature of sounds, ITDs and ILDs are not equally effective at all frequencies (Moore, 2003). For low-frequency tones, ITDs provide effective and unambiguous information about the location of the sounds. However, for higher-frequency sounds, ITDs provide ambiguous cues. For sinusoids, the physical cues of ILDs should be most useful at high frequencies, while the cues of ITDs should be most useful at low frequencies. The idea that sound localization is based on ILDs at high frequencies and ITDs at low frequencies has been called the "duplex theory." The minimum audible angle (MAA) for sinusoidal signals presented in the horizontal plane as a function of frequency has been investigated previously (Mills, 1958). The resolution of auditory space is measured in terms of the MAA, which is defined as the smallest detectable difference between the azimuths of two identical sources of sound. Performance worsens around 1500-1800 Hz. This is consistent with the duplex theory, which states that ITD differences above 1500 Hz between the two ears are ambiguous cues for localization, while ILDs up to 1800 Hz are small and do not change much with azimuth (Moore, 2003). Physiological correlates of the localization performance related to ITDs is still unclear.

ITDs can be measured by the interaural cross-correlation function (IACF) between two sound signals received at both the left and right ears. Whether there exist physiological processes that correspond to IACF processes is an important question, and answers have generally been sought in utilizing the so-called coincidence, or cross-correlation model for the evaluation of ITD first proposed by Jeffress (1948). Numerous theories of the binaural system rely on a coincidence detector or cross-correlator to act as a comparator element for signals arriving at both ears (e.g., Webster, 1951; Sayers & Cherry, 1957; Jeffress et al., 1962; Osman, 1971; Colbum, 1977; Lindemann, 1986; Joris et al., 1998). IAC can also be measured by the IACF. The width of the sound image changes according to the IAC (Licklider, 1948; Kurozumi & Ohgushi, 1983; Ando & Kurihara, 1986; Blauert & Lindemann, 1986). When sounds are delivered dichotically, the sound image varies with the IAC of the sound. If the IAC is high, the sound image is fused and occupies a narrow region. As the IAC decreases, the sound becomes more diffuse. Localization performance has been previously measured as a function of the degree of IAC (Jeffress et al., 1962; McEvoy et al., 1991; Zimmer & Macaluso, 2005), and the results showed that localization performance decreases slowly as the IAC is reduced especially below IAC $\approx$ 0.2.

Stimuli with ITDs have frequently been used in AEP and AEF studies of sound localization, and the processes underlying sound source localization have been analyzed (Ungan et al., 1989; McEvoy et al., 1990; Sams et al., 1993; McEvoy et al., 1993; 1994). The amplitude of N1m has been found to decrease with decreasing contralaterally-leading ITD (McEvoy et al., 1993; Sams et al., 1993). Magnetoencephalographic (MEG) research has benefited from the recent development of headphone-based 3D-sound technology, including head-related transfer functions, which are digital filters capable of reproducing the filtering effects of the pinna, head, and body (Palomäki et al., 2000; Fujiki et al., 2002; Palomäki et al., 2002; 2005). This research has found that the amplitude and latency of the N1m exhibits directional tuning to the sound location, with the amplitude of the right-hemisphere N1m being particularly sensitive to the amount of spatial cues in the stimuli. However, the processes underlying sound localization performance in the human auditory cortex have not been analyzed yet.

Therefore, in order to clarify the processes underlying basic binaural hearings in human auditory cortex, we investigated the physiological counterparts of binaural auditory filter bandwidth as a function of frequency and localization performance related to ITD, frequency, and IAC by AEFs.

2. Estimation of binaural auditory filter bandwidth

Some psychological experiments have examined whether monaural and binaural conditions have the same auditory filter bandwidths, and differences between the monaural and binaural conditions have been found (e.g., Kollmeier & Holube, 1989; Holube et al., 1998). However, there is little evidence of the physiological correlates of the auditory filter bandwidths under binaural listening conditions. Here, physiological counterparts to the binaural auditory filter bandwidth in the human auditory cortex were examined by AEFs. We tried to estimate the binaural auditory filter bandwidth as a function of frequency based on the amplitudes of the N1m components, which is prominent, robust, and controlled by the physical aspects of the stimulus (Näätänen & Picton, 1987).

The tone frequencies used in this experiment, f1 and f2, were geometrically centered on 125, 250, 500, 1000, 2000, 4000, and 8000 Hz. Frequency separations (f2-f1) were set at 2-160% of the center frequency. The higher frequency tone (f2) was presented to the right ear and the lower frequency tone (f1) was presented to the left ear. The duration of the stimuli used during the experiments was 0.5 s, including cosine rise and fall ramps of 10 ms. Participants were presented with stimuli dichotically at a sound pressure level (SPL) of 60 dB through insert earphones (Etymotic Research ER-2, Elk Grove Village, Illinois, USA) with 29-cm plastic tubes and eartips inserted into the ear canals. SPLs of all stimuli were checked with an ear simulator (Brüel & Kjaer Ear Simulator Type 4157, Naerum, Denmark).

Eight right-handed participants (22-37 years) took part in the experiment. All had normal audiological status and no history of neurological diseases. Informed consent was obtained from each participant after the nature of the study was explained. The study was approved by the Ethics Committee of the National Institute of Advanced Industrial Science and Technology (AIST).

AEFs were recorded using a 122-channel whole-head MEG system (Neuromag-122™; Neuromag Ltd., Helsinki, Finland) in a magnetically shielded room (Hämäläinen et al., 1993). Seven experimental sessions, each with a different center frequency, were carried out. In each session, stimuli were presented in a randomized order with an interstimulus interval selected at random from 1.0 to 1.5 s. To maintain a constant level of vigilance, participants were instructed not to pay attention to sounds but to concentrate on a self-selected silent movie projected on a screen in front of them. Magnetic data were sampled at 400 Hz after being band-pass-filtered between 0.03 and 100 Hz, and then averaged approximately 100 times. Responses were rejected if the magnetic field exceeded 3000 fT/cm in any channel. The averaged responses were digitally filtered between 1.0 and 30.0 Hz. The mean amplitude of the pre-stimulus period of the 0.2 s was used as the baseline level.

Source analysis based on the model of a single moving equivalent current dipole (ECD) in a spherical volume conductor was applied to the measured field distribution. Source estimates were based on a subset of 40-44 channels in the latency range of 70-130 ms over each left and right temporal hemisphere. ECDs were found separately for the left and right hemisphere data using a least-squares search (Hämäläinen et al., 1993). The amplitudes and latencies of the dipole with the maximal goodness of fit were defined as the N1m amplitudes and latencies for further analysis. Only dipoles with a goodness of fit of more than 80% were included in further analyses. The dipole location and orientation were determined in a head-based coordinate system with the origin set to the midpoint of the medial-lateral axis (x-axis) between the entrances of the left and right ear canals. The posterior-anterior axis (y-axis) was positioned through the nasion and the origin, and the inferior-superior axis (z-axis) was positioned through the origin perpendicular to the x-y plane.

Clear N1m responses were observed in both the right and left temporal regions in all participants with all stimuli (Fig. 1). The N1m latencies were not significantly affected by frequency separation and hemisphere with all center frequencies.

When the frequency separation was less than 10-20% of the center frequency, the N1m amplitude was independent of the frequency separation. When the frequency separation was more than about 10-20% of the center frequency, the N1m amplitude increased with increasing frequency separation (Fig. 2). Thus, N1m amplitudes show CB-like behavior under dichotic conditions. Regarding the increase in N1m amplitude above the CBW of the dichotically presented two-tone frequencies, Yvert et al. (1998) showed that the N1m amplitude increased with increasing frequency separation when the frequency separation

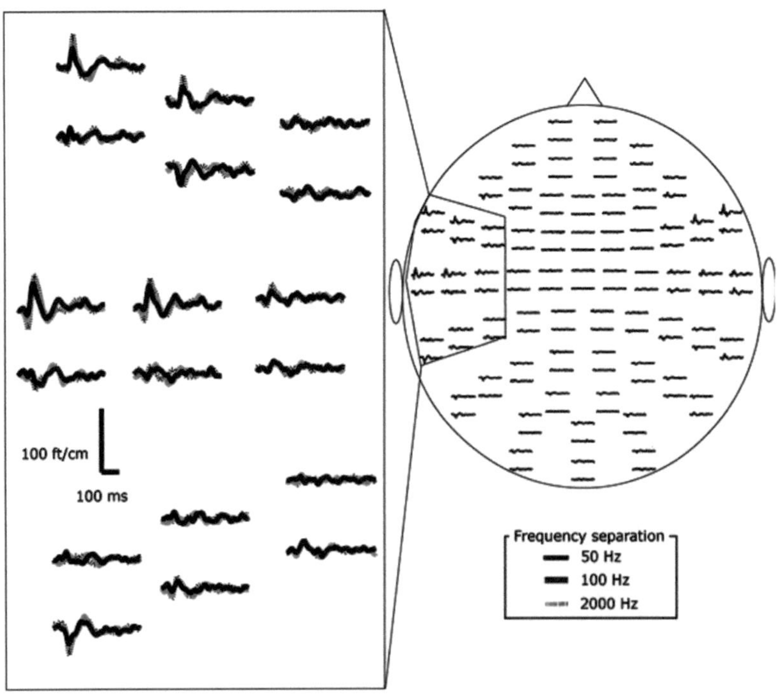

Fig. 1. Typical waveforms of AEFs in response to dichotically presented two-tones with different frequency separations from 122 channels in one subject. The center frequency was 1000 Hz. The waveforms of the AEFs have different frequency separations.

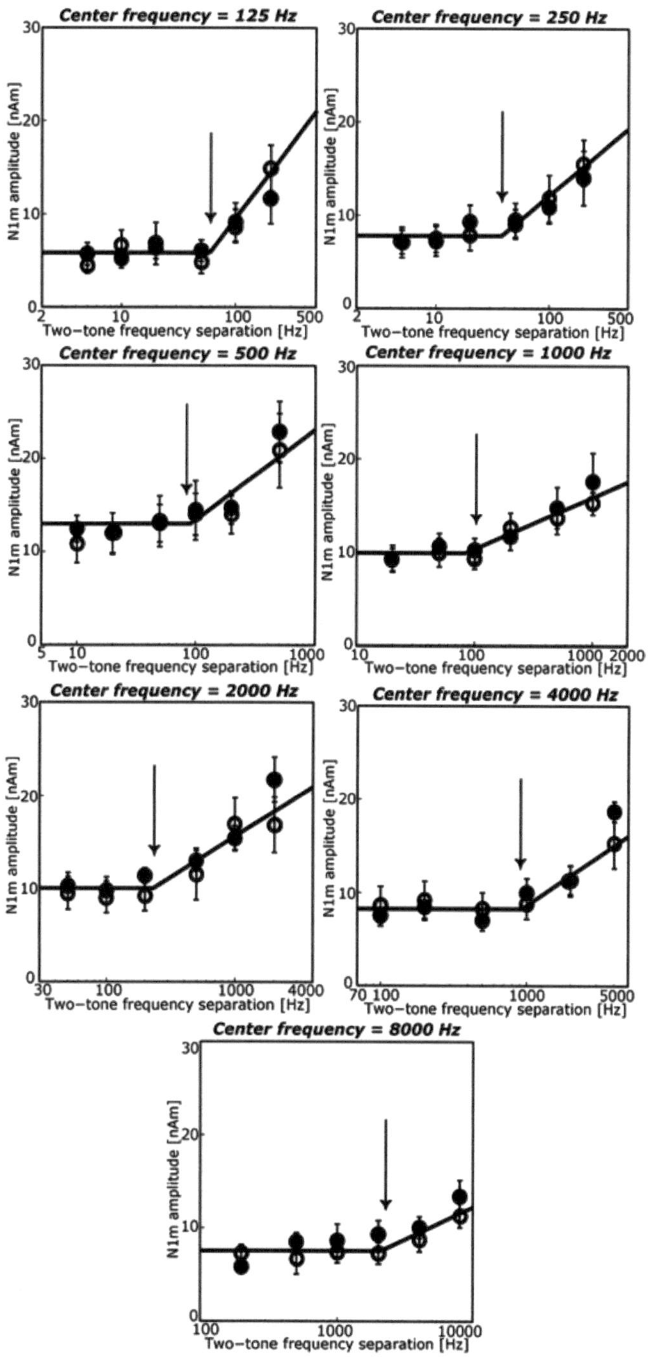

Fig. 2. Mean N1m amplitudes (± SEMs) from the right (●) and left (○) hemispheres as a function of the frequency separation. The data have been fitted with the best combination of two straight lines, one of zero slope for narrow frequency separations, and one of non-zero slope, by the method of least squares. The intersection estimates the critical bandwidth.

was more than 25% of the center frequency, which is consistent with the present finding. These results indicate that each tone stimulates both left and right hemispheres, and that the overall spectrum of the binaural stimulus becomes broader as the interaural frequency difference increases. This in turn reduces the interference between ipsilateral and contralateral pathways (binaural interaction) and activates many neurons in the auditory cortex.

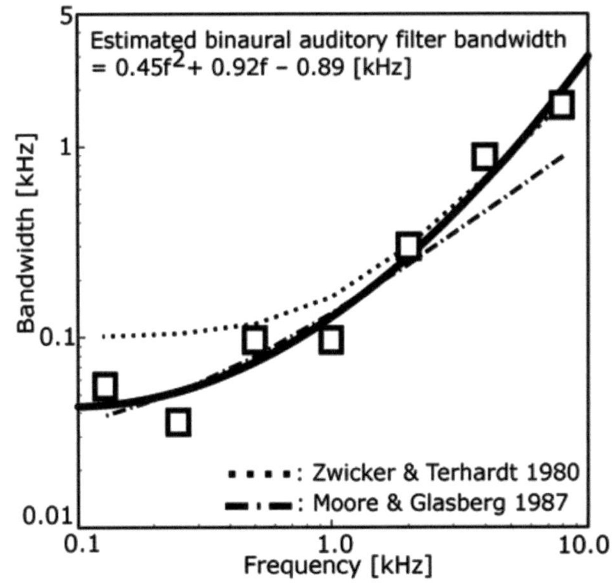

Fig. 3. The symbols (□) indicate the estimates of the binaural auditory filter bandwidth from the N1m amplitudes at various center frequencies. The curve fitted to the data is specified by the equation in the figure. For comparison, the dotted line and dash-dot line show the the monaural CB function (Zwicker & Terhardt, 1980) and equivalent rectangular bandwidth of the auditory filter (Moore & Glasberg, 1987), respectively.

We estimated the binaural auditory filter bandwidth by fitting the N1m amplitude as a function of frequency separation with the best combination of two straight lines as shown by the arrows in Fig. 2 in each center frequency. The averaged N1m amplitude from the left and right hemispheres was used for this fitting, because the main effect of hemisphere on the N1m amplitude was not significant. The estimated binaural critical bandwidth was approximately 10-20% of the center frequency and fitted to an equation (Fig. 3). The resulting function was $0.45f^2 + 0.92f - 0.89$ (Fig. 3). For comparison, the dotted line and dash-dot line show the estimated monaural auditory filter bandwisth (Zwicker & Terhardt, 1980; Moore & Glasberg, 1987). For the diotic condition, the effects of frequency separation of a two-tone complex and a three-tone complex on the AEFs have also been examined when the center frequency was 1000 Hz (Soeta & Nakagawa, 2006a). The auditory filter bandwidth was estimated in a similar way to that used in this study; the estimated auditory filter bandwidth was 153 Hz for a two-tone complex and 236 Hz for a three-tone complex. For the monaural condition, Sams & Salmelin (1994) investigated the frequency tuning of the human auditory cortex by masking tones using continuous white-noise maskers with frequency notches at the tone frequencies. The estimated auditory filter bandwidth for 1000 and 2000 Hz tones were 247 and 602 Hz, respectively. The reasons for these differing bandwidths are unclear. One factor might be the influence of a different presentation of the stimulus; that is, dichotic, diotic and monaural presentation. Additionally, different spectra or temporal shapes of the stimulus may have contributed to the discrepancies. Finally, different participants may have contributed to the discrepancies.

All eatimated ECDs were located at or near the Heschl's gyrus or planum temporale. The effects of frequency separation on the ECD locations of the N1m in each hemisphere and each center frequency were statistically analyzed by a repeated-measure ANOVA. While this analysis yielded some significant main effects of frequency separation for some of the dipole dimensions with a center frequency of 125 and 8000 Hz, none of these significant effects was replicated among center frequencies. It has been suggested that there is a hierarchy of pitch processing in which the center of activity moves away from the primary auditory cortex as the processing of music and speech proceeds, and the early stage of processing depends on core areas bilaterally; that is, pitch processing is largely symmetric in the hierarchy up to and including lateral Heschl's gyrus (Patterson et al., 2002; Zatorre et al., 2002; Hickok & Poeppel, 2004). In the present study, hemispheric differences in the latency and amplitude of the N1m were not observed. This might indicate that binaural frequency selectivity is symmetric up to the primary auditory cortex, including core areas of the auditory cortex such as Heschl's gyrus and planum temporale.

3. Estimation of localization performance related to ITD and frequency

For low-frequency tones, ITD provide effective and unambiguous cue for sound localization. For higher frequency sounds, however, ITD provide ambiguous cues. For pure tones, ITDs are only helpful when localizing sounds with frequencies less than 1500 Hz (Mills, 1958). The wavelength of the sound is about twice the distance between the two ears at these frequencies. Phase cues for tones with shorter wavelengths are ambiguous since after the first cycle of the wave, it is unclear which ear is leading or lagging. The present study aimed to evaluate responses related to the localization performance of ITDs, AEFs elicited by pure tones with different ITDs and frequencies were analyzed.

The stimuli used in this study were pure tones (sinusoidal sounds) of 800 and 1600 Hz. The ITD is an effective cue for sound localization when the frequency of the pure tone is 800 Hz, though it is not an effective cue for sound localization when the frequency of the pure tone is 1600 Hz (Mills, 1958). The stimulus duration used in the experiment was 500 ms, including rise and fall ramps of 10 ms. Stimuli were presented binaurally to the left and right ears through plastic tubes and earpieces inserted into the ear canals. All signals were presented at 60 dB SPL, and the ILD was set to 0 dB.

Ten right-handed participants (22-37 years) took part in the experiment. They all had normal audiological status and no history of neurological diseases. Informed consent was obtained from each participant after the nature of the study was explained. The study has been approved by the ethics committee of the National Institute of Advanced Industrial Science and Technology (AIST).

AEFs were recorded using a 122-channel whole-head MEG system in a magnetically shielded room (Hämäläinen et al., 1993). Two experimental sessions, each with a different frequency (800 or 1600 Hz), were conducted. In each session, combinations of a reference stimulus (ITD = 0.0 ms) and left-leading test stimuli (ITD = 0.1, 0.4, 0.7 ms) were presented alternately at a constant 1.5 s interstimulus interval. Usually, ITDs range from 0 ms for a sound at 0° azimuth (for a sound straight ahead) to about 0.7 ms for a sound at 90° azimuth (directly opposite one ear). To maintain a constant vigilance level, the participants were instructed to concentrate on a self-selected silent movie that was being projected on a screen in front of them and to ignore the stimuli. The method of MEG data analysis, that is, the latency, amplitude and ECD location of the N1m component, was the same way that we did in the previous experiment.

All the stimuli elicited prominent N1m responses in both the left and right hemispheres, with the near-dipolar field patterns, indicating sources in the vicinity of the auditory cortex of each hemisphere. The N1m latencies were not significantly affected by ITD and hemisphere in both frequencies (Fig. 4).

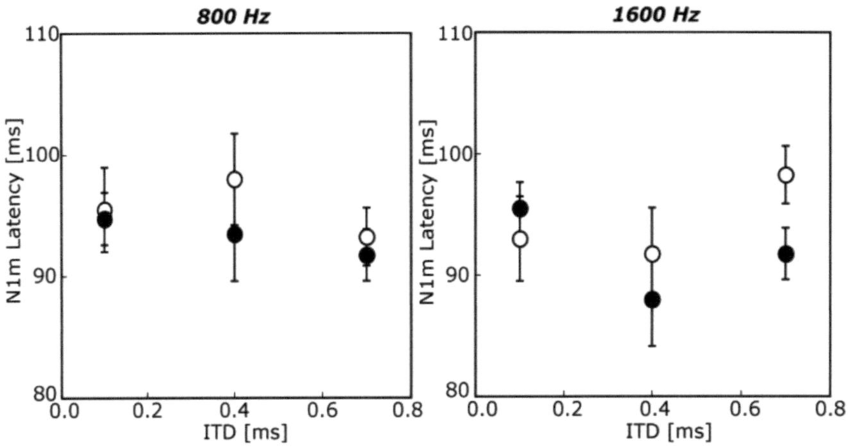

Fig. 4. Mean N1m latencies (± SEMs) as a function of the ITD from the right (●) and left (○) hemispheres.

Figure 5 shows the N1m amplitude as a function of ITD. When the frequency of the pure tone was 800 Hz, the N1m amplitude increased with increasing ITD. The main effect of the ITDs was significant ($P < 0.005$). This result is consistent with previous findings (McEvoy et al., 1993; Sams et al., 1993; Palomäki et al., 2005). The main effect of the hemispheres on the N1m amplitude was not significant. There were no significant interactions between the ITDs and hemispheres. When the frequency of the pure tone was 1600 Hz, the main effect of the ITDs was not significant. Humans can detect ITDs only up to 1500 Hz (Mills, 1958). When an ITD is conveyed by a narrowband signal such as a tone of appropriate frequency, humans may fail to derive the direction represented by that ITD. This is because they cannot distinguish the true ITD contained in the signal from its phase equivalents that are ITD + nT, where T is the period of the stimulus tone and n is an integer. This uncertainty is called phase-ambiguity.

Whether brain activity correlates with participants' localizations has been previously assessed using functional magnetic resonance imaging (fMRI) (Zimmer & Macaluso, 2005), with the results indicating that better localization performance is associated with increased activity both in Heschl's Gyrus (possibly including the primary auditory cortex) and in posterior auditory regions that are thought to process the spatial characteristics of sounds and generate the N1m components. Therefore, the present results indicate that localization performance could be reflected in N1m amplitudes.

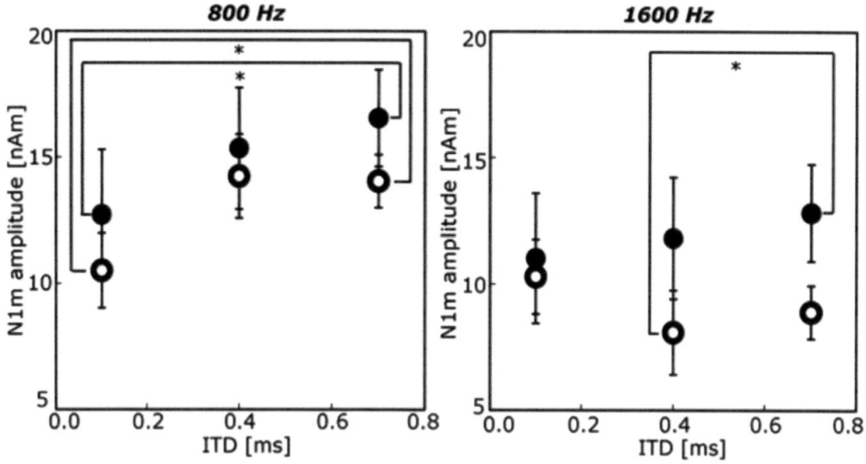

Fig. 5. Mean N1m amplitudes (± SEMs) as a function of the ITD from the right (●) and left (○) hemispheres. Asterisks indicate statistical significance (*$P<0.05$; Post hoc Newman-Keuls test).

There was a tendency that the N1m amplitudes in the right hemisphere were larger than those in the left hemisphere, although a significant effect was only found when the frequency of the stimulus was 1600 Hz ($P < 0.05$). The previous studies indicated that the N1m amplitude was significantly larger for stimuli presented with contralaterally-leading ITDs than for those with ipsilaterally-leading ITDs (McEvoy et al., 1993; 1994; Palomäki et al., 2000; 2002; 2005). These agree with our findings.

It has been found that the participant does not merely use the sound signals perceived at a given moment, but also makes a comparison with stored stimulus patterns in localization of a sound source (Plenge, 1974). The spectral cues generated by the head and outer ears vary between individuals and have to be calibrated by learning, which most probably takes place at the cortical level (Rauschecker, 1999). It has been reported that auditory training might develop enhanced auditory localization by using AEP (Munte et al., 2003). Three of the ten participants had increasing N1m amplitudes clearly with increasing ITDs in the right hemisphere even when the frequency of the stimulus was 1600 Hz. This might indicate that the effects of ITDs on N1m amplitudes depend on the individual, which is related to learning, training and so on.

The location of the ECDs underlying the N1m responses did not vary as a function of ITD in agreement with the previous results (McEvoy et al., 1993; Sams et al., 2003). Stimuli presented with different ITDs may excite somewhat different neuronal populations, though the cortical source location of the N1m did not vary systematically as a function of ITD. Therefore, we may conclude that the present data do not show an orderly representation of ITDs in the human auditory cortex that could be resolved by MEG.

4. Estimation of localization performance related to ITD and IAC

The detection of ITD for sound localization depends on the similarity between the left and right ear signals, namely IAC. Human localization performance deteriorates with decreasing IACs. The psychological responses to ITDs in relation to IACs have been obtained in humans (Jeffress et al., 1962; McEvoy et al., 1991; Zimmer & Macaluso, 2005), and the neurophysiological responses have been limited to animal studies (e.g., Yin et al., 1987; Yin & Chan, 1990; Albeck & Konishi, 1995; Keller & Takahashi, 1996; Saberi et al., 1998; D'Angelo et al., 2003; Shackleton et al., 2005). The present study aimed to evaluate the effects of ITDs of noises with different IACs on the AEF. In order to evaluate responses in the auditory cortex related to the ITDs and IACs of the sound, the AEFs elicited by noises with different ITDs and IACs were analyzed.

Bandpass noises were employed for acoustic signals. To create bandpass noises, white noises, each of 10 s duration, were digitally filtered between 200 and 3000 Hz (Chebychev bandpass: order 18). The IACF between the sound signals received at each ear $f_l(t)$ and $f_r(t)$ is defined by

$$\Phi_{lr}(\tau) = \frac{1}{2T} \int_{-T}^{+T} f_l'(t) f_r'(t+\tau) dt, \tag{1}$$

where $f_l'(t)$ and $f_r'(t)$ are obtained after passing through the A-weighted network, which approximately corresponds to ear sensitivity (Ando et al., 1987; Ando, 1998). The normalized IACF is defined by

$$\phi_{lr}(\tau) = \frac{\Phi_{lr}(\tau)}{\sqrt{\Phi_{ll}(0)\Phi_{rr}(0)}}, \tag{2}$$

where $\Phi_{ll}(0)$ and $\Phi_{rr}(0)$ are the autocorrelation functions at $\tau = 0$ for the left and right ear, respectively. The IAC is defined as the maximum of the IACF. The IAC of the stimuli was controlled by mixing in-phase diotic bandpass and dichotic independent bandpass noises in appropriate ratios (Blauert, 1983). The frequency range of these noises was always kept the same. The stimulus duration used in the experiment was 0.5 s, including rise and fall ramps of 10 ms, which were cut out of a 10 s long bandpass filtered noise with varying IAC and ITD. For stimulus localization, two cues were available to participants: envelope ITD and ongoing ITD. In this experiment, the envelope ITD was zero for all stimuli, and the ongoing ITD was varied, as shown in Fig. 6. Here, "envelope" refers to the shape of a gating function with 10-ms linear ramps at the onset and offset. Stimuli were presented binaurally to the left and right ears through plastic tubes and earpieces inserted into the ear canals. To check the frequency characteristics of the stimuli, stimuli were measured with an ear simulator. Figures 7 and 8 show examples of the power spectrum and the IACF of some of the stimuli measured. All signals were presented at 60 dB SPL, and the ILD was set to 0 dB.

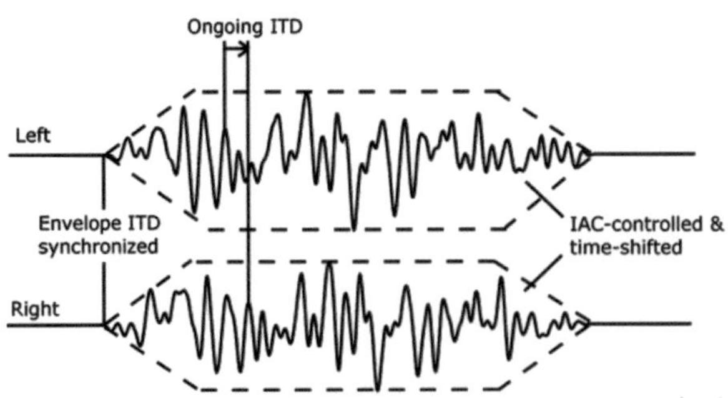

Fig. 6. Illustration of the stimuli used in the experiments. The fine structure (IAC controlled) of the stimulus was interaurally delayed, while the envelopes were synchronized between the ears.

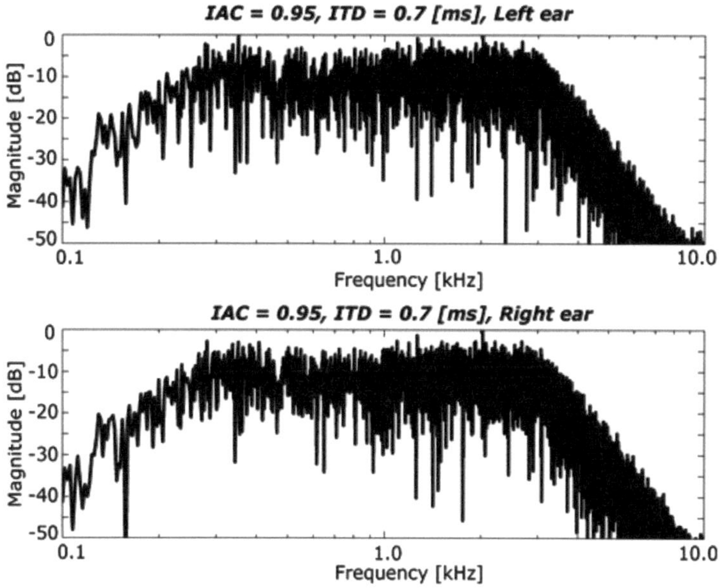

Fig. 7. Power spectrums of the stimuli used in the experiments.

Ten right-handed participants (22-35 years) took part in the experiment. They all had normal audiological status and none had a history of neurological disease. Informed consent was obtained from each participant after the nature of the study was explained. The study was approved by the Ethics Committee of the National Institute of Advanced Industrial Science and Technology (AIST).

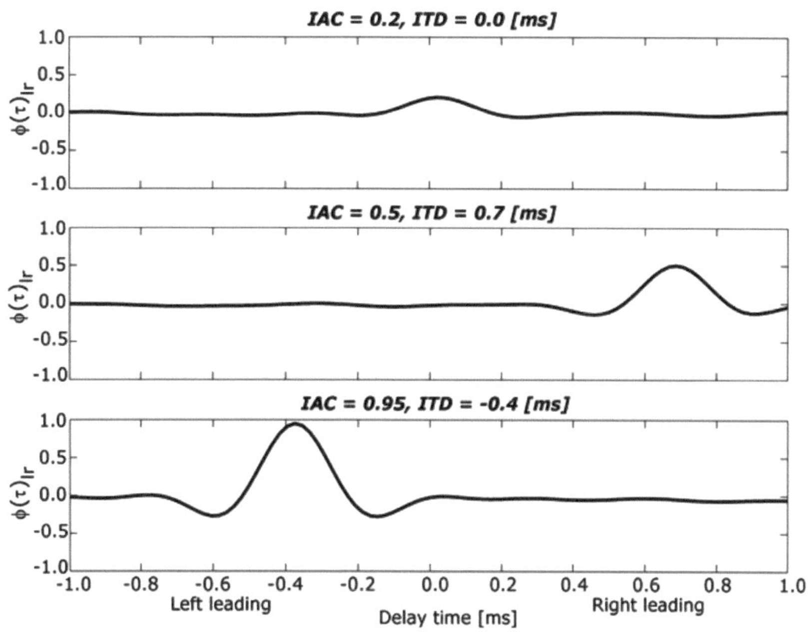

Fig. 8. IACFs of some of the stimuli used in the present study.

AEFs were recorded using a 122-channel whole-head MEG system in a magnetically shielded room (Hämäläinen et al., 1993). Combinations of a reference stimulus (IAC = 0.0) and test stimuli were presented alternately at a constant interstimulus interval of 1.5 s. Auditory evoked responses are affected by the preceding stimulus IAC (Ando et al., 1987; Chait et al., 2005). In order to reduce the effect of the IAC of the preceding stimulus, stimulus were alternated with the reference stimulus. The ITD of the test stimuli were 0, ±0.1, ±0.4, and ±0.7 ms, which had the IAC of 0.95 or 0.5. Two experimental sessions, each had right or left leading ITDs, were carried out. In order to maintain a constant vigilance level, the participants were instructed to concentrate on a self-selected silent movie that was being projected on a screen in front of them and to ignore the stimuli. The method of MEG data analysis was the same way that we did in the previous experiment.

All the stimuli elicited prominent N1m responses in both the left and right hemispheres, with near-dipolar field patterns (Fig. 9). Figures 10 show the N1m latency as a function of ITD. The N1m latency was not influenced by the ITDs. There was a tendency that the N1m latencies in the right hemisphere were shorter than those in the left hemisphere in the case of right-leading stimuli. That is, ipsilaterally localized stimuli produced shorter latencies in the case of right-leading stimuli. This result is consistent with previous findings (McEvoy et al., 1994; Palomäki et al., 2005).

Figures 11 show the N1m amplitude as a function of ITD. When the IAC of the stimulus was 0.95, the effect of ITD on the N1m amplitude was significant. The N1m amplitude increased with increasing ITD in the right hemisphere in the case of a left-leading stimulus and in both the left and right hemispheres in the case of a right-leading stimulus. This result is consistent with previous findings (McEvoy et al., 1993; Sams et al., 1993; Palomäki et al., 2005). The N1m amplitude increased slightly with increasing ITDs in the hemisphere contralateral to the ITDs when the IAC of the stimulus was 0.5; however, the main effect of ITDs on the N1m amplitude was not significant. Localization performance worsens with decreasing IACs (Jeffress et al., 1962; McEvoy et al., 1991; Zimmer & Macaluso, 2005); therefore, the present results indicate that localization performance is reflected in N1m amplitudes. Put another way, there is a close relationship between the N1m amplitudes, ITDs, and IACs of the stimuli.

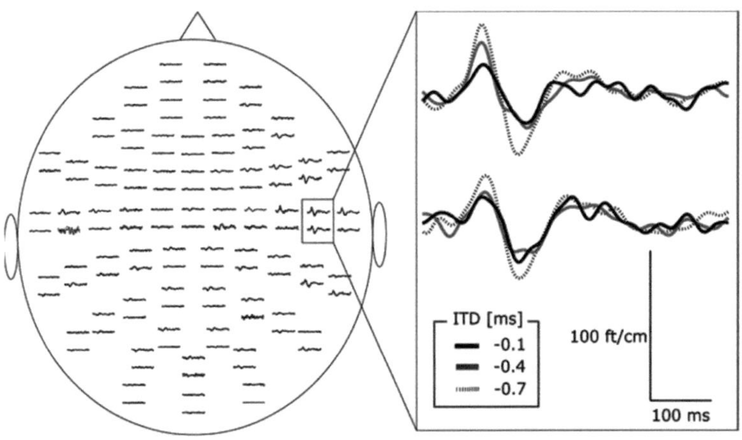

Fig. 9. Typical waveforms of AEFs from 122 channels in a subject when the IAC of the stimulus was 0.95.

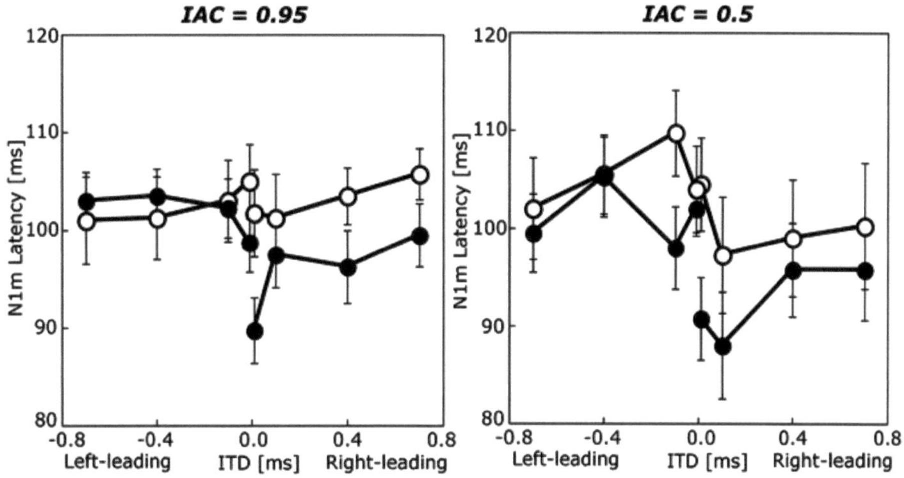

Fig. 10. Mean N1m latencies (± SEMs) as a function of the ITD from the right (●) and left (○) hemispheres.

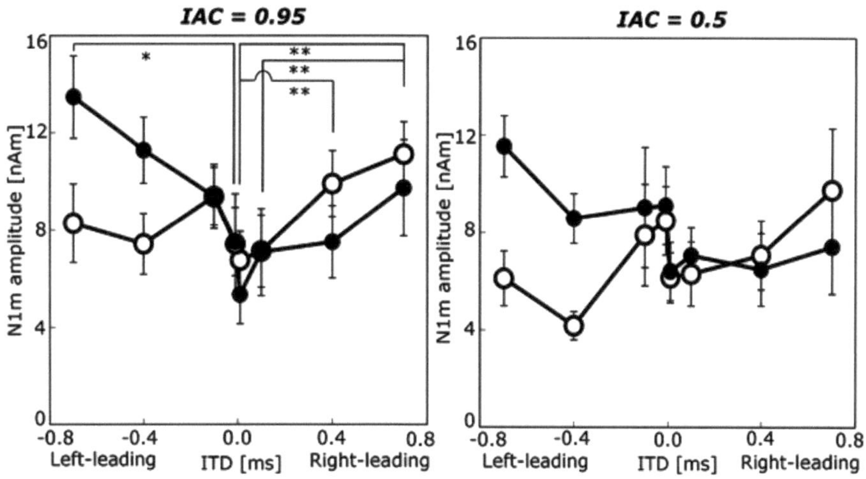

Fig. 11. Mean N1m amplitudes (± SEMs) as a function of the ITD from the right (●) and left (○) hemispheres. Asterisks indicate statistical significance (*P<0.05, **P<0.01; Post hoc Newman-Keuls test).

The effects of ITD and IAC on brain activity have recently been investigated using fMRI (Zimmer & Macaluso, 2005). The results showed that activity in Heschl's gyrus increased with increasing IAC and activity in posterior auditory regions also increased with increasing IAC, primarily when sound localization was required and participants successfully localized sounds. It was concluded that IAC cues are processed throughout the auditory cortex and that these cues are used in posterior regions for successful auditory localization. The activity in posterior regions might affect our findings of the N1m amplitude.

The right hemisphere dominance of the human brain in spatial processing has previously been reported (Burke et al., 1994; Butler, 1994; Ito et al., 2000; Kaiser et al., 2000; Palomäki et al., 2000; 2002; 2005). When the head-related transfer functions, ITD, and ILD were varied, the N1m amplitude in the right hemisphere was larger than that in the left hemisphere (Palomäki et al., 2002; 2005). In our study, the N1m amplitude in the right hemisphere was larger than that in the left hemisphere only in the case of a left-leading stimulus. However, the effects of ITDs on the right hemisphere were significant, with the N1m amplitude increasing with increasing ITD in the right hemisphere in the case of both left- and right-leading stimuli. These may indicate the right hemisphere dominance in spatial processing. The pattern of the right-hemisphere dominance observed in the current study is strikingly similar to that found in a previous fMRI study on the processing of sounds localized by ITDs (Krumbholz et al., 2005).

Figure 12 shows the averaged ECD locations in the left and right hemispheres. The ECD locations did not show any systematic variation across participants as a function of the ITDs or IACs. The location of the ECDs underlying the N1m responses did not vary as a function of ITD or IAC, a finding in agreement with previous MEG results (McEvoy et al., 1993; Sams et al., 1993; Soeta et al., 2004). As for fMRI, similarly, little evidence exists for segregated representations of specific ITDs or IACs in auditory cortex (Woldorff et al., 1999; Maeder et al., 2001; Budd et al., 2003; Krumbholz et al., 2005; Zimmer & Macaluso, 2005). Stimuli with different ITDs or IACs may excite somewhat different neuronal populations, although the cortical source location did not differ systematically as a function of ITD or IAC. Therefore, we conclude that the present data do not show an orderly representation of ITD or IAC in the human auditory cortex that can be resolved by MEG.

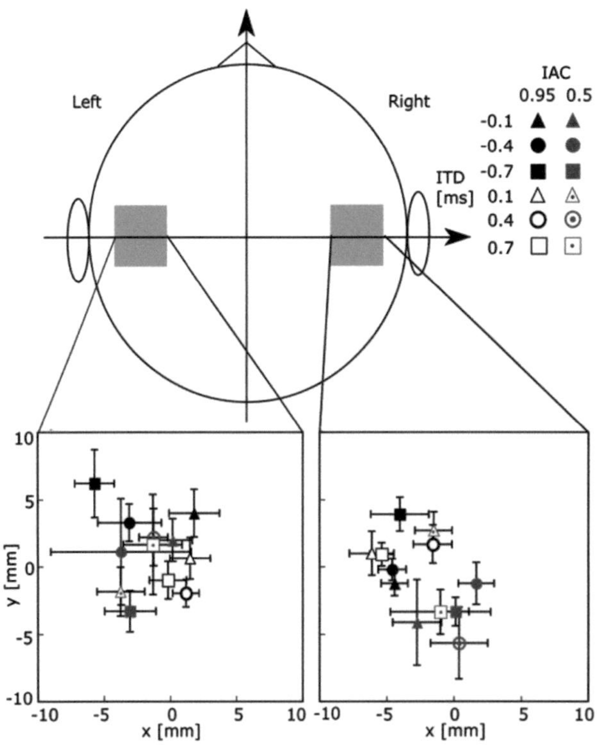

Fig. 12. Mean ECD location (± SEM) of all subjects in the left and right temporal planes. The ECD locations were normalized within each subject with respect to the position of ITD = 0.0 ms.

Recently it has been suggested that ITDs may be coded by the activity level in two broadly tuned hemispheric channels (McAlpine et al., 2001; Brand et al., 2002; McAlpine & Grothe, 2003; Stecker et al., 2005). The present study showed that the N1m amplitude varies with the ITD; however, the location of the ECDs underlying the N1m responses did not vary with the ITD. This could suggest that different ITDs are coded non-topographically but by response level. Thus, the current data seem to be more consistent with a two-channel model (McAlpine et al., 2001; Brand et al., 2002; McAlpine & Grothe, 2003; Stecker et al., 2005) rather than a topographic representation model (e.g., Jeffress, 1948).

5. Conclusion

We tried to estimate binaural auditory filter bandwidth as a function of frequency and localization performance related to ITD, frequency, and IAC by the response in human auditory cortex. First, in order to estimate binaural auditory filter bandwidth, two tones with different frequency separations and center frequencies, which were presented dichotically to the left and right ears, were used as the sound stimuli and AEFs were evaluated. The results indicated that the N1m amplitudes are approximately constant when the frequency separation is less than 10-20% of the center frequency; however, the N1m amplitudes increase with increasing frequency separation when the frequency separation is greater than 10-20% of the center frequency (Soeta & Nakagawa, 2007; Soeta et al., 2008). These results indicate that binaural auditory filter bandwidth is approximately 10-20% of the center frequency. The estimated binaural auditory filter bandwidth is roughly consistent with the estimated monaural auditory filter bandwidth by psychological experiment (Zwicker & Terhardt, 1980; Moore & Glasberg, 1987). Second, in order to identify the physiological correlates of the localization performance related to ITD and frequency, the AEFs in response to ITDs of pure tone with different frequency were examined. The results indicated that the N1m amplitudes increase with the ITDs when the frequency of the pure tone is 800 Hz; however, the N1m amplitudes do not vary with the ITDs when the frequency of the pure tone is 1600 Hz (Soeta & Nakagawa, 2006b). The results indicate that localization performance related to ITD and frequency is reflected in N1m amplitudes because ITDs provide effective and unambiguous information for low-frequency tones; however, ITDs provide ambiguous cues for higher-frequency tones. Finally, in order to identify the physiological correlates of the localization performance related to ITD and IAC, the AEFs in response to ITDs of bandpass noise with different IACs were examined. When the IAC is 0.95, the N1m amplitudes significantly increase with increasing ITD; however the effect of ITD on the N1m amplitudes is not significant when the IAC is 0.5 (Soeta & Nakagawa, 2006c). The results suggest that localization performance related to ITD and IAC is also reflected in the N1m amplitudes because human localization performance deteriorates with decreasing IACs. The results of two experiments related to localization performance suggest that ITDs are coded non-topographically but by response level.

6. References

Albeck, Y. & Konishi, M. (1995). Responses of neurons in the auditory pathway of the barn owl to partially correlated binaural signals. *J. Neurophysiol.*, Vol. 74, 1689-1700.

Ando, Y. & Kurihara, Y. (1986). Nonlinear response in evaluating the subjective diffuseness of sound fields. *J. Acoust. Soc. Am.*, Vol. 80, 833-836.

Ando, Y.; Kang, S. H. & Nagamatsu, H. (1987). On the auditory-evoked potential in relation to the IACC of sound field. *J. Acoust. Soc. Jpn. (E)*, Vol. 8, 183-190.

Ando, Y. (1998). *Architectural acoustics: Blending sound sources, sound fields, and listeners*, AIP Press Springer-Verlag, New York.

Blauert, J. (1983). *Spatial hearing: The psychophysics of human sound localization*, The MIT Press, Cambridge.

Blauert, J. & Lindemann, W. (1986). Spatial mapping of intracranical auditory events for various degrees of interaural coherence. *J. Acoust. Soc. Am.*, Vol. 79, 806-813.

Brand, A.; Behrend, O.; Marquardt, T.; McAlpine, D. & Grothe, B. (2002). Precise inhibition is essential for microsecond interaural time difference coding. *Nature*, Vol. 417, 543–547.

Budd, T. W.; Hall, D. A.; Gonçalves, M. S.; Akeroyd, M. A.; Foster, J. R.; Palmer, A. R.; Head, K. & Summerfield, A. Q. (2003). Binaural specialisation in human auditory cortex: an fMRI investigation of interaural correlation sensitivity. *Neuroimage*, Vol. 20, 1783 - 1794.

Burke, K. A.; Letsos, A. & Butler, R. A. (1994). Asymmetric performances in binaural localization of sound in space. *Neuropsychologia*, Vol. 32, 1409–1417.

Burrows, D. L. & Barry, S. J. (1990). Electrophysiological evidence for the critical band in humans: middle-latency responses. *J. Acoust. Soc. Am.*, Vol. 88, 180–184.

Butler, R. A. (1994). Asymmetric performances in monaural localization of sound in space. *Neuropsychologia*, Vol. 32, 221–229.

Chait, M.; Poeppel, D.; Cheveigne, A. & Simon, J. Z. (2005). Human auditory cortical processing of changes in interaural correlation. *J Neurosci.*, Vol. 25, 8518–8527.

Colburn, H. S. (1977). Theory of binaural interaction based on auditory-nerve data. II Detection of tones in noise. *J. Acoust. Soc. Am.*, Vol. 61, 525-533.

D'Angelo, W. R.; Sterbing, S. J.; Ostapoff, E. M. & Kuwada, S. (2003). Effects of amplitude modulation on the coding of interaural time differences of low-frequency sounds in the inferior colliculus. II. neural mechanisms. *J. Neurophysiol.*, Vol. 90, 2827-2836.

Fujiki, N.; Riederer, K. A. J.; Jousmäki, V.; Mäkelä, J. P. & Hari, R. (2002). Human cortical representation of virtual auditory space: differences between sound azimuth and elevation. *Eur. J. Neurosci.*, Vol. 16, 2207–2213.

Hämäläinen, M. S.; Hari, R.; Ilmoniemi, R. J.; Knuutila, J. & Lounasmaa, O. V. (1993). Magnetoencephalography - theory, instrumentation, and applications to noninvasive studies of the working human brain. *Rev. Mod. Phys.*, Vol. 65, 413-497.

Hickok, G. & Poeppel, D. (2004). Dorsal and ventral streams: a framework for understanding aspects of the functional anatomy of language. *Cognition*, Vol. 92, 67–99.

Holube, I.; Kinkel, M. & Kollmeier, B. (1998). Binaural and monaural auditory filter bandwidths and time constants in probe tone detection experiment. *J. Acoust. Soc. Am.*, Vol. 104, 2412–2425.

Itoh, K.; Yumoto, M.; Uno, A.; Kurauchi, T. & Kaga, K. (2000). Temporal stream of cortical representation for auditory spatial localization in human hemispheres. *Neurosci. Lett.*, Vol. 292, 215–219.

Jeffres, L. A. (1948). A place theory of sound localization. *J. Comp. Physiol. Psychol.*, Vol. 41, 35-39.

Jeffress, L. A.; Blodgett, H. C. & Deatherage, B. H. (1962). Effects of interaural correlation on the precision of centering a noise. *J. Acoust. Soc. Am.*, Vol. 34, 1122-1123.

Joris, P. X.; Smith, P. H. & Yin, T. C. (1998). Coincidence detection in the auditory system: 50 years after Jeffress. *Neuron*, Vol. 21, 1235–1238.

Kaiser, J.; Lutzenberger, W.; Preissl, H.; Ackermann, H. & Birbaumer, N. (2000). Right-hemisphere dominance for the processing of sound-source lateralization. *J. Neurosci.*, Vol. 20, 6631–6639.

Keller, C. H. & Takahashi, T. T. (1996). Binaural cross-correlation predicts the responses of neurons in the owl's auditory space map under conditions simulating summing localization. *J. Neurosci.*, Vol. 16, 4300-4309.

Kollmeier, B. & Holube, I. (1989). Auditory filter bandwidths in binaural and monaural listening conditions. *J. Acoust. Soc. Am.*, Vol. 92, 1889–1901.

Krumbholz, K.; Schönwiesner, M.; von Cramon, D. Y.; Rübsamen, R.; Shah, N. J.; Zilles, K. & Fink, G. R. (2005). Representation of interaural temporal information from left and right auditory space in the human planum temporale and inferior parietal lobe. *Cereb. Cortex*, Vol. 15, 317-324.

Kurozumi, K. & Ohgushi, K. (1983). The relationship between the cross-correlation coefficient of two-channel acoustic signals and sound image quality. *J. Acoust. Soc. Am.*, Vol. 74, 1726-1733.

Licklider, J. C. R. (1948). The influence of interaural phase relations upon masking of speech by white noise. *J. Acoust. Soc. Am.*, Vol. 20, 150-159.

Lindemann, W. (1986). Extension of a binaural cross-correlation model by means of contralateral inhibition, I: Simulation of lateralization of stationary signals. *J. Acoust. Soc. Am.*, Vol. 80, 1608-1622.

Maeder, P. P.; Meuli, R. A.; Adriani, M.; Bellmann, A.; Fornari, E.; Thiran, J. P.; Pittet, A. & Clarke, S. (2001). Distinct pathways involved in sound recognition and localization: a human fMRI study. *Neuroimage*, Vol. 14, 802–816.

McAlpine, D.; Jiang, D. & Palmer, A. R. (2001). A neural code for low-frequency sound localization in mammals. *Nature Neurosci.*, Vol. 4, 396–401.

McAlpine, D. & Grothe, B. (2003). Sound localization and delay lines - do mammals fit the model? *Trends Neurosci.*, Vol. 13, 347-350.

McEvoy, L.; Picton, T.; Champagne, S.; Kellett, A. & Kelly, J. (1990). Human auditory evoked potentials to shifts in the lateralization of noise. *Audiology*, Vol. 29, 163-180.

McEvoy, L. K.; Picton T. W. & Champagne, S. C. (1991). The timing of the processes underlying lateralization: psychophysical and evoked potential measures. *Ear Hear.*, Vol. 12, 389-398.

McEvoy, L.; Hari, R.; Imada, T. & Sams, M. (1993). Human auditory cortical mechanisms of sound lateralization: II. Interaural time differences at sound onset. *Hear. Res.*, Vol. 67, 98-109.

McEvoy, L.; Mäkelä, J. P.; Hämäläinen, M. & Hari, R. (1994). Effect of interaural time diferences on middle-latency and late auditory evoked magnetic fields. *Hear. Res.*, Vol. 78, 249-257.

Mills, A. W. (1958). On the minimum audible angle. *J. Acoust. Soc. Am.*, Vol. 30, 237-246.

Moore, B. C. J. (2003). *An introduction to the psychology of hearing*. New York: Academic Press.

Moore, B. C. J. & Glasberg, B. E. (1987). Formulae describing frequency selectivity as a function of frequency and level, and their use in calculating excitation patterns. *Hear. Res.*, Vol. 28, 209-225.

Munte, T. F.; Nager, W.; Beiss, T.; Schroeder, C. & Altenmuller, E. (2003). Specialization of the specialized: electrophysiological investigations in professional musicians. *Ann. N.Y. Acad. Sci.*, Vol. 999, 131-139.

Näätänen, R. & Picton, T. (1987). The N1 wave of the human electric and magnetic response to sound: a review and an analysis of the component structure. *Psychophysiol.*, Vol. 24, 375–425.

Osman, E. (1971). A correlation model of binaural masking level differences. *J. Acoust. Soc. Am.*, Vol. 50, 1494-1511.

Palomäki, K.; Alku, P.; Mäkinen, V.; May, P. & Tiitinen, H. (2000). Sound localization in the human brain: neuromagnetic observations. *Neuroreport,* Vol. 11, 1535–1538.

Palomäki, K.; Tiitinen, H.; Mäkinen, V.; May, P. & Alku, P. (2002). Cortical processing of speech sounds and their analogues in a spatial auditory environment. *Cogn. Brain Res.*, Vol. 14, 294-299.

Palomäki, K.; Tiitinen, H.; Mäkinen, V.; May, P. J. C. & Alku, P. (2005). Spatial processing in human auditory cortex: The effects of 3D, ITD, and ILD stimulation techniques, *Cogn. Brain Res.*, Vol. 24, 364-379.

Patterson, R. D.; Uppenkamp, S.; Johnsrude, I. S. & Griffiths, T. D. (2002). The processing of temporal pitch and melody information in auditory cortex. *Neuron*, Vol. 36, 767–776.

Plenge, G. (1974). On the differences between localization and lateralization. *J. Acoust. Soc. Am.*, Vol. 56, 944-951.

Rauschecker, J. P. (1999). Auditory cortical plasticity: a comparison with other sensory systems. *Trends Neurosci.*, Vol. 22, 74-80.

Saberi, K.; Takahashi, Y.; Konishi, M.; Albeck, Y.; Arthur, B. J. & Farahbod, H. (1998). Effects of interaural decorrelation on neural and behavioral detection of spatial cues. *Neuron*, Vol. 21, 789–798.

Sams, M.; Hämäläinen, M.; Hari, R. & McEvoy, L. (1993). Human auditory cortical mechanisms of sound lateralization: I. Interaural time differences within sound. *Hear. Res.*, Vol. 67, 89-97.

Sams, M. & Salmelin, R. (1994). Evidence of sharp frequency tuning in the human auditory cortex. *Hear. Res.*, Vol. 75, 67-74.

Sayers, B. M., & Cherry, E. C. (1957). Mechanism of binaural fusion in the hearing of speech. *J. Acoust. Soc. Am.*, Vol. 29, 973-987.

Shackleton, T. M.; Arnott, R. H. & Palmer, A. R. (2005). Sensitivity to interaural correlation of single neurons in the inferior colliculusof guinea pigs. *J. Assoc. Res. Otolaryngol.*, Vol. 6, 244-259.

Soeta, Y.; Hotehama, T.; Nakagawa, S.; Tonoike, M. & Ando, Y. (2004). Auditory evoked magnetic fields in relation to the inter-aural cross-correlation of bandpass noise, *Hear. Res.*, Vol. 196, 109-114.

Soeta, Y.; Nakagawa, S. & Matsuoka, K. (2005). Effects of the critical band on auditory evoked magnetic fields. *NeuroReport*, Vol. 16, 1787-1790.

Soeta, Y. & Nakagawa, S. (2006a). Complex tone processing and critical band in human auditory cortex. *Hear. Res.*, Vol. 222, 125-132.

Soeta, Y. & Nakagawa, S. (2006b). Effects of the frequency on interaural time difference in the human brain. *NeuroReport*, Vol. 17, 505-509.

Soeta, Y. & Nakagawa, S. (2006c). Auditory evoked magnetic fields in relation to interaural time delay and interaural correlation. *Hear. Res.*, Vol. 220, 106-115.

Soeta, Y. and Nakagawa, S. (2007). Effects of the binaural auditory filter in the human brain. *NeuroReport*, Vol. 18, 1939-1943.

Soeta, Y.; Shimokura, R. & Nakagawa, S. (2008). Effects of the center frequency on binaural auditory filter bandwidth in the human brain. *NeuroReport*, Vol. 19, 1709-1713.

Stecker, G. C.; Harrington, I. A. & Middlebrooks, J. C. (2005). Location coding by opponent neural populations in the auditory cortex. *PLoS Biol.*, Vol. 3, 520-528.

Ungan, P.; Sahinoglu, B. & Utkuçal, R. (1989). Human laterality reversal auditory evoked potentials: stimulation by reversing the interaural delay of dichotically presented continuous click trains. *Electroenceph. Clin. Neurophysiol.*, Vol. 73. 306-321.

Webster, F. A. (1951). The influence of binaural masking level differences. *J. Acoust. Soc. Am.*, Vol. 50, 1494-1511.

Woldorff, M. G.; Tempelmann, C.; Fell, J.; Tegeler, C.; Gaschler-Markefski, B.; Hinrichs, H.; Heinze, H. & Scheich, H. (1999). Lateralized auditory spatial perception and the contralaterality of cortical processing as studied with functional magnetic resonance imaging and magnetoencephalography. *Hum. Brain Mapp.*, Vol. 7, 49-66.

Yin, T. C.; Chan, J. C. & Carney, L. H. (1987). Effects of interaural time delays of noise stimuli on low-frequency cells in the cat's inferior colliculus. III. Evidence for cross-correlation, *J. Neurophysiol.*, Vol. 58, 562-583.

Yin, T. C. & Chan, J. C. (1990). Interaural time sensitivity in medial superior olive of cat. *J. Neurophysiol.*, Vol. 64, 465-488.

Yvert, B.; Bertrand, O.; Pernier, J. & Ilmoniemi, R. J. (1998). Human cortical responses evoked by dichotically presented tones of different frequencies. *NeuroReport*, Vol. 9, 1115-1119.

Zatorre, R. J.; Belin, P. & Penhune, V. B. (2002). Structure and function of auditory cortex: music and speech. *Trends Cogn. Sci.*, Vol. 6, 37–46.

Zerlin, S. (1986). Electrophysiological evidence for the critical band in humans. *J. Acoust. Soc. Am.*, Vol. 79, 1612–1616.

Zimmer, U. & Macaluso, E. (2005). High binaural coherence determines successful sound localization and increased activity in posterior auditory areas. *Neuron,* Vol. 47, 893-905.

Zwicker, E. & Terhardt, E. (1980). Analytical expression for critical-band rate and critical bandwidth as a function of frequency. *J. Acoust. Soc. Am.,* Vol. 68, 1523-1525.

22

Processing of Binaural Information in Human Auditory Cortex

Blake W. Johnson
Macquarie Centre for Cognitive Science, Macquarie University, Sydney Australia

1. Introduction

The mammalian auditory system is able to compute highly useful information by analyzing slight disparities in the information received by the two ears. Binaural information is used to build spatial representations of objects and also enhances our capacity to perform a fundamental structuring of perception referred to as 'auditory scene analysis' (Bregman, 1990) involving a parsing of the acoustic input stream into behaviourally-relevant representations. In a world that contains a cacophony of sounds, binaural hearing is employed to separate out concurrent sound sources, determine their locations, and assign them meaning. In the last several years our group has studied how binaural information is processed in the human auditory cortex, using a psychophysical paradigm to elicit binaural processing and using electroencephalography (EEG) and magnetoencephalography (MEG) to measure cortical function.

In our psychophysical paradigm listeners are posed with monaurally identical broadband sounds containing a timing or level disparity restricted to a narrow band of frequencies within their overall spectra. This results in the perception of a pitch corresponding to the affected frequency band, concurrent with, but spatially separated from, the remaining background (Yost, 1991). The illusion of "hearing out" (termed "dichotic pitch") has a close analogy in the visual system, where retinal disparities in random dot stereograms can be used to achieve the "seeing out" of a shape displaced in depth from a random background (Julesz, 1971).

Using EEG and MEG to measure brain activity in human listeners, we have found that the hearing out of dichotic pitches elicits a sequence of auditory cortical responses over a time window of some 150-400 ms after the onset of a dichotically-embedded pitch. In a series of experiments (Johnson et al., 2003; Hautus & Johnson, 2005; Johnson et al., 2007; Johnson & Hautus, 2010) we have shown that these responses correspond to functionally distinct stages of auditory scene analysis. Our data provide new insights into the nature, sequencing and timing of those stages.

2. Dichotic pitch paradigm

Dichotic pitch is a binaural unmasking phenomenon that is theoretically closely related to the masking level difference (MLD), and involves the perception of pitches from stimuli that contain no monaural cues to pitch (Bilsen, 1976; Cramer & Huggins, 1958). Dichotic pitch can be produced by presenting listeners with two broadband noises with interaurally identical amplitude spectra but with a specific interaural lag over a narrow frequency band (Dougherty et al., 1998). The interaurally-shifted frequency band becomes perceptually segregated from the noise, and the resulting pitch has a tonal quality associated with the centre frequency of the dichotically-delayed portion of the spectrum. Because the stimuli are discriminable solely by the interaural lag but are otherwise acoustically identical, the perception of dichotic pitch must ultimately depend upon the binaural fusion of interaural time differences (ITDs) within the central auditory system. The phenomenon of dichotic pitch demonstrates that the human auditory system applies its exquisite sensitivity for the fine-grained temporal structure of sounds to the perceptual segregation, localization, and identification of concurrently-presented sound sources.

Fig. 1 shows how dichotic pitches can be generated using a complementary filtering method described by Dougherty et al. (1998). Two independent broadband Gaussian noise processes, 500-ms in duration are digitally constructed, in this case with a sampling rate of 44,100 Hz. One noise process is bandpass filtered with a centre frequency of 600 Hz and 3-dB bandwidth of 50 Hz using a 4th-order Butterworth filter with corner frequencies of 575 and 625 Hz (Fig. 1: middle panels). The other noise process is notch filtered using the same corner frequencies as the bandpass filter (Figure 1: left panels). The sum of the filter functions for the notch and bandpass filters is equal to one for all frequencies.

The bandpass-filtered noise process is duplicated and, to produce the dichotic-pitch stimuli, one copy of the noise process is delayed by 500 μs. Control stimuli contain no delay. The notch and bandpass filtered noise processes are recombined, producing two spectrally flat noise processes, which are again bandpass filtered (4th-order Butterworth) with corner frequencies of 400 and 800 Hz (Fig. 1: right panels). All stimuli are windowed using a cos2 function with 4-ms rise and fall times. In our laboratory auditory stimuli are generated on two channels of a 16-bit converter (Model DAQPad 6052E, National Instruments, Austin, Texas, USA). Programmable attenuators (Model PA4, Tucker-Davis Technologies, Alachua, Florida, USA) set the binaural stimuli to 70 dB SPL prior to their delivery via earphones (In our lab, Etymotic insert earphones Model ER2 or ER3, Etymotic Research Inc., Elk Grove Village, Illinois, USA). For sequences of stimuli, a jittered interstimulus (offset to onset) interval is drawn from a rectangular distribution between 1000 and 3000 ms.

Comparable dichotic pitch perceptions can be elicited using interaural level (ILD) rather than timing differences. To produce ILD dichotic pitch, the relative amplitude of the two bandpass noises is adjusted to increase the level in one channel while reducing the level in the other, and the same is done for the two notched noises. The two noises for each channel are combined as for the ITD stimuli.

Fig. 2 illustrates some of the perceptions that can be evoked by dichotic pitch stimuli presented via earphones. Control stimuli (top row) contain an interaural time disparity (ITD) that is uniform over the entire frequency spectrum of the noise stimuli and results in a single percept of noise (represented as ###) lateralized to the side of the temporally leading ear. Dichotic pitch stimuli (bottom row) contain interaural disparities that are oppositely directed for a narrow notch of frequencies (e.g. 575-625 Hz) versus the remainder of the frequency spectrum. These stimuli evoke a perception of two concurrent but spatially separated sounds lateralized to opposite sides: a dichotic pitch (represented as a musical note) with a perceived pitch corresponding to the centre frequency binaurally delayed notch

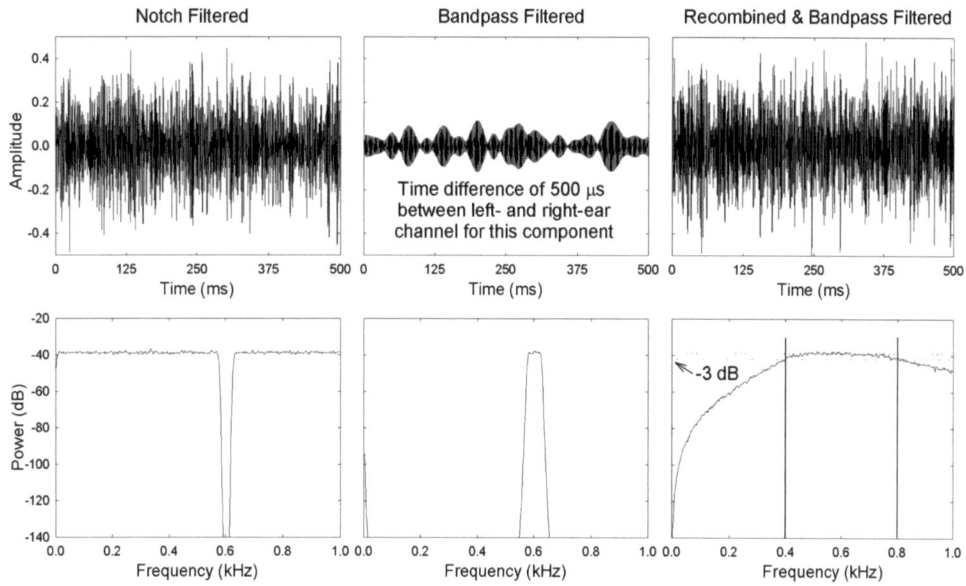

Fig. 1. Temporal and spectral representations of dichotic pitch stimulus. From Johnson et al., (2003) with permission.

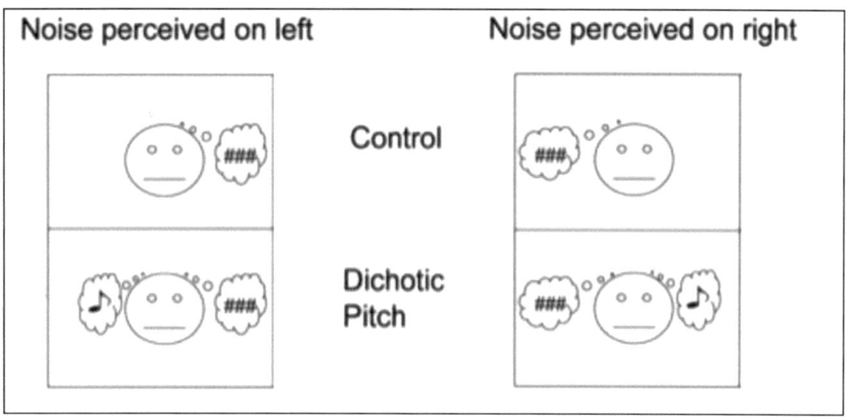

Fig. 2. Experimental stimuli and percepts of a listener. Adapted from Johnson and Hautus (2010) with permission.

(600 Hz) and a background noise corresponding to the remainder of the noise spectrum. From the point of view of an auditory researcher, the dichotic pitch paradigm has a number of features that make it useful for probing the workings of the central auditory system:

1. For experiments with dichotic pitch the control stimulus is simply one that has a uniform interaural disparity over its entire frequency range. Since the control and dichotic pitch stimuli are monaurally identical, any differences in perception or measured brain activity can be confidently attributed to differences in binaural processing;

2. Interaural disparities are first computed at the level of the medial superior olive in the brainstem (Goldberg & Brown, 1969; Yin & Chan, 1990) so perception of dichotic pitch can be confidently attributed to central rather than peripheral processes;
3. The perception of dichotic pitch depends on the ability of the auditory system to compute, encode, and process very fine temporal disparities (microseconds) and so provides a sensitive index of the temporal processing capabilities of the binaural auditory system. Consequently, dichotic pitch has been used to study clinical disorders such as dyslexia, that are suspected to involve central problems in auditory temporal processing (Dougherty et al., 1998);
4. The overall perceptual problem posed by dichotic pitch - that of separating a behaviourally relevant sound from a background noise or, more generally, that of segregating concurrent sound objects - is of considerable interest to those interested in how, and by what mechanisms, the brain is able to accomplish this important structuring of perception (Alain, 2007; Bregman, 1990).

Before proceeding to review experimental studies, we digress in the next section to describe for non-specialists the two main technologies used to measure auditory brain function in these studies, namely electroencephalography (EEG) and magnetoencephalography (MEG) and to introduce some terminology pertinent to these techniques.

3. EEG and MEG for measuring central auditory function

The methodologies for measuring brain function merit some consideration in any review of empirical studies, since the choice of method determines the type of brain activity measured (e.g. neuroelectric versus hemodynamic responses) and the spatial and temporal resolution of the measurements. These factors have a large impact on the types of inferences that can be derived from measured brain data.

Roughly speaking, EEG and MEG are the methods of choice when temporal resolution is an important or paramount requirement of a study. The reason for this is that the electromagnetic fields measured by these techniques are directly and instantaneously generated by ionic current flow in neurons. In contrast, positron emission tomography (PET) and functional magnetic resonance imaging (fMRI) techniques measure the indirect, and temporally sluggish, metabolic and hemodynamic consequences of neuronal activity. Consequently PET and fMRI have inherently coarse temporal resolutions, on the order of one to many seconds. EEG and MEG are often described as having "millisecond" temporal resolution, but this is a technical limit imposed by the sampling capabilities of analogue-to-digital converters: by the nature of the measurements EEG and MEG can theoretically track ionic currents as fast as they occur. In practice though there are a number of additional limitations to the temporal capabilities of EEG and MEG: for example, time series are typically averaged over spans of tens or even hundreds of ms to improve the reliability of measurements and to reduce the dimensionality of the data. Even so, EEG-MEG are the methods of choice when one studies brain events that change rapidly and dynamically over time. For example, EEG-MEG techniques have long been an essential tool of psycholinguists studying the brain processes associated with language (Kutas et al., 2006).

The very properties that confer a high temporal resolution to impart fundamental limits on spatial resolution and EEG-MEG are generally considered inferior to PET and fMRI for localizing brain events in space. For both techniques the algebraic summation of electromagnetic fields limits their ability to resolve concurrent and closely-spaced neuronal events. MEG has certain advantages over EEG in this regard because magnetic fields are not altered by conductive anisotropies and inhomogeneities. There are also advantages conferred to MEG by the fact that it is relatively less sensitive to distant sources and to neuronal sources in the crests of gyri (since these are oriented radially to the skull their magnetic fields do not exit the head). This lack of sensitivity is advantageous because MEG measurements present a relatively simpler picture of brain activity for researchers to interpret: simply put, there are fewer contributing brain sources that must be disentangled from measurements of fields on the surface of the head.

EEG-MEG measurements are typically carried out in event-related experimental designs in which stimuli are presented repeatedly (tens to hundreds or thousands of trials) and measurements are averaged over repeated trials to increase the signal-to-noise ratio of brain responses. In the case of EEG averaged signals are referred to as event-related potentials (ERPs) or evoked potentials (EPs). ERPs recorded on the surface of the head are often averaged across subjects as well to produce "grand-averaged" ERPs. In the case of MEG averaged signals are referred to as event-related magnetic fields (ERFs) but these are typically not analyzed as grand averages. This is because the higher spatial resolution of MEG means that it is not reasonable to assume that a given MEG sensor will record the same configuration of brain activations from subject to subject. For this reason MEG data is typically rendered into "source space" by computing the brain sources of the surface-recorded data, before performing descriptive and inferential statistics. Source analysis of EEG data is also possible and this is increasingly done by researchers. However the EEG source analysis problem is somewhat more complicated because of the need to specify the resistive parameters of the various tissue compartments of the head and brain. A final but essential piece of EEG-MEG nomenclature pertains to the naming of landmarks within ERP-ERF time series. ERP averages are presented as voltage deflections over time and deflections are named according to their polarity and latency (for example, "P100" may refer to a positive deflection at a latency of 100 ms after stimulus onset) or polarity and relative timing in a sequence (for example, P1-N1-P2 refers to a sequence of a positive and a negative and another positive deflection). ERP-ERF are also roughly subdivided into "middle" (about 20-70 ms) and "late" latency responses (greater than 80 ms or so). ERPs contain a third class of "early" (less than 10 ms) responses generated in CN VIII and the auditory brainstem. Because of the distance, MEG sensors are relatively insensitive to the sources of these early responses. Although this approach will not be further discussed in this review, we note in passing that it is also informative to analyse the frequency content of EEG and MEG signals and these are computed as "event-related spectral perturbations" (ERSPs).

4. Brain responses to dichotic pitch: the ORN

4.1 Passive listening conditions

Fig. 3 illustrates brain responses to dichotic pitch and control sounds, recorded with EEG from healthy adult subjects in a "passive" listening experiment (Johnson et al., 2003). In this experiment participants were instructed to attend to an engaging video viewed with the soundtrack silenced while they ignored experimental stimuli presented via insert earphones. Prior to the EEG recording session all subjects underwent a psychophysical screening procedure to ensure they could detect dichotic pitch (hereafter, "DP").

The left column of Fig. 3 shows ERPs averaged over 400 trials of each stimulus type and grand averaged over a group of 13 subjects, and recorded from electrodes placed at a frontal midline location on the head and at two lateral positions about 4 cm to the left and right of the midline. ERPs are plotted as voltage time series over a time base of -100 ms before stimulus onset to 500 ms after stimulus onset. Voltages are plotted with negative up (a convention used in many EEG labs), and ERPs evoked by DP stimuli are overlayed on top of ERPs to control stimuli.

For both types of stimuli ERPs are characterized by positive-negative-positive sequence of responses labelled P1, N1 and P2, and with peak latencies of 76 ms, 108 ms and 172 ms respectively, typical of late cortical responses to a variety of acoustic stimuli with abrupt onsets. The DP and control ERPs begin to diverge just prior to the peak of the P2 component at a latency of about 150 ms, with the DP waveform becoming more negative than the control ERP. The differences between DP and control responses are best visualized in the subtraction waveforms in the centre column of Fig. 3, showing that the amplitude difference is maximal by about 210 ms with large differences persisting until about 280 ms, after which amplitude differences decline sharply and the two ERPs show similar amplitudes again by 380 ms latency. In this early study we referred to the difference wave simply as a "late negativity" or LN component.

As can be surmised from the fairly similar responses obtained at electrode sites as much as 8 cm apart, the ERPs have an extensive spatial distribution. In the right hand column of Fig. 3 the amplitude distribution of ERPs is shown as isovoltage contours on a schematic top view of the head. In the head schematic the dots represent electrode positions and the topographic maps are based on a fairly dense spatial sampling of 128 recording electrodes. The maps of the LN component (bottom row of contour maps) shows that this component has a broad distribution centred over the frontal midline of the head. This reinforces a point made in the preceding section about the relatively coarse spatial resolution of the EEG. However the EEG does quite a good job of localizing neural processing of DP in time: the difference waveforms demonstrate a robust phase of neural processing -- specific to DP -- that occurs some 150-350 ms after stimulus onset. As the two stimuli employed in this study were discriminable solely by a dichotic delay (in the DP stimulus) but were otherwise acoustically identical, we can be confident that the late cortical LN wave reflects neural processing that is dependent on binaural fusion within the brain. These results confirm that the late cortical ERPs are highly sensitive to the binaural processes underlying the perception of DP, and suggest that these may be a useful electrophysiological tool for assessing the binaural processing capabilities of the central auditory system.

4.2 Active listening conditions

While the LN response to DP is clearly based on binaural processing, it closely resembles an ERP response associated with a perceptual structuring based on a monaural cue - the inharmonicity of one component of a complex sound composed of multiple harmonics. Alain et al. (2002) measured ERPs from subjects presented with sounds containing tuned or mistuned harmonics. In two different listening conditions the subjects either actively attended to the sounds and indicated their perceptions (a single sound versus two sounds) with a button press, or ignored the acoustic stimuli while watching a silent movie. The perception of a mistuned harmonic as a separate sound was associated with a negative wave that has a peak latency of about 160 ms, which these authors termed the "object-related negativity" (ORN). The ORN was elicited in both active and passive listening conditions, while a later P400 wave was elicited by the mistuned harmonic stimuli only when subjects

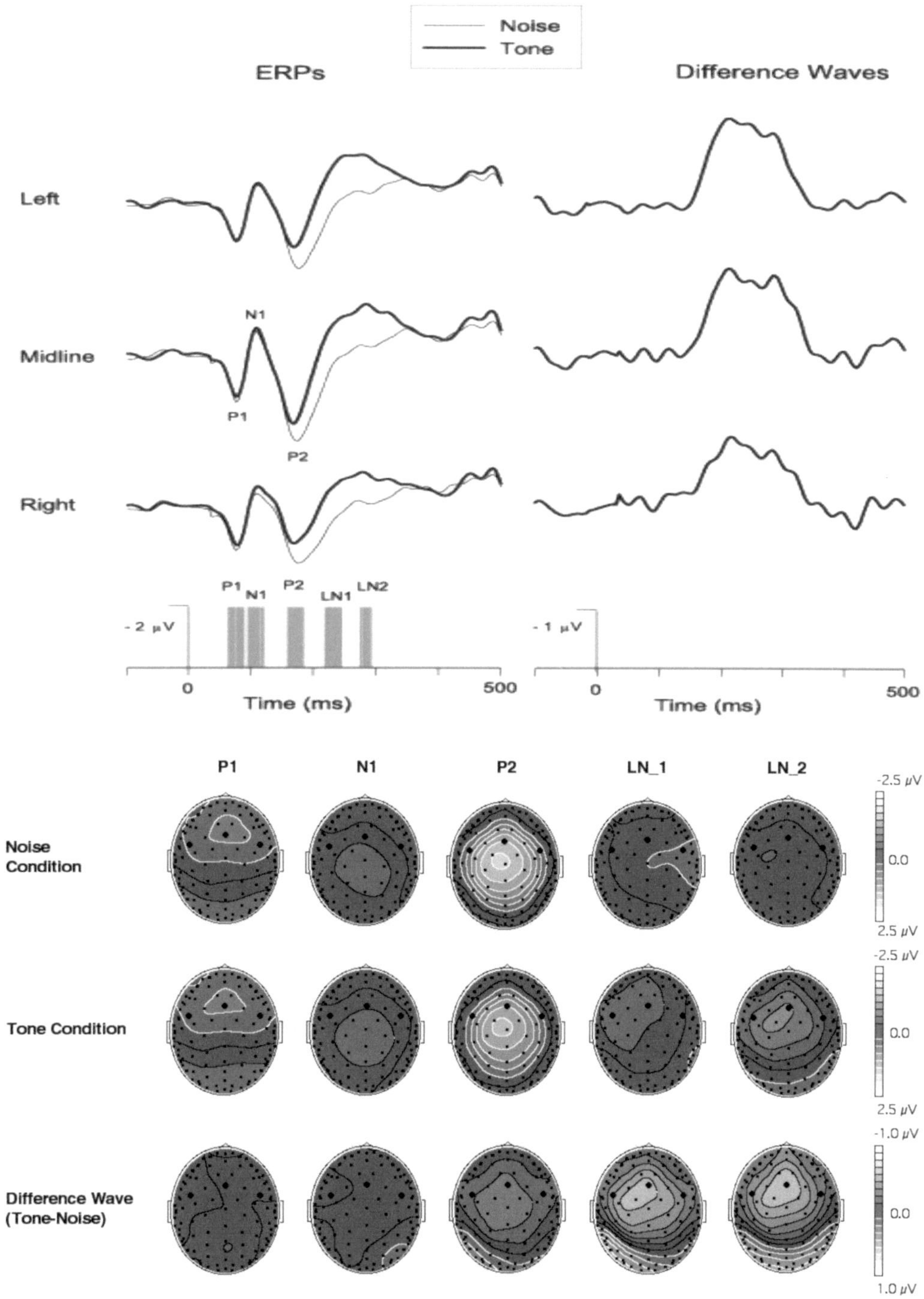

Fig. 3. Grand averaged ERPs in passive listening condition. "Tone" refers to dichotic pitch stimulus, "Noise" refers to control stimulus. Large dots indicate electrode positions for waveforms shown at top. Adapted from Johnson et al. (2003) with permission.

actively attended to these stimuli. The authors concluded that the two sequential components support a two-stage model of auditory scene analysis. In this model, complex sounds are first parsed based on an automatic process, reflected by the ORN and involving a detection of a mismatch between the harmonic template extracted from the incoming stimulus and the harmonic frequency expected based upon the fundamental of the complex sound. The second stage, indexed by the P400, is invoked only when subjects are actively attending to the stimuli and seems to reflect controlled processes responsible for identification of stimuli and selection of behavioural responses.

While the acoustic stimuli are radically different, the overall perceptual problem posed by the sounds in the DP paradigm used by us and the mistuned harmonic paradigm used by Alain et al. (2002) is the same: that of breaking a complex sound wave into components that correspond to two concurrent perceptual objects. Indeed, we found that under active listening conditions the mechanisms described in Alain et al.'s (2002) two-stage model are also deployed for the perception of DP.

Fig. 4 shows DP and control ERPs recorded under two passive listening conditions: one in which DP and control stimuli were randomly interleaved (P-R) and one in which they were presented in uniform blocks (P-B); and an "active" listening condition (A-R) in which listeners were required to actively attend to stimuli on each trial and to indicate with a button press whether a (randomly interleaved) DP or control stimulus had been presented. The ORN was elicited by DP stimuli in all three listening conditions, while a P400 response was elicited only in the active condition.

Fig. 5 is an instructive summary of the ERP data from this experiment because it clearly shows that successive segments of the ERPs show quite distinctive behaviours as a function of stimulus and listening conditions. The N1 component is modulated by attention but not stimulus type, showing a generalized increase in amplitude (i.e. greater negativity) when actively attending. The ORN component shows a generalized attention effect and a main effect of stimulus type (i.e. it is more negative when processing DP stimuli). Finally the P400 is manifest as an interaction between listening condition and stimulus type, because it is elicited only by DP stimuli and only when listeners are required to actively discriminate stimuli. These distinctive functional profiles clearly indicate that the three components reflect distinct stages of auditory processing; and conversely, that ERPs are capable of localizing different stages of processing in time.

These results show that the perception of DP is associated with two cortical processing stages - indexed by the ORN and the P400 - that are functionally comparable to those elicited by the mistuned harmonic stimuli used by Alain et al. (2002). Since the physical compositions and critical cues of these two classes of sounds are radically dissimilar, it is reasonable to conclude that these processing events are more related to the overall perceptual problem posed by both sounds: that of partitioning a complex acoustic input stream into two distinct perceptual objects. Alain et al. (2002) proposed that the ORN component indexes a transient automatic mismatch process between the harmonic template extracted from the incoming stimulus and the harmonic frequency expected based upon the fundamental of the complex sound. In the case of dichotic pitch, however, a neural mismatch response must be generated on the basis of location, since the noise processes by definition have no harmonic structure to match to. This suggests that the ORN reflects the activity of fairly general mechanisms of auditory stream segregation that can broadly utilize a range of cues to parse simultaneous acoustic events.

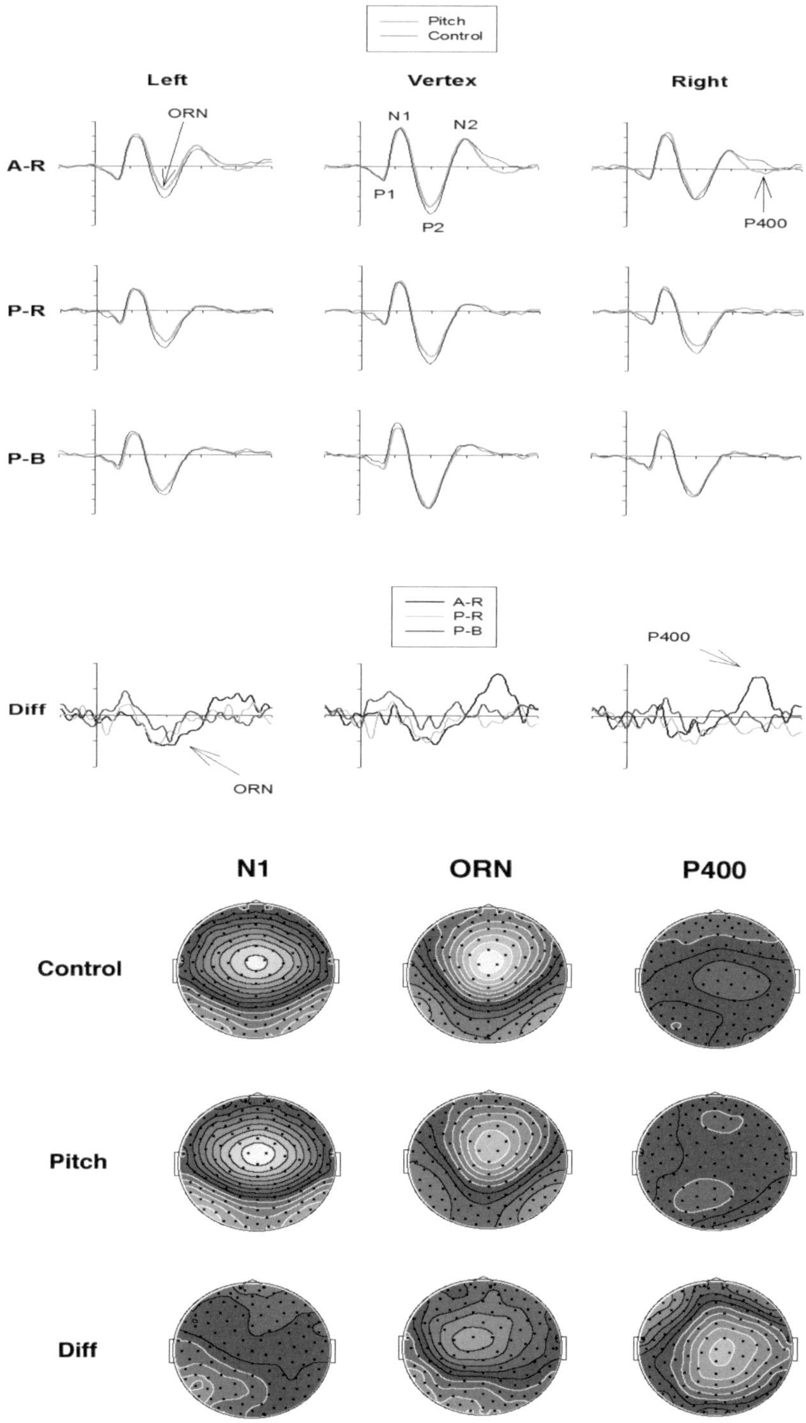

Fig. 4. Grand averaged ERPs under three listening conditions. Adapted from Hautus and Johnson (2005) with permission.

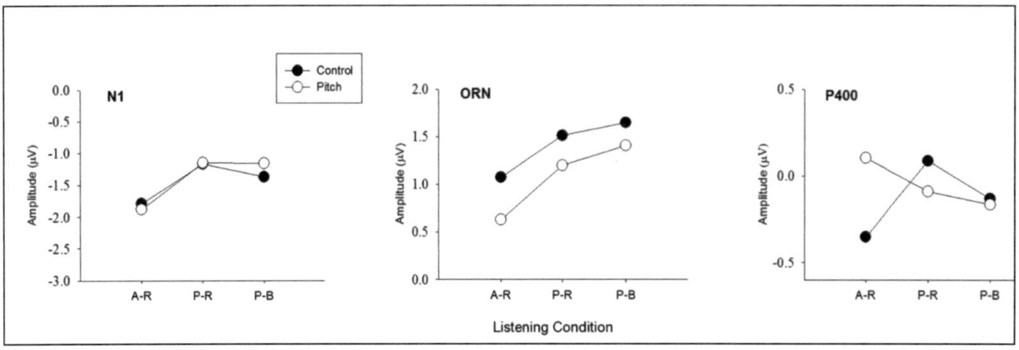

Fig. 5. Functional profiles of ERP components. Adapted from Hautus and Johnson (2005) with permission.

4.3 Cortical generators of the ORN

Consistent with what one would expect of mechanisms that play such a basic role in auditory perception, the results discussed in the preceding section, and those of previous researchers (Alain & Izenberg, 2003; Alain et al., 2002) suggest that the ORN indexes relatively automatic processes that function independently of attention. Such automaticity points to relatively low levels of the auditory system, and indeed physical modelling of the generators of the auditory P1-N1-P2 complex indicates that these waves are best modelled by current sources in or near the primary auditory cortices (Picton et al., 1999; Scherg et al., 1986). More evidence for the involvement of primary auditory cortex in stream segregation comes from a study by Dyson and Alain (Dyson & Alain, 2004), who reported that the first reliable effect of mistuning harmonic stimuli was a modulation of the amplitude of the Pa peak of the middle latency auditory evoked response. There is good evidence from convergent sources that the Pa wave is generated in primary auditory cortex (Liegois-Chauvel et al., 1991; McGee et al., 1992; Pantev et al., 1995; Scherg & Von Cramon, 1986), and the early latency of this response (about 32 ms) is consistent with the preattentive processing of acoustic cues to auditory stream segregation.

As noted previously, MEG is relatively most sensitive to cortical generators that are oriented tangentially to the surface of the head; consequently the auditory cortices of the supratemporal plane are well-visualized with MEG. Fig. 6A (left column) shows the distribution of AEF amplitude measured with MEG sensors (indicated as dots) and projected onto the surface of the head. In comparison to the EEG topographic maps described in previous sections, the MEG maps show some obvious differences. First, while the AEP fields were maximal over the superior surface of the head, the AEF fields are maximal over temporal regions. Second, the AEF fields have a much more focal distribution than the AEP fields. The first point of difference can be reconciled by appreciating that both electrical and magnetic fields are generated by populations of tangentially-oriented pyramidal cells in the supratemporal plane. A population of synchronously activated neurons can be well approximated by an "equivalent current dipole" (ECD), represented as the coloured ball and sticks in Fig. 6B (left column). The ball of the ECD represents the "centre of gravity" of a region of activated cortex, while the stick points in the direction of positive current flow. Accordingly, this ECD will generate an electric field with a negative polarity on the superior surface of the head and a positive polarity below the supratemporal plane. The magnetic field generated by the same ECD will follow the "right hand rule" of electromagnetism: If current is flowing in the direction of your extended right thumb, then magnetic flux will circle the ECD in the direction of the right fingers. For the ECD of Fig. 6B (left column) this will result in the magnetic field pattern of Fig. 6A (left column): on the left hemisphere positive flux emerging from the posterior temporal region (shown as red) and negative flux re-entering the head in the anterior temporal region (shown as blue). Following this logic it is clear why the opposite flux pattern is seen over the right hemisphere. The second point of difference, the more focal distribution of the AEFs, is due the fact that magnetic fields are not subject to the smearing effects of conductive inhomogeneities: the brain, cerebrospinal fluid, skull and scalp are transparent to magnetic fields.

These considerations indicate that AEFs can be quite effectively modelled with very simple models of both the brain sources and the physical characteristics of the brain and head. With such computational models hundreds of channels of surface-recorded MEG data can be rendered into a simple configuration of brain sources that nonetheless capture quite significant dynamics of brain function. Fig. 6A (right column) shows grand-averaged source waveforms from the bilateral dipole model of AEFs elicited by broadband noise. The source waveforms show that the responses of the left and right hemispheres are distinctively different: the right hemisphere has a much reduced middle latency M50 (equivalent to the P1 AEP) response but a larger amplitude M100 (equivalent to the N1 AEP) with an earlier

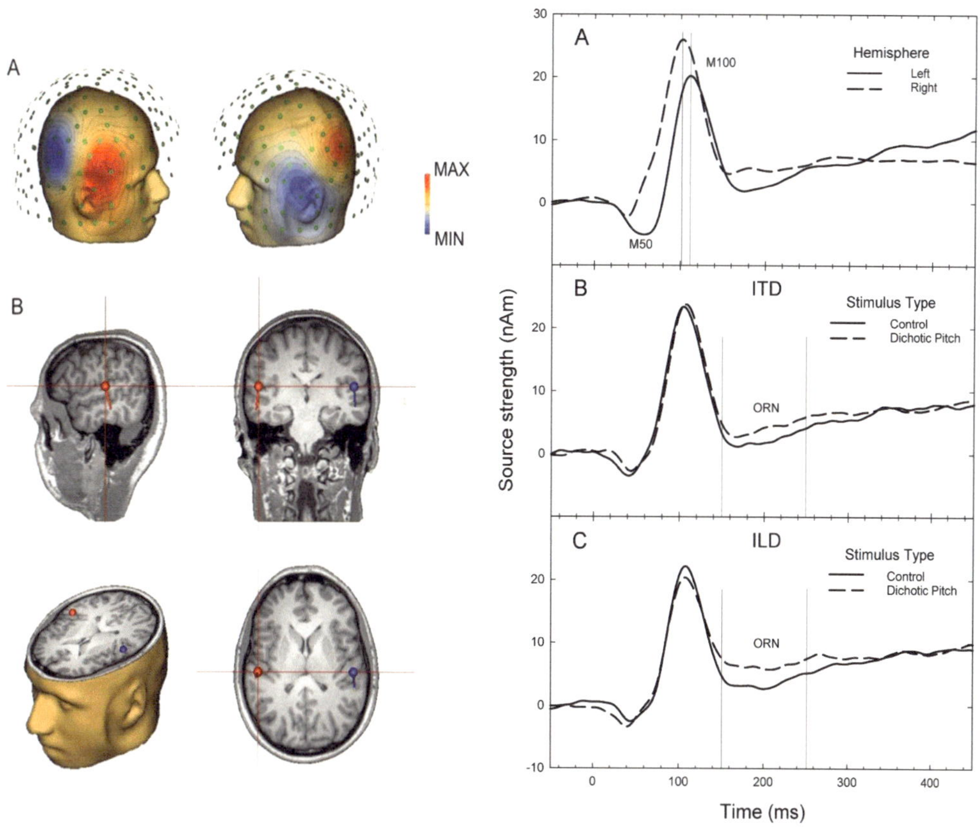

Fig. 6. ORN measured with MEG. Adapted from Johnson and Hautus (2010) with permission.

peak latency. Fig. 6B and C (right column) show the MEG source waveform version of the ORN elicited by interaural disparities in the left hemisphere. The timing of the MEG ORN coincides with the timing of the EEG ORN described previously.

5. Differential processing of interaural timing and level differences

If the ORN reflects the activity of fairly general mechanisms of auditory stream segregation that can draw on a range of cues to parse simultaneous acoustic events, we reasoned that it might be useful in helping to address an unresolved issue in binaural hearing concerning the processing of ITD and ILD cues in the auditory cortex (Johnson and Hautus, 2010). It has long been appreciated that ITDs are the dominant cues for low frequency sounds, while ILDs dominate for high frequency sounds, the so-called duplex theory of audition (Rayleigh, 1907). Since the formulation of the duplex theory, researchers have suggested the existence of separate neural processing channels for ITDs and ILDs. Indeed, there are several lines of physiological evidence for independent processing mechanisms from unit recordings in the auditory brainstem of animals (Phillips & Brugge, 1985; Smith et al., 1993; Yin & Kuwada, 1984) and also from surface recordings of auditory brainstem responses in humans (Pratt et al., 1997).

It would seem a logical requirement for the auditory system to eventually pool the spatial information extracted from disparate cues into a common code that can be used to solve the broad perceptual problems posed for auditory scene analysis. However it remains unclear when, or even if, information from ITDs and ILDs may be combined into a common code for spatial lateralization (Schroger, 1996; Ungan et al., 1997). On the one hand, psychophysical studies have shown that lateralization to one ear induced by one cue can be precisely counterbalanced by the complementary cue leading at the other ear, according to a systematic "trading ratio" (Hafter & Jeffress, 1968; Harris, 1960). This suggests that information from the two cues is eventually merged at some stage of the central nervous system. On the other hand, the trade-off between ITD and ILD does not seem to be complete: Listeners report that they experience distinctly different types of sound lateralization "images" for the two types of cues (Hafter & Carrier, 1972).

This suggests that segregation may be maintained, at least to some degree, to the level of conscious perception. Indeed, evidence from the cat (Phillips, 1993) and from brain-damaged human patients (Yamada et al., 1996) indicates that separate representations of ITDs and ILDs exist at the level of the auditory cortex. Further, an EEG study has reported different surface topographies for the circa 100 ms latency N1 event-related potential (ERP) component elicited by ITDs and ILDs, indicating spatially separated neural representations at a relatively late stage of processing in the cerebral cortex (Ungan et al., 2001). MEG recordings show that the two cues have independent effects on the amplitude of the M100, the magnetic counterpart of the N1 (Palomäki et al., 2005). A similar finding for the mismatch negativity (MMN) component of the ERP (Schroger, 1996) suggests at least partially independent cortical processing of timing and level cues up to 200 ms in latency.

Taken together, these studies provide good evidence for segregation of ITDs and ILDs to quite late stages of auditory processing. However, they shed little light on when (if ever) ITDs and ILDs might be incorporated into a common signal that could mediate perceptually relevant phenomena. We measured auditory brain function with MEG to determine if the ORN might represent a stage when information from the two cues is merged into a common code for auditory scene analysis (Johnson & Hautus, 2010).

5.1 Segregated processing of ITD and ILD cues

The experiment was a 2 x 2 x 2 design with variables location cue type (ITD or ILD), stimulus type (control or dichotic pitch), leading ear for background noise (noise perceived on left or noise perceived on right). Each location cue type could result in four possible percepts: a single noise on the left or right (control stimuli), or concurrent background noises and dichotic pitches (dichotic pitch stimuli), with the background noise perceived on the right or left and the dichotic pitch perceived on the opposite side.

The results showed that ITD and ILD cues elicit distinctive hemispheric patterns of activation during the M100 time window. Fig. 7 (top row) shows that left lateralized control sounds elicited larger amplitude M100s in the contralateral hemisphere, while right-lateralized sounds elicited similar amplitude responses in both hemispheres. In contrast, both left and right lateralized sounds for ILD cues elicited similar patterns of activation (larger amplitude on the right). Interestingly, the same stimulus x hemisphere interaction was obtained for ITDs only when the left-lateralized DP stimuli were compared to the right lateralized control stimuli. This is a striking result because the bulk of the stimulus energy of the left lateralized DP stimulus (the noise) was perceived on the right. We interpreted this pattern in terms of a stronger right-hemisphere bias for spatial information pitted against a left hemisphere bias for timing information: the left hemisphere holds its own if sounds originate solely from the opposite hemispace: the right hemisphere wins the tug of war if any sounds are present in the left hemispace (Johnson and Hautus, 2010).

While a unilateral ITD cue can engage the left hemisphere if presented in the right hemispace, no such effect was obtained for ILD cues (Figure 5, right). This finding reinforces the interpretation that it is timing information, contained in the ITD representation but not the ILD representation, that is crucial for engaging the left hemisphere. The greater capacity of ITD cues to activate the left hemisphere cannot be attributed to a greater salience of this cue, since our behavioural data showed that the ILD cues were in fact more detectable, despite our efforts at loudness and laterality matching. These results therefore support the hypothesis that, during the time window of the M100 response, ITD and ILD cues are processed in at least partially distinct channels of the auditory cortex (Wright & Fitzgerald, 2001), aimed at differentially elaborating the specific kinds of spatial information represented in each channel.

Our results are supported by an EEG topographic mapping study by Tardiff et al. (Tardif et al., 2006). These authors also found that ITD/ILD differences were primarily in terms of hemispheric lateralization, although they reported somewhat different hemispheric patterns than ours, with bilateral responses to ILD cues and left-lateralized responses to ITD cues. The differences between studies may be attributable to differences in EEG and MEG recording methodologies, but in any case both studies support a greater involvement of the left hemisphere in processing ITD cues than for ILD cues. An involvement of the left hemisphere in sound localization is supported by a recent study of patients with right or left hemisphere brain damage (Spierer et al., 2009). While these authors found a prominent right hemispheric dominance for auditory localization, significant spatial deficits were also observed in patients with strictly left hemisphere damage. These authors concluded, as do we, that right hemispheric dominance is better conceptualized as an asymmetrical activation of both hemispheres, rather than an absolute role for the right hemisphere.

5.2 Pooling of spatial information for auditory scene analysis

For both ITD and ILD location cues during a time window of 150-250 ms, sources in both hemispheres showed a prominent increase in amplitude (ORN) for sounds containing a

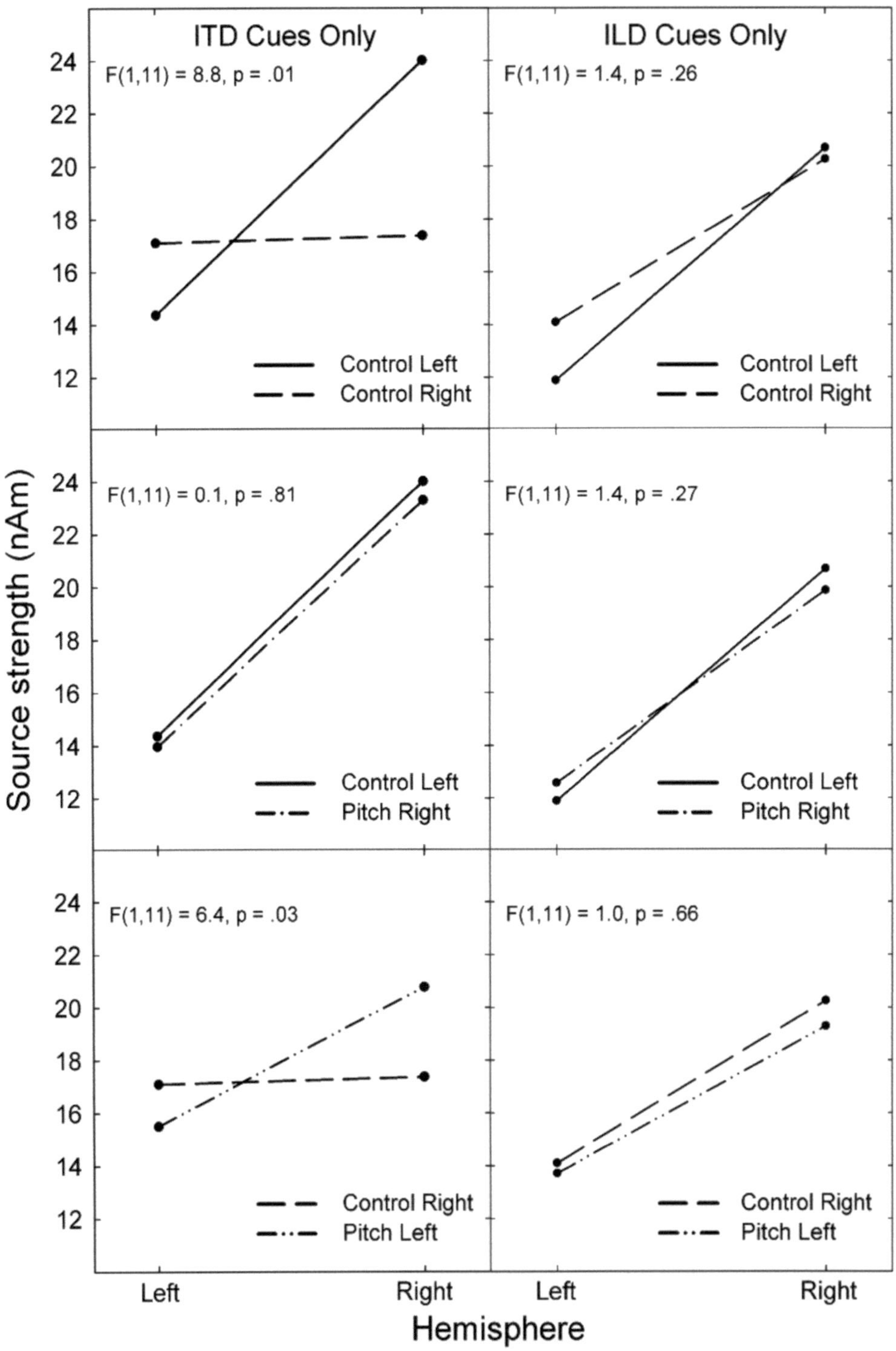

Fig. 7. Distinctive hemispheric activation patterns for ITD and ILD cues. Adapted from Johnson and Hautus (2010), with permission.

dichotic pitch, in comparison to sounds that contained no binaurally-embedded pitch (Fig. 6 right panel, B & C). The finding that the ORN is elicited to ILD cues to sound location further strengthens the contention that ORN mechanisms are able to broadly draw on information derived from a variety of cues: that is, the generation of the ORN is not constrained to the operation of highly specific processes such as a spectral template matching mechanism. The mean time window of the ORN effect was essentially identical for both ITD and ILD cues. In contrast to the M100 component, the ORN component was not itself modulated in amplitude by location cues, nor was its lateralization influenced by the type of location cue. This negative result stands in contrast to previous findings for the M100 and later components (Schroger, 1996; Tardif et al., 2006). Therefore, the ORN can potentially place an important and novel temporal boundary on the extent to which ITD and ILD cues, or the information derived from them, remain segregated in separate channels of the cerebral cortex, since information that can be used to parse a pitch from a background noise must have been extracted from both cues prior to the generation of the ORN. Further investigations are required to address these important theoretical issues. We note also that the ORN component was superimposed upon brain activity in the same time window that was strongly modulated by cue type (i.e. there was a main effect of cue type but no interaction with stimulus type). It seems that the brain may continue to process information in independent streams even after spatial information has been extracted to support auditory scene segregation.

In summary, these results show a strong modulation of interhemispheric activity by ITD cues, but only when these cues are presented unilaterally from the right hemispace. These data support the interpretation of a relatively strong right hemisphere bias for spatial information in conjunction pitted against a relatively weaker left hemisphere preference for timing information. The hemispheric biases are large in comparison to the modest contralateral bias exhibited at the population level in primate auditory cortex (Werner-Reiss & Groh, 2008). In contrast, ILD cues lack the capacity of ITDs to engage the left hemisphere, presumably because their cortical representations lack the timing information that is preferentially processed in that hemisphere. Finally, spatial information that is common to both ITD and ILD cues seems to be extracted prior to the ORN time-window for use by the cerebral mechanisms of auditory scene segregation.

6. Sequential processing of ITDs for sound source segregation and spatial localization

In everyday life, ITDs play a fundamental role in two basic aspects of auditory perception. First, they serve as the primary cues for localization of low frequency sounds in azimuthal space (Blauert, 1997). Second, ITDs are one of a set of Gestalt-style grouping cues employed by the auditory system to sort and parse sound mixtures produced by concurrently active sound sources (Drennan et al., 2003) and therefore play a role in a structuring of perception referred to by Bregman (1990) as "auditory scene analysis." This analysis allows us (for example) to attend to the voice of one speaker among the babble of many others at a cocktail party.

Considered individually these two perceptual roles for ITDs have been well studied. However, the relationship between these fundamental processes has received scant attention from auditory scientists. Intuitively one may suppose that both perceptual results may be achieved by the same stage of processing, since identifying the locations of two temporally concurrent (but spatially disparate) sounds would seem to automatically result in their spatial separation. However there are both empirical and theoretical reasons to believe that auditory scene analysis and spatial localization are achieved in distinct steps of auditory processing. Neuropsychological evidence comes from a recent report of a patient with an ischemic lesion of right temporo-parieto-frontal areas (Thiran & Clarke, 2003). This patient exhibited a severe impairment in localizing sounds in everyday life and was entirely unable to use ITD cues to localize sounds in a spatial release from masking task and several diotic tasks requiring spatial localization from ITDs. Despite her profound 'spatial deafness' she was nonetheless able to use ITDs to segregate concurrent sound sources.

Recent theoretical views of auditory perception also point to unique neural mechanisms for auditory segregation and localization. For example, Griffiths and Warren (2002) have suggested that the segregational processes of auditory scene analysis are accomplished by a 'computational hub' of the auditory system, which they suggest is located in the planum temporale. These authors proposed that the neurons of this region employ an algorithm akin to independent component analysis (Bell & Sejnowski, 1995) to separate the components of complex acoustic signals. On this view, the parsed outputs from this stage of processing are subsequently routed via anatomically and functionally separate pathways to higher order destinations for segregated processing of auditory attributes including object identity and spatial position (Rauschecker & Tian, 2000). In line with the neuropsychological evidence described above, this model suggests a sequence of processing in which binaural information in complex sounds is initially employed for sound segregation, followed by an analysis which results in the perceptual elaboration of sound location.

We attempted to address this issue by measuring ERPs in experimental conditions that required listeners to extract dichotically-embedded pitches lateralized to the right or left of auditory space. Following the sequential model of ITD processing outlined previously we predicted that the ORN, as a marker of an early step in auditory scene analysis, should be relatively unaffected by variations of location. On the other hand we expected location-specific processing to be manifest in relatively later components of the auditory ERP. A second objective was to determine if the neural processing of ITDs is influenced by the nature of the listening task. To test this, we compared DP ERPs obtained in a task that specifically required listeners to locate the stimuli in auditory space (localization task), to those from a task that could be performed without actually computing locations (detection task). We predicted that the relatively low-level and automatic processing indexed by the ORN should be relatively unaffected by task demands. However the relatively higher level and more controlled levels of auditory scene analysis, suggested to be indexed by the P400 component, could be expected to be more heavily influenced by the goals of the behavioural task.

6.1 The ORN and perceptual segregation

The experiment consisted of three conditions based on two different yes-no tasks: detection and localization.

Detection: Participants listened to a random sequence of control stimuli and DP stimuli. The a priori probability was 0.5. Within a block of presentations, the DPs were located consistently to one side of auditory space - either the right or the left. Thus, the detection task consisted of two conditions (detect-right and detect-left). The participant indicated on a button-box whether the stimulus was a dichotic pitch or a control stimulus.

Localization: (We note that spatial judgements of sounds perceived intracranially are properly referred to as "lateralization" judgements. We prefer to use localization in this context to avoid confusion when referring to hemispheric lateralization of neural activity.) Participants listened to a sequence of DP stimuli (no control stimuli) with the pitches randomly located with equal probability on either the left or right side of auditory space. The participant indicated on a button-box whether the pitch stimulus presented was located to their right or left. Across the two conditions there were three classes of stimuli, each with two types: detection DPs (left and right) and detection controls (for left and right pitches); localization DPs (left and right).

Fig. 8 shows that a robust ORN response was elicited in all experimental conditions. The finding that the ORN was not modulated by location nor by task supports the interpretation that this is a relatively automatic response to DPs.

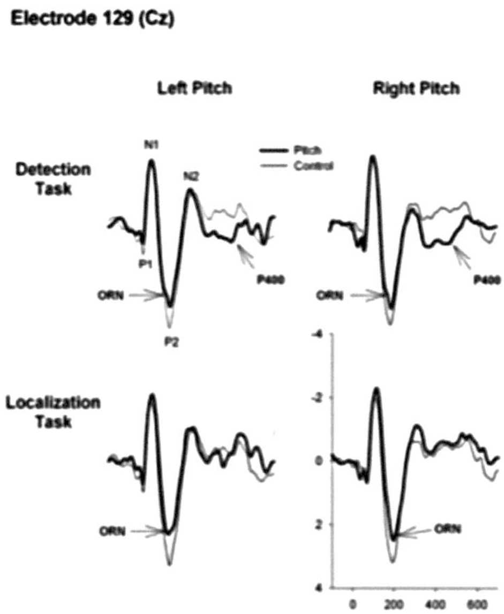

Fig. 8. Grand averaged ERPs for detection and localization tasks. Adapted from Johnson et al. (2007) with permission.

6.2 Location specific processing and the N2 component

Fig. 9 shows an amplitude modulation at lateral electrode sites which we termed "N2" because it overlaps during the time window of the vertex (Cz) N2 peak labelled in Fig. 8. We note that our N2 modulation is sustained several hundred milliseconds beyond the N2 peak (see Fig. 5) and has maximal amplitude distribution at lateral temporal sites contralateral to perceived pitch location. The most salient characteristic of the lateral temporal N2 component is its sensitivity to the spatial attributes of dichotic pitch, suggesting that this component reflects a location-specific phase of neural processing. We found a similar N2 contralateralization for lateralized pitches whether these were presented in the detection task, or if listeners were required to localize the pitches, suggesting that the spatial attributes of the stimuli were processed to some extent regardless of whether accurate task performance actually required a computation of spatial position.

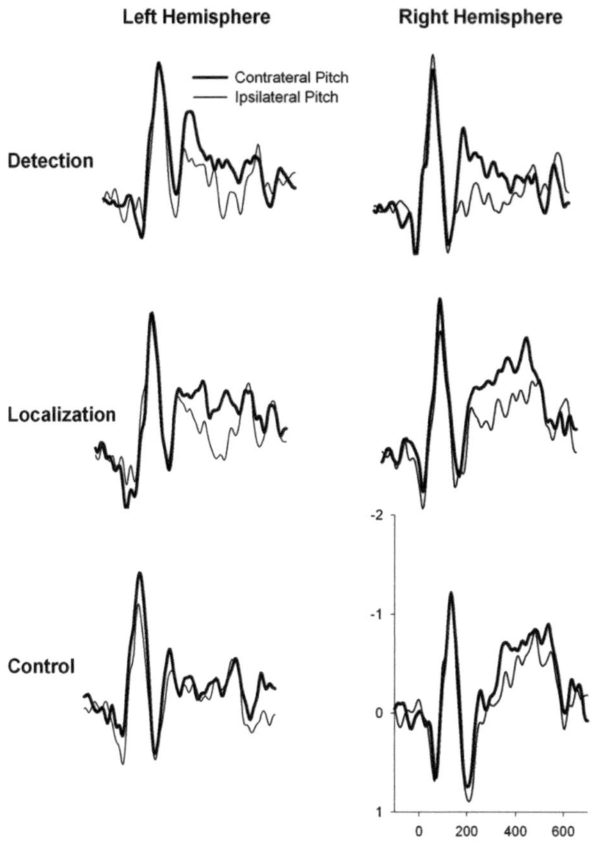

Fig. 9. Grand averaged ERPs recorded at lateral temporal electrodes on the left and right hemispheres. From Johnson et al. (2007) with permission.

6.3 The P400 and perceptual context

The ORN and N2 components were robustly evoked by ITDs in both tasks. In contrast, the P400 - a prominent feature of the detection ERPs - was entirely absent from the localization ERPs (Fig. 8). Previous studies have shown that, as with the ORN, the P400 component is tightly linked to concurrent sound segregation, but unlike the ORN is elicited only when listeners are required to actively attend and generate behavioural responses to sounds (Alain et al., 2002; Hautus & Johnson, 2005). As such, it seems clear that it indexes a controlled, rather than automatic, level of processing associated with the identification of concurrent sound sources and selection of a behavioural response (Alain et al., 2002).

The lack of a P400 component in the localization task shows that this component is not an obligatory consequence of auditory processes involved in concurrent sound segregation. Since both tasks require perceptual segregation of, and active responses to, concurrent sound sources, why do physically identical dichotic pitches elicit a P400 in the detection tasks but not in the localization task? One possibility is suggested by the fact that in the detection task the dichotic pitch stimuli were interleaved with control stimuli that did not contain any binaural cues, while in the location task all stimuli contained binaural cues. This

suggests that the P400 may be strongly influenced by the perceptual context in which the sounds are presented.
To test this possibility we performed a second order analysis in which we re-averaged ERPs based on the type of sound that immediately preceded a given sound. The results supported the conjecture that the P400 to DP is highly sensitive to the perceptual context in which the sounds are presented. Maximal P400 amplitudes were elicited when the dichotic pitch was presented after a control stimulus, but were much reduced in amplitude when the dichotic DP followed another DP. No P400 was elicited when the sounds in a sequence were all DPs (the localize condition). This pattern of results suggests that the P400 component may index a change in the number of perceptual objects in the acoustic environment. This appears to be a unidirectional phenomenon, since no P400 is obtained when the change is from a DP to a control stimulus. Thus, the P400 may be specifically sensitive to a change from one object (the centrally located noise) to two objects (the noise plus a lateralized pitch).

6.4 ERP components as indices of sequential processing

The sequence of processing events revealed by the results of this ERP study are summarized by the functional profiles of Fig. 10. The earliest, N1 component exhibits no modulation by experimental variables (cf. Fig 5, where attentional factors modulated the N1). The ORN was elicited by ITDs regardless of location or task. In contrast, the later N2 response (250-350 ms) was strongly contralateralized to the perceived location of a DP. Finally, DP stimuli in the detection task elicited a P400 at a latency of 400-500 ms, but this response was entirely absent from ERPs elicited by identical stimuli in the localization task. The sequence of cortical processing events shown here supports the prediction that operations associated with spatial localization of sounds are functionally distinct from, and are preceded by, operations associated with the perceptual segregation of those sounds. This functional-temporal dissociation is consistent with a model of auditory system functioning that suggests that the components of auditory information are initially separated on the basis of Gestalt-style auditory grouping cues (Bregman, 1990) including location (specified here by ITDs) and inharmonicity. This initial structuring of the acoustic waveform is considered a crucial computational step in auditory perception (Griffiths & Warren, 2002), which feeds parsed outputs to higher cortical areas for perceptual elaboration of key sound features including spatial position and object identity (Rauschecker & Tian, 2000).

7. Conclusion

Nature has elected to provide most of its creatures with a pair of ears rather than economizing with a single hearing sensor (Schnupp and Carr, 2009). Binaural hearing -- like stereo vision -- analyses informational disparities between two sensors to build spatial representations of objects and the underlying neural computations in the two modalities may employ common algorithms (Wagner, 2004). Both stereo vision and binaural hearing serve to enrich our perceptions of the world. In the auditory modality these improvements enhance our capacity to perform a fundamental structuring of perception referred to as 'auditory scene analysis' (Bregman, 1990), involving a parsing of the acoustic input stream into behaviourally-relevant representations. In a world that contains a cacophony of sounds, binaural hearing is employed to extract out a single sound source, determine its location,

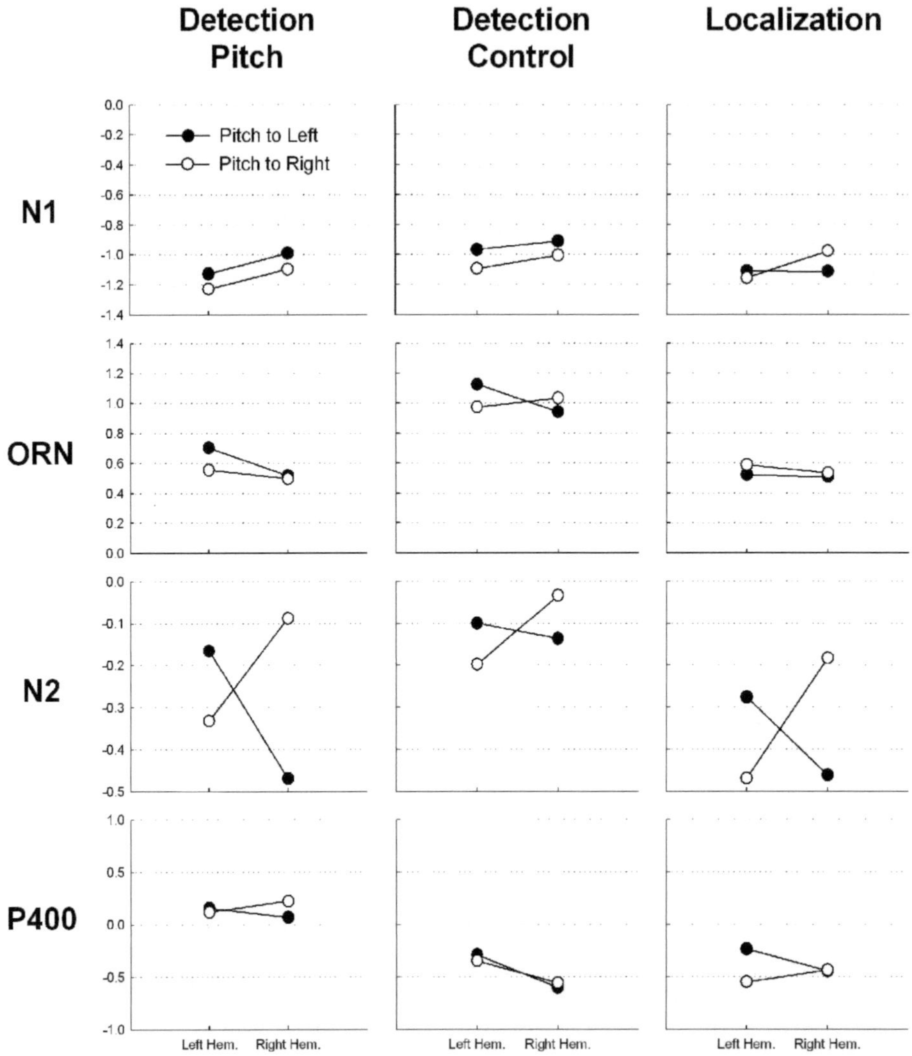

Fig. 10. Functional profiles for ERP components, plotted for left and right hemispheres. From Johnson et al. (2007) with permission.

and assign it meaning (Erikson and McKinley, 1997, p. 722). In the last several years our group has studied how binaural information is processed to these ends in the human auditory cortex, using a psychophysical paradigm to elicit binaural processing and using electroencephalography (EEG) and magnetoencephalography (MEG) to measure cortical function.

The dichotic pitch paradigm has a number of features that have made it useful for probing the workings of the binaural auditory system. Interaural disparities are first computed at the level of the brainstem so the perception of DP can be confidently attributed to central rather than peripheral processes. Further, the overall perceptual problem posed by DP -that of segregating concurrent sound objects - is of considerable interest to those interested in how, and by what mechanisms, the brain is able to accomplish this important structuring of perception. The experiments reviewed here show that EEG and MEG responses to DP consist of a sequence of auditory cortical responses that provide important markers of a number of functionally distinct stages of auditory scene analysis in the human brain.: (1) The M100 ERF seems to reflect the operation of right-hemispheric mechanisms for analysis of spatial information pitted against left hemisphere mechanisms for analysis of timing information; (2) The ORN ERP and ERF reflect the operation of fairly automatic and generalized brain mechanisms for auditory scene segregation. The ORN mechanisms can broadly draw on information about scene analysis from a variety of acoustic cues, including inharmonicity, ITDs, and ILDs. As such, the ORN appears to represent a stage of auditory processing that draws on information extracted from disparate cues into a common code that can be used to solve the broad perceptual problems of auditory scene analysis. (3) The P400 ERP is an electrophysiological signpost of a later, more controlled stage of processing, involving identification and generation of a behavioural response. This stage is highly dependent on the task and context in which stimuli are presented. (4) The N2 ERP recorded at lateral sites over the temporal lobes is highly sensitive to the spatial attributes of dichotic pitch, suggesting that this component reflects a location-specific phase of neural processing. The N2 has not been observed in MEG responses, likely because the generators have a radial orientation that the MEG is relatively less sensitive to than EEG.

Future work can leverage these electrophysiological markers to gain clearer insights into clinical conditions in which one or more of these important central processing stages may have gone awry. For example, psychophysical studies have reported that DP detection is significantly impaired in individuals with developmental dyslexia compared to normal readers (e.g. Dougherty et al., 1998). A current study in our laboratory is measuring concurrent EEG-MEG responses to DP in dyslexic and normal reading children (Johnson et al., submitted), to determine if auditory processing deficits in reading impaired children can be localized to one or more of the processing stages delineated in studies of healthy adults.

8. Acknowledgements

The MEG work described in this chapter was supported by Australian Research Council Linkage Infrastructure Equipment and Facilities Grant LEO668421. The author gratefully acknowledges the collaboration of Professor Stephen Crain, the Kanazawa Institute of Technology and Yokogawa Electric Corporation in establishing the KIT-Macquarie MEG laboratory.

9. References

Alain, C. (2007). Breaking the wave: effects of attention and learning on concurrent sound perception. *Hearing Research*, 229, 1-2., (July 2007) 225-236, 0378-5955 (Print).

Alain, C., & Izenberg, A. (2003). Effects of attentional load on auditory scene analysis. *Journal of Cognitive Neuroscience*, 15, 7, 1063-1073. 0898-929X (Print) 1530-8898 (Electronic)

Alain, C., Schuler, B. M., & McDonald, K. L. (2002). Neural activity associated with distinguishing concurrent auditory objects. *Journal of the Acoustical Society of America*, 111, 990-995, 0001-4966 (Print) 1520-8524 (Electronic).

Bell, A. J., & Sejnowski, T. J. (1995). An information-maximization approach to blind separation and blind deconvolution. *Neural Computation*, 7, 6, 1129-1159, 0899-7667 (Print).

Bilsen, F. A. (1976). Pronounced binaural pitch phenomenon. *Journal of the Acoustical Society of America,* 59, 2, 467-468, 0001-4966 (Print)

Blauert, J. (1997). *Spatial hearing: The psychophysics of human sound localization,* MIT Press, 0-262-02413-6 , Cambridge, MA.

Bregman, A. (1990). *Auditory scene analysis: The perceptual organization of sound,* MIT Press, 0-262-52195-4 , Cambridge, MA.

Cramer, E., & Huggins, W. (1958). Creation of pitch through binaural interaction. *Journal of the Acoustical Society of America,* 30, 413-417, 0001-4966 (Print) 1520-8524 (Electronic).

Dougherty, R. F., Cynader, M. S., Bjornson, B. H., Edgell, D., & Giaschi, D. E. (1998). Dichotic pitch: a new stimulus distinguishes normal and dyslexic auditory function. *Neuroreport,* 9(13), 3001-3005, 0959-4965 (Print) 1473-558X (Electronic).

Drennan, W. R., Gatehouse, S., & Lever, C. (2003). Perceptual segregation of competing speech sounds: the role of spatial location. *Journal of the Acoustical Society of America,* 114, 2178-2189, 0001-4966 (Print).

Dyson, B. J. & Alain, C. (2004). Representation of concurrent acoustic objects in primary auditory cortex. Journal of the Acoustical Society of America, 115, 280-288, 0001-4966 (Print).

Erikson, M., & McKinley, R. (1997). The intelligibility of multiple talkers separated spatially in noise. In R. Gilkey & T. Anderson (Eds.), *Binaural and Spatial Hearing in Real and Virtual Environments* (pp. 701-724). New Jersey, 13: 978-080581654, Lawrence Erlbaum.

Goldberg, J. M. & Brown, P. B. (1969). Response of binaural neurons of dog superior olivary complex to dichotic tonal stimuli: some physiological mechanisms of sound localization. *Journal of Neurophysiology,* 32, 613-636, 0022-3077 (Print) 1522-1598 (Electronic).

Griffiths, T. D., & Warren, J. D. (2002). The planum temporale as a computational hub. *Trends in Neurosciences,* 25, 348-353, 0166-2236.

Hafter, E. R., & Carrier, S. C. (1972). Binaural interaction in low-frequency stimuli: the inability to trade time and intensity completely. *Journal of the Acoustical Society of America,* 51, 6, 1852-1862, 0001-4966 (Print).

Hafter, E. R. & Jeffress, L. A. (1968). Two-image lateralization of tones and clicks. *Journal of the Acoustical Society of America,* 44, 2, 563-569, 0001-4966 (Print).

Harris, G. (1960). Binaural interactions of impulsive stimuli and pure tones. *Journal of the Acoustical Society of America,* 32, 685-692, 0001-4966 (Print).

Hautus, M. J., & Johnson, B. W. (2005). Object-related brain potentials associated with the perceptual segregation of a dichotically embedded pitch. *Journal of the Acoustical Society of America,* 117, 275-280, 0001-4966 (Print).

Johnson, B. W., Hautus, M., & Clapp, W. C. (2003). Neural activity associated with binaural processes for the perceptual segregation of pitch. *Clinical Neurophysiology,* 114, 2245-2250, 1388-2457 (Print) 1872-8952 (Electronic).

Johnson, B. W., & Hautus, M. J. (2010). Processing of binaural spatial information in human auditory cortex: neuromagnetic responses to interaural timing and level differences. *Neuropsychologia,* 48, 2610-2619, 0028-3932 (Print) 1873-3514 (Electronic).

Johnson, B. W., Hautus, M. J., Duff, D. J., & Clapp, W. C. (2007). Sequential processing of interaural timing differences for sound source segregation and spatial localization:

evidence from event-related cortical potentials. *Psychophysiology*, 44, 541-551, 0048-5772 (Print) 1540-5958 (Electronic).

Johnson, B.W., McArthur, G., Hautus, M., Reid, M., Brock, J., Castles, A., Crain, S. (submitted). Development of lateralized auditory brain function and binaural processing in children with normal reading ability and in children with dyslexia.

Julesz, B. (1971). *Foundations of Cyclopean Perception*, University of Chicago Press, 0-262-10113-0, Chicago.

Kutas, M., Van Petten, C. & Kluender, R. (2006). Psycholinguistics Electrified II: 1994-2005. In: *Handbood of Psycholinguistics*, M. Traxler & M. Gernsbacher (Eds.) (2nd ed.), (659-724), Elsevier, 0-12-369374-8, New York.

Liegois-Chauvel, C., Musolino, A., & Chauvel, P. (1991). Localization of the primary auditory areas in man. *Brain*, 114, 139-153, 0006-8950 (Print) 1460-2156 (Electronic).

McGee, T., Kraus, N., Littman, T., & Nicol, T. (1992). Contribution of the medial geniculate body subdivision to the middle latency response. *Hearing Research*, 61, 147-152, 0378-5955 (Print).

Palomäki, K. J., Tiitinen, H., Mäkinen, V., May, P. J., & Alku, P. (2005). Spatial processing in human auditory cortex: the effects of 3D, ITD, and ILD stimulation techniques. *Cognitive Brain Research*, 24, 364-379, 0926-6410 (Print).

Pantev, C., Bertrand, O., Eulitz, C., Verkindt, C., Hampson, S., Schuierer, G., et al. (1995). Specific tonotopic organizations of different areas of the human auditory cortex revealed by simultaneous magnetic and electric recordings. *Electroencephalography and Clinical Neurophysiology*, 94, 26-40, 0013-4694 (Print) 0424-8155 (Electronic).

Phillips, D., & Brugge, J. (1985). Progress in the neurobiology of sound direction. *Annual Review of Psychology*, 36, 245-274, 0066-4308 (Print) 1545-2085 (Electronic).

Phillips, D. P. (1993). Representation of acoustic events in the primary auditory cortex. *Journal of Experimental Psychology: Human Perception and Performance*, 19, 203-216, 0096-1523 (Print) 1939-1277 (Electronic).

Picton, T. W., Alain, C., Woods, D. L., John, M. S., Scherg, M., Valdes-Sosa, P., et al. (1999). Intracerebral sources of human auditory-evoked potentials. *Audiology & Neurotology*, 4, 64-79, 1420-3030 (Print).

Pratt, H., Polyakov, A., & Kontorovich, L. (1997). Evidence for separate processing in the human brainstem of interaural intensity and temporal disparities for sound lateralization. *Hearing Research*, 108, 1-8, 0378-5955 (Print) 1878-5891 (Electronic).

Rauschecker, J. P., & Tian, B. (2000). Mechanisms and streams for processing of "what" and "where" in auditory cortex. *Proceedings of the National Academy of Sciences of the United States of America*, 97, 11800-11806, 0027-8424 (Print).

Rayleigh, L. J. (1907). On our perception of sound direction. *Philosophical Magazine (Series 6)*, 13, 74, 214-232, 1941-5982 (Print).

Scherg, M., Vajsar, J., & Picton, T. W. (1986). A source analysis of the late human auditory evoked potentials. *Journal of Cognitive Neuroscience*, 1, 326-355, 0898-929X (Print) 1530-8898 (Electronic).

Scherg, M., & Von Cramon, D. (1986). Evoked dipole source potentials of the human auditory cortex. *Electroencephalography and clinical Neurophysiology*, 65, 344-360. 0013-4694 (Print).

Schnupp, J., & Carr, C. (2009). On hearing with more than one ear: lessons from evolution. *Nature Neuroscience*, 12(6), 692-697. 0022-3077.

Schroger, E. (1996). Interaural time and level differences: Integrated or separated processing? *Hearing Research,* 96, 191-198, 0378-5955 (Print) 1878-5891 (Electronic).

Smith, P. H., Joris, P. X., & Yin, T. C. (1993). Projections of physiologically characterized spherical bushy cell axons from the cochlear nucleus of the cat: evidence for delay lines to the medial superior olive. *Journal of Comparative Neurology,* 331(2), 245-260, 0021-9967 (Print).

Spierer, L., Bellmann-Thiran, A., Maeder, P., Murray, M. M., & Clarke, S. (2009). Hemispheric competence for auditory spatial representation. *Brain,* 132(Pt 7), 1953-1966. 1460-2156 (Electronic).

Tardif, E., Murray, M. M., Meylan, R., Spierer, L., & Clarke, S. (2006). The spatio-temporal brain dynamics of processing and integrating sound localization cues in humans. *Brain Research,* 1092, 161-176. 0006-8993 (Print).

Thiran, A. B., & Clarke, S. (2003). Preserved use of spatial cues for sound segregation in a case of spatial deafness. *Neuropsychologia,* 41, 1254-1261. 0028-3932.

Ungan, P., Yagcioglu, S., & Goksoy, C. (2001). Differences between the N1 waves of the responses to interaural time and intensity disparities: scalp topography and dipole sources. *Clinical Neurophysiology,* 112, 485-498, 1388-2457 (Print) 1872-8952 (Electronic).

Ungan, P., Yagcioglu, S., & Ozmen, B. (1997). Interaural delay-dependent changes in the binaural difference potential in cat auditory brainstem response: implications about the origin of the binaural interaction component. *Hearing Research,* 106, 66-82, 0378-5955 (Print) 1878-5891 (Electronic).

Wagner, H. (2004). A comparison of neural computations underlying stereo vision and sound localization. *Journal of Physiology Paris,* 98, 135-145. 0928-4257 (Print).

Werner-Reiss, U., & Groh, J. M. (2008). A rate code for sound azimuth in monkey auditory cortex: Implications for human neuroimaging studies. *Journal of Neuroscience,* 28, 3747-3758. 0270-6474.

Wright, B. A., & Fitzgerald, M. B. (2001). Different patterns of human discrimination learning for two interaural cues to sound-source location. *Proceedings of the National Academy of Sciences USA,* 98, 12307-12312, 0027-8424 (Print).

Yamada, K., Kaga, K., Uno, A., & Shindo, M. (1996). Sound lateralization in patients with lesions including the auditory cortex: comparison of interaural time difference (ITD) discrimination and interaural intensity difference (IID) discrimination. *Hearing Research,* 101, 173-180, 0378-5955 (Print).

Yin, T., & Kuwada, S. (1984). Neuronal mechanisms of binaural interaction. In G. Edelman (Ed.), Dynamic *Aspects of Neocortical Function*. New York, 0471805599, Wiley.

Yin, T. C., & Chan, J. C. (1990). Interaural time sensitivity in medial superior olive of cat. *Journal of Neurophysiology,* 64, 465-488. 1522-1598 (Electronic), 0022-3077 (Print).

Yost, W. A. (1991). Thresholds for segregating a narrow-band from a broadband noise based on interaural phase and level differences. *Journal of the Acoustical Society of America,* 89, 838-844. 0001-4966 (Print).

23

The Impact of Stochastic and Deterministic Sounds on Visual, Tactile and Proprioceptive Modalities

J.E. Lugo, R. Doti and J. Faubert
Visual Psychophysics and Perception Laboratory,
School of Optometry, University of Montreal,
C.P. 6128 succ. Centre Ville, Montréal,
Quebéc, H3C3J7
Canada

1. Introduction

Stimulus localization and particularly directional hearing can be considered as methods for investigating neural activity and they have proven to be useful tools for research in physiology and psychology. Human directional hearing techniques have been reflected upon way back by Von Békésy in Austrian forests [1]. For example, he observed that some of the roads took a perfectly straight course through deep, dark woods. He could not imagine how such straight roads had been cut through the forest when the usual optical methods used by road surveyors would seem to be useless in this case. Further some of these roads were very old and probably built before the introduction of the theodolite. Many of these roads were laid out by an acoustic method. How did they do it? A man stationed at the starting point noted the direction of the sound produced by someone at the other end blowing a horn. The first man then walked toward the sound source, marking the threes on the way. It turned out that this method produced a straight line from start to finish [1]. From this observation Békésy was motivated to perform a series of studies on stimuli localization not limited to hearing but also to vibration sensations on the skin, electrical pulses on the tongue and odors through the nose as well. Strikingly, his results showed an underlying ubiquitous mechanism present in the different stimuli localization modalities. For instance, the effect on localization of the time delay between two stimuli on the skin, the tongue, the two nostrils in the nose and the two ears, presented the same dynamics [2-4]. These results were quite exciting because it showed that, in humans, the senses work similarly for stimuli localization although the basic underlying neural pathways are not the same.

It was this kind of general principle on stimuli localization that motivated us in the search for more general principles related to how senses interact to generate multisensory perceptions but with a special emphasis on auditory stimulation. This is known as multisensory integration and its study is very important because it is the foundation of how humans bind all the information coming from the senses to generate a coherent percept. We began by studying something that we called cross-modal stochastic resonance. This consists in the concurrence of a threshold, a subthreshold stimulus present in one sense and noise at different amplitudes entering through another sense. What we found was that the same auditory noise can enhance the sensitivity of tactile, visual and propioceptive system responses to weak signals. Specifically, we showed that the effective auditory noise significantly increased tactile sensations of the finger, decreased luminance and contrast visual thresholds and significantly changed EMG recordings of the leg muscles during posture maintenance [5]. We also found that in all the cases the interactions follow the same sort of physical dynamics. Moreover, we unveil that the same result is obtained if we use auditory deterministic sounds instead of auditory noise [6] to enhance tactile sensations. We further demonstrated that we could use tactile noise and enhance visual detection [7] or use visual deterministic signals to enhance tactile detection [6]. These surprising results guided us to propose that these multisensory integration interactions can be explained under the same general principle that we call the Fulcrum principle.

In this chapter we present material emerging from our own research experience concerning human perception in general with emphasis in auditory interactions. We introduce in an accessible way a non-linear mathematical model supporting our hypothesis, and we provide experimental results and conclusions. We also propose that the Fulcrum principle may have numerous implications in a number of neurobiological alterations such as autism, aging and age-related neurodegenerative disorders and ADHD. We conclude by presenting to the readers with what we consider could be the next hurdles in this area, and the main points that we think should be emphasized in future work.

2. Multisensory Integration: MI

A general description: MI is a non linear process that binds information from all the participating sensory stimuli. The original approach shows that MI results from the brain's capacity for integrating information originating from more than a single sensory stimulus. Here we would like to present the two stimuli conditions allowing us to introduce the mathematical model.

The first aspect involves the concept of Signal **Coherence**, and the second important aspect is the **Sense Threshold** for those signals [8]. **Coherence** is intended to be the propriety that gives the signal a continuous and repetitive harmonic shape. A signal involves the concept of evolution in the time domain, harmonic shape implies the same amplitude at regular time intervals, and very importantly, the same amount of energy transferred per unit of time [9].

If we have more than one stimulus applied to a big surface interface, we can split this concept in two: **Temporal Coherence** (frequency) and **Spatial Coherence** (front- wave)

Temporal Coherence: when we consider the coexistence of more than one stimulus signal, the coherence associated with this compound stimulus is the correlation (proportional correspondence) between the evolutions in the time domain for both signals (together). When the signals are periodic this represents the same ***frequency spectrum content*** and results in the same ***bandwidth*** (BW) [10]. In the case of a pure tone, we would have only one frequency component in the signal spectrum.

Spatial Coherence: if for a fixed point in space along the signals pass the superposition of these simultaneous signals presents Temporal Coherence, we say that signals have ***spatial coherence***. The *front -wave* of this compound signal preserves the shape along its pass (when traveling along an ideal non dispersive mean).

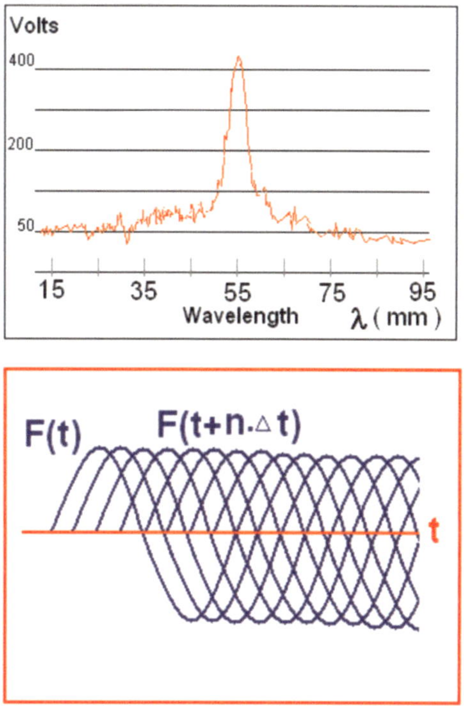

Examples of periodic signals

So, depending on the intensity and characteristic of the stimulus signal we can have different situations. For instance, for a given perceptual threshold level we can have: **supra-threshold** (perceived signal), or **sub-threshold** (not perceived) stimuli. Depending on the stability and consistency of the signal stimuli we can have **deterministic signals** (coherent or not) or **stochastic signals**.

Periodic signals means a fixed ***frequency spectrum content*** or a fixed ***bandwidth*** (BW).For a pure tone, we have a ***narrow*** frequency spectrum

Deterministic signals always present a limited bandwidth or a ***repetitive pattern***. They can be described and recreated without error along the time domain. We know the evolution of the instantaneous energy transferred trough these signals.

Stochastic signals represent a random pattern and a very large bandwidth. We can establish the limits of their characteristics (amplitude or BW), but we do not know in advance their evolution along the time domain. We know the mean energy transferred trough these signals. A good example of a Stochastic Signal is **White Noise** [11].

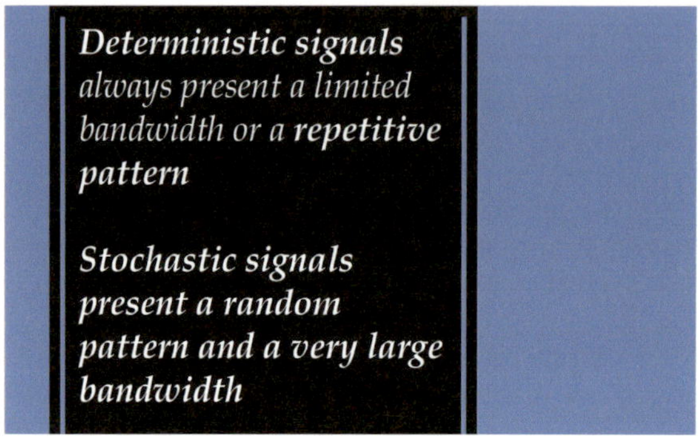

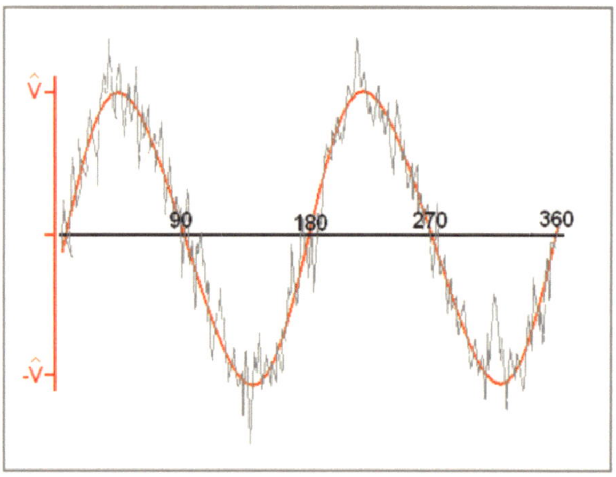

Example of a deterministic signal with noise

Because of the ***random instantaneous frequency content*** compared with a pure tone, we call it NOISE. As its ***Frequency Spectrum*** extends from zero Hz to infinite, we call it WHITE (in analogy with the visible spectrum and the eyes perception of the ***white light***).

3. The Inverse-effectiveness law

So far we defined the MI as the complex way in which our brain binds the different sensory stimuli that contributes to create a ***phantom image*** of the real world outside its perceptual limits. This image is the ***only reality we have***. Researchers tried to define the human sensory stimulus span from threshold to ceiling. They tested humans applying deterministic stimuli signals to the different senses. This generated normalized thresholds for auditory, tactile, visual, etc.

Here we find the first cue in reference to MI: it was determined that if two weak (close to threshold level) stimuli are applied together, the presence of the additional stimulus facilitates perception. And this happens for an elastic ***temporal coincidence***. But, this perceptual improvement is not possible if one of the stimuli is clearly supra-threshold. This is known as: **Inverse-Effectiveness Law** [12]. This means that perceptual enhancement takes place trough the MI mechanism when we apply: weak, supra-threshold, deterministic and coincident signals to the subject. However, there is an MI phenomenon that cannot be described by the inverse-effectiveness rule: cross-modal SR.

4. Stochastic resonance

Stochastic resonance (SR) [13] is a nonlinear phenomenon whereby the addition of noise can improve the detection of weak stimuli. An optimal amount of added noise results in the maximum enhancement, whereas further increases in noise intensity only degrade detection or information content. The phenomenon does not occur in linear systems, where the addition of noise to either the system or the stimulus only degrades the measures of signal quality. The SR phenomenon was thought to exist only in stochastic, nonlinear, dynamical systems but it also exists in another form referred to as 'threshold SR' or 'non-dynamical SR'. This form of stochastic resonance results from the concurrence of a threshold, a subthreshold stimulus, and noise. These ingredients are omnipresent in nature as well as in a variety of man-made systems, which accounts for the observation of SR in many fields and conditions. The SR signature is that the signal-to-noise ratio, which is proportional to the system's sensitivity, is an inverted U-like function of different noise levels. That is, the signal-to-noise ratio first is enhanced by the noise up to a maximum and then lessened. The SR phenomenon has been shown to occur in different macro [14], micro[15] and nano physical systems [16]. From the cyclic recurrence of ice ages, bistable ring lasers, electronic circuits, superconducting quantum interference devices (SQUIDs) and neurophysiological systems [17] such as receptors in animals. Several studies have suggested that the higher central nervous system might utilize the noise to enhance sensory information [13]. SR studies in humans can be divided in unimodal SR (signal and noise enter the same sense) [18,19], central SR (signal and noise enters in similar local receptors and later mix in the cortex) [20] and behavioral SR (similar to central SR but its effect is observed in one sense and then enacted in the behavior of the subjects) [21]. Before the SR principle was proposed, Harper [22] discovered what we currently would call crossmodal stochastic resonance while studying the effect of auditory white noise on sensitivity to visual flicker. Recently a similar result [23] has been found where auditory noise produces SR when subthreshold luminance stimuli are present. However what has not been explored is the extension of these interactions in humans. New results show that the noise induces large scale phase synchronization of human-brain activity associated with behavioral SR [24]. It is shown that both detection of weak visual signals to the right eye and phase synchronization of electroencephalogram (EEG) signals from widely separated areas of the human brain are increased by addition of weak visual noise to the left eye. These results imply that noise-induced large-scale neural synchronization may play a significant role in information transmission in the brain. Interestingly SR can be seen as a synchronization-like phenomenon between two energy states of a physical system for example [25]. Furthermore, the synchronization-like phenomenon plays a key role in the enhancement of the signal-to-noise ratio in SR. Therefore, we can hypothesized that if the noise induced large scale phase synchronization in different areas of the cortex and peripheral systems with dynamics similar to SR, the crossmodal SR would be a ubiquitous phenomenon in humans because it involves different cortical areas and peripheral systems. Consequently under the same auditory noise conditions, the crossmodal SR should be present among tactile, visual and proprioceptive sensory systems, for instance.

5. Facilitating and excitatory stimulus

In order to outline a ***synoptic scheme*** that represents the basis of some experiments that we have performed, we introduce another two concepts. First, **Excitatory Stimulus**: signal applied to the sense that we want to study. Second, **Facilitating Stimulus:** signal applied simultaneously to the same subject, intended to trigger the MI mechanism in a way that facilitates the perception of the Excitatory Stimulus. When both, ***facilitating* and *excitatory*** signals act as stimuli of the same sense (auditory, tactile, visual stimulus, etc) we have *Uni-modal* ***Interactions (U.M).*** When each one of these signals act in different senses (for instance ***excitatory*: *tactile***; and ***facilitating*: *auditory***) we have *Cross-modal* ***Interactions (C.M).*** Either of the precedent cases are part of the general *Multi-modal* Interactions model.

6. Crossmodal interactions paradigms and the sensory threshold enhancement

On the basis of what was presented so far, it is possible to combine those elements to create the experiments that allow us to explore ***human perception*** and outline a plausible model. All of them allow a positive response from the subject under test, by the action of the facilitating stimulus, when the excitatory stimulus is Sub threshold. This means an improvement of the human perception. Examples of multimodal interactions that have been tested so far are:

1. Excitatory: **Tactile – Deterministic- threshold** **E:T-D- T**
   Facilitating: **Auditory or Visual -Deterministic – threshold** **F: AoV-D-T**
2. Excitatory: **Tactile – Deterministic- Sub threshold** **E:T-D-ST**
   Facilitating: **Auditory - Stochastic - Supra threshold** **F: A-S- SST**
3. Excitatory: **Visual – Deterministic- Sub threshold** **E:V-D-ST**
   Facilitating: **Auditory - Stochastic - Supra threshold** **F: A-S- SST**
4. Excitatory: **Propioception – Deterministic- Sub threshold** **E:P-D-ST**
   Facilitating: **Auditory - Stochastic - Supra threshold** **F: A-S- SST**
5. Excitatory: **Visual – Deterministic- Sub threshold** **E:V-D- ST**
   Facilitating: **Tactile - Stochastic - Supra threshold** **F: T-S- SST**
6. Excitatory: **Tactile – Deterministic- Sub threshold** **E:T-D-ST**
   Facilitating: **Auditory - Deterministic - Supra threshold** **F: A-D- SST**
7. Excitatory: **Tactile – Deterministic- Sub threshold** **E:T-D-ST**
   Facilitating: **Visual - Deterministic - Supra threshold** **F: V-D- SST**

We observe that **1** is a cross modal example of the *Inverse Effectiveness Law* **(IEL)**. These kinds of examples have been studied massively and they are well documented on the literature [12]. **2 to 5** belong to the *Multi modal Stochastic Resonance* **(MmSR)** and **6** and **7** belong to the *Multi modal Deterministic Resonance* (MmDR). In what follows we will explain more in detail these multimodal interactions.

Excitatory: **Tactile – Deterministic- Sub threshold** **E:T-D-ST**
Facilitating: **Auditory - Stochastic - Supra threshold** **F: A-S- SST**

In the first series of experiments we studied the effects of auditory noise on tactile sensations in three subjects. Tactile vibrations were delivered to the middle finger of the right hand of the subjects at a frequency of 100Hz and were asked to report the tactile sensation. If they felt the signal they had to click on a yes button or on a no button otherwise (yes-no paradigm). Each subject was tested twice for every auditory noise and baseline condition. In all the experiments were the facilitating signal was auditory the normalized thresholds were computed as follows: once the absolute threshold was obtained for different auditory noise conditions, their values were divided by the absolute threshold measured for the baseline condition. Figure 1 (left column) shows the normalized tactile thresholds for three subjects and it is clear that, as the noise level increased, the threshold decreased reaching a minimum and then increased in a typical SR signature fashion. In general we found that the subject's minimum peaks are not always localized at a specific noise level but within a band centered at 69± 7 dBSPL. Can the above results be explained only on the bases of SR theory? Can one potentially rule out an explanation based on attention/arousal? If the noise creates a more interesting/arousing condition than the baseline condition, all neural systems could be correspondingly more excitable, not because the noise facilitates a resonance like behaviour but because the auditory noise nonspecifically boosts neural excitability. However, the Yerkes- Dodson law demonstrates an empirical relationship between arousal and performance [26]. Such relationship is task dependent. For instance, in a simple task the relationship between arousal and performance is linear and only in a difficult task this relationship becomes curvilinear (inverted u-shape similar to SR). Since a yes-no procedure with vibration thresholds would be considered a very simple task, we would not expect an inverted ushape between the noise level and tactile sensitivity if the mechanism involved in these interactions was only arousal. That was not the case as Fig. 1 clearly shows a curvilinear relationship. In order to further explore the notion of possible attention effects we performed an additional experiment on sixteen subjects where we used two different auditory stimuli plus the baseline condition. One stimulus was a specific auditory noise condition as described above, and another was a 3D-like sound. Both sounds had an intensity of 69 dBSPL and the 3D sound contained frequencies in a similar range as the auditory noise (between 100 Hz up to 19 kHz). The 3D sound gave the impression of very close movements near, up and down, and around the subjects' head resulting in a very strong attention getting sound sequence. If our previous results were only a result of attention modulation created by the sound intensity, we should expect that for, the 3D auditory condition, the tactile thresholds would be lower in most people because this sequence had strong attention modulation properties and the noise level we chose was the same as the averaged peak noise level we measured in the first experiment that generated the lowest tactile thresholds. An alternative hypothesis is that this attention-producing stimulus would not influence or maybe even hinder tactile performance. On the other hand, we did expect the auditory noise condition to generate lower tactile thresholds given that we chose the averaged peak noise level that generated the lowest thresholds in the previous experiment. Each subject was tested twice for every condition in randomized order. Fig. 1 (right column, top) shows the normalized tactile thresholds for the 3D sound and baseline conditions. Eight subjects augmented significantly their thresholds comparatively to baseline condition, four subjects lessened theirs thresholds and in other four subjects the threshold values remained unchanged. Fig. 1 (right column, middle) shows the normalized tactile thresholds for the auditory noise and baseline condition. Twelve subjects significantly lessened their thresholds, only two subjects increased their thresholds and another two subjects had unchanged threshold values. Fig. 1 (right column, bottom) shows the group average of the normalized tactile threshold for the three conditions. The average group sensitivity increased significantly (with respect the baseline) in the presence of noise ($p<0.001$) while no significant change was found for the 3Dlike sound ($p = 0.72$). It is clear from these experimental controls, that the noise effects on tactile sensations are not due to

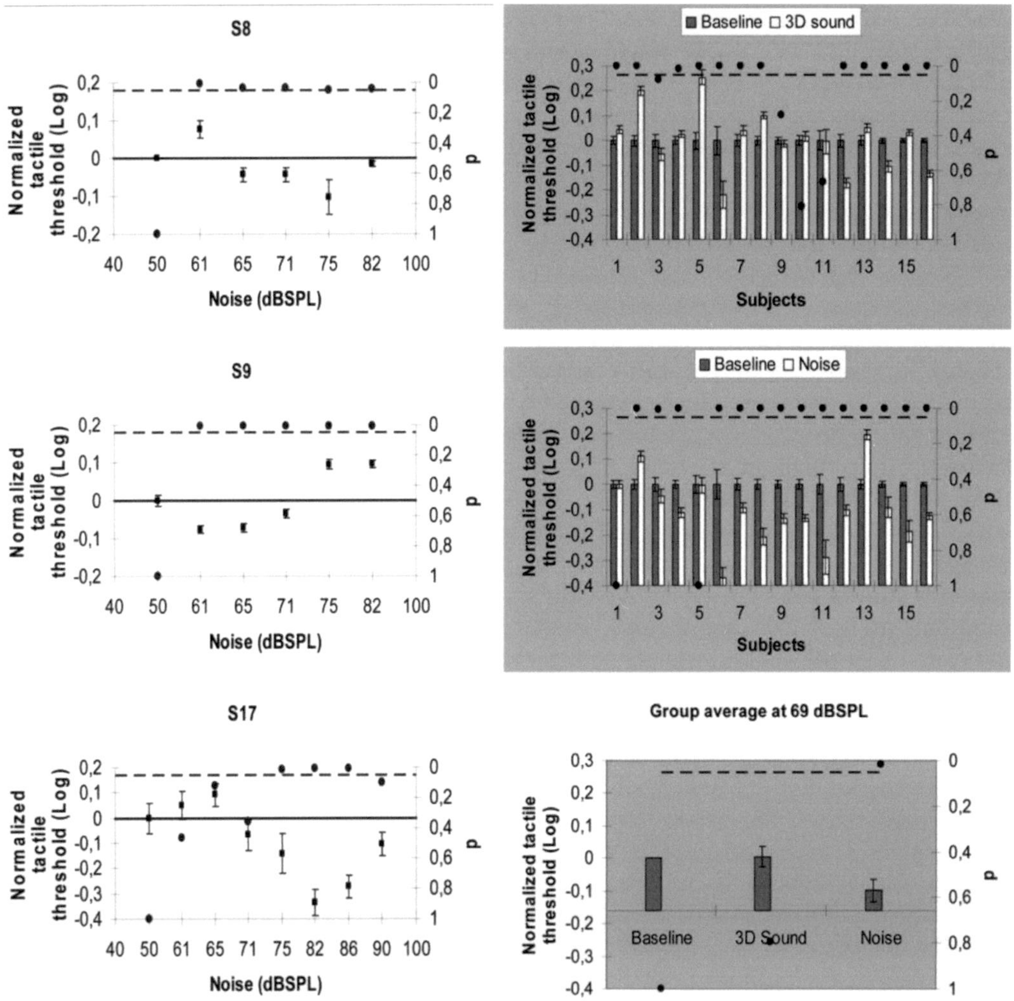

Fig. 1. Interactions between auditory noise and tactile signals. (Left column) normalized tactile threshold changes with the noise level in three subjects. (Right column, top) normalized tactile thresholds of sixteen subjects when the 3D sound level was fixed at 69 dBSPL. (Right column, middle) normalized tactile thresholds of sixteen subjects when the noise level was fixed at 69 dBSPL. (Right column, bottom) Group average results for three conditions: baseline, 3D sound and noise. The average group threshold decreased significantly in the presence of noise (p,0.001) and no significant change was found for the 3D-like sound (p = 0.72). In all the graphs the no-noise condition is taken as baseline; the black dots indicate pvalues (right y-axis) and the broken line represents the 5% significance level. Error bars correspond to one standard error.

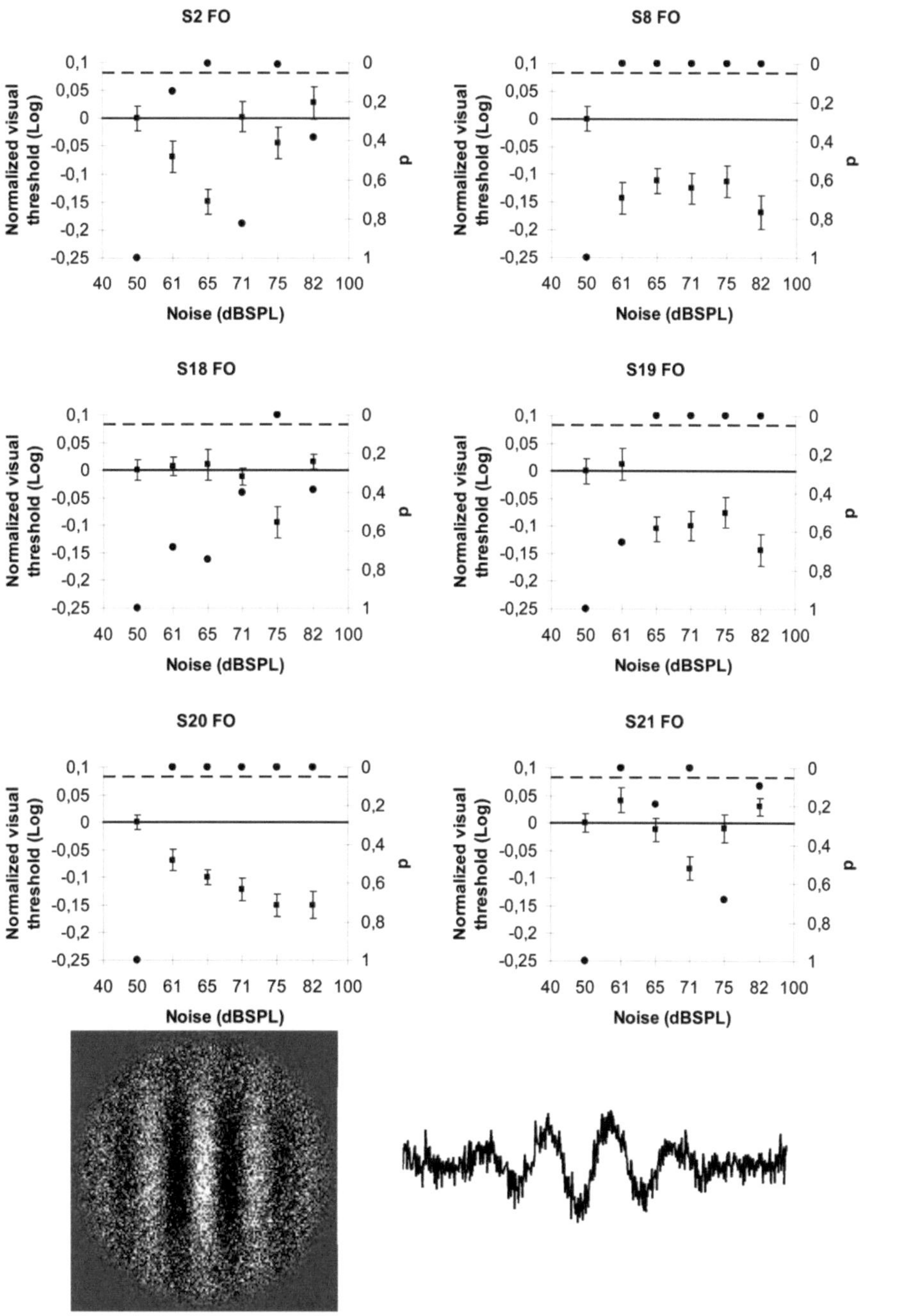

Fig. 2. Interactions between auditory noise and first order visual signals. Normalized visual threshold changes with the noise level in sixth subjects for luminance modulated (first order) stimuli. In all the graphs the no-noise condition is taken as baseline; the black dots indicate pvalues (right y-axis) and the broken line represents the 5% significance level. Error bars correspond to one standard error. In the last row an example of the first order stimulus is displayed.

attention/ arousal effects but result from the way the brain processes the energy (and probably the frequency) content of noise and signal.

Then we studied auditory-visual interactions. In previous work [22,23] only visual stimuli classified as first order stimuli were used. We wanted to evaluate the effect of SR on an additional visual attribute called second order processing. For first order stimuli, the local luminance spatial average varies throughout the stimulus while the local contrast remains constant. In second order stimuli, known to be processed by separate mechanisms and assumed to be more complex to process, the local spatial luminance average remains constant but the local contrast varies throughout the stimulus [27,28].

Excitatory: **Visual - Deterministic- Sub threshold** **E:V-D-ST**

Facilitating: **Auditory - Stochastic - Supra threshold** **F: A-S- SST**

In the second series of experiments, we studied whether auditory noise can facilitate luminance-modulated (first order) stimuli detection in six subjects. To evaluate visual thresholds, we used a two-alternative forced choice paradigm. In a two-alternative forced choice paradigm, the subject is presented two choices and must pick one (even if the observer thinks he/she did not see the stimulus), which produces a more stringent control of observer criteria than a yes/no response. Here the observers had to discriminate between vertical or horizontal luminance-modulated stimuli (LM) defined sinusoidal gratings [27,28]. We measured the LM thresholds for six auditory conditions (baseline plus five noise levels) in a random order. Five thresholds (5 separate staircases) were established for each condition and averaged. Fig. 2 shows the normalized visual LM thresholds for six subjects. As in our previous auditory-tactile experiments, the visual threshold profiles of the observers varied as a function of the different auditory noise levels demonstrating a typical SR function with zones of threshold values significantly different from the control condition. The SR average peak for our data was 75 ±3 dBSPL for LM stimuli. Previous reports show an average value of 70±2.5 dBSPL for visual flicker detection [22] and a value of 73.8±15.5 dBSPL for a luminance-defined stimulus [23].

In the third series of experiments, we studied whether auditory noise can facilitate contrast-modulated (second order) stimuli detection. With the same procedure as above, the observers had to discriminate between vertical or horizontal contrast-modulated stimuli (CM) defined sinusoidal gratings [27,28]. We measured the CM thresholds for six auditory conditions (baseline plus five noise levels) in a random order. Five thresholds (5 separate staircases) were established for each condition and averaged. Fig. 3 shows examples of the normalized visual CM thresholds for the same six subjects. As in our previous auditory-visual experiments, the visual CM threshold profiles of the observers varied as a function of the different auditory noise levels demonstrating a typical SR function with zones of threshold values significantly different from control. The SR average peak was found at 70±2 dBSPL for CM stimuli. Clearly both peaks are inside the same experimental region and there is no significant difference between them meaning that within the experimental accuracy we have used both SR mechanisms are similar.

Excitatory: **Propioception - Deterministic- Sub threshold** **E:P-D-ST**

Facilitating: **Auditory - Stochastic - Supra threshold** **F: A-S- SST**

In the fourth series of experiments we evaluated electromyography (EMG) responses of the subject's leg muscles during posture maintenance with different auditory noise conditions.

Recent evidence has demonstrated that tactile stimulation of the foot with noise could increase postural stability by acting on the somatosensory system and that noise can induce

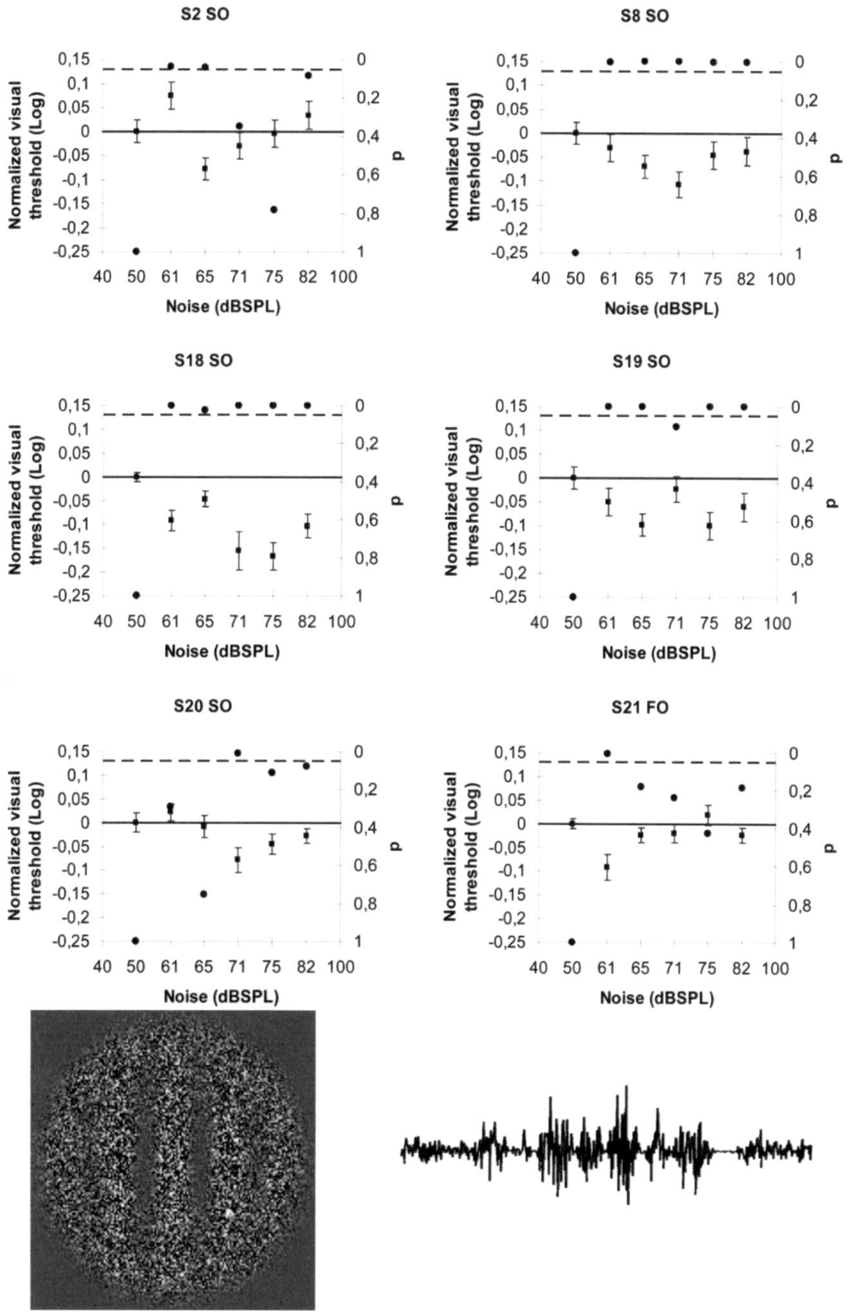

Fig. 3. Interactions between auditory noise and second order visual signals. Normalized visual threshold changes with the noise level in sixth subjects for contrast modulated (second order) stimuli. In all the graphs the no-noise condition is taken as baseline; the black dots indicate p-values (right y-axis) and the broken line represents the 5% significance level. Error bars correspond to one standard error. In the last row an example of the second order stimulus is displayed.

transitions in human postural sway [29-31]. Four subjects were asked to stand with their feet aligned one in front of the other and touching like in a tightrope position. For all conditions (the baseline plus five noise levels) we have measured the EMG activity of each subject three times in a randomized order. In figure 4 (left column) we show the averaged EMG power spectrum density as a function of noise intensity in four subjects. The right column of figure 4 shows the normalized power of the EMG activity in the same four subjects with different noise levels and the baseline. The EMG activity refers to the muscle's activity during posture maintenance. In this context a less stable posture represents more activity of the muscles related to this task. Again the SR signature was observed by using similar noise levels as the tactile and visual experiments and surprisingly, the subject's averaged peak 74±4 dBSPL lies in the same experimental range found in our previous experiments.

Excitatory: **Visual - Deterministic - Sub threshold** **E:V-D- ST**
Facilitating: **Tactile - Stochastic - Supra threshold** **F: T-S- SST**

In a sixth series of experiments, we applied different tactile noise intensity levels plus a baseline (no tactile noise) in randomized order (Figure 5) in 7 healthy subjects [7]. This randomized order of sessions assured that the observed effects are not simply due to a generalized modulation in attention/arousal. We maintained the intensity of the continuous tactile input noise constant for each session and varied it between sessions. We have measured absolute first order visual (in arbitrary units) thresholds and then normalized. Normalized visual thresholds were computed as follows: once the absolute threshold was obtained for different tactile noise conditions, their values were divided by the absolute threshold measured for the baseline condition. The experiments took place in a dark room for vision testing. The tactile noise was presented by means of a specific designed transferred signal spectrum actuator (TSSA) that converted the auditory signal spectrum energy into mechanical signal spectrum energy. The subjects held the TSSA against their right internal metacarpus. The tactile noise has a cut-off frequency around 1kHz. We found that tactile noise also facilitated first order stimuli perception in 5 subjects similar to the auditory noise case (the tactile noise may be was out of range to show facilitation in the other two subjects).

We decided to explore if facilitating deterministic signals can induce changes on the perception in a similar fashion as in the stochastic experiments [6]. In this case we used electrical signals that were delivered to the right calf (gastrocnemius medial head) of different subjects (fig.6). With the right electrical signal amplitude, the signal was not perceived but the electrical activity in the muscle it was measurable with electromyography (EMG) electrodes. If the subjects were presented a noticeable sound or a visible pip at the same time their muscles received the electrical signal, their muscular EMG response was amplified. Furthermore, the dynamic of these interactions was similar to the precedent stochastic case. In order to obtain individual tactile thresholds the signal amplitude started out at a low level so that it could not be detected, then the amplitude was gradually increased until the subjects reported that they were aware of it. This is known as the ascending threshold. Then signal amplitude started out at a high level so that it was perfectly detected, then the amplitude was gradually decreased until the subjects reported that they were not aware of it, this is the descending threshold. The absolute threshold was the average of both thresholds. After the data were collected, the power spectral density (PSD) of each EMG measurement was obtained. To calculate the normalized PSD for each condition, $\Psi_N(\omega)$ (where ω is the frequency in hertz), we divided the PSD at the suprathreshold level by the corresponding PSD at the subthreshold level on each trial and then averaged across trials.

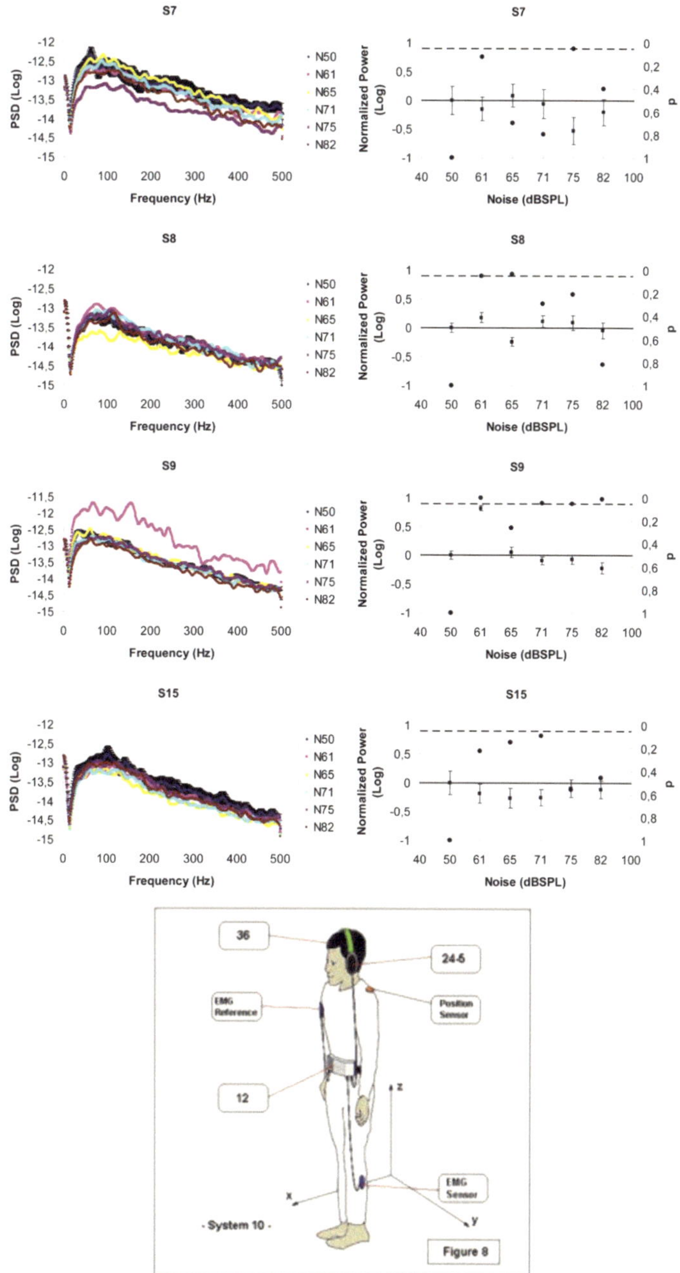

Fig. 4. Interactions between auditory noise and propioceptive signals. (Left column) average EMG power spectral densities as a function of noise level in four subjects for the tightrope posture position. For clarity only the baseline condition shows error bars (one standard error). (Right column) normalized power in four subjects. Again, the no-noise condition is taken as baseline; the black dots indicate p-values (right y-axis) and the broken line represents the 5% significance level. Error bars correspond to one standard error. In the last row an example on how the experiments were done is displayed.

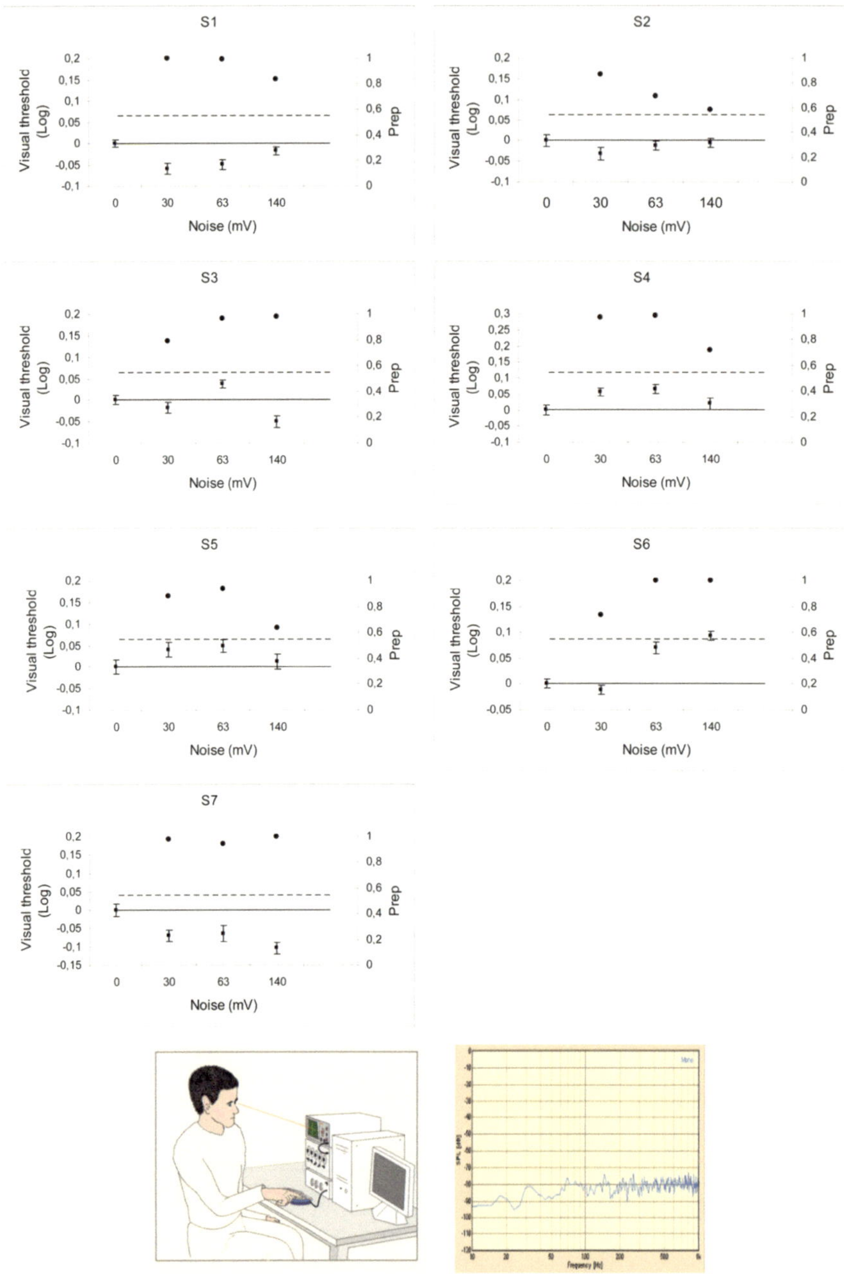

Fig. 5. Interactions between tactile noise and first order visual signals. Normalized visual threshold changes with the noise level in seven subjects for luminance modulated (first order) stimuli. In all the graphs the no-noise condition is taken as baseline; the black dots indicate the probability to replicate the same result (right y-axis) and the broken line represents the 50% chance level. Error bars correspond to one standard error. (Last row) shows an example on how the experiments were done and the effective tactile noise Fourier spectral density.

The normalized PSD was used to calculate the integral signal-to-noise ratio (integral SNR), defined as follows:

$$Integral.SNR = \int_{-\infty}^{\infty} \Psi_N(\omega)\Theta d\omega \Big/ \int_{-\infty}^{\infty} \Theta d\omega \tag{1}$$

where Θ is a step function that equals zero when $\Psi_N(\omega) < 1$ and equals one otherwise. On each trial, we obtained two paired measurements: the EMG for a tactile stimulus at a subthreshold level with a fixed amplitude (1.5% below threshold) and the EMG for a tactile stimulus that was presented concurrently with a stimulus in another modality, depending on the experiment. Every EMG measurement lasted 30 s, and the order of the paired measurements within each trial was randomized to ensure that the observed effects were not simply due to a generalized modulation in attention or arousal

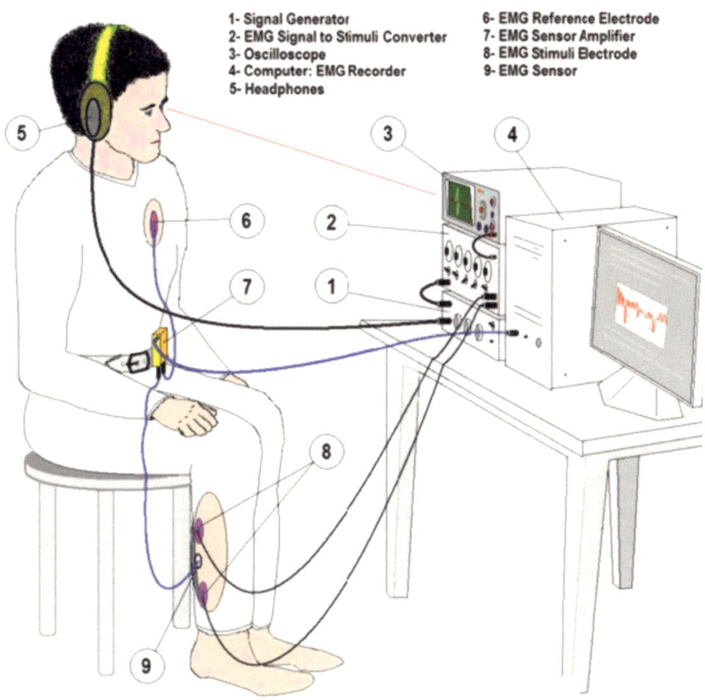

Fig. 6. Experimental lay-out for all the procedures related to deterministic signals described in the text, including the nine components of the experiment set-up.

Excitatory: **Tactile – Deterministic- Sub threshold** **E:T-D-ST**
Facilitating: **Auditory - Deterministic - Supra threshold** **F: A-D- SST**
The auditory stimuli were presented binaurally by means of a pair of high-precision headphones. We evaluated first the subjects' hearing from 250 Hz to 8 kHz using an audiometer; these evaluations were conducted in a 6-ft by10-ft double-wall audiometric sound suit that met the American National Standards Institute (Standard 3.1-1991) for permissible ambient noise levels (in one-third-octave bands) for testing in free-field conditions with headphones. During the experimental trials, all subjects were seated and

were asked to listen to the sound in the headphones and report when they first felt a tactile sensation. Once the subjects reported a change in tactile sensation, the EMG measurements started. The electrical amplitude signal was set to a subtreshold level (1.5% below threshold) and the auditory signal had a fixed amplitude of 9 mV (peak voltage). Figure 7a shows an example of the normalized power spectral density PSD. The enhancement ranges between 3% and 9% for all the subjects (fig. 7b).

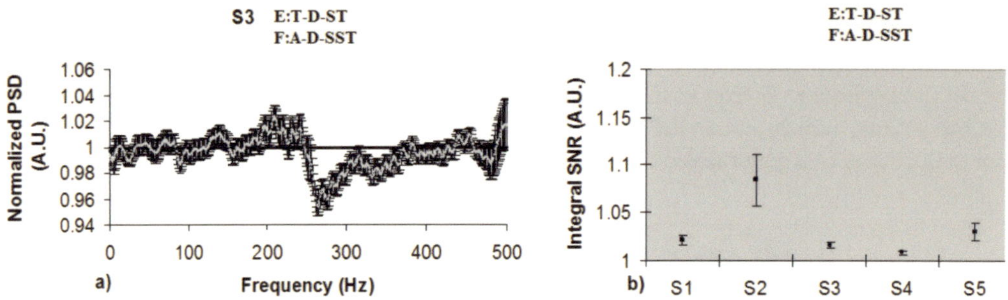

Fig. 7. a) An example of the normalized PSD of subject S3 for tactile-auditory interactions of deterministic signals. The grey line represents the mean and the black bars indicate one standard error. b) The graph in the second column shows the integral SNR for five subjects.

Excitatory: **Tactile – Deterministic- Sub threshold E:T-D-ST**
Facilitating: **Visual - Deterministic- Supra threshold F:V-D- SST**

Second, we investigated how tactile perception and the corresponding EMG activity were affected when the amplitude of the tactile stimulus was subthreshold (1.5% below threshold) and a suprathreshold visual stimulus was presented concurrently. The biphasic visual signal (Component 3) was displayed on an oscilloscope (Kikusui COS6100) and looked like a dot expanding to a line, first up and then down. All subjects were seated 45 cm from the oscilloscope screen and were asked to look at the screen and report when they first felt a tactile sensation. Once the subjects reported a change in tactile sensation, the EMG measurements started. The visual stimulus augmented tactile perception and the corresponding EMG activity. When we introduced the visual stimulus, the EMG activity increased correspondingly, primarily in frequencies between 290 and 380 Hz (Fig. 8a displays the EMG results from 1 subject). Figure 8b shows the integral SNR for all subjects, which ranged from approximately 1.03 (increase of 3% relative to baseline) to 1.1 (increase of 10%).

Can we explain the results of the last two experiments in terms of MI? The first condition for MI, temporal synchronicity, was satisfied in our experiments, because the two stimuli were presented at the same time. However, because the visual and auditory stimuli were suprathreshold and the tactile stimuli were subthreshold, the inverse-effectiveness rule seems not to be applicable to this case (greatest multisensory-mediated effects are generally seen when the individual stimuli are both weak in eliciting a response on their own). Therefore, we predicted (a) that visual or auditory noise also enhances tactile sensations, and (b) that there is a particular intermediate level of visual or auditory stimulation at which tactile-visual or tactile-auditory MI is optimally enhanced. We tested these predictions in the next two experiments by using auditory stimuli only.

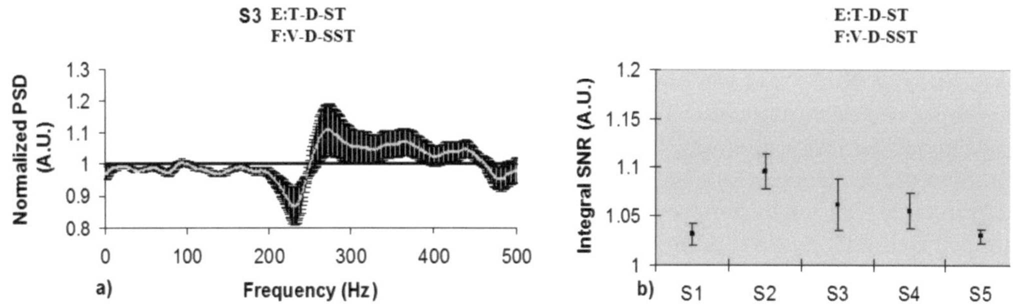

Fig. 8. a) An example of the normalized PSD of subject S3 for tactile-visual interactions of deterministic signals. The grey line represents the mean and the black bars indicate one standard error. b) The graph in the second column shows the integral SNR (left y-axis) for all subjects.

First, we tested tactile-auditory interaction using auditory noise instead of a deterministic auditory signal. In this experiment, we tested only the 3 subjects whose results for tactile-auditory interactions were similar. In this experiment, a subthreshold tactile stimulus (1.5% below threshold) was presented concurrently with a clearly audible noise stimulus (rather than the deterministic auditory stimulus). The amplitude of the white-noise signal was fixed at a value of 9 mV (peak voltage) and it has an effective acoustic noise spectrum (ENS).We estimate that the ENS upper bound is around 15 kHz.

Figure 9a indicates that the auditory noise enhanced tactile sensations because, on average, the EMG signal increased when the auditory noise was present. In addition, the integral SNR (see Fig. 9b) ranged from 1.05 (increase of 5% relative to baseline) to 1.10 (increase of 10%; similar to the range of tactile-visual SNRs), indicating that the energy transfer of the auditory noise was bigger than the energy transfer of the deterministic auditory signal. These differences in energy transfer could have been due to the fact that the frequency content was larger in the auditory noise signal than in the auditory deterministic signal. This would imply that the frequency content, and not just the energy content, is important in inducing transitions in tactile perception.

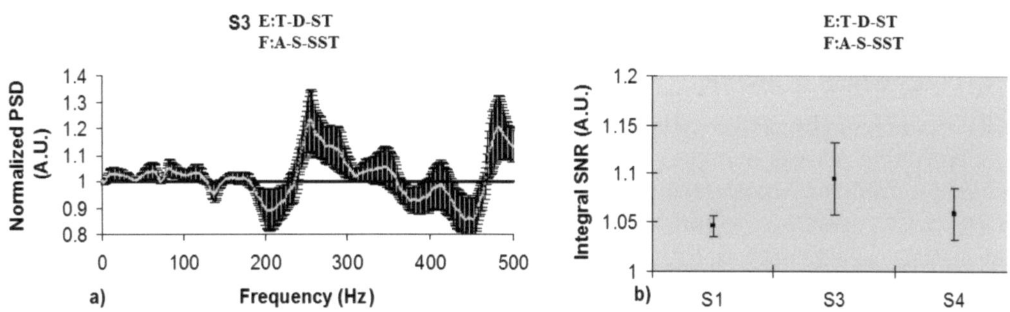

Fig. 9. a) An example of the normalized PSD of subject S3 for tactile interactions with auditory noise. The grey line represents the mean and the black bars indicate one standard error. b) The graph in the second column shows the integral SNR (left y-axis) for three subjects.

Finally, in the last experiment, we tested tactile-auditory interaction using deterministic auditory signals with different amplitudes and measured EMG activation in 1 subject (S4). A different amplitude of the auditory signal was tested at each session. The six amplitudes were 0, 8, 12, 20, 30, and 300 mV (peak voltage) at the amplifier exit. To show the inverted-U-shaped function, we chose the upper limit to be 300mV.We kept the intensity of the continuous auditory stimulus constant within each session and varied the intensity (in random order) between sessions. The order of the paired measurements was randomized within each trial (as in the previous experiments), and the order of the sessions was also randomized; this randomization ensured that the observed effects were not simply due to a modulation in attention or arousal. Figure 10 demonstrates that as we increased the amplitude of the auditory stimulus, EMG activity increased, reached a maximum, and then decreased (inverted- U-shaped function). This implies that there is indeed a particular intermediate level of auditory stimulation at which tactile auditory MI is optimally enhanced. Surprisingly, the same pattern of results shown in Figure 10 has been demonstrated in systems that show SR, deterministic resonance, or both [32].

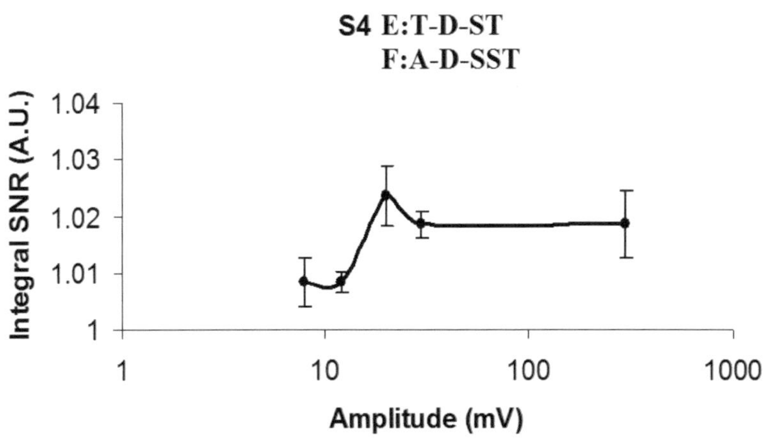

Fig. 10. Results for tactile-auditory interactions with deterministic auditory signals at different amplitudes. The integral signal-to noise ratio (SNR) of 1 subject (S4) is shown for the full frequency range of the electromyographic signal, from 0 through 500 Hz

7. The Fulcrum principle

SR was shown to be capable of improving sensitivity for sub threshold excitatory stimuli. At the beginning SR was thought as a local peripheral effect. But evidence has confirmed the ubiquitous influence of the facilitating stimulus by triggering a mechanism that involves the cortex acting upon the peripheral sensory activity. And this mechanism was shown for both, stochastic and deterministic supra threshold facilitating signals as part of a general principle for the stimuli interactions that could explain all the paradigms.

So, experiments 2 to 7 permitted to introduce a General Dynamics Model that involves the entire ***MI threshold enhancement***. This non-linear model that handles deterministic or stochastic facilitating signals has shown be useful for explaining all the paradigms and we therefore call it: The **FULCRUM** Principle. A fulcrum is one that supplies capability for action and we believe that this best describes the fundamental principle at work in these multisensory interactions. The principle can be summarized as follows: a subthreshold excitatory signal (entering in one sense) that is synchronous with a facilitation signal (entering in a different sense) can be increased (up to a resonant-like level) and then decreased by the energy and frequency content of the facilitating signal. As a result the sensation of the signal changes according with the excitatory signal strength. In this context, the sensitivity transitions represent the change from spontaneous activity to a firing activity in multisensory neurons. Initially the energy of their activity (supplied by the weak signals) is not enough to be detected but when the facilitating signal enters the brain, it generates a general activation among multisensory neurons, modifying their original activity. In our opinion, the result is an integrated activation that promotes sensitivity transitions and the signals are then perceived. In other words, the activity created by the interaction of the excitatory signal (for example tactile) and the facilitating signal (auditory noise) at some specific energy, produces the capability for a central detection of an otherwise weak signal.

8. Mathematical model for the Fulcrum

We can simulate neurons as natural devices with dynamics that consist of random low-amplitude motions (spontaneous neuronal activity) from which escapes occur at certain intervals [32]. The escapes are referred to as firings, and are associated with high amplitude bursts (spikes). We begin by proposing a similar bistable model for the response of neurons as in [32]

$$\ddot{x} = -V'(x) + \varepsilon\left[\gamma Cos(\omega_0 t) + \sigma G(t) - \beta\dot{x}\right], \tag{1}$$

Where x represents the neurons' amplitude activity, $\dot{x}$ is the neurons' amplitude activity velocity (how their activity changes with time), $V(x)$ is a double-well potential defined by a polynomial, ε is a perturbation parameter that may have a stepwise variation over x. $Cos(\omega_0 t)$ represents the excitatory weak signal, $G(t)$ is the facilitating signal and it can be a nearly white noise process or a deterministic one, γ, σ and β are adjustable parameters. The quantities between brackets represent excitatory, facilitating energy, and energy losses. Equation (1) can achieve simulations of neuronal time histories (with the appropriate parameter values) and it has solution with the qualitative features observed in the experiments described earlier. To achieve good neuronal time history simulations, the potential $V(x)$ must be asymmetric, which is deeper for $x > 0$ than for $x \le 0$ as shown in figure 11 (left column, top row).

Neuronal firing necessary condition

Associated with an unperturbed system ($\varepsilon = 0$ for all x) are the homoclinic orbits Γ^+ and Γ^- shown in figure 11 (left column, middle row). In order for the escapes to take place we require that the maximum total energy produced during the motion over an entire homoclinic loop will be bigger than zero. Suppose the motion takes place on the unperturbed system's homoclinic orbit. If the motion occurs over a small distance δx_h (h designates coordinates of the homoclinic orbit), then the maximum total energy is given by :

$$E_{tot} = E_{loss} + E_{exc} = -\varepsilon\beta\int_{-\infty}^{\infty}\dot{x}_h^2 dt + \varepsilon\int_{-\infty}^{\infty}\left\{\gamma Cos\left[\omega_0(t)\right] + \sigma G(t)\right\}\dot{x}_h dt\,. \tag{2}$$

The condition $\max(E_{tot}) > 0$ implies that the maximum of the second term between braces in equation (2) is larger than the first term. This implies that the energy of the system can drive the motion over the potential barrier and out of a potential well.

Fulcrum neuron firing condition

It is possible to show that the necessary condition for the Fulcrum to occurs [5], for the stochastic process $G(t)$, is

$$-\frac{4\beta\sqrt{\alpha}}{3} + \gamma S(\omega_0) + \sigma \sum_{k=1}^{N} a_k S(\omega_k) > 0\,, \tag{3}$$

where the constants a_k are related to the Fourier one-side spectral density. For a second harmonic signal $\sigma Cos(\omega_1 t)$ instead of white noise the conditions writes:

$$-\frac{4\beta\sqrt{\alpha}}{3} + \gamma S(\omega_0) + \sigma S(\omega_1) > 0\,. \tag{4}$$

where $S(\omega_j) = \left(\frac{2}{\alpha}\right)^{\frac{1}{2}} \pi\omega_j \operatorname{sech}\left\{\frac{\pi\omega_j}{2\sqrt{\alpha}}\right\}$ is known as the Melnikov scale factor. It is clear that if we want to optimize the energy transfer from the stochastic process $G(t)$ or deterministic process $\sigma Cos(\omega_1 t)$ then the spectral density of $G(t)$ needs to contain frequencies around the Melnikov scale factor maximum and the frequency ω_1 from the signal $\sigma Cos(\omega_1 t)$ must be centered at the Melnikov scale factor $S(\omega)$ peak as well. The central column in figure 11 shows the neurons spectrum amplitude as a function the noise intensity σ. As it is expected for low noise intensities the energy transfer from the noise to the signal is not enough to achieve the synchronization and as a result the spontaneous activity dominates and no firings occur. However as the noise intensity increases firings also increase up to a maximum peak, where the mean escape rate approximately equals the signal frequency. Beyond this point, random firings can occur at different frequencies meaning that the synchronized energy transfer from the noise to the signal is destroyed and the signal is embedded in the spontaneous activity. The insert (center column, middle row) shows the well-known SR inverse u-shape function and its maximum peak. Right column in figure 11, shows neuron firing histograms with their correspondent time histories. It is clear from Equations (3) and (4) that if we increase the energy losses we have to increase accordingly the excitatory energy to fulfill the fulcrum neuron firing condition always. This means that the energy transfer is always fixed no matter how long is the neuronal network.

9. Consequences of the Fulcrum principle

The first consequence is that signals in the peripheral nervous system can be modulated by crossmodal interaction at the central level as we have seen clearly from examples 6 and 7. What normally could be considered to be a simple, peripheral, and reflexive muscular reaction to a directly applied stimulus turns out be a multimodal function. No sense, even the most peripheral, works on its own [33]. Indeed, the energy and frequency content of the facilitating signal induces the transition in perception of the excitatory signal. However, we are not proposing that the sensory activity is only peripheral. Initially, the energy level of

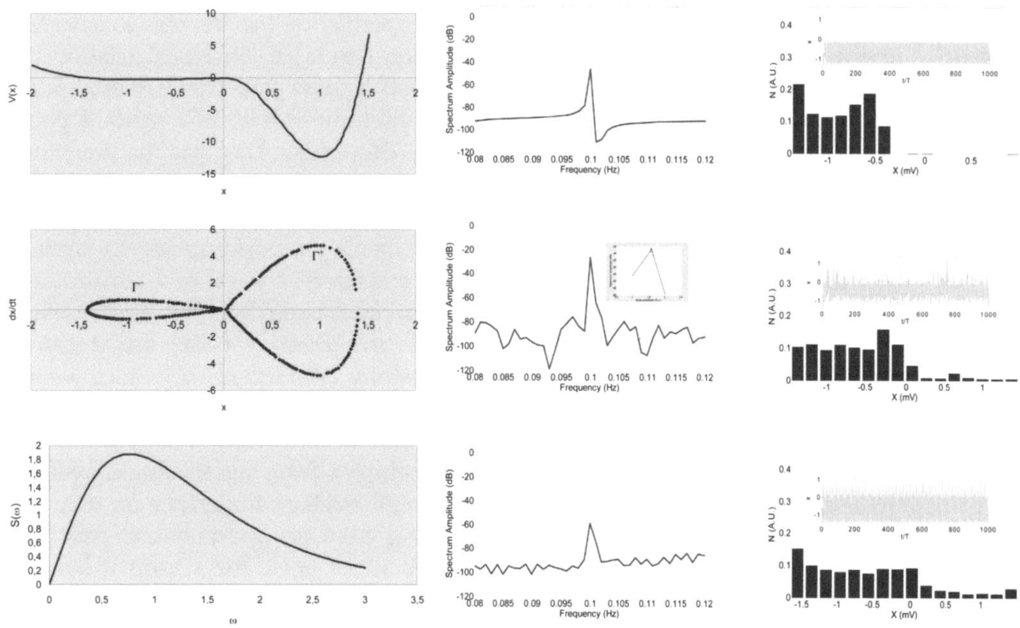

Fig. 11. Theoretical model for the fulcrum. (Left column, top row) Potential V(x); (Left column, middle row) Phase plane diagram showing homoclinic orbits; (left column, bottom row) Melnikov scale factor; (Center column) shows the neurons' spectrum amplitude as a function of the noise intensity. The insert (Center column, middle row) shows the well-known SR inverted u-shape function. Right column shows neuronal firing histograms with their corresponding time histories. T is the signal period and N means the probability to have certain neuronal activity levels.

the peripheral activity is not high enough to be detected by the central system; therefore, there is no interaction between central and peripheral systems at that time. When the facilitation signal enters the central system, it generates an activation that goes all the way back and modifies the original peripheral activity. The result is an activation that promotes resonance-like behavior, increasing the peripheral signal up to a level where it is perceived by the central system. This means that once the peripheral signal is perceived, the integration is represented not only at a central level, but also at a peripheral level. At some energy level of the facilitating stimulus, the peripheral activity reaches a maximum, and peripheral activity begins to decrease if the energy is increased further (see Fig. 10). Because the increase in peripheral activity comes from the way the brain processes the energy and frequency content of the facilitation signal in each individual, the nature of the signals (deterministic or stochastic) involved in the interactions is not important. If the facilitation signal has the right energy and frequency content, the phenomenon will occur. That is why deterministic signals (visual or auditory) and a stochastic signal (auditory noise) demonstrated the same effects in our experiments.

A second consequence is that these MI interactions do not follow the inverse-effectiveness rule, but are consistent with the fact that tactile, visual, and proprioceptive detection, and audiovisual comprehension of spoken words are substantially improved at an intermediate level of auditory noise [5,23,34-35].

A third consequence is the possibility to explain properties of the Yerkes-Dodson law dynamics under certain conditions. The **Yerkes–Dodson law** is an empirical relationship between performance and arousal, and it was developed in 1908 [26]. This law establishes that performance increases with physiological or mental arousal, but only up to a point. When the arousals become too high, performance decreases. The law is illustrated graphically as an inverted U-shaped curve which increases and then decreases with higher levels of arousal. Nonetheless, it is known that for simple or well learned tasks the relationship can be considered linear with improvements in performance as arousal increases. For difficult tasks however, the relationship between arousal and performance becomes inverse, with declines in performance as arousal increases. There has been research indicating that the correlation suggested by Yerkes and Dodson exists but a causal explanation of the correlation has not yet been successfully established [36]. Since we are using auditory noise, one might argue that 70 dBSPL (clearly audible) could be judged annoying by some people (although previous crosmodal SR claims have shown that this is the effective range [23]). Indeed sound annoyance is a complex thing and no single level can be pointed to as a threshold for it, there are reports of high levels of annoyance for very soft sounds indeed (e.g. 35dBA sound of a toilet flushing from an apartment above) [37]. Annoyance is defined by the context, and 70 dB SPL white noise for a normal hearing person could easily be construed as annoying under some conditions, for example if it were perceived to affect performance in an experiment where the participant wanted to do well. Indeed, subjects were exposed to white noises from 60-95 dB SPL during the experiment, so the noise could have been construed as interfering and annoying at all of the levels used, and could have caused arousal optimal for the task at around 70 dB. From these arguments one could possibly advance the hypothesis that the crossmodal effects are due to arousal. Arousal is a physiological and psychological state of being awake and represents physiological readiness for activity. Readiness or preparedness is the state of having been made ready or prepared for use or action. We argue that this classic definition of arousal cannot account for the crossmodal facilitation results presented here and elsewhere [23,5- 6] for several reasons. First our experimental conditions were all randomized and our subjects naïve, which would reduce the possibility of being specifically prepared for one condition or another. Second, we have shown that we can obtain similar dynamics with deterministic signals experimentally [6] and via modeling as well [5]. Given that the deterministic facilitation signals were simultaneously paired with the detection signal (no anticipation) we can also argue that the classic definition of arousal from noise would fall short at explaining these dynamics. Further, from the model we have developed it is clear that it is not the stochastic process that defines the noise (its uncontrollable nature) that makes the synchronization-like phenomenon occur. Instead it is the energy and frequency that are contained in the noise (or a harmonic signal) and the interaction between the excitatory and facilitation signals that makes the phenomenon possible and allows the subjects to improve perception. Another argument can be made against a simple arousal interpretation of our experiments. We found that the crossmodal SR effect was similar between luminance versus contrast-defined stimuli. It is well known that such stimuli require different processing levels where the contrast-defined stimuli are more complex to process [27,28] and are differentially affected by other factors such as attention, fatigue and learning. We would therefore have expected a greater and different effect of arousal on the contrast defined stimuli but we did not find this. Rather we found very similar results and this would be difficult to account with a simple arousal explanation. Nonetheless the fact that the subject's perception is enhanced by SR mechanisms might change the subjects' behavior if we would ask them to do a second task in parallel with the detection task such as in behavioral SR [38]. This implies that known behavioral effects induced by noise or other deterministic signals may have their origin at a lower level. We therefore propose that the fulcrum possibly explains properties of the Yerkes-Dodson law dynamics under certain conditions.

From unimodal SR studies it can be inferred that 70 dBSPL is much louder than the noise required for auditory SR [39–40]. This may make the SR label we have used here problematic. However the auditory unimodal SR works in a simpler architecture than the crossmodal SR [5], where larger neuronal networks are necessarily involved between modalities. Since the crossmodal architecture is vaster, and complex, one would expect more energy losses in such network and according with the model we have developed it is possible to have synchronized neuronal firings with these conditions. The aforementioned studies have shown that auditory unimodal SR happens between 5 dB [39] and 3–5 dB [40] below a point defined as noise threshold [40]. The noise threshold is the point where the noise hinders the signal detection and the sensitivity worsens to levels above threshold (the crossing point in the inverse u-shape curve). If we use this level as our reference instead of the SPL absolute scale (we will call this level the noise ceiling level that defines a ceiling decibel dBc) then we found that crossmodal SR threshold minima occur approximately in the same experimental range as the ones mentioned above. We found that for visual experiments the minima are localized at -6±1 dBc (first order) and -5±1 dBc (second order). In the proprioception experiments the minima occurs around -6±1 dBc and for tactile experiments at -8±1 dBc and for the experiment 6 with one subject (figure 10) -4 dBc. These results underscore the very important fact that independently of the unimodal or crossmodal interaction the energy transfer from signal plus noise is approximately fixed, which is the fourth consequence of the fulcrum. Note that for measuring the noise ceiling level we have used a similar approach than the one presented in [40].

Implications for autism: The fifth consequence is related to a better understanding of disorders such as **autism**, in which altered sensory processing often occurs that causes perceptual dysfunction, it causes problems with one or more sensory channels from the world to the brain. In a very general classification given in [41] the sensory channels are abnormal in one of the followings ways: Hyper: the sensory channel is too open and, as a result, too much stimulation gets in for the brain to be handled comfortably. Hypo: the sensory channel is not open enough and, as a result, too little of the stimulation gets in and the brain is deprived. White noise: the sensory channel creates its own stimulus because of faulty operation and, as a result, the message from the outside world is garbled or, in extreme cases, is overcome by the noise in the system. The broad autism classification can be qualitatively understood by using the neuron firing condition:

$$-\frac{4\beta\sqrt{\alpha}}{3} + \gamma S(\omega_0) + \sigma \sum_{k=1}^{N} a_k S(\omega_k) > 0$$

Let us assume in this case that the stochastic energy is due to the internal noise and the excitatory signal is deterministic. Therefore the Hyper type can be described by: small values of parameters β and α that make the energy losses small. This represents an internal noise energy level close to the energy losses level and a Melnikov scale factor that is very narrow (this is because α is small). With these conditions a very weak excitatory signal elicits neuronal firing and because of the Melnikov narrowness factor, the optimal condition for the firings is easily achieved (but only for a small frequency bandwidth). Note that if the frequency content of the excitatory signal is outside of this bandwidth then the classification will change to Hypo. Hypo condition: High values of parameters β and α that make the energy losses high. Therefore, the internal noise energy is lower than the energy losses. This represents a Melnikov scale factor that is very broad (this is because α is high). With these conditions a very strong excitatory signal is needed to elicit neuronal firing and, because the Melnikov factor broadens, more frequencies can induce the optimal condition for the firings with less selectively than the Hyper type. White noise condition: the stochastic energy is higher than both the energy losses and the excitatory signal combined. Then neuronal firings occur but they are mainly driven by the internal stochastic process. Note that nothing precludes that sensory channels present one or more classifications. That is, a person might be hyper or hypo for the different stimuli because optimal neuron firings depending on the frequency and energy content of the involved signals.

10. More experimental evidence for the Fulcrum Principle

The consequences of spinal cord injury or Parkinson's disease are not just a break in communication between neurons; a cascade of events occur that promote further neuronal degeneration, cell death and motor dysfunctioning [42]. Locomotion training is a very effective tool in neuronal degeneration rehabilitation. Besides regular locomotion exercises (or similar strategies) are associated with neuroprotective effects in different brain areas. Nevertheless, numerous patients are unable to do locomotion therapy and therefore the possibility of rehabilitation is reduced. Haas [43] has bypassed this problem by using vibratory stimulations, leading to reflex responses similar to reflex elicitations during human locomotion. He found that stochastic mechanical stimulations might be a useful method to counteract neuronal degeneration and to promote regenerative processes. His patients either stood up or sat down in a special chair and both legs were connected with two independently oscillating platforms. The platforms could oscillate with a mean frequency of 6 Hz and superimposed by random and stochastic influences which facilitate neuronal threshold crossing and enhance neuromuscular activity. Patients with Parkinson's disease and spinal-cord-injury patients that were stimulated regularly lead to significantly improved postural control and locomotion abilities. Interestingly, treated Parkinson's disease patients also showed reduced symptoms (tremor, rigidity) in the upper extremities. As improvements in manual coordination (for instance writing performance) were confirmed in further standardized experimental setting, it seemed unlikely that this vibratory stimulation affected only the muscle or exclusively the peripheral nervous system. That is, if only the lower limbs were excited, how could we explain improved writing performance?

The startle reflex is the response of brain and body to an unexpected stimulus, such as a loud noise, a flash of light, or a sudden movement near the face. The reaction includes physical movement away from the stimulus, a contraction of the muscles of the arms and legs, and often blinking. The startle reflex provides a unique tool for the investigation of sensorimotor gating and information processing. Neuner et al. presented the first MR study using a single trial approach with simultaneous acquired EMG and fMRI data on the human startle response [44]. The startle reflex was recorded from the right orbicularis oculi muscle. Electrodes for recording electromyographic activity of this muscle were fixed below the eye in midline and the outer canthus. The air puffs were delivery to the region below the left clavicula. They investigated the neural correlates for isolated air puff startle pulses (PA), prepulse–pulse inhibition (PPI), and prepulse facilitation (PPF), via air puffs onto the skin. PPI is a neurological phenomenon in which a weaker prepulse inhibits the human reaction to a subsequent strong startling stimulus. PPI is present in a vast number of species (from mice to human) and usually is measured through muscular reactions, which are normally diminished as a result of the nervous inhibition. The opposite reaction is known as prepulse facilitation (PPF). A common core network engaged by all three conditions (PA, PPI, and PPF) was identified. The network consists of bilateral primary and secondary somatosensory cortices, right insula, right thalamus, right temporal pole, middle cingulate cortex, and cerebellum. The cerebellar vermis exhibits distinct activation patterns between the startle modifications. It is differentially activated with the highest amplitude for PPF, a lower activation for PA, and lowest for PPI. The orbital frontal cortex exhibits a differential activation pattern, not for the type of startle response but for the amplitude modification. For pulse alone it is close to zero; for PPI it is activated. This is in contrast to PPF where it shows deactivation. In addition, the thalamus, the cerebellum, and the anterior cingulate cortex add to the modulation of the startle reflex. In summary, this research shows that peripheral activation, through somatosensory stimulation and measured with EMG techniques, correlates with central activation measured with fMRI techniques.

Fine-motor performance of the hand is more than important in our life and work. However, some injuries might damage the hand fine-motor performance. Previous studies show that fine-motor performance of the hand might be improved according to the mechanism of coactivation [45]. Lei Ai et al. tested if fine-motor performance could be enhanced by the presence of auditory noise. They used a pegboard which has 50 holes arranged in four rows and 50 metal pins placed in a container. In every trial subjects were asked to pick these 50 pins with their right hand, one by one, from the container and insert them into 50 holes on the peg-board. Every hole is inserted by one pin. If one pin is dropped during the transfer, they were instructed to pick the next pin from the container to insert the hole that they just failed. The dependent variable was the length of time required for completing the process of inserting all the pins. The less the time the more dexterous the hand is. The result was a U-shape function of the intensity of different levels of auditory noise showing that optimal auditory noise can largely improve the fine-motor performance.

Noise is typically conceived of as being detrimental to cognitive performance. However, a certain amount of noise can benefit performance. Soderlund et al. investigated cognitive performance in noisy environments in relation to a neurocomputational model of attention deficit hyperactivity disorder (ADHD) and dopamine [46]. They hypothesized that dopamine levels modulate how much noise is required for optimal cognitive performance. They experimentally examined how ADHD and control children responded to different encoding conditions, providing different levels of environmental stimulation. Participants carried out a high memory performance task and a low memory task. These tasks were done in the presence, or absence, of auditory white noise. They found that noise exerted a positive effect on cognitive performance for the ADHD group and deteriorated performance for the control group, indicating that ADHD subjects need more noise than controls for optimal cognitive performance.

11. Final remarks

The Fulcrum principle describes a ubiquitous process in humans related in how our peripheral and central systems use energy and frequency content of external and internal signals to modulate our perception of reality. We have seen that stochastic or deterministic sounds of one modality can facilitate perception of stimuli in another modality and underscore that they share the same dynamics. This and the fact that the energy transfer necessary for neuronal firings to occur is approximately constant for all the interactions presented here, implies that the central system also can modulate the peripheral system. Consequently, no single sense works alone. These interactions may be the basis for explaining certain aspects of arousal dynamics related with the Yerkes-Dawson law. At the same time they challenge us because they seemingly do not follow the inverse effectiveness law from the classic multisensory integration theory. We propose that the fulcrum principle also gives us a theoretical framework for a better understanding of autism an ADHD conditions. Finally, these results have obvious implications in developing methods for enhancing human performance in easy non-invasive ways. One possible application is with Parkinson disease as we have seen stochastic vibrations applied to lower limbs not only enhanced mobility and decreased tremors in the same anatomical part where the vibration was applied but on the upper limbs as well. We know from our results that stochastic sound should also work. Spinal cord, motor system, memory, ADHD and Alzheimer diseases may be treated with the same acoustic therapy that would be beneficial for everyone and in particular to the elderly. As we age, we depend more and more on multisensory perception and it has been recently suggested that, despite the decline in sensory processing that accompanies aging, the use of multiple sensory channels may represent an effective compensatory strategy to overcome uni-sensory deficits [47]. In summary, in the presence of any one sensory deficit or any neurobiological alteration the Fulcrum principle takes on a new and important meaning.

12. References

[1] Georg Von Békésy (1967), Sensory inhibition, Princeton University Press, Princeton NJ,

[2] Georg Von Békésy (1957), Sensations on the skin similar to directional hearing, beats, and harmonics of the ear, Journal of the Acoustical Society of America, 29: 489-501.

[3] Georg Von Békésy (1964), Rhythmical variations accompanying gustatory stimulation observed by means of localization phenomena, Journal of General Physiology, 47:809-825.

[4] Georg Von Békésy (1964), Olfactory analogue to directional hearing, Journal of Applied Physiology, 19: 369-373.

[5] Lugo E, Doti R, Faubert J (2008) Ubiquitous Crossmodal Stochastic Resonance in Humans: Auditory Noise Facilitates Tactile, Visual and Proprioceptive Sensations. PLoS ONE 3(8): e2860. doi:10.1371/journal.pone.0002860

[6] The final, definitive version of this paper has been published in Psychological Science, 19:989-997, 2008 by SAGE Publications Ltd./SAGE Publications, Inc., All rights reserved. ©

[7] Lugo J E, Doti R and Faubert J (2010), Effective tactile noise can decrease luminance modulated thresholds, Journal of Vision 10 doi:10.1167/10.7.857.

[8] Schwartz M (1977), Information, Transmission, Modulation, and Noise, Mc Graw-Hill book Co., New York, N.Y.

[9] Athanasios Papoulis (1959), SIGNAL ANALYSIS, Page 12 Chapter 1-2: Analog Signals and Systems, Mc Graw-Hill book Co., New York, N.Y.

[10] Athanasios Papoulis (1977), SIGNAL ANALYSIS, Page 184 Chapter 6-1: Properties of Band-limited Functions, Mc Graw-Hill book Co., New York, N.Y.,.
[11] Athanasios Papoulis (1977), SIGNAL ANALYSIS, Page 300, Chapter 9: Stochastic Processes, Mc Graw-Hill book Co., New York, N.Y.
[12] Stein B E, & Meredith M A (1993), The merging of the senses, Cambridge: MA MIT Press..
[13] Moss F, Ward L, Sannita W (2004) Stochastic resonance and sensory information processing:a tutorial and review of application. Clinical Neurophysiology 115:267-281.
[14] Benzi R, Stuera A, Vulpiani A (1981) J. Phys. A 14:L453.
[15] Simon A J, Libchaber A J (1992) Escape and Synchronization of a Brownian Particle. Physical Review Letters 68 (23):3375-3378.
[16] Badzey RL, Mohanty P (2005) Coherent signal amplification in bistable nanomechanical oscillators by stochastic resonance. Nature 437(7061):995-998.
[17] Ivey C, Apkarian A V, Chialvo D R (1998) Noise-Induced Tuning Curve in Mechanoreceptors. J. Neurophysiol. 79 (4):1879-1890.
[18] Simonotto E, Massimo R, Seife C, Roberts M, Twitty J, et al. (1997) Visual Perception of Stochastic Resonance. Physical Review Letters 78: 1186-1189.
[19] Collins J J, Imhoff T T, Grigg P (1997) Noise-mediated enhancements and decrements in human tactile sensation. Physical Review E 56: 923-926.
[20] Hidaka I, Nozaki D, Yamamoto Y (2000) Functional stochastic resonance in the human brain: Noise induced sensitization of baroreflex system. Physical Review Letters 85: 3740-3743.
[21] Kitajo K, Nozaki D, Ward LM, Yamamoto Y (2003) Behavioral stochastic resonance within the human brain. Physical Review Letters 90:218103.
[22] Harper D W (1979) Signal detection analysis of effect of white noise intensity on sensitivity to visual flicker. Percept. Mot. Skills 48: 791–798.
[23] Manjarrez E, Mendez I, Martinez L, Flores A, Mirasso C R (2007) Effects of auditory noise on the psychophysical detection of visual signals: Cross-modal stochastic resonance. Neuroscience letters 415: 231-236.
[24] Kitajo K, Doesburg S M, Yamanaka K, Nozaki D, Ward L M, et al. (2007) Noise-induced large-scale phase synchronization of human-brain activity associated with behavioural stochastic resonance. Europhysics Letters 80 (4):40009.
[25] Shulgin B, Neiman A, Anishchenko (1995) Mean Switching Frequency Locking in Stochastic Bistable Systems Driven by a Periodic Force. Physical Review Letters 75(23):4157-4160.
[26] Yerkes R M, Dodson J D (1908) The relation of strength of stimulus to rapidity of habit-formation. Journal of Comparative Neurology and Psychology 18: 459-482.
[27] Allard R, Faubert J (2006) Same calculation efficiency but different internal noise for luminance and contrast-modulated stimuli detection. Journal of Vision, 6(4): 322-334.
[28] Allard R, Faubert J (2007) Double dissociation between first- and second-order processing. Vision Research, 47(9): 1129-1141.
[29] Priplata A A, Niemi J B, Harry J D, Lipsitz L A, Collins J J (2003) Vibrating insoles and balance control in elderly people. The Lancet 362: 1123-1124.

[30] Priplata A., Niemi J B, Salen M, Harry J D, Lipsitz L A, et al. (2002) Noise-enhanced human balance control. Physical Review Letters 89:238101

[31] Eurich C W, Milton J G (1996) Noise-induced transitions in human postural sway. Physical Review E 54: 6681-6684.

[32] Simiu E. (2002) Chaotic Transitions in Deterministic and Stochastic Dynamical Systems. New Jersey: Princeton University Press. 178 p.

[33] Rosenblum L (2010), See What I'm Saying: The extraordinary Powers of Our Five Senses, New York London: W W Norton & Company Inc. page 270.

[34] Ross, L.A., Saint-Amour, D., Leavitt, V.M., Javitt, D.C., & Foxe, J.J. (2007). Do you see what I am saying? Exploring visual enhancement of speech comprehension in noisy environments. Cerebral Cortex, 17, 1147–1153.

[35]Faubert, J., Hahler, E.-M., Doti, R., & Lugo, J.E. (2007). Auditory noise can facilitate low level visual processing. Journal of Vision,7, Abstract 866a. Available http://journalofvision.org/7/9/866/

[36] Anderson KJ, Revelle W, Lynch MJ (1989). "Caffeine, impulsivity, and memory scanning: A comparison of two explanations for the Yerkes-Dodson Effect". Motivation and Emotion 13: 1–20.

[37] Sato S, Kitamura T, Ando Y (2004) Annoyance of noise stimuli in relation to the spatial factors extracted from the interaural cross-correlation function. Journal of Sound and Vibration 277:511-521.

[38] Kitajo K, Nozaki D, Ward LM, Yamamoto Y (2003) Behavioral stochastic resonance within the human brain. Physical Review Letters 90:218103.

[39] Zeng F-G, Fu Q-J, Morse R (2000) Human hearing enhanced by noise. Brain Research 869:251-255.

[40] Ries D T (2007) The influence of noise and level upon stochastic resonance in human audition. Hearing Research 228:136-143.

[41] Delacato C H, The ultimate stranger:The autistic child, Academic Therapy Publications, Novato California, pp.77.

[42] Horner P J, Gage F H (2000), Regeneration the damaged central nervous system. Nature 407, 963-970.

[43] Haas C T (2008), Vibratory Stimulation and Stochastic Resonance Therapy: Results from studies in Parkinson's disease and spinal cord injury, Technologies of Globalization Congress Darmstadt, Germany.

[44] Neuner, I., Stöcker, T., Kellermann, T., Ermer, V., Wegener, H. P., Eickhoff, S. B., Schneider, F. and Shah, N. J. (2010), Electrophysiology meets fMRI: Neural correlates of the startle reflex assessed by simultaneous EMG–fMRI data acquisition. Human Brain Mapping, n/a. doi: 10.1002/hbm.20965.

[45] Ai L, Liu J, Liu J (2009), Using auditory noise to enhance the fine-motor of human's hand due to cross-modal stochastic resonance, J. Proceedings of the 2009 2nd International Conference on Biomedical Engineering and Informatics, BMEI 2009 , art. no. 5305070.

[46] Söderlund G, Sikström S and Smart A (2007), Listen to the noise: noise is beneficial for cognitive performance in ADHD, Journal of Child Psychology and Psychiatry 48: 840–847.

[47] Laurienti PJ, Burdette JH, Maldjian JA, Wallace MT (2006) Enhanced multisensory integration in older adults. Neurobiology of Aging 27(8):1155-1163.

24

Discrete Damage Modelling for Computer Aided Acoustic Emissions in Health Monitoring

Antonio Rinaldi[1,2,3], Gualtiero Gusmano[1] and Silvia Licoccia[1]
[1]*Department of Chemical Science and Technology; University of Rome "Tor Vergata", Via della Ricerca Scientifica, 00133, Rome,*
[2]*Department of Safety Technologies, INAIL, Via Alessandria 220/e, 00189, Rome,*
[3]*ASSOINGE R&D, K. Doormanlaan 10 - 2283 AS, Rijswijk,*
[1,2]*Italy*
[3]*The Netherlands*

1. Introduction

This chapter is conceived as an essay on modern multiscale discrete damage modelling, providing a brief personal perspective about its foreseeable applications-implications for structural health monitoring purposes. In particular, it is argued that this sort of damage modelling could be potentially useful in damage detection by acoustic emissions (AE), which is a class of non-destructive techniques (NDT) used to capture damage evolution in a number of materials (e.g. from concrete systems such as bridges and beam elements to composites in aircraft components and pressure equipments) and from a number of external actions (e.g. sustained load, monotonic testing, fatigue, corrosion, etc.) (Biancolini & Brutti, 2006 ; Carpinteri & Lacidogna, 2008 ; Grosse & Ohtsu, 2008). With AE it is possible to "hear" the microcracking phenomenon and characterize the location and magnitude of a single microcrack (of size and "strength"[1] beyond certain thresholds) acting as an acoustic source. Hence, it is routinely possible to plot the released energy of each crack as a time series or to map them over a 2D spatial domain by counting and locating individual acoustic events in time. Yet the analysis of this type of output is not straightforward and major difficulties exist, let alone sensitivity issues of equipment, material dependence, and other practical issues. The scope of this discussion covers two issues of general interest:

1. the randomness of the AE signal,
2. the need for structure-property relations as companion to AE monitoring.

The first problem is rooted in the very same nature of the collected signal, which is a highly random time series that needs to be analyzed and interpreted. How to do that is a non-trivial task and remains an open research topic to date. Of course, elimination of the outer noises is one of the most concerned aspects in the applications and is usually achieved by simply setting a minimum cut-off threshold (low enough to retain all relevant information

---

[1] Here, the term "strength" alludes figuratively to the energy released by a microcrack during formation, which is linked to the amplitude of the collected signal and can largely differ between cracks of equal size (e.g. consider grain boundary microcracks with different orientation).

but above the noise level), by a band-pass filtering, or by a post-analysis of the data[2]. However, more importantly, even when the external noise could be filtered out completely, the AE data would retain a highly complex and random structure due to the inherent nature of the damage process and to material inhomogeneities, as depicted in Fig.1. A fundamental question, then, is whether the whole detected signal is essential to health monitoring and failure prediction, or criteria can be derived to discern what is relevant from what is not in the dataset. This is one main aspect to be explored later by discrete damage modelling. The development of such a filtering capability would have an impact, enabling not only greater understanding, but also the discard of the unwanted signals in favour of simplified time-series and optimal hardware usage (e.g. data storage, transmission facilities, longer monitoring period, etc.).

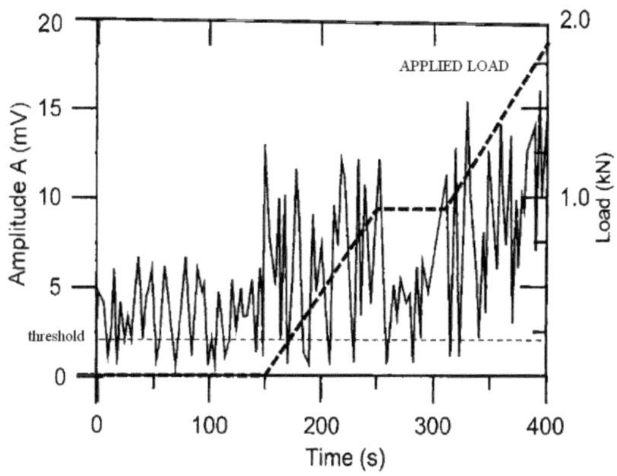

Fig. 1. Preliminary AE test for noise detection and quantification in a sintered ceramics loaded according to the dashed ramp. The cut-off threshold is indicated (after Palma & Mansur, 2003).

The second problem is the impossibility to correlate the AE output with the actual microstructure of the material without pairing AE with microscopy[3] to cross-correlate and perform a companion microstructural characterization. Macroscale observations (e.g. AE measures) render partial information that captures only the overall effect of microstructral phenomena happening at a much finer scale and not normally observable in field applications. It is nowadays well recognized that the material cannot be regarded as a black-box in the study of damage and strain localization phenomena (including failure analysis), which require thorough understanding of the structure-property relations. Likewise, the consideration of the deformation mechanisms active in a given material microstructure and triggered by a certain load configuration may indeed be crucial for the interpretation of AE signals and for the estimate of the current damage state. Unfortunately, with the exception of LOM and despite a few field trials for AFM (by INAIL(IT) private communication), it is

[2] Noises sometimes have similar frequency contents and amplitudes to AE signals, or sources of the noises are unknown. Then noises characteristics have to be estimated and modelled prior to measurement in order to separate the actual AE signals from raw data. The use of filters is very useful also in this respect, e.g. for determining the proper frequency range.

[3] For example, scanning electron microscopy (SEM), optical microscopy (LOM), atomic force microscopy (AFM), electron backscattered diffraction (EBSD), focus ion beam (FIB), etc.

not easy to perform these auxiliary microscopic investigations *in-situ* at present. As a matter of fact, in alternative to LOM for field metallurgy, industry relies on surrogate NDTs, such as the "replication" technique[4] (VGB, 1992; Rinaldi et al., 2010). Also in this context, discrete damage modelling seems to offer a possible route to overcome experimental limitations and establish the correct (micro)structure-property relationships of a desired material by means of accurate numerical simulations[5].

We should now move on to clarify how discrete damage modelling can be used to address the aforementioned AE problems, after a brief overview of AE facts. The outline of the discussion is as follows:

- Data collection, seismic similarity, and statistics in AE
- Discrete mechanical models for damage and fracture
- Lattice model highlights and AE
- Closing remarks

2. Data collection, seismic similarity, and statistics in AE

AE signals are electrical signals generated by fracture phenomena. After acquisition, the characteristics of AE parameters are used to infer fracture or physical phenomena. The following definitions (ref. ISO 12716 2001) refer to some popular signal parameters (Grosse & Ohtsu, 2008).

1. Hit: a signal that exceeds the set minimum threshold and causes a system channel to accumulate data;
2. Count: the number of times the waveform (signal) exceeds the given threshold within a hit;
3. Amplitude: a peak voltage of the signal waveform is usually assigned. Amplitudes are expressed on a decibel scale instead of linear scale.

The one waveform in Fig. 2 corresponds to one hit or to nine counts. "Hits" are the classical data used to show AE activity by means of the accumulated number n (parameter-based approach). Also "counts" can be employed (signal-based approach) to quantify the AE activity in place of "hits" but have several cons, as they require more acquisition capability, more consuming/sophisticated data analysis, and depend strongly on selected threshold and operating frequency[6]. For the sake of this discussion, the scope can be limited to

---

[4] Essentially a technique to collect a copy of the actual material surface in-situ for afterwards examination in laboratory. It is used for example for microstructural monitoring in low-allow carbon steels subject to creep.

[5] Structure-property relationships in solid mechanics represent a multidisciplinary research topic, sitting primarily at the crossroad of material science and engineering. Its widespread recognition and popularity is best witnessed by the growing number of scientists and engineers that have engaged in multiscale modeling of damage processes of all kinds and in all kinds of materials over the past ten years (check mechanics journals; JMPS, Acta Mater, Mech Mater.,etc.). The development of microstructure-based models is indeed a major trend in solid mechanics research. A great effort is also been directed in scaling laws able to predict the behaviour of materials in components of different size, trying to model the sample-size effects of damage, structural failure and other properties existing in real material systems. (Krajcinovic, 1996; Krajcinovic & Rinaldi, 2005, Carpinteri & Lacidogna, 2008).

[6] Approaches in recording and analyzing AE signals can be divided into two main groups: parameter-based (classical) and signal-based (quantitative) AE techniques. Both approaches are currently applied with success for different applications. Rapid developments in microelectronics over the last few

classical parameter-based approach with no loss of generality. In that case, the waveforms like Fig.2 are measured but only simpler parameters are stored, such as hits and corresponding amplitudes *vs.* time. As an example, Fig.3 shows actual AE monitoring in ceramics (Palma & Mansur, 2003) during controlled tensile test. One plot (left) correlates the amplitude and number of AE signals at each strain with the force response of the material specimen throughout the test. The second plot (right) reports the cumulative number of hits n vs. time, along with the force signal. Both pictures clearly render an increased activity, both in terms of signal amplitude and signal density (i.e. dn/dt) at well defined points, namely the end of the elastic regime and prior to failure. The force signal indicates the substantial loss of load bearing capability of the material in correspondence to these (transition) points.

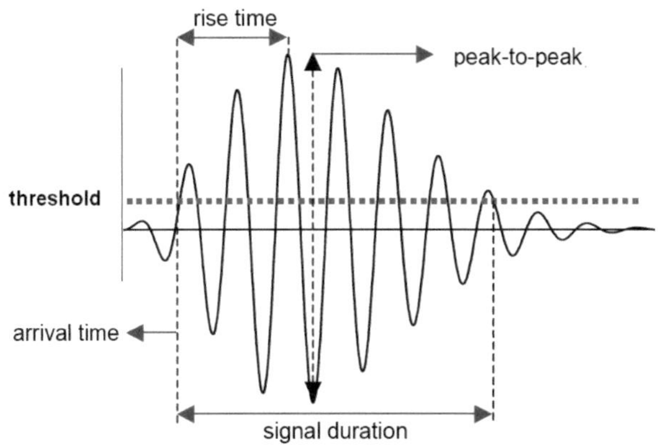

Fig. 2. Example of typical 'AE parameters' according to international and national standards (ASTM E610, 1982; Berger, 1977; DGZFP SE-3, 1991).

---

decades are largely responsible for the existence of two approaches. In the past, it was not possible to record and store a large number of signals over a sufficiently short period of time. Despite significant technical advances in recent years, it is still not possible to use signal-based techniques to monitor large structures and buildings. In addition, the relatively high financial costs and the time required to apply modern signal-based techniques are a sufficient reason for why parameter-based techniques are still popular. It should be emphasized that the discrepancies between the two approaches are becoming smaller and that devices intended for the classical AE technique are now able to store entire waveforms of the detected AE signals, even though this is not the primary function of these devices. Instead, applications using signal-based analysis techniques rely on equipment based on transient recorders, which facilitate the use of custom software tools to extract AE parameters for statistical analyses. In fact, in that case, not only "counts" but also other parameters can be chosen, e.g. "counts to Peak" (i.e. counts between the triggering time over the threshold and the peak amplitude, equal to four in Fig. 2), the arrival time (defined as the first crossing of a given amplitude threshold), the rise time (defined as the duration between the arrival time and the time where the maximum amplitude is recorded) and the duration (defined by the last crossing of a given amplitude threshold) (ref. ASTM E610 1982; CEN 1330-9 1999). The signal-based approach, so-called quantitative AE technique, record and store as many signals as possible after converting waveforms from analogue-to-digital (A/D) signals, which enables a comprehensive and time-consuming analysis of the data but usually only in a post-processing environment and not in real-time (Scruby 1985; Sachse and Kim 1987). The next generations of instruments will tend to be universal and adapt to different applications, capable either of recording waveforms if a signal-based approach is being taken or storing a large number of events n if a parameter-based approach is being taken requiring the statistical analysis of many events.

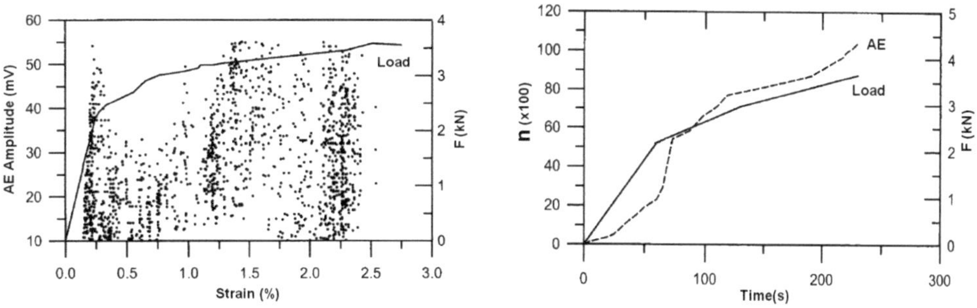

Fig. 3. (left) Correlation between AE amplitude and deformation in a sintered material. (right) Number of threshold crossings *n vs.* time. Damage localization points are seen as discontinuity in the force response and steep increase in AE activity (Palma & Mansur, 2003)

Another noteworthy feature of AE data is the similarity with seismology. AE and seismological techniques are very affine because they basically exploit the same concept but at a different scales. An AE signal is defined as the spontaneous release of localized strain energy in stressed material and, as such, it can be regarded as a form of microseismicity generated during the failure process as materials are loaded. AE transducers (sensors) placed on the materials surface sense and record this energy release due to microcracking, just like seismographers measure earthquakes. In turn, there is a well established theory connecting earthquakes and fracture processes in (micro)structural elements near failure (Mogi, 1967), which are both described by the Gutemberg-Richter law

$$Log\, N = a - b\, m \tag{1}$$

expressing the empirical relation between a certain magnitude m to the number N of events exceeding m in a earthquake (or a failure). By further assuming the magnitude to be related to the energy level as $m = Log\, E$, Eq.(1) can be rewritten as

$$Log\, N = a - b\, Log\, E \tag{2}$$

which implies a linear relation between N and E in a log-log plane (Fig.4a) and, equivalently, a power law relationship "$N \sim E^{-b}$" for the decay of number of seismic/microcracking events of larger energy. These equations state that major fracture events (leading to catastrophic failure) are expectedly preceded by many events of smaller

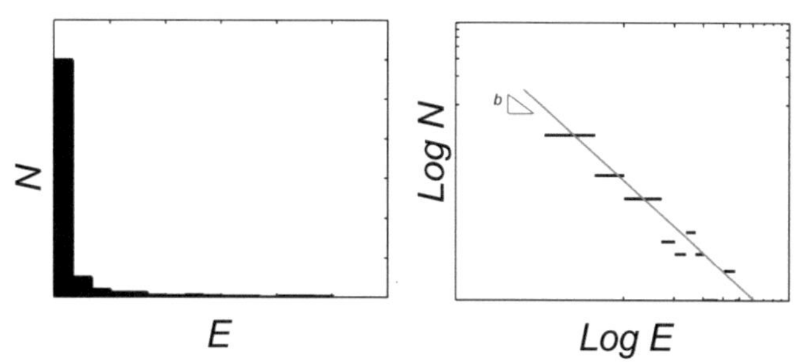

Fig. 4. Statistical distribution of microcracking events valid for seismic and AE data (left). The number of events N scale with energy (magnitude) according to a power law (right).

entity. For what said, this same framework and expectation apply to AE monitoring and explain why the AE technique is in principle suited to detect a failure at a very early stage, long before a structure completely fails.

3. Discrete mechanical models for damage and fracture

Now, discrete damage modelling is briefly addressed in consideration of the experimental facts just recalled. The importance of having a theoretical model to analyze AE measures was pointed out in the introduction, but no universal damage modelling has emerged to date. The goal of damage mechanics is to develop predictive models for damage tolerant design and failure prevention, just as AE monitoring. Damage models can continuum or discrete (Krajcinovic, 1996). Continuum models, which represent the mainstream tool in solid and structural mechanics, are very commonly used in industry but are unsuited in this case. Most continuum damage models are derived from micromechanics via one of the many homogenization or coarse graining techniques available. The "representative volume element" (RVE) is the traditional basic instrument of micromechanics to convert a disordered (i.e. randomly microcracked) material into an equivalent continuum model and, as depicted in Fig. 5, it represents the smallest specimen volume of disordered matter that can be considered as statistically homogeneous (and, hence, in thermodynamic equilibrium) under the action of nearly-uniform tractions at its boundaries. This formal definition simply means that a continuum model takes in consideration an idealized material that is mechanically equivalent to the real one and has properties obtained from averaging local micro-properties over the RVE domain. But this procedure, although numerically convenient, poses severe limitations and overheads, because details of the microstructure that are fundamental to the damage process (e.g. the grain and grain boundaries of a polycrystalline metal - Fig.5) are completely discarded. Further more, an RVE may not even exist sometimes, rendering the continuum approach ill-posed and not applicable. This is typically the case at the onset of damage localization and failure pointed out in Fig.3. In extreme synthesis, the good candidate model needs to have a resolution length of the same order of the relevant microstructure (e.g. l in Fig.5).

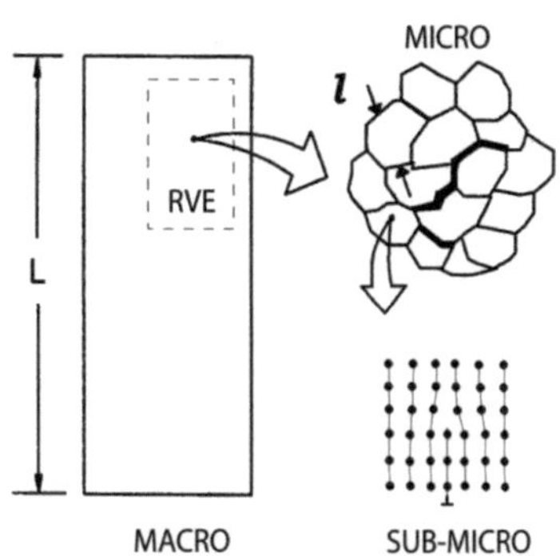

Fig. 5. Example of hierarchy of length scales associated to a damage problem in a polycrystalline material. The scope of damage mechanics does not conventionally entail sub-microscale for the estimate of residual life (i.e. unless nanoscale components of nanotechnology are involved) but necessitates direct consideration of the microscale.

In that regard, modern discrete damage modelling, also known as statistical damage mechanics (SDM), appears to be a better option to bridge such a theoretical gap. SDM is a new branch of damage mechanics (and more at large of solid mechanics) (Rinaldi, 2010). Unlike the continuum modelling (e.g. micromechanics but also non-local continuum theory), SDM is natively a multiscale approach, where discrete statistical models accurately reproduce the random microstructure of a material with a sufficient degree of detail, incorporating the relevant random microscale properties via statistical distributions. Such discrete models are applicable over the entire damage evolution, regardless of whether damage is homogeneously dispersed or localized in the material (following a transition), near and away from failure. SDM offers a fertile ground for the application of advanced statistical methods and non-standard mathematical method (e.g. fractal theory) to obtain innovative physically-based constitutive relations and damage theories that effectively reckon subtle aspects (e.g. such as sample-size effects, localization threshold, intrinsic variability of mechanical properties, and damage-induced anisotropy), which has important implications for the study of sound localization, as we shall see.

In damage mechanics, the modelling problem consists of determining the proper damage variable D that fully encapsulates the complexity of the stochastic damage process and is a random variable ranging from 0 in pristine conditions to 1 at failure. Damage is a weakening transformation of the microstructure that is driven by one or more external causes (i.e. quasi-static load, fatigue, corrosion, impact, etc. or a combination) and consists of microcracks formation, growth, and coalescence into a final fracture, which is perceived as a depletion of the elastic stiffness at the macroscale. The material response, as damage accumulates under the action of an increasing load σ, is generally expressed by

$$\sigma = E^{*}(D)\varepsilon_e = E_0(1-D)\varepsilon_e \tag{3}$$

where E_0 is the initial Young modulus, D is the damage parameter, ε_e is the elastic strain and E^* is the secant stiffness, both measurable during unloading in a ductile material. This relation states that the damage parameter is equal to the normalized loss of secant stiffness $D= \Delta E^*/ E_0$, which is the sum of each damage increment associated to the i-th event $\Delta D= \Delta E^*_i/ E_0$ (the stiffness decrement normalized to the pristine stiffness), such that after n random microcracks it is

$$D = \frac{\sum_{i=1}^{n} \Delta E_i}{E_0} \tag{4}$$

Of course the evaluation of Eq.(4) in not trivial, since the ΔE^*_i is unknown a priori and is a complicated random function of microstructure and applied load(or generalized action).

3.1 One dimensional SDM: fiber bundle model

The problem (4) has been investigated and solved exactly long time back (see my paper and reference therein) only for 1D fiber bundle models (FBM). As an example, consider the brittle FBM in Fig. 6 made of N parallel fibers (ideally representing actual fibers as well as springs, rods, bars, ropes, etc.) endowed with finite elasticity and connected to two transversal bus-bars loaded under tension. The disorder is typically quenched in the system by sampling the fracture displacements of each fiber from a given distribution $p_f(u)$, which is a characteristic attribute of the microstructure. If the fibers do not interact locally (e.g. limited cross-linking) and the end-bars are rigid, the rupture of a fiber (mimicking a microcrack) produces a "democratic" load redistribution over all extant fibers. When fibers have equal stiffness *k*, the force-displacement response of an instance bundle at the *n*-th rupture is given by

$$F = \sum_{i=1}^{N-n} f_i = \sum_{i=1}^{N-n} k u_i = K_0 \left(1 - \frac{n}{N_{TOT}} \right) u = K_0 \left(1 - D \right) u \quad (5)$$

which is the FBM counterpart of Eq.(3) in terms of force vs. displacement, with the Young's moduli being replaced orderly by the bundle stiffness K_0 and K in pristine and serviced conditions. The damage parameter is the mentioned order parameter, i.e. a random variable taking values from zero (in pristine state) up to 1 (at failure), and linking microscale disorder and macroscale structural degradation during the whole damage process. Because all fibers have equal applied displacement but fracture thresholds randomly sampled from $p_f(u)$, the expected value of D can be readily obtained at any damage state (sometimes analytically) as

$$D = \frac{n}{N_{TOT}} = \int_0^n \frac{dn}{N_{TOT}} = \int_0^u p_f(u) du = P(u) \quad (6)$$

Notably, the knowledge of D allows expressing the mean response of this class of FBM as

$$F = K_o \left(1 - D \right) \cdot u \quad (7)$$

As a numeric example of this model, consider the results in Fig. 6 (right) for the case of a uniform distribution p_f, where the damage curve (6) is a straight line and the force response (7) is a parabola.

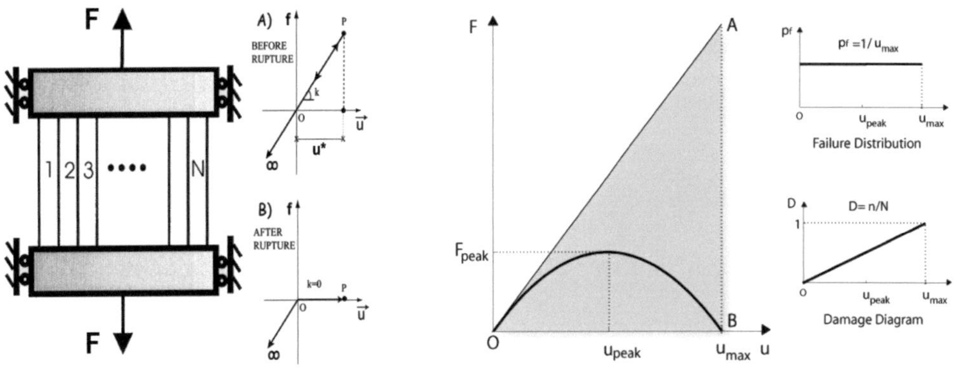

Fig. 6. FBM damage model and micro constitutive law of one bar (left), showing the tensile response prior and after rupture. Uniform strength distribution, corresponding damage curve, and average force response are shown on the right. The grayed area AOB represents the domain of all possible FBM responses.

3.2 Two dimensional SDM: lattice model

The above result is not only a rational mathematical model of intrinsic theoretical value, but has also several engineering applications (e.g. steel rope design, EN 12385-6:2004; EN 13414-3:2003; ISO 4101:1983). However, it only applies to 1-D structural systems that resemble a FBM and is of little usage for AE purposes. Most materials, despite their discrete nature, are multidimensional systems, with a high degree of interconnection between near-neighbour elements, e.g. polycrystalline or multiphase microstructures. Unfortunately, the damage process is much more complex in these systems and no rational theories have been formulated, with one notable exception being the 2-D lattice model in Fig. 7.

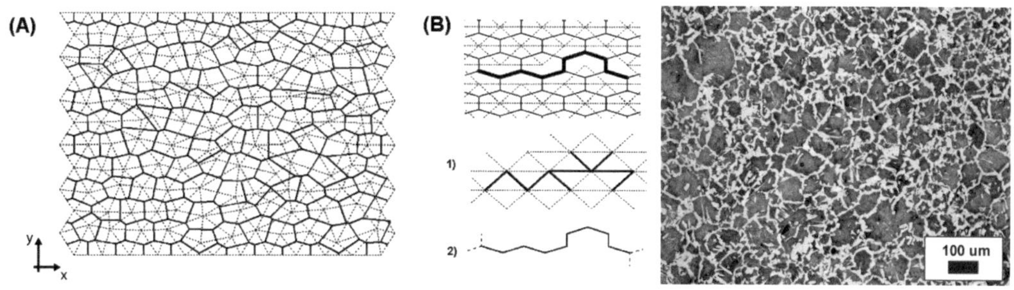

Fig. 7. (a) Sample lattice model obtained as the Delaunay network associated to a Voronoi froth approximating a polycrystalline microstructure. (b) Damage (microcracks) representation in Voronoi and Delaunay representations. An example of an actual network of ferrite (bright signal) framing pearlite grains (dark signal) in a C55 steel, as observed after metallographic attack (utmost right).

This mechanical lattice consists of a disordered spring network and provides a first order approximation of a polycrystalline microstructure (and an exact representation for actual as truss structure), where each spring represents a grain boundary (GB) normal to it in pristine condition. It has been investigated for decades to understand the physics of the damage mechanics underlying brittle failures (not just in brittle materials but in some ductile ones too) from inter-granular microcracking (Krajcinovic & Rinaldi 2005, Krajcinovic, 1996, and references therein). This model is the natural multidimensional extension of the FBM model from Fig. 6 but the damage process is different because of the local load redistribution effect and the geometrical disorder. In fact, when all springs have stiffness k and micro-strength sampled from a given $p_f(u)$ in strict similarity with the previous FBM, the rational model for the lattice subject to uniaxial load is demonstrably (Rinaldi & Lai , 2007 ; Rinaldi, 2009)

$$D(\varepsilon)=\frac{k}{E_0}\left(\frac{\ell}{L}\right)^2\left[\sum_{i=1}^{n(\varepsilon)}\left(1+\eta_p\right)\left(\frac{\varepsilon_i^*}{\varepsilon}\right)^2\right] \qquad (8)$$

Compared to Eq.(6), the damage parameter (8)depends on a number of extra parameters:

i. the ratio ℓ / L between the average grain size and the lattice overall dimension;
ii. the "strain energy" redistribution parameter η characteristic of the given microstructure and dependent on coordination number (i.e. the average number of grain boundaries of a grain), and orientation of the failed GBs with respect to the applied load;

iii. the kinematic parameter $\varepsilon^*/\varepsilon$ expressed by the ratio of the critical microstrain at spring failure (i.e. a microcrack forming at a grain boundary) over the corresponding macroscopic strain applied to the lattice (marked with a bar sign for clarity).

The fact that these variables are random may seem discouraging at first but they were demonstrated to actually exhibit a structure (Rinaldi, 2009), rendering the mathematical problem indeed tractable and allowing the formulation of approximate closed-form solutions of Eq.(8). The mathematical derivation and extensive discussion of each parameter is outside of the present scope and the interested reader is referred to the original scientific papers. Instead we shall focus on the aspects relevant to AE applications and to what is new in the SDM model, trying to keep math and technical jargon at a minimum.

4. Lattice model highlights and AE

The principal merit of the rationale model (8) is perhaps the disclosure of the "mathematical structure" of the brittle damage process, not just for the lattice problem that only served as a convenient setting for the proof. The problem of computing D in a higher dimensional system, i.e. most real materials, evidently requires the determination of several micro-variables, here η, $\varepsilon^*(\varepsilon)$, and $n(\varepsilon)$. Remarkably, this type of SDM models allows an unprecedented insight of the damage process at the microstructure level, which is one of the two main advocated limitations of AE in the introduction. To that end, some relevant results of the lattice model are illustrated in the remaining of this section. However, for the sake of argument, the concepts are discussed in the context of the "perfect" lattice example shown in Fig. 8, which consists of two classes of springs with orientation 0° or ±60° during a tensile test along 0°. The same figure (Fig.8(B)) reports the simulated tensile response σ *vs.* ε for an instance lattice, where the peak response at ε = 2.7 10^{-3} marks the damage localization, usually accompanied by a large microcracks avalanche (analogous to increased AE activity).

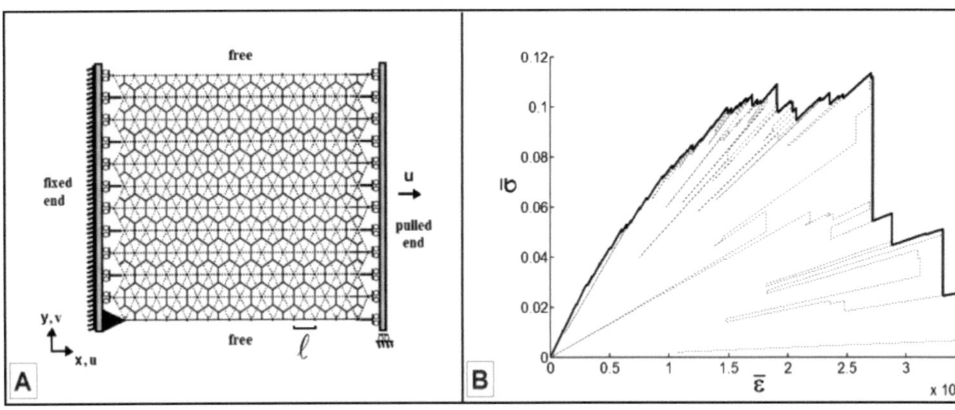

Fig. 8. (A) Perfect lattice with springs (GBs) orientated at 0° or ±60° during a tensile test along 0°. (B) Simulated lattice response from tensile test (stress values reflects an arbitrary numerical scale). Dotted lines relate to the formation of either isolated or avalanche of microcracks.

The first practical result is the clear demonstration of the non-linearity between the damage parameter D and the number of microcracks n. This is implicitly stated by Eq.(8) but is more readily verified by visual examination of the corresponding n and D data in Fig. 9 for the same tensile test in Fig.8(B). The marked difference of n *vs.* D is of consequence. Primarily, since n and D are not proportional, the damage parameter D cannot be deduced by a simple count of AE events as often attempted (i.e. n in Fig.3). Instead, such evaluation requires, as a prerequisite, that each AE event could be properly weighted to fit into a theoretical model similar to Eq.(8), after tailoring it for the material under consideration of course. We speculate that this might be somehow achieved practically by using the AE amplitude data to quantify the weights. Secondarily, Fig. 9 features a spectral decomposition of the n and D data into three components, each accounting for ruptures of springs with same orientation (recall that only 0° and ±60° are possible here). This breakdown of pooled data reveals that the horizontal springs in the perfect honeycomb lattice tend to break at a fastest pace and to contribute most to the damage parameter. Note in fact that, while diagonal ruptures happen (i.e. $n_{2,3} \neq 0$) since early in the damage process, they have a null effect in terms of damage (i.e. $D_{2,3} = 0$) and play a secondary role. After the transition at $\varepsilon = 2.7\ 10^{-3}$, the situation reverses and there is a crossover between n_1 that levels off and $n_{2,3}$ that rises, becoming dominant. This means that

- the importance of the springs (i.e. GBs in general) in the damage process heavily depends on their orientation relative to the load;
- the formation of (secondary) microcracks can be of minimal or negligible importance to D, such that these events can be classified as secondary;
- the relative importance of GBs with different orientation may change during the damage process, before and after damage localization.

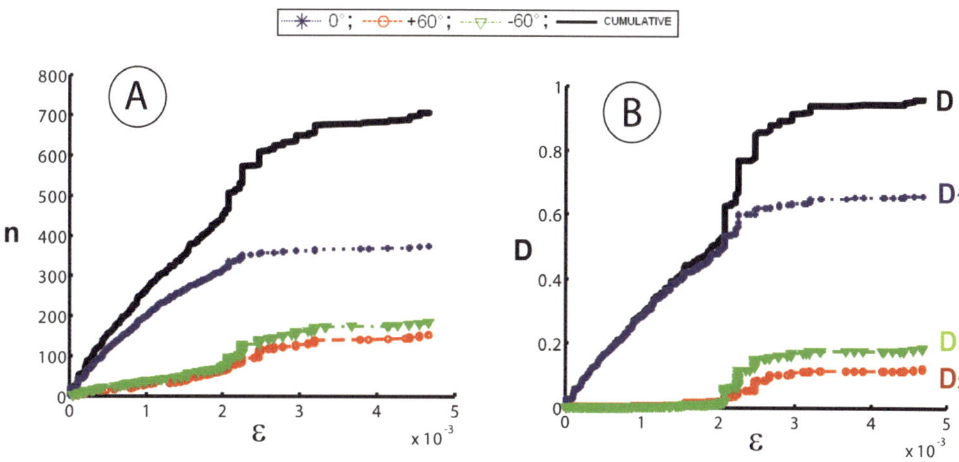

Fig. 9. (A) Cumulative microcracks n , as well as partition for GBs with orientation normal to 0° and ±60° for the tensile test in Fig.8B) (the cumulative curve is a typical AE output); (B) likewise, the damage parameter D and the spectral decomposition D_i. The comparison shows that only one type of GBs is relevant before damage (i.e. sound) localization.

These facts make immediately sense but are actually hard to quantify with classical modelling tools during cooperative phenomena, such as microcracks interaction at the onset of localization. This evaluation is also very hard experimentally and would require the advanced microscopy investigation (e.g. SEM, TEM, AFM, etc.) invoked in the introduction.

Next, consider the problem from another angle, by examining the simulation data shown in Fig. 10 about the critical strains series ε_p^* *vs.* ε of the *p*-th broken springs, presented both in aggregate form (A) and as partitioned into two groups (B), as per spectral decomposition. Monitoring ε_p^* during the simulation is a meaningful idea because it is a means of tracking the strain (and stress) fluctuations induced by damage in the lattice microstructure. Fig. 11(A) readily demonstrates that both the mean value and the scatter tend to increase progressively with ε (i.e. applied load) until the transition (load localization) is reached and a sudden burst occurs. This is fine and very interesting, also because this type of output, in the aggregate form, is very similar to the random signal from AE (ref. AE magnitude Fig.3(A)) - after all the energy released by a microcrack (spring here) is related to ε_p^{*2}. Yet, the aggregate form yields only a partial view of the microstructural phenomenon, as demonstrated by the spectral decomposition in Fig. 11(B). Then, it becomes very understandable that before the transition the rupture with higher ε_p^* (i.e. bearing more energy) corresponds almost exclusively to the horizontal springs, whereas afterwards large values of ε_p^* comes from springs of any orientation, which is consistent with the scenario drawn from Fig.9.

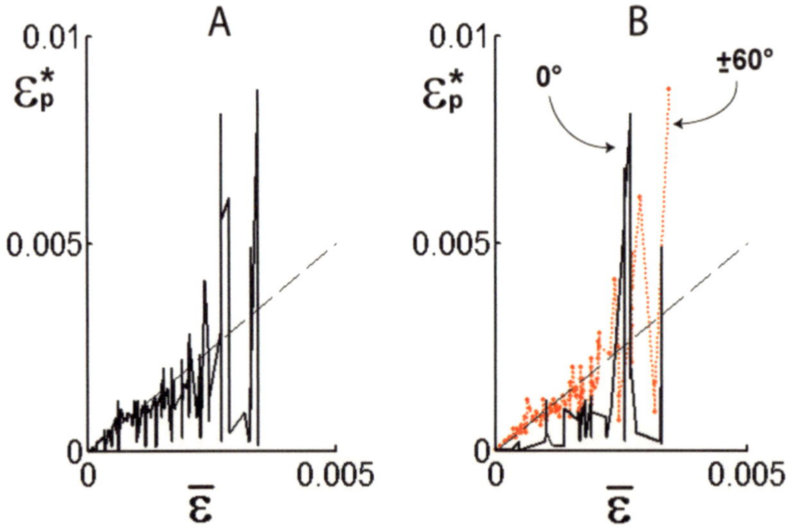

Fig. 10. Critical strains ε_p^* *vs.* ε of broken springs (i.e. GBs) subdivided in aggregate form (A) and partitioned into two groups (B), based on orientation relative to tensile axis. The peak response in Fig.8(B) has damage localization at $\varepsilon = 2.7\ 10^{-3}$, which happens with a large microcracks avalanche - a signature of the transition. As opposed to misaligned GBs, the GBs normal to the pulling action are more prone to damage before the localization because they carry most of the load and involve also stronger springs. After localization, damage formation involves GBs of any strength and orientation.

As far as the AE technique in polycrystalline materials, this result suggests that the whole AE signal may not be essential and that before sound localization (i.e. damage localization) it may possibly be filtered to extract the higher energy AE part that mostly governs the damage process, i.e. that part associated to GBs "favourably" oriented with the load and carrying large portions of strain energy, then released upon cracking. In other words, the present finding represents a potential basis to design a partition of AE data based on a microstructural interpretation of low and high energy events. At the same time, as far as failure prediction for field applications, the onset of damage localization could be detected by monitoring the spread in the AE amplitude signal, or in alternative by detecting rising trends in the low energy events, anticipating the cited crossover. By this viewpoint, modern discrete models theory seems like a viable route to device filters aimed at breaking the complexity of random AE signal and aiding in its interpretation.

As a last result of the section, we linger a little longer on the lattice problem to examine in greater detail the physical mechanism for the lattice transition in Figs. 9 and 10, a phenomenon observed phenomenologically in most brittle materials and failures. Based on our analysis, the damage localization at the onset of failure can be explained in terms of the stress amplification in the microstructure due to the local load redistribution induced by the previously accumulated microcracks. With reference to the perfect triangular lattice model in Fig.8, it can be shown that diagonal GBs would initially carry a near-zero stress until in pristine condition but, if one horizontal spring fails, this produces an overstraining influence that immediately raises the load level in the diagonals (inducing actually a strain-gradient). Fig. 11 shows graphically this effect in terms of percent strain perturbation on the ij-th extant spring between the i-th and j-th grains defined as

$$\%\ Strain\ Perturbation = \frac{\varepsilon_{(ij)} - \varepsilon_{(ij)}^{REF}}{\varepsilon_{(ij)}^{REF}} \times 100 \tag{9}$$

where $\varepsilon_{(ij)}^{REF}$ is the reference strain in pristine condition. The magnitude of the perturbation decays away from the damaged location but the maximum tensile perturbation induced on diagonal GBs is 10^3 % to 10^4 %, against the modest 20% of the horizontal springs. Such a remarkable magnification of the strain field is responsible for triggering the ruptures in the otherwise weakly loaded diagonal GBs. Eventually, as more microcracks form, the microcracking probability of unfavorably oriented GBs keeps increasing, to the point that the initial order in damage formation breaks down and a sudden transition ushers in a new mode, involving microcracking of GBs of any orientation. Of course this phenomenon is

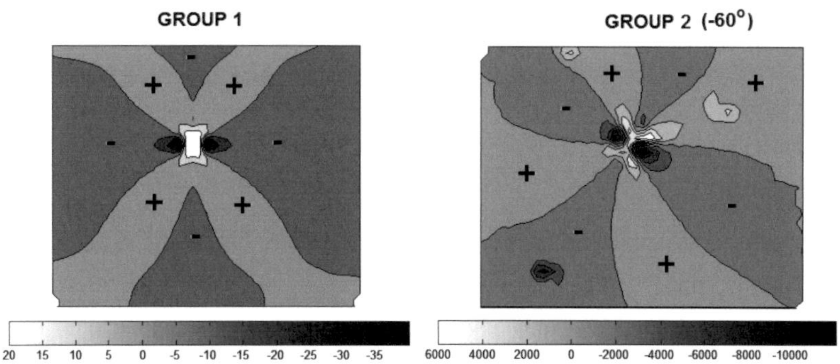

Fig. 11. Percent perturbation fields on horizontal (Group 1) and diagonal (Group 2) extant springs for a sample lattice with ~600 grains loaded as in Fig.8 and containing just one horizontal rupture. The magnitude of the perturbation on secondary spring is 1000-folds.

dependent on the loading direction, as the differential rupturing of GBs is tied to their orientation relative to the load. This is the root cause behind the damage-induced elastic anisotropy experienced by a damaged solid. The latter consists of the reduction of the elastic stiffness moduli only for the constants related to those GBs that participate to the damage process, leaving the elastic moduli in other directions only slightly affected. This is appreciated in Fig. 12, showing the different failure patterns for the same lattice from four uniaxial loading schemes, the ultimate evidence of the anisotropic damage evolution.

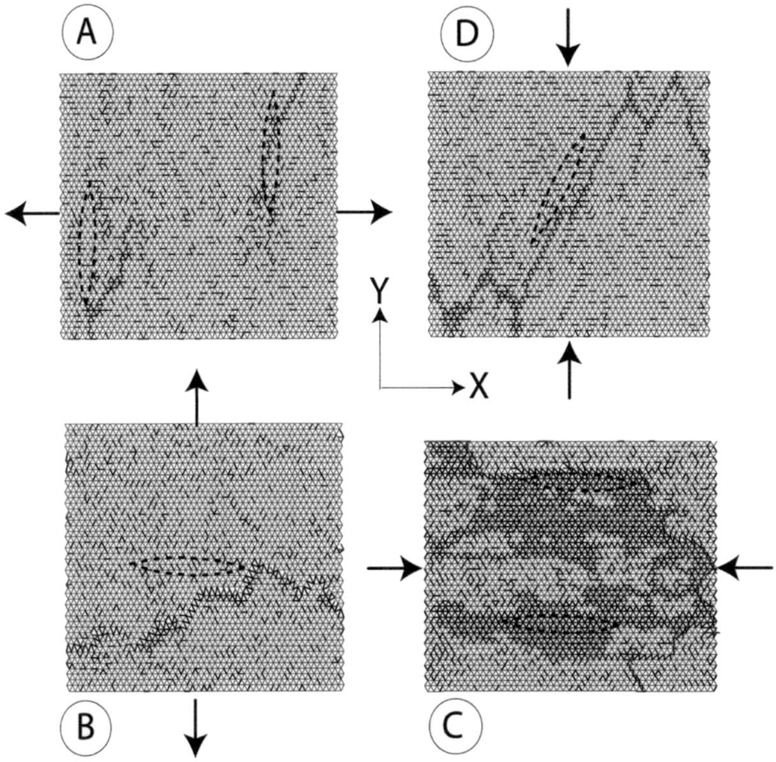

Fig. 12. Failure patterns for four load cases, revealing different failure modes. In agreement with experimental evidence on rock, concrete, and other brittle materials, tensile schemes are linked to cracks formation whether compressive loads produce shear banding and split (after Rinaldi, 2009).

5. Concluding remarks

Recent advances in discrete modelling were discussed in the context of AE monitoring. Starting from the limitations of AE stemming from the intrinsic randomness of AE data and from lack of knowledge/consideration of the microstructure, it was argued why SDM discrete modelling could become a companion tool for computer aided AE analysis. From the analysis of mechanical lattices we illustrated how SDM

1. can lead to an exact expression for the damage parameter, this proof-of-concept being a template to formulate physically-inspired damage models of D from parameter-based AE experimental data;

2. can capture the role of microstructural texture in the damage process and damage localization, demonstrating that knowledge of actual microstructure cross-correlate with AE signal, aiding its interpretation.

Thus, SDM is a powerful tool to look into structure-property relationships for damage and fracture. The featured analysis of the lattice model proved that the driving force in the fracture of heterogeneous matter resides in the stress amplification induced in the microstructure by the previously accumulated damage, following local load redistribution. This type of insight about the damage process could not be gained by classical continuum mechanics in such a straight forward manner. However, although the discussion supports the potential of the computational approach for damage assessment and AE structural monitoring, especially as far as the issues highlighted in the introduction, presently this remains a perspective, primarily because of the conceptual stage of the SDM theory for higher order structural system and calibration issues. Further research is on demand to validate these results on many real systems beyond lattice and customize them specifically for AE (field and lab) applications. On the other side there is a strong demand for modern computational tools for AE, which appear particularly welcome in consideration of the ever broadening range of AE applications that span from the determination of mechanical damage in metallic constructions (cracks, pits, and holes) to corrosion monitoring, from composites to concrete.

6. References

ASTM (1982) E610 - *Standard Definitions of Terms Relating to Acoustic Emission*. ASTM, 579-581

Berger, H. (Ed) (1977). *Nondestructive testing standards - a review*. Gaithersburg, ASTM, Philadelphia

Biancolini , M. E. ; Brutti, C. ; Paparo, G. & Zanini, A. (2006). Fatigue Cracks Nucelation on Steel, Acoustic Emissions and Fractal Analysis, *I. J. Fatigue*, 28, 1820-1825

Carpinteri, A. & Lacidogna, G. (Eds.) (2008). *Acoustic Emission and Critical Phenomena*, CRC Press, Boca Raton

DGZfP. MerkblattSE-3(1991) Richtlinie zur Charakterisierung des Schallemissionsprüfgerätes im Labor. Deutsche Gesellschaft für Zerstörungsfreie Prüfung. Recommendation SE-3

Grosse, C. U. & Ohtsu, M. (Eds) (2008). *Acoustic Emission Testing*. Springer-Verlag Berlin Heidelberg, ISBN 978-3-540-69895-1

Mogi, K. (1967). *Earthquakes and fracture*, Earthquakes Research Institute, Univ. Tokyo, Technophysics 5(1) .

Krajcinovic, D. (1996). *Damage mechanics*. North-Holland, Amsterdam, The Nederlands

Krajcinovic, D. & Rinaldi, A. (2005). Statistical Damage Mechanics - 1. Theory, *J.Appl.Mech.*,72, pp 76-85.

Palma, E.S. & Mansur, T.R. (2003). Damage Assessment in AISI/SAE 8620 Steel and in a Sintered Fe-P Alloy by Using Acoustic Emission *Journal of Materials Engineering and Performance* Volume 12(3), pp 254-260

Rinaldi, A. & Lai, Y-C. (2007). Damage Theory Of 2D Disordered Lattices: Energetics And Physical Foundations Of Damage Parameter. *Int. J. Plasticity*, 23, pp. 1796-1825

Rinaldi, A. (2009). A rational model for 2D Disordered Lattices Under Uniaxial Loading. *Int. J. Damage Mech.* Vol. 18, 3, pp 233-257

Rinaldi, A. (2011). Advances In Statistical Damage Mechanics: New Modelling Strategies, In: *Damage Mechanics and Micromechanics of Localized Fracture Phenomena in Inelastic Solids*, Voyiadjis G. (Ed.), CISM Course Series, Vol. 525, Springer, ISBN 978-3-7091-0426-2.

Rinaldi, A ; Ciuffa, F.; Alvino, A.; Lega, D.; Delle Site, C.; Pichini, E.; Mazzocchi, V. & Ricci, F. (2010). Creep damage in steels: a critical perspective: standards, management by detection and quasi-brittle damage modeling, In : *Advances in Materials Science Research. Vol.1*, ISBN 978-1-61728-109-9 (in print).

Sachse, W. & Kim, K.Y. (1987). Quantitative acoustic emission and failure mechanics of composite materials. *Ultrasonics* 25:195-203

Scruby, C.B. (1985). Quantitative acoustic emission techniques. *Nondestr. Test.* 8:141-210

VGB-tw 507 (1992) *Guideline for the Assessment of Microstructure and Damage Development of Creep Exposed Materials for Pipes and Boiler Components.* VGB, Essen

Part 6

Sound Localization in Animal Studies

25

Comparative Analysis of Spatial Hearing of Terrestrial, Semiaquatic and Aquatic Mammals

Elena Babushina and Mikhail Polyakov
Karadag Natural Reserve,
National Academy of Sciences of Ukraine
Ukraine

1. Introduction

The comparative analysis of own experimental researches of accuracy and mechanisms of orientation in acoustic space of the Black Sea bottlenose dolphins, north fur seals and dogs were accomplished depending on parameters and environment of sound distribution. From all of the probed representatives of marine mammals dolphins differ the most exact indexes of sound source localization (1.5-2°). Fur seals localization possibilities in water are substantially less to such the dolphins in 1.6-1.8 time in a horizontal plane and in 5-9 times, sometimes more in a vertical plane. The accuracy of localization of sound source by fur seals in a horizontal plane in air (3.5-5.5°) a few exceeds such in water (6.7-7.5°) and appeared substantially better, than for dogs (7-11°), but in 1.5-2 times worse, than in water in a vertical plane . The mechanisms of acoustic orientation depend on the type of animal, his ecology, parameters, conducting path, sound path environment, features of sound path structures. For all speeches the direction of acoustic signal arrival encoding is carried out by means of space-frequency filtration and interaural differences.

Peculiarities, accuracy and mechanisms of acoustic orientation of high level progress animals are investigated during many years by different authors using various methods. We have been already working on this theme over 30 years, in particular carrying out experimental researches (using behavioral response techniques operant conditioning with food reinforcement) of main characteristics of hearing, including space hearing, of aquatic and semiaquatic mammals - bottlenose dolphins and pinnipeds - big-eared and real seals (northern sea fur-seal and Caspian seals). The results of our researches and survey of summary of other authors' works are cited in our articles (Babushina, 1979, 1997, 1998, 1999, 2000, 2001 a, b, c; Babushina et al., 1991; Babushina & Polyakov, 2001, 2003, 2004; Babushina & Yurkevich, 1994 a, b; and others). All investigated mammals' representatives showed excellent abilities to take their bearings in space by means of hearing: to discover successfully acoustic signals, with high enough accuracy (but for every species - with its own one) to determine the place of the sound source, to define operating factors of signals, delicate structure of composite sounds, to use all functional possibilities of acoustic analyzer for solution of complicated experimental problems. Perhaps, at first it was a success for us to carry out multi-aspect, complex researches of peculiarities and mechanisms of mammals' acoustic orientation with different adaptive modifications of peripheral structures of hearing organ. It was determined, that main physical principles of sound source localization, using by a man, are just applied for other mammals according to anatomic, morpho-functional peculiarities of a concrete specie, its ecology.

In this work the range of investigated mammals is extended - the data of localization accuracy of the acoustic signals source by three dogs in the horizontal plane which we obtained are cited at first. The results are discussed in comparison with carried out analogical researches of dolphins and pinnipeds.

Localization of tonal signals source by dogs. The information about functional characteristics of acoustic analyzer of family doggy representatives are not numerous (Gorlinskiy & Babushina, 1985; Kalmykova, 1977; Goldberg & Brown, 1969; Issley & Gysel, 1975; Peterson et al., 1966, 1969). By the average data (on seven mammals) (Peterson et al., 1969) the range of dog's hearing is stretched from sound frequencies to 60 kHz with area of high sensitivity from 0.2 to 15 kHz. The most microphone potential was registered in uniform in magnitude response of area from 0.25 to 7 kHz.

At I. V. Kalmykova's work (Kalmykova, 1977) on dogs using the method of defensive conditioned reflexes lateralization of sound image was investigated in dichotic presentation of a series of clicks for the two signal levels - 60 and 20 dB above standard sound pressure level. Interaural minimum discernible differences in the intensity and time were found to be 2.2 dB and 75 ms, i.e. much higher than similar values for humans.

Investigation of localization abilities of dogs (mongrel, with erect ear shells) was carried out by the method of instrumental conditional reflexes with food reinforcement.

The dogs have elaborated a conditioned reflex to hold the original position, touching the tip of its nose, one of three (central) manipulator - rubber ball suspended at some distance from a line parallel to the plane of the emitters.

The distance from the middle base between acoustic meatuses to the plane of emitters location was at a frequency of 4 kHz, 1 m, at higher frequencies - 0.5 m. Head position at which both ears were in a plane parallel to the arrangement of the emitters at the same distance from central manipulator, was taken conformity with relevant zero azimuth. Two emitters were mounted at the height of acoustic meatuses of dogs at the same distance from the 0°-azimuth plane. During the experiment, the angle of signal arrival relative to the zero azimuth direction could vary from 45 to 3°. In the experiments, the signal was fed by alternately one of emitters in a random order. Animals were trained to touch with a paw the left or the right manipulator according to the direction of sound arrival. Each adequate reaction of the animal was accompanied by food reinforcement.

To study the limits of dogs localization abilities the azimuth of emitters decreased in increments of 10° from 45 to 15° and increments $5\text{-}1^{\circ}$ of 15° or less. An indicator of the dogs localization limit ability was the minimum detectable angle (MDA), equal to the azimuth of the emitter, corresponding to 75% level of positive reactions.

The limit values for azimuth localization by dogs of tone source parcels were measured for frequencies 4, 10 and 30 kHz. The choice of frequencies 4 and 30 kHz was due to their correspondence to tonal stimuli, which were presented in experiments with dolphins in the air environment (Babushina, 1979). The frequency 10 kHz was chosen as intermediate between two above signals. Rise time of the amplitude of tone parcels frequency 4 and 10 kHz was 20 ms, decay time - 25 ms. For signal with frequency 30 kHz corresponding values were within 20 mks. The duration of signals was 1 s.

The sound pressure level in the initial position of the animal reached 75, 88 and 65 dB (relative to 0.0002 dyn/cm^2) at frequencies of 4, 10 and 30 kHz, accordingly.

The work was done using standard radio measuring equipment. The experiments were performed on three dogs.

4 kHz			10 kHz			30 kHz		
α°	n	P(%)	α°	n	P(%)	α°	n	P(%)
3	107	61 ± 5.6	3	20	45 ± 1.3	6	36	53 ± 6.2
5	42	62 ± 8	6	76	64 ± 3.9	7	31	61 ± 14
6	188	68 ± 3.9	8	107	70 ± 8.9	8	53	77 ± 12
7	107	75 ± 8.2	10	136	71 ± 2	9	95	67 ± 2
8	101	63 ± 7	12	52	77 ± 11	10	63	84 ± 0.1
9	147	67 ± 3.1	15	149	82 ± 2.7	15	38	97 ± 2.5
10	474	73 ± 4	16	30	83 ± 6.2	25	88	83 ± 8
11	21	67 ± 8.7	20	37	84 ± 12	35	20	100
12	31	90 ± 9.7	25	10	100	45	20	100
15	258	79 ± 3.4	36	30	100			
20	54	81 ± 10.6	45	30	100			
25	67	97 ± 2.8						
35	38	100						
45	244	100						

Table 1. Localization of tonal signals source by dogs in the horizontal plane (averaged data). Symbols: α°– the azimuth of emitters; n – the number of tests; P (%) – - the percentage of positive reactions.

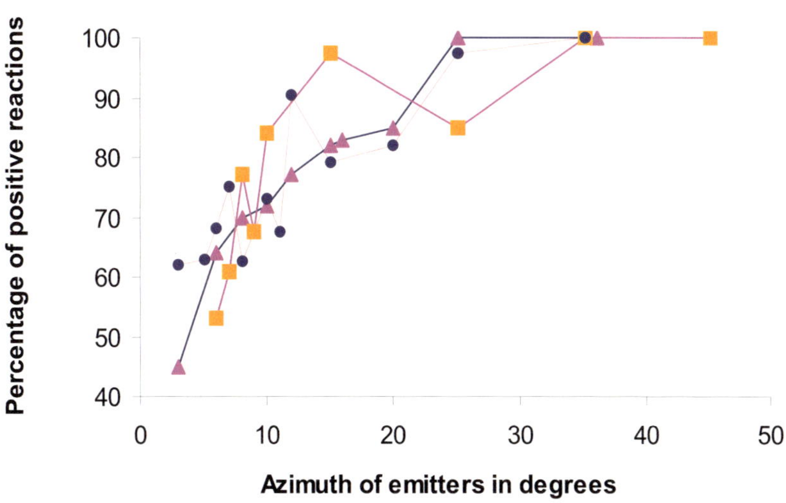

Pic. 1. The dependence of the percentage of positive reactions (P %) of dogs on the azimuth (α°) of tonal signals sources (averaged data).
Symbols: circles – 4 kHz, triangles – 10 kHz, squares – 30 kHz

Averaged results of experiments are presented in the table 1 and on figure 1. For signals with frequency of 4 and 10 kHz the data were averaged for three dogs, for the signal with

frequency of 30 kHz - in two animals, as one dog in further experiments did not participate. The data show that with decreasing of sources azimuth the share of positive reactions of animals decreases. Sustainable dependence of the percentage of positive reactions from the frequency of the signal at each given value of the azimuth emitters was not found. In pic. 1 one can see that the curves are close to one another. For frequencies 4 and 30 kHz there is a double crossing of the curves with the threshold level. Defined graphically the minimum detectable azimuth of emitters by animals was within the limits of 7-11° for the signal with frequency 4 kHz, 11° for the stimulus 10 kHz, 8-9.5° 0 for the signal with frequency 30 kHz, i.e. localization indices for the investigated monofrequency signals were similar.
Thus, the maximum perceived by dogs change of tone source azimuth in the frequency range 4-30 kHz (at the level of 75% of positive reactions) is not less than 7°.
Apparently, in the investigated frequency range the dogs oriented mainly on binaural differences in intensity of the stimulus. Similar values of limiting angles of localization obtained for different frequencies in our experiments with dogs suggest equal efficiency of binaural differences in intensity in all investigated frequencies (4-30 kHz).
Measuring of periferal hearing orientation in dogs (Gorlinskiy & Babushina, 1985) showed that with increase of frequency and the angle of the sound arrival the tendency to the growth of interaul differences is watched in the intensity of sound (Δ I), that increase efficiency of using Δ I in mechanism of source signal localization. The focus of auditory reception in dogs is provided at frequencies 0.5 and 1 kHz by acoustic properties of the animal's head, and over 1 kHz - spatial-frequency selective of external ear.
These experiments have allowed to understand the mechanisms which ensure a successful sound orientation, and revealed a crucial role towards the properties of ears in space hearing of terrestrial animals. The received material in conjunction with the analysis of other authors data suggests that the peripheral structures of the dogs auditory analyzer, like all mammals, not only terrestrial but also aquatic, decode acoustic space on the principle of directional frequency filtration. Of particular significance for the detection and localization of a sound source by dogs has the mobility of ears. In mid and high frequency ranges of sounds the turn of auricle influenced on the position of the maxima and the shape of the directivity patterns of reception. After the motor component of the orienting reaction the animal's head is turned to the sound source. Observed with the movement of auricles in a frontal position transfers maxima diagrams admission closer to the midline of a head. Steepening of the diagrams in this area along with some narrowing of focus, as well as increasing near the midline dog's head of a strictly monotonic function Δ I from the angle of sound arrival provide some optimization of processing acoustic information.
The values of the minimum perceptible by our experimental dogs azimuth of monofrequency signals source (7-11° in the researched frequencies range) are in good agreement with the results of experiments with dichotic presentation of the sound stimulus (Kalmykova, 1977). The 75% level of positive reactions in these experiments corresponded to binaural time differences equal to 75 microseconds, and the binaural difference in the intensity of 2.2 dB. As you can see, these values are significantly higher than the minimum values of Δ T (10 ms) and Δ I (0.5 dB) for a human obtained at a frequency 0.75 kHz (Casseday & Neff, 1973). At the same time binaural ΔT andΔ I for dogs compared with the corresponding values for the monkeys (60-180 ms and 6-10 dB, with 85% level of positive reactions) (Don & Starr, 1972) and slightly higher than the data for the cat (20 - 50 ms) (Masterton & Diamond, 1964; Masterton et al., 1967, 1968). In all experimental dogs interaural distance was in the range 9.5-11 cm. In such basis the binaural differences in

arrival time or phase of the signal become effective at a frequency of less than 1.7 kHz (the sound wavelength of more than double interaural distance), and only at a frequency 3.3 kHz the wavelength is comparable to the base (an average of 10.2 cm for the experimental dogs). For a human such a transition zone corresponds to the frequency 1.7 kHz, for the cat - 4 kHz.

Consequently, at a frequency of 4 kHz and above dogs were able to focus on the binaural difference in stimulus intensity, which probably took place. Somewhat smaller accuracy of localization of tonal sounds source by cats in the range of 2-8 kHz (Casseday & Neff, 1973) due, apparently, the size of the head, and consequently, less interaural distance, compared with dogs. High resolution of human auditory analyzer (1.5°) (Mills, 1958) at a frequency of 4 kHz to some extent also due to the size of the base.

High accuracy of definition of the ultrasound source direction (less than 1°) was found at bats (Gorlinsky, 1975, 1976). Based on the analysis of directional diagrams of receiving of ears of sharp-eared bats and the Greater Horseshoe Bat, as well as the results of localization experiments, the author concludes that neither the time nor phase binaural mechanisms can cannot explain such high localization ability of the animals. Only the assumption that the threshold of perception of interaural differences n the intensity in bats, like other mammals, does not exceed 1 dB, could satisfactorily explain the obtained data.

5 kHz			**20 kHz**			**120 kHz**		
α°	**n**	**P(%)**	**α°**	**n**	**P(%)**	**α°**	**n**	**P(%)**
2	241	68 ± 5.9	2	278	73 ± 6.1	1	8	37 ± 12
3	204	66 ± 6.1	3	200	73 ± 4.9	1.5	70	64 ± 6
4	66	70 ± 1.2	4	312	72 ± 5	2	462	80 ± 2.5
5	205	83 ± 1.2	5	206	89 ± 6.1	3	247	81 ± 4.7
45	30	100	10	20	100	4	76	80 ± 5
			45	30	100	5	35	85 ± 1.2
						6	400	83 ± 3.7
						15	20	100
						25	60	100
						35	60	100
						45	176	100

Table 2. Localization of tonal signals source by dolphins in the horizontal plane (averaged data). Symbols: α° – the angles between underwater sound transmitters; n – the number of tests; P (%) – - the percentage of positive reactions.

The dependence of correct responses percentage of two bottlenose dolphins on the angle between underwater sound transmitters for tonal signals is shown in the figure 2 and in the table 2 (Data on Babushina, 1979) for the comparison with the same values for the dogs (fig. 1 and table 1).

It was found by the experiment that animals which live in the water and have a rich set of adaptations of the specialized auditory analyzer (respectively ecology of concrete species), are able to orient successfully in an acoustic space, determine the direction on the sound source. To be efficient under water, the organ of hearing must be sensitive and capable of binaural analysis. In addition, the resonance frequency of the mechanical vibration system of the middle ear must be shifted under water (relative to that in air) to allow ultrasound

reception. The hearing system of aquatic and semiaquatic mammals possesses all these properties (Lipatov, 1978).
The good indexes of directional hearing in the water and in the air (in this environment - even better than dogs have) are found in different species of pinnipeds (Babushina, 1998; Babushina & Polyakov, 2004; Babushina & Yurkevich, 1994 a; Gentry, 1967; Moore, 1975; Möhl, 1964; Terhune, 1974). Let us draw attention on our own studies of northern fur seal space hearing.

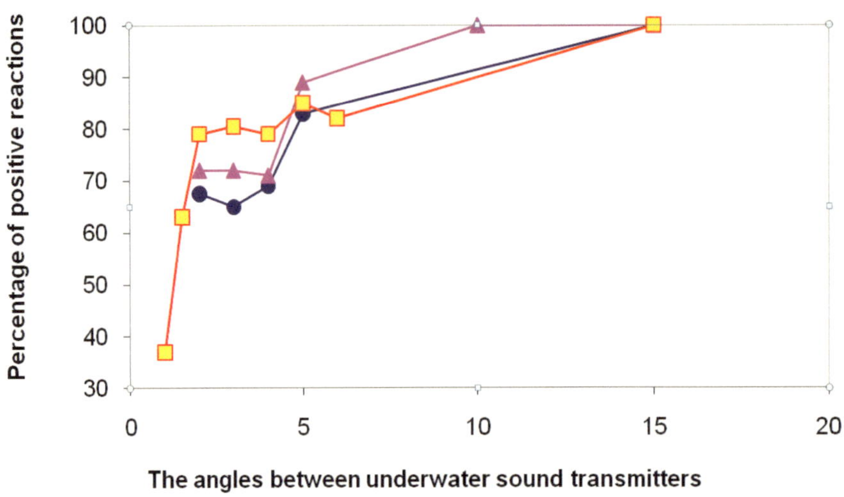

Pic. 2. The dependence of the percentage of positive reactions (P %) of two dolphins on the angles between underwater sound transmitters for tonal signals sources(averaged data). (Data on Babushina, 1979).
Symbols: circles - 5 kHz, triangles - 20 kHz, squares - 120 kHz.

Using the method of instrumental conditioned reflexes technique with food reinforcement, we investigated the accuracy of localization by northern fur seals of different sources of acoustic signals in the horizontal plane in the water and air environments (Fig. 2,3) (Babushina & Polyakov, 2004). Threshold limit values of angles were estimated (in all our experiments) on the level of 75% of positive reactions. In the frequency range 0.5-25 kHz the accuracy of localization by fur seals tone source pulses (duration 3-90 ms) is in the water 6.5-7.5°, in the air (for duration of the pulse 3-160 ms) - 3, 5-5,5° (better than dogs have). The source of noise pulses (bandwidth 1-20 kHz, duration 3 ms) is localized by fur seals in the water with accuracy 3°, continuous (duration 1 s), narrow-band (10% of the central frequency) and broadband (bandwidth 1-20 kHz) noises in the air - with an accuracy of 2-5° 0 and 4.5°, respectively. The data obtained allow to conclude that the signs used by fur seals in the localization of tonal pulses are likely to be equally effective for different frequencies (at least in the investigated frequency range). The source of broadband noise pulses, carrying a few signs of binaural localization, bears by seals at greater accuracy than a source of tonal pulses. Contrary to expectations, the significant increase of accuracy of localization in the air (as compared with the results for water) - about five times, according to the theory of binaural hearing, due to the decrease of sound velocity and, consequently, the increase of binaural time differences - not observed. Perhaps this is due to the change of the system resonances and transmission characteristics of the seal's external ear, slightly open in the air. In addition, the ears of the seal rolled into a tube and are oriented front to back, which also is not conducive to the directional auditory reception in the air.

A significant contribution to the study of space hearing is the study of localization capabilities of the animal in the vertical plane. According to our data (Babushina & Yurkevich, 1994 a), the accuracy of determine by fur seals the direction of arrival sound in the vertical plane in the water depends on the parameters of acoustic signals and amounts (peak angle, i.e. the angle between the upper and lower emitters at zero azimuth): 7-8° - for clicks (representing the reaction of the emitter to rectangular pulses of 0.5 ms and 1 ms), broadband noises (bandwidth 0.5-20 kHz), narrowband (10% of the central frequency) noise pulses with center frequencies 2-4 kHz, 12-20°- for continuous narrowband noises and noise pulses with central frequencies 5-20 kHz; 18 -20° - for tonal pulses with smooth fronts of amplitude variation.

The results showed that the accuracy of localization by fur seals of the source of acoustic signals in the vertical plane in the aquatic environment depends on their parameters such as in humans rises for the sounds with a complex spectra and is probably substantially reduced due to the presence of reverberation noise, especially for tonal pulses of long duration and high pulse repetition rates. Considerable difficulties which fur seals have (like humans) at localization in the vertical plane of monofrequency sounds source, due, apparently, and the absence of signals have to be, at least three frequency components (as shown in studies in humans and some terrestrial animals) with a certain ratio of the amplitudes. The deterioration of seal's localization abilities vertically with increasing center frequency of narrowband noise pulses is difficult to explain - in pinnipeds underwater sound reception provided the full range of conductive structures, the specific role of each of which encode the vertical coordinates of the source is still unclear.

The accuracy of localization by fur seals of the source sound vertically in the air at nonzero values of the emitters azimuth (27° -35°) is (peak angle): 14.5^{o} and 21°, respectively, for broadband and narrowband (with a center frequency of 5 kHz) noise pulses (Babushina, 1998). The source of narrowband noise pulses with center frequencies 2, 4, 10 kHz is localized by seal at the level of random selection (with the angles between the emitters - 22-30°).

It turned out that in the air the direction to the sound source in the medial (with zero values of the emitters azimuth) vertical plane, and at azimuth 90° the seal cannot define. The reason probably lies in the simple structure of the auricle seal which is devoid of typical for a human of many folds and ledges, which create for different angles the elevations of the sound source the complex combination of the diffraction pattern, interference, scattering, rounding, the resonances which significantly improve the accuracy of localization by a human the sound source in the vertical plane. In addition, the tubular shape of the seal auricle and its specific orientation (front to back) is also not conducive to the orientations of the auditory reception.

Perhaps some of these factors explain the inability of fur seals to locate the source of even complex sounds in the medial vertical plane in the air. At zero emitters azimuth variation with the change of frequency and angle of elevation caused by tissues of the head are minimal but increases with nonzero values of the azimuth (Searle et. al., 1975), which probably contributes to the determination of northern fur seal in the direction of the sound source in the vertical plane at non-zero azimuths emitters. However, it comes with a

noticeably less success than in humans (Altman, 1983, Altman et al, 1990). Perhaps the seal as a human also uses the additional binaural cues localization vertically through the light asymmetry of the ears.
These data indicate that the seal's ability to determine the direction to sound source in a vertical plane in the air depends on the parameters of acoustic signals, as well as in humans, rises for the sounds of complex spectra (containing much information about the coordinates of the sound source) and 1.5-2 times worse than in the water (which can be partly attributed to the different seal conductive channels in the water and in the air).
Localization opportunities and mechanisms of the dolphins directional hearing have been studied by many authors (Akopian et al., 1977; Bel'kovich & Dubrovsky, 1976; Bel'kovich & Solntseva, 1970; Voronov, 1978; Dyachenko et al., 1971; Zaitseva, 1978; Zaitseva et al., 1975; Ivanenko & Chilingiris, 1973, 1978, Korolev et al., 1973; Andersen, 1970; Dudok van Heel, 1959; Renaud & Popper, 1975 and others). Dolphins have several channels of sound conduction and this availability makes it difficult to study the mechanisms of space hearing in these mammals. Detailed review of works that have examined the features of sound conduction in marine mammals, is given in the article (Babushina, 2001). Auditory channal and the lower jaw which dolphin has in the aggregate with their surrounding tissues to a large extent form the direction of auditory reception (Purves & Utrecht, 1964). It was proved that in the formation of directional reception by dolphins of high frequency signals can take part different entities of soft and bone tissues, such as hypodermis of the lower jaw (Ravens, 1978; Stosman & Voronov, 1978; Stosman et al., 1978). Scanning movements of the head contribute to more precise analysis of the differences in the intensity and spectral pattern on the two receivers (Bel'kovich & Solntseva, 1970). Complex sounds, we can say "fall apart" by the conductive channels, interact with them, changing, creating a specific spectral pattern in the auditory centers, depending on the coordinates sound source.
Let us dwell on our own studies of space hearing of the Black Sea bottlenose dolphin

Tursiops truncatus p.

According to our data (Babushina, 1979), the limit angles of the localization by two bottlenose dolphins the source of acoustic signals in the horizontal plane in the water are as follows: for tonal pulses in duration of 1 s, frequency of 5, 20 and 120 kHz, respectively, 4,5, 4° and less than 2°, for pulses whose parameters vary within the limits of variability of echolocation
signals - $1.5\text{-}2^{\circ}$.
Investigation of limiting localization capabilities of bottlenose dolphins in the vertical plane showed high resolution of the auditory analyzer for both the tone and for pulsed signals (Babushina & Polyakov, 2008). The minimum detectable angle for the tone frequency of 5 and 20 kHz was within 2.5°. The magnitude of the limit angle for the stimulus frequency of 120 kHz was 2°, i.e. coincided with that for the horizontal plane. The maximum angle of localization of pulse click sequences (with a maximum energy at a frequency of 120 kHz, the duration of the pulses 20 ms, repetition rate 300 Hz) in the vertical plane was little more than 1.5°.
Changing of the intensity of the received stimuli spectrum with the change of place angle, perhaps, reports to the dolphin the primary key for sound localization in the vertical plane. Characteristics of conditioned reflex reactions generated by the dolphin do not preclude the possible movements of the head at the time of presentation of the signal. So, perhaps, the dolphin used the binaural-added information (both in the time and intensity) to determine

of sound source position. Dolphin, carrying out scanning reception by turning the head, can change the characteristics of its receiving filters, matching them with the test signals, and thus fulfill the optimal space-frequency filtering (Ayrapet'yants and others, 1973). Comparable values of limit localization angles of monofrequency source of acoustic signals by our dolphin in the vertical plane give rise to assume the existence of different, in a similar degree the effective cues for localization at different frequencies. This is consistent with data obtained by us in the localization of the various sounds by dolphins in the horizontal plane (Babushina, 1979), and with the results and hypotheses of localization mechanisms of other authors (Terhune, 1974; Moore, 1975; Renaud & Popper, 1975).

The results of our research of localization capabilities of the Black Sea bottlenose dolphins (*Tursiops truncatus p.*) in the horizontal and vertical planes are in very good agreement with experimental data Renaud D. and A. Popper (Renaud & Popper, 1975) obtained for Atlantic bottlenose dolphin (*Tursiops truncatus*). This is all the more interesting that, unlike our experiments, in the above-mentioned authors' work the animal's head was fixed. In addition, the localization in the vertical plane was investigated at the location of a dolphin on his side. As in our experiments, in the work (Renaud & Popper, 1975) it was investigated the accuracy of localization over a wide frequency range (6-100 kHz) for tonal signals and sequences of clicks with an energy spectrum similar to that of echolocation pulses of dolphins. There was no significant difference in the accuracy of localization in the horizontal and vertical planes. Thus, at frequencies of 30, 60 and 90 kHz limit vertical localization angles were, respectively, 2.5°, 3° and 3°. In this work the limit localization angles of the clicks sequences (with the parameters, similar to those of sonar signals) were 0.7 and 0.9°, respectively, in the vertical and horizontal planes. Based on the data obtained and the results of other researchers (Bullock et. al., 1968; McCormick et. al., 1970; Norris & Harvey, 1974) the authors suggested that at low frequencies the sound localization is carried out by the meatus, at frequencies around 20 kHz and above - through the lower jaw and, at frequencies above 20 kHz, the animals are guided by binaural differences in signals intensity. At the vertical localization the dolphin's low jaw was focused on one emitter, and the top - on the other one. In the experiments in the vertical plane (Renaud & Popper, 1975) the dolphin could not use binaural information as his head at the time of the signals supply was fixed. T. Bullock with co-authors (Bullock et.al, 1968) showed that the sounds coming through the jaw, cause in the lower colliculus midbrain responses greater magnitude than the sounds that pass through the dorsal part of the rostrum. These differences could be used by dolphin, believed to J. Renaud and A. Popper. Free animal (with movable head, as in our experiments) can, moreover, determine the vertical coordinates of the sound source by remembering on the short time parameters of the signals and comparing them with the characteristics of sounds in the other head position (Renaud & Popper, 1975).

Dolphins' localization abilities what we investigated exceed similar capabilities of semiaquatic animals - pinnipeds, especially in the vertical plane (when compared to the optimum for each frequency bands) (Babushina, 1998; Babushina & Polyakov, 2004; Babushina & Yurkevich, 1994 a). So, the best indicator of the "horizontal" localization by fur seals of broadband noise pulses (3°) is in 1.6-1.8 times less than localization abilities of a dolphin, measured in the horizontal plane at high frequencies (50-120 kHz), short pulses - 1-1.9°. In the vertical plane, similar differences increase substantially - the accuracy of localization by fur seals of tone source and a variety of complex sounds vertically in the water (Babushina & Yurkevich, 1994 a) in 5-9, sometimes more than once is inferior of localization dolphin's opportunities. Obviously, successful vertical localization requires a certain set of high-frequency components of the spectrum. Naturally, with such a task only complex dolphin auditory analyzer could easily handle, to a large extent formed by the evolution of echolocation function. One reason for the above differences in terms of space hearing, undoubtedly due to anatomical and functional differences in the dolphins' conductive structure and pinnipeds. In details it is outlined about conductive structures of pinnipeds in the work (Babushina & Yurkevich, 1994 a), showing all the major studies on this topic. Clearly, a large role in the auditory orientation belongs to the functional characteristics of the central sections of hearing organ of marine mammals.

For further study of the mechanisms of dolphins' directed auditory reception, as well as for comparison with terrestrial mammals from which they descended, we measured the accuracy of localization of acoustic signals by dolphins in the air (Babushina, 1979). According to our data (Babushina, 1986), the range of perception by dolphin of acoustic signals in the air ranges from 1 to 110 kHz with the greatest sensitivity to low frequencies (1-40 kHz). The lowest auditory thresholds ware recorded on frequency of 40 kHz (-44 dB relative. 1mkb). Dolphin worse hears the air sounds at 10-13 dB, when its alveary immersed in the water, compared with thresholds in the case when the whole head is in the air. The comparison of aerial and underwater audiograms of a bottlenose dolphin in the coordinates "intensity-frequency" has shown that the sensitivity of the dolphin's ear to the sounds in the air worsens by 30-60 dB (depending on frequency). For comparison (Babushina, 1997; Babushina et al., 1991): hearing sensitivity of pinnipeds to the underwater sounds at 15-20 dB exceeds the sensitivity in the air and only in 7-15 dB is inferior to that of dolphins in comparison at the best frequencies (for each species) of auditory perception. Northern fur seal hears in the water as good as the humans in the air; the sensitivity of hearing of seals to underwater sounds only in 7-10 dB below than the sensitivity of human hearing in the air. Comparing the curves of our dolphin hearing in the air and a human underwater in the frequency range of 0.125-8 kHz (Hollien & Brandt, 1969), we can say that the dolphin in the air at low frequencies hears much better than people in the water. From 0.125 to 2 kHz thresholds of human hearing in the water are equally high (about 70 dB relative to 0.0002 mcB) and up to 8 kHz is further increas by 12.5 dB. The difference between thresholds in two environments for a human at frequencies 0.25, 1 and 2 kHz is about 29 and 51 and 59 dB (relative to 0.0002 mcB), respectively.

As shown by studies of many authors, a human hears under water, mostly through bone conduction. In this paper, using the contact stimulation by tonal signals at the frequencies of 1 and 30 kHz it was showed that the thresholds corresponding to the bone structures and soft tissues of the human head differ only slightly (Soluha, 1973). On this basis, it was hypothesized that in aquatic environment sound conduction is realized by tissue structures - a distributed receiver about the size of 0.2 m. Researchers related the ability of the human organ of hearing to detect the direction of underwater sound signals and to locate their source mainly to the sound-conducting properties of the tympanic structures and to a lesser extent to bone conduction (Hollien, 1973 and others). However, not all phenomena could be explained. There were studies that reported the involvement of human skin in locating the source of underwater sound. For example, in Hollien's experiments (Hollien, 1973), human skin was found to possess sound-conducting properties: the subjects could sense underwater sound signals with foot, hand, or face skin.

Through mathematical calculations it was showed that human's hearing thresholds under water, at least, on sound frequency are defined as in the air by the passage of acoustic vibrations through external auditory channel (Lipatov, 1978).

There is evidence that people under water localize the sound source not worse than a dolphin in the air (as shown below), and almost in the same extent as semiaquatic mammals in the water. The literature cites a number of experimental data about fairly successful, especially at low frequencies, the localization by a human under the water of low frequency sound sources and broadband signals - 7-11° after exercise (Feinstein, 1973; Hollien, 1973). The authors suggest that binaural time differences are most informative for human and in underwater sound localization at low frequencies. Localization of the sound source under water can be provided by a number of mechanisms, as usual for the man - the air, and additional, caused by aquatic environment.

It turned out that in the air in a horizontal plane the source as of tone (4.5 and 28 kHz) and as pulse (with a carrier frequency of 20 kHz) signals is localized by a dolphin with the same accuracy consistent with the limit angle between the emitters of the order 20-21°.

By investigating the limits of localization capabilities of dolphins in the air (Babushina, 1979) the frequency of tonal stimuli was chosen on the basis of equality of the wavelength of tones in the air and water. Consequently, the sizes of the base (distance between the host signal structures) for each wavelength were identical in two environments. For pulsed stimulus temporal binaural differences are also the only factor, mixed in the water, and air. Thus, the localization conditions differ only in the speed of sound. Since the air has an advantage for the localization on the time parameter, it could be expected increase of accuracy of localization in this environment, similarly to the tests on the seals. However, experiments have shown the deterioration of the localization ability of a dolphin in the air, both for tone and for the pulse signals at about 10 times, compared with those for water. However, the obtained values of the limit localization angle by a dolphin the sound in the air is about 2 times differed from the corresponding minimum of values of the angle, as shown by dogs.

A dolphin has two independent acoustic receivers, as well as several sound-transmitting channels, due to complete isolation of the hearing organ from the vibration of skull bones it could be expected the best localization indexes in the air. The question is through what binaural mechanisms dolphins localize the source sound in the air, is still controversial. In the air at a frequency of 4.5 kHz is probably a manifestation of the shielding effect, certainly, in different extent with different possible dolphin's bases. So there is a reason to believe that binaural differences in intensity at two ears ware in our experiments the most informative. Concerning pathways of sound to the dolphin's ear in the air and their weight fraction in sound localization presently uniquely it is difficult to say.

Conclusion. The study of limit localization abilities of a human and animals so far showed that from the terrestrial nonecholocational mammals human auditory analyzer has the highest resolution on the angle of the arrival of low-frequency sound. Echolocational mammals, both terrestrial and marine, through a highly specialized auditory system with high accuracy can determine the direction of the ultrasound source. The principles of sound localization in space, first formulated for humans, are apparently applicable to other mammals. Animals studies have confirmed and even more showed significant role of external structures of the hearing organ in sound localization. As a result of experiments on individual representatives of terrestrial, semiaquatic and aquatic mammals proved that the external structure of the auditory system are actively transforming the audio stream, creating a sharp focus of the reception of acoustic oscillations. Thus, initiated the study of specific mechanisms of peripheral auditory analysis of acoustic space.

It is experimentally proved that the auditory system of mammals is an environmentally adapted, the perfect device to ensure the success orientation in space.

The results of experiments on dolphins, namely, comparable values of the limit angle of localization in the water of source of various signals in both the horizontal and vertical planes, the independence of accuracy of acoustic localization from change (in the studied range) parameters of pulse signals give rise to assume the existence of different, in a similar degree of effective binaural cues for localization at different frequencies.

Similar values of the limiting localization angle of source signals of different frequencies in the air allow the ability to save different localization cues and in this environment. Comparison of mammals' space-hearing in the unusual habitats for them showed that both terrestrial and aquatic mammals, possessing highly specialized, adapted to their own environment by the auditory system, can, however, to navigate rather successfully in the uncharacteristic acoustic spaces for them. Characteristically, the dolphin localizes the sound and ultrasound source in the air with an accuracy of about two times smaller to localization abilities of dogs. At the same time, space hearing of a man under water does not yield the same characteristics of the dolphin in the air. Indicators of space hearing of semiaquatic animals do not vary greatly in two environments.

Figuratively speaking, the evolution of disposition ordered equally rightly in the development and functional specialization of the auditory system, both terrestrial and aquatic mammals, providing them with some "margin", and dolphins, as secondary aquatic animals, some "balance" of the old features.

Nevertheless, a significant deterioration in hearing sensitivity of dolphins to ultrasound in a wide range of frequencies in the air, along with the deterioration of the order of 10 times of their limit of localization abilities, characterizes the habitat as the main factor determining the physiological capabilities of mammal.

As a result of experiments with pinnipeds it is showed that signals containing explicit temporal characteristics, and are localized much more successful than the signals that carry information only about the intensity or phase. Signs used by northern fur seals in the localization in the horizontal plane of tone pulses, probably (as in dolphin), are equally effective for a variety of frequencies. Source of complex sounds that carry a few signs of binaural localization, is bearing by seals at a higher accuracy than the source of tonal pulses. The accuracy of localization of fur seals source of acoustic signals in the vertical plane in the water and air depends on their parameters, as in humans and other animals, increases to sounds with complex spectra. Moreover, the vertical source of sound in the air is localized by fur seals in 1.5-2 times worse than in the water.

It was found that the air direction of the sound source in the medial (at zero values of azimuth emitters) vertical plane, as well as when an azimuth 90° seal cannot determine. Localization capabilities of a seal in aquatic environment substantially inferior to those of a dolphin in 1.6-1.8 times in the horizontal plane and 5-9, and sometimes more than once - in the vertical plane (when comparing the best rates in the optimum for each frequency ranges). Such differences in indexes of space-hearing explained as anatomical and functional differences of dolphins and pinnipeds conductive structures, as well as differences of characteristics of the central sections of hearing organ of marine mammals.

Of all investigated representatives of marine mammals, dolphins are distinguished by the most accurate analysis of the acoustic space.

Comparison of directional reception outside of the hearing of bats, dogs, dolphins and other animals revealed the presence of a single mechanism for all types of encoding mechanism of the direction of arrival of acoustic signals through space-frequency filtering and interaural differences. The nature of orientation of auditory reception of mammals is due to excitation by impact of environmental factors.

The similarity in the properties of sound-conducting structures in dolphins and pinnipeds viewed through different modes of acoustic information processing in these species opens new avenues for comparatively analyzing the mechanisms of hearing. The results of basic studies of hearing in animals capable of perfect assessment of their acoustic environments may be useful in solving various applied problems (Babushina, 2001 c).

We obtained the very interesting data on investigation of sound reception in marine mammals: effect of stimulus parameters end transmission pathways (Babushina, 2000).

Underwater audiograms of the northern fur seal *Callorhinus ursinus,* the Caspian seal *Pusa caspica,* and the Black Sea bottlenose dolphin *Tursiops truncatus* were determined in experiments with fully or partially submerged (head out of water) animals by the operant conditioning technique with food reinforcement. The partial submergence conditions (the pinnae isolated from the sound-transmitting medium) were used to assess the sound-conducting characteristics of marine mammals body tissues. In the Caspian seal, the detection thresholds for acoustic signals of different frequencies were also determined in the presence of broadband or narrowband noise maskers of varied center frequency. The effect of the masker depended on the medium (air or water) in which the signal and the masker propagated, and on the conditions of sound reception (pinnae under or above water). The aerial and underwater sound-conducting pathways were shown to be functionally interlinked in the Caspian seal. The masking effect of noise on its hearing depended on (I) whether the signal and the masker were aerial, conducted via the external ear; or underwater, conducted via the specialized structures of the head and via head and body tissues; (II) the sensitivities of the hearing system to the signal and the noise; and (III) the signal and noise spectra. The data obtained suggested that the seal body tissues like tissues of the bottlenose dolphin altered the amplitude and frequency characteristics of acoustic signals.

2. References

Akopyan, A.I.; Zaitseva, K.A.; Morozov, V.P. & Titov, A.A. (1977). The spatial orientation of the dolphin auditory system in the perception of signals once-term frequency in the noise, *IX Proc. Acoust. Conf., Section. N.,*M. pp. 9-12.

Altman, J.A. (1983).Localization of moving sound sources nick. L., *Science*, p. 176.

Altman, J.A.; Bibikov, N.G.; Vartanian, I.A.; Dubrovskyi; N.A., Ishchenko, S.M.; Konstantinov, A.I.; Makarov, A.K.; Movchan, E.V.; Radionova, E.A.; Telepnev, V.N.; Hachunts, S.A.; Shmigidina, G.N. & Shooplyakov, V.S. (1990). *Auditory system. L.*, Science, 620 p .

Andersen, S. (1970). Auditory sensitivity of the harbour porpoise Phocoena phocoena. *Investigation on Cetacea.* V. 2. , pp. 255-259.

Ayrapet'yants, E.Sh.; Voronov, V.A.; Ivanenko, Yu.V.; Ivanov, M.P.; Ordovsky, D.L.; Popov, V.V.; Sergeev, B.F. & Chilingiris, V.I. (1973). To physiology of Black See dolphin sonar system. *Journal Evolution. Biochem. and Physiology.* Vol. IX, No. 4., pp. 416-422.

Babushina, E.S. (1979). Localization by dolphin of sources of tone and pulse signals in the water and air., *Vestn. Leningr. Univ.*, No. 3. ,pp. 119-121.

Babushina, E.S. (1986).The sensitivity of the bottlenose dolphin hearing to sounds in the air. Sea mammals. - Abstracts. Dokl. IX Proc. soveshch. to study, protect and sustainable use of *marine mammals (Arkhangelsk, 9-11 September 1986)*. Archangelsk.,pp. 16-17.

Babushina, E.S.; Zaslavsky, G.L. & Yurkevich, L.I. (1991). Characteristics of hearing of northern fur seal in the water and air environments: auliogrammy, differential thresholds on frequency. *Biophysics*. Vol. 36, No. l., pp. 904-907.

Babushina, E.S. & Yurkevich, L.I. (1994 a). Localization of the sound source by fur seals of underwater sounds in the vertical plane. *Sensor Systems*. Vol. 8, No. 1., pp. 87-90.

Babushina, E.S. & Yurkevich, L.I. (1994 b). The directivity of the auditory reception of the northern fur seal in the horizontal plane in the water and air environments. Sensor Systems. Vol. 8, No. 1. , pp. 91-93.

Babushina, E.S. (1997). Underwater and air audiograms of Caspian seal. *Sensor systems*. - 1997. - Vol. 11., No. 2., pp. 101-106.

Babushina, E.S. (1998). Localization by northern fur seals (Callorhinus ursinus) the source of acoustic signals in the vertical plane in air stifling environment. *Sensor Systems*., Vol. 12, No. 4., pp. 444-451.

Babushina, E.S. (1999). Sound Reception of marine mammals depending on the parameters and pathways of the sound. *Biophysics*., Vol. 44, No. 6., pp. 1101-1108.

Babushina, E.S. (2000). Sound Reception of marine mammals depending on the parameters and sound conduction - Part 2. *Biophysics*. Vol. 45, No. 5., pp. 927-934.

Babushina, E.S. (2001 a). Auditory reception, peculiarities of acoustic orientation of marine mammals. In : *Karadag. History, Biology, Archaeology (Proceedings of the dedicated 85-th anniversary of the Karadag Biological Station. Vyazemsky TI)*. Simferopol., Sonata. ,pp. 230-235

Babushina, E.S. (2001 b). Auditory reception of marine mammals (Research in Karadagh dolphinarium) In:*Actual questions of innovation in ODL activities in countries with economies in transition (Proceedings of the international scientific-practical conference to the 80 anniversary of the National Academy of Sciences of Ukraine)*. Sympheropol. SONATA, pp. 51-52.

Babushina, E.S. (2001 c). Underwater sound condaction in mammals. *Biophysics*. Vol. 46, No. l., pp. 80-87.

Babushina, E.S. & Polyakov, M.A. (2001). Localization of the sum of acoustic signals by the northern seal in the air. *Biophysics*. Vol. 46, No. 3., pp. 557-562.

Babushina, E.S. & Polyakov, M.A. (2003). The frequency discrimination in the hearing of bottlenose dolphins and the northern fur seal depending on the parameters and sound pathways. *Biophysics*. Vol. 48, No. 2., pp. 332-336.

Babushina, E.S. & Polyakov M.A. (2004). Localization of sound sources by northern fur seals in the horizontal plane in the water and air. *Biophysics*. Vol. 49, No. 4., pp. 723-726.

Babushina, E.S. & Polyakov, M.A. (2008). Vertical Auditory Acuity of the Bottlenose Dolphin. *Biophysics*.,Vol. 53, No. 3., pp. 499-503.

Bel'kovich, V.M. & Dubrovsky, N.A. (1976). Sensor basics of orientation of cetaceans. Leningrad, pp. 204.

Belkovich, V.M. & Solntseva, G.N. (1970). Morphological and functional features of the dolphin's hearing organ. *Zoological magazine*. Vol. 49, No. 2., pp. 275-282.

Bullock, T.H.; Grinnel, A.D.; Ikesono, E.; Kameda, K.; Katsuki, Y.; Nomoto, M.; Sato, O.; Suga, N. & Yanagisawa, K. (1968). Electrophysiological studies of central auditory mechanisms in cetaceans. *Z. Vergl. Physiol*. Vol. 59, No. 2., pp. 117-156.

Casseday, J.H. & Neff, W.D. (1973). Localization of pure tones. *J. Acoust. Soc. Amer.* Vol. 54, No2., pp. 365-372.

Don, M. & Starr, A. (1972). Lateralization performance of squirrel monkey (Samiri Sciureus) to binaural click signals. *J. Neurophysiol.* Vol. 35, No. 4., pp. 493-500.

Dudok van Heel, W.H. (1959). Audio-direction finding in the perpoise (Phocoena Phocoena). *Nature.* Vol. 183, No. 4667., p. 1063.

Dyachenko, S.M.; Korolev, L.D.; Rezvov, R.N. & Chemodanov, B.K. (1971). Investigation of the ability of bottlenose dolphins to determine the direction to the source of the noise signal. *Transactions Acoust. Inst. M.*, Vol. 17. pp. 43-46.

Feinstein, S.H. (1973). Acuity of the human sound localization response underwater. *J. Acoust. Soc. Amer.* Vol. 53, No. 2., pp. 393-399.

Gentry, R.L. (1967). Underwater auditory localization in the California sea lion (Zalophus californianus). *J. Auditory Res.* Vol. 7., pp. 187-193.

Goldberg, J.M. & Brown, P.B. (1969). The response of binaural neurons of dog superior-olivary complex to dichotic tonal stimuli: some physiological mechanisms of sound localization. *J. Neurophysiol.* Vol. 32., pp. 613-636.

Gorlinsky, I.A. (1975). Features of the localization of ultrasound and role of the peripheral regions of the auditory analyzer in the first echolocational bats. *Abstract. Cand. Dis. L.*, pp. 14.

Gorlinsky, I.A. (1976). Features of the localization of ultrasound by Greater Horseshoe Bat in the vertical plane. *Vestn. Leningr. Univ.* No. 15., pp. 79-87.

Gorlinsky, I.A. & Babushina, E.S. (1985). About directivity of sound perseption with the external ear of dogs. Biophysics. Vol. XXX, No. 1., pp. 133-136.

Hollien, H. & Brandt, J.F. (1969). Effect of air bubbles in the external auditory meatus on underwater hearing thresholds. *J. Acoust. Soc. Amer.* Vol. 46, No. 2 (Part 2)., pp. 384-387.

Hollien, H. (1973). Underwater sound localization in humans. *J. Acoust. Soc. Amer.* Vol. 53, No. 5. ,pp. 1288-1295.

Iseley, T.E. & Gysel, L.W. (1975). Sound-source localization by the red fox. *J. Mammal.* Vol. 56, No. 2., pp. 397-404.

Ivanenko, Yu.V. & Chilingiris, V.I. (1973). Characteristics of receiving of dolphin sonar . *Ref. Proceedings. 8-th All-Union-SW. Acoust. Conf. M.*, Vol. 1., p. 126.

Ivanenko, Yu.V. & Chilingiris, V.I. (1978). The directivity of the monofrequency and pulse signals reception in dolphins. *Marine Mammals. - Abstracts. Proceedings. VII Proc. soveshch. (Simferopol).* , pp. 138-139.

Kalmykova, I.V. (1977). Study of sound localization in dogs during dichotic presentation of signals. *IX Proc. Acoust. Conf., Section. "T".* , pp. 33-35.

Korolev, L.D.; Lipatov, N.V.; Rezvov, R.N.; Savel'ev, M.A. & Flenov, A.B. (1973). Investigation of possibilities of radar apparatus dolphin passive location. *Ref. Proceedings. 8-th All-Union. Acoust. Conf. M.*, Vol. 1., pp. 125-126.

Lipatov, N.V. (1978). The functional role of the external auditory canal under water sea mammals. The results and research methods. , pp. 112-124.

Masterton, R.B. & Diamond, I.T. (1964). Effects of auditory cortex ablation on discrimination of small binaural time differences. *J. Neurophysiol.* Vol. 27., pp. 15-36.

Masterton, R.B.; Jane, J.A. & Diamond, I.T. (1967). Role of brain-stem auditory structures in sound localization. I. Trapezoid body, superior olive and lateral lemnis-cus. *J. Neurophysiol.* Vol. 30., pp. 341-359.

Masterton, R.B.; Jane, J.A. & Diamond, I.T. (1968). Role of brain-stem auditory structures in sound localization. II. Inferior colliculus and its brachium. *J. Neurophysiol.* Vol. 31. pp. 96-107.

McCormick, J.G.; Wever, E.G.; Palin, J. & Ridgway, S.H. (1970). Sound conduction in the dolphin ear. *J. Acoust. Soc. Amer.* Vol. 48, No. 6 (pt. 2)., pp. 1418-1428.

Mills, A.W. (1958). On the minimum audible angle. *J. Acoust. Soc. Amer.* Vol. 1930, No. 4., pp. 237-246.

Moore, P.W.B. (1975). Underwater localization of click and pulsed pure-tone signals by the California sea lion (Zalophus californianus). *J. Acoust. Soc. Amer.* Vol. 57, No. 2., pp. 406-410.

Möhl, B. (1964). Preliminary studies on hearing in seals. *Videsk. Medd. Fra Dansk Naturh. Foren.* Vol. 127. , pp. 283-294.

Norris, K.S. & Harvey, G.W. Sound transmission in the porpoise head. *J. Acoust. Soc. Amer.* Vol. 56, No. 2., pp. 659-664.

Peterson, E.A.; Pate, W.G. & Wruble, S. (1966). Cochlear potentials in the dog: 1. dIfferences with variation in exter-nal ear structure. *J. Auditory Res.* Vol. 6. , pp. 1-11.

Peterson, E.A.; Heaton, W.C. & Wruble, S. (1969). Levels of auditory respons in fissiped carnivores. *J. Mammal.* Vol. 50, No. 3.

Purves, P.E. & Utrecht, W.L. The anatomy and function of the ear of the bottlenose dolphin, Tursiops truncatus. *Beaufortia.* Vol. 3,No. 9., pp. 241-255.

Renaud, D.L. & Popper, A.N. (1975). Sound localization by the bottlenose porpoise Tursiops truncatus. *J. Exp. Biol.* Vol. 63, No. 3., pp. 569-585.

Searle, C.L.; Braida, L.D.; Cuddy, D.R. & Davis, M.F. (1975). Binaural pinna disparity: another auditory localization cue. *J. Acoust. Soc. Amer.* Vol. 57, No2., pp. 448-455. Soluha, B.V. (1973). The underwater sound localization by a man. *The Reports of AS of SSSR,* Vol. 213, No. 1, pp. 246-248.

Stosman, I.M. & Voronov, V.A. (1978) Features of forming directional reception of ultrasonic sound oscillations in the auditory analyzer of the dolphin-fotseny. *Problems of Neurophysiology. - Nervous system. L.* Vol. 20., pp. 73-78.

Stosman, I.M.; Voronov, V.A.; Ivanenko, Yu.V. & Miheev, A.G. (1978). The role of the anatomical structures of the head in the formation of directional reception of the dolphin. In : *Behavior and bioacoustics of dolphins.* , pp. 146-154.

Terhune, J.M. (1974). Directional hearing of a harbor seal in air and water. *J. Acoust. Soc. Amer.* Vol. 56, No. 6., pp. 1862-1865.

Voronov, V.A. (1978). Functional organization of a dolphin -phocoena hearing systems in emission and reception of ultrasound signals. *Abstract. Cand. Dis. L.,* p. 18.

Zaitseva, K.A.; Akopian, A.I. & Morozov, V.P. (1975). Immunity of dolphin's auditory analyzer as a function of the angle of presentation of the noise. *Biophysics.* Vol. 3. , pp. 519-521.

Zaitseva K.A. (1978). The role of spatial thrust of the dolphin auditory analyzer in the allocation of the signal from the noise. In: *Marine mammals. Results and research methods.,* pp. 99-105

26

Directional Hearing in Fishes

Richard R. Fay
Loyola University Chicago
USA

1. Introduction

Directional hearing and sound source localization by fishes has several related meanings that arise from our assumptions about localization by human beings and assumptions about the cognitive capacities of fishes. For human listeners, we assume that when we determine the position of a sound source in the space around us, we know in a cognitive sense where the source is located, we can point to it with some accuracy, and we can move directly toward it and remember where the source is. Furthermore, we have the capacity to segregate in perception and locate multiple, simultaneous sources that make up an auditory scene (Bregman, 1990). The quantitative measure of localization for human listeners is the minimum audible angle (MAA), usually defined as the minimum angular deviation (usually in azimuth) required for reliable discrimination between two source locations. In a MAA experiment, we simply assume that not only can we discriminate the difference between two source locations, but that we "know" the direction to both sources in an absolute sense. We also often make the reasonable assumption that most other animal species function the same way: that they too "know" where the sound sources are located.

But how well founded is this assumption in the case of fishes? This question arises for several reasons, including that fishes are thought not to use the same binaural acoustic cues as terrestrial animals, that the underwater environment makes source localization an exceedingly difficult, and sometimes impossible task, and that fishes may have few or any of the cognitive capacities required to "know" anything at all. In addition to these considerations, the history of research on source localization by fishes is contradictory and confusing.

This chapter summarizes the literature on sound source localization in fishes and concludes that the evidence for a localization ability is strong, but that the mechanisms of sound source localization remain a fascinating question and an essential mystery in need of further experimentation and theoretical analysis.

2. Earliest experiments

Sound source localization was first studied in the European minnow (*Phoxinus laevis*), by Reinhardt (1935) in a laboratory tank, and then by Karl von Frisch and Sven Dijkgraaf (1935) in a shallow lake (Lake Wolfgang, Germany). Von Frisch and Dijkgraaf pointed out that the dominant view of human azimuthal source localization was that the determination of minute interaural time differences (ITD - on the order of several microsceonds) was required. ITD processing seemed hardly imaginable for fish because effective inputs to the inner ears are separated by only millimeters, and because sound travels more than four times faster in water than in air. Finally, they emphasized that their minnow *Phoxinus* detects sound pressure indirectly via the swim bladder, a midline structure that fluctuates in volume (vibrates) in response to sound pressure and that would therefore stimulate both ears equally and simultaneously, regardless of source location. They were unable to demonstrate sound source localization by their method and reached the conclusion that it made sense that fish were not be able to locate sound sources for the reasons noted above, even though they thought that this conclusion would be displeasing to biologists. After all, of what use was the great auditory acuity of their fish (*Phoxinus*) if it could not recognize the location of a sound source (see also Pumphrey, 1950)?

3. First re-evaluation

The sound source localization question arose again with the work of Moulton on the directional tail-flip response of goldfish (a species, like *Phoxinus*, having the swimbladder intimately linked to the inner ears via a series of specialized bones - the Weberian ossicles). Moulton and Dixon (1967) conditioned goldfish using food reward to change the preferred direction of a naturally occurring tail-flip response to a sound source. When the saccular and lagenar (auditory) nerve was severed on one side, the conditioned animals flipped their tails as if the sound source was on the side of the intact nerve. Moulton and Dixon concluded that the goldfish behaved as if they had localized the source, and that both ears were necessary for the directional response.

Moulton and Dixon assumed that the directional responses they observed were initiated by the Mauthner cells (M-cells) of the lower brainstem (Furshpan and Furukawa 1962). It now seems questionable that the Mauthner cells, alone, were involved. The M-cells mediate reflex orienting responses (e.g., Canfield and Eaton, 1990), but are probably not responsible for localization capacities that we associate with sound source localization behaviors of the type investigated by von Frisch and Dijkgraaf (1935). Thus, fishes may have at least two pathways for directional hearing; a descending one for reflexive responses and an ascending one possibly mediating more intentional behaviors.

At about this time, Willem van Bergeijk (1964, 1967) had a great influence on this field, and he argued that hearing in fishes should be defined as sound pressure detection (via volume fluctuations of the swim bladder). Since pressure is a scalar quantity, without directionality, and since the swim bladders of most fishes impinge on both ears equally, there would be little or no possibility of directional hearing for fishes. As von Frisch and Dijkgraaf (1935) had argued before him, van Bergeijk reasoned that some other directional sensory system must be responsible for directional orientation behaviors. Van Bergeijk touted the "acoustico-lateralis" hypothesis that the lateral line system and the ears functioned together in hearing, and that only the lateral line was responsible for directional determination.

We now know that the otolith organs of the ears are exquisitely sensitive to oscillatory motion of the head and ears (i.e., acoustic particle motion), with saccular nerve fiber sensitivities to low-frequency displacements as small as 0.1 nanometers, root mean square (e.g., Fay 1984, Fay and Edds-Walton, 1997a). At 100 Hz, displacements of this magnitude accompany a propagating sound wave in the far field at 100 dB re: 1µPa. We now also know that van Bergeijk's (1967) assumption that ear-mediated hearing in fishes was a matter only of processing the sound pressure waveform using the swim bladder or other gas bubble acting as a pressure-to-displacement transformer, was essentially an error. If van Bergeijk were correct in his view, sound source localization mediated by the ears in the near- and far-fields would indeed be impossible for fishes. But it is also now widely believed that the otolithic ears of fishes function with great sensitivity in all species as if they were inertial accelerometers (de Vries, 1950; Dijkgraaf, 1960) responding directly to acoustic particle motion in all sound fields.

4. Discrimination experiments

4.1 Directional masking

Chapman (1973), Chapman and Johnstone (1974), and Hawkins and Sand (1977) investigated the effect of signal and masking noise source separation on the signal-to-noise ratio at signal detection threshold for haddock (*Melanogrammus aeglefinus*), and pollack (*Pollachius pollachius*). In general, fish were restrained in a free-field acoustic test range about 21 meters deep, and conditioned to detect tone signals in the presence of a noise masker using cardiac conditioning. Masked thresholds were highest (most masking occurred) when the signal and noise sources were separated by less than 10° azimuth or elevation, but that an 8-15 dB release from masking occurred when the sources were separated by 85° or more (up to 180°). These experiments and results are similar to those on human listeners investigating the binaural masking level difference (BMLD) (Hirsch, 1948) and the "cocktail party effect" (Cherry, 1953), and demonstrate that the directional aspects of hearing operate in fishes as well as human beings, and presumably other terrestrial animals. The peripheral mechanisms underlying these unmasking effects appear to be quite different in fishes and humans, but the consequences for hearing are similar: spatial resolution and filtering that promotes signal detection in noise.

4.2 Minimum audible angles and distance discrimination

A series of "heroic" experiments and theories of sound source localization in fishes were conceived by Hawkins, Chapman, Sand, Schuijf, and their colleagues, mainly in the 1970s (e.g., Schuijf et al. 1972, Chapman 1973, Chapman and Johnstone 1974, Schuijf 1975, Schuijf and Buwalda 1975, Hawkins and Sand, 1977, Schuijf and Hawkins, 1983). In the first psychophysical conditioning experiment on sound source localization, Schuijf et al. (1972) studied the Ballan wrasse (*Labrus berggylta*) using appetitive conditioning in a deep fjord near Bergen, Norway. Two sound sources were separated in azimuth and a conditioning trial consisted of a brief change in which loudspeaker broadcast the 115 Hz tone bursts. Positive responses were rewarded with a piece of food. The discriminations based on source location indicated that the fish detected that the sound came from a different loudspeaker, and this was assumed to result in the perception of a purely spatial change. The authors pointed out, however, that this experiment demonstrated the detection of a spatial change, but did not necessarily indicate that the wrasse correctly determined the locations of the sources. Any difference in the perception caused by switching between the two loudspeakers could have produced these results, and it is only an assumption that the difference in perception was of loudspeakers at different locations. Therefore, this kind of experiment represents a somewhat weak demonstration of sound source localization, and will always be open to alternative interpretations. In other experiments, Chapman and Johnstone (1974) found that azimuthal angular separations of 20° or more were required for the fish to discriminate between sources.

Schuijf (1975) demonstrated that cods could be conditioned to discriminate between different azimuthal source locations with a resolution of 22°, and that two, intact ears were necessary for this discrimination. The minimum audible angle (MAA) of 22° was determined using two- and four-alternative forced choice experiments. Schuijf recognized that the cods could possibly solve this problem by recognizing the identity of each sound projector through timbre difference cues, and solve the problem by associating a correct response location with each projector without being able to determine the actual locations of the sources. Hawkins and Sand (1977) measured the smallest discriminable change in elevation (about 16°). From earlier experiments on the microphonic potentials of the ear, Sand (1974) suggested that two ears seem to be required for azimuthal localization, but that elevation discrimination could be possible using only one ear. This hypothesis has not yet been tested, but is consistent with more recent physiological data on the peripheral encoding of directional information in *Opsanus tau*, the oyster toadfish (e.g. Fay and Edds-Walton, 1997).

These experiments are among the best evidence we have that sound source localization, as we think of it in human experience, is a capacity shared by fish, and additionally, that azimuthal discrimination requires binaural processing. But it must be kept in mind that this conculsion depends on the assumption that fish responded with respect to the actual locations of the sources and not some correlated cues that did not signal actual source location.

Schuijf and Hawkins (1983) studied the question of source distance determination in cod using cardiac conditioning. Two cod were able to discriminate between two sound sources at two distances, at both at 0° azimuth and elevation. This distance discrimination was interpreted to be based on the distance-dependent phase angle between sound pressure and acoustic particle motion within the nearfield of a sound source. It is also possible that the discrimination is based on processing the amplitude ratios between these two acoustic components rather than phase differences. The authors calculated that these ratio differences were less than 4 dB for their sources and that this difference was near the level discrimination threshold for cod, determined previously by Chapman and Johnstone (1974). Thus, this distance discrimination could be based on the processing of simultaneous amplitude ratios between pressure and particle motion. These observations are consistent with the hypothesis that these fish have truly three-dimensional directional hearing, but are not critical experiments in the sense of directly demonstrating that the fish could correctly locate the test sound sources.

5. The 'phase model' of directional hearing

Directional hearing in fishes is thought to depend upon the direct stimulation of the otolithic ears by acoustic particle motion impinging on the head (de Vries 1950, Dijkgraaf 1960). In this case, the axis of motion deflecting on the hair cell cilia could be determined by the pattern of hair cell activation over a population with diverse axes of best sensitivity. Hair cells are morphologically and physiologically polarized to respond best along one particular axis (Flock, 1964, 1965). All three otolith organs of fishes (saccule, lagena, and utricle) have different orientations in the head in most fish species, and within each organ, hair cells are oriented along various axes (e.g., Popper, 1977). In this way, directional hearing seems to be solved through the assumption that the pattern of neural activity across cell arrays could encode the axis of acoustic particle motion. This idea was called "vector detection" (Schuijf and Buwalda, 1975).

This conception assumed that one end of the axis of acoustic particle motion pointed directly at the sound source, that each auditory nerve fiber received input from only one hair cell or from a group of hair cells having the same directional orientation, and that this mode of stimulation was effective enough to operate at the sound levels usual for the species. The first assumption is valid only for monopole sound sources (e.g., a pulsating source fluctuating in volume), and not for dipoles or higher-order source types. The second assumption was not confirmed until the work of Hawkins and Horner (1981) on the directional response properties of saccular afferents in cod, and more recent work on other species (e.g., Fay 1984, Fay and Edds-Walton 1997, Lu & Popper, 1998). The third assumption of adequate sensitivity was tested indirectly in psychophysical experiments on sound detection by flatfishes without a swim bladder (Chapman and Sand 1974), indicating that displacement detection thresholds were as low as –220 dB re: 1 meter (less than 0.1 nm) at the best frequency of hearing (near 100 Hz). So, it is now thought that the axis of acoustic particle motion can be determined by looking across the population of primary otolith afferents for characteristic spatial patterns.

4.2 The 180° ambiguity problem

The concept of a 'vector detector' immediately suggested an important problem that remained to be solved. That is, while the particle motion axis could be determined by arrays of hair cells, this solution could not determine which end of the axis pointed toward the source or specified the direction of sound propagation. This is known as the "180° ambiguity problem" and has dominated most theoretical and empirical work on directional hearing in fishes since the mid 1970s. Schuijf (1975) and Schuijf and Buwalda (1975) outlined a possible solution to this problem. A determination of the phase angle between acoustic particle motion and sound pressure could resolve this ambiguity. Imagine an axis of particle motion that is from side-to-side. The source could be oscillating from side to side either on the left or right of the receiving animal to produce this axis of particle motion. However, if the sound is propagating from a source at the right, then leftward particle accelerations are coincident with rising pressure and leftward accelerations coincident with a falling pressure. This "phase model" of directional hearing requires that both the sound pressure and particle motion waveforms be encoded at the periphery, and that appropriate central computations take place using useful representations of their phase or timing relations.

Schuijf and Buwalda (1975) evaluated this theory experimentally. They were able to condition cods to discriminate between sound sources directly in front and directly behind the animals, and these directional choices could be reversed by manipulating the phase of sound pressure with respect to the phase of particle acceleration (180° phase shift) of a synthesized standing wave, just as the phase model predicted. This experiment was repeated and extended several times (e.g., van den Berg and Schuijf 1983, Buwalda et al. 1983), and represents the best evidence in support of the phase model for sound source localization by fishes.

A potential weakness of the phase model is its requirement that both sound pressure and acoustic particle motion be encoded separately at the periphery or segregated by central computations. In most unspecialized species with a swim bladder, this could possibly take place through one set of hair cells oriented so as to respond to re-radiated particle motion from the swim bladder (for the pressure-dependent component), and another set shielded from swim bladder signals that responded to direct particle motion stimulation. In Otophysi and other hearing specialist species, the lagena and utricle may also function as auditory organs (e.g., Wubbles and Schellart 1998) but do not receive swim bladder input (Coombs, et al., 2010). Rather, they respond with great sensitivity to acoustic particle motion as if they were inertial accelerometers (Fay 1984). However, for species without a swim bladder (or equivalent) such as elasmobranchs and flatfish, and for species without specializations for sound pressure detection, this dual encoding assumption is less likely to be valid.

Although not dealing directly with the 180° ambiguity question, Kalmijn (1997) has suggested an ethological explanation for sound source localization in fishes. He pointed out that a fish might not 'know' the location of any sound source, but could reach any sound source successfully simply by swimming in a direction that maintained a constant angle with the local axis of particle motion, which itself need not point to the sound source. Note that for this sort of mechanism to work, the sound source must be assumed to be broadcasting nearly continuously for a relatively long period of time, and that the receiver must be able to decide which direction along the pathway to take in approaching or avoiding the source.

6. Phonotaxis experiments

For many species of fish, males signal their breeding territory locations through advertisement calls that attract females of the species (Fine et al. 1977). It is presumed, and sometimes has been demonstrated, that females are able to localize these sources. Toadfish (family Batrachoididae) are the best studied family (e.g., Fish 1972, Gray and Winn 1961, Winn 1964). McKibben and Bass (1998) presented various continuous sounds mimicking advertisement calls to plainfin midshipman toadfish (*Porchthys notatus*) from one of two loudspeakers near the center of a 4-meter diameter tank (0.75 meter deep) and observed the responses of gravid females released within about 1 meter from the loudspeakers. For continuous tones and harmonic complexes with a fundamental frequency near 100 Hz (at about 130-140 dB re: 1 μPa), females were observed to exhibit phonotaxis, or a naturally occurring behavior of approaching the source of these stimuli. These and other (e.g., McKibben and Bass 2001) studies on this species also represent some of the clearest evidence available that fishes are able to locate sound sources. It is not known whether these animals were moving up an intensity gradient (klinotaxis), or approached the source using another search strategy (e.g., the constant-angle mechanism proposed by Kalmijn (1997)), or whether they had determined the source location at the time of initial release in the test arena.

6.1 New phonotaxis experiments on midshipman

Zeddies et al. (2010a) recently presented new phonotaxis observations on midshipman in the same arena used by McKibben and Bass (1998). In this case, the whole sound field was completely and quantitatively measured in terms of sound pressure and acoustic particle motion, and the pathways of approach to the source were videotaped. Female plainfin midshipman fish were collected by hand in the intertidal zone during the reproductive season on the same day as testing. For testing, a US Navy J9 sound projector was suspended from a beam in the center of the tank. An opaque plastic tarp was used as a screen and placed immediately in front of, but not touching, the sound projector to remove any visual cues that might affect sound source localization behavior. The playback signal consisted of a continuous tone at 90 Hz that was similar to the fundamental frequency of the male advertisement call (80-100 Hz; McKibben & Bass, 1998). The tone level at the calibration site was set at 130 dB (re 1 μPa).

The behavioral responses of the fish were recorded on videotape using a video recorder and a black-and-white camera mounted approximately 6 m above the tank's testing arena. The video records were digitized using video-to-DVD capture and recording software. The track taken by the fish was reconstructed using a frame-by-frame analysis of the digitized video records. The sound playback experiments were conducted at night between 21:00 and 2:00 h, and the water flow to the test tank was shut off during all tests. Water depth was adjusted to 50 cm for all tests.

Tests began with an individual fish being placed in a 30 cm diameter plastic mesh cylinder positioned approximately 109 cm from the sound source. Fish were then released by manually raising the cylinder. Tests were terminated when the fish swam to the perimeter of the testing arena or when the sound was turned off after a positive phonotaxic response. A positive response was recorded when a fish approached the sound source and then directly touched the speaker face or circled in front or under the sound projector. There were no observations of fish returning to the center of the tank after reaching the walls, and rarely, if ever, did a fish remain in the center of the tank once the speaker was turned off.

Pressure measurements were made with an eight-element array of miniature hydrophones forming a cube, 5 cm on a side. This arrangement permittted particle motion to be calculated in the x, y, and z directions by finding the pressure gradient between adjacent hydrophones. Pressure is a scalar quantity consisting of only a magnitude and particle motion (i.e. the displacement, velocity, and acceleration of the media due to an acoustic disturbance) is a vector, having both magnitude and direction. To properly interpret the phonotaxic pathways of the fish to the source, quantitative descriptions (maps) of the acoustic pressure and particle motion in the behavioral arena were obtained. Figure 1A shows a contour plot of the sound pressure field, and Fig. 1B is a vector plot of the acoustic particle motion in the arena. Both measurements confirm that the sound projector is essentially a monopole source, with an omnidirectional pressure field, and particle motion axes that point toward and away from the source.

Only gravid females containing ripe eggs showed phonotaxic responses to the hum-like playback tone of 90 Hz while spent females containing little or no eggs did not exhibit phonotaxic responses. The phonotaxic responses of the gravid females consisted primarily of straight to slightly curved tracks to the monopole sound source, as illustrated in Fig. 2A. Once at the sound source, the fish responded unambiguously by either directly touching the speaker face and/or circling in front or underneath the sound projector with prolonged active interest around the sound source. The majority of the tested gravid females (72.5%, 45 of 62) responded to the 90 Hz playback tone and localized the monopole sound source. In contrast, none of the gravid females in the control group (n = 59) released with the sound turned off swam toward the sound projector and made physical contact or showed active interest in the silent projector. Thus, these results confirm that gravid females exhibit robust phonotaxis with a high degree of directionality toward the source at initial release, and move along the axis of the particle motion vectors in a monopole sound field.

However, monopole sound sources are a special case, and Kalmijn (1997) has argued that most biological sound sources (such as swimming fish) are dipoles or higher order types. A dipole source is simply modeled as a translating or vibrating sphere that doesn't change shape or volume. The sound field created by a dipole is more complexly shaped than that produced by monopole sources. It is axisymmetric, with a relative sound pressure null at locations in the field that are perpendicular to the axis of source motion. At and surrounding these pressure null locations, particle motion vectors are oriented parallel to the axis of

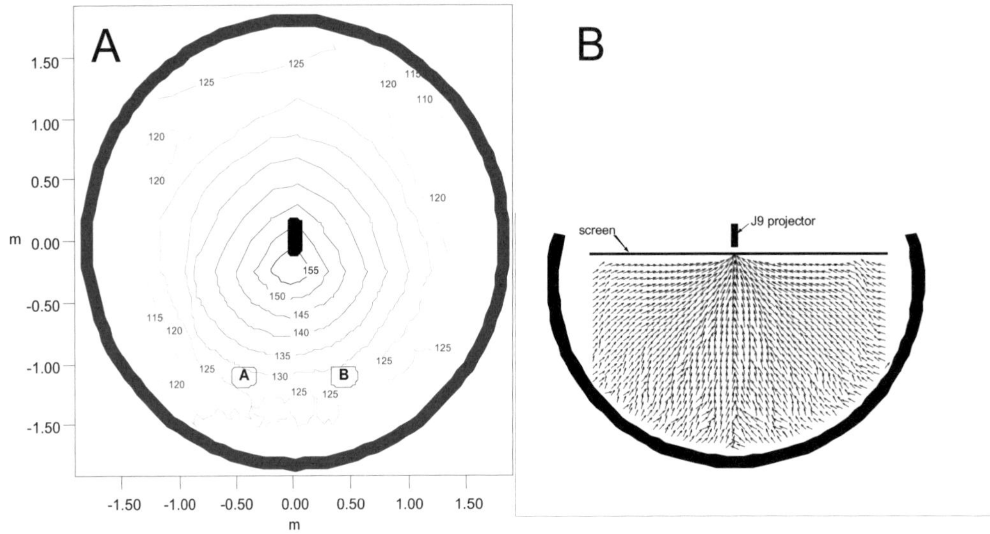

Fig. 1. A. Contour plot of sound pressure field in the test arena. A and B are alternative fish release sites. B. Particle motion field in the test arena calculated from pressure gradient measurements. The arrows on the vectors indicate the direction of increasing magnitude. Modified from Zeddies et al., 2010a.

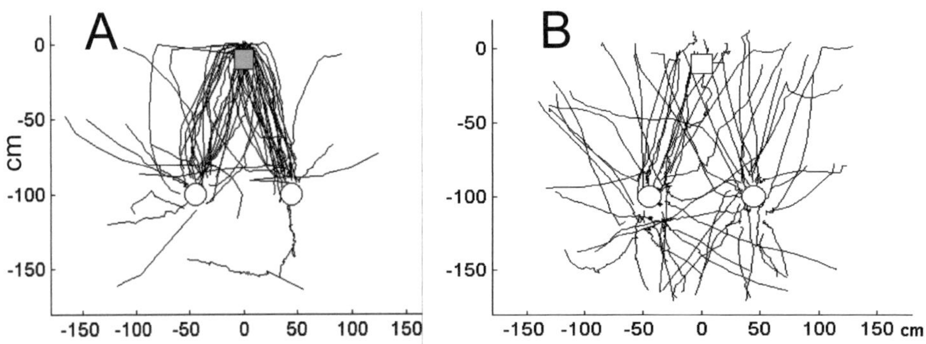

Fig. 2. A. Phonotaxis pathways for 45 gravid females that approached the source. B. Pathways for 59 gravid females without sound. Modified from Zeddies et al., 2010a.

source motion. Only on the axis of dipole vibration do particle motion axes point toward and away from the source. In other words, most particle motion vectors do not point toward and away from the source (as for monopoles), but are oriented at various angles to the axes that pass through the source.

What is the phonotaxic response of midshipman fish in a dipole field? In preliminary experiments on this topic (Zeddies et al., 2010b) a dipole source was created using two monopole sources back-to-back, creating a push-pull action. The test tank used and all procedures were essentially identical to the monopole experiments (Zeddies et al, 2010a). After measuring the dipole sound field, the pathways taken by 25 gravid female midshipman were recorded following release of the fish near the sound pressure null (nearly perpendicular to the source's axis of motion). The results showed that the pathways taken to the source were not straight lines to the source, but rather were curved, essentially following the axes of particle motion as experienced by the fish as it made its way to the source. Some fish initially swam left, and some swam right from the release site, but all of the responding fish swam parallel to the particle motion vectors to the source. Thus for dipole sources, fish can locate the source in the sense that they can eventually arrive at the source, but in this case, they do not and probably could not "know" where the source is. All they seem to know is what the axis of particle motion is at each position in the field in which they find themselves.

These observations on dipole sources add to our understanding of directional by fishes. First, these observations roughly correspond to Kalmijn's ethological scenario for approaching sources, but with the important exception that fish apparently don't use arbitrary but constant angles for approach, but rather seem to select a 0° approach angle with respect to the particle motion vectors. In general, these behaviors correspond to the predictions of the original 'vector detector' notions. Secondly, these observations raise issues with respect to the "180° ambiguity problem." When fish are released near a pressure null where the particle motion vectors are nearly perpendicular to a line to the source, turning right or left is equally effective; there is essentially no 180° ambiguity problem in the sense that there is no response that is more correct than another. As previously mentioned, about 50% of the released fish swim in each of the two correct directions as they approach the source. When released near the axis of dipole vibration, only one directed pathway is correct (taking the fish toward, not away from the source), and the 180° ambiguity problem has to be solved. But what is the difference between these two release sites that necessitates the solution to the problem at one, but not at the other? One possibility is that the particle motion intensity gradient contains information on the direction to the source at the on-axis release site, but not at the release site where the fish experiences a particle motion axis that is perpendicular to the line to the source. In other words, perhaps a detectable intensity gradient contributes to the solution of the "180° ambiguity problem."

7. Physiological studies

Peripheral and central neurophysiological studies of directional hearing in fishes have investigated the encoding of directional information in the primary afferents of the octaval nerve from the ears, and on these directional representations and computations in nuclei of the brainstem. The species investigated have included goldfish (*Carassius auratus*), toadfishes (*Opsanus tau* and *Porchthys notatus*), sleeper goby (*Dormitator latifrons*), rainbow trout (*Salmo gairdneri*), and Atlantic cod (*Gadus morhua*).

7.1 The periphery

Single unit studies on the peripheral encoding of directional information were first reported by Fay and Olsho (1979) and Fay (1981) for goldfish. Hawkins and Horner (1981) measured the first directional response patterns in recordings from the saccular and utricular nerve of the cod in response to whole-body oscillatory accelerations at various axes in the horizontal plane. They found that the response magnitude tended to vary according to a cosine-like function of vibration axis angle. Thus, each afferent studied apparently represented the presumed directionality of a single hair cell or group of hair cells having the same directional orientation. In other words, each hair cell orientation appeared to have a private line to the brain, a requirement of the notion of "vector detection" assumed by Schuijf (1975) as the first stages of the phase model. For the saccule, the best azimuthal axis of motion corresponded roughly with the horizontal-plane orientation in the head of the saccular organ and otolith. In utricular afferents, best azimuths varied widely, reflecting the diversity of hair cell orientations over the (horizontal) surface of the utriculus. Utricular best sensitivity was similar to that of the saccule, suggesting a possible role for the utriculus in directional hearing. It was noted that the phase angle at which afferents synchronized to the stimulus varied widely among the afferents and did not fall into two groups, 180° out-of-phase with one another. Fay and Olsho (1979) and Fay (1981) also reported a nearly flat distribution of synchronization angles among saccular and lagenar nerve units in goldfish. The phase model (and other related theories of directional hearing in fishes outlined above) assume that pressure and displacement "polarities" would be represented robustly in a bimodal distribution (two modes, 180° out-of-phase) of synchronization angles, as predicted by anatomical hair cell orientation maps for otolith organs (e.g., Dale 1976, Platt 1977, Popper 1977). The fact that phase-locking angles do not cluster in such a way (see also Fay and Edds-Walton 1997 for similar data on *Opsanus tau*) presents a problem for all current theories of sound source localization in fishes: Which neurons "represent" the phases of pressure or displacement waveforms that have to be compared to resolve the "180° ambiguity problem?"

Experiments on directional encoding in goldfish (Fay 1984, Ma and Fay 2002) and toadfish (Fay and Edds-Walton 1997a,b, Edds-Walton et al. 1999) have used a three-dimensional "shaker" system (Fay, 1984) to produce whole-body accelerations in both azimuth and elevation. Figure 3 illustrates typical directional response patterns (DRP) for saccular units of toadfish. These data can be summarized as follows:

1. Most saccular afferents respond in proportion to the cosine of the stimulus axis angle in azimuth and elevation, with a few exceptions (Fay and Edds-Walton 1997a). Thus, each afferent seems to represent the orientation of one hair cell, or a group of hair cells having the same directional orientation (Lu and Popper 1998). Some of the DRPs in Fig. 1 reflect the fact that primary afferents saturate at the highest levels, and therefore tend to loose directionality in these cases (e.g., unit H8 at the highest levels).
2. In the azimuthal plane, most saccular units of the left ear respond best to an axis approximately parallel with the saccular organ's orientation in the head (about -40°).
3. In the vertical plane, the best elevations among units correspond with the diversity of hair cell morphological polarizations on the saccular epithelium.
4. The best threshold sensitivity for these afferents is high: at 100 Hz, displacement at threshold is about 0.1nm. This is approximately the same amplitude of basilar membrane motion at behavioral detection threshold in mammals (Allen 1996).
5. Intracellular labeling shows that maps of anatomical hair cell orientation do not quantitatively predict physiological directionality (Edds-Walton et al. 1999). This is probably due to the simplifications of constructing two-dimensional maps of three-dimensional structures. Anatomical maps cannot substitute for physiological data in specifying the directional information transmitted to the brain by the octaval nerve.

Since best azimuths for the saccular afferents studied so far tend to cluster about the azimuthal angle in the head of the saccule and otolith (see also Sand 1974), the overall stimulation of the right and left saccules will tend to differ depending on the azimuth of particle motion. Theoretically, azimuth angle can be computed by comparing the summed

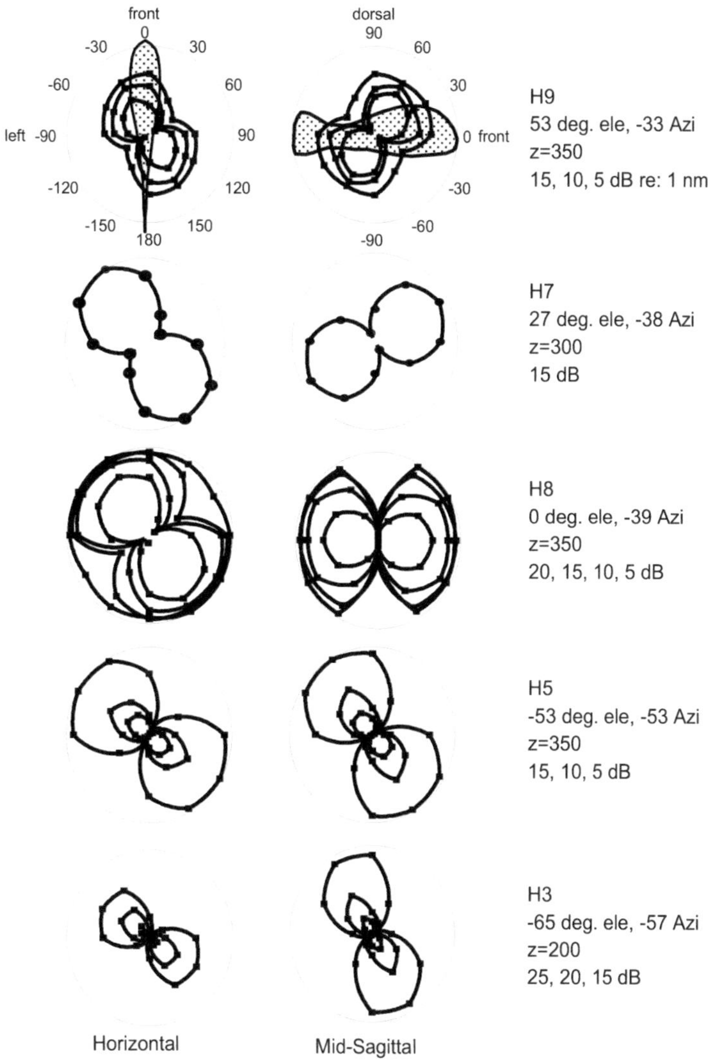

Fig. 3. Representative directional response patterns (DRP) for left the saccule of the oyster toadfish (*Opsanus tau*). The left column shows DRPs in the horizontal plane, and the right column shows DRSs in the mid-sagittal plane. The best azimuths (left) tend to cluster at about -40°. Best elevations (right) vary more widely, covering all elevations.
The keys associated with each pair of DRPs indicate the animal and unit ID, estimates of the best azimuth and elevation, the radial axis represented by the circle for each DRP, and the stimulus levels used.
Modified from Fay and Edds-Walton, 2000.

output of each saccule (e.g., through subtraction or common-mode rejection), but with several ambiguities. Sound source localization in azimuth seems to be a binaural process in fishes as it is in terrestrial animals. Edds-Walton (1998) has identified bilateral projections in a medullar nucleus (descending octaval nucleus), and there is physiological (Edds-Walton and Fay, 2009) and neural labeling data (Edds-Walton, et al., 2009) on *Opsanus tau* indicating that binaural processing occurs in the medulla. This is consistent with the observations of Moulton and Dixon (1967), Schuijf (1975), and Schuijf and Siemelink (1974) indicating that the information from the two ears is necessary for sound source localization in azimuth. Note, however, that binaural acoustic cues are probably not available to fishes; the binaural information derives from the inherent directionality of the ears that respond directly to acoustic particle motion. Fay and Edds-Walton (1997a) have observed that the phase angles at which the units synchronize to a stimulus vary with the effective stimulus level in non-spontaneous saccular afferents. This means that an interaural phase difference could represent response magnitude, giving rise to a robust interaural timing code for azimuth. Coding for elevation seems to be a different matter, however. The elevation of a sound source is represented within a sensory epithelium as the profile of activity across saccular afferents with different "best elevations" (see Fig. 3). There is a functionally similar hypothesis for determining elevation for mammalian listeners; this is the hypothesis that the spectral profile (pattern of activity over the length of the cochlear epithelium) as shaped by the frequency spectrum as filtered through the head-related transfer function (HRTF) (e.g., Wightman and Kistler 1993). In other words, it is hypothesized for both fishes and mammals, source elevation coded as a monaural profile of excitation over the surface of the sensory epithelium that encodes frequency for mammals, and elevation for fishes.

The directional responses of the auditory nerve have also been investigated for organs other than the saccule. Hawkins and Horner (1981) investigated utricular units in the cod and found them to be most sensitive in the horizontal plane with substantially cosine-like directional response patterns (DRP). Fay (1984) surveyed lagenar and utricular as well as saccular units in goldfish. All three otolith organs had a similar distribution of displacement thresholds (lowest thresholds near 0.1 nm at 140 Hz) and cosine-shaped DRPs. Lagenar and saccular units showed a wide distribution of best axes in elevation with a tendency to cluster in azimuth parallel to the orientation of the respective organs. In the experiments of Lu et al. (2003) on the lagena of the sleeper goby, DRPs deviated significantly from a cosine shape, showing more narrowly shaped DRPs than would be expected from hair cells, and best thresholds that were somewhat higher than saccular afferents from the same species. More broadly shaped DRPs could be explained by excitatory convergence from hair cells having different directional orientations (Fay and Edds-Walton 1997a), but narrowly shaped DRPs cannot be explained at present. The differences in sensitivity between lagenar and saccular units in the sleeper goby could possibly be related to the lagena's small size in most non-specialized species.

7.2 The auditory CNS

The representations of directional acoustic information in the brain have been studied in *Carassius auratus* by Ma and Fay (2002), *Opsanus tau* by Edds-Walton and Fay, and in *Salmo gairdneri* by Wubbles and Schellart. The major acoustic nuclei of the brainstem are the first-order descending octaval nucleus (DON), the higher-order secondary octaval population (SOP), and the torus semicircularis (TS) of the midbrain. Auditory responses of the SOP, thalamic, and other forebrain auditory nuclei have not been studied with respect to directionality.

Most of the single units recorded in the toadfish DON show simple directional preferences for the axis of whole-body acceleration. The occurrence of directionality in the DON (and other auditory nuclei) indicates that excitatory convergence from neurons having different directionality probably does not occur in the brain since the directional selectivity of the periphery is maintained by cells throughout the brainstem. The sensitivity, frequency response, and phase-locking of DON units are similar to those of saccular afferents, but the directional response patterns (DRP) of most units tend to be more directionally selective than saccular afferents. This increased selectivity has been termed "sharpening" (Fay and Edds-Walton 1999, Edds-Walton and Fay 2003). Figure 4 shows typically sharpened DRPs from the brainstem of toadfish along with a graphical representation of a simple model mechanism that could account for sharpening (Edds-Walton and Fay, 2003). The hypothesis is that a central cell receives excitatory input from one directional cell, and inhibitory input from another directional cell (possibly from the contralateral ear), both having cosine-like DRPs with different best axes in azimuth or elevation (Fay and Edds-Walton, 1999). This excitatory-inhibitory convergence appears to be a common interaction in the auditory brainstem, and it always results in some degree of directional sharpening, depending on the best axes and weights associated with each input. Recordings from the torus semicircularis (TS) of the midbrain (Fay and Edds-Walton, 2001, Edds-Walton and Fay, 2003) show similar

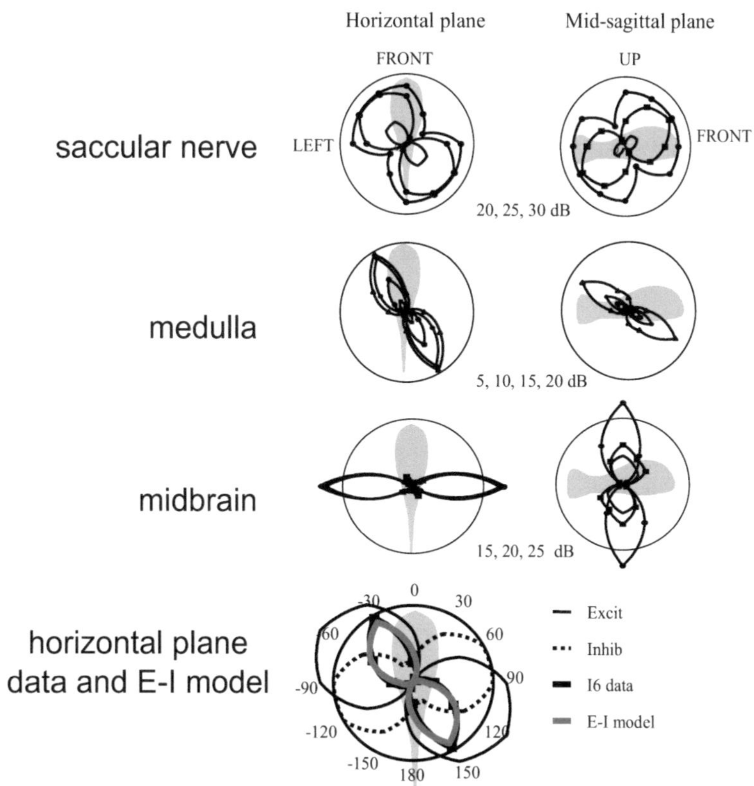

Fig. 4. Directional response patterns (DRP) for three representative cells of the saccular nerve, medulla (DON), and midbrain (TS) for three stimulus levels. Left column - horizontal plane. Right column - mid-sagittal plane (Modified from Edds-Walton and Fay, 200X). At bottom is shown a simple excitatitory-inhibitory model for sharpening the DRP along with data from unit I6 (see text) (modified from Fay and Edds-Walton, 2005).

unit sensitivity and frequency response as in the DON, but with dramatically reduced phase-locking, and directional sharpening that is augmented, on average (see Fig. 4). Edds-Walton (1998) has confirmed that binaural projections exist connecting the right and left DONs in the medulla, and Edds-Walton and Fay (2009) have confirmed physiologically that there are binaural interactions among some cells of the DON. In addition, Edds-Walton, et al. 2009 have confirmed that binaural computations involving excitatory and inhibitory inputs take place in the DON using the inhibitory neurotransmitter, GABA.

In the midbrain, directional auditory responses were found both in the nucleus centralis of the torus semicircularis (the "auditory" nucleus), and the nucleus ventrolateralis (the "lateral line" nucleus) in toadfish. In addition, many units recorded in both nuclei showed interactions of auditory and lateral line inputs (excitatory and inhibitory) (Fay and Edds-Walton 2001, Edds-Walton and Fay 2003). It is not known whether such bimodal interactions play a role in sound source localization, but source localization is likely a multi-modal function (Braun et al. 2003), and the lateral line system could play an important role close to the source (Weeg and Bass 2002). In general, the best axes for brainstem auditory units are more widely varied in best azimuth and elevation than the same distributions for saccular afferents.

The directional characteristics of TS units also have been studied in goldfish, a species specialized for sound pressure reception (Ma and Fay 2002). Most units recorded responded best to vertical vibration, as predicted by the vertical orientation of saccular hair cells in goldfish and other Otophysi. Thus, excitatory inputs to the TS appear to be primarily from the saccule in goldfish. Nevertheless, deviations from cosine directionality among unit DRPs (i.e., sharpening) were also observed in the goldfish TS, and could be accounted for by simple excitatory-inhibitory interactions as in toadfish. This suggests that sound source localization in Otophysi, if it occurs at all (see Schuijf et al. 1977), may be based on computations taking place elsewhere in the ascending auditory system where lagenar or utricular inputs could be used to help resolve the axis of acoustic particle motion. In any case, the representation of acoustic particle motion appears to be organized quite differently in the midbrains of goldfish and unspecialized species, corresponding to the anatomical differences between their respective saccules (essentially vertically oriented hair cells in goldfish and other Otophysi versus diverse orientations in most other species).

Wubbles, Schellart, and their colleagues have presented a series of studies on directional sound encoding in the midbrain (torus semicircularis or TS) of the rainbow trout (*Oncorhynchus mykiss*). Like the toadfish, this species is not specialized for sound pressure reception but is assumed to receive both direct motion as well as reradiated, pressure-dependent motion inputs from the swim bladder. Fish were stimulated in neurophysiological studies by whole-body acceleration at various angles in the horizontal plane using a vibrating platform that could be rotated to any angle (Schellart et al. 1995). Several important observations on directional representations were made:

1. About 44% of the units recorded were classified as directional (Wubbles and Schellart 1997).
2. Directional units were roughly mapped in the TS with the medial TS containing rostro-caudal orientations and the lateral TS containing cells with many different orientations (Wubbles et al. 1995).

3. The TS has a columnar organization with similar best axes of horizontal motion tending to be constant within vertical columns (Wubbles et al. 1995, Wubbles and Schellart 1998).
4. Some phase-locked units had phase angles of synchronization that did not vary with the stimulus axis angle (except for the expected 180° shift at one angle), while others showed a phase shift that varied continuously with stimulus angle over 360° (Wubbles and Schellart 1997).

Wubbles and Schellart concluded that those and other results strongly supported the phase model. They speculated that the rostro-caudally oriented units of the medial TS were channels activated by swim bladder-dependent motion input, while the diversely oriented units of the lateral TS represented direct motion input to the otolith organs. The utricle was thought to be the otolith organ supplying the direct motion-dependent input because of its horizontal orientation. The authors speculated that the units with synchronization angles independent of stimulus direction represented pressure-dependent swim bladder inputs while the units with variable synchronization phase angles represented direct motion inputs. Wubbles and Schellart (1997) then concluded that "…the phase difference between the(se) two unequivocally encodes the stimulus direction (0-360°)…" (i.e., solves the 180° ambiguity problem). This conclusion would be strengthened by a more clear and detailed explanation for the direction-dependent variation in synchronization angle shown by some units and by a testable theory for the final step that solves the 180° ambiguity.

8. Summary and conclusions

1. There are much data on the accoustical behaviors of several fish species that strongly suggest the capacity directional hearing and sound source localization. Most of these observations indicate the necessity that one or more otolith organs respond to acoustic particle motion.
2. The question of localization in the near- versus far-fields is no longer a critical issue because we now know that near field hearing does not imply that the lateral line system must be involved. The otolith organs respond directly to acoustic particle motion in both fields.
3. Most conditioning and psychophysical studies on the discrimination of sound source location provide evidence consistent with the hypothesis that fishes are able to locate sound sources in a way analogous to localization capacities of human beings and other tetrapods, both in azimuth and elevation. However, most of these studies fail to unequivocally demonstrate that fishes can actually perceive the location of sound sources.
4. An explanation for sound source localization behavior at the level of Mauthner cells and other reticulo-spinal neurons cannot serve to explain conditioning and discrimination learning phenomena with respect to source location.
5. All present accounts postulate that the process begins with the determination of the axis of acoustic particle motion by processing the profile of activity over an array of peripheral channels that directly reflect diverse hair cell and receptor organ orientations ("vector detection").
6. Neurophysiological studies on cells of the auditory nerve and brainstem are consistent with vector detection and show that most brainstem cells preserve and enhance the

directionality originating from otolith organ hair cells. Goldfish and other Otophysi present a clear problem for this view because there is little or no variation of hair cell directionality in the saccule or at the midbrain. This has lead to speculations that Otophysi use other otolith organs (lagena or utricle) in addition to the saccule for vector detection.

7. Vector detection leaves an essential "180° ambiguity" as an unsolved problem (Which end of the axis points to the source, or, in what direction is the sound propagating?). The "phase model" of directional hearing has been moderately successful in solving this ambiguity in theory and experiment. However, the 180° ambiguity is not the only ambiguity for sound source localization throughout the vertebrates. It is not certain that auditory processing, alone, must be able to solve this problem.
8. Although the phase model is successful in a general sense, it is difficult to apply in several important cases (i.e., for fishes without swimbladders, and for Otophysi) where effectively independent representations of the particle motion and pressure waveforms are required but are not evident.
9. Additional problems for vector detection and the phase model are that the axis of acoustic particle motion points directly at the source only for monopole sources, and that clear and unambiguous representations of waveform phase that could help in localization have not been observed in auditory nerve units (distributions of phase-locking angles tend to be uniform).
10. While there are behavioral and electrophysiological observations that are consistent with sound source localization in fishes, there are no examples of localization capacities in a single species that have a comprehensive theoretical explanation. Sound source localization in fishes remains incompletely understood.

9. References

Allen, J. (1996). OHCs shift the excitation pattern via BM tension, In: *Diversity in Auditory Mechanics,* Lewis, E.R.; Long, G.R.; Lyon, R.F.; Narins, P.M.; Steele, C.R. & Hecht-Poinar, E., (Eds.), pp. 167-175, World Scientific Publishers, Singapore

Berg, A.V. van den & Schuijf, A. (1983). Discrimination of sounds based on the phase difference between the particle motion and acoustic pressure in the shark *Chiloscyllium griseum. Proc. Roy. Soc. Lond. B,* 218, 127-134

Bergeijk, W.A. van (1964). Directional and nondirectional hearing in fish. In: *Marine Bioacoustics,* Tavolga, W.A., (Ed.), pp. 269-301, Pergamon Press, London

Bergeijk, W.A. van (1967). The evolution of vertebrate hearing, In: *Contributions to Sensory Physiology,* Vol. 2, Neff, W.D., (Ed.), pp. 1-49, Academic Press, New York

Braun, C.; Coombs, S. & Fay, R. (2002). Multisensory interactions within the octavolateralis systems: What is the nature of multisensory integration? *Brain Behav & Evol,* 59, pp. 162-176

Bregman, A.S. (1990). *Auditory Scene Analysis. The Perceptual Organisation of Sound,* MIT Press, Cambridge

Buwalda, R.J.A.; Schuijf, A. & Hawkins, A.D. (1983). Discrimination by the cod of sounds from opposing directions. *J Comp Physiol, A,* 150, pp. 175-184

Canfield, J.G. & Eaton, R.C. (1990). Swim bladder acoustic pressure transduction initiates Mauthner-mediated escape, *Nature,* 347, pp. 760-762

Chapman, C.J. (1973). Field studies of hearing in teleost fish. *Helgolander wiss Meeresunters,* 24, pp. 371-390

Chapman, C.J. & Sand, O. (1974). Field studies of hearing in two species of flatfish, *Pleuronectes platessa* (L.) and *Limanda limanda* (L.) (Family Pleuronectidae). *Comp Biochem Physiol,* 47, pp. 371-385

Chapman, C.J. & Johnstone, A.D.F. (1974). Some auditory discrimination experiments on marine fish. *J Exp Biol,* 61, pp. 521-528

Cherry, E.C. (1953). Some experiments on the recognition of speech, with one and with two ears. *J Acoust Soc Am,* 25, pp. 975-979

Dale, T. (1976). The labyrinthine mechanoreceptor organs of the cod (*Gadus morhua* L. (Teleostei: Gadidae). *Norw J Zool,* 24, pp. 85-128

Dijkgraaf, S. (1960). Hearing in bony fishes. *Proc Roy Soc, B,* 152, pp. 51-54

Edds-Walton, P.L. (1998). Anatomical evidence for binaural processing in the descending octaval nucleus of the toadfish (*Opsanus tau*). *Hear Res,* 123, 41-54.

Edds-Walton, P.L.; Fay, R.R. & Highstein, S.M. (1999). Dendritic arbors and central projections of auditory fibers from the saccule of the toadfish (*Opsanus tau*). *J Comp Neurol,* 411, pp. 212-238

Edds-Walton, P. & Fay, R.R. (2003). Directional selectivity and frequency tuning of midbrain cells in the oyster toadfish, *Opsanus tau. J Comp Physiol,* 189, pp. 527-543

Edds-Walton, P. and Fay, RR (2005) Sharpening of Directional Responses along the Auditory Pathway of the Oyster Toadfish, *Opsanus tau. J. Comp Physiol,* 191, 1079-1086.

Edds-Walton, P.; Holstein, G.M, & Fay, R. (2009) γ-Aminobutyric acid is a neurotransmitter in the auditory pathway of toadfish, *Opsanus tau. Hear. Res.* 262, 45-55.

Fay, R.R. (1981). Coding of acoustic information in the eighth nerve, In: *Hearing and Sound Communication in Fishes,* Tavolga, W.; Popper, A.N. & Fay, R.R., (Eds.), pp. 189-219, Springer-Verlag, New York

Fay, R.R. (1984). The goldfish ear codes the axis of acoustic particle motion in three dimensions. *Science,* 225, pp. 951-954

Fay, R.R..; Coombs, S.L. & Elepfandt, A. (2002). Response of goldfish otolithic afferents to a moving dipole sound source. *Bioacoustics,* 12, pp. 172-173

Fay, R.R. & Edds-Walton, P.L. (1997a). Directional response properties of saccular afferents of the toadfish, *Opsanus tau. Hear Res,* 111, pp. 1-21

Fay, R.R. & Edds-Walton, P.L. (1997b). Diversity in frequency response properties of saccular afferents of the toadfish (*Opsanus tau*). *Hear Res,* 113, pp. 235-246

Fay, R.R., and Edds-Walton, P.L. (1999). Sharpening of directional auditory responses in the descending octaval nucleus of the toadfish (*Opsanus tau*). *Biol. Bull,* 197, 240-241.

Fay, R.R. and Edds-Walton, P.L. (2000). Directional encoding by fish auditory systems. *Philosophical Transactions of the Royal Society London.* B, 355, 1281-1284.

Fay, R.R., and Edds-Walton, P.L. (2001). Bimodal units in the torus semicircularis of the toadfish (*Opsanus tau*). *Biol. Bull,* 201, 280-281.

Fay, R.R. & Olsho, L.W. (1979). Discharge patterns of lagenar and saccular neurones of the goldfish eighth nerve: Displacement sensitivity and directional characteristics. *Comp Biochem Physiol,* 62, pp. 377-386

Fine, M,; Winn, H. & Olla, B. (1977). Communication in fishes. In: *How Animals Communicate,* Sebeok, T., (Ed.), pp. 472-518, Indiana University Press, Bloomington

Fish, J.F. (1972). The effect of sound playback on the toadfish. In: *Behavior of Marine Animals,* Volume 2, Winn, H.E. & Olla, B.L. (Eds.), pp. 386-434, Plenum Publ. Corp, New York

Flock, Å. (1964). Structure of the macula utriculi with special reference to directional interplay of sensory responses as revealed by morphological polarization. *J Cell Biol* 22, pp. 413-431

Flock, Å. (1965). Electron microscopic and electrophysiological studies on the lateral line canal organ. *Acta Oto-laryngol Suppl,* 199, pp. 1-90

Frisch, K. von & Dijkgraaf, S. (1935). Can fish perceive sound direction? *Z vergl Physiol,* 22, pp. 641-655

Furshpan, E.J. & Furukawa, T. (1962). Intracellular and extracellular responses of the several regions of the Mauthner cell of the goldfish. *J Neurophysiol,* 25, pp. 732-771

Gray, G.A. & Winn, H.E. (1961). Reproductive ecology and sound production of the toadfish, *Opsanus tau. Ecology,* 42, pp. 274-282

Hawkins, A.D. & Horner, K. (1981). Directional characteristics of primary auditory neurons from the cod ear, In: *Hearing and Sound Communication in Fishes,* Tavolga, W.N.; Popper, A.N. & Fay, R.R., (Eds.), pp. 311-328, Springer-Verlag, New York

Hawkins, A.D. & Sand, O. (1977). Directional hearing in the median vertical plane by the cod. *J Comp Physiol,* 122, pp. 1-8

Hirsh, I.J. (1948). The influence of interaural phase on interaural summation and inhibition. *J Acoust Soc Am,* 20, pp. 536-544

Kalmijn, A.J. (1997). Electric and near-field acoustic detection, a comparative study. *Acta Physiol Scand,* 161, Suppl 638, pp. 25-38

Lu, Z.; Song, J. & Popper, A.N. (1998). Encoding of acoustic directional information by saccular afferents of the sleeper goby, *Dormitator latifrons. J Comp Physiol, A,* 182, pp. 805-815

Lu, Z.; Xu Z. & Buchser, W.J. (2003). Acoustic response properties of lagenar nerve fibers in the sleeper goby, *Dormitator latifrons. J Comp Physiol, A,* 189, pp. 889-905

Ma, W-L. & Fay, R.R. (2002). Neural representations of the axis of acoustic particle motion in the nucleus centralis of the torus semicircularis of the goldfish, *Carassius auratus. J Comp Physiol, A,* 188, pp. 301-313

McKibben, J.R. & Bass, A.H. (1998). Behavioral assessment of acoustic parameters relevant to signal recognition and preference in a vocal fish. *J Acoust Soc Am,* 104, pp. 3520-3533

McKibben, J.R. & Bass, A.H. (2001). Effects of temporal envelope modulation on acoustic signal recognition in a vocal fish, the plainfin midshipman. *J Acoust Soc Am,* 109, pp. 2934-2943

Moulton, J.M. & Dixon, R.H. (1967). Directional hearing in fishes. In: *Marine Bio-acoustics,* Vol. II, Tavolga, W.N., (Ed.), pp. 187-228, Pergamon Press, New York

Platt, C. (1977). Hair cell distribution and orientation in goldfish otolith organs. *J Comp Neurol,* 172, pp. 283-297

Popper, A.N.; Salmon, A. & Parvulescu (1973). Sound localization by the Hawaiian squirrelfishes, *Myripristis berndti* and *M. argyromus*. *Anim Behav,* 21, pp. 86-97

Pumphrey, R.J. (1950). Hearing. *Symp Soc Exp Biol,* 4, pp. 1-18

Reinhardt, F. (1935). Uber Richtungswharnehmung bei Fischen, besonders bei der Elritze (*Phoxinus laevis* L.) und beim Zwergwels (*Amiurus nebulosus* Raf.). *Z. vergl Physiol,* 22, pp. 570-603

Sand, O. (1974). Directional sensitivity of microphonic potentials from the perch ear. *J exp Biol,* 60, pp. 881-899

Schellart, N.A.M. Wubbels, R.J.; Schreurs, W. & Faber, A. (1995). Two-dimensional vibrating platform in nm range. *Med Biol Eng Comp,* 33, pp. 217-220

Schuijf, A. (1975). Directional hearing of cod (*Gadus morhua*) under approximate free field conditions. *J Comp Physiol, A,* 98, pp. 307-332

Schuijf, A. (1981). Models of acoustic localization. In: *Hearing and Sound Communication in Fishes,* Tavolga, W.N., Popper, A.N. & Fay, R.R., (Eds.), pp. 267-310, Springer-Verlag, New York

Schuijf, A. Baretta, J.W. & Windschut, J.T. (1972). A field investigation on the deiscrimination of sound direction in *Labrus berggylta* (Pisces: Perciformes). *Netherl J Zool,* 22, pp. 81-104

Schuijf, A. & Buwalda, R.J.A. (1975). On the mechanism of directional hearing in cod (*Gadus morhua*). *J Comp Physiol, A,* 98, pp. 333-344

Schuijf, A. & Hawkins, A.D. (1983). Acoustic distance discrimination by the cod. *Nature,* 302, pp. 143-144

Schuijf, A. & Siemelink, M. (1974). The ability of cod (*Gadus morhua*) to orient towards a sound source. *Experientia,* 30, pp. 773-774

Schuijf, A. Visser, C.; Willers, A. & Buwalda, R.J. (1977). Acoustic localization in an ostariophysine fish. *Experientia,* 33, pp. 1062-1063

Vries, H.L. de (1950). The mechanics of the labyrinth otoliths. *Acta Oto-Laryngol,* 38, pp. 262-273

Weeg, M.S. & Bass, A.H. (2002). Frequency response properties of lateral line superficial neuromasts in a vocal fish, with evidence for acoustic sensitivity. *J Neurophysiol,* 88, pp. 1252-1262

Wightman, F. & Kistler, D. (1993). Sound localization. In: *Human Psychophysics,* Yost, W.A.; Popper, A.N. & Fay, R.R. (Eds.), pp. 155–192, Springer-Verlag, New York

Winn, H.E. (1964). The biological significance of fish sounds. In: *Marine Bioacoustics,* Tavolga, W.N. (Ed.), pp. 213-231, Pergamon Press, New York

Wubbels, R.J. & Schellart, N.A.M. (1997). Neuronal encoding of sound direction in the auditory midbrain of the rainbow trout. *J Neurophysiol,* 77, pp. 3060-3074

Wubbels, R.J. & Schellart, N.A.M. (1998). An analysis of the relationship between the response characteristics and toporgaphy of directional- and non-directional auditory neurons in the torus semicircularis of the rainbow trout. *J exp Biol,* 201, pp. 1947-1958

Wubbles, R.J. Schellart, N.A.M. & Goossens, J.H.H.L.M. (1995.) Mapping of sound direction in the trout lower midbrain. *Neurosci Lett,* 199, pp. 179-182

Zeddies, D. Fay, R.; Alderks, P.; Shaub, K. & Sisneros, J. (2010a). Sound Source localization by the plainfin midshipman fish (*Porichthys notatus*). *Journal of the Acoustical Society of America*, 127, pp. 3104-3113

Zeddies, D. Fay, R.; Alderks, P.; Acob & Sisneros, J. (2010b). Sound source localization of a dipole by the plainfin midshipman fish (*Porichthys notatus*). *Journal of the Acoustical Society of America*, 127, pp. 1886 (Abstract)

Frequency Dependent Specialization for Processing Binaural Auditory Cues in Avian Sound Localization Circuits

Rei Yamada and Harunori Ohmori
Kyoto University
Japan

1. Introduction

Localizing sound sources is essential for survival of animals. It enables animals to avoid danger, or to catch their prey. The differences of sound information between two ears, those of interaural time and level difference (ITD and ILD), are important cues for sound source localization. The minimum resolvable angle of sound source separation is less than 30° along the horizontal plane in many species (cat, Casseday & Neff, 1973; rat, Masterton et al., 1975; songbirds, Klump et al., 1986; Park & Dooling, 1991; Klump, 2000), and in some species the resolution is extremely high. In human and in barn owl, the resolvable angle is as small as 1° (Mills, 1958; Knudsen & Konishi, 1979). ITD and ILD cues depend on the head size of animals and are quite small, particularly in small-headed animals. Thus processing of these cues may need specialization of individual neurons and neural circuits. The time and level information of sounds are captured in the cochlea, transformed to trains of action potentials in the auditory nerve fibers, and then transmitted to auditory nuclei in the brainstem. In the brainstem, time and level information are extracted in the cochlear nucleus and then transmitted in parallel pathways which are specialized to process ITD and ILD cues separately (Fig. 1A, indicating the auditory brainstem circuit in birds) (Sullivan & Konishi, 1984; Takahashi et al., 1984; Takahashi & Konishi, 1988; Warchol & Dallos, 1990; Moiseff & Konishi, 1983; Yin, 2002). Furthermore, in the auditory system, neurons are tuned to a specific frequency of sound (characteristic frequency, CF), and ITD and ILD cues are processed by each CF neuron (Brugge, 1992; Klump, 2000). Recently, a series of studies in the chicken have revealed several frequency dependent specializations in ITD coding pathway (Kuba et al., 2005; Yamada et al., 2005; Kuba et al., 2006). These specializations include the type and the density of ion channels, and their subcellular localization. Furthermore, recent observations in mammals and birds indicate that time and level information are not processed independently but rather cooperatively to enhance the contrast of interaural difference cues even at the first stage of processing of these cues in the brainstem auditory nuclei (Brand et al., 2002; Nishino et al., 2008; Sato et al., 2010). In this chapter, we will first summarize what is known about the neural specializations that enable the preciseness of coincidence detection of synaptic inputs, which is central to process the ITD. And then, we will review observations on how the interaction of time and level information of sounds modulates the processing of each ITD and ILD cue.

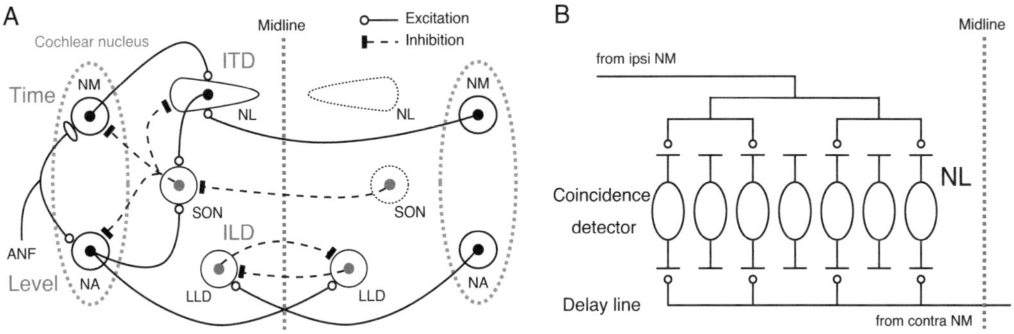

Fig. 1. (A) Schematic diagrams of the auditory brainstem circuits for processing ITD and ILD in birds. (B) Modification of Jeffress model incorporating features of NL of the chick. The contralateral projections from NM to NL form delay lines, while NL neurons act as coincidence detectors of bilateral excitatory inputs. When the sound source moves toward more contralateral locations, spikes from contralateral NM will arrive at NL faster, and bilateral spikes arrive simultaneously at the NL neuron located more laterally.

2. Specialization of ITD coding neurons

Extraction of ITDs in birds is explained on the classical Jeffress model (Jeffress, 1948), which requires delay lines and an array of coincidence detectors (Fig. 1B). Delay lines delay the arrival time of action potential to the coincidence detectors, while the coincidence detectors fire maximally when they receive synaptic inputs simultaneously from both ears. These two elements allow each ITD to be encoded as the place of neuron in the neuronal array. In birds, ITDs are processed in the nucleus laminaris (NL, Fig. 1A) (Konishi, 2003), which is a homologue of the mammalian nucleus of the medial superior olive (MSO). NL is innervated bilaterally from the nucleus magnocellularis (NM). NM extracts fine temporal information of sounds from auditory nerve fibers. In the chicken, the projection fibers from contralateral NM to NL form delay lines (Young & Rubel, 1983; Carr & Konishi, 1988), while NL neurons act as coincidence detectors of bilateral synaptic inputs (Fig. 1B) (Carr & Konishi, 1990; Overholt et al., 1992). Sensitivity to ITDs is extremely high in NL neurons. *In vivo* single-unit studies in the barn owl NL showed that the half-peak width of the ITD tuning curve varies with the CF of neurons, and reaches about 0.1-0.2 ms at 3-7 kHz (Carr & Konishi, 1990; Fujita & Konishi, 1991). This sharpness of ITD tuning of NL neurons should underlie the resolution of a microsecond order of ITDs in the barn owl (Moiseff & Konishi, 1981) and should be determined by the coincidence detection of NL neurons. The cellular mechanism of coincidence detection in NL neurons was studied *in vitro* (Kuba et al., 2003). Experiments were made in brainstem slices of the posthatch chick of P3-P11 at the body temperature of birds (40°C). Under the whole-cell recording, EPSPs were evoked in NL neurons by electrical stimuli applied to both sides of projection fibers from NM, while the time interval between the two stimuli (Δt) was varied (Fig. 2A). The EPSPs were summated to generate an action potential as the interval of two stimuli decreased. The probability of firings peaked at Δt of 0 ms (Fig. 2A and B), and the half-peak width of the coincidence detection curve (time window) was 0.4 ms (Fig. 2B), which is comparable to that observed in the barn owl NL *in vivo* (Carr & Konishi, 1990). What cellular mechanisms underlie to achieve such a high accuracy of coincidence detection?

The acceleration of EPSP time course is essential for the accurate coincidence detection (Kuba et al., 2003) by limiting the time window for the summation of bilateral EPSPs. NL neurons reduce their input resistance extensively by activating several membrane conductances at the resting membrane potential (Reyes et al., 1996; Trussell, 1999; Kuba et al., 2002; Kuba et al., 2003). Among them, the most important is the conductance of low-threshold K^+ current (I_{KLT}). I_{KLT} is mediated by subtypes of voltage-gated K^+ channels, Kv1.1 and 1.2, and in particular, Kv1.2 channels are predominant in the NL (Fukui & Ohmori, 2004; Kuba et al., 2005). Developmentally, I_{KLT} increases nearly fourfold around the hatch, and becomes the dominant conductance at resting potential in NL neurons (Kuba et al.,

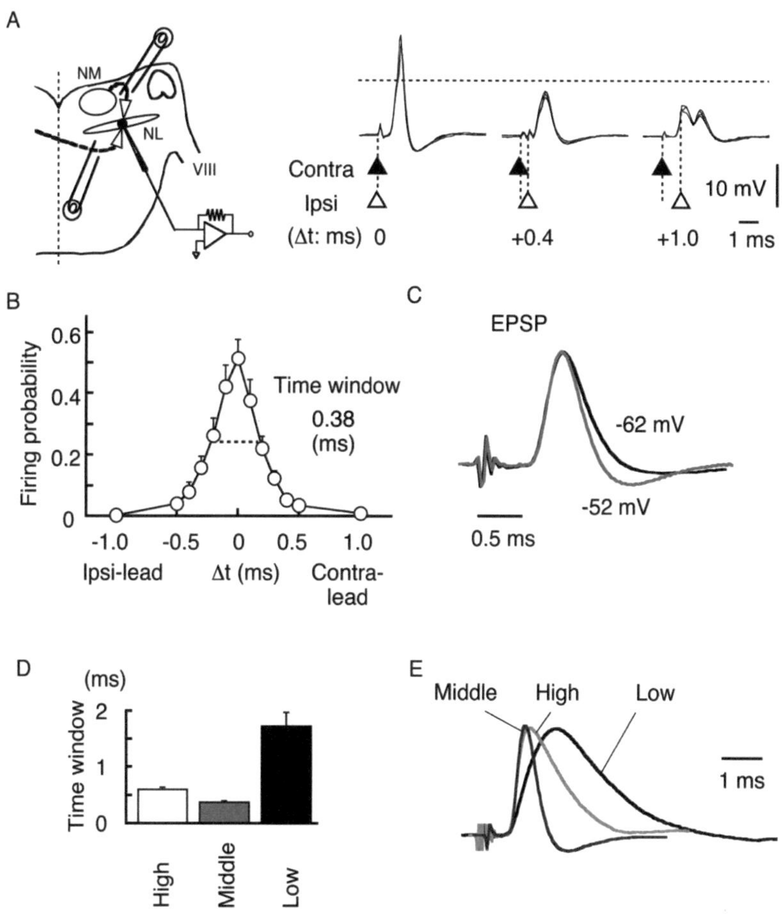

Fig. 2. Rapid EPSP time course is essential for coincidence detection (from Kuba et al., 2003; Kuba et al., 2005). (A) Bilateral EPSPs are evoked at different time intervals (Δt). Spikes are generated when Δt is small. (B) Probability of spike generation as a function of Δt. The time window is indicated by the horizontal broken line. (C) EPSPs from the same NL neurons at different holding potentials. EPSP is accelerated with membrane depolarization (from -62 to -52 mV). Data are from middle CF neurons. (D) Time window of coincidence detection at each CF. (E) EPSPs from each CF are normalized and superimposed. EPSP is fastest and coincidence detection is the most accurate at middle CF.

2002). Moreover, it is activated near the resting membrane potential with rapid kinetics (-60 mV; Rathouz & Trussell, 1998). I_{KLT} is activated by a small membrane depolarization and accelerates the falling phase of EPSP. Consequently, EPSP has a fast time course as fast as EPSC at the resting membrane potential, and is even faster than EPSC with a small membrane depolarization (Fig. 2C). These findings indicate that I_{KLT} plays a crucial role in shortening the time window of coincidence detection to submillisecond order. Recently, a similar developmental increase of I_{KLT} has been reported to shape the EPSPs in the mammalian MSO neurons (Scott et al., 2005).

3. Frequency specific expression of I_{KLT}

Although the range of audible frequencies varies among species, precision is the highest in the middle frequencies in most avian species; thus the acuity of azimuthal sound source localization depends on the sound frequency (Klump, 2000). NL is organized tonotopically; the CF of neurons is high in the rostro-medial (high CF) NL and decreases monotonically to the caudo-lateral (low CF) NL (Rubel & Parks, 1975). ITDs are determined separately by frequency-specific NL neurons. The coincidence detection is dependent on the frequency region of NL (Kuba et al., 2005), and their time window of coincidence detection was the smallest at the middle CF neurons, closely followed by the high CF neurons, and was the largest at the low CF neurons (Fig. 2D). Thus the acuity of coincidence detection is the highest in the middle CF NL neurons.

The EPSP time course is the fastest in the middle CF NL neurons (Fig. 2E). The size of I_{KLT} conductance is the largest at the middle CF. The expression of Kv1.2 channels is the highest in the middle CF neurons, followed by the high CF neurons, and is the lowest in the low CF neurons (Kuba et al., 2005). These observations indicate that the high level of Kv1.2 expression accelerates the EPSPs and determines the tonotopy of the coincidence detection in NL. Thus, the dominant expression of Kv1.2 may underlie the high resolution of sound localization in the middle frequency range in avian species (Klump, 2000).

4. HCN channel

Hyperpolarization-activated cation current (I_h) is another major conductance activated at the resting membrane potential in NL neurons (Kuba et al., 2002). I_h has slow activation and deactivation kinetics, and has the reversal potential positive to the resting membrane potential (-50 to -20 mV) (Pape, 1996). These allow I_h to accelerate the EPSPs in two ways. First, it works as a shunting conductance to shorten the membrane time constant. Second, it depolarizes the resting membrane potential and activates I_{KLT}. Thus, I_h contributes to improve the coincidence detection.

I_h is mediated by HCN (hyperpolarization-activated and cyclic nucleotide-gated) channels and four channel subtypes have been described (HCN1 ~ 4) with different rates of activation and different sensitivities to cyclic nucleotides (Santoro & Tibbs, 1999). Expressions of HCN1 and HCN2 are demonstrated in NL neurons and the level of expression varies along the tonotopic axis (Yamada et al., 2005). HCN1 is expressed highest at the low CF and decreases toward the high CF NL region, while HCN2 is evenly distributed along the tonotopic axis. What is the functional significance of this CF-dependent expression of HCN channels? HCN1 channels have a more positive activation voltage than HCN2 channels (Santoro & Tibbs, 1999). Because of the predominant expression of HCN1 channels, I_h

conductance shortens the membrane time constant and improves the coincidence detection in the middle-low CF NL neurons. In contrast in high CF neurons, the I_h conductance is rather small at the resting potential because HCN2 channels are activated at more negative membrane potentials than the resting level. HCN2 channels are more sensitive to $[cAMP]_i$ than HCN1 channels are, and the increase of $[cAMP]_i$ shifts the voltage-dependence of activation to a positive direction (Ludwig et al, 1998; Santoro et al., 1998; Santoro & Tibbs, 1999). This makes it possible for the high CF neurons to increase the I_h conductance at the resting potential through the elevation of $[cAMP]_i$ (Fig. 3A) (Yamada et al., 2005). Monoamine or acetylcholine is known to modulate I_h by regulating $[cAMP]_i$ (DiFrancesco et al., 1986; DiFrancesco & Tromba, 1988a,b; Bobker & Williams, 1989). In NL, noradrenaline elevates $[cAMP]_i$ and increases the I_h conductance, depolarizes the membrane and accelerates the EPSPs (Fig. 3B). Thus, the acuity of coincidence detection is enhanced by noradrenaline via the modulation of I_h in the high CF neurons (Fig. 3C). A small depolarization of the membrane by the current injection enhanced the coincidence detection almost to the same extent as that caused by depolarization by noradrenaline. This indicates that the noradrenergic effect on the coincidence detection is mediated by the membrane depolarization through the activation of I_{KLT} conductance.

These results raise the possibility that coincidence detection is under sympathetic control. An interesting observation was made in the barn owl (Knudsen & Konishi, 1979). The accuracy of sound source localization was tested by using either a short sound of 75 ms long or a long sound of 1 s long. There was no difference in the error of localization at the initial stage of head orientation whether the test sound stimulus was short or long and whether the sound was a broadband noise or a pure tone; perhaps barn owl measures the ITD at the onset of sound. However, adjustment of the head orientation at the end of a long sound stimulus clearly improved in the middle-high CF ranges (6-8 kHz) (Figure 3 of Knudsen & Konishi, 1979). This improvement might be related to the sympathetic activity when the

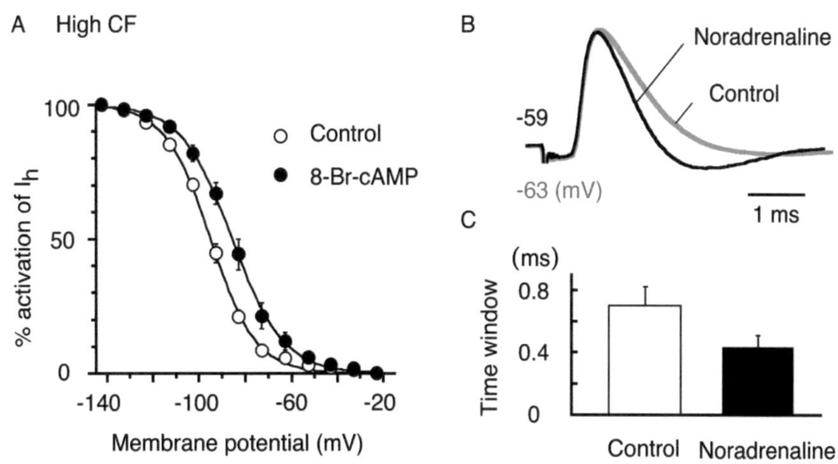

Fig. 3. Enhancement of coincidence detection by noradrenaline at high CF NL neurons (from Yamada et al., 2005). (A) Voltage-dependent activation curve of I_h at high CF. Membrane permeable analogue of cAMP (8-Br-cAMP) shifts the voltage dependence of I_h positively (filled circles). Noradrenaline depolarized the membrane potential, accelerates EPSP (B), and improves coincidence detection (C) at high CF.

animal was exposed to a long sound stimulus. However, the expression pattern of HCN channel subunits has not been examined in owls.

The CF-specific ITD information is integrated across frequencies at higher order nuclei to create an auditory space map (Konishi, 2003). Therefore, the noradrenergic enhancement of coincidence detection in the high CF NL neurons should increase the resolution of sound source localization. Neurons in the nucleus locus ceruleus send noradrenergic projections to almost all regions of the brain (Jones & Moore, 1977), and activities of these neurons are increased during a high arousal state (Aston-Jones & Bloom, 1981). This may suggest that noradrenergic systems are effective to increase the resolution of sound localization when animals are listening carefully to the sounds.

5. Specialization of action potential initiation site along the tonotopic axis

NL neurons are also specialized along the tonotopic axis in initiating action potentials in the axon. The axon initial segment has a high density of Nav channels (Catterall, 1981), and is the site of action potential initiation in many neurons (Mainen et al., 1995; Luscher & Larkum, 1998). However, the electron-microscopic studies indicated that the axon initial segment of NL neurons is myelinated in the chicken and the barn owl (Carr & Boudreau, 1993). Since the myelination was not observed in low-frequency NL neurons (below 1 kHz), they considered that the myelinated initial segment could be a consequence of adaptation for accurate binaural processing of high frequency sounds. This raises questions as to the location and role of action potential initiation site in NL neurons.

The distribution of Nav channels was studied in NL of the chicken (Kuba et al., 2006), and found that Nav1.6 channels are expressed and clustered in the axon, while they are almost absent in the soma. The distribution is different tonotopically, and in the high CF neurons, the cluster of Nav1.6 channels is located at some distance from the soma (50 μm) and stretches a short segment of the axon (10 μm), while it is located closer to the soma (5 μm) and is extended much longer segment (25 μm) in the low CF neurons. Thus, the site of action potential initiation is displaced more distant from the soma as the CF of neurons becomes higher. Consistently, the somatic amplitude of action potentials is small in the high CF NL neurons.

The CF-specific distribution of Nav channels ensures the acuity of coincidence detection. In the high CF neurons, the higher rates of synaptic inputs temporally summate and generate a plateau depolarization at the soma. This depolarization inactivates Nav channels and impedes the generation of action potentials, and consequently reduces the ITD sensitivity of the neuron. A distant localization of Nav channels from the soma may reduce the level of depolarization and the level of inactivation electrotonically. A computer simulation predicted that a distant localization of Nav channels enables the processing of ITD with a high peak-trough contrast (the contrast of firing rate between the peak and trough of the ITD tuning curve) in the high CF neurons.

6. Sound level dependent inhibition modulates the ITD tuning in NL

Processing of ITDs in NL *in vivo* is affected by sound loudness. Loud sound was expected to reduce the peak-trough contrast by simulation (Dasika et al., 2005). However, the peak-trough contrast was maintained rather at high sound pressure level in the barn owl (Pena et al., 1996). They proposed that inhibition from the superior olivary nucleus (SON) controls

ITD tuning in NL, rendering it tolerant to sound pressure level. The level information of sound is extracted in the nucleus angularis (NA), which is another subdivision of cochlear nucleus (Fig. 1A). The SON receives excitatory inputs from the NA and makes an inhibitory innervation to NA, NM, and NL in a sound-level-dependent manner (Lachica et al., 1994; Yang et al., 1999; Monsivais et al., 2000; Burger et al., 2005; Fukui et al, 2010). By recording single unit activity in NL of chicken *in vivo*, the ITD tuning in NL is found being controlled by both the frequency and level of sounds (Nishino et al., 2008). In the following discussion, best frequency (BF) is used as an alternative to CF. BF is the sound frequency at which the neuron generates spikes at the highest rate, while CF is the frequency at which neurons are driven at the lowest level of sound.

The peak-trough contrast of ITD tuning in the low BF units (BF lower than 1 kHz) became larger as the sound became louder, and was maintained high even at the loudest sound levels (Fig. 4A). After electrical lesion of the SON, the peak-trough contrast of ITD tuning curve collapsed at loud sound levels in the low BF NL neurons (Fig. 4B). In contrast, the peak-trough contrast of the middle-high BF units (higher than 1 kHz) was maximized at the intermediate sound pressure level and was practically lost when a loud sound was applied, which was similar to that of the low BF units after the lesion of SON. Furthermore, the level dependence of peak-trough contrast of middle-high BF neurons was not different from the control after the lesion of SON. These observations demonstrated that the BF dependence of level-dependent ITD tuning reflects the BF dependence of SON control on NL. The pattern and density of the SON projection to NL is correlated with this BF dependent effect of the SON. The GABAergic projection from SON to NL is robust in the low BF region of the nucleus and is less prominent towards the high BF region (Nishino et al., 2008). Therefore, the dense inhibitory projection from SON to NL is concluded to regulate the ITD tuning in NL.

The computer simulation that is based on a NEURON model reproduced a level dependence of ITD tuning in NL neurons (Nishino et al., 2008). The simulation further showed that without balance in the bilateral excitation, the peak-trough contrast of ITD tuning lost tolerance to the loud sounds. The SON inhibition might also play a role in maintaining the balance of excitation from NM on the two sides (Dasika et al., 2005).

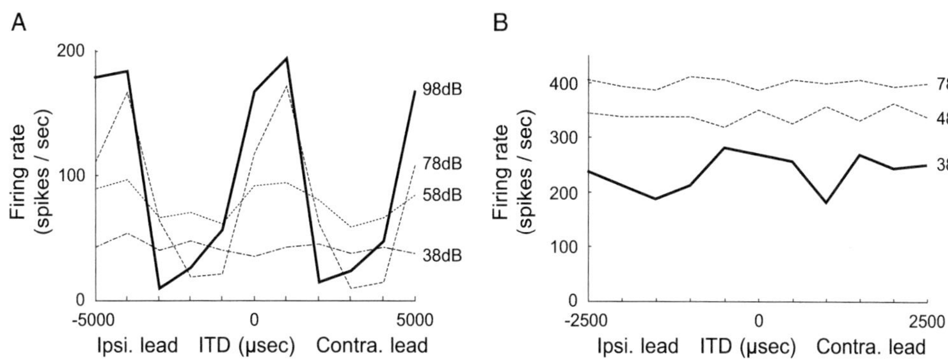

Fig. 4. ITD tuning to a pure-tone sound stimulus of low BF unit in NL (from Nishino et al., 2008). (A) ITD tuning curves from a low BF NL unit (200 Hz) with different sound pressure levels. The solid line indicates the ITD tuning curve of best peak-trough contrast. (B) ITD tuning curves from a low BF NL unit (400 Hz) after SON lesion.

7. ILD coding

In birds, the interaural level differences ILDs are processed in the dorsal lateral lemniscal neurons (LLD). The LLD receives excitatory inputs from the contralateral NA and inhibitory inputs from the ipsilateral NA via the contralateral LLD (Manley at al., 1988; Takahashi & Konishi, 1988; Mogdans & Knudsen, 1994; Konishi, 2003). Therefore, LLD neurons are excited by contralateral sound and inhibited by ipsilateral sound, and encode ILDs by comparing the sound level between two ears (Fig. 5A). However, small head diameter of the animal and the limited audible frequency range (< 4kHz) may limit the physiological relevant range of ILD to about ±5 dB or narrower in the chicken. By recording single unit activity in NA and LLD of chicken *in vivo*, the neural activity in these neurons was found being affected by the interaural phase difference (IPD), which is a frequency-independent formula of ITD, through acoustic interference across the interaural canal that connects the middle ears of the two sides in birds (Sato et al., 2010).

The firing of the NA unit increased monotonically not only by the ipsilateral sound but also by the contralateral sound, whereas the sensitivity was lower (about 15 dB) with the contralateral sound. Activity in the NA is affected by strong contralateral sound through the interaural canal, an air-filled connection between the two middle ear cavities (Fig. 5A). During the binaural sound stimulus, the interaction of contralateral sound shows IPD dependence (Fig. 5B). Increasing the level of out-of-phase (IPD = 180°) contralateral sound monotonically increased the firing rate of the NA neurons, whereas increasing the in-phase (0°) sound produced a local minimum (dip-ILD) and then increased the firing rate, and the depth of the dip was affected by the IPD (Fig. 5B). According to the NA activity, the LLD unit is strongly modulated by the IPD. LLD neurons are activated by contralateral NA activity and are inhibited by ipsilateral NA activity. Therefore, the firing activity of LLD neurons is high at negative ILDs (ipsi < contra) and declines to positive ILDs (ipsi > contra). Fig. 5C shows a unit that exhibited a low firing rate when the sound level was not different in two ears. The firing activity was nearly absent at 0 dB to positive ILDs, demonstrating a strong ipsilateral inhibition on this unit. Another unit (Fig. 5D) fired robust even when the sound to the ipsilateral ear was loud (positive ILDs). The ipsilateral inhibition may not be strong in this unit. The rate-ILD relationship varied with the IPD in both units, and the firing rate was lowest for the in-phase sound (0° IPD, thick solid lines), and the rate increased in most cases when IPD was included, to some extent.

In the open field, any displacement of the sound source from the midline must cause a correlated change in both the level and phase of sounds between two ears. When the sound source is presented at the midline, the ILD is 0 dB and IPD is 0°. A sound source displacement towards the contralateral ear generates negative ILD and positive IPD in the binaural sounds (by definition), and towards ipsilateral ear generates positive ILD and negative IPD (Fig. 5C and D). With any IPD, the firing rate of most units increases (Fig. 5C and D). Therefore, the responsiveness of the LLD units to small changes of ILD, namely the slope of rate-ILD relationship, is increased toward the cotralateral ear (negative ILD) and decreases toward the ipsilateral ear (positive ILD) corresponding to the respective displacement of the sound source from the midline.

Consequently, the modulation of neuronal activity by IPD enhances the responsiveness of LLD neurons to the contralateral field. Any particular dependence of this enhancement on the BF was not found; however the sample numbers were small and most recordings were made in high-BF LLD units.

A simple model is proposed to explain the interaural coupling effects and IPD modulation of LLD activity (Sato et al., 2010), and concluded that the modulation of neuronal activity by IPD increases the sensitivity of LLD neurons to the contralateral field, and may improve the processing of small ILD cues.

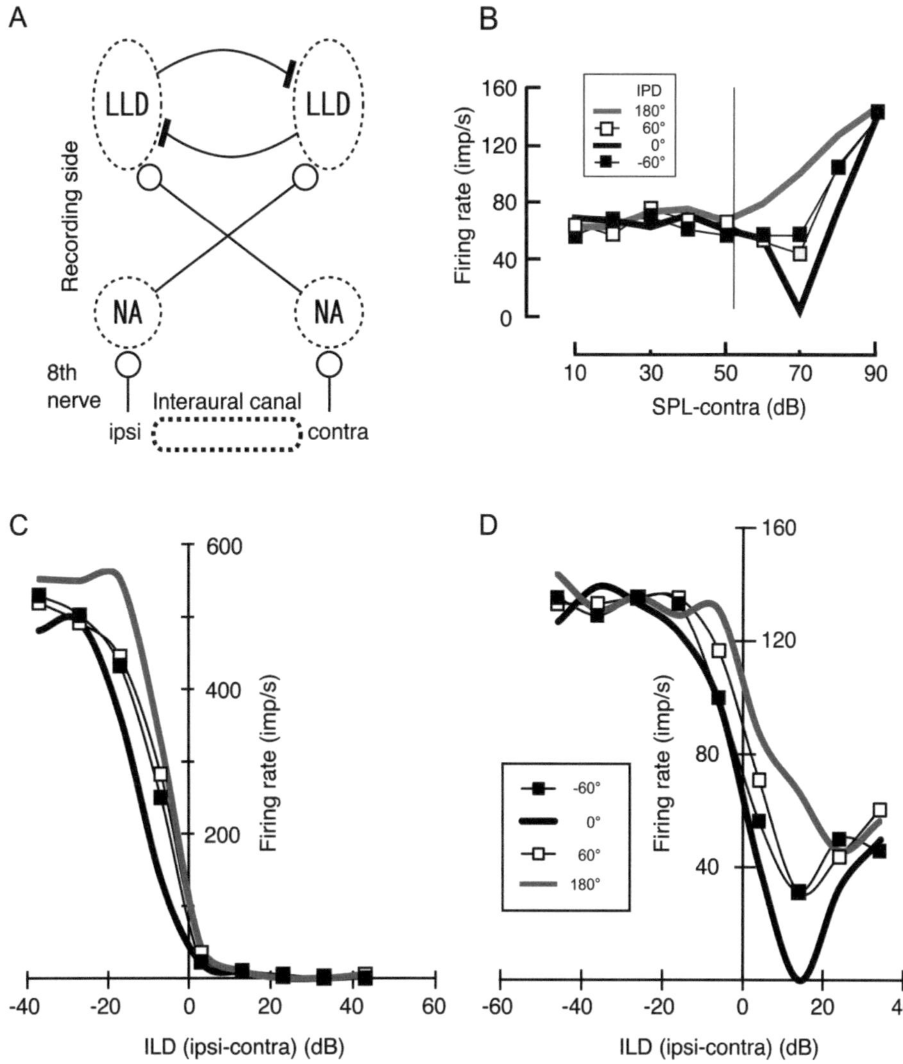

Fig. 5. IPD modulates the neural activity in the NA and LLD (from Sato et al., 2010). (A) Schematic diagrams to show the ILD processing circuit in the brainstem. Open circles indicate excitatory projections and filled bars indicate inhibitory projections. (B) The firing rate of NA unit (BF 200 Hz) as a function of contralateral sound pressure level (SPL). Ipsilateral sound (52 dB SPL) is constant at 20 dB above the threshold. A vertical thin line indicates the 0 ILD in this unit. Binaural sound is presented at four IPDs, as indicated by the different symbols. (C and D) IPD-dependence of rate-ILD relationship of two typical LLD units. The inset indicates IPDs applied to both (C) and (D).

8. Comparison to mammals

MSO neurons have several morphological and biophysical features common to NL neurons (Oertel, 1999; Trussell, 1999). These include bipolar dendrites (Scheibel & Scheibel, 1974), rapid time course of EPSCs (Smith et al., 2000), and large conductance of I_{KLT} and I_h (Smith, 1995; Svirskis et al., 2002). Furthermore, channel molecules underlying the synaptic and membrane conductances are also common between MSO and NL (Parks, 2000; Rosenberger et al., 2003; Koch et al., 2004), suggesting that the two structures share some common mechanisms for enhancing the coincidence detection of binaural excitatory inputs. However, no tonotopic specializations have been reported in the morphological and biophysical features in MSO. This might be related to the limited frequency range that mammals use for the ITD extraction (below 1.5 kHz; Heffner & Heffner, 1988). Nevertheless, more thorough studies need to be conducted in MSO along the tonotopic axis.

Single unit recordings from the MSO of gerbils revealed that glycinergic inhibition improved ITD processing for low-frequency sound (Brand et al., 2002). Suppression of inhibition by the iontophoretic application of strychnine increased the firing rate of MSO neurons and shifted the peak of ITD tuning curves from contralateral-leading ITD to 0 ITD. They concluded that precisely timed inhibition from the contralateral ear via the medial nucleus of the trapezoid body (MNTB) precedes the excitatory input from that side and creates an effective delay in the excitatory response, which is essential for ITD coding (Brand et al., 2002). The cell in MNTB is a relay neuron, which receives excitatory input from contralateral globular bushy cells in the anteroventral cochlear nucleus, and projects ipsilaterally to MSO and lateral superior olive (LSO) (Spangler et al., 1985; Adams & Mugnaini, 1990; Cant & Hyson, 1922). The MNTB neurons are also sensitive to the sound level (Tollin & Yin, 2005). In fact, the ITD tuning of MSO neurons could be maintained even at loud sound (Pecka et al., 2008). It has also been shown that the processing of ILD in LSO, which is a homologue of the LLD in birds, depends critically on timing; the timing of contralateral inhibition through MNTB has to be matched with ipsilateral excitation (Finlayson & Caspary, 1991; Smith et al., 1993; Joris & Yin, 1995; Tollin & Yin, 2005). These evidences suggest that also mammals may use the time and level information of sounds cooperatively to extract ITD and ILD cues.

9. Conclusion

We reviewed here how the ITD and ILD cues are precisely processed basing on the *in vitro* and *in vivo* researches conducted in the chicken auditory brainstem. In the ITD coding circuit, NL neurons show several functional as well as morphological refinements along the tonotopic axis to enhance the coincidence detection at each frequency. In particular, the expression of channel molecules is highly organized in NL neurons to regulate auditory coincidence detection across frequencies. We need to know further how the subcellular localization of these molecules contributes to the computation of neurons and also to the behavior of animals. In addition, new evidences suggest that time and level information of sounds are used not independently but rather cooperatively to improve the processing of both ITD and ILD cues. Interaural difference cues can be small, particularly for an animal with a small head. Both mammals and birds may use similar strategies to compensate the small interaural difference cues for the accurate sound source localization.

10. References

Adams, J.C. & Mugnaini, E. (1990). Immunocytochemical evidence for inhibitory and disinhibitory circuits in the superior olive. *Hear. Res.*, 49, 281-298,

Aston-Jones, G. & Bloom, F.E. (1981). Activity of NE-containing locus coeruleus neurons in behaving rats anticipates fluctuations in the sleep-waking cycle. *J. Neurosci.*, 1, 876-886,

Bobker, D.H. & Williams, J.T. (1989). Serotonin augments the cationic current I_h in central neurons. *Neuron*, 2, 1535–1540,

Brand, A.; Behrend, O.; Marquardt, T.; McAlpine, D. & Grothe, B. (2002). Precise inhibition is essential for microsecond interaural time difference coding. *Nature*, 417, 543-547,

Brugge, J.F. (1992). An overview of central auditory processing, In: *The mammalian auditory pathway: Neurophysiology*, Popper, A.N. & Fay, R.R., (Ed.), 1-33, Springer-Verlag, New York

Burger, R.M.; Cramer, K.S.; Pfeiffer, J.D. & Rubel, E.W. (2005). Avian superior olivary nucleus provides divergent inhibitory input to parallel auditory pathways. *J. Comp. Neurol.*, 481, 6-18,

Cant, N.B. & Hyson, R.L. (1992). Projections from the lateral nucleus of the trapezoid body to the medial superior olivary nucleus in the gerbil. *Hear. Res.*, 58, 26-34,

Carr, C.E. & Konishi, M. (1988). Axonal delay lines for time measurement in the owl's brainstem. *Proc. Natl. Acad. Sci. USA*, 85, 8311-8315,

Carr, C.E. & Konishi, M. (1990). A circuit for detection of interaural time differences in the brain stem of the barn owl. *J. Neurosci.*, 10, 3227-3246,

Carr, C.E. & Boudreau, R.E. (1993). An axon with a myelinated initial segment in the bird auditory system. *Brain Res.*, 628, 330-334,

Casseday, J.H. & Neff, W.D. (1973). Localization of pure tones. *J. Acoust. Soc. Am.*, 54, 365-372,

Catterall, W.A. (1981). Localization of sodium channels in cultured neural cells. *J. Neurosci.*, 1, 777-783,

Dasika, V.D.; White, J.A.; Carney, L.H. & Colburn, H.S. (2005). Effects of inhibitory feedback in a network model of avian brain stem. *J. Neurophysiol.*, 94, 400-411,

DiFrancesco, D.; Ferroni, A.; Mazzanti, M. & Tromba, C. (1986). Properties of the hyperpolarizing-activated current (If) in cells isolated from the rabbit sino-atrial node. *J. Physiol.*, 377, 61-88,

DiFrancesco, D. & Tromba, C. (1988a). Inhibition of the hyperpolarization-activated current (if) induced by acetylcholine in rabbit sino-atrial node myocytes. *J. Physiol.*, 405, 477-491,

DiFrancesco, D. & Tromba, C. (1988b). Muscarinic control of the hyperpolarization-activated current (if) in rabbit sino-atrial node myocytes. *J. Physiol.*, 405, 493-510,

Finlayson, P.G. & Caspary, D.M. (1991). Low-frequency neurons in the lateral superior olive exhibit phase-sensitive binaural inhibition. *J. Neurophysiol.*, 65, 598-605,

Fujita, I. & Konishi, M. (1991). The role of GABAergic inhibition in processing of interaural time difference in the owl's auditory system. *J. Neurosci.*, 11, 722-739,

Fukui, I. & Ohmori, H. (2004). Tonotopic gradients of membrane and synaptic properties for neurons of the chicken nucleus magnocellularis. *J. Neurosci.*, 24, 7514-7523,

Fukui, I.; Burger, R.M.; Ohmori, H. & Rubel E.W. (2010). GABAergic inhibition sharpens the frequency tuning and enhances phase locking in the chicken nucleus magnocellularis neurons. *J. Neurosci.*, 30, 12075-12083,

Heffner, R.S. & Heffner, H.E. (1988). Sound localization and use of binaural cues by the gerbil (*Meriones unguiculatus*). *Behav. Neurosci.*, 102, 422-428,
Jeffress, L.A. (1948). A place theory of sound localization. *J. Comp. Physiol. Psychol.*, 41, 35-39,
Jones, B.E. & Moore, R.Y. (1977). Ascending projections of the locus coeruleus in the rat. II. Autoradiographic study. *Brain Res.*, 127, 25-53,
Joris, P.X. & Yin, T.C. (1995). Envelope coding in the lateral superior olive. I. Sensitivity to interaural time differences. *J. Neurophysiol.*, 73, 1043-1062,
Klump, G.M.; Windt, W. & Curio, E. (1986). The great tit's (*Parus major*) auditory resolution in azimuth. *J. Comp. Physiol.*, 158, 383-390,
Klump, G.M. (2000). Sound localization in birds. In: *Comparative Hearing: Birds and Reptiles*, Dooling, R.J.; Fay, R.R. & Popper, A.N. (Eds.), 249-307, Springer, New York
Knudsen, E.I. & Konishi, M. (1979). Mechanisms of sound localization in the barn owl (*Tyto alba*). *J. Comp. Physiol.*, 133, 13-21,
Koch, U.; Braun, M.; Kapfer, C. & Grothe, B. (2004). Distribution of HCN1 and HCN2 in rat auditory brainstem nuclei. *Eur. J. Neurosci.*, 20, 79-91,
Konishi, M. (2003). Coding of auditory space. *Annu. Rev. Neurosci.*, 26, 31-55,
Kuba, H.; Koyano, K. & Ohmori, H. (2002). Development of membrane conductance improves coincidence detection in the nucleus laminaris of the chicken. *J. Physiol.*, 540, 529-542,
Kuba, H.; Yamada, R. & Ohmori, H. (2003). Evaluation of the limiting acuity of coincidence detection in nucleus laminaris of the chicken. *J. Physiol.*, 552, 611-620,
Kuba, H.; Yamada, R.; Fukui, I. & Ohmori, H. (2005). Tonotopic specialization of auditory coincidence detection in nucleus laminaris of the chick. *J. Neurosci.*, 25, 1924-1934,
Kuba, H.; Ishii, T.M. & Ohmori, H. (2006). Axonal site of spike initiation enhances auditory coincidence detection. *Nature*, 444, 1069-1072,
Lachica, E.A.; Rubsamen, R. & Rubel, E.W. (1994). GABAergic terminals in nucleus magnocellularis and laminaris originate from the superior olivary nucleus. *J. Comp. Neurol.*, 348, 403-418,
Ludwig, A.; Zong, X.; Jeglitsch, M.; Hofmann, F. & Biel, M. (1998). A family of hyperpolarization-activated mammalian cation channels. *Nature*, 393, 587–591,
Luscher, H.R. & Larkum, M.E. (1998). Modeling action potential initiation and back-propagation in dendrites of cultured rat motoneurons. *J. Neurophysiol.*, 80, 715-729,
Mainen, Z.F.; Foerges, J.; Huguenard, J.R. & Sejnowski, T.J. (1995). A model of spike initiation in neocortical pyramidal neurons. *Neuron*, 15, 1427-1439,
Manley, G.A.; Koppl, C. & Konishi, M. (1988). A neural map of interaural intensity differences in the brain stem of the barn owl. *J. Neurosci.*, 8, 2665-2676,
Masterton, B.; Thompson, G.C.; Bechtold, J.K. & RoBards, M.J. (1975). Neuroanatomical basis of binaural phase-difference analysis for sound localization: a comparative study. *J. Comp. Physiol. Psychol.*, 89, 379-386,
Mills, A.W. (1958). On the minimum audible angle. *J. Acoust. Soc. Am.*, 30, 237-246,
Mogdans, J. & Knudsen, E.I. (1994). Representation of interaural level difference in the VLVp, the first site of binaural comparison in the barn owl's auditory system. *Hear. Res.*, 74, 148-164,
Moiseff, A. & Konishi, M. (1981). Neuronal and behavioral sensitivity to binaural time differences in the owl. *J. Neurosci.*, 1, 40-48,

Moiseff, A. & Konishi, M. (1983). Binaural characteristics of units in the owl's brainstem auditory pathway: precursors of restricted spatial receptive fields. *J. Neurosci.*, 3, 2553-2562,

Monsivais, P.; Yang, L. & Rubel, E.W. (2000). GABAergic inhibition in nucleus magnocellularis: implications for phase locking in the avian auditory brainstem. *J. Neurosci.*, 20, 2954-2963,

Nishino, E.; Yamada, R.; Kuba, H.; Hioki, H.; Furuta, T.; Kaneko, T. & Ohmori, H. (2008). Sound-intensity-dependent compensation for the small interaural time difference cue for sound source localization. *J. Neurosci.*, 28, 7153-7164,

Oertel, D. (1999). The role of timing in the brain stem auditory nuclei of vertebrates. *Annu. Rev. Physiol.*, 61, 497-519,

Overholt, E.M.; Rubel, E.W. & Hyson, R.L. (1992). A circuit for coding interaural time differences in the chick brainstem. *J. Neurosci.*, 12, 1698-1708,

Pape, H.C. (1996). Queer current and pacemaker: the hyperpolarization-activated cation current in neurons. *Annu. Rev. Physiol.*, 58, 299–327,

Park, T.J. & Dooling, R.J. (1991). Sound localization in small birds: absolute localization in azimuth. *J. Comp. Psychol.*, 105, 125-133,

Parks, T.N. (2000). The AMPA receptors of auditory neurons. *Hear. Res.*, 147, 77-91,

Pecka, M.; Brand, A.; Behrend, O. & Grothe, B. (2008). Interaural time difference processing in the mammalian medial superior olive: the role of glycinergic inhibition. *J. Neurosci.*, 28, 6914-6925,

Pena, J.L.; Viete, S.; Albeck, Y. & Konishi, M. (1996). Tolerance to sound intensity of binaural coincidence detection in the nucleus laminaris of the owl. *J. Neurosci.*, 16, 7046-7054,

Rathouz, M. & Trussell, L.O. (1998). Characterization of outward currents in neurons of the avian nucleus magnocellularis. *J. Neurophysiol.*, 80, 2824-2835,

Reyes, A.D.; Rubel, E.W. & Spain, W.J. (1996). In vitro analysis of optimal stimuli for phase-locking and time-delayed modulation of firing in avian nucleus laminaris neurons. *J. Neurosci.*, 16, 993-1007,

Rosenberger, M.H.; Fremouw, T.; Casseday, J.H. & Covey, E. (2003). Expression of Kv1.1 ion channel subunit in the auditory brainstem of the big brown bat, *Eptesicus fuscus*. *J. Comp. Neurol.*, 462, 101-120,

Rubel, E.W. & Parks, T.N. (1975). Organization and development of the brain stem auditory nuclei of the chicken: tonotopic organization of N. magnocellualris and N. laminaris. *J. Comp. Neurol.*, 164, 411-434,

Santoro, B.; Liu, D.T.; Yao, H.; Bartsch, D.; Kandel, E.R.; Siegelbaum, S.A. & Tibbs, G.R. (1998). Identification of a gene encoding a hyperpolarization-activated pacemaker channel of brain. *Cell*, 93, 717-729,

Santoro, B. & Tibbs, G.R. (1999). The HCN gene family: molecular basis of the hyperpolarization-activated pacemaker channels. *Ann. N.Y. Acad. Sci.*, 868, 741–764,

Sato, T.; Fukui, I. & Ohmori, H. (2010). Interaural phase difference modulates the neural activity in the nucleus angularis and improves the processing of level difference cue in the lateral lemniscal nucleus in the chicken. *Neurosci. Res.*, 66, 198-212,

Scheibel, M.E. & Scheibel, A.B. (1974). Neuropil organization in the superior olive of the cat. *Exp. Neurol.*, 43, 339-348,

Scott, L.L.; Mathews, P.J. & Golding, N.L. (2005). Posthearing developmental refinement of temporal processing in principal neurons of the medial superior olive. *J. Neurosci.*, 25, 7887-7895,

Smith, A.J.; Steven, O. & Forsythe, I.D. (2000). Characterization of inhibitory and excitatory postsynaptic currents of the rat medial superior olive. *J. Physiol.*, 529, 681-698,

Smith, P.H.; Joris, P.X. & Yin, T.C. (1993). Projections of physiologically characterized spherical bushy cell axons from the cochlear nucleus of the cat: evidence for delay lines to the medial superior olive. *J. Comp. Neurol.*, 331, 245-260,

Smith, P.H. (1995). Structural and functional differences distinguish principal from nonprincipal cells in the guinea pig MSO slice. *J. Neurophysiol.*, 73, 1653-1667,

Spangler, K.M.; Warr, W.B. & Henkel, C.K. (1985). The projections of principal cells of the medial nucleus of the trapezoid body in the cat. *J. Comp. Neurol.*, 238, 249-262,

Sullivan, W.E. & Konishi, M. (1984). Segregation of stimulus phase and intensity coding in the cochlear nucleus of the barn owl. *J. Neurosci.*, 4, 1787-1799,

Svirskis, G.; Kotak, V.; Sanes, D.H. & Rinzel. J. (2002). Enhancement of signal-to-noise ratio and phase locking for small inputs by a low-threshold outward current in auditory neurons. *J. Neurosci.*, 22, 11019-11025,

Takahashi, T.; Moiseff, A. & Konishi, M. (1984). Time and intensity cues are processed independently in the auditory system of the owl. *J. Neurosci.*, 4, 1781-1786,

Takahashi, T.T. & Konishi, M. (1988). Projections of nucleus angularis and nucleus laminaris to the lateral lemniscal nuclear complex of the barn owl. *J. Comp. Neurol.*, 274, 212-238,

Tollin, D.J. & Yin, T.C. (2005). Interaural phase and level difference sensitivity in low-frequency neurons in the lateral superior olive. *J. Neurosci.*, 25, 10648-10657,

Trussell, L.O. (1999). Synaptic mechanisms for coding timing in auditory neurons. *Annu. Rev. Physiol.*, 61, 477-496,

Warchol, M.E. & Dallos, P. (1990). Neural coding in the chick cochlear nucleus. *J. Comp. Physiol. A*, 166, 721-734,

Yamada, R.; Kuba, H.; Ishii, T.M. & Ohmori, H. (2005). Hyperpolarization-activated cyclic nucleotide-gated cation channels regulate auditory coincidence detection in nucleus laminaris of the chick. *J. Neurosci.*, 25, 8867-8877,

Yang, L.; Monsivais, P. & Rubel, E.W. (1999). The superior olivary nucleus and its influence on nucleus laminaris: a source of inhibitory feedback for coincidence detection in the avian auditory brainstem. *J. Neurosci.*, 19, 2313-2325,

Yin, T.C. (2002). Neural Mechanisms of Encoding Binaural Localization Cues in the Auditory Brainstem. In: *Integrative Functions in the Mammalian Auditory Pathway*, Oertel, D, (Ed), 99-159, Springer-Verlag, New York

Young, S.R. & Rubel, E.W. (1983). Frequency-specific projections of individual neurons in chick brainstem auditory nuclei. *J. Neurosci.*, 3, 1373-1378,

28

Highly Defined Whale Group Tracking by Passive Acoustic Stochastic Matched Filter

Frédéric Bénard[1], Hervé Glotin[2] and Pascale Giraudet[3]
[1,2]*Systems & Information Sciences Laboratory (LSIS - UMR 6168 USTV&CNRS), Université du Sud-Toulon-Var*
[3]*Department of Biology, Université du Sud-Toulon-Var*
France

1. Introduction

In this paper, we compare two low cost time-domain tracking algorithms based on passive acoustics. The problem consists in tracking an unknown number of sperm whales (Physeter catodon). Clicks are recorded on two datasets of 20 and 25 minutes on an open-ocean widely-spaced bottom-mounted hydrophone array. The output of the method is the track(s) of the Marine Mammal(s) (MM) in 3D space and time. Firstly, we briefly review studies of the Stochastic Matched Filter (SMF) detector and its performances with a reflected click cancellation, the Teager-Kaiser-Mallat (TKM) filtering, the source separation methods and the main characteristics of MM signals. Then, we propose a real-time algorithm for MM transient call localization. We also recall the Cramér-Rao Lower Bound (CRLB) Kay (1993) and the confidence ellipses theory to predict the reachable accuracy and compare it to the tracking results. In Section 3 we show and compare results of track estimates with results from specialized teams and compare SMF versus TKM localization. Then, the system is evaluated with the confidence ellipses on the trajectories. Finally, we discuss on the possible dynamic behavior of the whale that these localizations offer, like hunting and foraging strategies.
This paper deals with the 3D tracking of MM using a widely-spaced bottom-mounted hydrophone array in deep water. It focuses on sperm whale clicks. There were previous algorithms developed in the state of the art Giraudet & Glotin (2006a;b); Morrissey et al. (2006); Nosal & Frazer (2006) but none of them has satisfying results for multiple tracks and most of them are far from being real-time. Our main goal is to build a robust and real-time tracking model, despite ocean noise, multiple reflected clicks, imprecise sound speed profiles, an unknown number of MM, and the non-linear time-frequency structure of most MM signals. Background ocean noise results from the addition of several noises: sea state, biological noises, ship noise and molecular turbulence. Propagation characteristics from an acoustic source to an array of hydrophones include multipath effects (and reverberations, Fig. 1), which create secondary peaks in the Cross-Correlation (CC) function that the generalized CC methods cannot eliminate. In Caudal & Glotin (2008b); Glotin et al. (2008), we gave an extension of Giraudet & Glotin (2006b) that shows multiple tracking using TKM. Here we improve this model using SMF which also allows an efficient Inter-Click-Interval and reflected click removal process.

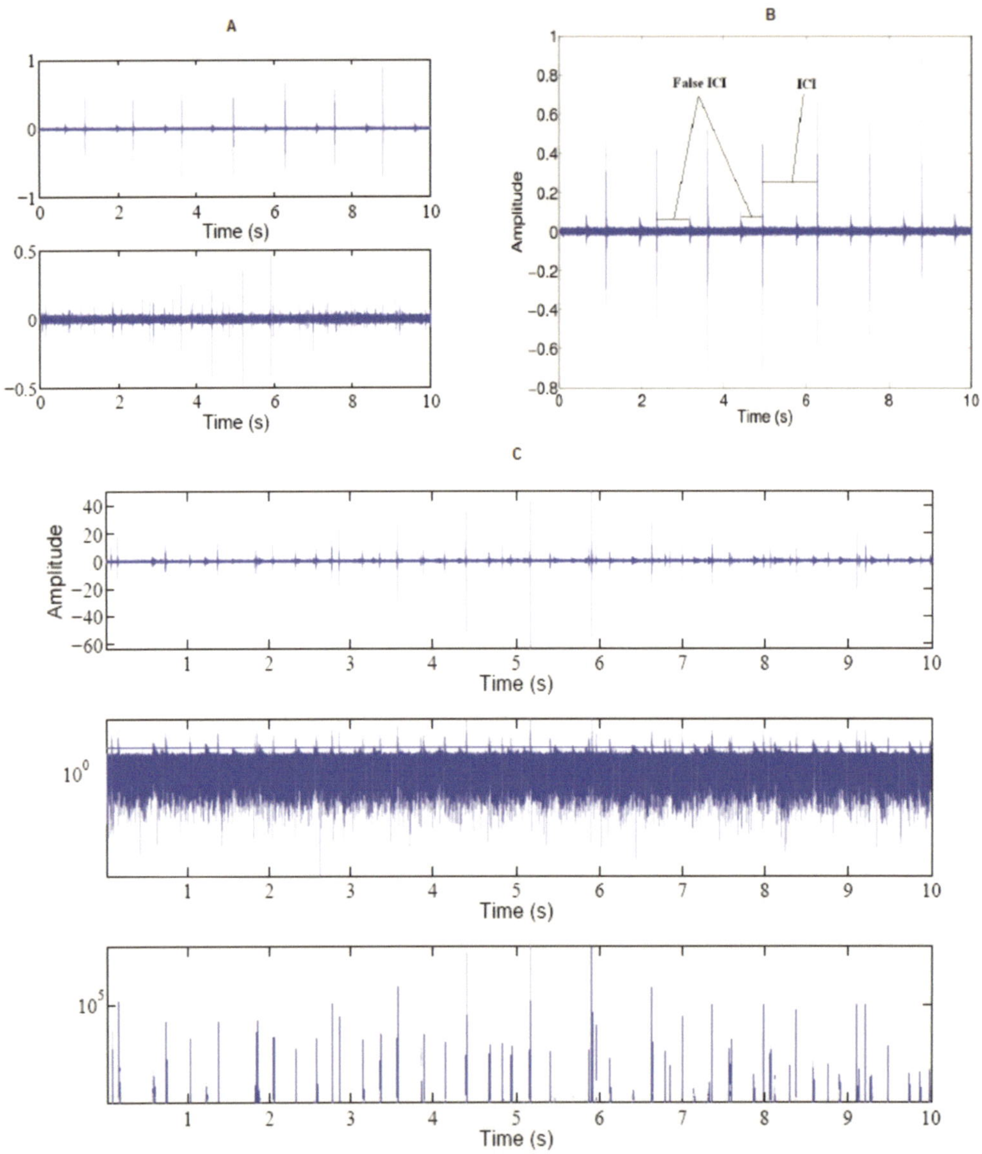

Fig. 1. (A): on the top, a raw signal from dataset2 (D2) and hydrophone 7 (H7) during the first 10 s of recording, containing 7 clicks and their reflected click. At the bottom, 10 s from dataset 1 (D1), hydrophone 1 (H1) containing several (4) simultaneous emitting whale clicks and the reflected clicks. (B): a click train with reflected click from a single sperm whale. We can see an Inter-Click-Interval (ICI), and two false ICI between direct and reflected clicks. (C): Example of a raw multipe whales' signal on H1 (10 s) (top) and the corresponding $\Lambda(x)$ presented in paragraph 3.2 with the threshold in a log-scale (middle) and the thresholded signal (bottom).

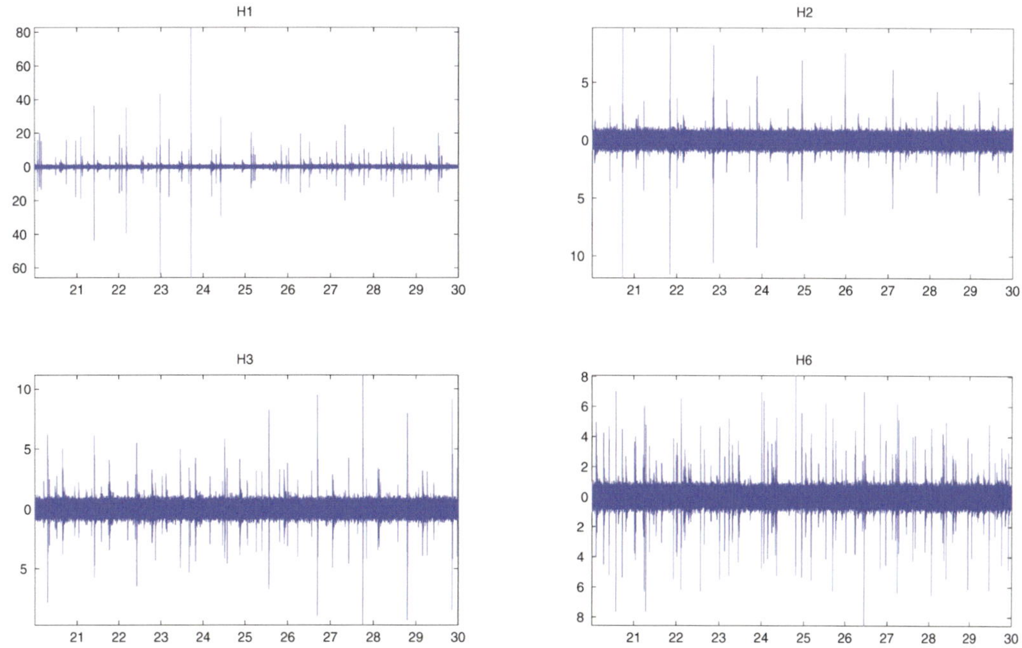

Fig. 2. H1 to H6: plots of a 10 sec samples of raw signals from the four hydrophones of the dataset D1 (time in sec).

2. Hydrophone array characteristics

2.1 Records Settings

D	Hydro	Dist (m)	X (m)	Y (m)	Z (m)
D1	H 1	5428	18501	9494	-1687
	H 2	4620	10447	4244	-1677
	H 3	2514	14119	3034	-1627
	H 4	1536	16179	6294	-1672
	H 5	3126	12557	7471	-1670
	H 6	4423	17691	1975	-1633
D2	H 7	1518	10658	-14953	-1530
	H 8	4314	12788	-11897	-1556
	H 9	2632	14318	-16189	-1553
	H 10	3619	8672	-18064	-1361
	H 11	3186	12007	-19238	-1522

Table 1. Hydrophones positions: Dist=Distance to the barycenter of the set. (H4 and H5 are out of order)

The signals are records of March 2002 from the ocean floor (about 1500 m) near Andros Island - Bahamas (Tab.1), provided with celerity profiles. Datasets are sampled at 48 kHz and contain MM clicks and whistles, background noises like distant engine boat noises. Dataset1 (D1) is recorded on hydrophones 1 to 6 during 20 min (see Fig.2 for a sample view) while the dataset 2 (D2) is recorded on hydrophones 7 to 11 with 25 min length. We will use a constant sound speed with $c = 1500ms^{-1}$ or a linear profile with $c(z) = c_0 + gz$, where z is the depth,

$c_0 = 1542ms^{-1}$ is the sound speed at the surface and $g = 0.051s^{-1}$ is the gradient Caudal & Glotin (2008b). Sound source tracking is performed by continuous localization in 3D using Time Delays Of Arrival (TDOA) estimation from four hydrophones (Tab.1).

2.2 Cramér-Rao lower bound from the hydrophone array geometry

For each hydrophones array, the Cramér-Rao Lower Bound (CRLB) provides the maximum accuracy for the estimation of any source position. Considering a constant sound speed profile, the function model of the Time Delay Of Arrival (TDOA) is defined by:

$$s(\theta) = \frac{1}{c_s}\left[||X_i - \theta|| - ||X_j - \theta||,\ ||X_i - \theta|| - ||X_k - \theta||,\ ||X_i - \theta|| - ||X_l - \theta||\right]^T, \quad (1)$$

where $||\ ||$ denotes the euclidian norm, X_i is the hydrophone i vector coordinate, θ is the unknown parameters vector $[x\ y\ z]^T$ and c_s the celerity. Here $i = 1, j = 2, k = 3, l = 4$. Thus, considering the TDOA noise as a Gaussian process and B its variance-covariance matrix, the Fisher Information matrix is:

$$I_\theta = \nabla_\theta s(\theta) B^{-1} \nabla_\theta^T s(\theta). \quad (2)$$

Then, the CRLB is $B_\theta = I_\theta^{-1}$. The solution error ellipses are contours of constant value of the inner product $\theta I \theta$.

We compute the CRLB (in meter) in the space (x,y,z) and plot the values for both datasets (Fig.3). We consider that the standard deviation of the noise is equal to the quantification noise with a sampling frequency of 480 Hz. The main dependencies of the bounds are the noise and the array configuration. In figures 3.A to F, the CRLB on y and z is shown for a depth of 500 m, and is just about the same for a depth of 1000 m as shown in figures 3.G-H.

3. Filters design

3.1 Teager-Kaiser-Mallat filtering

A sperm whale click is a transient increase of signal energy lasting about 20 ms (Fig.1). Therefore, we use the Teager-Kaiser (TK) energy operator Kandia & Stylianou (2006) on the discrete data:

$$\Psi[x(n)] = x^2(n) - x(n+1)x(n-1), \quad (3)$$

where n denotes the sample number. Considering the raw signal $s(n)$ (sample n of the raw signal) as:

$$s(n) = x(n) + u(n), \quad (4)$$

where $x(n)$ is the signal of interest (click), $u(n)$ is an additive noise defined as a process realization considered Wide Sense Stationary (WSS) Gaussian during a short time. By applying TK to $s(n)$, $\Psi[s(n)]$ is:

$$\Psi[s(n)] \approx \Psi[x(n)] + w(n), \quad (5)$$

where $w(n)$ is a random gaussian process Kandia & Stylianou (2006). The output is dominated by the clicks energy. Then, we reduce the sampling frequency to 480 Hz by the mean of 100 adjacent bins to reduce the variance of the noise. We apply the Mallat's algorithm

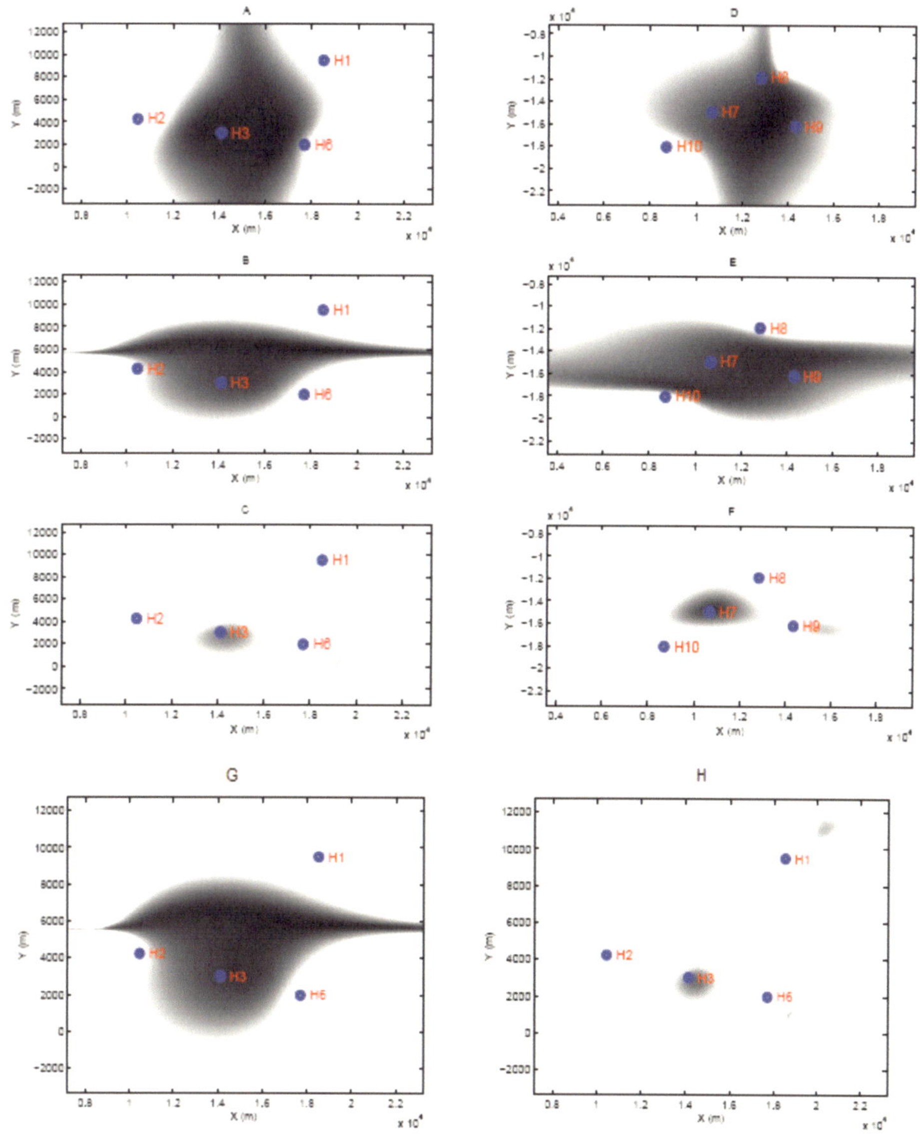

Fig. 3. CRLB values scaled in gray colors with a plan view: black means a null CRLB and white a CRLB≥10 m. For the figures A to F, a depth of 500 m was chosen. (A): CRLB on x values, dataset D1. (B): y, D1. (C): z, D1. (D): x, D2. (E): y, D2. (F): z, D2. (G): CRLB on y axis, plan view, dataset D1, depth=1000 m. (H): CRLB on z axis, plan view, dataset D1, depth=1000 m.

Mallat (1989) with the Daubechies wavelet Adam et al. (2005) (order 3) into the process. The signal is denoised by a universal thresholdingDonoho (1995). This filtering step is very fast and without any parameter. The Fig.5-b shows the filtered signal of multiple emitting MM (Fig.5-a). The click detection Kandia & Stylianou (2006) is not a concern. The purpose is to ameliorate the TDOA estimation, but the reflected clicks are not removed. The TK operator can be used for detection but need an empirical threshold, thus we need a more adaptive method. The purpose of this paper is not a full comparison between the performances of two detectors, but to offer two methods for the whale tracking and the performance relative to these methods.

3.2 The Stochastic Matched Filter

The SMF, which is a filtering method, is employed here for detection. The clicks and sea noise are considered as gaussian stochastic process with 0-mean. Considering a stochastic process s (of length N), the covariance matrix is $E(ss^T) = A$. We also consider an additive, centered and independent noise b with the variance-covariance matrix B. Those processes are not correlated to each other and the matrices are supposed positive and full rank. The SMF theory Courmontagne & Chaillan (2006); Juennard (2007) says that the linear filter h of length N that maximizes the Signal to Noise Ratio (SNR) is the eigen vector solution of:

$$Bh = \lambda Ah, \tag{6}$$

associated to the greatest eigen value λ_0. Thus, we are looking for the eigen values and vectors of $B^{-1}A$. The function used for the detection is:

$$\Lambda(x) = [h^T x]^2, \tag{7}$$

with x a N-length window. Denoting ρ as the SNR gain after filtering, and $\tilde{M}$ a matrix normalized by its trace, we have $\rho = \frac{h^T \tilde{A} h}{h^T \tilde{B} h}$. We work on windows of 20 ms which correspond to a click mean length. A is computed with an average of 1000 sperm whale clicks, and B is calculated directly from the hydrophone signals. After h is calculated for each channel, we are able to filter the signal with one bin of shifting. Thus, we obtain $\Lambda(x)$ (Fig.1.C) with a threshold chosen considering the performance that maximizes the synthetic Receiver Operating Characteristic (ROC).

We compute the ROC for a SNR of -10 dB which corresponds to an emitting whale at about 5 km from the receptor. SNR (dB) is computed with $10log_{10}(\frac{P_s}{P_b})$ with $P_s = \frac{s^T s}{N}$ and $P_b = E(\frac{b^T b}{N})$. The detection rate for 1% false alarm rate of the SMF is 49%. To the contrary of the TKM filter, as we use SMF as a detector, we have a date for each probable click, and thus we can eliminate false detections like reflected clicks.

3.3 Reflected click cancellation

In order to generate robust estimates of the TDOA we avoid relying completely on correlation based techniques. In reality, the SMF detector often detects the reflected clicks present after each click (Fig.1.B). To remove them, we work on each detection date given by the SMF in a channel, considering it as a reflected click or a click, and we discriminate the direct paths and the reflected arrivals. Since the multi-path arrivals pass through the surface layer and are reflected from the sea air interface, they are subject to significant surface reverberation. This causes a temporal elongation (Fig.4.C, White et al. (2006)). Dan Ellis and al. Halkias & Ellis (2006) proposed a cancellation method based on frequential propreties. Here we propose for a simpler temporal discrimination. Therefore, on the raw signal, we smooth each potential click and calculate the sum of the normalized envelope (Fig.4.D). Consequently, the results from direct and reflected paths are significantly different.

A relatively crude threshold allows one to distinguish the majority of reflected signals from the direct arrivals. There is no demand for highly accurate discrimination; subsequent Giraudet & Glotin (2006b) delay estimation algorithm performs well as long as the majority of events surviving discrimination correspond to direct arrivals.

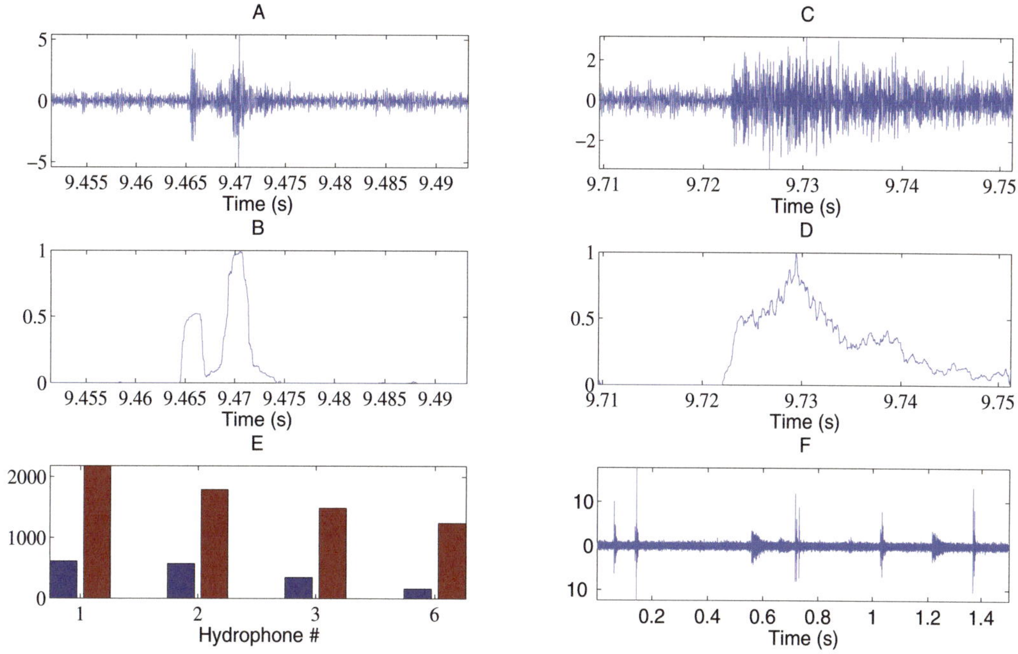

Fig. 4. A direct click (A) with its absolute normalized smoothed envelope (B) compared to a reflected click (C) and its envelope (D). (E): number of reflected clicks (bar on the left) and clicks (bar on the right) detected for each hydrophone in D1 (H4 and H5 are out of order). (F): Raw signal (H1) showing clicks and reflected clicks with several emitting whales.

hydrophone #	7	8	9	10	11
Reflected clicks detected	602	488	473	467	439
Clicks detected	1355	1232	1147	1058	1032
Real number of clicks	1378	1378	1378	1378	1378

Table 2. SMF + reflected click cancellation statistics for dataset 2.

hydrophone #	1	2	3	6
Reflected clicks detected	609	481	245	176
Clicks detected	2129	1894	1461	1293

Table 3. SMF + reflected click cancellation statistics for dataset 1. We do not know yet the exact number of clicks in D1.

The Tab.2-3 summarizes the reflected clicks and clicks detected in both datasets. All clicks in D2 are manually detected. We see that the number of clicks detected with the SMF is varying with the hydrophone because of the different SNR on each hydrophone, and that the reflected clicks are partly rejected (considering one reflected click per click).

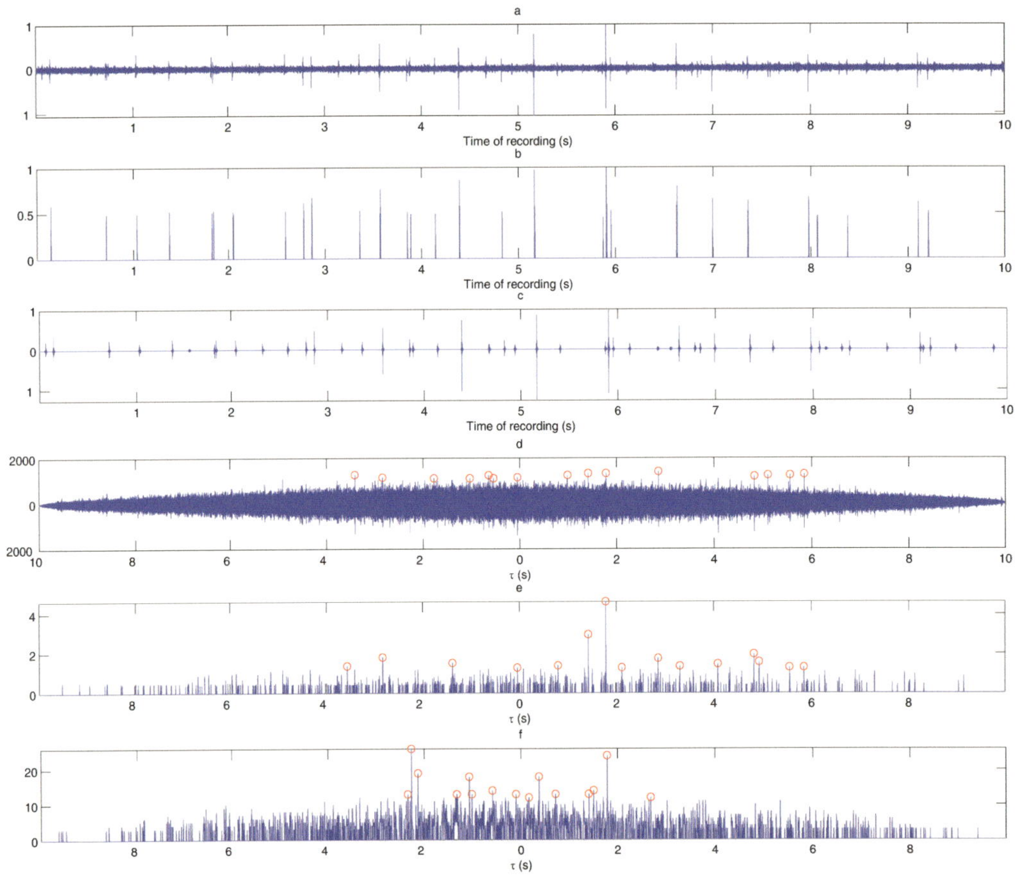

Fig. 5. (a): raw signal for dataset D1 of H1 during the first 10s of recording showing multiple emissions. (b): (a) after TKM filtering. (c): (a) after the SMF. (d): CC between (a) and corresponding raw signal chunk of H3. (e): idem than (d) but with (b). (f): idem than (d) but with (c). Circles on the top of some peaks correspond to the 15 maximum peaks (that are used for localization).

4. Robust and Fast localization method

4.1 Rough TDOA estimation and selection

Either after TKM or after [SMF + reflected click cancellation], we process rough TDOA estimation and selection. The Fig.5-a,c are respectively a raw multiple emitting MM signal, and the corresponding complete filtered signal. First, TDOA estimates are based on MM click realignment only. Every 10 s, and for each pair of hydrophones (i, j), the difference between times t_i and t_j of the arrival of a click train on hydrophones i and j is referred as $T(i, j) = t_j - t_i$. Its estimate $\widetilde{T}(i, j)$ is calculated Giraudet & Glotin (2006a;b) by CC of 10 s chunks (2 s shifting) of the filtered signal for hydrophones i and j. We keep the 15 highest peaks on each CC to determine the corresponding $\widetilde{T}(i, j)$. The filtered signals give a very fast rough estimate of TDOA (precision $\pm$ 2 ms). The Fig.5.d shows the CC with the raw signal, and the Fig.5.e,f with respectively the TKM filtered signal and the SMF. The number of common peaks between TKM, SMF and raw filtering methods are in Tab.4. At most, only 23% of the TDOA are common between the 2 filters, the 23% corresponding to the TDOA of the high SNR clicks. Without any filtering, CC generates spurious delays estimates and the tracks are not correct. Finally, thanks to the $\widetilde{T}$ transitivity constrained system described in Giraudet & Glotin (2006b), we keep $\widetilde{T}$ triplets coming from the same source.

4.2 Localization with a constant and a linear profile

Thanks to the measured delays and an acoustic model based on a constant or a linear sound speed profile, the least squares cost function determines the MM positions using a multiple non-linear regression with the Gauss-Newton method Giraudet & Glotin (2006b) (Levenberg-Marquardt technique Marquardt (1963)). The residuals are approximatively following a Chi-square distribution with $Nc - d$ degrees of freedom, which is noted X^2_{Nc-d}, where Nc is the number of hydrophones couples considered and d the number of unknowns, that are the coordinates (x, y, z). The position is accepted if the residual is inferior to a threshold x, that is calculated solving $P = prob(X^2_{Nc-d} > x)$ with $P = 0.01$ (we keep 99% of the estimates).

5. Results

5.1 Tracking comparisons

In this section, we give the tracks results for TKM and SMF for D1 and D2 dataset. For the dataset D2 (Fig.6), a constant and a linear sound speed profile were used like in Caudal & Glotin (2008b) and the results are similar to those of Morrissey's Morrissey et al. (2006) and Nosal's Nosal & Frazer (2006) methods. The diving profile underlines a bias of about -70 m between the linear and the constant profiles results, emphasizing the importance of the chosen profiles. Moreover with the linear sound speed, the results are about the same as those of Morrissey and Nosal, who used profiles corresponding to the period and place of the recordings. The results on D1 are shown in Fig.7,8,9, and are for a linear sound speed profile. The TKM method lets appear an artifact whale (yielding to 5 whales), which is due to the reflected clicks produced by the whale with the (+) symbol which are eliminated thanks to the reflected click removal with the SMF method (without the reflected click removal, the same virtual whale appears in the SMF results) but we can not apply a reflected click removal on TKM because it is not used as a detector. We thus localize 4 MM with the SMF method. The number of positions estimated for each method is in Tab.4. The confidence regions are computed for both datasets with a Monte Carlo method. The ellipses maxima (30 m) fit with MM length (20 m), which is acceptable.

In the SMF method, there are much more estimated positions because the SMF, detects partically all the clicks and even the ones with a bad SNR. This last method depends on the direction and the distance of the whale to the hydrophone. The signals correlation for the TDOA estimation are thus binary, and do not contain the information of distance, whereas the Teager-Kaiser method just enhances the correlation, where the signal is filtered without detection and thus low click energy results in small correlation, compared with the high click correlation. This is why in the 3D region, when the whale is in the opposite direction and/or emits small energy clicks, the SMF method returns more positions. The result for the left whale (represented with '+' in Fig. 7) in the multiple whales case is a good illustration. In the first trajectory part, the whale is off-axis with all the hydrophones, so the SNR is low on pratically

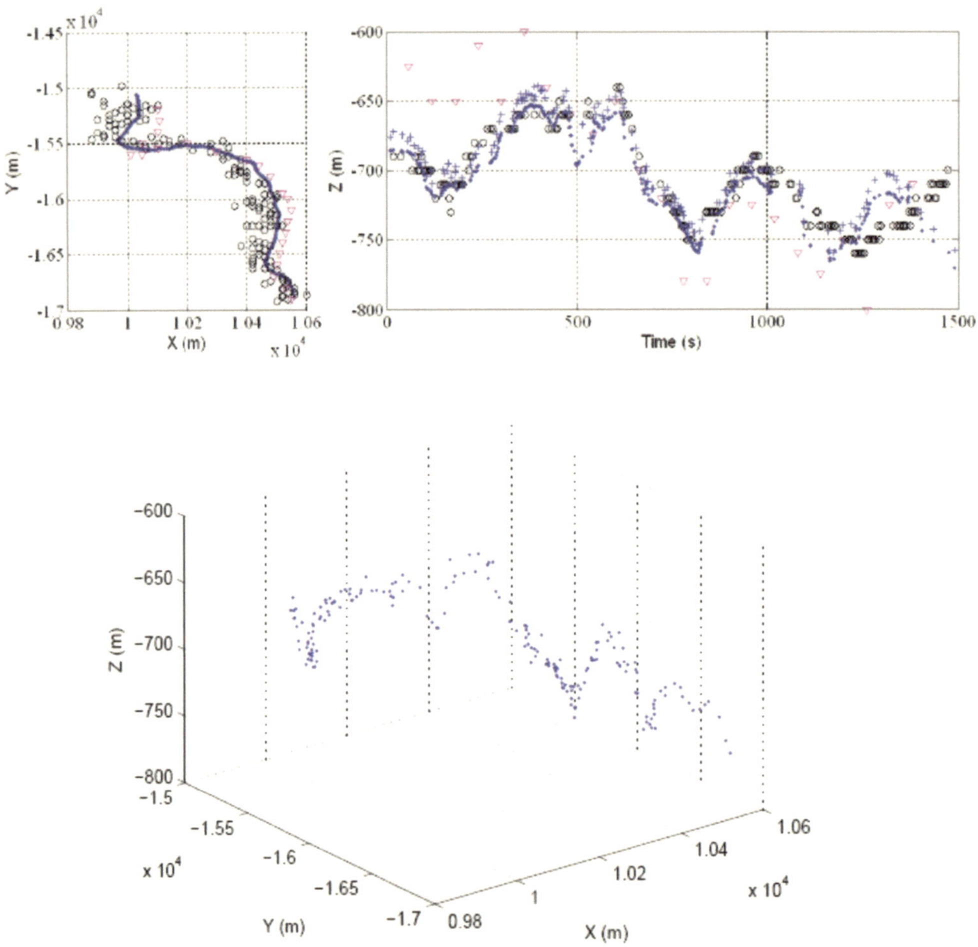

Fig. 6. Top: plan view and diving profile of the MM in D2, our estimates with a linear profile and the SMF method (.) or TKM (+); and estimates of Morrissey's (∇) and Nosal's methods (o). The results with a constant profile underline a bias of about -70 m Caudal & Glotin (2008b). Bottom: 3D plot of the trajectory in D2.

all the hydrophones, and thus, TDOA extraction and localization are difficult with the TKM method (SMF is much better). In the last part, the whale is on-axis with 2-3 hydrophones, then the TDOA and positions estimation are facilitated thanks to the high SNR and then the TKM method generates more positions per time sample relatively to the first trajectory part.

To summarize the confrontation, SMF produces more positions than TKM because it is a more efficient detector. But it does not modify the accuracy of the position estimation, as shown below in section 5.2.

5.2 Confidence region analyses

To compute the ellipses, we apply a Monte Carlo method and a gaussian distribution noise with the standard deviation described above. For each $\widetilde{T}$ realization, the source position is calculated. We deduce the variance and the mean for each position to plot the confidence

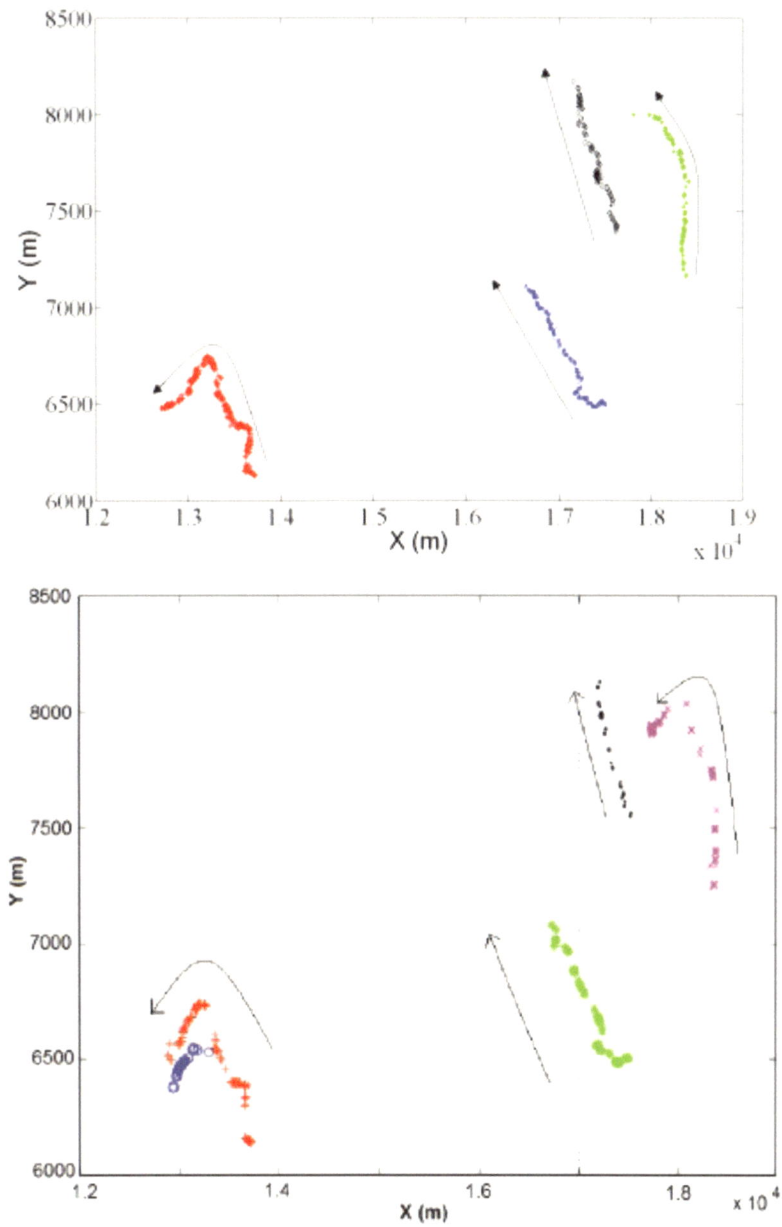

Fig. 7. On the top, plan view of several whale tracks in D1 with the SMF method. Each symbol corresponds to one of the 4 whales. The arrows show the directions. On the bottom, plan view in D1 with the TKM method. Each symbol corresponds to one of the 5 whales. Compared to the SMF, one false whale appears due to the lack of reflected click removal for TKM. Another great difference is the high definition of the SMF track compared to TKM ones, as discribed in Tab.4. A 3D plot of the trajectores with the SMF method can be seen on Fig.8.

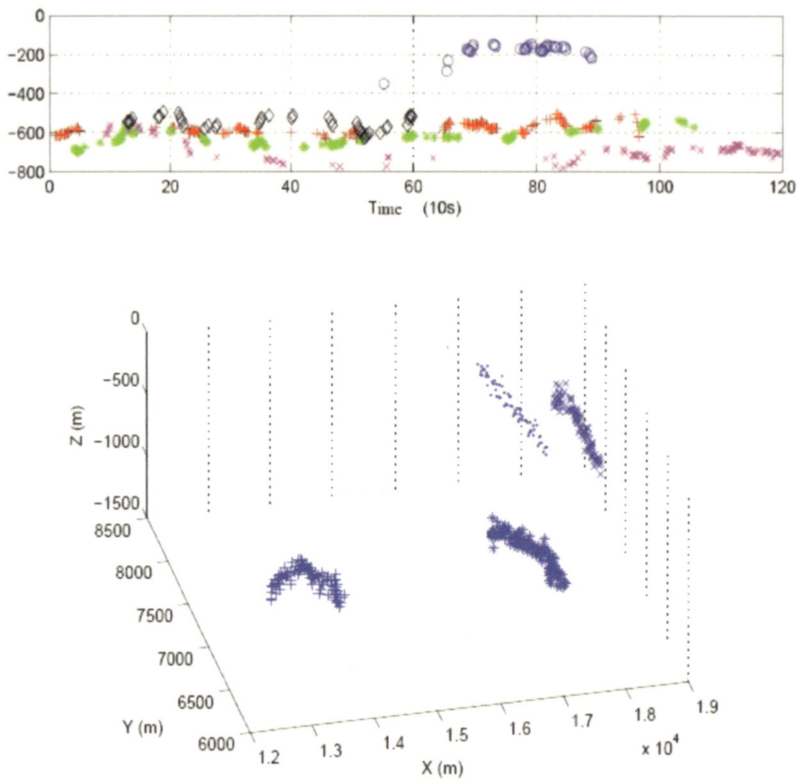

Fig. 8. Top: averaged diving profile in D1 with TKM method. Each symbol corresponds to one whale in Fig.7 ((+), (o), (hexagon), (diamond)). The SMF results in section 5.2 demonstrate that the 'o' is a reflected click of the '+' whale (cf Fig.7). Bottom: 3D plot of the trajectories in D1.

% of common TDOA	Raw	TKM	SMF
Raw	5/13	3%	2%
TKM	12%	103/57	14%
SMF	10%	23%	387/143

Table 4. In diagonal, the number of positions obtained in D1/D2. Above the diagonal of this table, we give the % of common TDOA in D1, and under it, the % of common TDOA in D2.

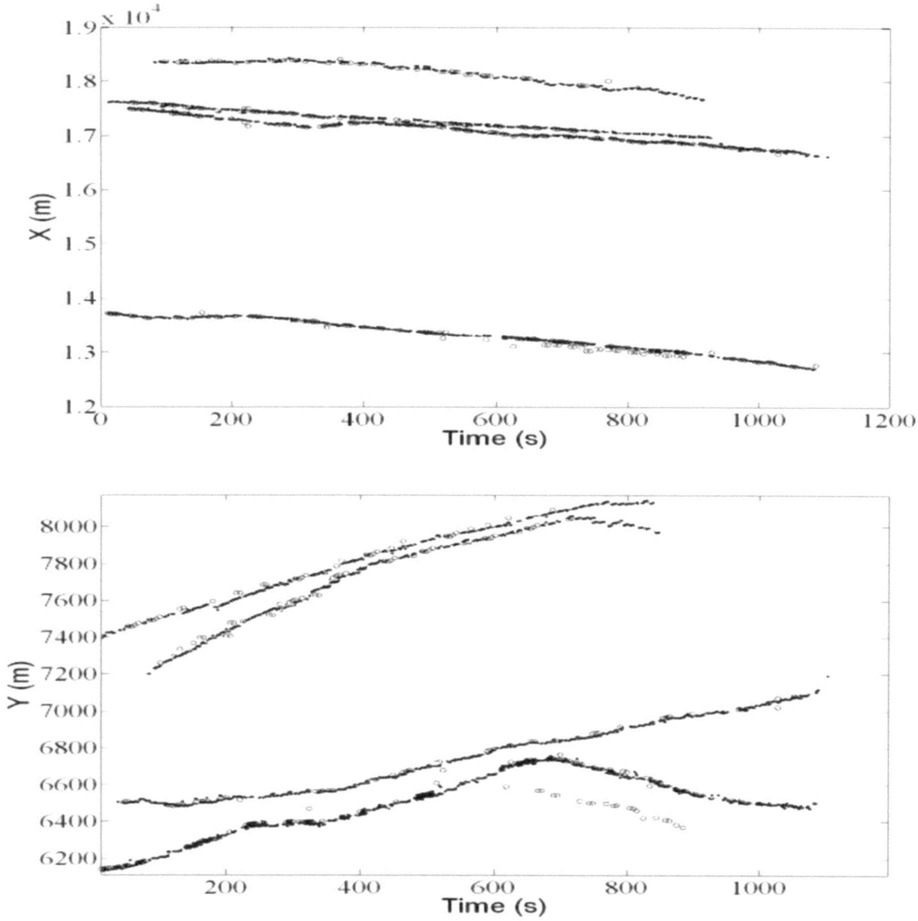

Fig. 9. Comparisons between SMF tracks (.) and TKM tracks (o) in D1, demonstrating that SFM outperforms TKM. X(t) (Y(t)) coordinates are on the top (bottom). We can see that the number of positions is far larger with the SMF than TKM (see Tab.4).

regions with a confidence level of 0.95, which means that there is 0.95 probability for the whale to be in the ellipse centered on the position. In dataset D2, the mean values of the confidence intervals on X, Y, Z axes are about 18, 16 and 30 m (Fig.10). The results confirm that the errors on the vertical axis are meaningfully higher than the other axes because the distance between each hydrophones in this direction is smaller (the maximum difference on the Z axis between hydrophones is 200 m). As estimated by the CRLB analysis in section 2.2, the farthest whales in dataset D1 from the hydrophones array center have a larger uncertainty with an error of about 20 to 30 m on X and Y axes (Fig.10), while the whales close to the center (Fig.10) exhibit an error of about 10 to 20 m like for D2 (Fig.7). These uncertainties are reasonable according to the sperm whale length (20 to 30 meters).
Comparing the CRLB with the ellipses, we can see the correlation between maximum accuracy and confidence regions. Figures 3-C and 3-F show the accuracy on z axis is larger than 10 m

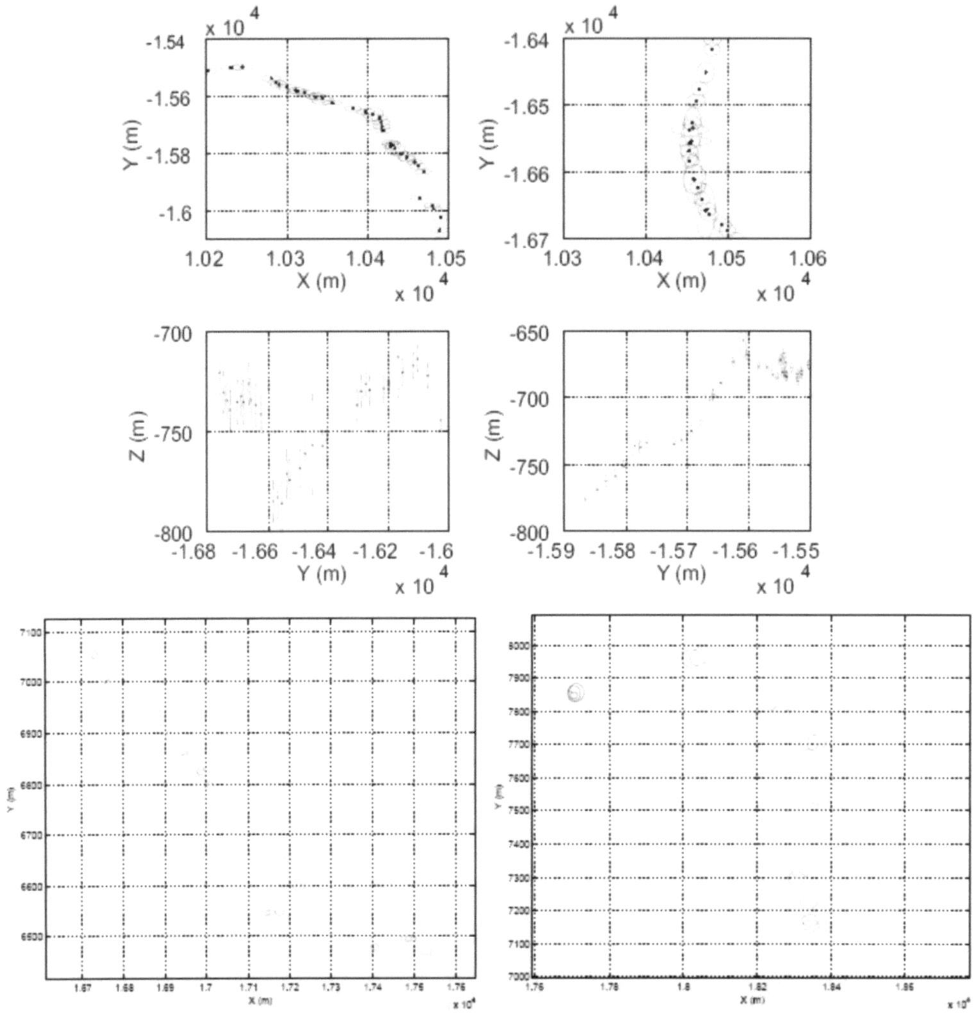

Fig. 10. Top: confidence region projection on (X,Y) and (Z,Y) axes for different trajectories in D2 with the SMF. Bottom left: confidence region projection on X and Y for the whale in the middle ('*' symbol in Fig.7), D1 dataset, trajectory with the SMF. Bottom right: confidence region projection on X and Y for the whale in the right (x symbol), D1 dataset, trajectory with the SMF.

for both datasets in the tracking regions, which is consistent with the ellipses results, but in the case of D1, the diving profile estimation is not suitable, mainly due to the z-component of the CRLB (Fig.3.H). The CRLB (in D1, Fig.3.D-E) also explains that the farthest whale has larger confidence regions. But for both datasets, the CRLB on x and y is far inferior to 10 m inside the array whereas our ellipses are about 10 to 20 m, which is maybe caused by other parameters involved in, like the approximated celerity profile.

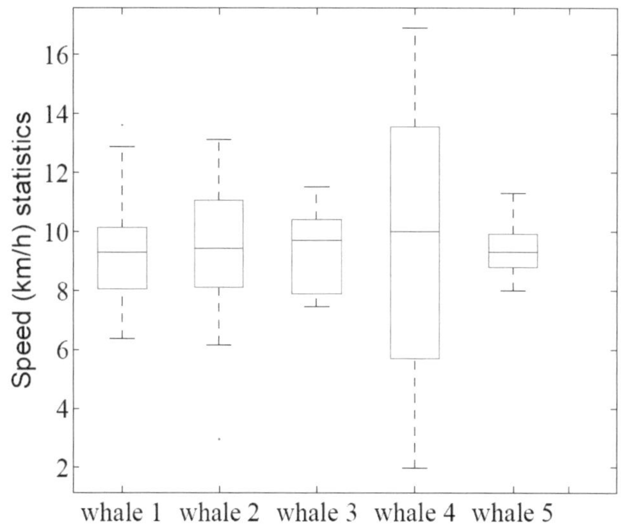

Fig. 11. Speed (averaged on 30 s windows) statistics on the whole set for each whale in dataset D1 (whales 1 to 4, numeroted from left to right) and dataset D2 (whale 5). The central line of the box is the median of speed and the lower and higher lines are the quartiles. The whiskers show the extent of the speed (the average speed is about 9 km/h, likely as regular sperm whale).

6. Perspectives on whale behavior analyses

Our method gives a real-time multi tracks of a whale group. The Fig.11 shows the speed profile for each whale. These localizations allow to use the true TDOA and to label the signal. This leads to a precise Inter-Click-Interval (ICI) extraction Caudal & Glotin (2008a). Other features can also be extracted thanks to this localization, such as the whale speed, the energy of each click, distance to a given hydrophone and head's angles with a given hydrophone. These features would give some relevant informations on the whale's behavior when hunting prey at depth. It is admitted that sperm whales make a slow pitch movement and created a faster pitch or yaw movement in synchronization with the clicking activity. The literature Laplanche et al. (2005),Laplanche et al. (2006) suggested that sperm whales would, at depth, make an asymmetric scan of the surrounding water and that during the search phase, sperm whales would methodically scan a cone-shaped mass of water when searching for prey. This scan would suggest that each sperm whale click is generated to aim in a specific direction and at a specific range. Sperm whales would move physically to change the click beam direction, and control level and ICI to change the click target range. Authors in Laplanche et al. (2005) then pointed out a correlation between click level variations and ICI. The hypothesis explaining such a correlation would be click level control: sperm whales would click slowly at a high source level and faster at a lower source level. We are currently analysing the dependencies of all these results that will validate or note this model.
The Fig.12 summarizes different possible features computation in our framework, and their connections with our tracking algorithm. Actually, these measurements are correlated with each other and offer a new large research field for whale behavior analysis.

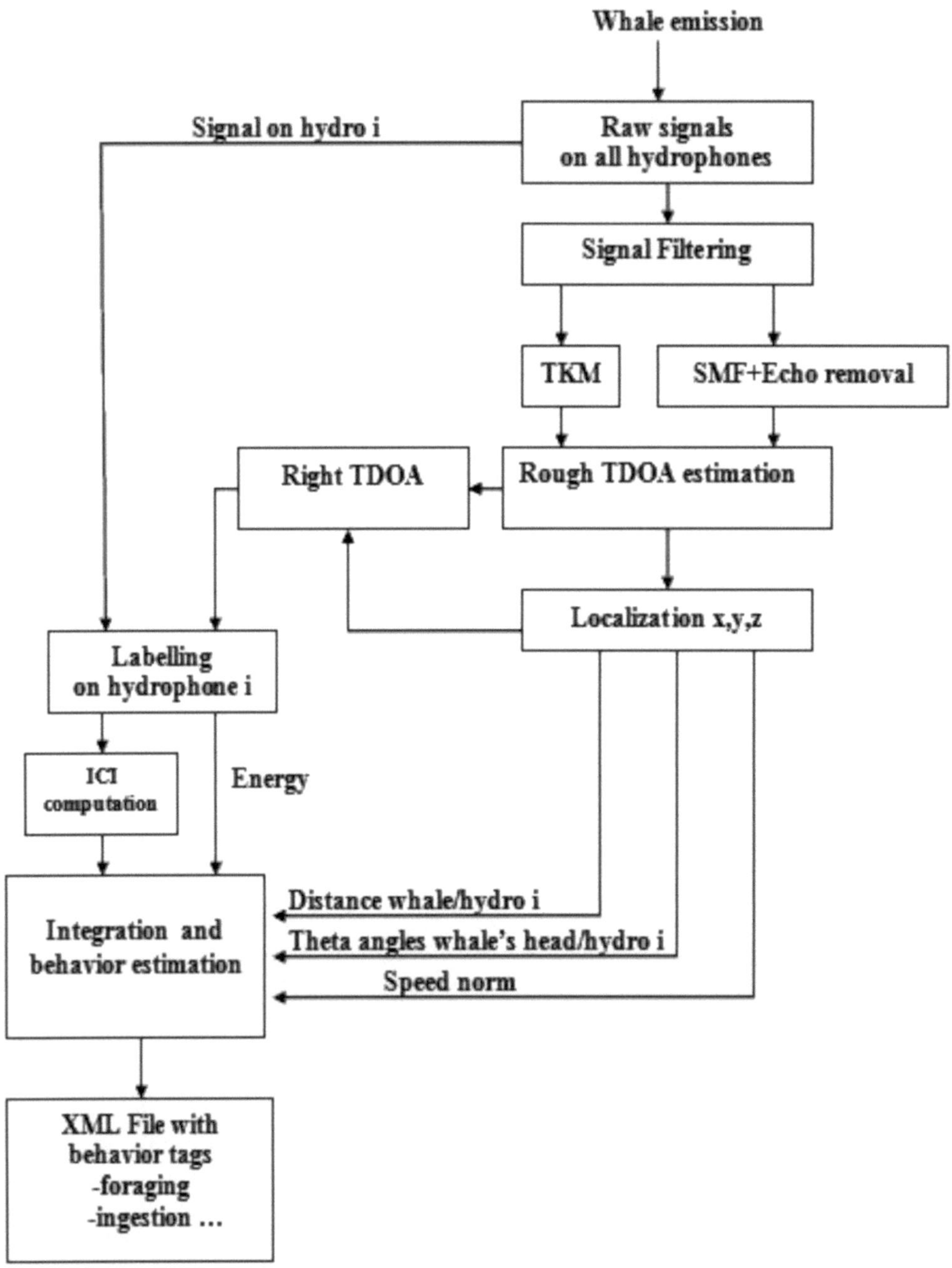

Fig. 12. Block scheme of the possible features extraction.

7. Conclusion

We present in this paper a method using a real-time algorithm for tracking one or multiple emitting sperm whales, in order to analyse whale behavior and to index hydrophones audio files. All the derived features (speed, direction changes) are likely according to the cetologists knowledge. Part of the results presented in this paper are in video demonstrations on http://glotin.univ-tln.fr/PIMC/DEMO.

The localization step is run in real-time on a standard laptop, and works for one or multiple emitting sperm whales. The depth results with a constant speed contain a bias error due to the refraction of the sound paths from the MM to the receivers which a linear speed corrects. Another way to tackle the speed profile issue Glotin et al. (2008) would be to estimate it as a fourth unknown in the regression. However, SMF perfoms better than TKM method (see Tab.4, 103 positions estimated for TKM versus 387 for SMF in D1). The SMF provides a simple way for detecting the sperm whales with good performance, where all the thresholds are learned online and no parameters are needed considering the database. In D2, results indicate that only one sperm whale was emitting in the area, as also analysed by MorrisseyMorrissey et al. (2006) and NosalNosal & Frazer (2006), but not in real-time. Moreover, according to ROSA Lab estimationHalkias & Ellis (2006) based on click clustering, and also preliminary results from the SOEST (School of Ocean and Earth Science and Technology), the number of MM for each 5 min chunks on D1 (Tab.5) is similar to ours. The localization accuracy is computed via the CRLB and the confidence ellipses are correct considering the MM length. The depth error is mainly due to the low precision reachable with the CRLB considering the array configuration.

Our method allows to label the signal and then will be used to extract features presented in Fig.12. Other characteristics such as inter-pulse-interval (a click is composed of multiple pulses), which contain informations about the whale's pitch and yaw behavior. The features discussed above would allow us to analyse whale's foraging and hunting behavior with a good resolution. Finally, we can index the files thanks to the features extracted. Therefore, a XML structure can be generated and include some behavior tags for a rapid access to the acoustic data. Our method offers facilities for robust online passive acoustics behavior studying of clicking MM in open ocean. More research will be conducted to reveal high level features (ICI, behavior, hunting) from the tracks.

5 min chunks	0-5	5-10	10-15	15-20
ROSA Lab	4.3	5.3	4	3.6
PIMC TKM	4	4	4	3
SOEST	4	4	4	2
PIMC SMF	4	4	4	2

Table 5. Counting number estimations of whales in D1. First row is the five minutes chunks of D1, second is the averaged number of whales estimations from ROSA Lab, third and fifth are our estimations (PIMC TKM and SMF). The estimates of the SOEST lab Hawaii are given by personnal communication from E.M.Nosal (preliminary results from Bellop model).

8. Acknowledgments

We thank the Atlantic Undersea Test And Evaluation Center (AUTEC) and the Naval Undersea Warfare Center (NUWC) for having provided the dataset. This research was conducted within the international sea *pôle de compétitivité* at Toulon-France, through "Plateform for Integration of Multimodal Cetacean data (PIMC)". Part of this work is funded by the "Conseil régional Provence-Alpes-Côte d'Azur" France, and Chrisar Software Inc.

Another part of this work has been previously pattented[1]. We thank as well O.Adam, R.Morrissey and E.Nosal for the use of their work.

9. References

Adam, O., Lopatka, M., Laplanche, C. & Motsch, J.-F. (2005). Sperm whale signal analysis: Comparison using the autoregressive model and the wavelets transform, *International Journal of Information Technology* 2: 1–8.

Caudal, F. & Glotin, H. (2008a). Automatic inter-click-interval (ici) and behavior estimation for one emitting sperm whale., *PASSIVE 08 IEEE*.

Caudal, F. & Glotin, H. (2008b). Multiple real-time 3d tracking of simultaneous clicking whales using hydrophone array and linear sound speed profile, *ICASSP IEEE* p. 4p.

Courmontagne, P. & Chaillan, F. (2006). The adaptive stochastic matched filter for sas images denoising, *OCEAN 2006* pp. 1–6.

Donoho, D. L. (1995). De-noising by soft thresholding, *IEEE Trans. IT* 41: 613–627.

Giraudet, P. & Glotin, H. (2006a). Echo-robust and real-time 3d tracking of marine mammals using their transient calls recorded by hydrophones array, *ICASSP IEEE*.

Giraudet, P. & Glotin, H. (2006b). Real-time 3d tracking of whales by echo-robust precise tdoa estimates with a widely-spaced hydrophone array, *Applied Acoustics* 67: 1106–1117.

Glotin, H., Caudal, F. & Giraudet, P. (2008). Whales cocktail party: a real-time tracking of multiple whales, *International Journal Canadian Acoustics* 36(1): 141–147.

Halkias, X. & Ellis, D. (2006). Estimating the number of marine mammals using recordings from one microphone, *ICASSP IEEE*.

Juennard, N. (2007). *Acoustic submarine detection and localisation of very high energy particules*, Thèse de doctorat, University of Toulon.

Kandia, V. & Stylianou, Y. (2006). Detection of sperm whale clicks based on the teager-kaiser energy operator, *Applied Acoustics* 67: 1144–1163.

Kay, S. (1993). *Fundamentals of statistical signal processing*, Prentice Hall, PTR.

Laplanche, C., Adam, O., Lopatka, M. & Motsch, J.-F. (2005). Measuring the off-axis angle and the rotational movements of phonating sperm whales using a single hydrophone, *Journal of Acoustical Society of America.* 119: 4074–4082.

Laplanche, C., Adam, O., Lopatka, M. & Motsch, J.-F. (2006). Male sperm whale acoustic behavior observed from multipaths at a single hydrophone, *Journal of Acoustical Society of America.* 118(5): 2677–2687.

Mallat, S. (1989). A theory for multiresolution signal decomposition: The wavelet representation, *IEEE Transaction on Pattern Analysis and Machine Intelligence* 11: 674–693.

Marquardt, D. W. (1963). An algorithm for least-squares estimation of nonlinear parameters, *SIAM Journal on Applied Mathematics* 11(2): 431–441.

Morrissey, R., Ward, J., DiMarzio, N., Jarvisa, S., & Moretti, D. (2006). Passive acoustics detection and localization of sperm whales in the tongue of the ocean, *Applied Acoustics* 62: 1091–1105.

Nosal, E. & Frazer, L. (2006). Delays between direct and reflected arrivals used to track a single sperm whale, *Applied Acoustics* 62: 1187–1201.

White, P., Leighton, T., Finfer, D., Powles, C. & Baumann, O. (2006). Localisation of sperm whales using bottom-mounted sensors, *Applied Acoustics* 62: 1074–1090.

[1] H. Glotin, P. Giraudet, F. Caudal at Inst. Nat. de la Propriété Intellectuelle, nb 07/06162, (2007).

29

Localising Cetacean Sounds for the Real-Time Mitigation and Long-Term Acoustic Monitoring of Noise

Michel André, Ludwig Houégnigan, Mike van der Schaar, Eric Delory, Serge Zaugg, Antonio M. Sánchez and Alex Mas
Laboratory of Applied Bioacoustics, Technical University of Catalonia, 08800 Vilanova i la Geltrú, Barcelona
Spain

1. Introduction

Noise can have a detrimental effect on cetaceans, as well as on other marine animal species. It can cause stress and increase risk of mortality by interfering with their use of sounds in communication (social behaviour and reproduction) and in navigation (echolocation or bio-sonar to orientate and look for food). Acoustic overexposure, e.g. in areas of heavy shipping, seismic surveys, military exercises, offshore windmills or gas/oil exploration, can lead to hearing loss. While temporary threshold shift (TTS) represents a reversible hearing loss over time, a permanent threshold shift (PTS) results in non-reversible lesions in mammal ears, i.e. a permanent hearing loss caused by long term and/or intense exposure. Although the impact of low to mid frequency (<5kHz) acoustic pollution from the above mentioned human marine activities with regard to cetacean disorientation and death remains poorly understood, available evidence is strongly suggestive of some negative direct or indirect effects: There is an increasing mortality rate from shipping collisions, and cetacean mass strandings after military maneuvers have also been recently related with the use of active sonar, both suggesting that some populations may already be suffering from acoustic impact (i.e. TTS, PTS or blast injuries). The control of noise impact on the marine environment constitutes a scientific challenge and requires a dynamic analysis of the situation based on the parallel development of applied solutions to balance human interests and the conservation of marine species. This objective implies the ambitious synthesis of many advanced acoustic technologies that must be designed to monitor the real-time presence of determined cetacean populations in conflictive areas.

Many cetacean species can be identified by their specific calls. The recording of these signature acoustic signals can reveal their presence in monitored areas. Since sound propagates efficiently in water, the detection range of these signals can be quite large, exceeding 100 km in favourable conditions for low-frequency calls far above visual detection methods. This acoustic potential to non-intrusively detect and monitor cetacean species in their environment gave rise to Passive Acoustic Monitoring (PAM) techniques, for which research is very active. The localisation of whales from their sounds in their habitats was initiated in the 1970s. This was rapidly applied to tracking whales over large distances.

Advances in electronics, computers and numerical analysis now make this PAM technology more accessible and affordable to small research budgets. Various systems have been used, including radio-linked systems, drifting buoys, and arrays of autonomous recorders for versatile and long-term deployments. The goal of such PAM systems, is the continuous mapping of presence and distribution of whales over ocean basins and assessing their densities, sometimes in quasi real-time. Their performance in effectively accomplishing these tasks depends on the characteristics of the targeted cetacean acoustic signals, the environment, the type of equipment used, its deployment and configuration. This performance may significantly vary from case to case.

However, in any case, PAM's success first depends on the capacity to detect and isolate the target signals from the rest of the sounds in which they are imbedded, especially for distant sources and low signal to noise ratios (SNR). The acoustic signal source level, propagation loss, and local background levels determine detection ranges. Moreover, cetacean sounds vary considerably in time-frequency, from infrasonic calls of baleen whales to ultrasonic clicks of toothed whales, and in amplitudes among species and within a species' vocal repertoire. The ocean noise level also exhibits considerable variability in space and time, in response to fluctuating natural sources, such as wind, ice, rain, sounds produced by various organisms and anthropogenic sources such as shipping. Sound speed structures over the water column can focus sounds from distant sources into sound channels. The 3D spatial arrangements of the sources and the hydrophones are therefore relevant to the PAM configuration.

The Laboratory of Applied Bioacoustics of the Technical University of Catalonia has developed PAM solutions to prevent ship collisions with sperm whales but also to detect, classify and track acoustic sources for the long-term study of noise effects on the marine environment at deep-sea underwater observatories. The first system, called WACS (Whale Anti-Collision System) is a passive system designed to monitor the presence of individual cetaceans or objects and transmit the real-time information of their movements to any ship concerned in preventing possible collisions or harmful operations in areas of interest. The WACS original concept is to instrument a safety corridor for marine mammals, within which cetaceans can be detected, classified, localised and their positions notified to vessels using the corridor to permit timely course alterations. WACS is based on passive acoustics detection and ocean ambient noise to locate and identify the marine mammals present in the survey area. WACS integrates two inter-correlated systems: one 3D listening system called Loc3D which allows the 3D detection of the underwater sound sources (distance, azimuth and elevation), and an azimuthal location system (locAz) of the non vocalizing marine mammals by the spatio-temporal contrast produced by the ambient noise in the survey area. The latter is fundamental, because many of the cetacean species (e.g. sperm and probably beaked whales) are silent while at or near the surface or produce very low frequency sounds difficult to detect above background noise (e.g. most baleen whales species). A simulation tool for 3D acoustic propagation was designed where wideband 3D curved ray solution of the wave equation is implemented. This tool was developed to simulate a bi-static solution formed of an arbitrary number of active acoustic sources, an illuminated object, and a receiver all positioned in 3D space with arbitrary bathymetry. The software recreates the resulting sound mixture of direct, reverberated and echoed signals arriving at the array sensors for any array configuration and any number of sources. One object can be placed in the water column and its impact on the acoustic field at the receiver is resolved. The software simulations set bounds as for the concept viability. Detection and bearing estimates could be evaluated for vocalising sperm whales.

In addition to the development and use of PAM techniques for mitigation and prevention of ship collisions, the challenge to assess the large-scale influence of artificial noise on marine organisms and ecosystems requires long-term access of this data. Understanding the link between natural and anthropogenic acoustic processes is indeed essential to predict the magnitude and impact of future changes of the natural balance of the oceans. Deep-sea observatories have the potential to play a key role in the assessment and monitoring of these acoustic changes. ESONET is a European Network of Excellence of 12 deep-sea observatories that are deployed from the Arctic to the Gulf of Cadiz (http://www.esonet-noe.org/). ESONET NoE provides data on key parameters from the subsurface down to the seafloor at representative locations and transmits them in real time to shore. The strategies of deployment, data sampling, technological development, standardisation and data management are being integrated with projects dealing with the spatial and near surface time series. LIDO (Listening to the Deep Ocean environment, http://listentothedeep.com) is one of these projects that is allowing the real-time long-term monitoring of marine ambient noise as well as marine mammal sounds in European waters.

In the frame of ESONET and the LIDO project, vocalising sperm whales were detected offshore the port of Catania (Sicily) with a bottom-mounted (around 2080m depth) tetrahedral compact array intended for real-time detection, localisation and classification of cetaceans. Various broadband space-time methods were implemented and permitted to map the sound radiated during the detected clicks and to consequently localise not only sperm whales but also vessels. Hybrid methods were developed as well which permit to make space-time methods more robust to noise and reverberation and moderate computation time. In most cases, the small variance obtained for these estimates reduces the necessity of additional statistical clustering. Consistent tracking of both sperm whales and vessels in the area have validated the performance of the approach.

The development of these techniques that we present here represent a major step forward the mitigation of the effects of invasive sound sources on cetaceans and monitoring the long-term interactions of noise.

2. The sperm whale sonar

Sperm whales are known to spend most of their time foraging and feeding on squids at depths of several hundreds of meters where the light is scarce. While foraging, sperm whales produce a series of acoustic signals called 'usual clicks'. The coincidence of the continuous production of usual clicks together with the associated feeding behaviour has led authors to suppose that those specific signals could be involved in the process of detecting prey. Because the usual click has known acoustic signal features differing from most of the described echolocation signals of other species, there has long been speculation about the sperm whale sonar capabilities. While the usual clicks of this species were considered to support mid-range echolocation, no physical characteristics of the signal had, until very recently, clearly confirmed this assumption nor had it been explained how sperm whales forage on low sound reflective bodies like squid. The recent data on sperm whale on-axis recordings have shed some light on those questions and allowed us to perform simulations in controlled environments to verify the possible mid-range sonar function of usual clicks during foraging processes (André et al., 2007, 2009).

Research on the acoustic features of sperm whale clicks is well documented, but the obtained quantitative results have varied substantially between publications. Only recently have the intricate sound production mechanisms been addressed with reliable quantitative data (Møhl et al., 2003; Zimmer et al., 2005).

Source level and directionality

In 1980 Watkins reported a source level (SL) of 180 dB *re* 1μPa-m and suggested that clicks were rather omnidirectional (Watkins, 1980), whereas recent results from Møhl et al. estimate this source level to be as high as 223 dB*peRMS* *re* 1μPa-m with high directionality (Møhl et al., 2003). Morphophysiological observations on the unusual shape and weight of the sperm whale nose are in clear agreement with the hypothesis of its highly directional and powerful sonar function, supported by Møhl's results. Goold & Jones (1995) recorded clicks from both an adult male and female and measured a shift to higher frequencies of the main spectral peaks, from 400 Hz to 1.2 kHz, and 2 kHz to 3 kHz, though they noticed that this shift was rather unstable. Spectral contents of clicks as a function of body size and, most importantly, animal orientation information could help to explain this difference in received levels. The almost ubiquitous lack of animal heading information at click recording time in published material makes results hardly usable for a reliable 3D model. To date, Møhl et al. (2003) and Zimmer et al. (2005) are the only studies that provide sufficient calibrated material to produce a correct model. The reported 15 kHz centroïd frequency and apparent source levels higher than 220 dB*RMS* *re* 1μPam corroborate the fact that most previously published click levels and characteristics certainly stemmed from off-axis recordings or unsuitable recording bandwidth. Sperm whale click source level and time-frequency characteristics can be predicted by inferring a threedimensional model, which is based upon well-known physics principles, such as the direct relationship between the size of the sound production apparatus and its directionality (Tucker & Glazey, 1966).

Click time-frequency characteristics

Acoustic recordings of distant sperm whales have often revealed the multi-pulsed nature of their clicks, with interpulse intervals that may be related to head size or more specifically the distance between the frontal and distal air sacs situated at both ends of the spermaceti organ (Alder-Frenchel, 1980). While the utility of this multipulsed pattern is unclear, Møhl et al. (2003) have shown that one single main pulse appears for on-axis recordings. They suggest that the radiated secondary pulses are acoustic clutter resulting from the on-axis main pulse generation. This clearly advocates that the animal orientation must be known in order to create a 3D click time-frequency model from recorded sound. These multiple pulses are found in the upper half of the received click spectrum while on-axis recordings reveal a centroïd frequency of 15 kHz and a monopulse pattern (Figure 1). On recordings we performed in the Canary Islands from whales of unknown orientation, more than six secondary pulses could at times be observed. A continuous low frequency part (below 1 kHz), which does not seem to follow a repetitive pattern and may last more than 10 ms, has also been documented (Goold & Jones, 1995; Zimmer et al., 2003). Proper time-frequency modelling from recorded clicks should therefore account for animal instantaneous distance, heading and depth, and environmental conditions with sufficient space-time resolution. To our knowledge, no other report fulfils these requirements. Yet, our aim here will not be to model an even near-perfect click generator, but a system that is in agreement with our current knowledge.

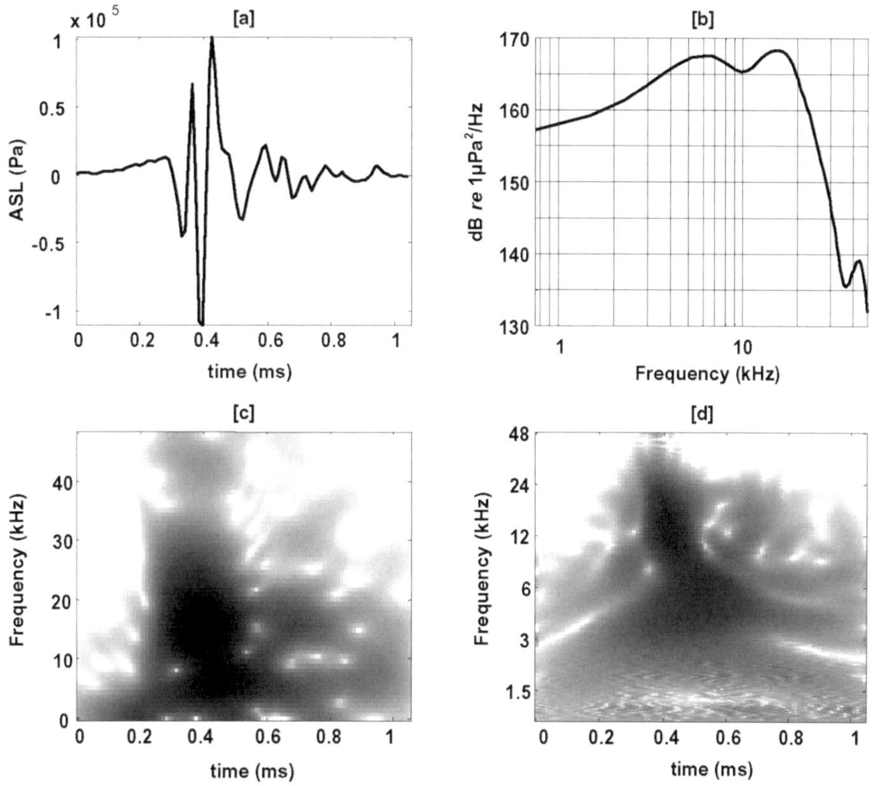

Fig. 1. This monopulse click was recorded near on-axis from an adult sperm whale off Andenes (B. Møhl et al., 2003). Sampling rate is 96 kHz. (A) Waveform, apparent source evel in μPa; (B) the received power spectral density by averaged periodogram, continuously on 32-sample windows, Hamming weighted; (C) continuous spectrogram, Hanning weighted, calculated on 128 pts-zero-padded FFT windows of 32 samples; (D) click scalogram by Meyer continuous wavelet transform envelope. (C) and (D) greyscales span 180–230 dB *re* 1µPa2/Hz, apparent source level.

Temporal patterns of click series

Sperm whale clicks were also chosen as a possible source for this work for the known steadiness of the click production rates. The obvious advantage is the possibility for the monitoring system to search the environment for steady and coherent responses, as a means of raising the detection thresholds and, as a result, reducing false alarm rates. Sperm whale clicks are mostly sequential and interclick-intervals (ICIs) rarely exceed 5 s. Most commonly encountered are the so-called 'usual clicks', which are produced a few seconds after the feeding dive starts and end a few minutes before surfacing. ICIs of usual clicks span 0.5 to 2 s. Clicks of ICI lower than 0.1 s are called rapid clicks, and those of ICI higher than a few seconds are called slow clicks. Creaks are series of clicks with a much higher repetition rate, as high as 200 s-1, and are believed to be used for sonar and foraging exclusively. Sperm whales are also known to produce 'codas', defined as short sequences (1–2 s) of clicks of irregular but geographically stereotyped ICIs (Pavan et al., 2000; van der Schaar & André,

2006). A more elaborate form of ICI analysis performed on usual clicks showed that the ICI may follow a rhythmic pattern that could be used as a signature by individuals of the same group. This pattern is a frequency modulation of the click repetition rate of usual clicks (André & Kamminga, 2000).

3. Ambient noise imaging to track non-vocalising sperm whales

Sound propagates in water better than any other form of energ, thus cetaceans have adapted and evolved integrating sound in many vital functions such as feeding, communicating and sensing their environment. In areas where marine mammal monitoring is a concern, detection and localization can therefore be efficiently achieved by passive sonar, but provided that the whales are acoustically active. When near or at the surface, where they may remain for 9 to 15 min between dives (André, 1997), sperm whales (*Physeter macrocephalus*) are known to stop vocalizing (Jaquet et al., 2001). Not discarding the possibility of deploying static active sonar solutions that would scan the high-risk areas, the concern that whales are highly sensitive to anthropogenic sound sources (Richardson et al., 1995) has motivated the search for alternative passive means to localize them. The whale anti-collision system (WACS) is a passive sonar system to be deployed along maritime routes where collisions are a concern for public safety and cetacean species conservation (André et al., 2004a,b; 2005). The WACS will integrate a three-dimensional localization passive array of hydrophones and a communication system to inform ships, in real-time, of the presence of cetaceans on their route. To detect silent whales, alternatives to conventional passive methods should be explored in order to avoid or complement active sonar support. In the present case, i.e. a group of sperm whales consisting of silent and vocal individuals, using the latter's highly energetic clicks might prove effective as illuminating sources to detect silently surfacing whales. Ambient noise imaging (ANI) uses underwater sound just as terrestrial life forms use daylight to visually sense their environment. Instead of filtering the surrounding ocean background noise, ANI uses it as the illuminating source and searches the environment for a contrast created by an object underwater (Potter et al., 1994; Buckingham et al., 1996). Although ANI is fraught with technical difficulties and has been validated, to date, at relatively short ranges, it opens new insights into acoustic monitoring solutions that are neither passive nor active in the strict sense. The solution introduced in this paper is conceptually based on both ANI and multi-static active solutions, where the active sources are produced by surrounding foraging sperm whales at greater depths (from 200 m downwards), which vocalize on their way down and at foraging depths (Zimmer et al., 2003), and in reported cases, likely on their way up until a few minutes before surfacing (Jaquet et al., 2001). The full analysis can be found in Delory et al., 2007.

A comparable approach was introduced for the humpback whale (*Megaptera novaeangliae*) off eastern Australia (Makris & Cato, 1994; Makris et al., 1999). In this study, if the solution were to be applied for monitoring purposes, it would be difficult to implement due to the need for near real-time shallow water propagation modelling as humpback whale vocalizations' spectra peaks are at rather low frequencies and as a result happen to be severely altered in the shallow water waveguide. This may prevent correct pattern matching between the direct and reflected signals unless accurate modelling techniques are applied. Comparatively, sperm whales' vocalizations spectra are considerably wider, higher in frequency, and of greater intensity. Their transient nature also makes received signals less prone to overlaps. Furthermore, our interest is in the propagation of these clicks in deep water and at relatively shorter distances, where the wave propagation problem is more tractable than for shallow water and long distances. These differing characteristics motivated us to revisit this passive approach and test the efficiency of using deep diving sperm whale clicks as a source to illuminate silent whales near the surface. Amongst numerous constraints, a prerequisite for sperm whale clicks to be used as active sources is that acoustically active whales should be close and numerous enough to create a repeated detectable echo from silent whales. The chorus created by these active whales should occur day and night and possibly all year long. Hence the following demonstration relies on the condition that whales are foraging in a group spread over not more than a few Squire kilometres and where a substantial amount of them are present within that range. Such a scenario has been observed consistently in the Canary Islands (André, 1997) and in the South Pacific (Jaquet et al, 2001), where sperm whales tend to travel and forage in groups of around ten adults, mostly female, spread over several kilometre distances with a separation on the order of one kilometre between individuals. In addition to the above, a substantial amount of information on temporal, spectral and directional aspects of the sources is essential (see section 2).

The essential information is that we can rely upon a high click repetition rate that may generate better estimates in a short time period. We believe that simulations that would implement all known types of click temporal patterns would probably not add significant information at this phase of the study. Consequently, our demonstration will contemplate usual clicks only. As a result, in a simulation where a given group of sperm whales are clicking in chorus, each individual will be assigned an ICI sampled from a uniform probability density function on the [0.5;2] second interval.

In order to evaluate the possibility of detecting and localizing silent whales near the surface using other conspecifics' acoustic energy, information on sperm whale acoustics was analysed and computed to create a simulation framework that could recreate a real-world scenario. Amongst other modules, a piston model for the generation of clicks is described that accounts for the data available to date (Delory et al, 2007). The modelled beam pattern supports the assumption that sperm whale clicks may be good candidates as background active sources. A sperm whale target strength (TS) model is also introduced that interpolates the sparse data available for large whales in the literature.

3D simulation of sperm whale wave sound

3D simulation of wave propagation from source-to-receiver and source-to-object-to-receiver in the bounded medium is implemented by software that we designed based on a ray-tracing model. This well documented and thoroughly utilised method provides good approximation of the full wave equation solution when the wavelength is small compared to water depth and bathymetric features. As seen above, whale TS and click spectra curves prompted our approach only for frequencies above 1 kHz, i.e. a 1.5 m wavelength, a value far smaller than any other physical scale in the problem.

Bathymetry and sound speed profile

Bathymetric data between the islands of Gran Canaria and Tenerife (Canary Islands, Spain) were obtained with a SIMRAD EM12 multibeam echo-sounder and provided by S. Krastel, University of Bremen, Germany. The bathymetric map horizontal resolution is 87 m. Sound speed profile was estimated by salinity, temperature and pressure measurements up to 1000 m applied to Mackenzie's equation, and from 1000 m to the ocean bottom (>3000 m at many locations) by linear extrapolation and increasing pressure, while considering temperature and salinity constant, because no deeper data were available to us. The resulting profile was close to typical North Atlantic sound speed profiles found in the literature.

Boundaries

The operating mechanisms at the surface and seafloor boundaries were incorporated through their physical characteristics. Sea surface effects were limited to reflection loss, reflection angle and spectral filtering. Surface reflection loss was estimated by the Rayleigh parameter, as a function of the acoustic wavelength and the root-meansquare amplitude of surface waves. Angles of reflection were determined by the Snell law, whereas neither surface nor bottom scattering were modelled. Sea-floor effects were limited to reflection loss and reflection angle.

Other parameters

An arbitrary number of acoustically active whales and one passive object defined by a 3D TS function were arbitrarily positioned in the three dimensions. All active whales were assigned a different and arbitrary waveform, the spectral information of which was estimated and affected the absorption parameter as well as the source radiation pattern. To test the efficiency of arbitrary hydrophone arrays, beamforming was processed at the receiver location by mapping direction of arrival into phase delays and recreating the sound mixture at all sensors. To ease the implementation and testing of the ray solution, a graphical user interface was created under Matlab and called *Songlines*.

Implementation

We first delimited a 5 km×5 km square area around the monitoring point, located at 40 m depth, half-way between Tenerife and Gran Canaria islands (Canary Islands, Spain), where 8 clicking whales of 10 m size are pseudo-randomly positioned between a depth of 200 m and 2000 m, with the condition that animals maintained a minimum distance of 1 km between each other. One silent whale was at 100 m depth and at a controlled distance from the monitoring point of 1000 m. All whales travel in the same direction at a 2-knot horizontal speed and random elevation. Inter-click intervals, radiation patterns and maximum intensities were set according to the above sections. The simulation setup described above was run 200 times with all active whales randomly repositioned with 1000 m minimal inter-individual separation and the silent whale being 1000 m away from the buoy. This amounted to a total of 1600 simulations, each calculating the resulting signals at the buoy stemming from one vocal and one silent whale. For each click produced in a simulation the following information was stored: whale position (vocal and silent), on-axis click sound pressure level, piston model diameter, environmental conditions (wave height, reflection ratio at the bottom, ambient noise level and type), ray angular tolerance, azimuth and elevation of the whale, levels, bearings and delays of the reverberated clicks arriving at the buoy. Every click produced by a single whale created 12 paths of measurable arrival levels at the buoy (see Figure 2): three from its source to the buoy (direct, surface- and bottom-reflected); three to the silent whale, each producing another three paths to the buoy. Consequently, the signal at the buoy was altered 9 times by the silent whale.

Results

Figure 3 shows the distribution of the received levels at the buoy from rays reflected by the silent whale. The number of echoes represents those received out of the 72 reflected rays (8

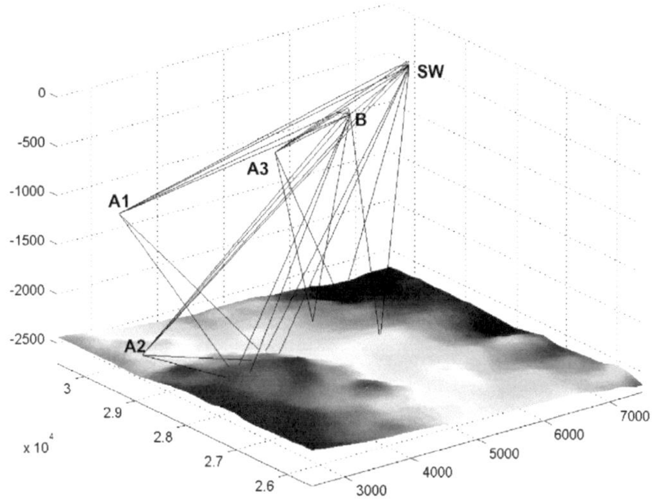

Fig. 2. 3D representation of rays with bottom, surface and object reflections with varying bathymetry resulting from our simulation software *Songlines*. A1–3, 3 vocal whales; SW, silent whale at 100 m depth; B, monitoring buoy, here located half-way between Gran Canaria and Tenerife Island (km 28) on the maritime channel. Ray paths account for vocal whale to buoy, vocal whale to non-vocal whale, silent whale to buoy, and their respective bottom and surface reflection paths. All dimensions are in metres.

clicks create 3 paths to the silent whale, each resulting in another 3 paths to the buoy) for each scenario. Signal level distribution is centred on sea-state 1 background noise level (1–30 kHz) with a right-hand side tail decreasing until seastate 3 background noise level. As sea-states are rarely below 2, especially in the Canary Islands, a first conclusion is that techniques to increase the SNR must be applied to ensure reasonable detection rates. These techniques could build upon the following observations:

1. The fact that clicks are to be repeated on an average of 1 click per second and per whale, implies that the silent whale is likely to be illuminated at least at this rate, and in the rather conservative case that only one whale is a contributing source. Integrated on a 10 s window, the coherent addition of the silent responses is to increase the SNR by at least 10 dB.
2. A beam-formed phased array would increase the SNR, with the additional benefit of resolving bearing information of the silent whale. Moreover, the broadband nature of the signals of interest here permits the use of sparse arrays of high directionality because frequency-specific grating lobes do not add up coherently in space. This technical scenario was simulated with Songlines. A 4 m-diameter ring array of 32 omni-directional hydrophones was beam-formed in the time-domain on one typical scenario, under the same control parameters as above. The silent whale was positioned 100 m deep and 1500 m away from the antenna. The software also allowed recreating the full waveforms resulting from the multi-path propagation of clicks to the buoy. Each whale produced a click at a random ICI taken from a uniform distribution in the 0.5–1 s interval during a 25 s period. Whales were separated by at least 1 km and repositioned every 5 s according to a group horizontal speed of 2 knots. The rest of the simulation settings remained unchanged. Results are presented in Figure 3.

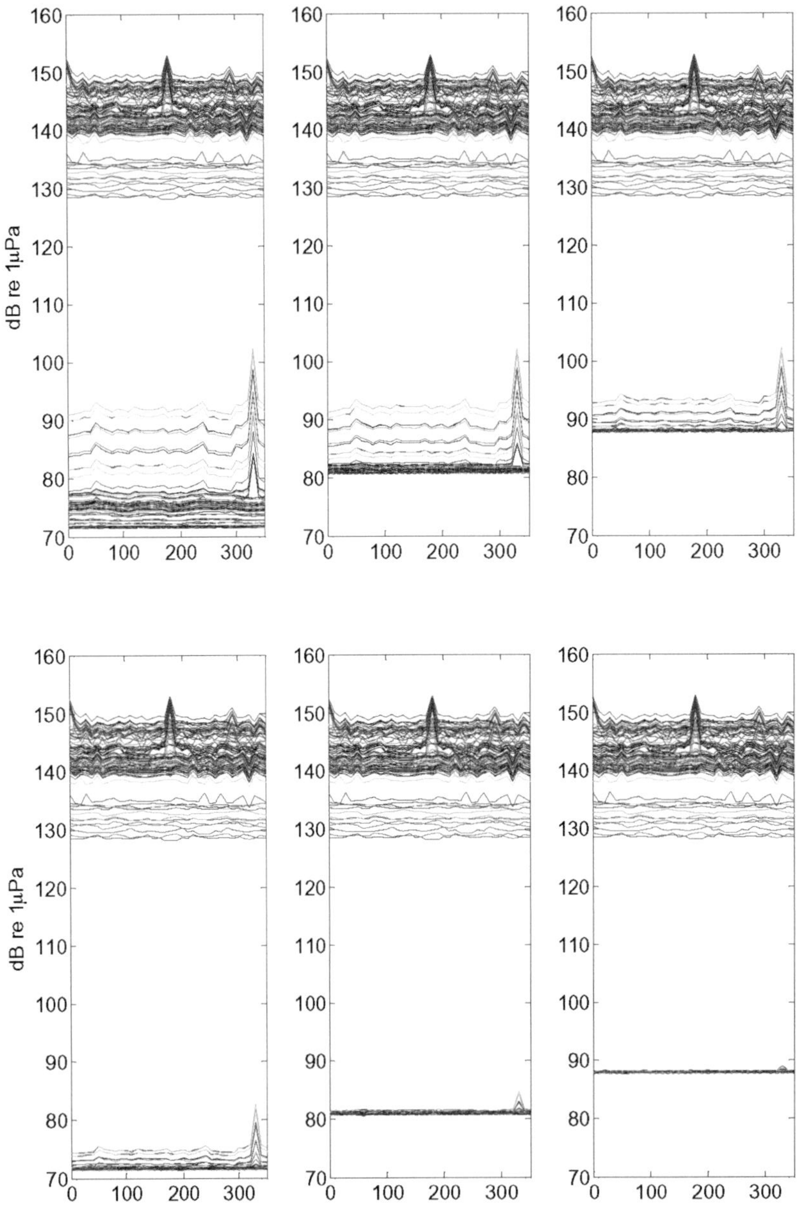

Fig. 3. Received levels on the 32 time-based beam-formed beams of a Ø4m-32-sensor-antenna for sea state 1, 3 and 6 (left to right) and three passive-active whale types of orientation: from top to bottom: whale angle of view is near beam aspect, and tail-aspect (see text). Array DI is 12 dB (see text). The simulated silent whale is at 330° azimuth, 100 m depth, 1100 m horizontal distance from the buoy. The cumulated plot results from a 25-s period with 8 whales clicking at depth (see text). Total number of clicks was 189. Beams are altered by the direct and reverberated paths from the vocal whales' clicks directly to the buoy (90 dB and over).

3. Matched filtering using pre-localized sources could raise the SNR in cases when sea-state and the resulting greater noise levels and reverberations alter the detection rates. However, as clicks are highly directional, matched filtering in the case of sperm whales may not always perform as expected as both source signal and reverberated replicas tend to differ when the source heading changes. As seen in the previous section on click time–frequency characteristics, both time and frequency contents are angle-dependent. As this angle is random to the receiver in most cases, the hypothesis of a deterministic signal is not fulfilled and thus matched filtering would not be optimal. It is also likely that matched filtering would be less efficient at greater ranges, where signals are more distorted. According to Daziens (2004), sperm whale clicks matched filtering was indeed outperformed by an energy detector for ranges greater than 3000 m. In fact, the latter outperformed matched filtering only for sperm whale click detection. Detection ranges were then nearly doubled as compared to matched filtering, for the same source level, detection and false-alarm probabilities, of 50% and 1% respectively. In our case, as the two-way propagation (source to silent whale to receiver) results in greater attenuation and distortion than those resulting from a one-way propagation of the same distance, it is expected that the energy detector will outperform matched filtering.

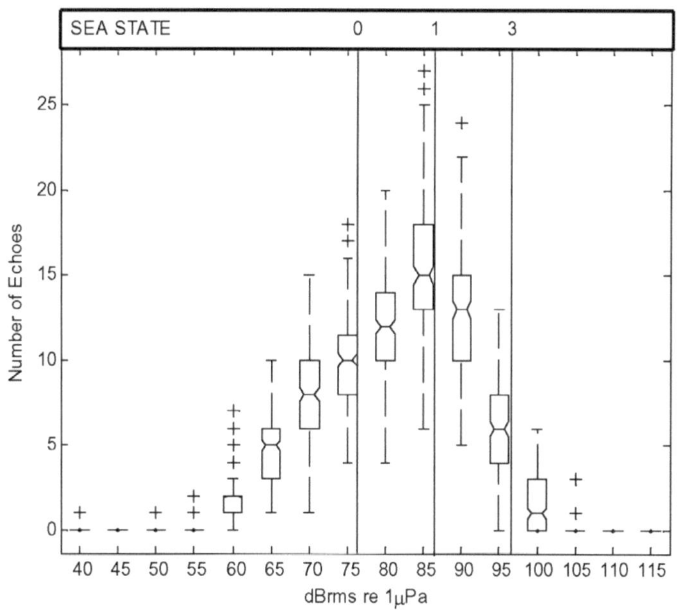

Fig. 4. Statistical plot of the simulated received RMS levels of clicks reflected on a silent whale located at 1000m distance from the buoy (see text for details on simulation settings). Ordinates represent the median number of contributing clicks per simulation drawn from 200 simulations (each simulation includes 8 vocal whales clicking once). Also plotted are lines at the lower quartile and upper quartile values. The whiskers are lines extending from each end of the box to show the extent of the rest of the data. Outliers are data with values beyond the ends of the whiskers. Notches over and below median values are medians' 95% confidence intervals. Sea-states 0 to 3 and above noise levels in the 1-30 kHz bandwidth are represented (calculated from Urick, 1996).

4. In view of the above, which advises a simplistic preprocessing method based on beam-forming and signal energy, we plotted the received signal intensity distributions from 25 ms time-intervals in Figure 4 (no background noise, no beam-forming) and Figure 5 (with background noise and beam-forming). Figure 4 shows that the resulting probability density function is bimodal, where the low-level mode represents the click energy reverberated from the silent whale, and the high-level mode, centred above 120 dB, stems from the click direct, surface and bottom reflected energy at the receiver. We anticipate that simultaneous occurrence of these two modes on a limited number of beams could prove robust for a decision stage.

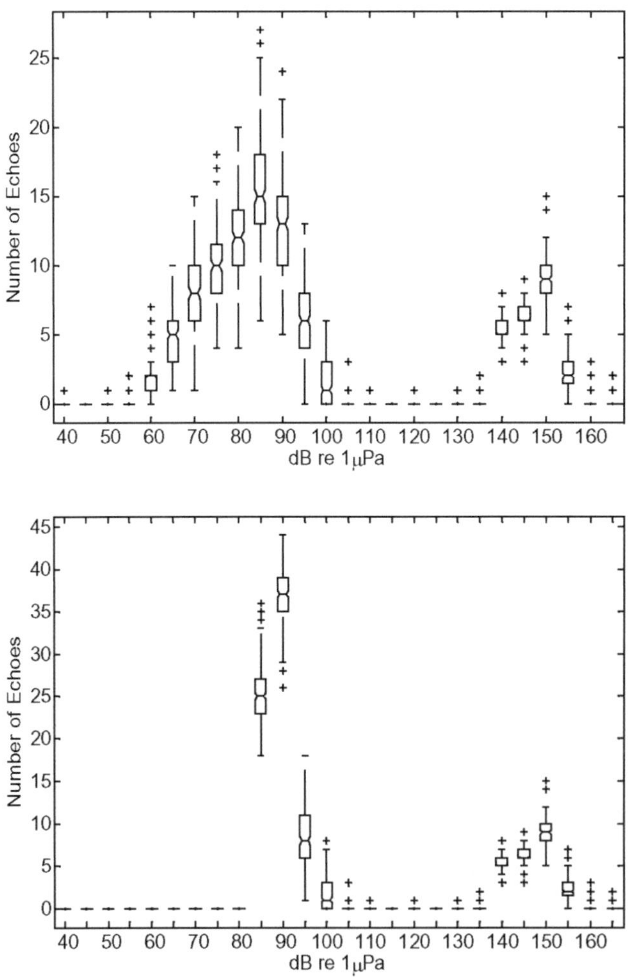

Fig. 5. Distribution of direct, surface, bottom-reflected and silent-whale reverberated clicks. The top figure is the level-expanded version of Figure 4, which highlights the bimodal aspect of the received level distribution. The bottom figure represents the resulting distribution at sea-state 1 with an omni-directional receiver. The same results are obtained on one beam for sea-state 3 after beam-forming with the antenna described in the text.

4. Space–time and hybrid algorithms for the passive acoustic localisation of sperm whales and vessels

The prominent approach, described in the previous section, for the passive acoustic localisation of cetaceans is based on the estimation and spatial inversion of time differences of arrival of an emitted signal at spatially dispersed sensors, which form an array. A second class of methods, space–time methods, originated from underwater applications such as sonar and found valuable applications in other fields such as the analysis of seismic waves or digital communications. In the latter, a significant amount of research has been devoted to space–time methods leading to powerful developments over the last 20 years. This approach has indeed shown to provide more accurate results than TDOA-based methods (Krim & Viberg, 1996). By maximising the mutual information between the source signal and array out- put, space–time methods achieve reduced variance in position estimates. Furthermore they offer simple means for the localisation of multiple simultaneously radiating sources. While the case of narrowband signals is well documented, the application of space–time methods to broadband signals, such as those emitted by sperm whales, only recently found satisfying developments in terms of complexity and accuracy (Dmochowski et al., 2007). These broadband developments could be imported and largely benefit the localisation of cetaceans: they indeed outperform TDOA-based methods even with a similar small number of sensors, a performance, which increases in harsher conditions with high levels of noise and reverberation. It is not the intention of this paper to thoroughly compare TDOA-based and space–time methods: this is an evaluation, which requires fairness and constant updates. Rather, this paper aims to illustrate the interest of developing an alternative frame concerning localisation, which may be well suited for certain array configurations. It will present the newly developed and challenging principles behind these methods and the results they can achieve for the passive acoustic localisation of multiple sperm whales and vessels. The principles which underlie the increased robustness of space–time methods will be recalled, and remarks are made concerning other interesting results which can be obtained via these methods such as broadband beam pattern estimation and dynamic estimation of attenuation factors. The full description of the approach can be found at Houégnigan et al., 2010.

A promising new class of hybrid localisers is introduced and its abilities for the localisation of sperm whales are shown. An important achievement of these hybrid localisers, in the case of compact arrays, is the reduction of the necessary processing time for results equivalent to those obtained for space–time methods. All of the developments to follow are intended to be included in a real-time developed at the Laboratory of Applied Bioacoustics (LAB) of the Technical University of Catalonia, for the passive monitoring of cetaceans from deep-sea observatories (http://listentothedeep.com).

4.1 General frame of the technical developments

Propagation Model

In this paper a compact array and real far-field sources are under consideration, far beyond the Rayleigh limit (Ziomek, 1995). The main focus is on the quality of bearing estimation provided by space-time methods and hybrid methods rather than on their range estimation capabilities, even though high-resolution space-time estimates of range could be obtained under certain conditions (Dmochowski et al., 2007). The model moreover focuses on

broadband sound, hence throughout this paper when reference is made to "cetaceans" this actually only refers to cetaceans producing broadband sound; note that the developments are valid for all types of broadband sounds, which includes some vessel sounds.
A three-dimensional array of M sensors is assumed. Due to propagation, each sensor receives attenuated, phased and noisy versions of the signal *s* emitted by a cetacean at spherical position $\underline{\mathbf{r}}_{\mathbf{s}} = [r_s\ \Theta_s\ \Phi s]$. The coordinates of $\underline{\mathbf{r}}_{\mathbf{s}}$ respectively represent range, azimuth and elevation.
The signal $x_i\ (t)$ received at the ith sensor at instant *t* is modelled as:

$$x_i(t) = \alpha_i(\underline{r}_s) \cdot s\left(t - \tau_{j,i}(\underline{r}_s)\right) + v_i(t)\,, \tag{1.1}$$

where v_i represents the additive noise at sensor *i*, which may include background and propagation noise, reverberation, and electronic noise. If sensor *j* is taken as the reference sensor, the ith signal can be expressed by using the propagation delay $\tau_{j,i}(\underline{r}_s)$ which is related to the path difference between the signals received at sensors *j* and *i*. Each x_i is thus modelled as a noise-corrupted phased and attenuated by distance (term $\alpha_i(\underline{r}_s)$) and version of the signal **s** emitted by the cetacean or broadband sound source.

4.2 Methods for the localisation of cetaceans

Methods based on Time Differences of Arrival (TDOA)

To understand the hybrid methods presented below, it is necessary to understand some aspects of TDOA-based methods (see section 3), but also to compare them to space-time methods.
The basic principle behind TDOA-based methods is that the time differences of arrival between the signals received at each sensor are related to the propagation path and the position of the estimated source. Hence TDOA-based methods feature two main steps: firstly time-delay estimation (TDE), and secondly a time-space inversion which consists in forming the position of the radiating source from the group of estimated TDOA related to the array geometry.

Limits of TDOA-based methods

The estimated time-delays between two noisy signals are themselves corrupted with broadband noise. Generalised Cross Correlation can improve estimation but this may not be sufficient. Each of the noisy estimates is then used in a time-space inversion phase and participates in the construction of a location estimate strongly affected by noise. This is a severe a priori hindrance that causes anomalies and high variance in the localisation results even if sophisticated statistical post-processing is applied. Combining all the sensors at disposal and not using only pairs could yield a strong noise reduction: space-time and hybrid methods precisely carry out such a beneficial processing. Indeed, the distinction between the spatial propagation of the signal emitted by cetaceans as opposed to the supposedly incoherent nature of noise offers powerful means of spatial separation.

Space-time methods

Several space-time methods were implemented for the localisation of cetaceans. The space-time terminology covers beamformers, spatial spectral estimators, and more generally methods based on the processing of a spatial observation vector estimated at various time instants. Space-time methods construct a spatial spectrum by virtually steering the array in various directions and estimating the received power (in some cases only a power-like index is estimated). When steered in the direction of a source the power received by the array and the signal-to-noise ratio will be maximised, and hence the spectrum will exhibit a high peak, whereas in directions where no sound or only low-power incoherent noise is radiated the received power will be weak and therefore the spatial spectrum will be relatively flat. Another way to interpret space-time methods and in particular spatial spectral estimators is to link them to frequency estimation; indeed these methods do extract information concerning a spatial frequency: the *wavenumber*. There exists a strong theoretical link between spatial frequency estimation and the more familiar temporal frequency estimation to the point that many methods moved from one domain to the other over the last decades (Johnson, 1982).

Power estimation

A power $P(\theta_k, \phi_k)$ is received when the array is steered in the direction (θ_k, ϕ_k). Steering is concretely achieved by delaying each signal according to the theoretical delays observed at each sensor for a waveform coming from direction (θ_k, ϕ_k). One sensor is to be chosen as reference.

Hence when only one source is present its estimated bearing $(\hat{\theta}_S, \hat{\phi}_S)$ is given by:

$$\left\langle \hat{\theta}_S, \hat{\phi}_S \right\rangle = \underset{k}{\operatorname{argmax}} \left(P(\theta_k, \phi_k) \right) \tag{2.2}$$

Multiple sources can be located by searching for multiple peaks in the spatial spectrum. The accuracy and resolution of the spatial spectrum is related to the way the calculation of power is carried out. In this paper, the general frame for power calculation is based on the estimation of a spatial correlation matrix and on various spatial estimators, which function as spatial filters.

Derivation of the spatial correlation matrix

The spatial correlation matrix (SCM) carries information about the correlation between the signals received at the sensors and the phase and amplitude differences between them. Other names may be encountered in literature such as space-time covariance matrix, spatio-spectral correlation matrix or spectro-temporal covariance matrix, but the same spatial second order statistics is always meant.

The SCM noted as $\Re$ is defined by:

$$\Re = E\left\{ x x^H \right\}, \tag{2.3}$$

where $E\{\ \}$ denotes mathematical expectation and where H indicates Hermitian conjugation.

In practice the signals' finite nature only permits an estimation of $\Re$. Estimation is made more difficult by short duration signals like some of those emitted by cetaceans. In a discrete frame, the most widely used estimate of $\hat{\Re}$ can be expressed as:

$$\hat{\Re} = \frac{1}{N_S} \sum_{n=1}^{N_S} z_n z_n^H, \tag{2.4}$$

where N_S is the number of samples corresponding to the signal, where z_n is a spatial observation vector at instant n.
$\widehat{\Re}$ should not be confused with the cross-correlation function $R_{x_i x_j}$ as presented in section (2.1.2), this will be important for the hybrid methods presented in 2.3.
At instant n, i.e. for the nth sample acquired by the array, the observation vector is given by $z_n = [x_1(n)\ x_2(n)\ \dots\ x_M(n)]^T$, (2.5).

Derivation of the steered spatial correlation matrix

The steered spatial correlation matrix $\widehat{\Re}(\theta_k,\phi_k)$ is the spatial correlation matrix associated with the array when it is virtually steered in the direction (θ_k,ϕ_k) to estimate the power received by the array from that particular direction. Steering in the direction (θ_k,ϕ_k) is done by adequately delaying the received signals with regard to a chosen reference sensor. The observation vector z_n then transforms to $z_n^{(k)}$ and $\widehat{\Re}(\theta_k,\phi_k)$ can then be expressed as :

$$\widehat{\Re}(\theta_k,\phi_k) = \frac{1}{N_S}\sum_{n=1}^{N_S} z_n^{(k)} z_n^{(k)H}\ , \quad (2.6)$$

For example if the jth sensor is chosen as a reference, the expression of $z_n^{(k)}$ is given by:

$$z_n^{(k)} = [\,x_1(n-\delta_{j1}^{(k)})\ \ x_2(n-\delta_{j2}^{(k)})\ \dots x_M(n-\delta_{jM}^{(k)})\,]^T, \quad (2.7)$$

where $\delta_{jm}^{(k)}$ represents the theoretical delay in samples between the signals at the jth and m[th] sensor for a far field source radiating from direction (θ_k,ϕ_k). Note that this process may suffer slight limitations from the sampling frequency since the computable delay in samples and the actual delay for direction (θ_k,ϕ_k) do not exactly match.

Spal Spectral Estimator	Power estimate	Theoretical Spectral resolution and accuracy	Computation time
Steered Response Power (SRP or Bartlett)	$P(\theta_k,\phi_k) = w^T \cdot \widehat{\Re}(\theta_k,\phi_k)\cdot w$	+ (lowest)	+ (shortest)
Capon (Minimum Variance) [15]	$P(\theta_k,\phi_k) = \dfrac{1}{w^T \cdot \left(\widehat{\Re}(\theta_k,\phi_k)\right)^{-1} \cdot w}$	++	++
Eigenvalue decomposition (EIG)	$P(\theta_k,\phi_k) = \lambda_{\max}(\theta_k,\phi_k)$	+++	+++
MuSiC [14]	$P(\theta_k,\phi_k) = \dfrac{1}{w^T \cdot \widehat{\Pi}_{(\theta_k,\phi_k)} \cdot w}$	++++ (highest)	++++ (longest)
Other estimators: ESPRIT, Root-MusiC Propagator…[16]	…	…	…

Table 1. Description of a few spatial spectral estimators

where $w = [1^1 ... 1^m ... 1^M]$, $\lambda_{\max}(\theta_k, \phi_k)$ denotes the maximum eigenvalue of $\widehat{\Re}(\theta_k, \phi_k)$, and $\widehat{\Pi}_{(\theta_k, \phi_k)}$ denotes the noise subspace of $\widehat{\Re}(\theta_k, \phi_k)$.

Based on the matrix defined in (2.6) we present in table 1 various spatial spectral estimators used to obtain our results (see below). EIG, Capon, and MuSiC are often referred to as high-resolution algorithms, and MuSiC is also labelled as subspace-based.

Hybrid spatial spectral estimation

The newly defined and developed hybrid methods are composed of three steps related both to space-time methods and TDOA-based methods.

Step 1: Calculation of the generalised cross-correlation for all pairs of sensors

Note that using other functions than GCC at this step may bring other interesting results.

Step 2: Construction of a Steered hybrid SCM $\widehat{\Re}_{hyb}(\theta_k, \phi_k)$ based on the generalised cross-correlation functions.

There exists a clear mathematical relationship between the cross correlation and the hybrid SCM such that the element $\widehat{r}_{ij}$ on the ith line and j[th] column of $\widehat{\Re}_{hyb}(\theta_k, \phi_k)$ is given by:

$$\widehat{r}_{ij} = \widehat{R}_{x_i x_j}(\delta_{ij}^{(k)}), \tag{2.8}$$

$\widehat{R}_{x_i x_j}$ represents the estimated generalised cross-correlation function between the signals at the ith and j[th] sensor. The use of $\delta_{ij}^{(k)}$ follows from Eq. (2.7). The operation in Eq (2.8) selects realisable delays within the cross-correlation functions and repositions the temporal second-order statistics in a spatial frame.

Step 3: Space-time power estimation

Space-time power estimation can be conducted based on the steered hybrid covariance matrix $\widehat{\Re}_{hyb}(\theta_k, \phi_k)$. The power estimators presented in table (2.2) can be re-used simply by replacing $\widehat{\Re}(\theta_k, \phi_k)$ by $\widehat{\Re}_{hyb}(\theta_k, \phi_k)$.

Nomenclature of hybrid methods

The name of a hybrid method will be composed of two parts: firstly the type of spatial power estimator used and secondly the type of GCC filter used. For example, SRP-SCOT corresponds to a SCOT filter applied to the Cross-Correlation function at step 1 and a Steered Response Power at step 3. Similarly MuSiC-ROTH corresponds to a ROTH filter applied to the Cross-Correlation function in the first phase and a MuSiC Power Estimation in the third phase. When no filtering is done, a standard Cross-Correlation function is used and the hybrid method is almost equivalent to the corresponding space-time method except that the estimated SCM remains hybrid with regard to its construction. In that case we would write for example SRP-hybrid or MuSiC-hybrid to differentiate them from the classical space-time SRP and MuSiC. In the case presented here hybridisation typically consists in going from a temporal second order statistics to a spatio-temporal second-order statistics.

Note that some methods developed by other authors are very close to the class of hybrid methods. This is notably the case of the SRP-PHAT algorithm developed by Griebel and Brandstein (2001). Developed mostly for conference settings with high reverberation it uses firstly the generalised cross-correlation with a PHAT filter and secondly a steered response power approach to localise speakers. However the method is obviously derived in a different manner and its authors class it as TDOA-based (DiBiase et al., 2001). Indeed, it does not rely on steered correlation matrices, which would have permitted to relate the spatial and temporal second order statistics and which would formally place their estimator in the hybrid group. To our knowledge, the first technical equivalent of a hybrid method was presented by Dmochowski et al, in 2007, who introduced the parameterised spatial correlation matrix, a powerful framework which inspired the hybrid steered SCM.

Final methodical remarks

The space-time and hybrid approaches presented here are well suited for far-field cetacean localisation and in particular for broadband cetacean sound. Typically a relatively small number of widely spaced sensors are featured while some cetaceans emit sound with a proportionately high frequency content, which may yield spatial aliasing. Spatial aliasing is a well known but poorly studied phenomenon caused by the relation between the aperture of the array and the wavelengths present in the signal.
The philosophy behind the methods presented here is, as in most TDOA-based methods, to treat the broadband signals received as truly broadband, and not as an artificial composition of narrowband components. This permits to gain accuracy, to mitigate the effects of spatial aliasing and to reduce processing time. In order to implement this time approach for broadband cetacean sound, a simple time-derived spatial correlation matrix is computed. Sophisticated frequency derivations of the SCM (Wang & Kaveh, 1985) do exist but they may have difficulties in coping with real-time requirements. Furthermore, given the spatial dimensions of most arrays deployed underwater, the frequency approach is likely to be heavily corrupted by spatial aliasing, which will then affect the accuracy of cetaceans' localisation.

A Short Presentation of the datasets and material

In the frame of the NEMO collaboration (Neutrino Mediterranean Observatory) for neutrino detection (Riccobene, 2009), more than 2000 hours of multichannel recordings were gathered. An underwater station was installed 25 km East of the port of Catania (Sicily) at approximately 2000 m depth. The station was equipped with four hydrophones working in a frequency band, which is sufficiently large (from 36 Hz to 43 kHz) for the detection, classification and localisation of vocalising cetaceans. The average distance between the sensors was 2.5m. Data was acquired at a sampling rate of 96 kHz. Vocalising sperm whales were detected with an algorithm for the real-time detection of impulsive sounds, which provided an estimation of the onsets and offsets of the sperm whale clicks (Zaugg et al, 2010).
Information from these datasets was extracted to estimate the beampattern and to perform localisation. The calculations were run under Matlab on a desktop with a 2.8 Ghz Pentium IV with limited memory which explains some relatively high calculation (Houégnigan et al, 2010).

4.3 Results

Determination of the beam pattern of the array

The beam pattern represents the variation of intensity or sound pressure level received as the direction of arrival varies, range being fixed. This is valuable information concerning the capability of the array to localise sources. The beam patterns presented in figures 6.1 and 6.2, respectively based on SRP and EIG, demonstrate that the array possesses good spatial separation capabilities with regard to bearing even with only four sensors and is not strongly affected by sidelobes, grating lobes and aliasing. A broadband sperm whale click of average energy was selected from the available data sets as a representative reference source. The traces and maxima, in the beampatterns 6.1 (left) and, even more clearly, in 6.1 (right), are related to the power received by the array. This power is itself related to the path difference between the sensors for a particular angular position of the source. The simplest maxima, yet not the most obvious, occurs at the borders of the spectra when the elevation is at 0° or 180°, i.e. when the source is pointing towards the array from above or from below. This position minimizes the path difference between three hydrophones (i.e. those with cartesian coordinate z=0 in the tetrahedron) and maximizes the power received by the whole array. Given the regular form of the array (the array is almost tetrahedral in shape but not exactly) it is clear that the power received will be invariant by rotation or by certain movements. This is verified by the six other maxima, which can be found in the pattern. There is a clear symmetry among them due to the choice of an azimuth varying from 360° and not just 180°. In the same way, traces can be explained by considering the array geometry and how the DOA of the source influences the path difference and power received. The 9 traces observed (6 traces appear at constant azimuth and 3 traces oscillate with azimuth in a manner reminiscent of a sine wave) show us that certain positions of the source create invariance of the power received, this power being relatively high. In these cases only the power received between pairs of hydrophones is actually maximized and thus only the path difference between pairs of hydrophones is minimized. There are clearly more ways of maximizing the power received for pairs than for triplets of sensors and this explains the extension of the traces and their number. On the whole, the traces observed are strongly dependent on the array geometry in the sense that they follow all the spatial positions, which maximize the power received (or minimize the path difference) in pairs of sensors. With EIG, spectral lines appear much sharper and spatial regions are much more clearly separated in terms of power than with SRP. For localisation, this implies less ambiguity in the estimation through clearer and narrower peaks.

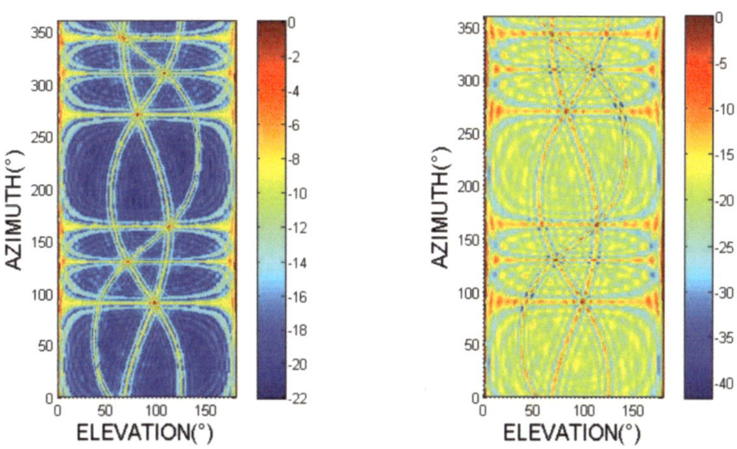

Fig. 6.1. Broadband beam pattern for a broadband click computed through SRP (left); broadband beam pattern for a broadband click, computed through EIG (right). Colour scale indicates average output power in dB.

Click-by-click localisation

Click-by-click localisation assumes that each click in a sequence contains information concerning the position of a vocalising sperm whale. Hence applying various spatial spectral estimators to a unique click can give an indication concerning their performance. Among the numerous 5 minutes duration datasets at disposal, the dataset recorded on 14th August 2005 from 3pm to 3.05pm was chosen. In this short sequence 819 impulsive sounds were detected and classified as sperm whale clicks. The localisation procedure was run for the methods presented above. In order to compare the localisation capabilities of those methods a single click of average energy, the 40th in the sequence, was selected. The processing of this click was also used to assess processing time. This will permit to decide on the choice of a suitable algorithm for real-time tracking.

Via Space-time methods

Figure 6.2 present the spatial distribution of power received for the selected click for space-time methods. A one-degree resolution was used for the computation of the spectra. There is a clear similarity between them, with spectral lobes, which are characteristic of the array, the strongest of which should converge towards the putative source location. The located source appears without ambiguity as a sharp peak within a dense zone of high power in figures 6.2 (left) and 6.2 (right), respectively for the SRP and EIG algorithms. The spatial spectra for MuSiC and Capon are not presented here since they provided inconsistent location estimates. The Capon spatial spectrum appeared extremely noisy with many secondary peaks while the MuSiC spectrum was obviously less noisy but did not have a clear unique peak. The circles, which appear in 6.2 and 6.3 are artefacts in the construction of the spatial spectrum. Spectral lines other than circles are actually observed in different positions of the spectrum when the source is at a different position. However, these artefacts are not appearing randomly: in the same way as for the beam pattern, spectral lines appear in correlation with the position of the source and the geometry of the array. This is comparable to frequency estimation where the spacing between the sampling points (sampling rate) constraints the spectrum as much as the spectral content of the signal. Here, the placement of the sensors in the array operates a sampling of space, which has an influence on the spatial spectrum.

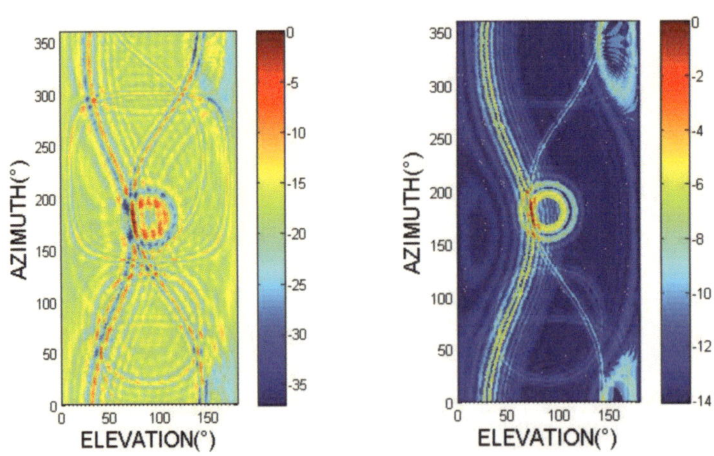

Fig. 6.2. Localisation of a broadband click computed with SRP (left); localisation of a broadband click computed with eigenanalysis spatial spectral estimation (right). Estimated position: $(\hat{\theta}_s, \hat{\phi}_s) = \{176^\circ, 74^\circ\}$. Color scale indicates average output power in dB.

Via Hybrid methods

Figure 6.3 and 6.4 present the spatial distribution of power received for the selected click for the hybrid methods, which were implemented. A one-degree resolution was used for the computation of the spectra. In figure 3.7 a side view of the spatial spectra (corresponding to elevation against power) is shown which permits to evaluate the number of side lobes, the separation between signal and noise for hybrid MuSiC and to visualise a narrow localisation peak, which is not obvious from 3.6. There is clearly a similarity between the hybrid spectra and the spectra obtained with space-time methods.

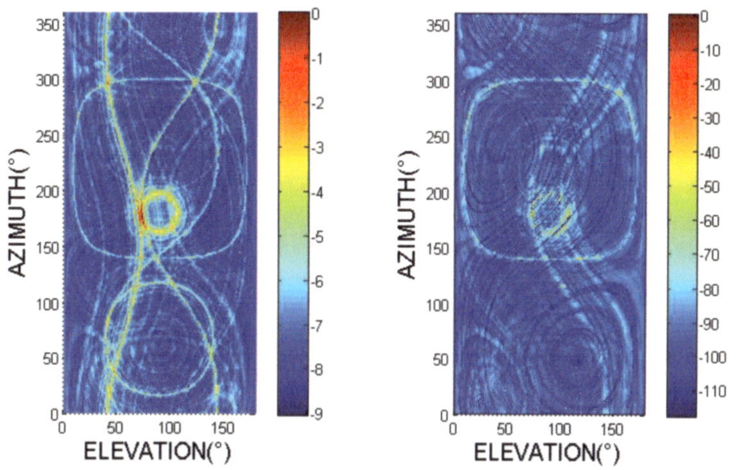

Fig. 6.3. Performance of SRP-ROTH (left); performance of MUSIC-SCOT (right), colour scale indicates average output power in dB.

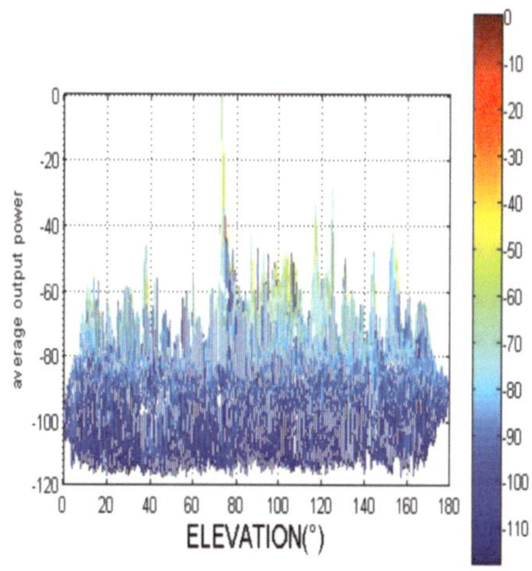

Fig. 6.4. Performance of MUSIC-SCOT, (elevation only), colour scale indicates average output power in dB.

The located source appears clearly as a sharp peak within the red-coloured zone in figure 6.3, respectively for SRP-ROTH and MuSiC-SCOT. The hybrid EIG algorithm failed to give results, which could compare with its non-hybrid version, it featured large spectral lines of high power which could not correspond to a real scenario and therefore it is not included here. The performance achieved by SRP-ROTH was very similar to that obtained for the non-hybrid EIG, with a reduced processing time (Houégnigan et al, 2010). With SRP-SCOT various high amplitude secondary peaks appeared which was not the case was for SRP-ROTH.

The Capon and Music methods did seem to perform more reliably when hybridised. They could isolate a main peak, which reduced ambiguity as figure 6.4 shows for MuSiC-SCOT. MuSiC-SCOT and MuSiC-ROTH in particular did achieve a powerful separation of signal (peak) and noise (lower power zones) as could be expected from the (non-hybrid) theory of MuSiC (Schmidt, 1986). The localisation obtained for the hybridised versions of Capon permitted to achieve a consistent localisation but figures are not presented for conciseness. Several secondary peaks appeared for Capon-SCOT but they were not yet problematic; they were not present for Capon-ROTH. In general ROTH hybrids seemed to provide the most reliable localisations.

Tracking of sperm whales and boats

Repeating the localisation procedure for each of the impulsive sounds detected in a 5-minute window allowed to track the movement of emitting sources classified as sperm whales or boats.

Track 1: Dataset 18th August 2005, 10 pm

Besides some isolated locations, which may be anomalies or simply scarcely vocalising sperm whales, two main tracks can be isolated with a clear separation in azimuth and elevation against time. The first one is found close to $(\theta_1, \varphi_1) = \{160°, 60°\}$ and the second one close to $(\theta_2, \varphi_2) = \{200°, 55°\}$.

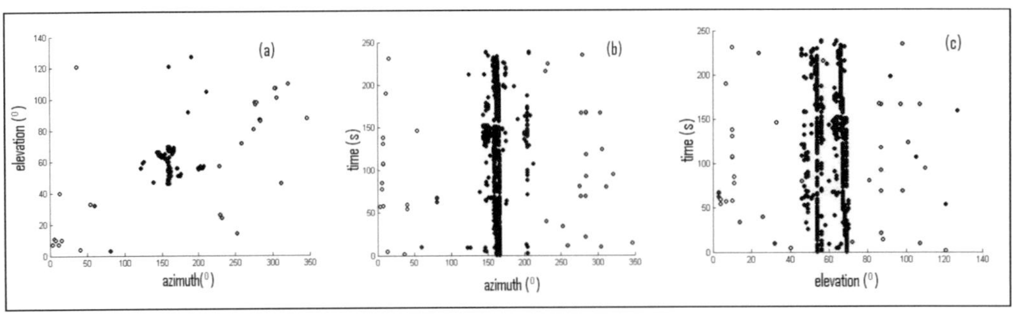

Fig. 6.5. Sperm whale tracking, 18th August 2005, 10pm

Track 2: Dataset 09th August 2005, 09 pm

776 sperm whale clicks were taken into account for localisation. There are two main clusters of points with sound sources moving around $(\theta_1,\varphi_1)=\{80°,50°\}$ and $(\theta_2,\varphi_2)=\{290°,30°\}$ and some more isolated clicks. The second cluster may contain several closely spaced animals but on the whole at least two vocalising mammals can be numbered in this sequence. The mammal corresponding to the first cluster has a very clear pattern of decreasing elevation

and azimuth in time. The second cluster is less obvious; there could be two animals close to each other. Further clustering and disentanglement of click series could be useful to obtain a better separation. From (c), elevation seems to indicate that there could be more than just two animals in the second cluster indeed elevation normally varies very little at large distances whereas well separated values of elevation (>5°) were found at a particular instant in time. Since the particular geometry makes it also less sensitive to small variation in azimuth, there could well be more than one animal in that cluster.

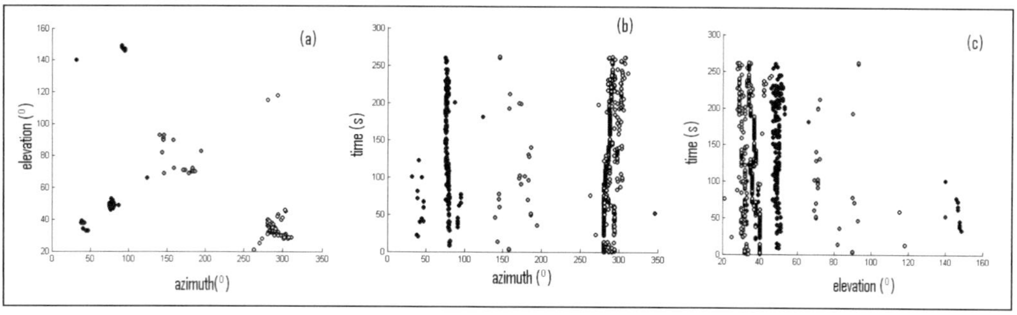

Fig. 6.6. Sperm whale tracking, 09th August 2005, 09pm

Track 3: Dataset 18th August 2005, 11 pm

760 sperm whale clicks were taken into account for localisation. One animal is clearly localised around (θ_1, φ_1) = {110°, 45°} and features a relatively stable elevation and a decreasing azimuth.

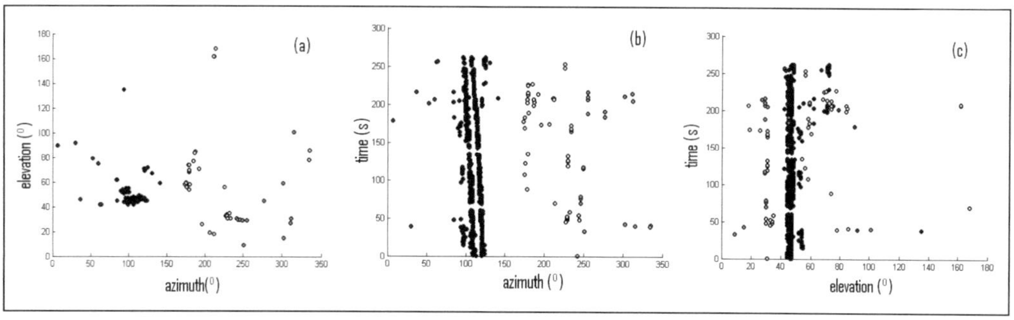

Fig. 6.7. Sperm whale tracking, 18th August 2005, 11pm

Track 4: Dataset 09th August 2005, 02 am

701 impulsive sounds were taken into account for localisation. An experienced operator aurally identified them as being shipping impulsive sounds. Contrary to the tracking of sperm whales, the tracking of boats features a clear evolution of DOA during the available five minutes. This seems to confirm the fact that boats are localised since their speed is expected to be much faster than that of sperm whales. The first cluster around (θ_1, φ_1) = {100°, 65°} corresponds to a source which starts to radiate around 150s. It features a slow but clear increase of both azimuth and elevation. The second cluster around (θ_2, φ_2) = {275°, 45°} corresponds to a source which radiates regularly during the 5 minutes of recording. It features a fast decrease of azimuth and a fast increase of elevation.

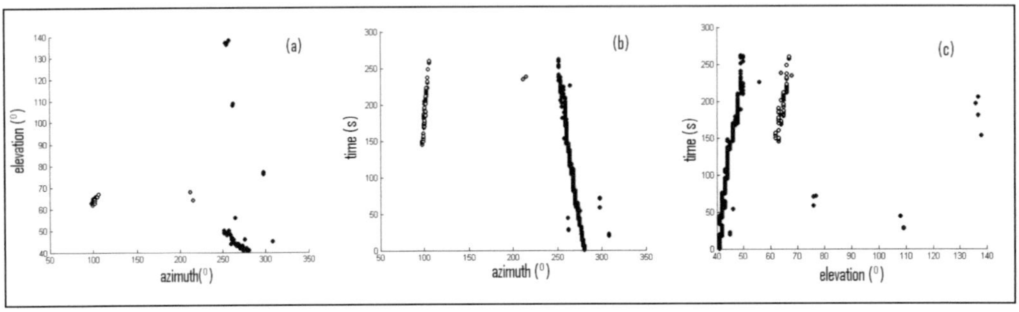

Fig. 6.8. Vessel tracking, 09th August 2005, 02am

4.3 Discussion

Discussion on click-by-click localisation

For space-time methods, two main reasons could explain the poor performance of the Capon and MuSiC algorithms which theoretically perform better than SRP. Firstly, both of these methods are extremely sensitive to the possible misestimation of the SCM (Krim & Viberg, 1996). SCM is in particular difficult to estimate correctly for short duration signals. This problem appeared to be partly solved by hybrid methods. Secondly, these methods are sensitive to the amplitude mismatch caused by unknown differences in sensitivity of the hydrophones. This could be corrected for but only at the expense of additional computations, which are not developed here for conciseness. These corrections would also add to the respective processing times. Some of the hybrid methods presented in the next section seemed to demonstrate that this problem could be solved with reasonable processing times.

Among the space-time methods, the SRP algorithm, even though it is less sophisticated, seems to be the best compromise between accuracy and processing time (Houégnigan et al, 2010). Considering that calculations could be carried out much faster in parallel and on a dedicated computation platform, considering also commonly observed inter-click intervals for sperm whales between 0.5 and 2 seconds and the pauses been sequences of clicks (Wahlberg, 2002), an SRP implementation could be well-suited for real-time applications.

The interest of hybrid methods seems manifest from the results, which can be compared to those obtained with space-time methods while requiring less processing time. For example, SRP-ROTH was comparable to the non-hybrid EIG but took about a third of its time. In general the hybrid methods presented here (many other filters -and thus other hybrids- could be considered) are also extremely profitable to the simple SRP algorithm. An interpretation of these results based on the nature of the filters used in the Generalised Cross-Correlation can be done.

(1) By filtering the signals, hybrids seem to construct a better estimation of the spatial correlation matrix. This estimate is not necessarily closer to the real spatial correlation matrix but more likely this estimate is more adapted to the nature of the spatial spectral estimators to which it is associated. Each of the estimators indeed uses a particular balance of noise and signal to achieve localisation, which is affected by the filters used for the Generalised Cross-Correlation.

The bad performance of the hybridised EIG, which relies on signal information by estimating the highest eigenvalue, is a sign that the signal components obtained after filtering are incorrect. The SCOT and ROTH filters, by taking into account coherence,

blindly enhance spectral regions with high energy which may contain both noise and signal. The signal estimated by EIG-SCOT or EIG-ROTH is hence probably not only signal but a mixture of signal and noise, which leads to localisation errors. A filter more adapted to that algorithm could be imagined. On the contrary, since MuSiC and Capon rely on noise estimation, they location estimation is improved. Obviously, even though some noisy components are labelled as signal, the remaining noise components, those with low energy, are likely to contain less signal and hence to improve MuSiC and Capon.
(2) The filters have to be well adapted to the ambient noise present and to the levels of reverberation. It was for example noticed that the PHAT filter, frequently used for human speaker localisation was not well suited for data from the NEMO deep-sea observatory. More adaptive pre-filters could also be created. In general, in the use of hybrids one should be aware of the effects of each filter and of the modus operandi of each spectral estimator with regard to the spatial correlation matrix.
(3) The signals received on hydrophones resulting from broadband sound emitted by cetaceans are corrupted by noise and may feature an important dynamic range across the frequency bands, which makes estimation more difficult. Indeed, the contribution of weak signal components is likely to be underestimated whereas they could provide valuable information. Pre-whitening can reduce this dynamic range and is one of the capabilities of the SCOT and ROTH filters.

4.4 Discussion on tracking

Although they cannot be confirmed by sightings, the estimated tracks were consistent with what can be expected from a sperm whale. In five-minute sequences the bearing may not change drastically given the expected slow speed of sperm whales (such as 0.2 to 2.6 m/s observed in Wahlberg, 2002) but coherent evolution of azimuth and elevation with time can be reconstructed. Track 6 shows that the localisation of vessels performs consistently without even proceeding to clustering. For sperm whales, additional clustering may add consistency to the results displayed but might as well discard valid isolated clicks. Already, the spatial separation abilities permitted to proceed to an estimation of the minimal number of vocalising mammals. The developed methods would benefit from being trained on data using a known moving source; this would permit to assess more precisely their performances.

5. Conclusions

5.1 Ambient noise imaging to track non-vocalising sperm whales

For a given and well-characterized signal, detection probabilities mostly depend on the background noise level. Before attempting the implementation of our passive approach in a specific area, it should be noted that ambient noise level statistics are the most limiting factor. We inferred from the literature that, in the band of interest, noise level was around 90 dB*rms re* 1μPa for sea-state 1 and a 1–30 kHz bandwidth. From our simulation results, energy-based detection thresholds would work until 1000 m. Nonetheless, each increase of 6 dB in background noise level, which is far from unusual, would half the detection range, as most propagation spreading is spherical in our case, and would make the system unreliable due to the dependency on weather conditions. Advanced post-processing of the received low-level signals was not studied. The inherent spatio-temporal nature of sperm whale acoustics and behaviour requires the use of either stochastic or determinist signal processing to further increase the SNR. Statistical methods for ANI have been thoroughly studied in shallow water (Potter & Chitre, 1996, 1999), but due to the numerous contextual differences, especially the limited number of active sources, it is likely that a stochastic approach would not be appropriate in our case. On the other hand, a determinist approach founded on proper modelling of source angular variability could prove robust. Among other well-documented methods, passive 3D localization of active sperm whales could then provide triggering information to coherently sum up the silent whale's response and increase the SNR and compensate for the ambient noise variability.

The reported multi-pulse structure of (most probably) offaxis clicks was not simulated, due to our incapacity to infer a model of its three-dimensional properties. We hence limited our study to the propagation of the first main pulse. Yet, including this feature would not impact upon the received levels except in the rare cases of constructive or destructive overlaps. The greatest impact would more likely be on the 'fillup' of the time–space window with more high-energy pulses at the monitoring point, which may handicap the search for low level echoes in background noise. It is generally reported that the secondary pulses are rarely more than two or three and only appear at frequencies higher than 4 to 5 kHz (see Figure 1). The whole signal duration may then increase to 20 msec which results in a maximum 20×8×2=320 msec time period. This is one-third of the search time window, for 8 vocal whales and taking direct, surface and bottom reflected signals to the buoy into account, at a rate of 1click/whale/s.

In the usual case, detection rates would not be drastically altered. This paper would not be complete without a note on false alarm rates and how they would impact on a vessel's decision, as detectable echoes from the surface may often come from different sources, like a densely concentrated group of fish. At-sea experiments and real recordings may provide the relevant information to discriminate these other types of objects, e.g. by incorporating their monitored spatio-temporal and behavioural characteristics. Scattering was only modelled by surface and bottom reflection coefficients being altered depending on sea-state and bottom type, respectively. As a result, our scattering model only affects specular rays. Reverberation, e.g. nonspecular rays back-scattered from surface, bottom or deep scattering layers was not mentioned nor simulated. When propagating through a deep scattering layer, direct rays from source to target could also reach the receiver with interference scattered from the deep layer, attenuated by 40 to 50 dB (Jensen et al., 2000). Such attenuation could differ when deep scattering layers are at lower depth at night-time. During daytime, such layers tend to be at greater depths and would be further attenuated due to propagation loss. In either case, the resulting reverberations may interfere with the low-level echoes from silent whales. Similarly, modelling of surface and bottom scattering would provide important information on the interferences from the reverberated sources as a function of sea-state and time, since no detection will be possible if these are omnipresent, even for low scattering strengths. Even though we have shown that signals echoed from silent whales could be detectable at only low sea-states, when surface scattering may become negligible, bottom scattering strength could constantly interfere with and increase noise to critical levels. In this work, simulations accounted for a given number of vocalizing whales, each producing one direct, one surface reflected and one bottom-reflected ray to the receiver and to one silent whale, which in turn radiated the corresponding echoes modelled by one direct, one surface-reflected and one bottom-reflected ray to the receiver. In fact, these 12 resulting rays represent only one part of the real signal at the receiver, as all vocalizing whales would also scatter energy from other whales' clicks. In addition, simulations were limited to allow only one bottom and one surface reflection. Multiple reflections from vocalizing whales' clicks would originate weak signals of a similar order of magnitude as the simulated silent whale's echoes and should be discarded as well. So far, we have not studied how adding these additional scatterers and pathways could alter the current results, as the objective of this work was to study whether a signal excess from a silent whale near the surface could be measured. The raised ambiguity and false-alarm rates due to unpredicted and more complex pathways would probably call for a more advanced detector. As the primary task of the WACS is to localize active whales using an array of receivers, the resulting information could be used to perform forward modelling of the arrival structure, and then to compare this with observations to identify the anticipated replica arrivals. Echoed signals from silent whales could then be detected by a band-limited energy detector. In future work the authors hope to be able to simulate the same scenario with an unlimited number of reflections and enable back-scattering from active whales so that more complex detectors and matched field methods can properly be evaluated.

While this study is restricted to sperm whales, the ANI approach might progressively extend to wider possibilities, as large baleen whales passing through a wide pod of sperm whales are also to be detected, probably with higher contrast in the case of species such as fin and blue whales. Most large baleen whales only produce very low frequency sounds (most of the energy remains below 100 Hz) that reverberate in a complex way in the SOFAR channel and mix with all types of low-frequency sources summing up to great sound pressure levels (Potter & Delory, 1998). As a direct consequence, designing a permanent solution for passive localization of these whales is a difficult task and furthermore can be performed only with very wide aperture bottom-mounted arrays. The low and, at times, negative signal-to-noise ratios at relatively short range from the whales have motivated the specific development of advanced signal processing algorithms that have not yet been implemented and still need further development (Delory & Potter, 1999; Delory et al., 1999). We believe that our approach could be an alternative worth considering in areas where sperm whale populations are geographically dense and stable over time. Furthermore, this method would have to be a complementary component of a more complex system like the previously described WACS in order to be viable and useful.

In conclusion, the results provided quantitative information as regards the implementation of a passive approach using sperm whale clicks as illuminating sources. Received levels are centred on ambient noise levels for low sea-states, motivating the use of beam-forming to raise signal levels and extract bearing information. Validation of the method introduced in this paper is essential before advanced signal enhancement techniques can be properly evaluated, leading to the prior necessity of performing experiments in the field. From a broader perspective, as permanent passive techniques based on natural acoustic energy would be probably less costly and less prejudicial to cetaceans than conventional active solutions the authors believe that they merit further investigation.

5.2 Space–time and hybrid algorithms for the passive acoustic localisation of sperm whales and vessels

This paper presented space-time estimators in a broadband frame and introduced novel hybrid methods. These developments could benefit the localisation of cetaceans emitting broadband sound, e.g. sperm whales, and can also be used for the localisation of vessels emitting broadband sound. When hybridised, basic space-time algorithms such as SRP were improved and performed as consistently as more sophisticated high-resolution estimators such as MuSiC or EIG. It was observed that hybrid high-resolution algorithms did improve the robustness of space-time high-resolution methods provided that pre-filtering is consistent with their treatment of noise and signal. All of these developments are always considered in the scope of a real-time processing frame in which at least SRP and hybrids would fit.

SRP-based localisation of series of detected impulsive sounds pre-classified as sperm whales or shipping noise permitted to construct consistent tracks of the radiating sources. The good spatial separation achieved by the algorithm permitted to obtain a first estimate of the number of sperm whales. For boats, the clarity of the tracks was higher while both the variance and the number of anomalies seemed drastically reduced probably due to the regular and mechanical aspect of sound radiation. In all cases, but especially for sperm whales, the developed methods would benefit from being tested on longer data sets, which may confirm the consistency of the track by displaying more complex motion patterns.

Future work will focus on the validation of the methods with known sound sources and on range estimation. Promising simulations showed that information on phase and amplitude contained in the spatial correlation matrix could permit an estimation of range. This approach may though require well-calibrated and equivalent sensors. Such an engineering requirement is highly constrained by technical costs; however it should be considered that such costs could be balanced by the gain of more precise results in localisation and enhanced capabilities for de-noising and sound enhancement. This may in turn be used to improve detection and classification. Another approach could consist in considering reflections for passive ranging. The association of the use of reflections and space-time or hybrid methods will be a challenging task and could bring powerful developments. Future work shall also include the influence of sound speed variation via an adequate insertion of sound speed profiles.

6. References

Alder-Frenchel, H.S. (1980). Acoustically derived estimate of the size distribution for a sample of sperm whales, Physeter catodon, in the western north Atlantic. *Canadian Journal of Fisheries and Aquatic Sciences*, 37, 2358–2361.

André, M. (1997.) El cachalote, Physeter macrocephalus, en las Islas Canarias. PhD Thesis. Universidad de Las Palmas de Gran Canaria.

André, M. & Kamminga, C. (2000). Rhythmic dimension in the echolocation click trains of sperm whales: a possible function of identification and communication. *Journal of the Marine Biological Association of the United Kingdom*, 80, 163–169.

André, M., Delory, E. & van der Schaar, M. (2004a). A passive acoustic solution to 3D whale monitoring. Brest, France: Sea-Tech Week.

André, M., Delory, E. & van der Schaar, M. (2004b). A passive mitigation solution to the effects of human-generated sound on marine mammals. *London: Policy on Sounds and Marine Mammals: an International Workshop.*

André, M., Delory, E., van der Schaar, M. & Castell, J.V. (2005). On the possibility of detecting and tracking echolocating whales by passive acoustics and ambient noise imaging. La Rochelle, France: *Workshop on Active Sonar*, 19th Conference of the European Cetacean Society.

André, M., Johansson, A.T., Delory, E., van der Schaar, M. (2007). Foraging on squids: the sperm whale mid-range sonar. *Journal of the Marine Biological Association of the United Kingdom*. 2007, 87, 59–67 .

André, M. (2009). The sperm whale sonar: monitoring and use in mitigation of anthropogenic noise effects in the marine environment. *Nucl. Inst. Meth. Phys. Res. A* 602, 262.

Buckingham, M.J., Potter, J.R. & Epifanio, C.L. (1996). Seeing underwater with background noise. *Scientific American*, 274, 40–44.

Daziens, J.M. (2004). Assessing the performance of omni-directional receivers for passive acoustic detection of vocalizing odontocetes. Monterey, *California: Naval Postgraduate School.*

Delory, E. & Potter, J.R. (1999). Transient, tonal, and background noise filtering with wavelet and cosine transforms. *Journal of the Acoustical Society of America*, 105, 1106.

Delory, E., Potter, J.R., Miller, C. & Chiu, C.-S. (1999). Detection of blue whales A and B calls in the northeast Pacific Ocean using a multi-scale discriminant operator. Maui, Hawaii: .

Delory, E., André, M., Navarro Mesa, J.-L., Van der Schaar, M. (2007). On the possibility of detecting surfacing sperm whales at risk of collision using others' foraging clicks. *J. Mar. Biol. Ass. U.K.* (2007), 87, 47–58.

DiBiase, J. H., Silverman, H.F. and Brandstein, M.S. (2001). Microphone Arrays: Techniques and Applications, *Springer-Verlag*, 2001, pp. 157–180.

Dmochowski, P., Benesty, J. and Affes, S. (2007). Direction of arrival estimation using eigenanalysis of the parameterized spatial correlation matrix, in *Proc. IEEE ICASSP*, 2007, pp. I-1--I-4.

Goold, J. C. & Jones, S. E. (1995). Time and frequency domain characteristics of sperm whale clicks. *Journal of the Acoustical Society of America*, vol. 98, pp. 1279-1291.

Griebel, S.M. & Brandstein, M.S. (2001). Microphone array source localization using realizable delay vectors, in Proc. *IEEE Workshop on Applications of Signal Processing to Audio and Acoustics*, 2001.

Houégnigan L., Zaugg, S., van der Schaar, M., André, M. (2010). Space-time and hybrid algorithms for the passive acoustic localisation of sperm whales and vessels. *Appl Acoust.*, doi:10.1016/j.apacoust.2010.05.017.

Jaquet, N., Dawson, S. & Douglas, L. (2001). Vocal behavior of male sperm whales: why do they click? *Journal of the Acoustical Society of America*, 109, 2254–2259.

Johnson, D.H. (1982). The application of spectral estimation methods to bearing estimation problems, *Proc. IEEE*, vol. 70, 1982, pp. 1018-1028.

Krim, H. & Viberg, M. (1996). Two decades of array signal processing research: The parametric approach", *IEEE Signal Process. Mag.* 13, 1996, pp. 67–94.

Makris, N.C. & Cato, D.H. (1994). Using singing whales to track nonsingers. *Journal of the Acoustical Society of America*, 96, 3270.

Makris, N.C., Lai, Y.-S. & Cato, D.H. (1999). Using broadband humpback whale vocalizations to locate nonvocal whales in shallow water. *Journal of the Acoustical Society of America*, 105, 993.

Møhl, B., Wahlberg, M., Madsen, P.T., Heerfordt, A. & Lund, A., (2003). The monopulsed nature of sperm whale clicks. *Journal of the Acoustical Society of America*, 114, 1143–1154.

Pavan, G., Hayward, T.J., Borsani, J.F., Priano, M., Manghi, M., Fossati, C. & Gordon, J. (2000). Time patterns of sperm whale codas recorded in the Mediterranean Sea 1985–1996. *Journal of the Acoustical Society of America*, 107, 3487–3495.

Potter, J.R., Buckingham, M.J., Deane, G.B., Epifanio, C.L. & Carbone, N.M. (1994). Acoustic daylight: preliminary results from an ambient noise imaging system. *Journal of the Acoustical Society of America*, 96, 3235.

Potter, J.R. & Chitre, M. (1996). Statistical models for ambient noise imaging in temperate and tropical waters. *Journal of the Acoustical Society of America*, 100, 2738–2739.

Potter, J.R. & Delory, E. (1998). Noise sources in the sea and the impact for those who live there. Singapore: Acoustic and Vibration, Asia' 98.

Potter, J.R. & Chitre, M. (1999). Ambient noise imaging in warm shallow seas; second-order moment and model-based imaging algorithms. *Journal of the Acoustical Society of America*, 106, 3201– 3210.

Riccobene, G. (2009). NEMO Collaboration. Long-term measurements of acoustic background noise in very deep sea, *Nucl. Instr. and Meth. A.* 604 S149-S157.

Richardson, W.J., Greene, C.R., Malme, C.L. & Thomson, D.H. (1995). Marine mammals and noise. San Diego: *Academic Press*.

Schmidt, R. (1986). Multiple emitter location and signal parameter estimation, *IEEE Transactions on Antennas and Propagation*, Vol.34, No. 3, March 1986, pp:276 - 280.

Tucker, D.G. & Glazey, B.K. (1966). Applied underwater acoustics. London: *Pergamon Press Ltd.*

Urick, R.J. (1996). Principles of Underwater Sound. Slough, UK: Peninsula Publishing.

Van der Schaar, M. & André, M. (2006). An Alternative Sperm Whale (*Physeter macrocephalus*) Coda Naming Protocol. *Aquatic Mammals* 32(3), 370-373.

Wahlberg, M. (2002). The acoustic behaviour of diving sperm whales observed with a hydrophone array". *Journal of Experimental Marine Biology and Ecology* 281(1-2), pp. 53-62.

Wang, H. & Kaveh, M. (1985). Coherent signal-subspace processing for the detection and estimation of angles-of-arrival of multiple wideband sources, *IEEE Trans. Acoust. Speech Signal Process* 33 4, August 1985, pp. 823–831.

Watkins, W. A. (1980). Sperm whale clicks, in *Animal Sonar Systems*, R.-G. B. a. J. F. Fish, Ed. New York: Plenum, 1980, pp. 283-290.

Serge Zaugg, S., van der Schaar, M., Houégnigan, L., Gervaise, C., André, M. (2010). Real-time acoustic classification of sperm whale clicks and shipping impulses. *Appl Acoust*, doi:10.1016/j.apacoust.2010.05.005.

Zimmer, W.M.X., Tyack, P.L., Johnson, M.P. & Madsen, P.T., (2005). Three-dimensional beam pattern of regular sperm whale clicks confirms bent-horn hypothesis. *Journal of the Acoustical Society of America*, 117, 1473–1485.

Ziomek, L.J. (1995). Fundamentals of Acoustic Field Theory and Space-Time Signal Processing, *CRC Press*, 1995.

30

Sound Localisation in Practice: An Application in Localisation of Sick Animals in Commercial Piggeries

Vasileios Exadaktylos[1], Mitchell Silva[1], Sara Ferrari[2], Marcella Guarino[2] and Daniel Berckmans[1]
[1]*Measure, Model and Manage Bioresponses (M3-BIORES), Department of Biosystems, Katholieke Universiteit Leuven, Kasteelpark Arenberg 30, 3001 Heverlee*
[2]*Department of Veterinarian Sciences and Technologies for Food Safety, Faculty of Veterinary Medicine, University of Milano, Via Celoria 10, 20133 Milan*
[1]*Belgium*
[2]*Italy*

1. Introduction

Application of sound localisation algorithms requires a good description of the problem to be solved, detailed specifications and the choice of the algorithms that are most suited for that specific application. For the correct choice of the localisation procedure, the objectives of the study or application need to be clearly defined. These objectives are most of the times conflicting (e.g. high accuracy and low computational complexity) and often the choice of the method is not unique. Additionally, the localisation method can be part of a bigger objective (i.e. the application for which the localisation algorithm will be used) and as such, it must be able to interact with the other components of the bigger project. Previous research in relation to localisation of animal vocalisations has focused on localising animals in the wild (e.g. Hayes et al., 2000; Thomas et al., 2002) in order to mainly study animal behaviour. This chapter describes the steps that were followed in relation to sound localisation in commercial piggeries.

More specifically, the objective of this study is to monitor respiratory diseases in commercial piggeries (Fig 1). Similar to the effect on humans, respiratory diseases in pigs result in coughing and in a different sound of coughing due to the different response of the respiratory system when contacting different pathogenic agents. In humans, an experienced physician can identify over 100 different respiratory diseases based on the sound timbre (Korpáš et al., 1996). In animals, veterinarians use a similar approach to detect sick animals when they enter a farm. Their initial impression over the herd is based on visual and auditory observation when they collect information about the welfare, health and productive status of the animals. In this direction, Marx et al. (2003) have studied pig vocalisations related to pain, while Manteuffel et al. (2004) and Schön et al. (2004) have employed vocalisation analysis in livestock farms as a measure of welfare. Similarly, considerable research has been conducted in the characteristics of pig coughing (e.g. Ferrari et al., 2008), the effect of environmental noise on the cough-frequency features (Van Hirtum & Berckmans, 2003a), in identification of pig coughing based on continuous recordings (e.g. Van Hirtum & Berckmans, 2001) and algorithms have been developed for automatic detection of coughs (e.g. Van Hirtum & Berckmans, 2003b, Exadaktylos et al., 2008a, Exadaktylos et al., 2008b).

Fig. 1. A pen with pigs in a commercial piggery

1.1 The problem

Feeding the world with quality assured food remains a significant challenge for the food supply chain within which meat production plays an important role. As countries become more affluent and the world's population continues to rise, demand for meat and other livestock products has grown substantially, according to the Food and Agriculture Organisation (FAO). To be able to satisfy this higher demand for meat products, global animal food production is undergoing a major transformation in the last decades. According to FAO, global meat production was 200 million metric tons in 1999 and an increase of 25% is expected until 2015. Furthermore, in the next 17 years world food production is expected to increase by 62% to feed the world. The highest increase in the 17 years is in meat (42%) (2nd CIGR Conference on Agricultural Research, Iguassu Falls city, Brazil, 2008). This expansion will occur but at the same time with supply, industry will have to deal with concerns over animal and public health from livestock farming and also animal welfare. Finally, this expansion has an important environmental impact.

FAO warns that the risk of disease transmission from animals to humans will increase in the future due to human and livestock population growth, dynamic changes in livestock production, the emergence of worldwide agro-food networks and a significant increase in mobility of people and goods. In a review of 1407 species of human pathogens, 58% were broadly classified as zoonotic (Woolhouse and Gowtage-Sequeria, 2005), defined by the World Health Organization (WHO) as those diseases and infections which are naturally transmitted between (other) animals and man. Of 177 of 1407 human pathogens that were identified as "emerging", 130 (73%) were zoonotic.

Over the last century, there has been a shift away from livestock production as a highly localised enterprise, where animals were typically born, fattened and slaughtered in the same region. The number of live animals traded for food quintupled in the 1990s, where more than one billion were moved across borders in 2005 (FAO, 2007). Transport of animals from different herds or flocks is ideally suited for spreading disease (FAO, 2002), e.g. the spread of the highly pathogenic avian influenza virus H5N1 in Southeast Asia and the spread of swine influenza viruses in the US.

Crowding of greater numbers of animals into smaller spaces has been identified as a critical factor in the spread and maintenance of disease on the farms (Delgado et al., 2003). In those farms when the environment is inadequate, diseases evolve as endemics at considerable cost and many pathogens are zoonoses (e.g. Strept suis, swine flu, etc. in pigs). Those endemic multifactorial diseases are the major target for drug use in livestock and poultry production. The amount of manure produced by intensive animal husbandry creates a challenge to maintain hygienic standards. Industrialisation of animal production may lead not only to greater animal-to-animal contact, but also to increasing animal-to-human contact, particularly when production facilities border urban areas (Murphy, 1998). Furthermore, Hamscher et al. (2003) demonstrated that antibiotics can also be spread to the environment by dust. Besides environmental effects, there is concern whether the widespread use of antibiotics in animals exacerbates the rising incidence in human pathogens. In the U.S., the society expends US$ 30 billion (more than €20 billion) per year due to the cumulative effects of antimicrobial resistance (Centner, 2003).

In modern pig houses (see Fig 2), animals are grouped in separate compartments ranging in size from 70 animals up to 1000 pigs. Compartments are separately controlled regarding climate control, manure storage, feed supply, etc. Within the same compartment, animals are grouped in pens containing mostly 10 to 16 animals per pen. Pens are separated by 1 meter high walls so that animals have physical contact within one pen but limited contact with animals from neighbouring pens although they are in the same compartment.

These differences in size and the big number of animals per compartment, allow no punctual and individual animal monitoring. The sound analysis approach that we present in this chapter is able to automatically identify cough signals from a continuous sound registration. In this case, the number of coughing incidents will provide some information about the general respiratory health status of the animals in the compartment. However, it will not give any information as to where the coughs are coming from. Respiratory diseases are not only frequent in piggeries (appearing at least once during every growth period of about 130 days), they are also spreading fast within the group and therefore if we know where a disease is originating in a compartment, the veterinarian may decide to take action in a selective way by managing or treating only those animals indicated in the hazard area.

This selective treatment of animals can have multiple benefits both in the short as well as in the long run. Fewer antibiotics will be used that directly translates to a decreased cost for the farm manager. Healthy animals will not be unnecessarily treated and therefore the meat quality will not be affected by any medicament residuals. Extensive use of antibiotics also

results in fast mutation rate of bacteria. This poses a huge threat to the livestock industry in case the antibiotics fail to evolve as fast. Therefore, fewer and 'on time' use of antibiotics may result in reduction of the mutation rate of bacteria and as a consequence reduction of the chances that antibiotics become ineffective.
Economy of scale, transforms this problem of source localisation to a vital one for the pig-farming industry.

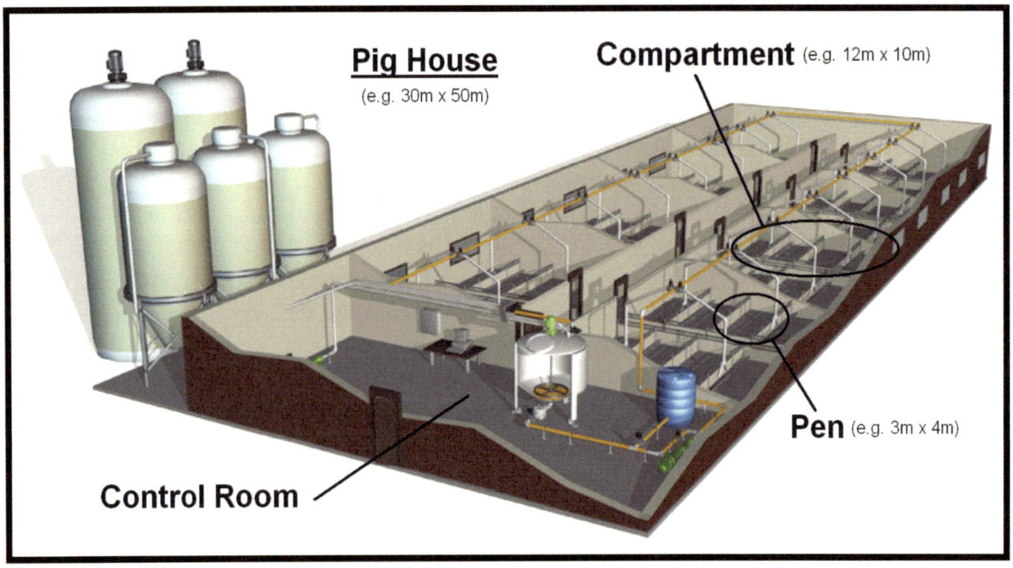

Fig. 2. Schematic overview of a typical layout of a commercial pig house (house > compartment > pen). Schematic courtesy of FANCOM bv. (http://www.fancom.com)

1.2 The proposed solution – objective

In order to identify where a disease is occurring within a pig compartment (and more specifically in which pen in a compartment), a system was developed that consists of a real-time sound extraction algorithm, a classification algorithm and a sound localisation algorithm. All three components of the system need to work in synergy in order for the desired outcome, which is the correct localisation of a sick animal in a commercial piggery. The real-time constraint is imposed by efficiency (the farmer or the veterinarian needs to be informed as soon as there is a health issue in a compartment), technical (the data flow is high and makes it impossible to store all the data or communicate it to the veterinarian) and economical (in order to be financially viable for a pig farm, such a system cannot cost more than 0.5€/pig) issues.

1.3 Additional restrictions

In intensive commercial piggeries, several housing topologies exist which vary from small compartments housing 70 pigs up to large compartments housing up to 1000 animals, with various ways to split the compartments from each other according to EU directives.
Furthermore, a number of EU directives (e.g. directive 2001/88/EC of 23/10/2001) exist for the amount of space for live animals ranging from $0.15m^2$ for young piglets (<10kg) up to $1m^2$ for larger pigs (>110kg), but this does not imply anything about the topology of the

compartment. The variation in size will also affect the number of the necessary microphones in order to cover the complete space. In addition, the room acoustics will change with time as the animals grow older in cases where the animals are kept in the same space during the growing period, because the pigs will occupy a bigger proportion of the compartment.
Finally, different wall material (can range from steel to wood or concrete) and the type of floor (fully or partially slatted, straw-bedded, concrete, etc., see EU directive 2001/93/EC for the conditions that the floor needs to fulfil) affect the acoustics of the acquired sounds (e.g. reverberation). Moreover, this variation in size and in the number of housed animals affects the number of the microphones that are necessary to cover the complete space. Consequently, for a system to be used in practice, the localisation system must be easily adaptable to different compartment topologies, construction materials and the number of microphones used.

2. Method

To fulfil the specifications that were presented, a practical localisation algorithm has been developed that is based on the Time Difference Of Arrival (TDOA) of the signal in multiple microphones (see Fig. 7 for different microphone arrangements). The algorithm accounts for the noisy environment and the uncertainty in the exact time of arrival of the signal at each microphone. The TDOA extraction algorithm is fully automatic (which is necessary for the real-time application of the system) and has been implemented and applied in field experiments.
More specifically, individual sounds are extracted using an algorithm based on the Hilbert Transform. Using the cough classification algorithm, each sound is detected as cough or not. For the sounds that have been detected as cough, each instance from different microphones is subsequently compared and the TDOA is estimated. Finally, the localisation algorithm is based on a weighted sum function that results in an inverse probability matrix for the position of the sound.
Below, each of the three steps is described in detail.

2.1 Extraction of individual sounds

In a pig compartment, the level of the recorded sound can vary considerably. In general, sound of higher intensity will be recorded during the day. Additionally, during feeding (in cases where feed is not provided ad libitum) the intensity of the sounds increases considerably due to movement, competition and the urge of pigs to eat. Also, episodes of increased sound intensity occur when someone is entering the compartment (e.g. Moura et al., 2008). To account for these characteristics of the recordings in a pig compartment, a 2-minute window is used for the analysis. More specifically, sound is continuously recorded and is stored in parts of two minutes for each of the microphones used. Then, each group of the recordings is processed.
It should also be taken into account that environmental noise is constantly present in the compartment. For example, low frequency ventilation noise is almost constantly present as well as social vocalisations of the animals, while the sound of the motors for the feeding system also appears periodically during the day. Since previous research has shown that for the cough identification algorithm, the frequency band 0.1-10 kHz is of the most importance (e.g. Exadaktylos et al., 2008a), low frequency ventilation noise can be eliminated by filtering. In this regard, the recording is initially filtered using a 10th order Butterworth filter

with a passband of 0.1-10 kHz. Depending on the classification approach (e.g. Exadaktylos et al., 2008b), the characteristics of the filter can vary. However, this does not affect the performance of the sound extraction algorithm presented here.
The sound extraction algorithm is using the energy envelope (Oppenheim et al., 1999) of the recording. The calculation of the energy envelope of the recording is done using the Hilbert transform (Oppenheim et al., 1999). The Hilbert transform of a discrete time signal s[k] is defined as:

$$H\{s[k]\} = \sum_{n=-N/2}^{N/2} s[n-k]h[n]\sin^2\left(\frac{n\pi}{2}\right) \tag{1}$$

where $h[k]=\frac{2}{k\pi}$, for $k=\pm1,\pm2,\dots,\pm\frac{N}{2}$, h[0]=0 and it introduces a 90° phase shift to the original signal. The procedure that is followed to calculate the energy envelope is summarised in the following

1. Calculate the energy of the recorded signal by taking its absolute value
2. Calculate the Hilbert transform of the energy as described in Eqn. 1
3. Add of the energy signal and the Hilbert transform
4. Calculate the square root of the above summation
5. Calculate the moving average of the square root

For the results presented below a moving average with N=100 is used that has empirically shown to provide a good trade-off between smoothing and phase shift.
Subsequently, a threshold is defined and the part of the continuous recording that exceeds this threshold is considered to be a sound. As mentioned above, variations in the sound intensity are expected throughout the day. To account for these, the threshold is automatically chosen for every recording as the average of the energy envelope of the continuous recording for each microphone.
Next, Fig. 3 shows three instances of the described procedure. It can be observed that just after 0.3 s of the beginning of the recording, a very small segment of the signal (that is not part of the sound that needs to be extracted) exceeds the identified threshold while after 0.49 s a very small segment (that is part of the sound) is below the threshold. To correct these possible errors, a minimum length of a sound has been set (0.25 s) along with the distance between two consecutive sounds (0.05 s). If two sounds are closer, then they are considered to be a single sound.

2.1 Estimation of the Time Difference of Arrival (TDOA)

After the individual sounds have been extracted, the Time Difference Of Arrival (TDOA) of the sound at the different microphones needs to be estimated. Fig. 4 presents the same sound as has been received by the different microphones where also the TDOA can be observed.
The energy envelope of the signal will be used to estimate the TDOA for each microphone. To get a clearer view of the starting point of a signal, the energy envelope is normalised to have values between 0 and 1. The normalised energy envelope of the same signal as has been received at different microphones is presented in Fig. 5. Defining a threshold as the minimum average value of the energy envelope signals, the arrival time of the signal at each microphone is defined as the time that the energy envelope is exceeding the threshold. Then the TDOA can easily be estimated.

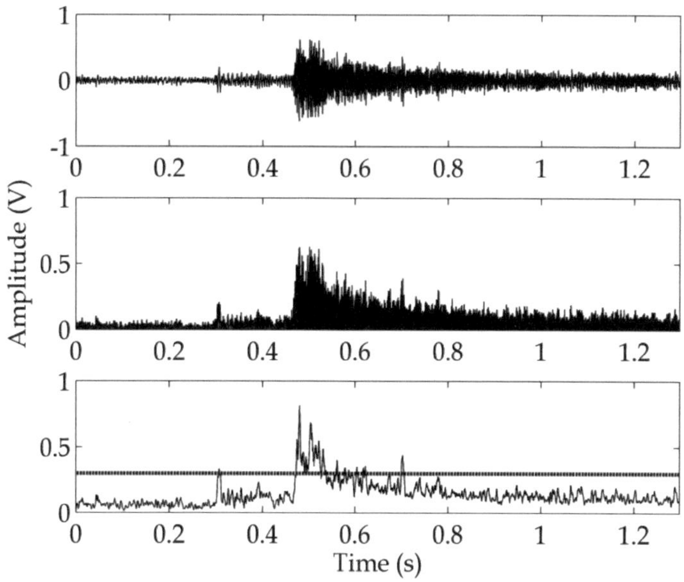

Fig. 3. Three steps in the procedure for extracting a single sound from a continuous recording. The initial/filtered signal (top), the energy of the signal (middle) and the energy envelope (bottom). The automatically chosen threshold is also shown as a horizontal dotted line (bottom)

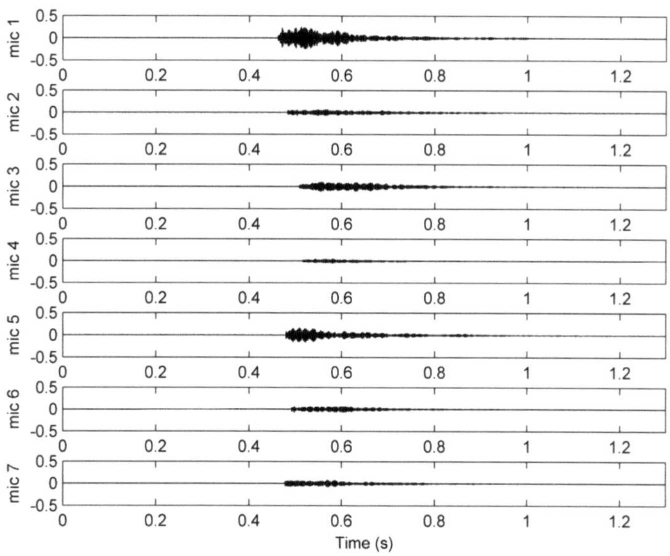

Fig. 4. The recordings of the same signal at 7 microphones in a pig compartment. The amplitude of the signal is in Volts

2.1 Sound localisation

Having estimated the TDOA of the signal at each microphone, the position P that the sound has originated from will be calculated.

By multiplying the TDOA with the speed of sound (343.4 m/s at 20°C), a distance d_τ (distance of the time delay) is calculated. Let us define $d_{p,1}$ and $d_{p,2}$ the distance between the source and microphones 1 and 2 respectively. For the perfect case where the times of arrival have been exactly identified, d_τ equals the absolute difference between $d_{p,1}$ and $d_{p,2}$. However, the simplicity of the algorithm for estimating the TDOA would not result in a difference that is exactly zero.

To estimate the probability that a sound originates from a specific point, the test field is divided into a grid with the necessary resolution. In this case, we have a resolution of 0.1m that is considered adequate of our application since the objective is to identify the pen in which the sick animal is located and treat all the animals in the same pen (and maybe those of the neighbouring pens) and not only the sick animal. Even if the algorithm provides more accurate results, the movement of the pigs makes the additional accuracy unnecessary. Then the following weight is calculated for every point of the grid:

$$w_{(k.l)} = \sum_{i=1}^{n-1} \sum_{j=i+1}^{n} \left| \left(d_{(k,l),i} - d_{(k,l),j} \right) - d_{t(i,j)} \right| \tag{2}$$

where $w_{(k,l)}$ is the weight at position (k,l), $\left(d_{(k,l),i} - d_{(k,l),j}\right)$ is the difference in distance between position (k,l) and microphones i and j, $d_{\tau,(k,l)}$ is the time delay distance between the signal as it arrived at microphones i and j, and n is the total number of microphones that are used. A visual 3D representation of the normalised w matrix is given in Fig. 6. In this graph, the inverse probability that the sound has originated from a point is given. The lowest point identifies the point in the grid from which the sound has most probably originated.

In an ideal context, the above algorithm is simplified to finding the position P in which the weight w_P is zero. However, this would require an exact estimation of the TDOA. In contrast, this algorithm can produce accurate results (with an accuracy of about 1 m) without an exact estimation of the TDOA. This can also be seen in Fig. 6 by examining the shape of the 3D curve. If there was a very accurate estimation of the TDOA, the curve would have the value zero at its lowest point and monotonically increase as it goes away from that point. However, in this particular example, this is the case only for an area around the global minimum. Going further, the curve has local minima and maxima that contradict the ideal expected behaviour.

3. Results

3.1 Localisation of triangle sounds

To test the accuracy of the developed algorithm, triangle sounds were produced in a pig compartment with known dimensions and microphone positions as in Fig. 7. Triangle sounds are used for the development and tuning of the algorithm because they are sharp enough and provide the best case that could occur in practice. Later, the algorithm is tested in a real situation.

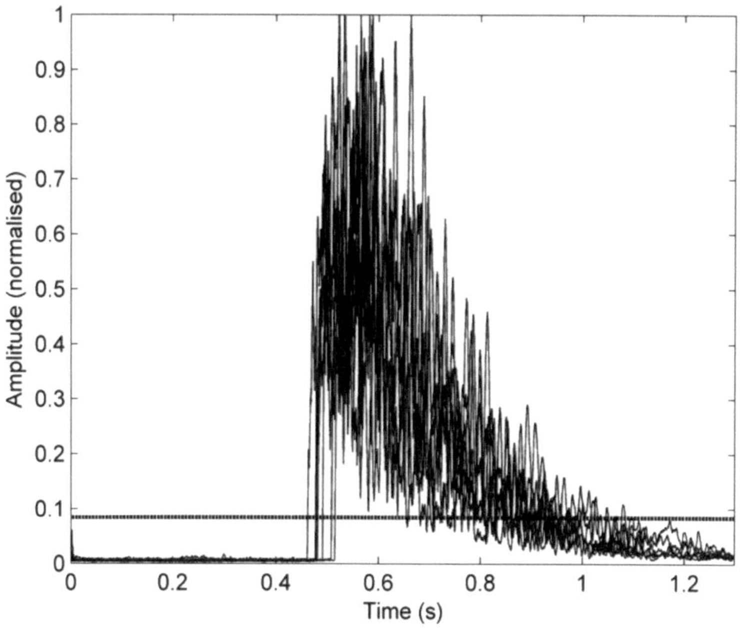

Fig. 5. Normalised energy envelope of a sound received by different microphones (solid lines). The automatically detected threshold is also shown (horizontal thick dotted line)

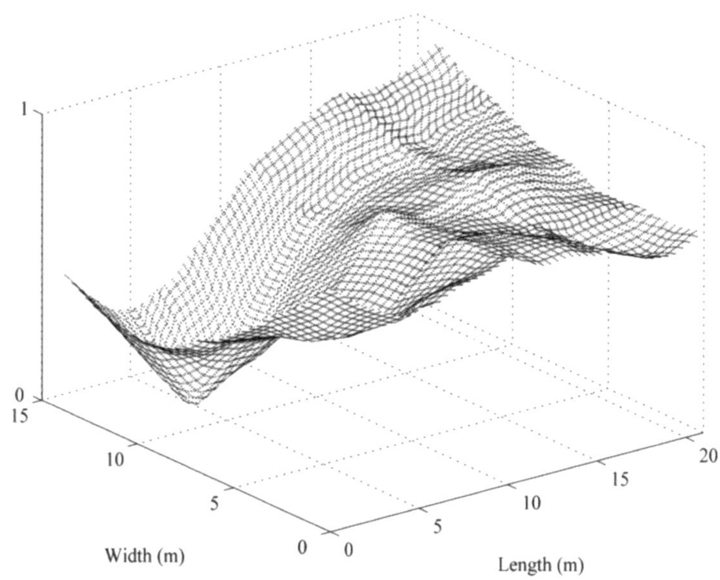

Fig. 6. Visual representation of the weight at every point of the grid. The lowest value on the graph is the estimated position from which the sound originated

For Experiments 1 and 3, six microphones were positioned against the walls at a height of 2 m, while Experiment 1 had an additional microphone hanged in the middle of the compartment at the same height. In Experiments 2 and 5, only a fraction of the compartment was covered with the microphones either hanged or placed against the wall at a height of 2 m. In Experiment 4, the microphones were hanged 2 m above the ground above the pens. Finally, in Experiment 6, two microphones were focused on half of the compartment while two more were placed over the corridor collecting sounds from the complete compartment. All experiments had a duration of 15 minutes with a total recording of 41 triangle sounds at known positions.

Each continuous recording was processed with the algorithms described above, the individual sounds were automatically extracted and location of the sound was estimated. In all cases, the real position of the triangle sound was within the area that is defined by the microphone positions. An overview of the average error is given in Table 1, while Table 2 presents the detailed results for Experiment 1.

It can be seen that the maximum error in all the experiments is less than 2 m and the average less than 1 m. This level of accuracy is considered adequate for this specific application because the vet only needs to know in which pen sick animals are, so that only animals in that pen and maybe in the two neighbouring pens are treated. Animal movement makes it impossible to identify the exact animal that is coughing because a possible alarm can only be provided after a significant increase in the number of coughing sounds. With the current accuracy of the algorithm, it is possible that a cough is identified in a neighbouring pen than the one it has occurred. This is still acceptable because in practice most probably pigs in neighbouring pens will also be treated since most of the time two adjacent pens share the same trough and animals are in close contact.

	Maximum Error (m)	Average Error (m)	Standard Deviation (m)
Experiment 1	1.98	0.49	0.44
Experiment 2	1.70	0.74	0.52
Experiment 3	1.41	0.78	0.28
Experiment 4	0.92	0.70	0.19
Experiment 5	0.60	0.34	0.16
Experiment 6	1.08	0.75	0.22

Table 1. Overview of the performance of the localisation algorithm for the different experiments

3.2 Localisation of real coughs in a livestock house

Using the microphone topology of Experiment 1, sound was recorded using 7 microphones for 3h. During the recording period an expert was using audio-visual observation being in the livestock house to observe the compartment and identified the pens in which every one of the coughs occurred. Subsequently the same expert listened to all the recordings from which 19 cough attacks, counting for a total of 179 individual coughs, were extracted. The number of individual coughs in a cough attack varied from 2 up to 23 coughs.

The sound extraction and localisation algorithm was then applied to the extracted cough attack signals. The individual coughs were automatically extracted and localised. The position of the cough attack was then defined as the average of the position of each of the coughs.

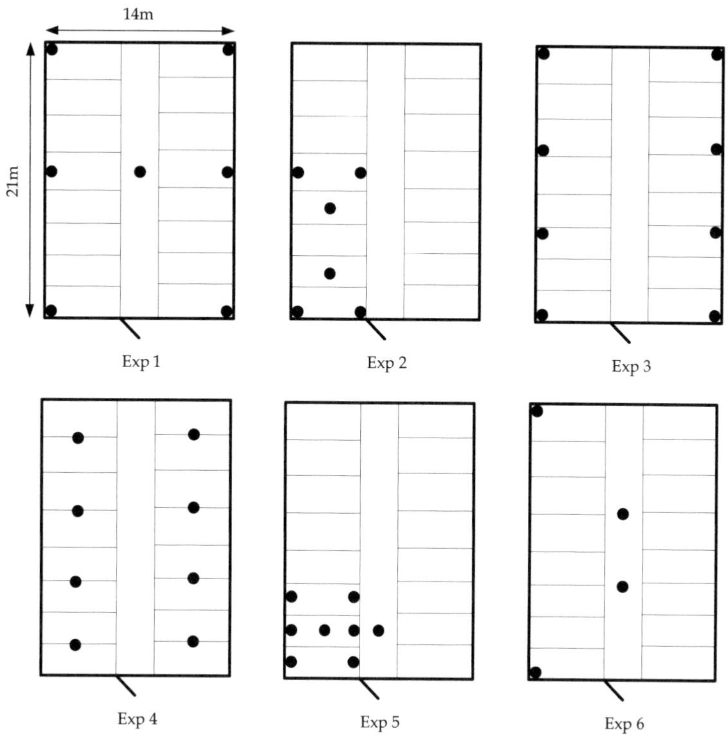

Fig. 7. Top view planimetry of multiple boxes for a swine building with dimensions 14x21m. Microphone positions are indicated with dots for each of the conducted experiments

Sound	Real (x,y) coordinates (m)	Estimated (x,y) coordinates (m)	Error (m)
1	(10.8,1.3)	(10.0,1.9)	1.0
2	(10.8,3.9)	(10.8,3.9)	0.0
3	(10.8,6.5)	(12.1,5.9)	1.4
4	(10.8,9.2)	(11.0,9.0)	0.3
5	(10.8,11.8)	(11.3,11.9)	0.5
6	(10.8,14.4)	(11.4,15.0)	0.8
7	(10.8,17.0)	(10.7,17.0)	0.1
8	(10.8,19.7)	(11.2,19.5)	0.4
9	(3.2,1.3)	(0.1,0.1)	3.3
10	(3.2,19.7)	(3.1,19.2)	0.5
11	(3.2,17.0)	(2.0,16.6)	0.6
12	(3.2,14.4)	(3.0,14.6)	0.3
13	(3.2,11.8)	(2.7,11.7)	0.5
14	(3.2,9.2)	(0.1,0.1)	9.6
15	(3.2,6.5)	(2.7,6.3)	0.5
16	(3.2,3.9)	(2.7,4.1)	0.5

Table 2. Detailed results for Experiment 1. The real and estimated positions of each triangle sound are shown along with the localisation error in meters.

The expert that was observing the compartment during the recording identified 1 cough attack in pen number 4, 3 in pen number 5, 6 in pen number 8, 1 in pen number 15 and 8 cough attacks in pen number 16. The automatic localisation result is visualised in Fig. 8 were the black stars depict cough attacks that were estimated to originate from the pen that they actually did and white stars represent cough attacks that were estimated to originate from a neighbouring pen.

The above suggests that 3 cough hazards may exist in the compartment, 2 at the sides of the compartment below the windows (pens 8 and 16) and 1 in the middle of the compartment (around pen number 5). From this experiment, it can be concluded that a good estimate of the areas where coughing animals are located can be estimated using the proposed algorithm.

3.3 Fully automatic identification and localisation of pig coughs

As already mentioned above, the objective of the development of this localisation algorithm is to identify the locations of coughing pigs in a pig compartment. To achieve this, the sound extraction algorithm presented above is coupled to a cough identification algorithm and if the acquired signal is a cough then the presented localisation algorithm is used to estimate the position that the sick cough originated from.

This fully automatic algorithm has been applied to the recordings of the previous subsection. Using the algorithm of Exadaktylos et al. (2008b), about 50% of the manually labelled cough sounds were correctly identified by the algorithm and subsequently the result is visualised in Fig. 9, where the dark areas identify potential cough hazards. Although this level of accuracy is fairly low for identification standards, it does present an improvement with the current situation where real-time monitoring is impossible due to the large number of animals per compartment and the big number of compartments in commercial intensive pig farming. Furthermore, it is expected that sick animals will cough repetitively and therefore the system will be able to detect a hazard. Comparison of Fig. 9 with Fig. 8 shows that 2 out of the 3 hazards have been correctly identified. The reason for

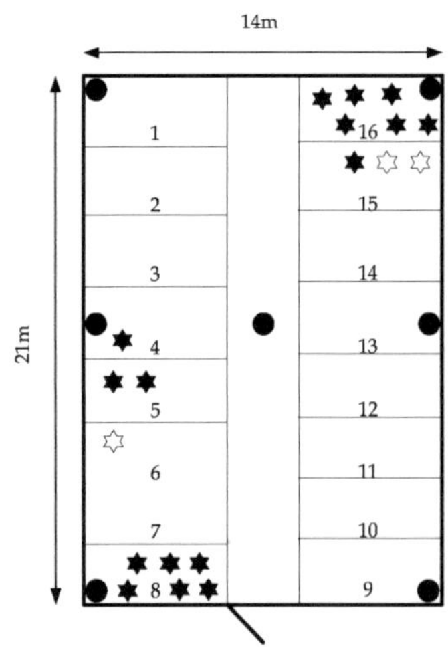

Fig. 8. Localisation result for cough attacks that were manually identified and automatically localised. Black stars show the cough attacks that were localised in the exact pen. White stars show cough attacks that were identified coming from a neighbouring pen.

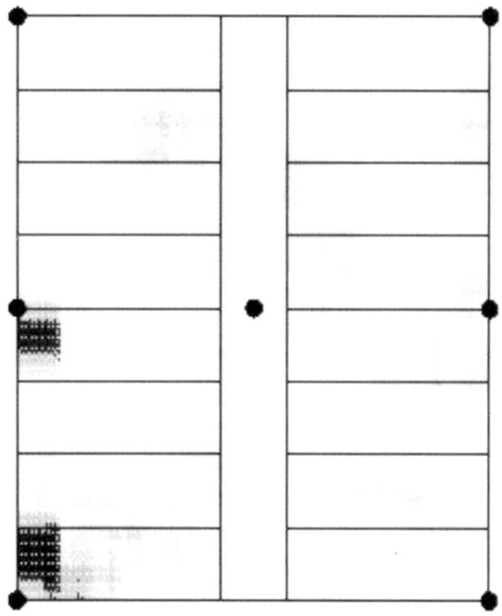

Fig. 9. Result of the combined cough identification and localisation algorithms. The dark areas identify potential cough hazards

not identifying the cough hazard near pen number 16 can be either due to the identification or the localisation algorithm.

In general, coughs are expected to occur repeatedly if an animal is sick or a disease is spreading. Therefore, the relatively low identification ratio (50%) can still be used for practical application of the system. It is claimed that application of the described system can provide a good and quick overview of the respiratory health status in animal housing (Fig. 9) that can lead to better management of the herd.

4. Shortcomings and future research

In the present chapter, we have presented the development process for a specific localisation algorithm application. As mentioned above, there are a number of choices to be made in the process about the different components of the system. Clearly, a different approach could include a more sophisticated TDOA detection algorithm and a less robust localisation algorithm. Our choice was based on the fact that many different practical issues (e.g. different building material) would require long calibration procedures for a very accurate TDOA estimation. The simplicity and robustness of our approach should still prove itself under different building conditions. Furthermore, sound deflection and reverberation was not taken into account in this study and is one of the key elements that should be further tested. Techniques to deal with reverberation have been developed (e.g. Marro et al., 1998; Gustafsson et al., 2003) and it is expected that if necessary can be integrated in our system.

The redundancy in the number of microphones used is an acceptable cost for research purposes. However, it may be very expensive in a real commercial setup where the cost needs to be kept below 0.5€/pig. To maintain the algorithm performance and reduce the number of microphones used, existing techniques for improving signal quality can be used. More sophisticated filtering or beam-forming (e.g. Krim & Viberg, 1996) are two options.

However, further development of the localisation algorithm should be performed in parallel with the cough identification algorithm since the individual blocks of the system are coupled. Any distortion or alteration of the signal must be linked with the rest of the steps in the system.

5. Conclusion

The present chapter has presented a system that can be used for continuous automatic health monitoring in commercial piggeries. Coughing is the main symptom of respiratory problems in pigs. In order to develop a monitoring system, a cough identification algorithm has been previously developed. In order to identify the pen in which a sick pig is located, a localisation algorithm has also been developed.

The harsh environment of a commercial piggery, along with the differences among the different piggeries requires a simple and robust localisation algorithm that can be individually adapted for the building topology and acoustics. This work has presented the development process, starting from concept, defining the specifications, and finally developing an algorithm for this specific application.

By adapting the identification algorithm, the application of this specific localisation algorithm can be extended to monitor respiratory health and welfare issues in livestock production beyond pigs, such as cattle and poultry.

6. References

Centner, T.J. (2003). Regulating concentrated animal feeding operations to enhance the environment. *Environmental Science & Policy*, Vol. 6, No. 5, (October 2003) 433-440, ISSN: 1462-9011.

CEC, Commission of the European Communities (2000). White paper on food safety. CEC Brussels, 12 January 2000 COM (1999) 719 final, 52 PP.

Delgado, C.L.; Narrod, C.A. & Tiongco, M.M (2003). Policy, technical, ad environmental determinants and implications of the scaling-up of livestock production in four fast-growing developing countries: a synthesis. International Food Policy Research Intitute.

Elbers, A.RW.; Stegeman, J.A. & De Jong, M.C.M. (2001). Factors associated with the introduction of Classical Swine Fever virus into pig herds in the central area of the 1997/98 epidemic in the Netherland. *Veterinary Record*, Vol. 149, No. 13 (September 2001) 377-382, ISSN: 0042-4900.

Exadaktylos, V.; Silva, M.; Aerts, J.-M.; Taylor, C.J. & Berckmans, D. (2008a). Real-time recognition of sick pig cough sounds. *Computers and Electronics in Agriculture*, Vol. 63, No. 2, (October 2008) 207-214, ISSN: 0168-1699.

Exadaktylos, V.; Silva, M.; Ferrari, S.; Guarino, M.; Taylor, C.J.; Aerts, J.-M. & Berckmans, D. (2008b). Time-series analysis for online recognition and localization of sick pig (Sus scrofa) cough sounds. *Journal of the Acoustical Society of America*, Vol. 124, No. 6, (December 2008) 3803-3809, ISSN: 0001-4966.

Farm Animal Industrial Platform (FAIP) (2003). Farm Animal Breeding in Europe 2003. 8 pages.

Ferrari, S.; Silva, M.; Guarino, M. & Berckmans, D. (2008). Analysis of cough sounds for diagnosis of respiratory infections in intensive farming. *Transactions of the ASABE*, Vol. 51, No. 3, (May-June 2008) 1051-1055, ISSN: 0001-2351.

Food and Agriculture Organization of the United Nations (FAO) (2002). Improved animal health for poverty reduction and sustainable livelyhoods. FAO animal production ad health paper 153.

Food and Agriculture Organization of the United Nations (FAO) (2007). FAO-STAT, Rome, Italy.

Gustafsson, T.; Rao, B. & Trivedi, M. (2003). Source localization in reverberant environments: Modeling and statistical analysis. *IEEE Transactions on Speech and Audio Processing*, Vol. 11, No. 6, (November 2003) 791-803, ISSN: 1063-6676.

Hamscher, G.; Pawelzick, H.T.; Sczesny, S.; Nau, H. & Hartung, J. (2003). Antibiotcis in dust originating from a pig-fattening farm: a new source of health hazard for farmers. *Environmental Health Perspectives*, Vol. 111, No. 13, (October 2003) 1590-1594, ISSN: 0091-6765.

Hayes, S.; Mellinger, D.; Croll, D.; Costa, D. & Borsani, J. (2000). An inexpensive passive acoustic system for recording and localizing wild animal sounds. *Journal of the Acoustical Society of America*, Vol. 107, No. 6, (June 2000) 3552-3555, ISSN: 0001-4966.

Korpáš, J.; Sadloňová, J. & Vrabec, M. (1996). Analysis of the cough sound: an overview. *Pulmonary Pharmacology*, Vol. 9, No. 56, (October 1996) 261-268, ISSN: 0952-0600.

Krim, H. & Viberg, M. (1996). Two decades of array signal processing research - The parametric approach. *IEEE Signal Processing Magazine*, Vol. 13, No. 4, (July 1996) 67-94, ISSN: 1053-5888.

Manteuffel, G.; Puppe, B. & Schön, P. (2004). Vocalization of farm animals as a measure of welfare. *Applied Animal Behaviour Science*, Vol. 88, No. 1-2, (September 2004) 163-182, ISSN: 0168-1591.

Marro, C.; Mahieux, Y. & Simmer K.U. (1998). Analysis of noise reduction and deverberation techniques based on microphone arrays with postfiltering. *IEEE Transactions on Speech and Audio Processing*, Vol. 6, No. 3, (May 1998) 240-259, ISSN: 1063-6676.

Marx, G.; Horn, T.; Thielebein, J.; Knubel, B. & von Borell, E. (2003). Analysis of pain-related vocalization in young pigs. *Journal of Sound and Vibration*, Vol. 266, No. 3, (September 2003) 678-698, ISSN: 0022-460X.

Moura, D.J.; Silva, W.T.; Naas, I.A.; Tolon, Y.A.; Lima, K.A.O. & Vale, M.M. (2008). Real-time computer stress monitoring of piglets using vocalization analysis. *Computers and Electronics in Agriculture*, Vol. 64, No. 1, (November 2008), 11-18, ISSN: 0168-1699.

Murphy, F.A. (1998). Emerging zoonoses. *Emerging Infectious Diseases*, Vol. 4, No. 3, (July-September 1998) 429-435, ISSN: 1080-6059.

OIE. 2005. *Terrestrial Animal Health Code*. 14th ed. Paris.

Oppenheim, A.; Schafer, R. & Book J. (1999). *Discrete-Time Signal Processing*, Prentice-Hall, ISBN: 978-0137549207, Upper Saddle River, NJ.

Schön, P; Puppe, B. & Manteuffel G. (2004). Automated recording of stress vocalisation as a tool to document impaired welfare in pigs. *Animal Welfare*, Vol. 13, No. 2, (May 2004) 105-110, ISSN: 0962-7286.

Silva, M.; Ferrari, S.; Costa, A.; Aerts, J.-M.; Guarino, M. & Berckmans, D. (2008). Cough localization for the detection of respiratory diseases in pig houses. *Computers and Electronics in Agriculture*, Vol. 64, No. 2, (December 2008) 286-292, ISSN: 0168-1699.

Thomas, R.; Fristrup, K. & Tyack, P. (2002). Linking sounds of dolphins to their locations and behavior using video and multichannel acoustic recordings. *Journal of the Acoustical Society of America*, Vol. 112, No. 4, (October 2002) 1692-1701, ISSN: 0001-4966.

Van Hirtum, A. & Berckmans D. (2001). Fuzzy approach for improved recognition of pig coughing from continuous registration. *Proceedings of Noise and Vibration Engineering* Vols. 1-3, pp. 1535-1541, ISBN: 90-73802-76-8, 1st International ISMA Workshop on Noise and Vibration in Agricultural and Biological Engineering, Leuven, Belgium, 13-14 September 2000.

Van Hirtum, A. & Berckmans D. (2003a). Considering the influence of artificial environmental noise to study cough time-frequency features. *Journal of Sound and Vibration*, Vol. 266, No. 3, (September 2003) 667-675, ISSN: 0022-460X.

Van Hirtum, A. & Berckmans D. (2003b). Fuzzy approach for improved recognition of citric acid induced piglet coughing from continuous registration. *Journal of Sound and Vibration*, Vol. 266, No. 3, (September 2003) 677-686, ISSN: 0022-460X.

Windhorst, H.W. (2004). Dynamics in global trade. *Pig Progress*, Vol. 20, No. 9 (September 2004) 36-39, ISSN: 0169-4405.

Windhorst, H.W. (2006). Regional patterns of livestock and poultry production in Europe. In: *Livestock Production and Society*, R. Geers and F. Madec (Ed.) 21-34, Wageningen Academic Publishers, ISBN: 90-76998-89-2, Wageningen, The Netherlands.

Woolhouse, M.E. & Gowtage-Sequiria, S. (2005). Host range and emerging and reemerging pathogens. *Emerging Infectious Diseases*, Vol. 11, No. 12, (December 2005) 1842-1847, ISSN: 1080-6059.